Thermal Physics

Kinetic Theory, Thermodynamics and Statistical Mechanics

Second Edition

About the Authors

S. C. Garg, PhD (University of Delhi), is Professor of Physics at School of Sciences, Indira Gandhi National Open University (IGNOU), New Delhi. Earlier he was Pro Vice Chancellor of IGNOU and Founder Director of Centre for Learning and Teaching, National University of Lesotho, Roma, Southern Africa. Before joining IGNOU in 1989, he was with the Department of Physics, Hindu College, University of Delhi for about 12 years.

With more than three decades of teaching experience in India and abroad, Prof. Garg has co-authored four textbooks in physics at the undergraduate and Masters levels. He has published several research papers in national and international journals and guided several students for their PhD degree. Having been a member of academic councils and management boards of several universities in the India and abroad, he has wide experience in planning and development of curricula and learning materials for different groups of learners. His areas of research are Theoretical Reactor Physics and Physics Education.

R. M. Bansal, PhD (University of Delhi), is Associate Professor, Department of Physics, Hans Raj College, University of Delhi. He has more than three decades of teaching experience and has been a member of several committees tasked to revise physics courses. His areas of research are Theoretical Solid State Physics and Neutron Transport.

C. K. Ghosh, PhD (University of Calcutta), is Director, National Centre for Innovations in Distance Education, Indira Gandhi National Open University (IGNOU), New Delhi. He has held several important positions in Student Support Services Division of IGNOU. Before joining IGNOU in 1989, he was with the Department of Physics, Scottish Church College, Kolkata. He has been the nodal officer of the Integrated MSc-PhD programme in physics and astrophysics under IGNOU-IIA collaboration. He has co-authored two books in physics and published several research papers.

Thermal Physics
Kinetic Theory, Thermodynamics and Statistical Mechanics

Second Edition

S.C. Garg
Professor of Physics
Former Pro Vice Chancellor of IGNOU
School of Sciences, Indira Gandhi National Open University, Delhi

R.M. Bansal
Associate Professor
Hans Raj College, University of Delhi, Delhi

C.K. Ghosh
Director
National Centre for Innovations in Distance Education
Indira Gandhi National Open University, Delhi

McGraw Hill Education (India) Private Limited
NEW DELHI

McGraw Hill Education Offices
New Delhi New York St Louis San Francisco Auckland Bogotá Caracas
Kuala Lumpur Lisbon London Madrid Mexico City Milan Montreal
San Juan Santiago Singapore Sydney Tokyo Toronto

McGraw Hill Education (India) Private Limited

Published by McGraw Hill Education (India) Private Limited
P-24, Green Park Extension, New Delhi 110 016

Thermal Physics: Kinetic Theory, Thermodynamics and Statistical Mechanics, 2e

ISBN (13): 978-1-25-909761-4
ISBN (10): 1-25-909761-7

Vice President and Managing Director: *Ajay Shukla*

Head—Higher Education Publishing and Marketing: *Vibha Mahajan*
Publishing Manager—SEM & Tech Ed.: *Shalini Jha*
Asst Sponsoring Editor: *Smruti Snigdha*
Editorial Researcher: *Amiya Mahapatra*
Manager—Production Systems: *Satinder S Baveja*
Copy Editor: *Preyoshi Kundu*
Sr Production Manager: *P L Pandita*

Asstt. General Manager—Marketing (Higher Ed.): *Vijay Sarathi*
Sr Product Specialist: *Sachin Tripathi*
Sr Graphic Designer—Cover: *Meenu Raghav*

General Manager—Production: *Rajender P Ghansela*
Manager— Production: *Reji Kumar*

Typeset at The Composers, 260, C.A. Apt., Paschim Vihar, New Delhi 110 063, and printed at Gopal Jee Enterprises, 190/5 Main Road, Maujpur, Delhi 110053

Cover Printer: SDR

RCXYCRLODLDRR

Dedicated to
Prof. L. S. Kothari

CONTENTS

FOREWORD TO THE FIRST EDITION

Physics is subtle, yet simple. The fundamental laws of physics encompass a large variety of physical, technological and natural phenomena. Physics influenced our material well-being as well as our philosophical outlook. Physics is therefore rightly regarded as natural philosophy.

In the last century, the time gap between a new discovery and its practical applications was measured in decades. It has now been reduced to a few years. The development of this close link between science and technology heavily depends on the availability of well-trained scientific and technical minds in a country. This is facilitated if the curriculum is so designed that it develops a sensitivity towards nature, and students are able to see the principles of physics in operation. Due to these varied demands on a teacher of physics, good books play a very important role in this endeavour. It is my considered view that this book on thermal physics satisfies many of these requirements and is therefore a valuable addition to our store of knowledge.

Thermal physics is one of the most fascinating courses taught to our undergraduate students. The authors of this volume present the subject matter on kinetic theory, thermodynamics and statistical mechanics, spread over thirteen chapters, in lucid and simple language. This book would also prove useful to those studying chemistry, engineering, technology or any other science course which requires a knowledge of thermal physics.

The best way to teach and learn a subject is to develop a healthy attitude towards problem solving. The numerous solved examples included should help in clarifying the concepts which are introduced and discussed in the text. A large number of problems given within and at the end of each chapter would help the students become active participants in the teaching learning process.

One of the major deficiencies in the higher education system in India is the paucity of well written textbooks which are within the reach of a majority of students. Any effort to improve the quality of higher education will not succeed until this requirement is met. The authors have succeeded in meeting a felt need. I am sure, this book will be a great source of help to students of physics in the years to come.

V C Kulandai Swamy
Vice Chancellor
Indira Gandhi National Open University
New Delhi

PREFACE TO THE SECOND EDITION

About the Book

When we began to develop the manuscript of the first edition of this book in 1990, we took a conscious decision to make it interactive so that students could be encouraged to participate in the learning process. This was based on the realization that an average Indian student was not comfortable with problem solving or shied away from participating in the classroom discussion. It was a challenging task and quite unlike all other available books on this subject. Yet in this experiment, we chose to swim against the tide and were fortunate to receive support from the editorial team of McGraw Hill Education (India) Private Limited. It is a matter of great satisfaction that our approach has been accepted by the peers as well as the students.

The book has grown out of our lectures delivered to our undergraduate students and we have continued to use it for different batches and groups of learners. In addition, useful feedback from the peers coupled with the research team of McGraw Hill Education (India) Pvt. Ltd. currently associated with this project also helped us to collate the perceived gaps. In so doing, we have included many new topics and rewritten a few chapters to further improve the quality of content. For instance, Chapter 11 has been added on Radiation and Chapters 12–15 on the Statistical Physics have been almost completely rewritten to facilitate better understanding of the subject. New topics such as measurement of temperature and temperature scales, power and refrigeration cycles, and liquefaction of gases and applications of low temperatures have been added in Chapters 4, 6 and 10. Moreover, many and varied new solved examples and unsolved problems with graded difficulty levels have been interspersed in the text. A conscious effort has been made to assist learners to apply acquired theoretical knowledge to real-life applications, phenomena and problems. We hope that these changes will extend the coverage and enhance the utility of the book further.

Latest learning theories suggest that use of access devices— learning objectives, in-text questions, practise problems and summary—should give strategic as well as adequate opportunities, information and instruction to aid understanding and make learning an enjoyable experience for students.

New to this Edition

- Change in the title of the book from *Thermal Physics* to *Thermal Physics—Kinetic Theory, Thermodynamics and Statistical Mechanics*
- A new chapter on Radiation has been added as Chapter 11
- Chapters 12 to 15 have been rewritten to facilitate better understanding among students
- New topics such as measurement of temperature and temperature scales, power and refrigeration cycles, and liquefaction of gases and applications of low temperatures have been added in Chapters 4, 6 and 10

- Discussion of Joule–Thomson Effect has be added in Chapter 3 as an implication of interatomic interactions between the molecules of a real gas.
- 200 new solved numericals
- New and enhanced pedagogical features include Learning Objectives and Summary at the end of each chapter.

Salient Features

- Comprehensive coverage of the syllabi of all major universities and the new UGC syllabus.
- Addition of a new chapter of Radiation and new topics such as Liquefaction of gases, Power and Refrigeration cycles and applications of low temperatures in the new edition.
- The treatment of the subject matter is rigorous while still retaining a student-friendly approach.
- Strong in-text pedagogical with numerous figures, solved examples and practise sets to support learning.
- Rich pedagogical devices include Learning Objectives, Summary and Mnemonics.
  252 solved examples
  149 practise problems with answers
  107 unsolved exercises with answers
  180 Illustrations

Acknowledgments

We would like to express our appreciation to our students who taught us how to teach better and all those peers who spared their invaluable time to share their suggestions. Our special thanks are due to the editorial team of McGraw Hill Education (India) Pvt. Ltd. and our families for their support.

S. C. Garg
R. M. Bansal
C. K. Ghosh

Feedback

Every possible care has been taken to develop an error-free book. Yet some omissions might have escaped our attention. We shall be grateful to the readers for bringing the same to our notice by writing to us at *prof1951@gmail.com*, *bansal_rm@yahoo.co.in* and *ckghosh@ignou.ac.in*. Suggestions are most welcome to further improve the content and presentation of the book.

Publisher's Note

Don't forget to write to us! We are always open to new ideas (the best ideas come from you!). You may send your comments to *tmh.sciencemathsfeedback@gmail.com* (don't forget to mention the title and author's name in the subject line).

Piracy-related issues may be reported as well!

PREFACE TO THE FIRST EDITION

The foundations of the subject of thermodynamics were laid by Joule in 1840 when he established that heat is intimately connected to motion. It developed on a postulatory basis without reference to the microscopic details of matter and soon blossomed into a powerful science, which dealt with systems in stable equilibrium. With the growth in our understanding of the nature and behaviour of constituents (electrons, atoms and/or molecules) of a system, Boltzmann showed, by supplementing statistical methods with the laws of (classical) mechanics, that the laws of thermodynamics have sound theoretical basis. The advent of quantum mechanics added a new dimension to the subject. Today, statistical mechanics together with quantum mechanics, lays the foundation of modern physics. In fact, thermodynamics and statistical mechanics are of vital importance in chemistry, biology, metallurgy, material science and engineering.

In its present state, thermal physics is one of the most fascinating courses taught to undergraduate science and engineering students in India and abroad. There are many books of good quality on the subject. Some of these are so costly that not many students can afford them. Moreover, while teaching this course to our undergraduate students, we could never recommend a textbook which presented lucid treatment of all the topics in one cover. Our book is a sincere effort in this direction. The language is simple and all care has been taken to make the book self-contained so that it is suitable for self-study as well.

The structure of the book has evolved out of classroom experience. The book contains thirteen chapters which can be classified into three parts. The first part deals with kinetic theory of gases which postulates that atoms/molecules are the building blocks of every gas and that they move about randomly. In **Chapter 1**, we begin by deriving expression for the pressure exerted by a perfect gas on the walls of the container. A detailed account of Maxwell–Boltzmann distribution is also given in this chapter. **Chapter 2** deals with mean free path, viscosity, thermal conductivity, diffusion, Brownian motion and radom walk under the title, Transport Phenomena. The random walk problem has been discussed in detail because of its importance. Moreover, students do not find its discussion in any existing book on thermal physics. The discussion in these two chapters pertains to a perfect gas. But in practice, perfect gas behaviour is hardly observed. In **Chapter 3**, we have discussed behaviour of real gases within the framework of van der Waals' equation of state.

The second part of the book, spread over eight chapters, is devoted to thermodynamics. There are two distinct approaches to learn this subject. The traditional approach is phenomenological; based on some abstract postulates derived from experience. In the statistical approach, firm theoretical basis of thermodynamics is demonstrated by relating the properties of bulk systems to the behaviour of their elementary constituents. One can argue for and against both alternatives. But we have preferred the historical sequence of events—the formal postulatory approach. The basic concepts of thermodynamics have been introduced in **Chapter 4**. The first law of thermodynamics and its applications to diverse phenomena are discussed in **Chapter 5**. The second law of thermodynamics forms the subject matter of **Chapter 6**. This law is intimately connected to entropy. The students will learn that entropy is the most effective vehicle for communicating observations on natural processes. Keeping this in view, a separate chapter is devoted to the discussion of entropy. The working tools of thermodynamics—Maxwell's relations are derived in **Chapter 8** and their applications in widely differing processes/systems are also discussed here. The students will learn of their use in the study of phase transitions, which apart from their immense technological importance, involve a lot of good physics. Every thermodynamic system has an inherent tendency to approach equilibrium. To describe the general condition of equilibrium, we must know free energy functions which constitute a fountainhead of vital information about the thermodynamic behaviour of a system. These are discussed in **Chapter 9**. **Chapter 10** is devoted to the production of low temperatures. Here we have described Joule-Thomson effect and adiabatic demagnetisation in great detail and also outlined a few other techniques. The third law of thermodynamics and its consequences, including unattainability of absolute zero, are also discussed in this chapter.

The third part, spread over three chapters, is devoted to statistical mechanics. **Chapter 11** serves to introduce basic concepts and tools used to study the behaviour of a system, using statistical mechanics. The relation connecting thermodynamics with statistical mechanics, and expressions for the Maxwell–Boltzmann, Fermi-Dirac and Bose-Einstein distribution functions are also derived. These distribution functions suggest that partition function is the most important, effective and convenient mathematical device to unfold information about a system. The evaluation of partition function for computing heat capacities of gases and solids as well as lasers and negative temperatures is illustrated clearly in **Chapter 12**. All these topics are presented in the most lucid and coherent form. In **Chapter 13** we have discussed the important applications of ideal F-D and B-E distributions including white dwarf stars and liquid helium.

We believe that learning efficiency of students is increased when they are active participants in the course. The best way to ensure this is to make them do numerical problems and pose application-oriented, reason-based questions. With this in mind, we have included more than one hundred 'in-text' and a hundred and fifty end-of-chapter practice exercises. More than one hundred solved examples are also given. We feel that if a student does all these thoroughly he or she would both appreciate and enjoy the subject better.

We wish to express sincere thanks to Mr Sunder Singh of the School of Sciences, IGNOU, New Delhi, for his help at various stages in the preparation of the manuscript.

AUTHORS

1 IDEAL GASES: ELEMENTARY KINETIC THEORY AND MAXWELLIAN DISTRIBUTION

Learning Objectives

In this chapter, you will learn how to

- list the basic assumptions of kinetic theory of gases;
- derive expression for pressure exerted by an ideal gas in terms of microscopic properties of the molecules;
- interpret temperature in terms of molecular properties and obtain ideal gas laws starting from the expression for pressure;'
- derive expressions for Maxwell distribution function of molecular velocities as well as speeds;
- obtain expression for average speed, most probable speed and root mean square speed of molecules;
- discuss indirect evidences in favour of Maxwell's distribution law for gases; and
- describe methods for direct experimental verification of Maxwell's distribution law.

1.1 INTRODUCTION

Matter is made up of atoms.* A few atoms bind together to form a molecule.+ A molecule is the smallest unit that can exist in a free state, and possesses all the basic properties of the substance concerned. Molecules move continuously and this motion becomes more and more chaotic on increased thermal excitation, but is constrained due to cohesive forces between molecules. We know that cohesive force is maximum in solids and almost negligible in gases. That is why a solid has a definite shape** and size but a gas does not.

In their efforts to obtain information about thermal properties of different states of matter, the earliest physicists, particularly Bernoulli, Clausius, Maxwell, van der Waals and Jeans focussed their attention on the gaseous state and developed a mathematical formulation of the kinetic theory. This emphasis seems to be due to the fact that gaseous systems are relatively simple to analyse and understand. Moreover, it helps illustrate all essential physical mechanisms and basic concepts. Towards the beginning of the 19th century, the kinetic theory gained recognition and acceptability of wider scientific community, with

---

*This concept was proposed by Democritus in 5th century B.C.

+ The existence of molecules was proposed by Avogadro in 1811.

**It has been found that an orderly arrangement of molecules can even be seen in some liquids. This however exists over an extremely short range ($\approx$ 10 nm).

indirect as well as direct experimental evidences to support it. In particular, Brownian motion put this theory on a very firm footing.

Broadly speaking, kinetic theory of gases is based on two basic postulates: (i) Matter is made up of identical molecules and (ii) Thermal energy can be identified with molecular motion. Accordingly, a few assumptions are made about the nature of molecules, interactions among the atoms and their movement. To describe this motion, simple laws of mechanics are used. This helps us to obtain elegant explanations of the laws governing the behaviour of ideal gases in equilibrium. Today, we understand the gaseous state, almost fully. A detailed discussion of the kinetic theory of gases forms the subject matter of this chapter.

In kinetic theory, we apply the laws of (classical) mechanics to each individual molecule. As you may recall from your school physics classes, a gas consists of an extremely large number of identical molecules and it is not possible to keep track of the positions and velocities of them all. Therefore, to obtain useful results, say for the pressure exerted by a gas, we resort to statistical averaging. Within the framework of this theory, we can also study the transport processes like viscosity, conduction, diffusion, effusion and Brownian motion. You will learn about these in detail in Chapter 2.

In Sec. 1.2, you will learn the assumptions of kinetic theory of gases. We have given justifications for their appropriateness as far as possible. In Sec. 1.3, you will learn to derive expression for pressure exerted by a gas on the walls of a container and use it to correlate temperature with molecular properties. In particular, you will discover that *absolute zero* is that temperature at which all molecular motion ceases to exist. You will use this expression for pressure to deduce gas laws and calculate the magnitude of molecular speeds for different gases. These help us to explain the observed distribution of gases in our atmosphere. Here we have also discussed the law of equipartition of energy. You will learn to express heat capacities of ideal monatomic, diatomic and polyatomic gases in terms of the gas constant, R. The ratio of heat capacities of a gas can be determined from their experimentally observed values and this proved to be a valid mechanism to test the validity of kinetic theory of gases in its initial years.

When Maxwell began to study the problems of kinetic theory of gases, he was aware of the work of Clausius. In his characteristically novel and profound work published in 1860, he provided the correct answer for the distribution of molecular velocities and introduced statistical concepts in kinetic theory of gases. His theory marked the beginning of a new era in physics; it formed the basis of modern theory of statistical mechanics and led to statistical interpretation of the laws of thermodynamics. Using the theory of probability, Boltzmann put Maxwellian theory on firm mathematical foundations. For this reason, distribution function for molecular velocities is also referred to as *Maxwell–Boltzmann distribution function*. In Sec. 1.4, you will learn to derive the expressions for Maxwell–Boltzmann distribution function for molecular velocities as well as speeds. Molecular distribution of velocities helps us to estimate the number of molecules having velocities in a particular range. The knowledge of Maxwell distribution function will then be used to obtain expressions for average speed, most probable speed and root mean square speed of molecules. The energy distribution of Maxwellian gas will be discussed in Sec.1.5. In science, no theory is accepted till such time that it is directly verified. In case of Maxwell distribution law, indirect evidences came much before direct measurements. In Sec. 1.6, you will learn about indirect evidences. The first direct proof of Maxwell's law was given by Stern in 1920. His technique was subsequently modified by Zartman and Ko. A detailed discussion of these and other more significant experiments is given in Sec. 1.7.

To develop the understanding of elementary kinetic theory, stating its basic assumptions with reference to an ideal gas forms a convenient starting point.

1.2 Basic Assumptions of Kinetic Theory

The basic assumptions made by the early contributors to the subject of kinetic theory of ideal gases are:

1. A gas consists of a very large number of identical molecules. As we now know, this assumption is quite justified. The number of molecules in a kilo-mole is 6.03×10^{26}. (This is commonly known as *Avogadro's number*.) From experimental evidences, we know that 1 cm^3 of an ideal gas at standard temperature and pressure (STP)* contains 3×10^{19} molecules. (This number is about four billion times the population of the world as of now.)
2. The gas molecules can be regarded as point masses. Experiments show that the diameter of a gas molecule is about $2 - 3 \times 10^{-10}$ m. The distance between any two neighbouring gas molecules at STP, on an average, is about 3×10^{-9} m, which is an order of magnitude bigger than their diameter.
3. The gas molecules are in a state of constant random motion. This is shown in Fig. 1.1. (In fact, the motion of gas molecules resembles the motion of honeybees disturbed from their hive.) In essence, it means that molecules of an ideal gas can move in all possible directions and all positions are equally probable. The support for this assumption came in the form of Brownian motion.

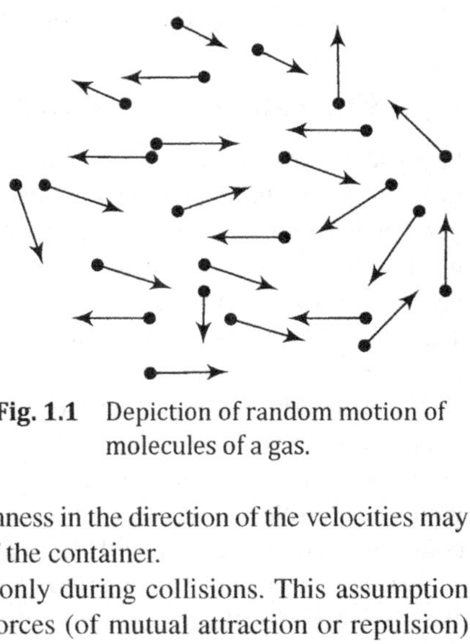

Fig. 1.1 Depiction of random motion of molecules of a gas.

4. In the absence of any external force field, the molecules are distributed uniformly in the container. It means that an ideal gas behaves as an isotropic medium. In practise, however, some randomness in the direction of the velocities may arise because of irregularity in the walls of the container.
5. The molecules of a gas experience force only during collisions. This assumption implies that there are no intermolecular forces (of mutual attraction or repulsion) between the molecules and the walls of the container. That is, molecule of a gas can be thought of as moving about freely unaware of the presence of other molecules. In other words, the molecules of an ideal gas possess only kinetic energy.

* **Standard conditions for temperature and pressure (STP)** refer to commonly agreed sets of conditions under which experimental measurements are reported. Internationally, two standards are in use: International Union of Pure and Applied Chemistry (IUPAC) and the National Institute of Standards and Technology (NIST). The IUPAC's standard refers to a temperature of 0 °C (273.15 K) and an absolute pressure of 100 kPa (0.986 atm), while NIST's standard is a temperature of 20 °C (293.15 K) and an absolute pressure of 101.325 kPa (1 atm).

6. The molecules of a gas behave as rigid, perfectly elastic hard spheres. It means that molecules neither lose energy nor deform in shape when they collide amongst themselves or with the walls of the container. (Its mechanical equivalence can be thought of as collisions between glass marbles.)
7. The duration of collision is negligible compared to the time interval between successive collisions.
8. All molecules do not move with the same speed. That is, there is a spread of molecular speeds ranging from zero to infinity*. An indirect justification for this assumption lies in the finite width of spectral lines.
9. The gravitational potential energy does not in any way affect the motion of gas molecules. This assumption is quite justified since the magnitude of gravitational force is 10^{-43} N, which is much less than the molecular force whose magnitude is about 10^{-13} N for normal separation between two molecules.

You now know the basic assumptions which provided the pedestal for development of kinetic theory of gases. Based on these assumptions, we will now obtain an expression for pressure exerted by the molecules of a gas on the walls of a container. This expression will then be used to obtain gas laws and some other useful results. The correctness of its various predictions put faith in the kinetic theory model in its formative years. However, some of these assumptions were modified in the light of experimental evidences and to conform our understanding of various phenomena to observed results.

1.3 Pressure Exerted by an Ideal Gas

Suppose that μ kilo-mole of a gas are confined in a cubical container of side L, as shown in Fig. 1.2a. If the total number of particles in the container is N, we can say that the number density $n = N/V$ molecules per m^3, where $V = L^3$ is volume of the cube. (We may mention that volume, not shape, of the container plays a role and we have chosen cubic geometry for mathematical convenience.) We assume that mass of each molecule is m. The gas molecules move randomly (Assumption 3) and at times these will collide with the walls of the container.

In general, the gas molecules move with different velocities (Assumption 8). But their number is large and to simplify mathematical derivation, we subdivide these into G energy groups $(1,2,3,\ldots G)$, each group having n_i $(i = 1,2,3,\ldots G)$ molecules per m^3 and moving with average velocity $\mathbf{v}_i$. In the absence of an external field, these molecules will exhibit complete chaos and collide with the walls of the container.

Let us first consider the motion of molecules in group 1 moving with velocity $\mathbf{v}_1$. We resolve $\mathbf{v}_1$ into three mutually orthogonal components v_{1x}, v_{1y} and v_{1z} along the x, y and z-axes and parallel to the sides of the cube, as shown in Fig. 1.2(b). Using the basic knowledge of vectors, you can write

$$v_1^2 = v_{1x}^2 + v_{1y}^2 + v_{1z}^2. \tag{1.1}$$

* Strictly speaking, at ordinary temperatures and pressures, neither of these limits is justified. The upper limit is not justified in view of the theory of relativity since no material particle can have velocity greater than the velocity of light. In any case, the relativistic effects are not important for molecular velocities. Similarly, a gas molecule does not cease to have motion until absolute zero. However, for mathematical convenience, we will keep the lower and upper limits as zero and infinity and there would not be much of a loss of generality.

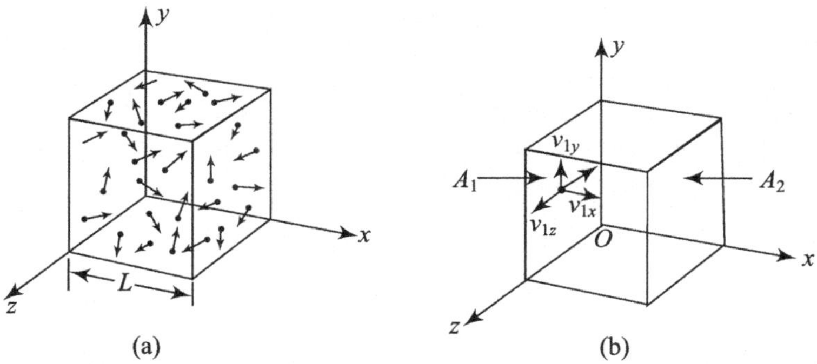

Fig. 1.2 (a) Velocity vectors of the molecules of a gas in a cubical container. The molecular motion is random and (b) Resolution of velocity of a molecule into components.

We now consider motion of the n_1 molecules in group 1 along any one of the three axes and the expression for pressure so obtained could then be generalised for other directions and groups of molecules. For simplicity, let us consider the motion of one of these molecules, which we identify as 'tagged' molecule, along the x-axis, which is normal to the faces A_1 and A_2 of the cube.

The (initial) momentum of the tagged molecule of group 1 at face A_1 moving along x-axis will be mv_{1x} normal to the face of the cube. It moves to the face A_2 at $x = L$ and makes an elastic collision. As a result, the molecule rebounds without any loss of momentum (Assumption 7). So, its momentum after the collision will be $-mv_{1x}$. (The negative sign signifies that the tagged molecule is moving in $-x$ direction.) Therefore, the change in momentum of the molecule at the face $x = L$ is $mv_{1x} - (-mv_{1x}) = 2\, mv_{1x}$. After rebounding, the tagged molecule will travel back to face A_1 and strike it again after travelling a distance $2L$ across the cube. That is, the tagged molecule covers a distance $2L$ before it strikes the same face of the cube again.

The time between any two successive collisions of the tagged molecule with face A_1 will be $\Delta t = 2L/v_{1x}$. Hence, the momentum imparted to face A_1 by a molecule per second along x-axis, i.e., the rate of change of momentum of a molecule is given by $2\dfrac{mv_{1x}}{\Delta t} = \dfrac{mv_{1x}^2}{L}$. The total momentum transferred per second by all the molecules of group 1 moving along x-axis will be $\dfrac{mn_1 v_{1x}^2}{L}$. You may now recall that the rate of change of momentum gives force (Newton's second law of motion). So we can say that the force exerted by all the molecules in group 1 on face A_1 while moving along the x-axis is given by $f_{1x} = \dfrac{mn_1 v_{1x}^2}{L}$.

The force per unit area defines pressure. Since group 1 molecules moving along the x-axis exert force f_{1x} over a surface area L^2, the expression for pressure exerted by all such group 1 molecules can be written as

$$p_{1x} = \frac{f_{1x}}{L^2} = mn_1 \frac{v_{1x}^2}{L^3}. \tag{1.2}$$

Let us pause for a while and reflect on the discussion so far. You will realise that we have almost achieved the goal we had set for ourselves. What remains to be done is to consider the motion of all other groups moving along the x-axis and generalise the results so obtained for molecules moving along the y- and z-directions.

The pressure communicated to the wall at $x = L$ by the molecules of groups 2,3,...G moving along the x-axis, by analogy, can be written as

$$p_{2x} = \frac{m}{L^3} n_2 v_{2x}^2$$

$$p_{3x} = \frac{m}{L^3} n_3 v_{3x}^2$$

$$\vdots$$

$$p_{Gx} = \frac{m}{L^3} n_G v_{Gx}^2$$

Hence, the total pressure exerted by all the molecules moving along the x-axis is

$$p_x = \sum_i p_{ix} = \frac{m}{L^3} \sum_{i=1}^{G} n_i v_{ix}^2 \ . \tag{1.3}$$

We now define the average value of v_x^2 as

$$\overline{v_x^2} = \frac{\sum_{i=1}^{G} n_i v_{ix}^2}{\sum_i^G n_i} = \frac{\sum_i^G n_i v_{ix}^2}{N} \ .$$

Here the bar indicates the average over the total number of molecules in the container. Using this result in Eq. (1.3), we can write

$$p_x = \frac{m}{L^3} N \overline{v_x^2} \tag{1.4}$$

Proceeding further, we note that when the gas is in equilibrium, the molecules move entirely randomly and all directions of motion are equally probable. Physically, it means that gas molecules do not have any preferred direction of motion (Assumption 4). So we can write

$$\overline{v_x^2} = \overline{v_y^2} = \overline{v_z^2} = \frac{1}{3}\overline{(v_x^2 + v_y^2 + v_z^2)} = \frac{\overline{v^2}}{3} , \tag{1.5}$$

where $\overline{v^2}$ is the mean square velocity of all the molecules in the gas,

$$\overline{v^2} = \frac{n_1 v_1^2 + n_2 v_2^2 + n_3 v_3^2 + \ldots + n_G v_G^2}{n_1 + n_2 + n_3 + \ldots + n_G} = \frac{\sum n_i v_i^2}{N}$$

On combining Eqs. (1.4) and (1.5), we can write*

$$p = \frac{1}{3} mn \overline{v^2} = \frac{1}{3V} mN \overline{v^2}$$

or

$$pV = \frac{1}{3} mN \overline{v^2} \tag{1.6}$$

where $n = N/V$ is number density and $V = L^3$ is volume of the container. This is an important result; it relates macroscopic properties (pressure p and volume V) with microscopic properties (mass m and mean square velocity $\overline{v^2}$) of individual molecules making up the system.

For one mole of a gas, N is Avogadro's number, N_A, and we can write

$$pV = \frac{1}{3} mN_A \overline{v^2} = \frac{1}{3} M \overline{v^2} \tag{1.7}$$

where M is the molecular weight of the gas. We know that 1 kmol of a substance has 6.023×10^{26} molecules and if a machine counts 10^9 molecules per second, it will take about 19 billion years to count them all. This is nearly three times the age of the earth.

We can rewrite Eq. (1.6) in a slightly different form as

$$p = \frac{1}{3} mn \overline{v^2} = \frac{1}{3} \rho \overline{v^2} \tag{1.8}$$

where n is number density and ρ is density of the gas.

If we define root mean square speed as

$$v_{\text{rms}} = \sqrt{\overline{v^2}} \tag{1.9}$$

the expression for pressure exerted by a gas on the walls of a container can also be written as

$$p = \frac{1}{3} mnv_{\text{rms}}^2 \tag{1.10}$$

Note that in this elementary derivation, intermolecular forces as well as finite size of gas molecules have been ignored. Before we give a more rigourous proof (Example 1.5) of Eq. (1.8), we would like you to go through the following example.

Example 1.1 One mole of a gas is contained in a cube of side 0.2 m. If these molecules, each of mass 5×10^{-26} kg, move with translational speed 483 ms^{-1}, calculate the pressure exerted by the gas on the sides of the cube.

Solution: The change in the momentum of the gaseous molecule between any two successive collisions with a wall of the container will be

$$\Delta p_x = 2mv_x = 2 \times (5 \times 10^{-26} \text{ kg}) \times (483 \text{ ms}^{-1}) = 4.83 \times 10^{-23} \text{ Ns}$$

* We have derived the expression for pressure exerted by a gas by considering a cubical container. Since only V occurs in Eq. (1.6), the shape of the container has no role in determining the magnitude of pressure exerted on it. It means that this result is valid for any type/shape of container.

The time interval between successive collisions on the same face

$$\Delta t = \frac{2L}{v_x} = \frac{2 \times 0.2\,\text{m}}{483\,\text{ms}^{-1}} = 8.3 \times 10^{-4}\,\text{s}$$

Hence, the rate of change of momentum of one molecule

$$\frac{\Delta p_x}{\Delta t} = \frac{4.83 \times 10^{-23}\,\text{Ns}}{8.3 \times 10^{-4}\,\text{s}} = 0.582 \times 10^{-19}\,\text{N}$$

Therefore, the total force exerted by all the molecules of the gas on a wall is

$$f_x = (0.582 \times 10^{-19}\,\text{N}) \times (6 \times 10^{23}) = 3.49 \times 10^4\,\text{N}$$

Hence, average pressure exerted by all the molecules of the gas on the walls of the container

$$p = \frac{3.49 \times 10^4\,\text{N}}{3 \times 4 \times 10^{-2}\,\text{m}^2} = 2.9 \times 10^5\,\text{Nm}^{-2}$$

You will recall that one atmospheric pressure is nearly 1×10^5 Nm^{-2}. So one mole of the gas contained in a cube of side 0.2 m exerts pressure nearly 3 atm on its sides. Yet we do not observe any change in the shape of a container.

You now know that though we started from a purely mechanical picture of a gas as a collection of randomly moving molecules, we have obtained an expression for pressure in terms of the microscopic properties of individual molecules. Let us now discover kinetic interpretation of temperature and learn to use Eq. (1.8) to deduce gas laws.

1.3.1 Kinetic Interpretation of Temperature

To seek kinetic interpretation of temperature, we first rewrite Eq. (1.8) as

$$pV = \frac{2}{3}N\left(\frac{1}{2}m\overline{v^2}\right) = \frac{2}{3}N\varepsilon, \tag{1.11}$$

where $\varepsilon = \frac{1}{2}m\overline{v^2}$ signifies the average kinetic energy of a molecule.

It is instructive to compare this equation with the ideal gas equation:

$$pV = \mu RT \tag{1.11a}$$

where $\mu = \frac{N}{N_A}$ is the number of kilo-mole of the gas, T is absolute temperature and R is *kilo-molar gas constant*. Its value is 8314 Jkmol^{-1}K^{-1} = 1.987 kcal kmol^{-1} K^{-1}. N_A denotes Avogadro number and its presently accepted value is 6.023×10^{26} molecules per kilo-mole.

Using Eqs. (1.11) and (1.11a), we can connect the average kinetic energy of a molecule to temperature through the relation

$$\frac{2}{3}\mu N_A \varepsilon = \mu RT$$

so that

$$\varepsilon = \frac{3}{2}\frac{R}{N_A}T = \frac{3}{2}k_B T, \tag{1.12}$$

where $k_B = R/N_A$ is *Boltzmann constant*. Its value is 1.38×10^{-23} JK^{-1}. You can easily convince yourself that kinetic energy of a molecule will be 6.21×10^{-21} J (= 0.04 eV) at 300 K and 1.6×10^{-19} J at 7730 K. Note that Eq. (1.12) relates a purely mechanical quantity—the average kinetic energy of a molecule—to temperature. This is a big step as it connects molecular and macroscopic viewpoints through Boltzmann constant. (In Chapter 12, you will learn that Boltzmann constant appears in the relation that bridges the statistical and the thermodynamic viewpoints for a system in equilibrium). Equation (1.12) assigns a completely new and deeper meaning to temperature. In particular, we note that

1. Temperature is linearly proportional to average (kinetic) energy of molecules.
2. At a given temperature, the (kinetic) energy of the molecules of all gases, irrespective of the differences in their masses, will be constant and the same. It implies that motion of gas molecules will become more lively and vigorous at higher temperatures and vice versa.
3. As mentioned earlier, at 300 K, $\varepsilon \approx 6 \times 10^{-21}$ J or 0.04 eV. This energy is too small to increase the internal energy of colliding molecules, even if all this could be absorbed in a head-on collision between them. This justifies the assumption that gas molecules behave as perfect hard spheres without any internal structure.
4. A*t absolute zero of temperature, the gas molecules will be devoid of all motion***.** So kinetic theory predicts that at absolute zero, all molecules will behave as if they were frozen in space (In Chapter 14 on Fermi-Dirac statistics, you will learn that due to quantum effects, the energy is finite even at absolute zero for some typical systems. This is known as *zero-point energy.*)

To test how far the model of gaseous state discussed in this section is reliable, let us use it to calculate the values of molecular speeds of different gases and compare these values with measured values. Let us discover the extent of agreement and usefulness of the model.

1.3.2 Root Mean Square Speed

To know how fast the molecules in a gas move, we rewrite Eq. (1.12) as

$$\frac{1}{2} m \overline{v^2} = \frac{3}{2} k_B T = \frac{3}{2} \frac{R}{N_A} T$$

so that

$$\overline{v^2} = \frac{3 k_B T}{m} = \frac{3 RT}{M}$$

where $M = mN_A$ is the molecular weight of the gas. It is important to mention here that when the molecules of a gas are in random motion, the velocity of a molecule averages out to zero. So we calculate the square of the velocity, which is non-zero, take its average and then obtain the square root. This gives root mean square speed.

The square root of mean square velocity is usually denoted as v_{rms} and is called *root mean square speed*:

$$v_{rms} = \sqrt{\overline{v^2}} = \sqrt{\frac{3 RT}{M}} = \sqrt{\frac{3 k_B T}{m}} \tag{1.13}$$

Note that v_{rms} is directly proportional to the square root of temperature of a gas. It means that the r.m.s. speed of a gas at 68.2 K will be one-half of its value at 273 K. At what temperature will r.m.s. speed of a gas be one-fourth of its value at 0°C? You should be able to compute it orally as ≈ 17 K.

The elementary kinetic theory further predicts that at a given temperature, r.m.s. speed of a gas molecule is inversely proportional to the square root of its molecular weight. It means that molecules of a lighter gas will move faster. For a few typical gases in our atmosphere, the values of v_{rms} at 300 K are given in Table 1.1. As may be noted, hydrogen molecules move faster than any other gas. This result has an interesting consequence for existence of life on our planet. Gases like hydrogen, helium, nitrogen and water vapour move upward in earth's atmosphere, whereas oxygen is available near earth's surface. Further, in your school mechanics classes, you have learnt that if an object acquires radially outward velocity greater than 11.2 km s^{-1}, it will escape from the influence of gravity of earth. ($v_e = \sqrt{2gR}$, where R is radius of the earth.) Since no gas molecule has speed greater than v_e, there is little chance of their escaping from the earth instantaneously. For Moon, Jupiter and the Sun, the values of escape velocity are 2.3 km s^{-1}, 60 km s^{-1} and 600 km s^{-1}, respectively. This explains why over a period of time, all gases have escaped from the surface of Moon, whereas even hydrogen is held back in the core of the Sun.

Table 1.1 *Values of v_{rms} for different gases at* 300 K

Gas	$v_{rms} = \sqrt{\frac{3k_BT}{m}}$ (ms^{-1})	Gas	$v_{rms} = \sqrt{\frac{3k_BT}{m}}$ (ms^{-1})
H_2	1934	O_2	483
He	1367	Ar	433
H_2O	615	CO_2	412
N_2	517		

We can also predict the presence of dust particles and pollutants such as carbon monoxide and oxides of nitrogen in atmospheric air, as also the greenhouse gases. These gases are emitted by factories which burn coal, vehicular/air traffic, air conditioners and refrigerators in our atmosphere. This highlights why emission standards for vehicles are being reviewed periodically and why we should get our vehicles regularly checked for pollution. In the upper layers of the atmosphere, these gases are breaking up ozone, which is so vital for sustaining life on this planet. (Ozone absorbs ultraviolet radiations generated in the outer space and stops these from entering the biosphere.)

In the early days of the formulation of kinetic theory, correct prediction of vertical distribution of gases in the atmosphere provided indirect but sound evidence in its favour and boosted the confidence of physicists in it. Another step in the positive direction was the deduction of various gas laws from the expression for pressure. Before you learn about these, we would like you to solve the following practise problem.

Problem 1.1 Dust particles suspended in a monatomic gas are in equilibrium with the gas at 300 K. If the mass of a suspended particle is 10^{-27} kg, calculate v_{rms}.

Ans: 3.5×10^3 ms^{-1}

1.3.3 Some Elementary Deductions from Kinetic Theory

Boyle's Law From Eq. (1.11), we recall that for a given mass of a gas

$$pV = \frac{2}{3} N \varepsilon$$

Since kinetic energy depends only on temperature, the right-hand side in the above relation will remain constant at a fixed temperature. Hence, we can write

$$pV = \text{constant} \tag{1.14}$$

Thus, the *pressure exerted by a given mass of a gas varies inversely with its volume, when temperature remains constant.* This is Boyle's law.

Gay-Lussac's Law From Eq. (1.14), we can also conclude that *at constant volume, pressure exerted by a given mass of a gas is directly proportional to its kinetic energy, i.e., temperature.* This is Gay–Lussac's law.

Charle's Law When pressure remains constant, Eq. (1.11) implies that *volume of a given mass of a gas increases linearly with kinetic energy, i.e., temperature.* This is Charle's law. These laws are graphically depicted in Fig. 1.3.

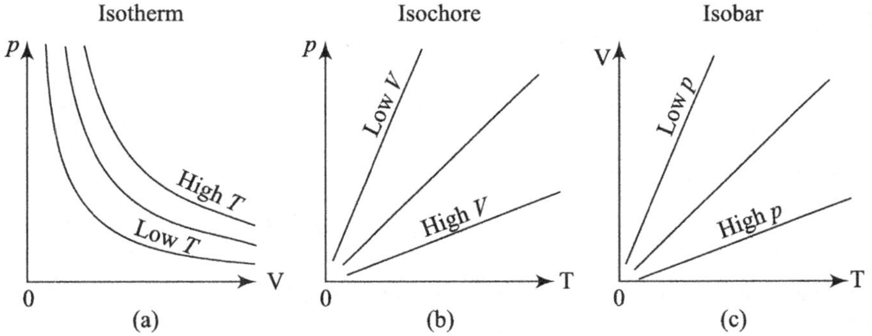

Fig. 1.3 Graphical depiction of (a) Boyle's law, (b) Gay–Lussac's law and (c) Charle's law.

We may mention that these laws were originally formulated based on experimental observations and are strictly valid only for ideal gases.

Avogadro's Law Avogadro's law states that *at constant temperature and pressure, equal volume of all gases contains the same number of molecules.* Let us consider two different gases at the same temperature and pressure. Then, from Eq. (1.8), we can write

$$p = \frac{1}{3} m_1 n_1 \overline{v_1^2} = \frac{1}{3} m_2 n_2 \overline{v_2^2}, \tag{1.15}$$

where $\overline{v_1^2}$ and $\overline{v_2^2}$ are respectively the mean squared speeds of the molecules of two gases. Recall that at constant temperature, the mean molecular energies of these gases will be equal. So we can write

$$\frac{1}{2} m_1 \overline{v_1^2} = \frac{1}{2} m_2 \overline{v_2^2}.$$

On combining this result with Eq. (1.15), we get

$$n_1 = n_2. \tag{1.16}$$

This is the mathematical statement of Avogadro's law. It may be pointed out here that Avogadro put forward this hypothesis while studying weights and proportions. So we can say that it had nothing to do with the motion of molecules. But the fact that the same result had been obtained from the theory of molecular motion proved a significant success for kinetic theory.

Before proceeding further, you should answer the following practise problem.

Problem 1.2 Graham's law of diffusion states that the rate of diffusion of a gas is inversely proportional to the square root of its density. Deduce this law from Eq. (1.8).

To get a feel of these results, you should go through the following example carefully.

Example 1.2 The number density of air at 1 atm = 1.013×10^5 Nm^{-2} and $T = 300$ K is 2.7×10^{25} m^{-3}. Calculate the number density at 1.33×10^{-4} Nm^{-2}, provided the temperature is kept constant.

Solution: From Eq. (1.8), we recall that

$$p = \frac{1}{3} mn \overline{v^2}$$

so that we can write

$$p_1 = \frac{1}{3} mn_1 \overline{v^2} \text{ and } p_2 = \frac{1}{3} mn_2 \overline{v^2}$$

Since temperature is kept constant, the value of mean square speed remains unchanged. Hence, we can write

$$n_2 = \frac{p_2 \times n_1}{p_1}$$

On inserting the given values, we get

$$n_2 = \frac{(1.33 \times 10^{-4} \text{ Nm}^{-2}) \times (2.7 \times 10^{25} \text{ m}^{-3})}{(1.013 \times 10^5 \text{ Nm}^{-2})} = 3.57 \times 10^{16} \text{ m}^{-3}$$

The elegance with which elementary kinetic theory was used to explain vertical distribution of gases in our atmosphere as well as the gas laws proved an important milestone for its further development. Also, its aesthetic appeal in that the motion of extremely large number of molecules can be described in terms of basic laws put faith in its profundity. In its progress, this theory was put to litmus test when its predictions about heat capacities of monatomic and diatomic gases were verified experimentally. We now discuss classical theory of heat capacities in some detail.

1.4 Classical Theory of Heat Capacities of Gases

When no external field is present, an ideal gas possesses only kinetic energy; there is no potential energy. For one kilo-mole of a gas, the total kinetic energy is given by

$$U = N_A \varepsilon = \frac{3}{2} N_A k_B T = \frac{3}{2} RT \tag{1.17}$$

This result implies that molar heat capacity, defined as the energy required to raise the temperature of one kilo-mole of an ideal gas by one kelvin, at constant volume is given by

$$C_V = \left(\frac{\partial U}{\partial T}\right)_V = \frac{3}{2} R = 2.98 \text{ kcal kmol}^{-1} \text{ K}^{-1} \tag{1.18}$$

Hence, molar heat capacity at constant pressure is given by

$$C_p = C_V + R = \frac{5}{2} R = 4.87 \text{ kcal kmol}^{-1} \text{ K}^{-1} \tag{1.19}$$

and the ratio of molar heat capacity at constant pressure to molar heat capacity at constant volume is

$$\gamma = \frac{C_p}{C_V} = \frac{5}{3} = 1.67 \tag{1.20}$$

Let us pause for a while and reflect on these results. We can conclude that *the molar heat capacities at constant pressure and at constant volume as also their ratio are the same and independent of temperature for all gases*. These predictions of elementary kinetic theory agree rather well with experimental results for monatomic gases (Table 1.2). However, for diatomic and polyatomic gases, we note that

1. The ratio of molar heat capacities decreases with increasing atomicity.
2. There are significant deviations between theoretical and measured values of molar heat capacities at constant volume.

Table 1.2 *kilo-molar heat capacities at constant volume for a few common gases at room temperature*

Gas	C_V (kcal kmol^{-1} K^{-1})	γ
Ar	2.98	1.67
He	2.98	1.66
H_2	4.88	1.41
O_2	5.03	1.401
N_2	4.96	1.404
Cl_2	6.15	1.360
CO_2	6.80	1.304
NH_3	6.65	1.310

To explain these differences between theoretical and experimental values, we need to introduce the concept of the degrees of freedom of the molecules of a gas.

Degrees of Freedom *The degree of freedom (d.f.) of a molecule is defined as the number of independent coordinates required to specify its position completely.* A point moving along a curved path has only one d.f, though two coordinates are required to specify its motion. This is because only one coordinate is independent; they have to satisfy the equation of the curve. For example, for every point in the path of an oscillating simple pendulum, the equation $x^2 + y^2 = \ell^2$ is satisfied. It means that x and y are not independent. Such an equation constitutes an equation of constraint. Similarly, an ant moving on a stretched string has only one d.f. as we can completely specify its motion/position with only one coordinate. However, when an ant moves on the floor, its degrees of freedom increase to two. And when a housefly flies, its motion is three-dimensional and we require three coordinates (x, y, z) to describe it. So, the number of degrees of freedom will be three. How many d.f. a randomly moving molecule of monatomic gas (like helium, argon, or krypton) has? If you are thinking that it has three translational degrees of freedom, you have visualised its motion correctly.

In addition to translational degree of freedom, a diatomic or a polyatomic molecule has a tendency to rotate (about fixed axes). Since the angular velocity of a rotating body (molecule in the instant case) can be resolved along three mutually perpendicular coordinate axes, you may expect a rigid diatomic molecule (H_2, O_2, N_2 ...) to have three rotational degrees of freedom. But the moment of inertia of a diatomic molecule about an axis along or parallel to the axis of the molecule (i.e., line joining the atoms) is very small and no rotation of the molecule as a whole is possible about it. So a linear diatomic molecule, in general, has only two rotational degrees of freedom. That is, a linear diatomic molecule will have a total of five degrees of freedom. (A non-linear molecule like H_2O has three rotational degrees of freedom.)

The number of d.f (f) of a molecule can be obtained from the general result

$$f = 3p - c, \tag{1.21}$$

where p is the number of particles constituting the system and c is the total number of constraints.

For a monatomic gas molecule, $p = 1$, and $c = 0$ so that $f = 3$. That is, a monatomic gas molecule has three degrees of freedom. These correspond to translational motion along three mutually perpendicular directions.

For a diatomic molecule, $p = 2$ and $c = 1$ as the distance between the atoms is fixed. So we have $f = 5$.

The number of degrees of freedom can also be defined as the total number of independent squared terms appearing in the expression of energy of a system. The general expression for the internal energy U of a gas may contain four separate terms corresponding to translational, rotational, vibrational and potential energies of the molecules. This is depicted in Fig. 1.4. In the language of mathematics, we can write

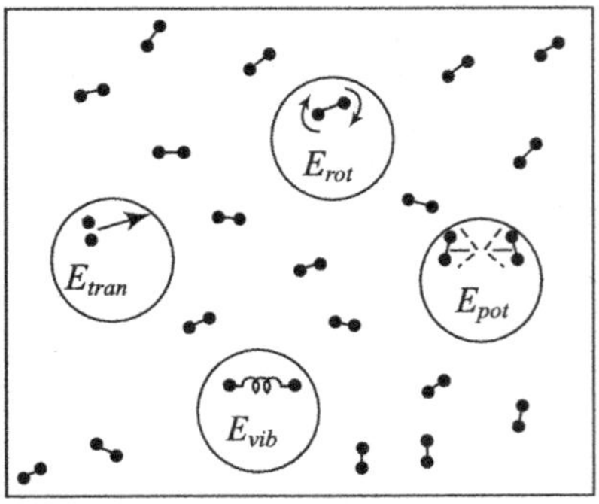

Fig. 1.4 Internal energy of a gaseous molecule.

$$U = E_{\text{trans}} + E_{\text{rot}} + E_{\text{vib}} + E_{\text{pot}} \tag{1.22}$$

The total energy of a monatomic gas is given by

$$E_{\text{trans}} = \frac{1}{2}m\left[\left(\frac{dx}{dt}\right)^2 + \left(\frac{dy}{dt}\right)^2 + \left(\frac{dz}{dt}\right)^2\right] \tag{1.23}$$

Note that energy of a monatomic gas molecule is only translational and accordingly there are three squared terms in Eq. (1.23) so that $f = 3$. This is consistent with Eq. (1.21).

Since a diatomic molecule can execute translational as well as rotational motion, there will be two additional terms to signify rotational motion about two mutually perpendicular axes:

$$E_{\text{rot}} = \frac{I_1 \omega_1^2}{2} + \frac{I_2 \omega_2^2}{2}, \tag{1.24}$$

where I_1 and I_2 are the moments of inertia about these axes and ω_1 and ω_2 are the corresponding components of angular velocity. Note that there cannot be any rotation about the line joining the atoms. So a diatomic molecule will have five degrees of freedom in all; $f = 5$. In analogy with monatomic molecule, we find that the internal energy of a diatomic molecule is $\frac{5}{2}RT$. Hence, for a diatomic molecule, molar heat capacity at constant volume and at constant pressure are given by

$$C_V = \frac{5}{2}R = 4.87 \text{ kcal kmol}^{-1} \text{ K}^{-1}. \tag{1.25a}$$

$$C_p = C_V + R = \frac{7}{2}R = 6.85 \text{ kcal kmol}^{-1} \text{ K}^{-1}. \tag{1.25b}$$

and the ratio of molar heat capacities for a diatomic molecule is

$$\gamma = \frac{C_p}{C_V} = \frac{7}{5} = 1.40. \tag{1.25c}$$

As may be noted, Eqs. (1.25a) and (1.25c) are in better agreement with observed results for nitrogen and oxygen.

The first sign of disagreement between theory and experiments came to the knowledge when γ was used to determine the number of degrees of freedom and the result was not, in general, an exact integer. Now refer to Fig. 1.5. It shows the temperature variation of heat capacity of hydrogen and chlorine. As may be noted, the measured value of C_V for hydrogen decreases to $\frac{3}{2}R$ at 20 K, whereas it increases to $\frac{5}{2}R$ at room temperature. Beyond 1000 K, it increases further. Physically we can say that at 20 K only translational modes are present whereas rotational modes begin to contribute at room temperature. At very high temperatures, vibrational modes are also excited and begin to contribute to the energy of hydrogen molecule. On the other hand, for chlorine, vibrational modes begin to contribute around 600 K.

Principle of Equipartition of Energy The general dependence of γ on f can be obtained from the *principle of equipartition of energy*. It states that *for a system in equilibrium at temperature T, the total energy is equally partitioned among the several*

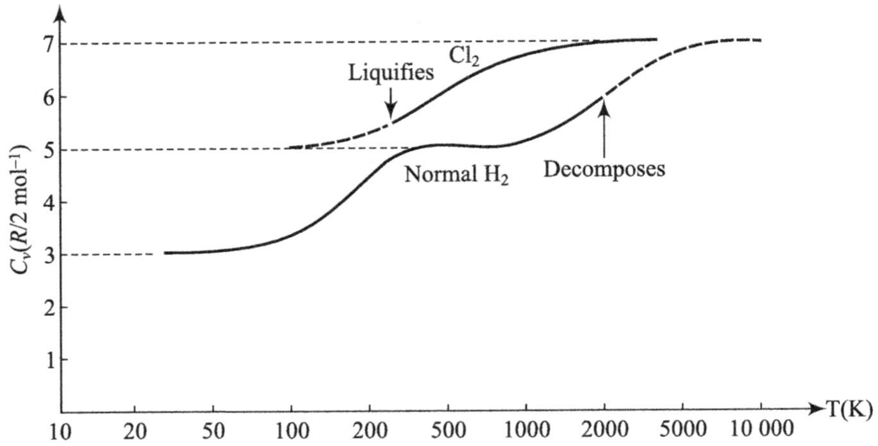

Fig. 1.5 Temperature variation of heat capacity of hydrogen and chlorine.

degrees of freedom and energy associated with each d.f. is equal to $\frac{k_B T}{2}$. You will learn the general proof of this result in Chapter 12. However, from Sec. 1.3.1, you may recall that translational kinetic energy of a molecule of a monatomic gas is given by

$$\varepsilon = \frac{1}{2} m \overline{v^2} = \frac{3}{2} k_B T$$

We also know that molecular motion is completely random. So when an isotropic distribution is in equilibrium, all three directions are equivalent and translational kinetic energy associated with each component of velocity of an ideal (perfect) gas molecule is one-third of its total translational kinetic energy, which is equal to $k_B T/2$. In other words, we can say that *energy is equally partitioned among the three components of velocity*. This is an important result of classical physics. In Sec. 1.4, we will show that it readily follows from Maxwell's velocity distribution law.

According to the principle of equipartition of energy, the internal energy of a system per kmol is given by

$$U = \frac{f}{2} k_B T N_A = \frac{f}{2} RT$$

so that

$$C_V = \frac{f}{2} R; C_p = C_V + R = R\left(\frac{f}{2} + 1\right)$$

and

$$\gamma = 1 + \frac{2}{f} \tag{1.26a}$$

You can convince yourself that Eqs. (1.20) and (1.25c) for monatomic and diatomic gases readily follow from this general result.

Since f is an indicator of atomicity of a gas, we can rearrange Eq. (1.26a) to obtain

$$f = \frac{2}{\gamma - 1} \tag{1.26b}$$

Now refer to Fig. 1.6, which shows a molecule of ammonia (NH_3), a polyatomic gas. In this case, E_{trans} and E_{rot} will contain three terms each, whereas E_{vib} will have six terms of the form $\frac{k\zeta^2}{2}$ corresponding to the possible vibrational modes along the lines joining the atoms. So if $f = 12$, $C_V = 6R$, $C_p = 7R$ and $\gamma = 1.17$, which differs significantly from the experimental value. However, if we assume that vibrational modes are not excited at room temperature, we can choose $f = 6$. This gives $\gamma = 1.33$, which agrees with the experimental value.

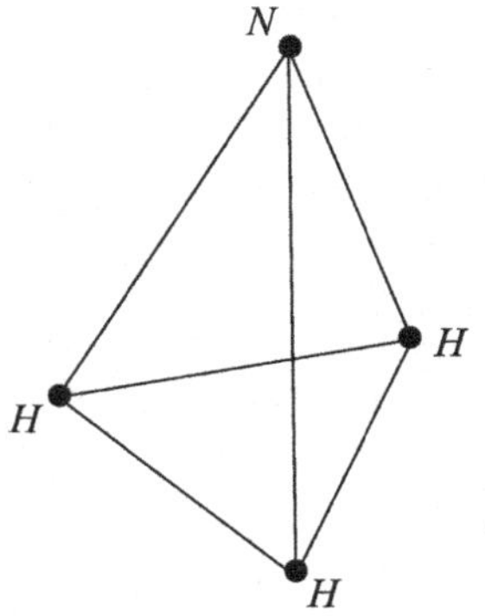

Fig. 1.6 An ammonia molecule.

The explanation of heat capacities of gases established the kinetic theory. However, its failure to explain the observed temperature dependence of heat capacity of diatomic gases presented a really serious challenge. A successful explanation of temperature variation of heat capacities of diatomic gases could be provided on the basis of quantum theories. You will learn these in Chapter 13.

So far we have referred to the root mean square speed of the molecules in a gas. We refrained from saying anything about the actual distribution of velocities among the molecules. In a gas, constant random motion of molecules makes them to collide against each other as well as against the walls of the container. This results in a continuous change both in magnitude and direction of their velocities. It implies that in a real system, even if all the molecules have the same velocity at a given time, we should expect that molecular collisions will result in a wide dispersion of molecular velocities at some later time. So the problem we wish to address now is: How many molecules will have velocities within a certain range? It is also interesting to know how this number changes with velocity, temperature and/or pressure. The correct answer to this more sophisticated problem was provided by Maxwell in his characteristically novel and profound work published in 1860.

Historically, while working out the solution, Maxwell introduced statistical concepts in kinetic theory of gases. His theory marked the beginning of a new era in physics; it formed the basis of modern theory of statistical mechanics and led to statistical interpretation of the laws of thermodynamics. Using the theory of probability, Boltzmann put Maxwellian theory on firm mathematical foundations. For this reason, distribution function for molecular velocities is also referred to as *Maxwell–Boltzmann distribution function*. We will now discuss it in some detail.

1.5 Distribution of Molecular Velocities in a Perfect Gas: Maxwell–Boltzmann Distribution Law

The derivation of the distribution law is based on a number of assumptions. These are in addition to the basic assumptions of elementary kinetic theory (Sec. 1.2). The assumptions are as follows:

Assumptions

1. In the equilibrium state, the molecules have complete randomness of direction and velocity.
2. There is no mass motion or convection current in the body of the gas. As long as there is no net force (or pressure gradient) acting on the system, this assumption is quite justified.

3. The probability of a molecule having x-velocity component, say after a large number of collisions, does not depend on the y- and z-velocity components. This assumption follows from random motion of gas molecules.
4. The probability that a molecule selected at random has velocity component in the given range is a function only of the magnitude of velocity component and the width of the interval.
5. The gas molecules have no vibrational or rotational energies. This assumption is quite justified because vibrational and rotational energies remain unchanged when gas molecules undergo collisions.

Derivation Consider a gas having N molecules enclosed in a vessel of arbitrary shape and moving randomly. To each molecule, we attach a vector, which represents its velocity in magnitude and direction as shown in Fig. 1.7a. We then transfer these vectors (not the molecules) to a common origin (Fig. 1.7b). In doing so, we use the property that *a vector remains unchanged when it is translated parallel to itself.* We represent these molecules on a velocity diagram with Ox, Oy and Oz as the coordinate axes as shown in Fig. 1.7c. Here $d\mathbf{v}$ is an infinitesimal volume element.

Note that each velocity vector will be defined by the coordinates of its end point. Let us denote the x, y and z components of the velocity $\mathbf{v}$ of a particle by v_x, v_y and v_z. Therefore, in terms of these three rectilinear components, we can write

$$v^2 = v_x^2 + v_y^2 + v_z^2 \tag{1.27}$$

The number of velocity vectors ending in element $d\mathbf{v} = dv_x\, dv_y\, dv_z$ gives the average number of molecules whose velocities lie between given limits $\mathbf{v}$ and $\mathbf{v} + d\mathbf{v}$ after a large number of collisions among identical molecules. It means that we have to calculate the number of molecules simultaneously having velocity components in the range v_x to $v_x + dv_x$, v_y to $v_y + dv_y$ and v_z to $v_z + dv_z$. Assumption (4) stated above implies that the fraction of molecules having velocity components in the range v_x to $v_x + dv_x$ is equal to $f(v_x)\, dv_x$. That is

$$\frac{dN_{v_x}}{N} = f(v_x)\,dv_x \quad \text{or} \quad dN_{v_x} = N f(v_x)\,dv_x \tag{1.28}$$

where dN_{v_x} is the number of molecules having velocity components in the range v_x and $v_x + dv_x$, N is the total number of molecules and f is an unknown function, which we have to

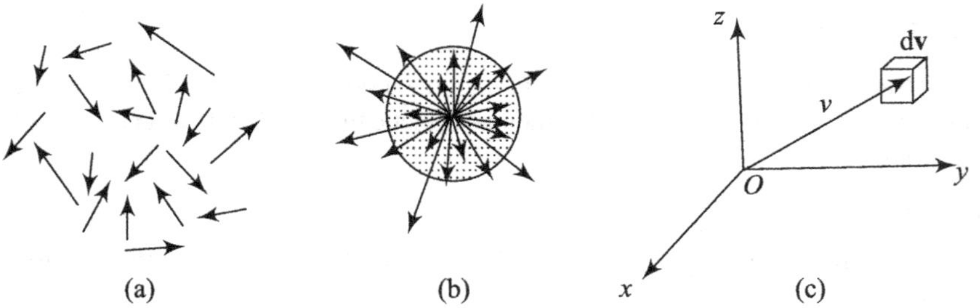

Fig. 1.7 (a) Gas molecules in random motion, (b) All velocity vectors transferred to a common origin and (c) A volume element $d\mathbf{v}$ in velocity space for a molecule of velocity $\mathbf{v}$.

determine. Mathematically, the ratio $\frac{dN_{v_x}}{N}$ denotes the probability* of finding a molecule with x-component of velocity in the range v_x to $v_x + dv_x$. Let us denote it by the symbol p_x.

Proceeding further, we note that existence of velocity component v_x does not in any way affect velocity components v_y and v_z, since these are mutually perpendicular and independent of each other. So we can write the expression for the probability that a molecule having velocity components in the range v_y to $v_y + dv_y$ and v_z to $v_z + dv_z$ as

$$p_y = f(v_y)\, dv_y \tag{1.29}$$

and

$$p_z = f(v_z)\, dv_z \tag{1.30}$$

Note that we have taken the same functional dependence in all three cases. It signifies that there is no preferred direction of motion of gas molecules.

Since the three perpendicular components of velocity are independent, we can express the probability for a molecule to simultaneously have velocity components in the range v_x to $v_x + dv_x$, v_y to $v_y + dv_y$ and v_z to $v_z + dv_z$ using the law of compound probabilities (According to the law of compound probabilities, the composite probability for independent events is equal to the product of the probabilities of individual events.) This gives

$$\frac{d^3 N_{v_x v_y v_z}}{N} = p_x p_y p_z = f(v_x)\, f(v_y)\, f(v_z)\, dv_x\, dv_y\, dv_z \tag{1.31}$$

Hence, the number of molecules simultaneously having velocity components lying between v_x to $v_x + dv_x$, v_y to $v_y + dv_y$ and v_z to $v_z + dv_z$ is

$$d^3 N_{v_x v_y v_z} = N f(v_x)\, f(v_y)\, f(v_z)\, dv_x\, dv_y\, dv_z \tag{1.32}$$

Note that all these $d^3N_{v_x v_y v_z}$ molecules lie in the small volume element $dv_x\, dv_y\, dv_z$. In Fig. 1.7c, these molecules are depicted as velocity points (A point representing a molecule with velocity components in the three coordinate directions is called *velocity point.*) Therefore, the density of velocity points, i.e., number density of gas molecules can be expressed as

$$\rho = \frac{d^3 N_{v_x v_y v_z}}{dv_x\, dv_y\, dv_z} \tag{1.33}$$

On combining Eqs. (1.32) and (1.33), we can write

$$\rho = Nf(v_x)\, f(v_y)\, f(v_z)$$

Since the velocity space has been assumed to be isotropic, the density of velocity points can be taken to be independent of inclination of **v** to the axes. So we can write

$$Nf(v_x)\, f(v_y)\, f(v_z) = \text{constant} = NF(\mathbf{v}) = NJ(v^2), \tag{1.34}$$

where F and J are some other function. Note that this equation holds for a fixed value of **v**, i.e., it is subject to the condition

$$v^2 = v_x^2 + v_y^2 + v_z^2 = \text{constant} \tag{1.35}$$

---

* The probability of an event is the ratio of the number of outcomes favourable to the event to the total number of outcomes. For example, the probability of getting a head in tossing a coin is one-half.

Physically, it means that after a large number of collisions, the distribution will be isotropic. Therefore, for fixed **v**, $J(v^2)$ is constant and $d[J(v^2)] = 0$.
In terms of dv_x, dv_y and dv_z, we can write

$$\frac{\partial f(v_x)}{\partial v_x} dv_x f(v_y) f(v_z) + f(v_x) \frac{\partial f(v_y)}{\partial v_y} f(v_z) dv_y + f(v_x) f(v_y) \frac{\partial f(v_z)}{\partial v_z} dv_z = 0 \quad (1.36)$$

On dividing Eq. (1.36) by $f(v_x) f(v_y) f(v_z)$, we obtain

$$\frac{1}{f(v_x)} \frac{\partial f(v_x)}{\partial v_x} dv_x + \frac{1}{f(v_y)} \frac{\partial f(v_y)}{\partial v_y} dv_y + \frac{1}{f(v_z)} \frac{\partial f(v_z)}{\partial v_z} dv_z = 0 \quad (1.37)$$

The differential form of Eq. (1.35), which expresses the condition under which v_x, v_y and v_z can vary while v^2 remains constant is

$$v_x dv_x + v_y dv_y + v_z dv_z = 0 \quad (1.38)$$

From this equation, it is clear that the differentials dv_x, dv_y and dv_z are not mutually independent; these can take any value but must satisfy Eq. (1.38). To relax this constraint, we use Lagrange's method of undetermined multipliers. In this method, the constraining relation is multiplied by a constant and the resultant expression is added to the constrained equation. In this case, we choose the undetermined multiplier to be $2B$. Accordingly, on multiplying Eq. (1.38) by $2B$ and adding the resultant expression to Eq. (1.37), we get

$$\left[\frac{1}{f(v_x)} \frac{\partial f(v_x)}{\partial v_x} + 2Bv_x\right] dv_x + \left[\frac{1}{f(v_y)} \frac{\partial f(v_y)}{\partial v_y} + 2Bv_y\right] dv_y$$

$$+ \left[\frac{1}{f(v_z)} \frac{\partial f(v_z)}{\partial v_z} + 2Bv_z\right] dv_z = 0 \quad (1.39)$$

Let us choose the constant B such that

$$\frac{1}{f(v_x)} \frac{\partial f(v_x)}{\partial v_x} + 2Bv_x = 0 \quad (1.40)$$

Then Eq. (1.39) reduces to

$$\left[\frac{1}{f(v_y)} \frac{\partial f(v_y)}{\partial v_y} + 2Bv_y\right] dv_y + \left[\frac{1}{f(v_z)} \frac{\partial f(v_z)}{\partial v_z} + 2Bv_z\right] dv_z = 0 \quad (1.41)$$

Of the three variables dv_x, dv_y and dv_z, we can consider any two variables as constant. So let us take dv_y and dv_z to be independent. Then, for finite values of dv_y and dv_z, Eq. (1.39) will be satisfied if the coefficients of these differentials in Eq. (1.41) vanish separately. That is,

$$\frac{1}{f(v_y)} \frac{\partial f(v_y)}{\partial v_y} + 2Bv_y = 0 \quad (1.42)$$

and

$$\frac{1}{f(v_z)}\frac{\partial f(v_z)}{\partial v_z} + 2Bv_z = 0 \tag{1.43}$$

To proceed further, we rewrite Eq. (1.40) as

$$\frac{df(v_x)}{f(v_x)} = -2Bv_x \, dv_x$$

This equation can be easily integrated to obtain

$$\ln f(v_x) = -\frac{2Bv_x^2}{2} + \ln A$$

where ln A is an integration constant.

We can rewrite this result as

$$f(v_x) = A \exp[-Bv_x^2] \tag{1.44}$$

where A and B are unknown constants. Note that the Lagrange's method of undetermined multipliers has helped us to discover the form of the function f; it is a decreasing exponential. But now we will have to deal with two unknown constants (A and B) rather than one unknown function (f).

On combining this result with Eq. (1.28), we can express the number of molecules having velocity components in the range v_x to $v_x + dv_x$ as

$$dN_{v_x} = NA \exp(-Bv_x^2)\, dv_x \tag{1.45a}$$

Proceeding further, we note that the differential equations satisfied by $f(v_y)$ and $f(v_z)$ are similar to that satisfied by $f(v_x)$, and by analogy, we can write

$$f(v_y) = Ae^{-Bv_y^2} \tag{1.45b}$$

and

$$f(v_z) = Ae^{-Bv_z^2} \tag{1.45c}$$

On combining Eqs. (1.45a), (1.45b) and (1.45c) we obtain the required expression for $d^3 N_{v_x v_y v_z}$:

$$d^3 N_{v_x v_y v_z} = NA^3 e^{-B\left(v_x^2 + v_y^2 + v_z^2\right)} dv_x \, dv_y \, dv_z = NA^3 e^{-Bv^2} dv_x \, dv_y \, dv_z \tag{1.46}$$

and number density

$$\rho = NA^3 \exp[-B(v_x^2 + v_y^2 + v_z^2)] = NA^3 \exp[-Bv^2] \tag{1.47}$$

This is the required expression for the *Maxwell's velocity distribution function.* In Fig. 1.8, we have plotted ρ versus Bv^2. As may be noted, the molecular number density decreases exponentially with velocity and its maximum value is NA^3, which corresponds to $v = 0$.

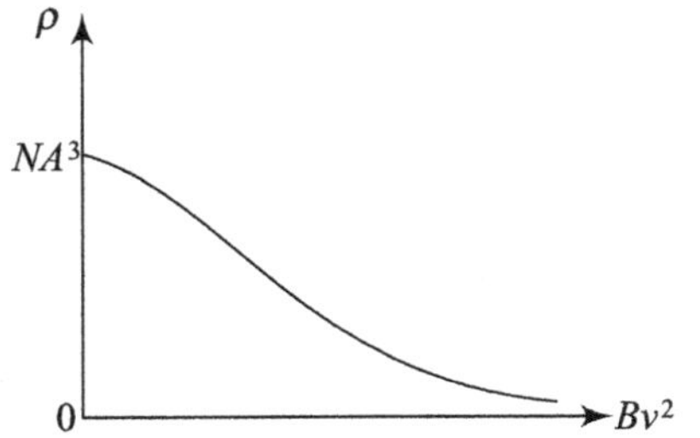

Fig. 1.8 Plot of Maxwell's velocity distribution function versus Bv^2.

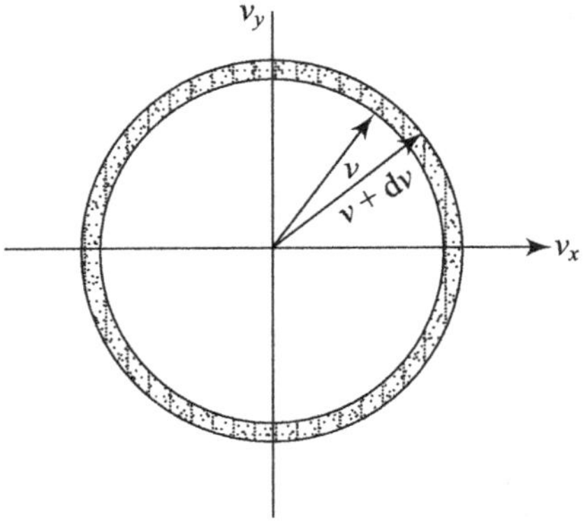

Fig. 1.9 A spherical shell of radius v and

1.5.1 Molecular Distribution of Speeds

Usually it is more convenient to express the distribution function in other forms since one is generally not interested in knowing the molecular distribution for individual velocity components. In particular, we wish to know the number of molecules having speeds in the range v to $v + \mathrm{d}v$. Since the distribution has been assumed to be isotropic, we do not expect gas molecules to show any preferential direction of motion. We can easily calculate this number by considering a spherical shell of radius v and thickness $\mathrm{d}v$ in the velocity space (see Fig. 1.9). The number of velocity vectors ending in such a spherical shell gives the required number.

The volume of the shell in velocity space will be $4\pi^2 v^2\,\mathrm{d}v$. In view of the geometry under consideration, it is more appropriate to express the volume element $\mathrm{d}v_x\,\mathrm{d}v_y\,\mathrm{d}v_z$ in terms of spherical polar coordinates. From your mathematical physics classes, you may recall that $\mathrm{d}v_x\,\mathrm{d}v_y\,\mathrm{d}v_z = v^2 \sin\theta\,\mathrm{d}\theta\,\mathrm{d}\phi\,\mathrm{d}v$. The limits of integration over θ vary from 0 to π and over ϕ from 0 to 2π. This gives

$$\mathrm{d}N_v = \mathrm{d}^3 N_{v_x v_y v_z} = \int_{\theta=0}^{\pi} \int_{\phi=0}^{2\pi} NA^3 e^{-Bv^2}\, v^2 \sin\theta\,\mathrm{d}\theta\,\mathrm{d}\varphi\,\mathrm{d}v$$

$$= NA^3 v^2 e^{-Bv^2}\left[(-\cos\pi + \cos 0)\,2\pi\right]\mathrm{d}v$$

$$= 4\pi NA^3 v^2 \exp[-Bv^2]\,\mathrm{d}v \tag{1.48}$$

The ratio $\dfrac{\mathrm{d}N_v}{\mathrm{d}v}$ determines the *Maxwellian distribution of molecular speeds.*

The qualitative shape of the Maxwellian distribution function as a function of molecular speed is shown in Fig. 1.10.

Note that

1. The shape of the curve is determined by the interplay of an increasing quadratic function and a decaying exponential.
2. For smaller values of v, the quadratic function dominates and induces the curve to rise. The exponential function begins to become important gradually.
3. Beyond a certain value of v, the exponential term begins to dominate and limits the maximum value of the function. At that point, the curve shows an inversion and decays exponentially thereafter.

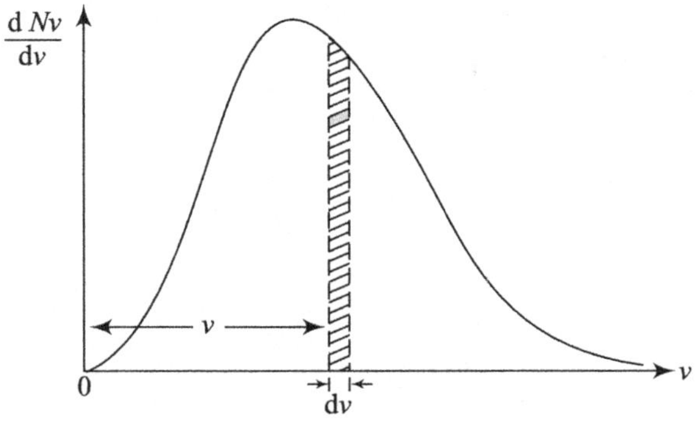

Fig. 1.10 A plot of Maxwellian distribution function versus molecular speed.

If we consider a strip of width dv, its area (shown by the shaded part) gives us the number of molecules with speeds between v and v + dv. As may be seen, the number of molecules with very small and very large speeds is extremely small. The area under the entire curve gives the total number of molecules in the gas.

As mentioned earlier, we introduced only one unknown constant (2*B*) to relax the constraint expressed by Eq. (1.38), but Eq. (1.48) for molecular distribution of speeds has two unknowns (*A* and *B*). Therefore, we must determine these to obtain the quantitative shape of the distribution curve. Let us do so now.

Determination of Constants To evaluate the unknown constants *A* and *B*, we first make use of the fact that if we integrate Eq. (1.48) for dN_v over all possible values of v from 0 to ∞, we will get the total number of molecules:

$$N = \int \mathrm{d}N_v = 4\pi N A^3 \int_0^\infty v^2 \exp[-Bv^2]\,\mathrm{d}v \tag{1.49}$$

Note that at ordinary temperatures, all molecules move with finite speed. However, for mathematical convenience, we will work within the limits 0 and ∞ without any reference to relativistic effects.

We use the standard integral

$$\int_0^\infty e^{-Bv^2} v^n\,dv = \frac{1}{2B^{((n+1)/2)}}\Gamma\left(\frac{n+1}{2}\right)$$

where $\Gamma(n)$ is gamma function. In Table 1.3, we have listed its values for a few values of n. For $n = 2$, the integral on the RHS of Eq. (1.49) can be expressed in terms of the gamma function as $\Gamma\left(\frac{n+1}{2}\right) = \Gamma\left(\frac{3}{2}\right) = \frac{\sqrt{\pi}}{2}$. Using this result, Eq. (1.49) for the total number of particles takes an elegant form,

$$N = 4\pi N A^3 \frac{1}{2B^{3/2}}\Gamma(3/2) = \frac{4\pi N A^3}{2B^{3/2}}\frac{\sqrt{\pi}}{2} \tag{1.50}$$

Table 1.3 *Values of gamma function integral*

n	$\Gamma(n)$
$\frac{1}{2}$	$\sqrt{\pi}$
1	1
$\frac{3}{2}$	$\frac{\sqrt{\pi}}{2}$
2	1

To apprise you of how the integral in Eq. (1.49) is evaluated, we introduce a change of variable by putting $Bv^2 = x$ so that $2Bv\,dv = dx$ or $v\,d\,v = \frac{1}{2B}dx$ and $v^2 d\,v = \frac{1}{2B^{3/2}}x^{1/2}\,dx$. However, the limits of integration will not change since $x = 0$ for $v = 0$ and $x = \infty$ for $v = \infty$. Hence, we can write

$$N = 4\pi NA^3 \frac{1}{2B^{3/2}} \int_0^\infty e^{-x} x^{1/2}\,dx \qquad \text{(i)}$$

The integral in this expression denotes a special function, known as gamma function:

$$\Gamma(n) = \int_0^\infty e^{-x} x^{n-1}\,dx$$

In this case, $n = 3/2$. Therefore, the integral in (i) is equal to $\Gamma(3/2) = (1/2)\Gamma(1/2) = \sqrt{\pi}/2$. In writing the last step, we have used the relation $\Gamma(n) = (n-1)\Gamma(n-1)$ and $\Gamma(1/2) = \sqrt{\pi}$. Hence, the expression for total number of particles in the gas takes a simple form:

$$N = 4\pi NA^3 \frac{1}{2B^{3/2}} \frac{\sqrt{\pi}}{2}$$

On simplification and arrangement of terms in Eq. (1.50), we can express A in terms of B as

$$A = \sqrt{B/\pi} \qquad (1.51)$$

We can determine B, either by calculating the average speed or the mean square speed $\overline{v^2}$ of a molecule, i.e., by relating B to the momentum or the average energy of a molecule. Here we will calculate v_{rms} and define $\overline{v^2}$ as

$$\overline{v^2} = \frac{\int_0^\infty v^2\,dN_v}{\int_0^\infty dN_v} \qquad (1.52)$$

On substituting for dN_v from Eq. (1.48), we get

$$\overline{v^2} = \frac{\int_0^\infty 4\pi NA^3 \exp[-Bv^2]v^2 . v^2 \, dv}{\int_0^\infty 4\pi NA^3 \exp[-Bv^2]v^2 \, dv} = \frac{\int_0^\infty v^4 \exp[-Bv^2] dv}{\int_0^\infty v^2 \exp[-Bv^2] dv}$$

By referring to the integral in the box, you can write

$$\overline{v^2} = \frac{\frac{1}{2B^{5/2}}\Gamma(5/2)}{\frac{1}{2B^{3/2}}\Gamma(3/2)} = \frac{2B^{3/2}\Gamma(5/2)}{2B^{5/2}\Gamma(3/2)}$$

Since $\Gamma(5/2) = (3/2)\Gamma(3/2)$, we get

$$\overline{v^2} = \frac{3}{2B} \tag{1.53}$$

From Eq. (1.12), we recall that the average kinetic energy of a molecule is $(3/2)k_BT$. Therefore, we can write

$$\frac{1}{2}m\overline{v^2} = \frac{3}{2}k_BT$$

On combining this result with Eq. (1.53), we can write

$$\frac{1}{2}m\overline{v^2} = \frac{1}{2}m . \frac{3}{2B} = \frac{3}{2}k_BT$$

so that

$$B = \frac{m}{2k_BT} \tag{1.54}$$

Before proceeding further, you should answer the following practise problem.

Problem 1.3 Verify Eq. (1.54) by starting with the definition of average speed as

$$\overline{v} = \frac{\int_0^\infty v \, dN_v}{\int_0^\infty dN_v}$$

On substituting this value of B in Eq. (1.51), we obtain the value of the constant A:

$$A = \sqrt{\frac{m}{2\pi k_BT}} \tag{1.55}$$

If we now substitute these values of A and B in Eq. (1.48), we can express the number of molecules in a Maxwellian gas having speeds in the range v to $v + dv$ as

$$dN_v = 4\pi N\left(\frac{m}{2\pi k_B T}\right)^{3/2} v^2 \exp\left[-\left(\frac{mv^2}{2k_B T}\right)\right]dv \tag{1.56}$$

Let us pause for a while and reflect as to what we have achieved so far. Equation (1.56) tells us that distribution of molecular speeds depends on the nature of the gas (through mass of the molecules) as well as the temperature. For a given gas at a fixed temperature:

1. The Maxwellian distribution function of molecular speeds, $f_v = \frac{dN_v}{dv}$, will be zero for $v = 0$ as well as $v = \infty$. This is consistent with the physical situation that molecular speeds can have only finite values.
2. For small molecular speeds, Maxwellian distribution function f_v increases quadratically with v.
3. As v increases, the exponential function, which is a decaying term, starts becoming important. Therefore, the curve representing the Maxwellian distribution function increases initially, attains a maximum value and then decreases exponentially. This is illustrated in Fig. 1.11 for three different temperatures.
4. As temperature increases, the curve becomes flatter and its peak value becomes smaller since the total area (number of molecules) under it has to remain constant. However, entire distribution is pushed towards the right including the peak value. This is because greater fraction of gas molecules becomes more energetic.
5. If we integrate $f(v)$ for all values of v from zero to infinity, we get unity. This means that the Maxwellian distribution is a true probability density function.

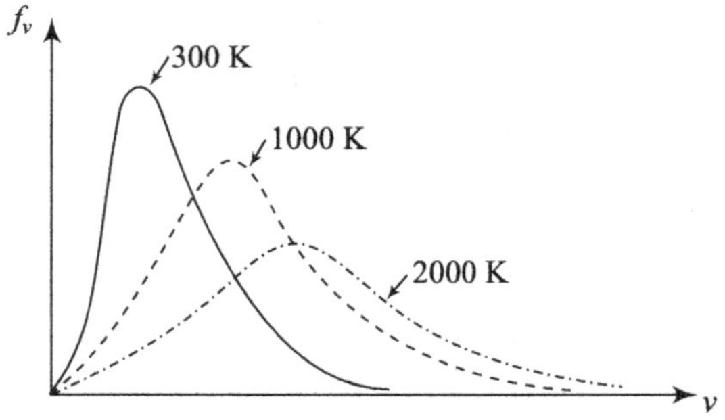

Fig. 1.11 Plot of Maxwell distribution function for molecular speeds at temperatures of 300 K, 1000 K and 2000 K.

Before proceeding further, you should answer the following practise problem.

Problem 1.4 Consider a mixture of molecules of different gases in complete equilibrium such that each type of molecule has the same Maxwellian distribution that it would have if other types of molecules were not present. Will the distribution of the mixture be also Maxwellian?

Since molecular distribution of speeds in a gas depends only on T and m, the mixture will have the same distribution as individual gases. That is, the distribution will be Maxwellian.

To help you grasp the ideas developed in this section and get a feel of the numbers involved, we would like you to go through the following example carefully.

Example 1.3 Calculate the probability that the speed of an oxygen molecule will lie between 100 ms^{-1} and 101 ms^{-1} at 200 K. The mass of oxygen molecule is 32 u. Take $k_B = 1.38 \times 10^{-23}$ JK^{-1} and $N_A = 6 \times 10^{26}$ kmol^{-1}.

Solution: From Eq. (1.56) we know that the probability of a molecule having speed in the range v to $v + dv$ is given by

$$f_v = 4\pi \left(\frac{m}{2\pi k_B T} \right)^{3/2} v^2 \exp\left[-\left(\frac{mv^2}{2k_B T} \right) \right] dv$$

The interval of speeds under consideration is $dv = 101 - 100 = 1$ ms^{-1} and $v = 100$ ms^{-1}.

The mass of an oxygen molecule, $m = \dfrac{32}{6.02 \times 10^{26}} = 5.31 \times 10^{-26}$ kg

At $T = 200$ K, $k_B T = (1.38 \times 10^{-23}\ \text{JK}^{-1}) \times 200\ \text{K} = 2.76 \times 10^{-21}$ J. On substituting these values in the expression for f_v, we get

$$f_v = 4\pi \times \left(\frac{5.31 \times 10^{-26}\ \text{kg}}{2\pi \times (2.76 \times 10^{-21}\ \text{J})} \right)^{3/2} \times (100)^2 \exp\left(-\frac{5.31 \times 10^{-26}\ \text{kg} \times \left(100\,\text{ms}^{-1}\right)^2}{2 \times (2.76 \times 10^{-21}\ \text{J})} \right) \times 1$$

$$= 12.57 \times 10^4 (3.06 \times 10^{-6})^{3/2} \exp(-9.6196)$$

$$= 12.57 \times 10^4 \times (5.35 \times 10^{-9}) \times 6.64 \times 10^{-5} = 4.47 \times 10^{-8}$$

1.5.2 Some Useful Deductions from Maxwell's Law

You now know that molecules of a gas move randomly with all velocities between zero and infinity. Therefore, we cannot characterise a Maxwellian gas with average velocity, which will be zero. For this reason, we consider average speed $\overline{v}$. Since Maxwell distribution function comprises an increasing quadratic function and a decaying exponential function, it should be possible to determine the value at which the function will have maximum value. It is referred to as the most probable speed, v_p. From Sec. 1.2, you may recall that energy of a molecule is defined in terms of root mean square speed, v_{rms}. (Most authors still choose to designate these as average velocity, root mean square velocity and most probable velocity, which is not correct strictly speaking.) It will be instructive to derive an expression for v_{rms} for a Maxwellian gas. An estimate of the values of these speeds gives us an idea about the nature of molecular distribution. We obtain expressions for these now.

Average Speed The average speed is defined as

$$\overline{v} = \frac{\int_0^\infty v\, dN_v}{\int_0^\infty dN_v} = \frac{1}{N} \int_0^\infty v\, dN_v \tag{1.57}$$

On substituting for dN_v from Eq. (1.56), we get

$$\bar{v} = \frac{1}{N} 4\pi N \left(\frac{m}{2\pi k_B T}\right)^{3/2} \int_0^\infty v^3 \exp\left[-\left(\frac{mv^2}{k_B T}\right)\right] dv$$

The value of the integral on the RHS of this expression is

$$I = \frac{1}{2(m/2k_B T)^2} \Gamma(2)$$

Since $\Gamma(2) = 1$, the expression for average speed simplifies to

$$\bar{v} = 4\pi \left(\frac{m}{2\pi k_B T}\right)^{3/2} \frac{1}{2\left(\frac{m}{2k_B T}\right)^2} = \sqrt{\frac{8k_B T}{\pi m}} = \sqrt{\frac{2.55 k_B T}{m}} \tag{1.58}$$

Root Mean Square speed You may recall that root mean square speed is defined as

$$v_{rms} = \sqrt{\overline{v^2}}$$

From Eq. (1.53), we recall that for a Maxwellian gas, mean square speed is related to constant B. On substituting the value of B from Eq. (1.54), we get

$$\overline{v^2} = \frac{3}{2B} = \frac{3k_B T}{m}$$

Hence,

$$v_{rms} = \sqrt{\overline{v^2}} = \sqrt{\frac{3k_B T}{m}} \tag{1.59}$$

Note that Eq. (1.59) obtained for a Maxwellian gas is same as Eq. (1.13) obtained for an ideal gas.

By comparing Eqs. (1.58) and (1.59) for average speed and root mean square speed, respectively, you will note that $v_{rms} > \bar{v}$. This can be seen from Table 1.4, where we have listed numerical values of root mean square speeds computed at STP for a few typical gases based on Eq. (1.59).

Most Probable speed Refer to Fig.1.10 again. The speed at which the Maxwellian distribution function f_v exhibits a maximum is known as the *most probable speed* and is denoted by the symbol v_p. To obtain an expression for most probable speed, we use the elementary knowledge of calculus: *At the maximum of a function, the first derivative with respect to the independent variable is zero and the second derivative is negative.* Therefore, we first calculate $\frac{df_v}{dv}$ using Eq. (1.56). This gives

$$\frac{df_v}{dv} = 4\pi N \left(\frac{m}{2\pi k_B T}\right)^{3/2} \frac{d}{dv}\left[v^2 \exp\left\{-\left(\frac{mv^2}{2k_B T}\right)\right\}\right]$$

$$= C\left\{2v\exp\left[-\left(\frac{mv^2}{2k_BT}\right)\right] + v^2\exp\left[-\left(\frac{mv^2}{2k_BT}\right)\right]\left(-\frac{2mv}{2k_BT}\right)\right\} \tag{1.60}$$

We now equate $\frac{df_v}{dv} = 0$ and solve for v. This value will define $v = v_p$.

Hence, $2v_p\exp\left[-\left(\frac{mv_p^2}{2k_BT}\right)\right] + v_p^2\exp\left[-\left(\frac{mv_p^2}{2k_BT}\right)\right]\left(-\frac{2mv_p}{2k_BT}\right) = 0$

We can rewrite it as

$$2v_p\exp\left[-\left(\frac{mv_p^2}{2k_BT}\right)\right]\left[1 - \frac{m}{2k_BT}v_p^2\right] = 0$$

For a non-zero value of v_p, this equality will hold only if

$$1 - \frac{m}{2k_BT}v_p^2 = 0$$

or

$$v_p^2 = \frac{2k_BT}{m}$$

Hence, the most probable speed of a Maxwellian gas is given by

$$v_p = \sqrt{\frac{2k_BT}{m}} \tag{1.61}$$

If you calculate the second order derivative of Maxwellian distribution function and substitute the value of most probable speed given by Eq. (1.61), you will find that $\frac{d^2 f_v}{dv^2} = -4$. We leave it as an exercise for you.

The numerical values of average speed, root mean square speed and most probable speed calculated for a few typical gases on the basis of Eqs. (1.58), (1.59) and (1.61), respectively at STP are given in Table 1.4. You will note that root mean square speed of a Maxwellian molecule is greater than its average speed, which, in turn, is greater than the most probable speed. You can easily convince yourself that $v_p : \overline{v} : v_{rms} : 1 : 1.128 : 1.224$.

Table 1.4 *Values of $\overline{v}$, v_{rms} and v_p for different gases at STP*

Gas	$\overline{v}$ (ms^{-1})	v_{rms} (ms^{-1})	v_p (ms^{-1})
H_2	1695	1838	1501
H_2O	567	615	502
N_2	455	493	403
Air	447	485	396
O_2	425	461	376
CO_2	362	393	321

To highlight the utility of the quantities discussed in this section, we now solve a few examples.

Example 1.4 For a Maxwellian gas, show that $\bar{v} \times \left(\overline{\frac{1}{v}}\right) = \frac{4}{\pi}$.

Solution: We define $\left(\overline{\frac{1}{v}}\right)$ as

$$\left(\overline{\frac{1}{v}}\right) = \frac{1}{N}\int_0^\infty \frac{1}{v}\,\mathrm{d}N_v$$

$$= 4\pi\left(\frac{m}{2\pi k_B T}\right)^{3/2} \int_0^\infty v \exp\left[-\left(\frac{mv^2}{k_B T}\right)\right]\mathrm{d}v$$

The value of the integral on the RHS of this expression is

$$I = \frac{1}{2(m/2k_B T)}\Gamma(1)$$

Since $\Gamma(1) = 1$, the expression for $\left(\overline{\frac{1}{v}}\right)$ takes the form

$$\left(\overline{\frac{1}{v}}\right) = 4\pi\left(\frac{m}{2\pi k_B T}\right)^{3/2} \frac{1}{2\left(\frac{m}{2k_B T}\right)} = \sqrt{\frac{2m}{\pi k_B T}}$$

Since $\bar{v} = \sqrt{\frac{8k_B T}{m\pi}}$, we get

$$\bar{v} \times \left(\overline{\frac{1}{v}}\right) = \sqrt{\frac{8k_B T}{m\pi}} \times \sqrt{\frac{2m}{\pi k_B T}} = \frac{4}{\pi}$$

Example 1.5 Starting from Maxwell's distribution function, obtain the expression for pressure exerted by the molecules of a gas on the walls of the container.

Solution Let us assume that the molecular number density is n. Suppose that of these molecules, a fraction $f(v)\,\mathrm{d}v$ has velocity components between v and $v + \mathrm{d}v$ and $f(v)$ is Maxwell's distribution function.

From elementary mathematics, we know that the solid angle included between the polar angles θ and $\theta + \mathrm{d}\theta$ is $2\pi \sin\theta\,\mathrm{d}\theta$. This is a fraction $(1/4\pi)\,2\pi \sin\theta\,\mathrm{d}\theta$, i.e., $(1/2)\sin\theta\,\mathrm{d}\theta$ of the total solid angle around a point. Hence, in unit volume, there are $(n/2)\sin\theta\,\mathrm{d}\theta\, f(v)\mathrm{d}v$ molecules, which approach the surface between angles θ and $\theta + \mathrm{d}\theta$ and having velocities in the range v and $v + \mathrm{d}v$. The projection of their velocity perpendicular to the surface will be $v\cos\theta$. Thus, the number of molecules hitting unit area of the container surface per unit time is $(n/2)\sin\theta\,\mathrm{d}\theta\, v\cos\theta f(v)\mathrm{d}v$.

Each of these molecules has a component of momentum $mv \cos \theta$ normal to the surface. So on rebound, the total exchange of momentum normal to the surface is $2mv \cos \theta$. Hence, the total rate of change of momentum per unit area per unit time communicated by these molecules is

$$(n/2) \sin \theta \, d\theta \, v \cos \theta f(v) \, dv (2mv \cos \theta) \quad \text{(i)}$$

But rate of change of momentum per unit area defines the pressure exerted by the molecules on the walls of the container. So total pressure is obtained by integrating the expression in (i) over all values of θ and v. This gives

$$p = mn \int_0^{\pi/2} \sin\theta \cos^2 \theta \, d\theta \int_0^{\infty} v^2 f(v) \, dv$$

We can easily show that

$$\int_0^{\pi/2} \sin\theta \cos^2 \theta \, d\theta = \frac{1}{3}$$

and

$$\int_0^{\infty} v^2 f(v) \, dv = \overline{v^2}$$

You may now like to know if this definition holds for any $f(v)$. Definitely not. We have got v^2 on the right-hand side only because $f(v)$ is a probability distribution function (see point 5 on page 1.26).

Hence, we obtain the desired expression for pressure exerted by the molecules on the walls of a container:

$$p = \frac{1}{3} mn \overline{v^2}$$

Example 1.6 Calculate the most probable speed, average speed and the root mean square speed for oxygen molecules at 300 K using the following data: $m(O_2) = 5.31 \times 10^{-26}$ kg and $k_B = 1.38 \times 10^{-23}$ JK^{-1}.

Solution: From Eqs. (1.58), (1.59) and (1.61), you will note that for calculation of all three speeds, we need the value of $\sqrt{\dfrac{k_B T}{m}}$. Therefore, let us calculate this first.

$$\frac{k_B T}{m} = \frac{(1.38 \times 10^{-23} \text{ JK}^{-1}) \times (300 \text{ K})}{5.31 \times 10^{-26} \text{ kg}} = 7.8 \times 10^4 \text{ m}^2 \text{ s}^{-2}$$

Hence,

$$\sqrt{\frac{k_B T}{m}} = 2.8 \times 10^2 \text{ ms}^{-1}$$

$\therefore$

$$v_p = \sqrt{2} \times 2.8 \times 10^2 = 395 \text{ ms}^{-1}$$

$$\overline{v} = \sqrt{2.55} \times 2.8 \times 10^2 = 446 \text{ ms}^{-1}$$

and

$$v_{rms} = \sqrt{3} \times 2.8 \times 10^2 = 484 \text{ ms}^{-1}$$

Example 1.7 Calculate the fraction of molecules of a gas within 1% of the most probable speed at STP. Will its value be same for all gases at all temperatures?

Solution: The fraction of molecules having speeds between v and $v + dv$ is given by

$$\frac{dN_v}{N} = 4\pi\left(\frac{m}{2\pi k_B T}\right)^{3/2} v^2 \exp\left[-\left(\frac{mv^2}{2k_B T}\right)\right] dv \quad \text{(i)}$$

The most probable speed is given by

$$v_p = \sqrt{\frac{2k_B T}{m}}$$

We can rewrite Eq. (i) in terms of the most probable speed as

$$\frac{dN_v}{N} = \frac{4}{\sqrt{\pi}}\frac{1}{v_p^3} \times v^2 \exp\left(-\frac{v^2}{v_p^{\,2}}\right) dv \quad \text{(ii)}$$

Since v varies within 1% of the most probable speed, we can take $v = v_p$ and $dv = 1.01v_p - 0.99v_p = 0.02v_p$. Hence, the fraction of molecules having speeds within 1% of the most probable speed is

$$\frac{dN_v}{N} = \frac{4}{\sqrt{\pi}}\frac{1}{v_p} \times (0.02\,v_p) \exp(-1)$$

$$= \frac{0.08}{\sqrt{\pi}} \times \exp(-1)$$

$$= 0.045 \exp(-1) = 0.0166.$$

We may therefore conclude that Maxwell's law predicts that

1. About 2% molecules will have speeds within 1% of the most probable speed.
2. The fraction of molecules is constant; independent of temperature as well as the nature of the gas.

Example 1.8 Calculate the temperature at which the root mean square speed of hydrogen and oxygen molecules will be equal to their escape velocities from the earth's gravitational field. Take the radius of the earth as 6400 km.

Solution: The average kinetic energy of a gas molecule of mass m and root mean square speed v_{rms} is

$$\frac{1}{2} m \overline{v^2} = \frac{3}{2} k_B T$$

We also know that the escape velocity from the surface of the earth is given by $v_{es} = \sqrt{2gR_0}$, where g is acceleration due to gravity and R_0 is radius of the earth.

The kinetic energy of a molecule moving with escape velocity will be $\frac{m}{2} v_{es}^{\,2}$ and we can write $\frac{3}{2} k_B T = mgR_0$

(a) *For hydrogen molecules*

$$T(H_2) = \frac{2}{3}\frac{m(H_2)\,gR_0}{k_B}$$

Here $m(H_2) = \frac{2}{N_A} = \frac{2}{6\times10^{26}}$ kg, $g = 9.8$ ms^{-2}, $R_0 = 6.4\times10^6$ m and $k_B = 1.38\times10^{-23}$ JK^{-1}.

Hence,

$$T(H_2) = \frac{2}{3}\times\left(\frac{2}{6\times10^{26}}\text{kg}\right)\times\left(\frac{(9.8\,\text{ms}^{-2})\times(6.4\times10^6\text{ m})}{1.38\times10^{-23}\text{ JK}^{-1}}\right) = 10.1\times10^3\text{ K}$$

(b) *For oxygen molecules*

$$T(O_2) = \frac{2}{3}\frac{m(O_2)\,gR_0}{k_B}$$

Since $m(O_2) = \frac{32}{6\times10^{26}}$ kg, we find that

$$T(O_2) = \frac{2}{3}\times\left(\frac{32}{6\times10^{26}}\text{kg}\right)\times\left(\frac{(9.8\,\text{ms}^{-2})\times(6.4\times10^6\text{ m})}{1.38\times10^{-23}\text{ JK}^{-1}}\right) = 161.6\times10^3\text{ K}$$

$$= 16T(H_2)$$

Example 1.9 Obtain an expression for the mean translational energy per degree of freedom for the molecules of a Maxwellian gas.

Solution: For molecules constrained to move along a particular direction, say x-axis, the distribution function is

$$f(v_x)\,dv_x = A\exp(-Bv_x^2)\,dv_x$$

where $A = \sqrt{B/\pi}$ and $B = \frac{m}{2k_BT}$.

The average kinetic energy of molecules moving along x- axis is

$$\overline{\varepsilon_x} = \frac{m}{2}\int_{-\infty}^{\infty} v_x^2 f(v_x)\,dv_x$$

On substituting for $f(v_x)$, the expression for average kinetic energy of molecules moving along x-axis takes the form

$$\overline{\varepsilon_x} = \frac{m}{2}\int_{-\infty}^{\infty} v_x^2 A\exp(-Bv_x^2)\,dv_x = \frac{m}{2}\times 2A\int_0^{\infty} v_x^2\exp(-Bv_x^2)\,dv_x$$

You may recall that this integral is $\frac{1}{2B^{3/2}}\Gamma(3/2) = \frac{\sqrt{\pi}}{4B^{3/2}}$. Hence, the expression for average kinetic energy of molecules moving along x-axis simplifies to

$$\overline{\varepsilon_x} = \frac{mA\sqrt{\pi}}{4B^{3/2}} = m\left(\frac{m}{2\pi k_B T}\right)^{1/2} \frac{\sqrt{\pi}}{4}\left(\frac{2k_B T}{m}\right)^{3/2} = \frac{k_B T}{2}$$

This result is consistent with the principle of equipartition of energy.

Example 1.10 Determine the temperature at which speeds of nitrogen gas molecules, $v_1 = 300\ \text{ms}^{-1}$ and $v_2 = 600\ \text{ms}^{-1}$ are associated with equal values of the Maxwellian distribution function f_v.

Solution: Let the temperature at which the two distributions (with v_1 and v_2) are same be T. Then, we can write

$$f_{v_1} = 4\pi N\left(\frac{m}{2\pi k_B T}\right)^{3/2} e^{-(mv_1^2/2k_B T)} v_1^2$$

and

$$f_{v_2} = 4\pi N\left(\frac{m}{2\pi k_B T}\right)^{3/2} e^{-(mv_2^2/2k_B T)} v_2^2$$

For $f_{v_1} = f_{v_2}$, we can write

$$e^{-(mv_1^2/2k_B T)} = \frac{v_2^2}{v_1^2} e^{-(mv_2^2/2k_B T)}$$

$$= \left(\frac{v_2}{v_1}\right)^2 e^{-(mv_2^2/2k_B T)}$$

Taking natural logarithm of both sides, we get

$$-\frac{mv_1^2}{2k_B T} = 2\ln\left(\frac{v_2}{v_1}\right) - \frac{mv_2^2}{2k_B T}$$

We can rewrite it as

$$\frac{m}{2k_B T}\left(v_2^2 - v_1^2\right) = 2\ln\left(\frac{v_2}{v_1}\right)$$

so that

$$T = \frac{m(v_2^2 - v_1^2)}{4k_B \ln\left(\frac{v_2}{v_1}\right)} = \frac{M(v_2^2 - v_1^2)}{4R\ln\left(\frac{v_2}{v_1}\right)}$$

$$= \frac{(0.028\ \text{kg mol}^{-1}) \times (36-9) \times 10^4\ (\text{ms}^{-1})^2}{4 \times (8.31\ \text{JK}^{-1}\ \text{mol}^{-1}) \times \ln(2)}$$

$$= \frac{0.028 \times 27 \times 10^4}{4 \times 8.31 \times 0.693}\ \text{K}$$

$$= \frac{75.6 \times 10^2}{23.03} \text{ K}$$

$$= 328 \text{ K}$$

Before proceeding further, you should answer the following practise problem.

Problem 1.5 Calculate the v_{rms} for neutrons and electrons at 300 K. Given $m_n = 1.675 \times 10^{-27}$ kg and $m_e = 9.11 \times 10^{-31}$ kg.

Ans: 2.723×10^3 ms^{-1} ; 6.700×10^4 ms^{-1}

So far we have confined our discussion to molecular distribution of velocities/speeds for a Maxwellian gas at a finite temperature. Another physical quantity of interest is energy of a gas molecule. Let us now learn about the energy distribution of a Maxwellian gas.

1.6 ENERGY DISTRIBUTION OF A MAXWELLIAN GAS

In pragmatic situations, we are frequently interested in the energy distribution function rather than the distribution of speeds. This can be easily obtained by recalling that the kinetic energy of a molecule of mass m moving with speed v is $E = (1/2)mv^2$. We use this relation to correlate the spread in speed with the spread in energy:

$$\mathrm{d}E = mv\mathrm{d}v$$

so that

$$\mathrm{d}v = (2mE)^{-1/2}\, \mathrm{d}E \tag{1.62}$$

We now assume that the number of molecules having kinetic energies in the range E to $E + \mathrm{d}E$ is the same as the number of molecules having speeds in the range v to $v + \mathrm{d}v$, i.e.,

$$\mathrm{d}N_E = \mathrm{d}N_v \tag{1.63}$$

On substituting for v^2 and $\mathrm{d}v$ in terms of energy, we get

$$\mathrm{d}N_E = 4\pi N \left(\frac{m}{2\pi k_B T}\right)^{3/2} \left(\frac{2E}{m}\right) \exp\left(-\frac{E}{k_B T}\right) \cdot \frac{\mathrm{d}E}{\sqrt{2mE}}$$

On simplification, we can write

$$\mathrm{d}N_E = 2N \left(\frac{E}{\pi}\right)^{1/2} \left(\frac{1}{k_B T}\right)^{3/2} \exp\left(-\frac{E}{k_B T}\right) \mathrm{d}E \tag{1.64}$$

This expression gives the number of molecules of a Maxwellian gas with energies in the range E and $E + \mathrm{d}E$. Hence, the *energy distribution function*, f_E, of a Maxwellian gas can be expressed as

$$f_E = \frac{\mathrm{d}N_E}{\mathrm{d}E} = 2N \left(\frac{E}{\pi}\right)^{1/2} \left(\frac{1}{k_B T}\right)^{3/2} \exp\left(-\frac{E}{k_B T}\right) \tag{1.65}$$

Note that at a given temperature, energy distribution function is a product of square root of energy and a decaying exponential. These features are depicted in Fig. 1.12 where we have plotted f_E as a function of E/k_BT.

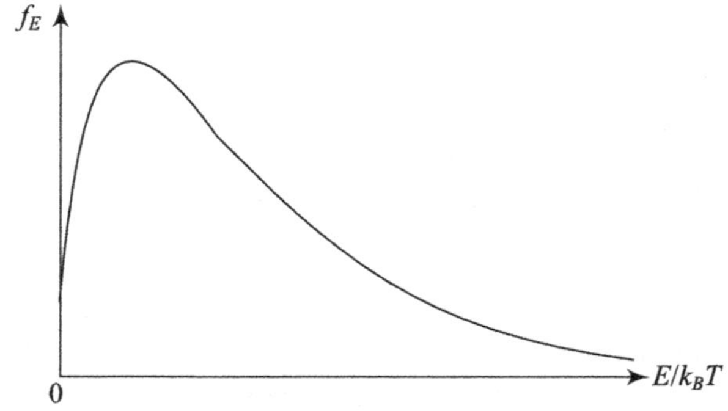

Fig. 1.12 Plot of Maxwellian distribution function for molecular energies.

Note that just like velocity distribution function, the energy distribution function also exhibits a maximum. The energy corresponding to the peak value is known as *most probable energy*, ε_p. Another physical quantity of great interest is *mean energy*. The expressions for $\bar{\varepsilon}$ and ε_p are derived in following example.

Example 1.11 Derive expressions for mean energy and most probable energy of the molecules of a Maxwellian gas.

Solution: The mean energy of a molecule in a Maxwellian gas is defined as

$$\bar{\varepsilon} = \frac{\int_0^\infty E dN_E}{\int_0^\infty dN_E}$$

On substituting for dN_E from Eq. (1.64), we get

$$\bar{\varepsilon} = 2\left(\frac{1}{\pi}\right)^{1/2}\left(\frac{1}{k_BT}\right)^{3/2}\int_0^\infty E^{3/2}\exp\left(-\frac{E}{k_BT}\right)dE$$

To evaluate the integral on the RHS, we introduce a change of variable by defining $x = E/k_BT$ so that $dx = (k_BT)^{-1}dE$. On substituting these results in the expression for mean energy, we get

$$\bar{\varepsilon} = 2\left(\frac{1}{\pi}\right)^{1/2} k_BT\int_0^\infty x^{3/2}\exp(-x)\,dx$$

This integral is, by definition, $\Gamma(5/2) = \frac{3\sqrt{\pi}}{4}$. Hence, the mean energy of a Maxwellian gas molecule is $\bar{\varepsilon} = \frac{3}{2}k_BT$. This expression shows that the mean energy of a molecule of an ideal monatomic gas at 273 K is 5.65×10^{-21} J. Similarly, hydrogen atoms will have to be heated to about 7.9×10^4 K to excite them to their first excited state, which occurs at 10.2 eV.

To calculate the most probable energy, we differentiate the expression for probability distribution with respect to energy and put the resultant expression equal to zero. That is

$$\frac{d f_E}{dE} = 0$$

This for $E = \varepsilon_p$, we can write

$$\frac{d}{dE}\left[E^{1/2} \exp(-E/k_B T)\right] = 0$$

or

$$\left[\frac{1}{2\varepsilon_p^{1/2}} - \frac{1}{k_B T}\varepsilon_p^{1/2}\right]\exp(-\varepsilon_p/k_B T) = 0$$

This equation will hold only if the term within the parentheses vanishes identically. This leads to the required result:

$$\varepsilon_p = \frac{k_B T}{2}$$

Note that

1. The mean energy of a Maxwellian gas molecule is three times its most probable energy.
2. The values of $\bar{\varepsilon}$ and ε_p do not follow from their definitions in terms of corresponding velocities.
3. At a given temperature, the mean energy is same for all gases, irrespective of their nature. At room temperature ($T = 300$ K), $\bar{\varepsilon} = 6.21 \times 10^{-21}$ J $= 0.0388$ eV.

Yet another physical quantity of interest is *momentum of a gas molecule*. You may like to express the Maxwellian distribution function in terms of momentum. The number of molecules whose momenta lie between p and $p + dp$ is given by

$$dN_p = \frac{4\pi N}{(2\pi m k_B T)^{3/2}} \exp\left[-\left(\frac{p^2}{2 m k_B T}\right)\right] p^2 \, dp \tag{1.66}$$

It is important to mention here that Maxwell's law of distribution of velocities proved a landmark in improving our understanding of the behaviour of ideal gases. But its direct experimental verification remained elusive for quite some time, though several indirect evidences were cited in its support. We will now discuss these in some detail.

1.7 Experimental Verification of Maxwell's Distribution Law

In science, no theory is accepted till such time that it is supported by experimental evidences. In case of Maxwell' distribution law, only indirect evidences were available till 1920. The finite width of spectral lines was cited as the first indirect evidence. Subsequently, Sir Owen W. Richardson qualitatively explained the behaviour of thermionic emission of electrons by assuming that electrons behave as Maxwellian gas. (The exact explanation though came when electrons were treated as fermions and Fermi–Dirac statistics was used.) This was also interpreted as evidence in favour of Maxwell's law. The first direct proof of Maxwell's law was given by Stern in 1920. His technique was subsequently modified by Zartman and Ko. You will now learn about these and other more elegant experiments.

1.7.1 Indirect Evidences

Finite Width of Spectral Lines While measuring the wavelength of spectral lines of mercury in your B.Sc Physics Laboratory, you must have used a spectrometer and a glass prism or a diffraction grating. Did you obtain fairly sharp lines? Even if so, these have finite width. We can understand their genesis as well as why their width is finite based on Bohr's theory and its underlying physical considerations.

We know that when energy is given to an atom in its ground state, electrons in its outermost orbit jump to permitted higher orbits. If this energy is not enough to make electrons leave the atom, they fall back to the lower energy states and the excess energy is emitted in the form of radiation. When a stationary atom is excited, the wavelength of the emitted radiation is determined only by the difference of the energies corresponding to the energy levels involved in transitions. Hence, the width of the spectral line should, in principle, be vanishingly small. That is, for a stationary atom, there is no broadening of the spectral line. But in actual practise, we observe that a spectral line has finite width. This can be understood if we assume that atoms move with finite velocities, obey Maxwell' distribution law of velocities and the frequency of emitted radiation changes on account of Doppler Effect.

For an atom at rest, the frequency of radiation emitted by it when it returns to its initial state is given by

$$f_0 = \frac{c}{\lambda_0} \tag{1.67}$$

If we assume that the atoms are moving freely, their frequency of radiation will be influenced by Doppler's effect. To calculate Doppler broadening, let us consider the motion of an atom along the x-axis. If it moves towards an observer with velocity v_x, the frequency of the spectral line will change to $f = f_0\left(1+\frac{v_x}{c}\right)$, where c is velocity of light. On the other hand, if an atom moves away from the observer, the emitted frequency will be $f_0\left(1-\frac{v_x}{c}\right)$. Since v_x can have any value in the range 0 to ∞, we can conclude that an atom can have all frequencies about f_0. It means that if Maxwell's law for distribution of molecular velocities holds, a spectral line can have infinite width. In practice, the number of particles having large speeds is rather small and the intensity of a spectral line falls off very rapidly about a central maximum.

We know that the frequency is inversely proportional to wavelength. Let us assume that $f_0\left(1\pm\frac{v_x}{c}\right)$ corresponds to the wavelength $\lambda_0 \mp x$. Then, we can write

$$\lambda_0 - x = \frac{c}{f_0\left(1+\frac{v_x}{c}\right)} = \frac{c}{f_0}\left(1+\frac{v_x}{c}\right)^{-1} \tag{1.68}$$

Using binomial expansion $\left[(1+x)^{-1} = 1 - x + \frac{x^2}{2} - \ldots\right]$ and retaining terms only up to first order in v_x/c, we can rewrite Eq. (1.68) as

$$\lambda_0 - x = \frac{c}{f_0} - \frac{v_x}{f_0}$$

so that

$$x = \frac{v_x}{f_0} \tag{1.69}$$

This result shows that spread of a spectral line is directly proportional to molecular velocity. We also know that intensity of a spectral line is proportional to the number of molecules, i.e., $I \propto N$. Since the number of particles having velocity components in the range v_x to $v_x + \mathrm{d}v_x$ is given by Eq. (1.45a), we can write

$$\mathrm{d}N_{v_x} = NA \exp\left(-\frac{mv_x^{\ 2}}{2k_B T}\right)\mathrm{d}v_x$$

so that intensity of spectral line along x-axis can be expressed as

$$I_x = I_0 \exp\left(-\frac{mv_x^2}{2k_B T}\right)$$

On combining this result with Eq. (1.69), we get

$$I_x = I_0 \exp\left(-\frac{mx^2 f_0^{\ 2}}{2k_B T}\right) \tag{1.70}$$

where $I_0 = NA$ denotes the maximum intensity of radiation emitted by an assembly of N stationary atoms.
Thus, we can express the intensity at half-width* as

$$\frac{I_0}{I_b} = 2 = \exp\left(\frac{mb^2 f_0^{\ 2}}{2k_B T}\right)$$

Taking natural logarithm of both sides, we get

$$\ln 2 = \frac{mb^2 f_0^2}{2k_B T}$$

so that the half width b

$$b = \frac{1}{f_0}\sqrt{\frac{2k_B T \ln 2}{m}} = \frac{\lambda_0}{c}\sqrt{\frac{2RT \ln 2}{M}} \tag{1.71}$$

where $M = mN_A$ is the molecular weight. We thus find that half-width of a spectral line is inversely proportional to the square root of the molecular weight of the substance emitting it. This implies that spectral lines in the hydrogen spectrum will be diffused, whereas cadmium and mercury lines should be sharp. (It is for this reason that heavier nuclei are used for precision work). This is in conformity with our observation and provided much needed indirect support to Maxwell's law.

* Actually, it signifies full-width at half-maximum.

You should now go through the following example.

Example 1.12 Calculate the Doppler broadening in sodium line having wavelength 589 nm at 400 K. Take $R = 8.31$ kJ kmol^{-1} K^{-1} and atomic weight of sodium is 23.

Solution From Eq. (1.71) we recall that half-width of a spectral line is given by

$$b = \frac{\lambda_0}{c}\sqrt{\frac{2\,RT\,\ln 2}{M}}$$

On substituting the given values, we get

$$b = \frac{(589\times10^{-9}\ \text{m})}{(3\times10^{8}\ \text{ms}^{-1})}\sqrt{\frac{2\times(8.31\times10^{3}\ \text{JK}^{-1}\text{kmol}^{-1})\times2.3\times0.3010\times(400\text{K})}{(23\ \text{kg kmol}^{-1})}}$$

$$= 8.78\times10^{-4}\ \text{nm}$$

Thermionic emission of electrons by metallic filaments When a metallic wire is heated continuously, electrons are emitted. By considering the electrons to behave as a Maxwellian gas, Richardson obtained an expression for current density. Here we just quote the result without giving any details:

$$j = CT^{1/2}\exp\left(-\frac{\phi}{k_B T}\right) \tag{1.72}$$

where ϕ is work function, T is the temperature of the filament and C is an arbitrary constant. This equation qualitatively explained the process of thermionic emission and provided indirect evidence in favour of Maxwell's law.

The first direct experimental measurement was made by Zartman and Ko. Let us discuss it now.

1.7.2 Direct Experiments

The first direct proof of Maxwell's law was given by Stern in 1920. His technique was subsequently modified by Zartman and Ko. In 1947, one of the most convincing and elegant experiments was performed by Estermann *et al.* A more precise measurement was reported by Miller and Kus in 1955. We now discuss these experiment in some detail.

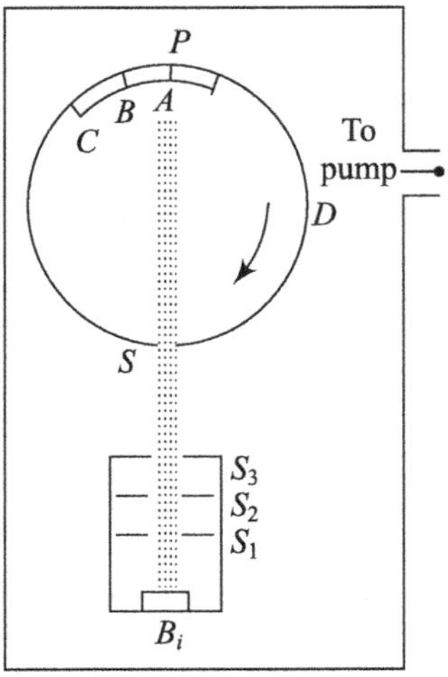

Fig. 1.13 A schematic diagram of the apparatus used by Zartman and Ko to verify Maxwell's law for

Zartman and Ko experiment The apparatus used by Zartman and Ko is illustrated in Fig. 1.13. A beam of bismuth molecules, produced in an oven, was collimated by a series of slits S_1, S_2, S_3. A glass plate P fixed inside a cylindrical drum, which can be rotated at a high speed, about an axis passing through its centre, was used to collect bismuth molecules. (Instead of the plate, a photographic film could also be placed). The entire apparatus was placed in an evacuated chamber. Note that the molecules could enter the drum only when the slit *S* crossed the molecular beam.

When the drum is stationary, the beam will strike the glass plate at a particular spot, *A*, say. As the drum is rotated clockwise, the glass plate moves towards the right and the faster moving molecules entering the cylinder will strike it to the left of *A*, the point of impact when the cylinder was stationary. Suppose that the slower molecules reach the plate between *B* and *C*. These features are illustrated in Fig. 1.14.

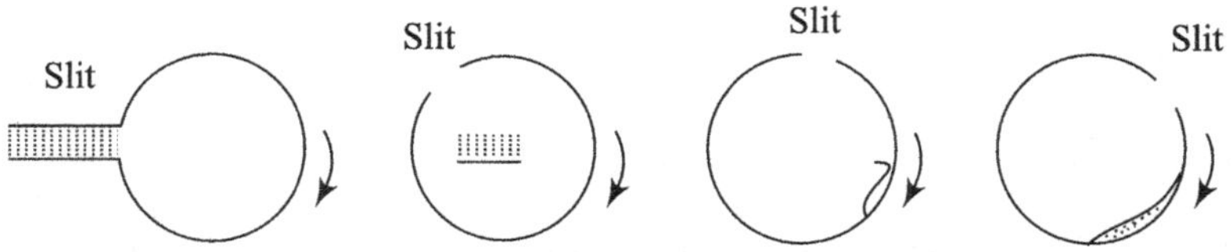

Fig. 1.14 Spread of deposit across the plate.

The density of deposit across the plate gives a measure of the velocity distribution of molecules. A schematic representation is shown in Fig. 1.15.

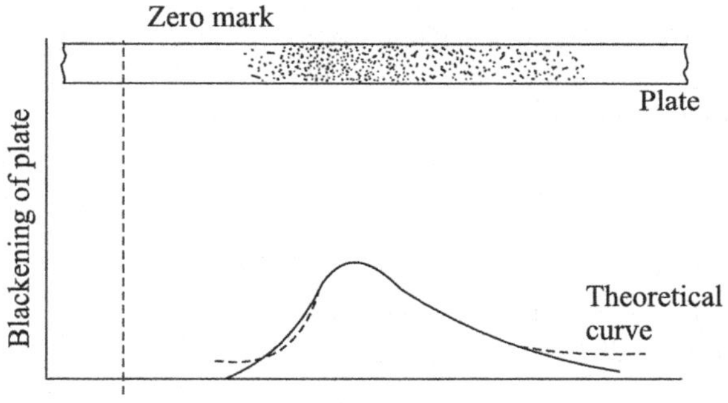

Fig. 1.15 Data on density of deposit obtained in the rotating drum experiment of Zartmann and Ko.

The results obtained by Zartman and Ko are shown in Fig. 1.16. As may be noted, the agreement between theoretical and experimental values is surprisingly good.

Estermann, Simpson and Stern Experiment Refer to Fig. 1.17. It depicts a schematic diagram of the apparatus used by Estermann *et al.* The atoms are deflected by gravity only, i.e., use is made of the free fall of the molecules in a beam. A molecular beam of cesium emerges from the oven placed in a long, highly evacuated chamber ($p \approx 10^{-8}$ mm of Hg). It is made to pass through the collimating slits and impinges

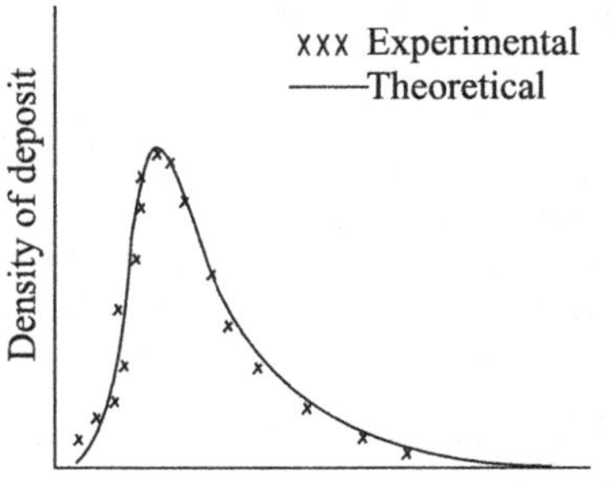

Fig. 1.16 Plot of the density of deposit as a function of molecular speed.

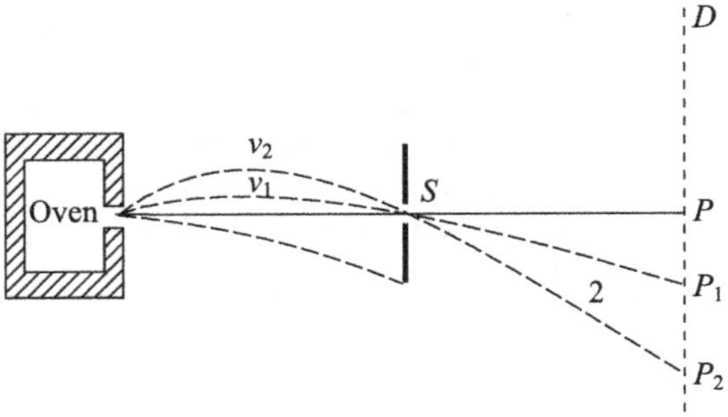

Fig. 1.17 A schematic diagram of the apparatus used by Estermann, Simpson and Stern.

on a hot tungsten wire *D*. The cesium atoms striking the wire get ionised and leave the wire as a positive ion. These are collected by a negatively charged cylinder surrounding the wire (not shown in the diagram). The current in the collecting cylinder gives a measure of the number of cesium atoms striking the detecting wire per unit time.

In the absence of gravitational field, all atoms emerging horizontally from the oven would pass through the slit *S*, travel without any deviation and strike the collector at *P*, irrespective of their velocities. However, due to gravitational field, each atom emerging horizontally would behave as a projectile and follow a parabolic path. Atoms proceeding along path 1 (speed v_1) would reach the collector at P_1 and those proceeding along path 2 (speed v_2) would reach the collector at P_2. The measurement of the ion current as a function of the vertical height of the collector gives a measure of the velocity distribution.

As may be seen from Fig. 1.18, the agreement between theoretical predictions and experimental results is exceedingly good.

In 1955, Miller and Kusch reported a still more precise measurement using a beam of thallium atoms. Their results show that the velocity distribution of thallium atoms agrees with the theoretical predictions to within 1%

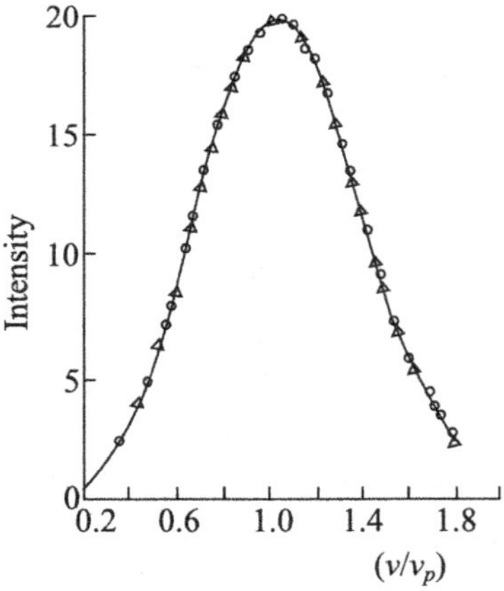

Fig. 1.18 Variation of ionisation current with (v/v_p).

Additional Examples

Example 1.13 At what temperature is the rms speed of hydrogen molecules equal to twice of that of oxygen molecules at 63°C? It is given that oxygen molecule is about 16 times heavier than a hydrogen molecule.

Solution: From Eq. (1.13), we can write

$$RT = \frac{1}{3} M v_{\text{rms}}^2$$

or
$$3R = \frac{M v_{\text{rms}}^2}{T}$$

It shows that $\frac{T}{M v_{\text{rms}}^2}$ is constant . If T is the required temperature of hydrogen, we can write

$$\frac{T}{M_{H_2} (2v_{\text{rms}})^2} = \frac{(273+63)\text{ K}}{M_{O_2} v_{\text{rms}}^2} = \frac{336\text{ K}}{16 M_{H_2} v_{\text{rms}}^2}$$

On simplification, we get

or
$$T = \frac{(4 \times 336)\text{ K}}{16} = 84\text{ K}$$

∴
$$\text{Required temp.} = (84 - 273)°\text{C} = -189°\text{C}.$$

Example 1.14 Calculate the rms speed of methane at 20°C and pressure of 5 atm. (Take 1 atm = 10^5 Nm^{-2}.)

Solution: Let the volume of the gas be V litre under the given conditions. We know that at STP, i.e., 0°C (≡ 273 K) and 1 atm pressure, the volume occupied by one mole of an ideal gas = 22.4 lit. Therefore, using the ideal gas equation of state, $\frac{pV}{T}$ = constant, we can write

∴
$$\frac{(1\text{ atm}) \times (22.4\text{ lit mol}^{-1})}{273\text{ K}} = \frac{(5\text{ atm}) \times V}{(20+273)\text{ K}}$$

or
$$V = \frac{293 \times 22.4}{273 \times 5}\text{ lit mol}^{-1}$$

We know that atomic weight of carbon is 12 and that of hydrogen is 1. So the mol. wt. of CH_4 = 12 + 1 × 4 = 16.

∴
$$\text{Density } \rho = \frac{16\text{ g mol}^{-1}}{\text{V lit mol}^{-1}}$$
$$= \frac{16}{\frac{293 \times 22.4}{273 \times 5}}\text{ g lit}^{-1}$$
$$= \frac{16 \times 273 \times 5}{293 \times 22.4 \times 10^3}\text{ g cm}^{-3}$$
$$= 3.328 \times 10^{-3}\text{ g cm}^{-3} = 3.328\text{ kg m}^{-3}.$$

Hence, the root mean square speed is given by

$$v_{\text{rms}} = \sqrt{\frac{3p}{\rho}} = \sqrt{\frac{3 \times (5 \times 10^5\text{ N m}^{-2})}{3.328\text{ kg m}^{-3}}} = \sqrt{45.07} \times 10^2\text{ ms}^{-1}$$
$$= 6.71 \times 10^2\text{ ms}^{-1}$$

Example 1.15 Calculate the number of molecules in **(a)** 1 litre of an ideal gas at 300 K and 1.5 atm pressure. The mean *K.E* of a molecule at 300 K is 4×10^{-11} J, and **(b)** a room of size 5 m × 4 m × 3 m (1 atm = 10^5 Nm^{-2}).

Solution: **(a)** From Eq. (1.8), we recall that pressure exerted by a gas is given by

$$p = \frac{1}{3} mn\overline{v^2}$$

On arranging terms, we can write

$$n = \frac{3p}{m\overline{v^2}} = \frac{\frac{3}{2}p}{\frac{1}{2}m\overline{v^2}} = \frac{\frac{3}{2}p}{\varepsilon}$$

where $\varepsilon = \frac{1}{2}m\overline{v^2}$ is mean energy. On substituting the values of p and ε, we get

$$n = \frac{1.5 \times 10^5 \text{ Nm}^{-2}}{4 \times 10^{-11} \text{ Nm}} = 3.75 \times 10^{15} \text{ m}^{-3}$$

To calculate the number density per litre, we write

$$n = \frac{3.75 \times 10^{15}}{1 \text{ m}^3} = \frac{3.75 \times 10^{15}}{10^6 \text{ cm}^3}$$

so that

$$n = 3.75 \times 10^{12} \text{ per } 1000 \text{ cm}^3$$

or

$$n = 3.75 \times 10^{12} \text{ per litre}$$

(b) Since one mole occupies 22400 cm^3, the number of molecules

$$N = \frac{60 \text{ m}^3}{(22.4 \times 10^{-3} \text{ m}^3)} \times 6.023 \times 10^{23} = 1.61 \times 10^{27}.$$

Example 1.16 1.29 litre of a perfect gas weighs 2.7×10^{-3} kg at 18°C and 1 atm pressure. Compute its rms speed. Use 1 atm = 10^5 Nm^{-2} and $R = 8.31$ J mol^{-1} K^{-1}.

Solution: We know that $v_{rms} = \sqrt{\frac{3k_BT}{m}} = \sqrt{\frac{3RT}{M}}$. Therefore, to determine v_{rms}, we must know M. To do so, we recall the equation of state of a perfect gas:

$$pV = \mu RT$$

$$\therefore \qquad \mu = \frac{pV}{RT}$$

and

$$M = \frac{m}{\mu} = \frac{mRT}{pV} = \frac{(2.7 \times 10^{-3} \text{ kg}) \times (8.31 \text{ J mol}^{-1}\text{K}^{-1}) \times (18 + 273)\text{K}}{(10^5 \text{ Nm}^{-2}) \times (1.29 \times 10^3 \times 10^{-6} \text{ m}^3)}$$

$\therefore$
$$M = 0.05061 \text{ kg mol}^{-1}$$

Hence,

$$v_{rms} = \sqrt{\frac{3RT}{M}} = \sqrt{\frac{3 \times (8.31 \text{ J mol}^{-1}\text{K}^{-1}) \times (291 \text{ K})}{0.05061 \text{ kg mol}^{-1}}} = 378.6 \text{ ms}^{-1}.$$

Example 1.17 **(a)** Two ideal monatomic gases at temperatures T_1 and T_2 are mixed so that there is no loss of energy. If the masses and the number of molecules of the two gases are m_1, m_2 and n_1, n_2, respectively, calculate the temperature of the mixture.
(b) Solve the above problem if the gases are polyatomic and the degrees of freedom of the molecules of the first and second gas respectively are f_1 and f_2.

Solution: **(a)** $K.E.$ of the first gas $E_1 = \frac{3}{2} n_1 k_B T_1$

and $K.E.$ of the second gas $E_2 = \frac{3}{2} n_2 k_B T_2$

$\therefore$ Total $K.E.$ of the molecules of the gases before mixing

$$E_1 + E_2 = \frac{3}{2} k_B (n_1 T_1 + n_2 T_2)$$

Let the temperature of the mixture be T after attaining equilibrium. Then, we can write

$$K.E. \text{ of the first gas} = \frac{3}{2} n_1 k_B T = E_1' \text{ (say)}$$

and
$$K.E. \text{ of the second gas} = \frac{3}{2} n_2 k_B T = E_2' \text{ (say)}$$

$\therefore$
$$\text{Total } K.E. = E_1' + E_2' = \frac{3}{2} k_B (n_1 + n_2) T$$

As there has been no loss of energy, we have

$$\frac{3k_B}{2}(n_1 + n_2) T = \frac{3k_B}{2}(n_1 T_1 + n_2 T_2)$$

or

$$T = \frac{n_1 T_1 + n_2 T_2}{n_1 + n_2}$$

(b) For polyatomic gases, whose molecules have f_1 and f_2 degrees of freedom, the expressions for E_1 and E_2 would be modified as

$$E_1 = \frac{f_1}{2} n_1 k_B T_1, \quad \text{and} \quad E_2 = \frac{f_2}{2} n_2 k_B T_2$$

$\therefore$
$$E_1 + E_2 = \frac{k_B}{2}(f_1 n_1 T_1 + f_2 n_2 T_2)$$

Again, after mixing, we can write

$$E_1' = \frac{f_1}{2} n_1 k_B T, \qquad E_2' = \frac{f_2}{2} n_2 k_B T$$

$$\therefore \qquad E_1' + E_2' = \frac{k_B T}{2}(n_1 f_1 + n_2 f_2)$$

Hence, conservation of energy leads us to

$$\frac{k_B T}{2}(n_1 f_1 + n_2 f_2) = \frac{k_B}{2}(f_1 n_1 T_1 + f_2 n_2 T_2),$$

or

$$T = \frac{f_1 n_1 T_1 + f_2 n_2 T_2}{n_1 f_1 + n_2 f_2}$$

Example 1.18 Use the result of Example 1.17 to work out the temperatures of mixture in the following cases:

(a) One gm mole of argon at 40°C is mixed with one gm mole of helium at 56°C.

(b) One gm mole of oxygen at 27°C is mixed with two gm mole of helium at 127°C.

Solution: **(a)** Since both gases are monatomic, we can write

$$T = \frac{n_1 T_1 + n_2 T_2}{n_1 + n_2}$$

Here, $T_1 = (40 + 273)$ K = 313 K, $T_2 = (56 + 273)$ K = 329 K and $n_1 = n_2 = 1$

$$\therefore \qquad T = \frac{1 \times 313 + 1 \times 329}{2}\text{K} = \frac{313 + 329}{2}\text{K} = \frac{642}{2}\text{K}$$

$$\therefore \qquad T = 321 \text{ K} = (321 - 273)°\text{C} = 48°\text{C}$$

(b) Since oxygen is diatomic and helium is monatomic, we have

$$T = \frac{f_1 n_1 T_1 + f_2 n_2 T_2}{n_1 f_1 + n_2 f_2}$$

Here, $f_1 = 5, f_2 = 3, n_1 = 1, n_2 = 2, T_1 = (27 + 273)$ K = 300 K, $T_2 = (127 + 273)$ K = 400 K

$$\therefore \qquad T = \frac{5 \times 1 \times 300 + 3 \times 2 \times 400}{1 \times 5 + 2 \times 3}\text{K}$$

$$= \frac{1500 + 2400}{11}\text{K} = \frac{3900}{11}\text{K} = 354.5\text{K}$$

$$= (354.5 - 273)°\text{C} = 81.5°\text{C}.$$

Example 1.19 10 g hydrogen is kept enclosed in a vessel at a temperature of 47°C. Calculate the quantity of heat required to increase the rms speed of its molecules by three-fold. Take $R = 8.31$ J mol^{-1} K^{-1}.

Solution: We know that

$$v_{\text{rms}} \propto \sqrt{T}$$

$$\therefore \qquad \frac{(v_{rms})_2}{(v_{rms})_1} = \sqrt{\frac{T_2}{T_1}}$$

or

$$\sqrt{\frac{T_2}{T_1}} = 3$$

$$\Rightarrow \qquad \frac{T_2}{T_1} = 9 \qquad \text{or} \qquad T_2 = 9T_1$$

Since

$$T_1 = 47°\text{C} = (47 + 273)\text{ K} = 320\text{ K},$$

it readily follows that

$$T_2 = 9 \times 320\text{ K} = 2880\text{ K}$$

Since the gas is kept in a closed vessel, it absorbs heat at constant volume. Therefore, we can write

$$Q = \mu C_V(T_2 - T_1) = \mu C_V(9T_1 - T_1) = 8\mu C_V T_1$$

$$\therefore \qquad Q = \frac{8m}{M} C_V T_1$$

where m is mass of the gas, and M is molecular weight. We can rewrite this expression as

$$\begin{aligned} Q &= \frac{8m}{M} \times \frac{5R}{2} \times T_1 \\ &= 20\frac{m}{M} \times R \times T_1 \\ &= 20 \times \frac{10\text{ g}}{2\text{ g mol}^{-1}} \times (8.31\text{ J mol}^{-1}\text{K}^{-1}) \times (320\text{ K}) \\ &= 10 \times 10 \times 8.31 \times 320\text{ J} \\ &= 265920\text{ J} \\ &= \frac{265920}{4.2}\text{ cal} = 63314\text{ cal.} \end{aligned}$$

Example 1.20 The measurement of v_{rms} of particles of mass 5.9×10^{-17} kg suspended in a liquid maintained at 27°C shows that they move with $v_{rms} = 1.45 \times 10^{-2}\text{ ms}^{-1}$. Calculate Avogadro's number N_A using the principle of equipartition of energy. Given $R = 8.314\text{ JK}^{-1}\text{ mol}^{-1}$.

Solution: Since the particles move in a liquid, they execute translational motion only. Therefore, from Eq. (1.12), we can recall that

$$\varepsilon = \frac{3R}{2N_A}T$$

In terms of v_{rms},
$$\varepsilon = \frac{1}{2}m(v_{\text{rms}})^2 = \frac{3RT}{2N_A}.$$

$\Rightarrow$
$$N_A = \frac{3RT}{m(v_{\text{rms}})^2}$$

On substituting the given values, we get

$$N_A = \frac{3\times(8.314\ \text{JK}^{-1}\text{mol}^{-1})\times(300\text{K})}{(5.9\times10^{-17}\ \text{kg})\times(1.45\times10^{-2})^2\ \text{m}^2\text{s}^{-2}}$$

$$= \frac{24.942\times300\ \text{J mol}^{-1}}{(5.9\times10^{-17}\ \text{kg})\times2.10\times10^{-4}\ \text{m}^2\text{s}^{-2}}$$

$$= \frac{7482.6\ \text{kg m}^2\text{s}^{-2}\text{mol}^{-1}}{12.39\times10^{-21}\ \text{kg}\text{m}^2\text{s}^{-2}}$$

$$= 6.03\times10^{23}\ \text{mol}^{-1}$$

Example 1.21 Calculate the value of v_x for which the probability of a molecule having x-velocity component falls to half of the maximum value.

Solution: The probability of a molecule having x-velocity component between $v_x + dv_x$ is given by

$$f(v_x)dv_x = \left(\frac{m}{2\pi k_B T}\right)^{1/2} \exp\left[-\frac{mv_x^2}{2k_B T}\right]dv_x$$

Obviously, the maximum value of $f(v_x)$ is

$$f_{\max}(v_x) = \left(\frac{m}{2\pi k_B T}\right)^{1/2}$$

Hence, for $f(v_x) = \dfrac{f_{\max}(v_x)}{2}$, we can write

$$\frac{1}{2}\left(\frac{m}{2\pi k_B T}\right)^{1/2} = \left(\frac{m}{2\pi k_B T}\right)^{1/2} \exp\left[-\frac{mv_x^2}{2k_B T}\right]$$

On simplification, we get

$$\frac{1}{2} = \exp\left[-\frac{mv_x^2}{2k_B T}\right]$$

On taking natural logarithm of both sides, we get

$$-\ln 2 = -\frac{mv_x^2}{2k_B T}.$$

$$\Rightarrow \qquad v_x = \sqrt{\frac{2k_B T}{m}\ln 2}$$

$$= \sqrt{1.386\frac{k_B T}{m}}$$

Example 1.22 Calculate the temperature at which **(a)** root mean square speed of oxygen molecules exceeds their most probable speed by 150 ms^{-1} and **(b)** the molecular distribution function for oxygen will peak for 400 ms^{-1}.

Solution: **(a)** We know that for a Maxwellian gas

$$v_{rms} = \sqrt{\frac{3k_B T}{m}}$$

and

$$v_p = \sqrt{\frac{2k_B T}{m}}$$

Hence,

$$v_{rms} - v_p = 150\ ms^{-1} = \sqrt{\frac{3k_B T}{m}} - \sqrt{\frac{2k_B T}{m}}$$

On squaring both sides, we get

$$22500\ m^2 s^{-2} = \frac{k_B T}{m}(\sqrt{3} - \sqrt{2})^2$$

$$= \frac{k_B T}{m}(5 - 2\sqrt{6})$$

Hence,

$$T = \frac{m \times (22500\ m^2 s^{-2})}{k_B\ (5 - 2\sqrt{6})}$$

Since

$$m = \frac{32\ kg}{6.023 \times 10^{-26}}$$

we get

$$T = \frac{(32\ kg) \times 22500\ (ms^{-1})^2}{6.023 \times 10^{26} \times (1.38 \times 10^{-23}\ JK^{-1}) \times (5 - 2\sqrt{6})}$$

$$= \frac{720}{8.312 \times (0.101)} \text{K}$$

$$= 858 \text{ K}$$

(b) The molecular function attains maximum value at $v_p = \sqrt{\frac{2k_B T}{m}}$. Therefore, by rearranging term, we can write

$$T = \frac{m v_p^2}{2k_B} = \frac{(32 \text{ kg}) \times (400 \text{ ms}^{-1})^2}{2 \times 6.023 \times 10^{26} \times (1.38 \times 10^{-23} \text{ JK}^{-1})}$$

$$= \frac{512 \times 10^4 \times 10^{-3}}{2 \times 8.312} \text{K}$$

$$= 308 \text{ K}$$

Example 1.23 Nitrogen molecules obey Maxwellian distribution law and their mean energy is 15.6×10^{-21} J. Calculate their mean speed.

Solution: The mean speed of a Maxwellian gas $\bar{v} = \sqrt{\frac{8k_B T}{m\pi}}$. Therefore, to determine $\bar{v}$, we must know $k_B T$. To this end, we recall that nitrogen is a diatomic gas and its molecules have five degrees of freedom. Hence, its mean energy $= \frac{f}{2} k_B T = \frac{5}{2} k_B T = 15.6 \times 10^{-21}$ J so that $k_B T = \frac{2}{5} \times 15.6 \times 10^{-21} \text{ J} = 6.24 \times 10^{-21}$ J.

The mass of nitrogen molecule $m = \frac{28}{6.023 \times 10^{26}}$ kg

$$\therefore \qquad \bar{v} = \sqrt{\frac{8 \times (6.24 \times 10^{-21} \text{ J}) \times (6.023 \times 10^{26})}{(28 \text{ kg}) \times 3.14}} = 5.85 \times 10^2 \text{ ms}^{-1}$$

Example 1.24 In a nuclear fusion reactor, deuterium nuclei fuse if their average energy is 0.72 MeV. Calculate the temperature required for nuclear fusion to occur.

Solution: The average kinetic energy

$$\varepsilon = \frac{3}{2} k_B T$$

$$\Rightarrow \qquad T = \frac{2}{3} \frac{\varepsilon}{k_B}$$

$$= \frac{2}{3} \times \frac{(0.72 \times 10^6 \text{ eV}) \times (1.6 \times 10^{-19} \text{ JeV}^{-1})}{1.38 \times 10^{-23} \text{ JK}^{-1}}$$

$$= 5.59 \times 10^9 \text{ K}$$

That is, nuclear fusion can occur at a temperature of about 5.6 billion kelvin. Such temperatures occur in the core of the Sun. Therefore, in producing energy by nuclear fusion, we have to reproduce conditions existing in the interior of the Sun. This anyway is a challenging task. But now we are very close to achieving it using extremely sophisticated technologies.

Example 1.25 14 g of N_2 and 4 g of He are mixed in a thermally insulated container. What will be the rms speeds of N_2 and He molecules after equilibrium is reached, if their rms speeds before mixing were respectively 500 ms^{-1} and 1500 ms^{-1}? The molecular weights of nitrogen and helium are respectively 28 and 4.

Solution: If we denote the degree of freedom of a gas molecule by f, then its average energy is given by

$$\varepsilon = \frac{f}{2} k_B T$$

Let v_{rms} be the rms speed at temperature T, then we can write

$$v_{rms}^2 = \overline{v^2} = \frac{3 k_B T}{m}$$

or

$$k_B T = \frac{1}{3} m \overline{v^2}$$

where m is mass of a molecule.

$$\therefore \qquad \varepsilon = \frac{1}{2} f \frac{1}{3} m \overline{v^2} = \frac{f}{6} m \overline{v^2}$$

If the gas has n molecules, then its total energy will be

$$E = n\varepsilon = \frac{f}{6} mn \overline{v^2} = \frac{f}{6} M \overline{v^2}$$

where M is total mass of n molecules.
Now, for nitrogen we have

$$f = 5 = f_1, \quad \text{say,} \quad M_N = 14\text{g}, \quad v_{rms} = 500 \text{ ms}^{-1} = v_1, \text{ say, and } M_1 = 28 \text{ g}.$$

Thus, the total energy before mixing is $E_1 = \dfrac{f_1}{6} M_N v_1^2$

For helium, we have

$$f = 3 = f_2, \quad \text{say,} \quad M_{He} = 4\text{g}, \quad v_{rms} = 1000 \text{ ms}^{-1} = v_2, \text{ and } M_2 = 4\text{g}.$$

Thus, the total energy of helium before mixing is $E_2 = \dfrac{f_2}{6} M_{He} v_2^2$

$\therefore$ The total energy of both the gases before mixing is

$$E_1 + E_2 = \frac{1}{6}\left(f_1 M_N v_1^2 + f_2 M_{He} v_2^2\right)$$

Now let the rms velocities after mixing be v_{11} and v_{22} for nitrogen and helium, respectively. So the total energy after mixing is given by

$$\frac{1}{6}\left(f_1 M_N v_{11}^2 + f_2 M_{He} v_{22}^2\right)$$

Using the principle of conservation of energy, we can write

$$\frac{1}{6}\left(f_1 M_N v_1^2 + f_2 M_{He} v_2^2\right) = \frac{1}{6}\left(f_1 M_N v_{11}^2 + f_2 M_{He} v_{22}^2\right)$$

or

$$\left(f_1 M_N v_1^2 + f_2 M_{He} v_2^2\right) = \left(f_1 M_N v_{11}^2 + f_2 M_{He} v_{22}^2\right)$$

On putting the values of various quantities, we get

$$\therefore \quad 5 \times (14\ \text{g}) \times (500\ \text{ms}^{-1})^2 + 3 \times (4\ \text{g}) \times (1000\ \text{ms}^{-1})^2 = 5 \times (14\ \text{g}) \times v_{11}^2 + 3 \times (4\ \text{g})\, v_{22}^2$$

or $$70\, v_{11}^2 + 12\, v_{22}^2 = 29.5 \times 10^6\ \text{m}^2\,\text{s}^{-2}$$

or $$35\, v_{11}^2 + 6\, v_{22}^2 = 14.75 \times 10^6\ \text{m}^2\,\text{s}^{-2} \qquad \text{(i)}$$

In the mixture, under equilibrium conditions, the molecules of both gases will be at the same temperature, say T. Therefore, we can write

$$\therefore \quad RT = \frac{1}{3} M_1 v_{11}^2 = \frac{1}{3} M_2 v_{22}^2$$

so that

$$\frac{v_{11}^2}{v_{22}^2} = \frac{M_2}{M_1} = \frac{4\,\text{g}}{28\,\text{g}} = \frac{1}{7}$$

$$\therefore \quad v_{22}^2 = 7\, v_{11}^2 \qquad \text{(ii)}$$

From Eqs. (i) and (ii), we get

$$35\, v_{11}^2 + 6 \times 7\, v_{11}^2 = 14.75 \times 10^6\ \text{m}^2\,\text{s}^{-2}$$

or $$77\, v_{11}^2 = 14.75 \times 10^6\ \text{m}^2\,\text{s}^{-2}$$

$$\therefore \quad v_{11}^2 = \frac{14.75}{77} \times 10^6\ \text{m}^2\,\text{s}^{-2}$$

$$\therefore \quad v_{11} = \sqrt{\frac{1475}{77} \times 10^4}\ \text{ms}^{-1} = 437.7\ \text{ms}^{-1}$$

From Eq. (ii), it readily follows that

$$v_{22} = \sqrt{7}\, v_{11} = 1158.0\ \text{ms}^{-1}$$

$\therefore$ v_{rms} for nitrogen and helium are respectively 437.7 ms^{-1} and 1158.0 ms^{-1}

Let us now sum up what you have learnt in this chapter.

SUMMARY

- The pressure exerted by the molecules of a gas on the walls of a container is given by $p = \frac{1}{3} mn\overline{v^2}$.
- The root mean square speed of a gas molecule is given by

$$v_{\text{rms}} = \sqrt{\overline{v^2}} = \sqrt{\frac{3RT}{M}} = \sqrt{\frac{3k_BT}{m}}.$$

- At absolute zero temperature, gas molecules are devoid of all motion.
- The law of equipartition of energy states that thermal energy is equally divided amongst various active degrees of freedom of a molecule and its magnitude is $k_BT/2$.
- For a monatomic gas, the molar heat capacity at constant volume, molar heat capacity at constant pressure and their ratio are given by $C_V = \frac{3R}{2}, C_p = \frac{5R}{2}$ and $\gamma = \frac{5}{3} = 1.67$. For a diatomic gas, the corresponding values are $C_V = \frac{5R}{2}$, $C_p = \frac{7R}{2}$ and $\gamma = 1.4$. The ratio of molar heat capacities decreases with increasing atomicity of gases and their ratios are independent of temperature.
- The distribution of molecular speeds is given by Maxwell's law:

$$dN_v = 4\pi N\left(\frac{m}{2\pi k_BT}\right)^{\frac{3}{2}} e^{-\left(\frac{mv^2}{2k_BT}\right)} v^2\,dv$$

- The root mean square speed, the average speed, and the most probable speed are respectively given by $v_{rms} = \sqrt{\frac{3k_BT}{m}}$, $\overline{v} = \sqrt{\frac{2.55\,k_BT}{m}}$ and $v_p = \sqrt{\frac{2k_BT}{m}}$. As may be noted, $v_{rms} > \overline{v} > v_p$.
- The number of molecules of a Maxwellian gas with energies in the range E and $E + \mathrm{d}E$ is given by

$$\mathrm{d}N_E = 2N\left(\frac{E}{\pi}\right)^{1/2}\left(\frac{1}{k_BT}\right)^{3/2} \exp\left(-\frac{E}{k_BT}\right)\mathrm{d}E$$

- The average energy of a molecule obeying Maxwell's law is $\overline{\varepsilon} = \frac{3}{2}k_BT$.
- The intensity of Doppler-broadened spectral line at half-width is given by

$$\frac{I_0}{I_b} = 2 = \exp\left(\frac{mb^2 f_0^2}{2k_BT}\right)$$

- The half-width of Doppler-broadened spectral line is given by

$$b = \frac{1}{f_0}\sqrt{\frac{2k_BT\ln 2}{m}} = \frac{\lambda_0}{c}\sqrt{\frac{2RT\ln 2}{M}}.$$

EXERCISES

1.1 Explain the following on the basis of kinetic theory of gases: (a) Thermal expansion, (b) Evaporation and the subsequent cooling of the residual liquid and (c) lower temperature in upper atmosphere though solar radiations incident directly raise its kinetic temperature to about 1000 K.

1.2 Estimate the time that a nitrogen molecule moving at rms will take in travelling across your lecture theatre of length 10 m. Assume that it does not undergo any collision. (*Ans:* $t \approx 0.02$ s)

1.3 A parallel beam of nitrogen molecules moving with velocity 400 ms^{-1} impinges on a wall at an angle of 30° to the normal. If $n = 9 \times 10^{24}$ m^{-3}, calculate the pressure exerted by the beam on the wall assuming that the molecules suffer elastic collisions. (*Ans:* $p \approx 1$ atm)

1.4 A gas composed of 10^6 carbon atoms has a Maxwellian velocity distribution at $T = 300$ K. Determine the number of atoms having velocities between **(a)** 100 ms^{-1} and 101 ms^{-1}, **(b)** 300 ms^{-1} and 301 ms^{-1}, and **(c)** 1500 ms^{-1} and 1501 m^{-1}. Express your results up to two significant figures. (*Ans:* (a) 83; (b) 613; (c) 82)

1.5 Show that the root mean square value of the Cartesian v_x of molecular velocity is equal to $\sqrt{p/\rho}$, which defines velocity of sound.

1.6 Calculate the temperature at which the root mean square velocity of hydrogen molecules will exceed their most probable velocity by 400 ms^{-1}. (*Ans:* 384K)

1.7 Express Maxwell velocity distribution function in terms of $u = v/v_p$.

$$\left(Ans:\ dN = \frac{4N}{\sqrt{\pi}} u^2 \exp(-u^2)\, du\right)$$

1.8 Calculate the ratio of the velocity of sound in a gas to the rms velocity of its molecules if the molecules are (i) monatomic and (ii) diatomic. (*Ans:* 0.75; 0.68)

1.9 At what temperature will the average speed of hydrogen molecules be the same as that of nitrogen molecules at 308 K? Take molecular weight of nitrogen as 28 and that of hydrogen as 2. (*Ans:* 22 K)

1.10 Starting from Eq. (1.71), calculate the Doppler broadening of hydrogen line having wavelength 486.1 nm at $T = 400$ K. Use $k_B = 1.38 \times 10^{-23}$ JK^{-1} and mass of hydrogen atom as 1.6725×10^{-27} kg. (*Ans:* 345.8×10^{-5} nm)

2 MEAN FREE PATH AND TRANSPORT PHENOMENA

Learning Objectives

In this chapter, you will learn how to

- define mean free path and derive an expression when all molecules move with the same average speed;
- describe the law of free paths and explain how it is used to measure mean free path;
- derive an expression for the coefficient of viscosity and discuss its temperature/pressure dependence;
- obtain expressions for thermal conductivity and coefficient of self-diffusion and relate these to the coefficient of viscosity and mean free path, which are measurable quantities;
- explain the significance of Brownian motion and obtain expressions for the mean square displacement on the basis of Einstein's and Langevin's theories;
- discuss examples of Brownian motion in physics; and
- describe Perrin's experiments and discuss their significance for kinetic theory of gases.

2.1 INTRODUCTION

In the previous chapter, we considered a gas in a state of equilibrium and calculated the pressure exerted by it on the walls of the container as well as the probability that a molecule has velocity components in a particular range. We know that if a system is not in the equilibrium state, it has inherent tendency to approach it. Therefore, a logical question arises: How does a gaseous substance, initially in a non-equilibrium state, attain equilibrium? Further, you have learnt that molecules of an ideal gas behave as point masses, which move randomly and obey Maxwell's velocity distribution law. We have also seen that even at room temperature, the molecules of oxygen move with very large speeds; $v_{rms} = 480\ ms^{-1}$. If this were true, a small amount of gaseous mass released in a large room from a vessel should spread throughout the room in no time. But this is not supported by common observation; when we open the lid of a perfume bottle in one corner of the room, the odour can be smelt at the other corner after a considerable time. This apparent paradox raised a serious question mark on the kinetic theory of gases in its early stages of development. In fact, prominent scientists were reluctant to accept the realities of the atoms and molecules. The German physicist, Wilhelm Ostwald regarded the atoms as a merely hypothetical conception that afforded a very convenient picture of matter. Similarly, Ernst Mach maintained that atoms and molecules must be treated as convenient fictions. However, Clausius was convinced about the molecular nature and gave a simple explanation to the question of time taken by the odour to spread in a room. He argued that gas molecules have finite size and as they move, they collide with one another and increase the diffusion time. (This amounts to modifying the basic assumption about the size of molecules.) It also implies that a gaseous system attains equilibrium through molecular collisions. You may now like to know: What is the average distance travelled by a molecule between successive collisions? What is the mechanism of these collisions? How frequently do these collisions occur. A careful consideration of these aspects leads us to the concept of *mean free path.*

In Sec. 2.2, we have obtained expressions for the mean free path under the assumption that all molecules move with the same average speed. You will notice that this will help us to get an insight into the basic physics of molecular collisions. (If we consider Maxwell's distribution law to calculate relative speeds, mathematical steps get quite involved. However, those of you who are interested in rigorous analysis, see Appendix 2.) The distribution of free paths and the experimental determination of mean free path are also discussed in this section.

The non-equilibrium state of a gas can arise due to its mass motion in a particular direction. Suppose that a gas moves with a flow velocity towards the right. In addition, if its molecules experience velocity gradient in a direction normal to the direction of flow, there is a net transport of momentum across an imaginary plane in the gas along the direction of mass motion. This is characterised by the *coefficient of viscosity*. You will note that in gases, unlike the case of liquids, random thermal motion associated with gas molecules gives rise to the phenomenon of viscosity. (In liquids, frictional force between successive layers causes viscosity.) In Sec. 2.3.1, you will learn to derive an expression for the coefficient of viscosity in terms of mean free path and discuss its temperature as well as pressure dependence.

When a gas endowed with mass motion experiences temperature gradient in a direction perpendicular to the direction of flow, gas molecules lead to net transport of thermal energy in the direction of flow. This gives rise to the phenomenon of thermal conduction, which is characterised by *thermal conductivity*. You will learn to correlate it with the coefficient of viscosity in Sec 2.3.2. When concentration gradient exists inside a gas normal to the direction of flow, there is net transport of matter (mass) in the direction of preferential motion. This is characterised by the *coefficient of diffusion*. You will learn to derive an expression for the coefficient of diffusion in terms of mean free path and discuss its temperature dependence in Sec. 2.3.3. These processes (viscosity, thermal conduction and diffusion) are collectively referred to as *transport phenomena*, since we observe net transfer of momentum, energy or matter.

In spite of visible success of kinetic theory model, there was no way to visualise it till the observations of English botanist Robert Brown on small pollen grains suspended in an aqueous solution. His experiments provided us a way to visualise the gaseous state. In fact, *Brownian movement* is taken as the first direct evidence in favour of kinetic theory of gases and the observations did put the theory on very firm foundations. In order to appreciate its importance, we have given a phenomenological treatment of Brownian motion due to Einstein and Langevin in Sec.2.4. In this section, we have also discussed a few interesting examples of Brownian motion. This is followed by a discussion of Perrin's experiments. The *random walk problem*, which provides models for numerous applications in physics, including Brownian motion, is discussed in Sec. 2.5.

2.2 Mean Free Path

You may recall that in deriving an expression for the pressure exerted by the molecules of an ideal (perfect) gas, we regarded gas molecules as point masses (Assumption 2, Sec.1.2) and ignored intermolecular collisions. This provided very satisfactory explanation of vertical distribution of gases in our atmosphere. We have also seen that even at room temperature, the root mean square speed of air molecules is about 480 ms^{-1}. It means that a given mass of a gas released in a large room from a vessel should spread throughout the room in no time. This however does not happen in practise; when we open the lid of a perfume bottle in one corner of the room, the smell reaches the other corner after a considerable time.

To explain the finite time taken by a gaseous mass in an open container to disperse, Clausius traced the discrepancy to the assumption of 'point' molecules, which is equivalent to ignoring intermolecular collisions. Clausius suggested that slow diffusion of a gas is due to finite, though small, size of gas molecules. He argued that when a gas molecule moves in the body of the gas under equilibrium conditions, it necessarily collides with other molecules it encounters in its path. As a result, its direction of motion changes frequently. This results in a series of zigzag paths as shown in Fig. 2.1. The path covered by a molecule between successive collisions is called a *free path*. Note that in Fig. 2.1, *AB*, *BC*, *CD*, etc., are all free paths.

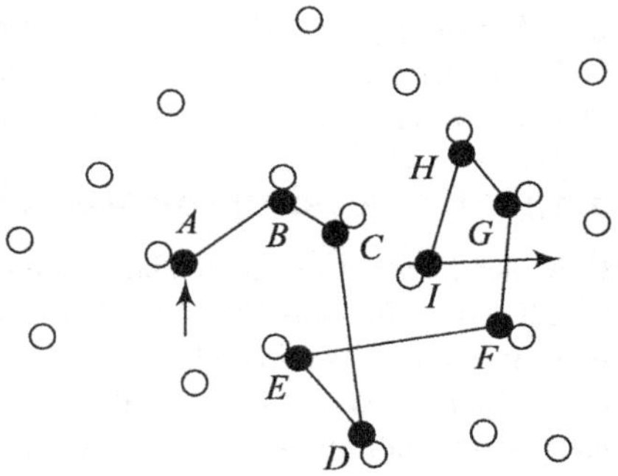

Fig. 2.1 The trajectory of a molecule moving in a gas: Molecular free paths.

To visualise the concept of free path, suppose that you are participating in a rally, which somehow turns unpleasant. The police resort to bursting tear gas shells to disperse the mob. While running for cover to safety, you are bound to collide with other people. In the course of your retreat, you are likely to change your direction quite frequently and randomly. A trace of your steps will resemble the motion of a gas molecule within the gas.

Let us examine Fig. 2.1 carefully. You will notice that free paths are of unequal length; some are short while others are long. The *average length of free paths is referred to as mean free path*. Thus, *mean free path is the average distance travelled by a molecule between two successive collisions*. We denote it by the symbol λ (pronounced as lambda). If λ_1, λ_2, λ_3, ... λ_N denote the successive free paths traversed in time t and N is the total number of collisions suffered in this period, we can write

$$\lambda = \frac{\lambda_1 + \lambda_2 + \lambda_3 + \ldots + \lambda_N}{N}$$

$$= \frac{\text{Total distance travelled}}{\text{Total number of collisions}}$$

If $\bar{v}$ is the average speed with which a molecule moves in the body of the gas, then we can write

$$\lambda = \frac{\bar{v}\,t}{N} \tag{2.1}$$

Under normal conditions, the mean free path is about 2×10^{-7} m, which is smaller than the wavelength of visible light (4×10^{-7} m – 6×10^{-7} m) but greater than the average intermolecular separation (3×10^{-9} m).

If $\tau\left(=\frac{t}{N}\right)$ denotes the mean time between two successive collisions, we can also write

$$\lambda = \bar{v}\,\tau = \frac{\bar{v}}{P_c} \tag{2.2}$$

Here $P_c = (\tau^{-1})$ denotes the collision frequency, which is a measure of the average number of collisions per second. You may now ask: How can we relate λ to the microscopic properties of a gas? To answer this question, we make an elementary calculation.

2.2.1 Expression for Mean Free Path

Consider a gas consisting of a large number of molecules, each of mass m and diameter d, which undergo random collisions. Mathematically speaking, we can say that the probability that a molecule suffers a collision in a small interval of time dt is independent of the history of past collisions. As a simple analogy, we can consider the throwing of a dice. In this case, the probability of getting a six does not depend on the preceding throw, where a six may or may not have appeared.

Expression for P_c We shall make this calculation in two simple situations: (i) When only one molecule is in motion and (ii) When all molecules are moving with an average velocity. We have referred to these as Zeroth and First order approximations, respectively. You will realise that essential physics involved in this phenomenon can be understood well through simple arguments.

Zeroth order approximation: Only one molecule in motion We assume that only one molecule moves with speed v in a sea of other molecules, which are assumed to be at rest, i.e., all molecules of the gas except one are frozen.* (It means that the relative speed of one molecule with respect to any other molecule is v).You can easily visualise that for a collision to take place, the centre-to-centre distance between the moving and a stationary molecule should be equal to the diameter d of a gas molecule as shown in Fig. 2.2(a). Note that the centre-to-centre distance at the time of collision between a moving and a stationary molecule will also be equal to d when the stationary molecules were shrunk to geometrical points and the radius of the moving molecule is increased to d as shown in Fig. 2.2(b). We can visualise this as if the moving molecule carries with it a circular disc of radius d. Therefore, in time t, it can be thought of as sweeping out a cylinder of cross sectional area πd^2 and length vt. This is illustrated in Fig. 2.2(c). (The quantity πd^2 is also termed as the *microscopic collision cross section* and is usually denoted as σ.) During time t, it will collide with all other molecules whose centres lie within a cylinder of volume $vt\,\pi d^2$.

If there are n number of molecules per unit volume, the number of molecules contained in the cylinder of volume $vt\,\pi d^2$ will be equal to $vt\,\pi d^2 n$. Note that this number is also equal

* As such, this is a purely hypothetical situation. Nevertheless, it brings out the essential physical ideas.

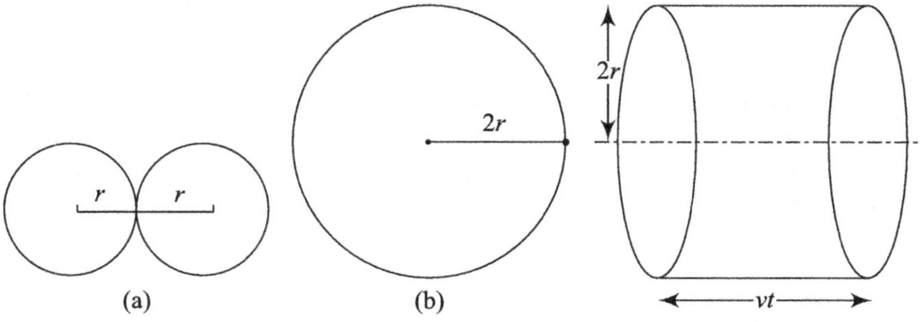

Fig. 2.2 (a) Collision of two molecules of radii r: Instantaneous representation, (b) Equivalent representation of collision depicted in (a) where radius of the moving molecule is increased to $d = 2r$ and the stationary molecule is shrunk a geometrical point and (c) Cylindrical volume swept out by a molecule of radius $2r$ and moving with speed v in time t.

to the number of collisions made by the moving molecule in time t. Hence, the number of collisions per second defines the collision frequency:

$$P_c = v\,\pi d^2\, n = n\sigma\, v \tag{2.3}$$

Substituting this result in Eq. (2.2), we get

$$\lambda = \frac{v}{P_c} = \frac{1}{\pi d^2 n} = \frac{1}{\sigma n}, \tag{2.4}$$

where σ is microscopic collision cross section. It is expressed in m^2. The quantity $n\sigma$ is called *macroscopic collision cross section.* Note that the unit of macroscopic collision cross section is m^{-1}, i.e., *reciprocal length*, not an area.

Equation (2.4) predicts that mean free path is inversely proportional to the macroscopic collision cross section, i.e., number density and second power of diameter of the molecule. It means that mean free path will be less for a denser and/or a heavier gas. This is in perfect agreement with common observation and lends support to the basic tenets of kinetic theory. You may now logically ask: What is the typical magnitude of mean free path in a gas at STP? The answer to this question is contained in the following example. Go through it carefully.

Example 2.1 A hydrogen molecule of radius 1.37×10^{-10} m moves with speed 1840 ms^{-1}. Calculate its (i) microscopic collision cross section, (ii) collision frequency and (iii) mean free path. Take $n = 3 \times 10^{25}\ m^{-3}$.

Solution:

(i) Since radius of a hydrogen molecule is 1.37×10^{-10} m and number density is $n = 3 \times 10^{25}\ m^{-3}$, we can calculate the microscopic collision cross section using the relation

$$\sigma = \pi d^2 = 4\pi \times (1.37 \times 10^{-10})^2\ m^2 = 23.6 \times 10^{-20}\ m^2$$

(ii) To calculate collision frequency and mean free path, we use Eqs. (2.3) and 2.4):

$$\begin{aligned} P_c &= \pi d^2 n v = n v \sigma \\ &= (3 \times 10^{25}\ m^{-3}) \times (1.84 \times 10^3\ ms^{-1}) \times (23.6 \times 10^{-20}\ m^2) \\ &= 1.3 \times 10^{10}\ s^{-1}. \end{aligned}$$

(iii) $\lambda = \dfrac{1}{n\pi d^2} = \dfrac{1}{(3\times 10^{25}\ \text{m}^{-3})\times(23.6\times 10^{-20}\ \text{m}^2)} = 14\times 10^{-8}\ \text{m} = 140\ \text{nm}$

Note that the number of collisions per second is of the order of 10^{10}, which is a very large number. For this reason, the path of molecule is made up of so many kinks and zigzags that it is almost impossible to follow its trajectory. Also note that λ is large compared to intermolecular distance, which is only a few angstroms ($\approx$ 3 nm).

Recall that using the ideal gas equation, we can express the pressure exerted by a gas in terms of its temperature as

$$p = nk_B T$$

On using this result to substitute for n in Eq. (2.4), the expression for mean free path takes the form

$$\lambda = \frac{k_B T}{\sigma p} \tag{2.5}$$

This result is very interesting; it predicts that mean free path is directly proportional to absolute temperature and inversely proportional to the pressure exerted by the gas on the walls of the container. It means that λ will increase as pressure decreases and/or temperature increases. Now suppose we reduce pressure to a very small value using a vacuum pump. According to Eq. (2.5), the mean free path can approach infinity. In actual practise, the value of λ can equal the dimensions of the container. This finds an interesting application in producing well-directed molecular beams for research in nuclear reactors and particle accelerators. For this reason, a proton is able to travel nearly 27 km in Hadron Collider Experiment conducted at CERN to search Higgs bosons.

As such, the analysis based on zeroth approximation seems rather artificial in that it does not account for the motion of all the molecules of a gas. Therefore, a slightly better representation of the actual situation will be to assume that all molecules move with the same velocity $\overline{v}$. Let us now discover how this will modify Eq. (2.4).

First order approximation: All molecules in motion Refer to Fig. 2.3. It shows the relative motion of two molecules moving with velocity $\overline{v}$ at an angle θ with respect to one another. Now the number of collisions will be determined by the magnitude of relative velocity between these molecules. Let us denote the relative velocity between these molecules as v_r. Then, we have to replace v in Eq. (2.3) with v_r and the expression for collision frequency modifies to

$$P_c = nv_r\sigma \tag{2.6}$$

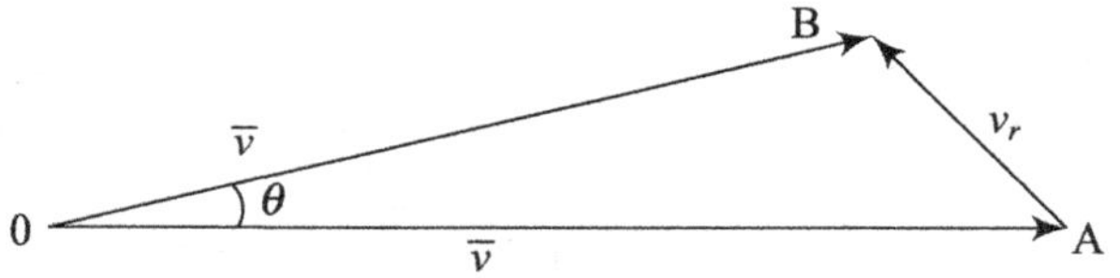

Fig. 2.3 Relative velocity of molecule 1 moving along *OA* with respect to molecule 2 moving along *OB*.

By referring to Fig. 2.3, we note that θ is the angle between *OA* and *OB*. Hence, we can write the magnitude of relative velocity of molecule 1 with respect to molecule 2 as

$$v_r = \sqrt{(\overline{v})^2 + (\overline{v})^2 - 2(\overline{v})^2 (\cos\theta)}$$

$$= \sqrt{2}\,\overline{v}(1-\cos\theta)^{1/2} = 2\overline{v}\sin\frac{\theta}{2} \tag{2.7}$$

In writing the last step, we have used the trigonometric relation $1 - \cos 2\theta = 2\sin^2\theta$. On replacing θ by $\theta/2$, we can write $\sin^2\left(\frac{\theta}{2}\right) = \frac{1}{2}(1-\cos\theta)$ so that $(1 - \cos\theta)^{1/2} = \sqrt{2}\sin\left(\frac{\theta}{2}\right)$.

Since all directions of velocity are equally probable, the mean relative *velocity* of any two molecules is obtained by averaging v_r as given by Eq. (2.7) over the solid angle, $d\Omega/4\pi$. Since $d\Omega = 2\pi\sin\theta\,d\theta$, with $0 < \theta < \pi$, we can write

$$\overline{v}_r = \frac{1}{4\pi}\int v_r\, d\Omega = \frac{1}{2}\int_0^{\pi} v_r \sin\theta\, d\theta$$

On substituting for v_r from Eq. (2.7), we get

$$\overline{v}_r = \overline{v}\int_0^{\pi} \sin\theta \sin\left(\frac{\theta}{2}\right) d\theta$$

If we now use the relation $\sin 2A = 2\sin A\cos A$, we can rewrite the expression for $\overline{v}_r$ as

$$\overline{v}_r = 2\overline{v}\int_0^{\pi} \sin^2\left(\frac{\theta}{2}\right)\cos\left(\frac{\theta}{2}\right) d\theta$$

To evaluate this integral, we introduce a change of variable by defining $x = \sin(\theta/2)$ so that $\cos(\theta/2)\,d\theta = 2\,dx$. Note that the limits of integration change to 0 and 1. Then the expression for $\overline{v}_r$ takes a very compact form

$$\overline{v}_r = 4\overline{v}\int_0^{1} x^2\, dx = \frac{4}{3}\overline{v} \tag{2.8}$$

On substituting this value of average relative velocity in place of v_r in Eq. (2.6), the expression for collision frequency in first order approximation modifies to

$$P_c = \frac{4}{3} n\sigma\,\overline{v} \tag{2.9}$$

On comparing this expression with that given in Eq. (2.3), we note that the collision frequency increases when motion of all the molecules is taken into consideration. This, of course, is expected physically since a molecule is more likely to suffer collision when all of them are in motion than in a collection of molecules at rest in their respective positions. On combining this result with Eq. (2.2) with $v = \overline{v}$, we find that the expression for mean free path modifies to

$$\lambda_{Cl} = \frac{3}{4n\sigma} = \frac{0.75}{n\sigma} = 0.75\frac{k_B T}{p\sigma}. \tag{2.10}$$

It may be mentioned here that this result was obtained by Clausius. (That is why we have put the subscript *Cl* with λ.) It shows that mean free path of the molecules in a gas decreases when all molecules are moving.

You will recall that in actual practise, we should have considered Maxwellian distribution of velocities while computing the average relative velocity. However, the mathematical calculations are quite involved and for the present purpose, it will be sufficient to quote the result. (Those of you who are interested in these details may refer to Appendix B. However, a compact alternative derivation is given in the box on page 2.10.):

$$\lambda_M = \frac{1}{\sqrt{2}\,n\sigma} = (0.707)\frac{1}{n\sigma} = (0.707)\frac{k_B T}{p\sigma} \tag{2.11}$$

Note that the ratio of mean free paths calculated on the basis of Maxwellian distribution of velocities and Clausius assertion of uniform velocity is 0.94. It means that there is a correction of about six per cent in the value of mean free path but the essential physics does not change. It suggests that replacing the Maxwell's distribution by a uniform velocity model is a fairly good approximation. You will also note that for point molecules ($d \to 0$) collision cross section $\sigma \to 0$ and mean free path $\lambda \to \infty$.

Now go through the following examples carefully to get a feel of the values involved in real situations.

Example 2.2 A gas is assumed to obey Maxwell's law for distribution of velocities. Calculate the **(a)** molecular diameter and **(b)** the number of collisions per unit distance if mean free path at STP is 2.85×10^{-7} m. Take $n = 3 \times 10^{25}$ m^{-3}.

Solution: From Eq. (2.11), we can write

$$\lambda = \frac{1}{\sqrt{2}}\frac{1}{\pi d^2 n}$$

(a) To determine d, we rewrite the expression for mean free path as

$$d^2 = \frac{1}{\sqrt{2}}\frac{1}{\pi \lambda n}$$

On substituting the given values, we get

$$d^2 = \frac{1}{\sqrt{2}} \times \frac{1}{3.14} \times \frac{1}{(2.85 \times 10^{-7}\text{ m})} \times \frac{1}{(3 \times 10^{25}\text{ m}^{-3})}$$

$$= \frac{1}{(37.97 \times 10^{18}\text{ m}^{-2})} = 2.63 \times 10^{-20}\text{ m}^2$$

so that $d = 1.62 \times 10^{-10}$ m $= 1.62$ Å

Note that the order of magnitude of mean free path is 10^3 times that of molecular diameter.

(b) Since number of collisions per unit distance is reciprocal of the mean free path, we can write

$$N_S = \frac{1}{\lambda} = \frac{1}{2.85 \times 10^{-7}\text{ m}} = 3.51 \times 10^6\text{ m}^{-1}$$

Example 2.3 ^{4}He gas is assumed to obey Maxwell's law for distribution of velocities. The molecular diameter of atoms of ^{4}He is 10^{-10} m. One mole of the gas occupies 20 litre at 20 K. Calculate the mean free path of the molecules, and the collision frequency. Given $R = 8.4$ Jmol^{-1}K^{-1} and $N_A = 6 \times 10^{23}$ mol^{-1}.

Solution: From Eq.(2.11), we recall that

$$\lambda = \frac{1}{\sqrt{2}}\frac{1}{\pi d^2 n}$$

Here $d = 10^{-10}$ m and $n = \dfrac{6\times 10^{23}\,\text{mol}^{-1}}{20\times 10^{-3}\,\text{m}^3\,\text{mol}^{-1}} = 3\times 10^{25}\,\text{m}^{-3}$. Hence,

$$\lambda = \frac{0.707}{(3.14\times 10^{-20}\,\text{m}^2)\times(3\times 10^{25}\,\text{m}^{-3})} = 7.5\times 10^{-7}\,\text{m}$$

The average speed of a molecule obeying Maxwellian distribution is given by

$$\bar{v} = \sqrt{\frac{2.55\,RT}{M}} = \left[\frac{2.55\times(8.4\,\text{Jmol}^{-1}\,\text{K}^{-1})\times(20\,\text{K})}{4\times 10^{-3}\,\text{kg mol}^{-1}}\right]^{1/2} = 3.3\times 10^2\,\text{ms}^{-1}$$

Hence, collision frequency is

$$P_c = \frac{\bar{v}}{\lambda} = \frac{3.3\times 10^2\,\text{ms}^{-1}}{7.5\times 10^{-7}\,\text{m}} = 4.4\times 10^8\,\text{s}^{-1}$$

The mean free time or average time between collisions is inverse of the collision frequency and in this case, it will be equal to

$$\tau = \frac{1}{4.4\times 10^8\,\text{s}^{-1}} = 2.3\times 10^{-7}\,\text{s}$$

Example 2.4 The mean free path of the molecules of a gas at a pressure p and temperature T is 3×10^7 m. Calculate the mean free path if **(a)** the temperature is doubled, and **(b)** the pressure is doubled.

Solution: From Eq. (2.11), we know that

$$\lambda = \frac{k_B T}{\sqrt{2}\pi d^2 p}$$

(a) When the temperature is doubled, we note that the mean free path will increase by a factor of two. Hence,

$$\lambda_{(p,2T)} = 2\,\lambda_{(p,T)} = 6\times 10^{-7}\,\text{m}$$

(b) When the pressure is doubled, the mean free path will be halved:

$$\lambda_{(2p,T)} = \frac{\lambda_{(p,T)}}{2} = 3\times 10^{-7}\,\text{m}$$

Example 2.5 The radius of argon atoms is 0.128 nm. Calculate their mean free path at 25°C and one atmosphere pressure. Given $k_B = 1.38 \times 10^{-23}$ JK^{-1}.

Solution: From Eq. (2.11), we recall that

$$\lambda = \frac{k_B T}{\sqrt{2}\pi d^2 p}$$

Here, $d = 2 \times 0.128 \times 10^{-9}$ m $= 0.256 \times 10^{-9}$ m, $T = 27°C = (273 + 27)$ K $= 298$ K, $k_B = 1.38 \times 10^{-23}$ JK^{-1} and $p = 1.013 \times 10^5$ Nm^{-2}. On substituting these values in (i), we get

$$\lambda = \frac{(1.38 \times 10^{-23}\ \text{JK}^{-1}) \times (298\ \text{K})}{1.414 \times 3.14 \times (0.256 \times 10^{-9}\ \text{m})^2 \times (1.013 \times 10^5\ \text{Nm}^{-2})}$$

$$= \frac{414 \times 10^{-10}\ \text{J}}{0.2948\ \text{N}} = 1.39 \times 10^{-7}\ \text{m}$$

Now you may like to solve the following problem.

Problem 2.1 Calculate the mean free path of the molecules of a gas of diameter 0.2 nm. Redo the calculations for molecules of diameter 0.4 nm. Take $n = 3 \times 10^{25}$ molecules m^{-3}. How does it compare with intermolecular separation at STP?

Ans: 1.88×10^{-7} m; 4.69×10^{-8} m

Problem 2.2 The mean free path of the molecules of a gas at 15°C is 6.28×10^{-8} m. If the radius of the molecule is 1.88 Å, calculate the pressure exerted by the gas. Also calculate the number of collisions suffered by a molecule in traversing one metre. Take $k_B = 1.38 \times 10^{-23}$ JK^{-1}.

Ans: 1.01×10^5 Nm^{-2}, 16×10^6 m^{-1}

ALTERNATIVE DERIVATION OF MAXWELL'S EXPRESSION FOR λ

Let us consider two molecules of masses m_1 and m_2 and diameters d_1 and d_2. When they collide, they can be considered as one body system having reduced mass $\mu = \dfrac{m_1 m_2}{m_1 + m_2}$ and the centre-to-centre distance $d = \dfrac{d_1 + d_2}{2}$.

Let the number densities of these gases be n_1 and n_2. We construct a cylinder of cross-sectional area πd^2 and length $\bar{v} = \sqrt{\dfrac{8 k_B T}{\mu \pi}}$, which defines average speed characterising Maxwellian molecules.

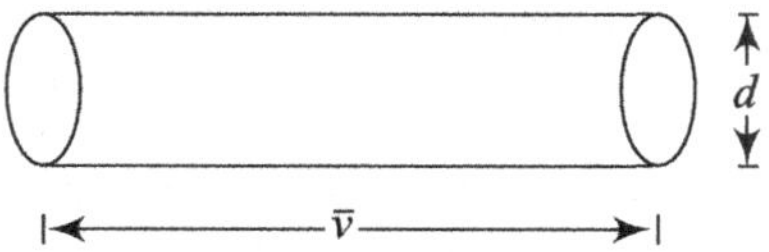

Fig. B.2.1

When one molecule of the first type moves, it will collide with all the molecules of the second type contained in the cylinder. The total number of such collisions is given by

$$f_{12} = n_2 \pi d^2 \bar{v} = n_2 \pi d^2 \sqrt{\frac{8 k_B T}{\pi \frac{m_1 m_2}{m_1 + m_2}}} = n_2 \pi d^2 \sqrt{\frac{8 k_B T}{\pi}} \left(\frac{1}{m_1} + \frac{1}{m_2} \right)^{1/2}$$

If all the molecules are identical, we can write $d_1 = d_2 = d$, $m_1 = m_2 = m$ and $n_1 = n_2 = n$ (say). Using these in the expression for f_{12}, which we redesignate as f, we get

$$f = n\pi d^2 \sqrt{2 \times \frac{8 k_B T}{m\pi}} = \sqrt{2}\pi n d^2 \bar{v}$$

Hence,

$$\lambda = \frac{\bar{v}}{f} = \frac{\bar{v}}{\sqrt{2}\pi n d^2 \bar{v}} = \frac{1}{\sqrt{2}\pi n d^2}$$

You now know that molecules undergo collisions very frequently. It means that distribution of free paths will be spread over a wide range. It is therefore instructive to calculate the distribution function for free paths. This forms the subject matter of the following subsection.

2.2.2 Distribution of Free Path

Consider a large number of molecules at a certain instant. As they travel, they will collide among themselves. We wish to estimate the number that have not made a collision at some later time. Let the number of molecules surviving a collision in travelling a distance x be N. If each molecule is allowed to travel a further distance dx, more collisions will occur. We assume that the number of collisions is proportional to the number of molecules N and the distance dx. That is, the number of molecules that undergo collisions will be proportional to Ndx. Since the number of molecules that do not make any collision decreases with increasing distance, we can write

$$\mathrm{d}N = -P_c N\mathrm{d}x \tag{2.12}$$

where collision probability P_c is the constant of proportionality.
We can rewrite this equation as

$$\frac{\mathrm{d}N}{N} = -P_c\,\mathrm{d}x$$

We can readily integrate it to obtain

$$N = N_0 \exp(-P_c x) \tag{2.13}$$

where N_0 is the number of molecules at $x = 0$.

This result shows that the number of molecules surviving a collision decreases exponentially. The probability that a gas molecule will cover a distance x without making any collision is

$$f(x) = \frac{N}{N_0} = \exp(-P_c x) \tag{2.14}$$

To evaluate the constant of proportionality P_c, we note that, by definition, mean free path is given by

$$\lambda = \frac{\int_0^\infty x\mathrm{d}N}{N_0} = \frac{-\int_0^\infty P_c N_0 x \exp(-P_c x)\,\mathrm{d}x}{N_0}$$

$$= \left[-x\exp(-P_c x) - \frac{\exp(-P_c x)}{P_c}\right]_0^\infty$$

Note that the first term within the parentheses vanishes identically at both limits; due to decaying exponential at the upper limit and the presence of x at the lower limit. However, the second term vanishes at the upper limit due to decaying exponential but it is non-zero at the lower limit; it has the value $1/P_c$:

$$\lambda = \frac{1}{P_c}$$

On using this result in Eq. (2.13), we obtain a compact form for the number of particles surviving collisions in travelling a distance x:

$$N(x) = N_0 \exp\left(-\frac{x}{\lambda}\right) \tag{2.15}$$

This equation is known as *survival equation*.

You will come across similar equations in other areas of physics. For example, radioactive decay obeys the equation $N(t) = N_0 \exp(-\Lambda t)$, where $N(t)$ is the number of atoms which survive disintegration for time t and Λ is disintegration constant. In optics, Biot's law describing the intensity of an incident beam after it has traversed a distance x in a medium exhibits exponential decay. Similarly, discharge of a capacitor in a R–C circuit is a decaying exponential.

Figure 2.4 graphically depicts survival equation. We have plotted N/N_0 as a function of x/λ. Note that the fraction of molecules with $x = \lambda$ is $e^{-1} \approx 0.37$.

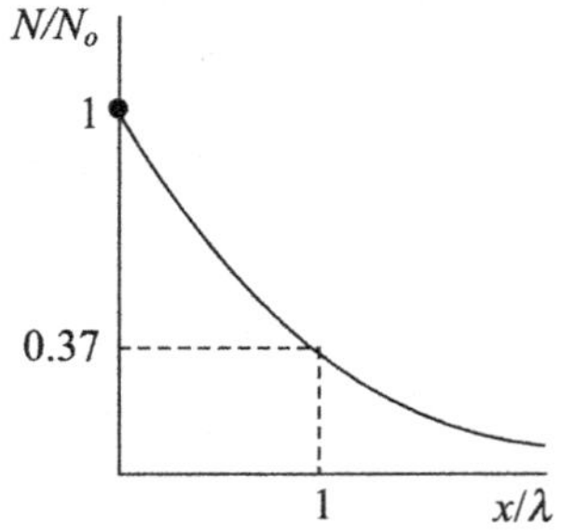

Fig. 2.4 Plot of Eq. (2.15): The survival equation.

If we take Maxwellian distribution of velocities into account, the calculations become somewhat involved but the form of the function remains unchanged. Therefore, we just quote the result, which is due to Jeans:

$$f(x) = \frac{N}{N_0} = \exp(-1.04\,x/\lambda) \tag{2.16}$$

We now solve an example to reinforce our discussion in this section.

Example 2.6 In a sample of 10^{26} gas molecules, each molecule is moving with the same speed. Calculate the fraction of molecules that will continue to travel undeflected after traversing distances of λ, 2λ, 5λ and 10λ.

Solution: The values of $f(x)$ are tabulated below for different distances traversed by the gas molecules and you can readily verify these using Eq. (2.15). To illustrate, for $x = \lambda$, we can write

$$f(x) = \exp(-1) = 0.3679$$

$\frac{x}{\lambda}$	1	2	5	10
$f(x)$	0.3679	0.1353	6.738×10^{-3}	4.540×10^{-5}

You will note that after traversing a distance of λ and 2λ, nearly 37% and 13.5% molecules respectively survive collisions. However, after covering a distance of 10λ, almost all molecules will have undergone a collision.

You may now like to answer a practise problem.

Problem 2.3 The mean free path of the molecules of a gas is 2×10^{-7} m. Calculate the probability that a molecule will travel 6×10^{-7} m without making a collision, if temperature is doubled.

Ans: 0.223

In 1920, Max Born put the law of free paths to test when he suggested a method for experimental determination of mean free path. Let us now learn about it.

2.2.3 Experimental Determination of Mean Free Path

A schematic diagram of the apparatus used by Born is depicted in Fig. 2.5a. It consists of a quartz tube, which can be evacuated to a desired level. Silver is heated at the lower end of the tube T. On evaporation, the silver vapours pass through the narrow slit S and then

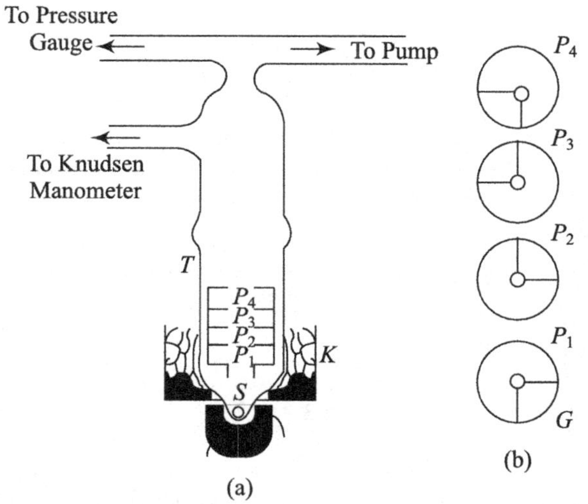

Fig. 2.5 (a) Schematic diagram of Born's appartus used to measure mean free path and (b) Arrangement of glass quadrants on brass discs.

passes through a circular hole in each of the four brass discs P_1, P_2, P_3, P_4. Each disc has a circular hole and carries a glass quadrant G (a glass plate in the shape of a quarter of circle) such that the apex of each quadrant lies at the centre of the hole. These are separated by 1 cm. The evaporated silver moves through these holes. Moreover, each quadrant is oriented in such a way that it is displaced through 90° relative to the preceding one. This arrangement is shown in Fig. 2.5b and ensures that each quadrant can receive only one-fourth of the incident beam.

The pressure in the quartz tube can be varied to any desired value using a vacuum pump. The discs are cooled by the cooling mixture and silver is deposited on the quadrants. The relative amount of silver deposited on a given quadrant is determined by photometric comparison.

Initially, the tube is completely evacuated. The mean free path will then be equal to the length of the chamber. It is because silver vapours will have no air/gas molecule to collide against. Let the density of the deposit on a quadrant under complete evacuation condition be d_{10}. Next we introduce the gas and determine the density of deposit on the same quadrant, after cleaning the initial deposition and replacing the brass disc back. Let us denote the density of deposit by d_1 and mean free path by λ. Then, using the law of free paths, we can write

$$d_1 = d_{10} \exp(-x_1 / \lambda) \tag{2.17}$$

where x_1 is the distance of the quadrant from S.
Similarly, for the quadrant at a distance x_2 from S, we can write

$$d_2 = d_{20} \exp(-x_2 / \lambda) \tag{2.18}$$

where d_2 and d_{20} are the densities of the deposit.
On dividing Eq. (2.17) by (2.18), we get

$$\frac{d_1}{d_2} = \frac{d_{10}}{d_{20}} \exp\left(\frac{x_2 - x_1}{\lambda}\right)$$

On rearranging terms, we can write

$$\lambda = \frac{(x_2 - x_1)}{\ln\left(\dfrac{d_1}{d_2}\dfrac{d_{20}}{d_{10}}\right)} \tag{2.19}$$

Using this equation, one can easily determine the value of mean free path. For the arrangement used by Born, $(x_2 - x_1) = 1$ cm. So once we know the density of the deposits, we can easily determine mean free path at a given temperature and pressure. Born's results are tabulated below:

p (mm)	λ (cm)	$p\lambda$
5.8×10^{-3}	1.7	9.86×10^{-4}
4.5×10^{-3}	2.4	1.08×10^{-3}
Atmospheric pressure	1.3×10^{-5}	9.88×10^{-4}

From the table, we note that the product $p\lambda$ is nearly constant. This is in conformity with the prediction of kinetic theory of gases and supports the law of distribution of free paths. So far we have assumed that the gaseous system is at uniform temperature and pressure.

And in the equilibrium state, there is no preferential direction of motion of any kind. This means that the rate of transfer of molecules across a given plane in the body of the gas exactly matches the number of molecules moving in the opposite direction. However, when the entire gas or a part of it moves as a whole in a particular direction, a gradient in velocity, temperature or concentration may arise. Then thermal molecular motion leads to what is commonly known as *transport phenomena*. We now discuss it in some detail.

2.3 TRANSPORT PHENOMENA

We now know that a gas molecule has a finite mass and is characterised by random thermal motion. Therefore, it possesses momentum as well as kinetic energy. So, while moving from one part of the container to another, it is a potential carrier of physical quantities such as matter, energy or momentum. When a gas is in equilibrium, there is no *net* transport of these physical quantities. However, when a gas is endowed with macroscopic motion in a particular direction, the following cases may occur singly or jointly:

1. Different parts of a gas move with different velocities This will give rise to relative motion to different layers of the gas. As a result, the molecules crossing from faster moving layers will transport greater momentum as compared to those originating from slower moving layers. Therefore, *across an imaginary plane, there will be a net transport of momentum in the preferential direction of motion.* This is characterised by the *coefficient of viscosity.*

It is important to realise that viscosity in gases arises due to the random thermal motion of molecules when it is endowed with macroscopic motion rather than a frictional force between any two adjacent layers, as in the case of liquids.

2. Different parts of a gas are at different temperatures In this case, the molecules of the gas will carry greater *thermal energy* from regions of higher temperature to regions of lower temperature and tend to establish thermal equilibrium. This leads to the phenomenon of *thermal conduction* and is characterised by *thermal conductivity.*

3. Different parts of a gas have different concentrations When molecules migrate from regions of higher concentration to the regions of lower concentration, there is net transport of matter (mass). This leads to the phenomenon of *diffusion* and is characterised by the *coefficient of diffusion.*

We thus find that *viscosity, conduction* and *diffusion* are bulk properties of gases and represent the *transport of momentum, energy* and *mass,* respectively. These are collectively categorised under the title of *transport phenomena.* These processes are of vital importance in physical sciences and find several important applications. We will now discuss these in some detail on the basis of the simple molecular model.

2.3.1 Viscosity: Transport of Momentum

We know that the property by virtue of which a fluid opposes the relative motion between the adjacent layers is known as *viscosity*. It is quantitatively expressed in terms of the coefficient of viscosity, η, which is defined as *the tangential force per unit area when a unit velocity gradient exists in a direction perpendicular to the direction of motion.* Mathematically, for small velocity gradient we can write

$$F = -\eta A \frac{du}{dy} \tag{2.20}$$

Here F denotes the viscous force acting on an area A and du/dy is the velocity gradient along the +ve y-direction. The negative sign signifies that the viscous force is directed against the velocity gradient. Let us consider a gas enclosed between two planes depicted as aa' and bb' in Fig. 2.6 and separated through a distance h. Suppose that the gas is endowed with mass motion so that the net flow is from left to right and a positive velocity gradient exists along the y-axis, which is perpendicular to the direction of flow. We consider a plane xx' located somewhere between aa' and bb'.

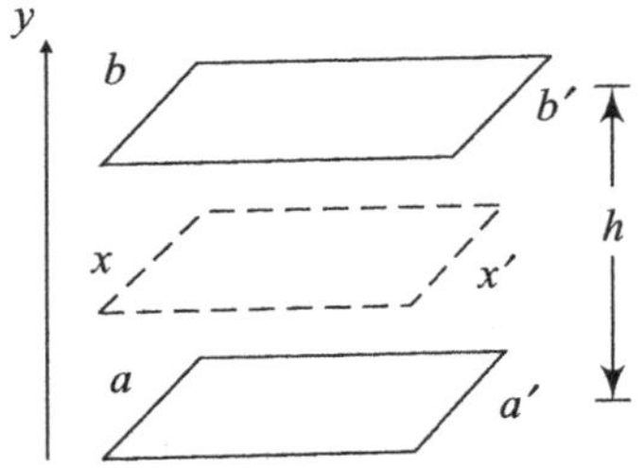

Fig. 2.6 Schematic representation of planes signifying velocity gradient in a gas.

Let the velocity of flow at xx' be u and the velocity gradient be $\frac{du}{dy}$. Because of thermal motion, gas molecules continuously cross the plane xx' both from above and below it. Since the velocity gradient is positive upwards, the gas molecules below xx' gain momentum whereas those above it lose momentum. That is, there will be a net transfer of momentum across the plane under consideration in the direction of mass motion. By Newton's second law of motion, we can correlate it to the coefficient of viscosity. Therefore, to derive an expression for the net rate of tranfer of momentum, we need to know (i) the number of molecules crossing the surface xx' and (ii) the average distance at which a molecule makes its last collision before crossing this surface.

To facilitate these calculations, it is convenient to assume that *each molecule acquires a flow velocity in the direction of mass motion characteristic of the height at which it made the last collision before crossing the surface xx'*. So, the velocity of a molecule at a distance δ from this surface will be $\delta \frac{du}{dy}$.

Calculation of Number of Molecules To calculate the number of molecules crossing the surface xx', we consider an element of volume dV at a distance r from an element of area dA in the plane xx' and in a direction making an angle θ with the normal to the plane as shown in Fig. 2.7.

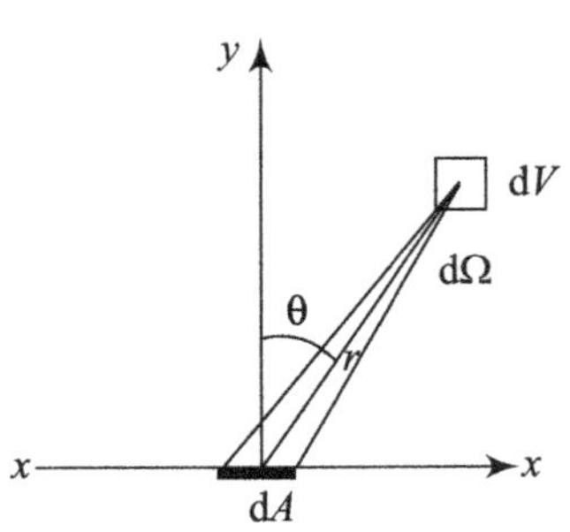

Fig. 2.7 Calculation of average number of molecules crossing an imaginary plane xx'.

Let n be the number density of molecules. Then the number of molecules in the volume element dV will be n dV. If P_c is the collision frequency, the total number of collisions occuring within dV in time dt will be $(1/2)P_c n$ dV dt. Note that the factor (1/2) has been introduced so that each collision is counted only once; each collision involves two molecules. Since each collision results in two free paths, the total number of free paths generated in dV in time dt will be $P_c n$ dV dt. These free paths start off in all 4π directions since all directions of molecular motion are equally probable. Therefore, the fraction of free paths, and hence molecules, heading towards dA is given by

$$\Delta n = \frac{d\Omega}{4\pi} P_c n dV dt$$

where $d\Omega$ is the solid angle subtended by dA at the volume element dV and is equal to $\frac{dA \cos\theta}{r^2}$. The number of molecules that will reach dA without making any further collisions is obtained using the survival equation. Hence, we can write

$$\Delta N = \Delta n \exp(-r/\lambda) = P_c\, n dV dt \frac{dA \cos\theta}{4\pi r^2} \exp(-r/\lambda)$$

In spherical polar co-ordinates, we can write $dV = 2\pi r^2 \sin\theta d\theta dr$. We substitute this expression in the above equation and obtain

$$\Delta N = \frac{1}{2} P_c\, n dA dt \sin\theta \cos\theta \exp\left(-\frac{r}{\lambda}\right) d\theta\, dr \tag{2.21}$$

To calculate the total number of molecules crossing dA in time dt from the half-space above xx' and from all directions, we integrate this expression over r from 0 to ∞, and over θ from 0 to $\pi/2$. This gives

$$N = \frac{1}{2} P_c\, n dA\, dt \int_0^\infty \exp\left(-\frac{r}{\lambda}\right) dr \int_0^{\pi/2} \sin\theta \cos\theta d\theta$$

You can readily verify that integration over θ gives (1/2) and integration over r yields λ. Hence,

$$N = \frac{\lambda}{4} P_c\, n dA\, dt$$

Since $P_c = \bar{v}/\lambda$, we find that the number of molecules crossing the plane xx' from either side per unit area per second is

$$\frac{N}{dA\, dt} = n_s = \frac{1}{4} n\bar{v} \tag{2.22}$$

Calculation of Average Height The mathematical steps involved in the calculation of average distance at which a molecule makes its last collision before crossing the imaginary *surface xx'* are somewhat involved. However, for mathematical ease, we first make a simplifying assumption and work in what we will refer to as Zeroth approximation.

Zeroth Order Approximation We assume that the average distance from the surface xx' at which a molecule makes its last collision before crossing it is equal to one mean free path, λ. The flow velocity of the gas molecules at this height from the surface xx' will be $u + \lambda \frac{du}{dy}$, where u is flow velocity at the xx'. Therefore, the momentum transported by a molecule moving with this velocity will be $m\left(u + \lambda \frac{du}{dy}\right)$. And the total momentum in the direction of the flow carried across the surface xx' per unit area per unit time by all the molecules crossing it in the downward direction from above after making their last collision at an average distance of λ from xx' is equal to the product of the number of particles and the momentum transported by each molecule. Hence, we can write

$$G^+ = \frac{1}{4} n\bar{v} m \left(u + \lambda \frac{du}{dy}\right)$$

The total momentum carried across the surface xx' per unit area per unit time by the molecules crossing it in the upward direction from below after making their last collision at an average distance of λ from xx' is

$$G^{-} = \frac{1}{4} n\,\bar{v}\,m\left(u - \lambda\frac{du}{dy}\right)$$

The net transport of momentum across surface xx' from below in the direction of mass motion per unit area per unit time, which is equal to the viscous force per unit area, is given by

$$G = G^{-} - G^{+} = -\frac{1}{2} m\,n\,\bar{v}\,\lambda\frac{du}{dy} \tag{2.23}$$

On comparing Eqs. (2.20) and (2.23) we find that the coefficient of viscosity of a gas in the zeroth approximation is given by

$$\eta_{Zero} = \frac{1}{2} m\,n\,\bar{v}\,\lambda = \frac{1}{2}\rho\,\bar{v}\,\lambda \tag{2.24a}$$

On combining this result with Eq. (2.10), we can write

$$\eta_{Zero} = \frac{3m}{8\sigma}\bar{v} \tag{2.24b}$$

Note that the coefficient of viscosity of a gas is directly proportional to the average speed $\bar{v}$ and hence $\sqrt{T}$ and inversely proportional to the second power of diameter. That is, the coefficient of viscosity of a gas is independent of pressure and varies as square root of its absolute temperature. This conclusion is in qualitative agreement with experimental findings.

First Order Approximation Let us now obtain an exact expression for the average distance at which a molecule makes its last collision before crossing the imaginary surface xx'. As before, we consider a volume element dV located at distance r and angle θ as shown in Fig. 2.7. If $y = r\cos\theta$ is the height of the volume element dV above xx', then the average distance at which a molecule makes its last collision before crossing xx' is given by

$$\bar{y} = \frac{\int_0^{\infty} y\,\Delta N}{\int_0^{\infty} \Delta N} \tag{2.25}$$

where ΔN denotes the number of molecules surviving collisions and the integrals are defined over the entire space above the plane xx'.

On substituting for ΔN from Eq. (2.21) in Eq. (2.25), we get

$$\bar{y} = \frac{\int_0^{\infty} r\exp\left(-\frac{r}{\lambda}\right)dr \int_0^{\pi/2} \sin\theta\cos^2\theta\,d\theta}{\int_0^{\infty} \exp\left(-\frac{r}{\lambda}\right)dr \int_0^{\pi/2} \sin\theta\cos\theta\,d\theta}$$

Note that

$$\int_0^\infty r \exp\left(-\frac{r}{\lambda}\right) dr = -\lambda e^{-r/\lambda}\Big|_0^\infty + \lambda \int_0^\infty e^{-r/\lambda}\, dr = \lambda^2$$

and

$\int_0^{\pi/2} \sin\theta \cos^2\theta \, d\theta = \int_0^1 x^2 \, dx = \frac{1}{3}$. Hence, we can write

$$\overline{y} = \frac{(1/3)\,\lambda^2}{(1/2)\,\lambda} = \frac{2\lambda}{3} \tag{2.26}$$

This result shows that, on an average, each molecule makes its last collision at two-third of the mean free path above or below the surface xx' before crossing it, rather than one mean free path. It means that the simple assumption made earlier was not so crude and we should not expect any major changes in our conclusions.

Let us now use Eq. (2.26) to calculate momentum and see how Eq. (2.24a) is modified. At a height $\left(\frac{2}{3}\right)\lambda$ above the plane xx', the flow velocity of the gas molecules will be $u + \frac{2}{3}\lambda\frac{du}{dy}$, where u is flow velocity at the plane xx'. The momentum transported by a molecule moving with this velocity will be $m\left(u + \frac{2}{3}\lambda\frac{du}{dy}\right)$. So, the total momentum in the direction of the flow carried across the surface per unit area per unit time by all the molecules crossing the surface xx' from above will be

$$G^+ = \frac{1}{4} n \overline{v} m \left(u + \frac{2}{3}\lambda\frac{du}{dy}\right)$$

Similarly, the total momentum carried across the surface xx' per unit area per unit time by the molecules crossing it in the upward direction from below will be

$$G^- = \frac{1}{4} n \overline{v} m \left(u - \frac{2}{3}\lambda\frac{du}{dy}\right)$$

Hence, the net transport of momentum across xx' from below in the direction of mass motion per unit area per unit time, which is equal to the viscous force per unit area, is given by

$$G = G^- - G^+ = -\frac{1}{3} m n \overline{v} \lambda \frac{du}{dy} \tag{2.27}$$

On comparing Eqs. (2.20) and (2.27), we note that the coefficient of viscosity of a gas is given by

$$\eta_F = \frac{1}{3} m n \overline{v} \lambda = \frac{1}{3}\rho \overline{v} \lambda = \frac{m\overline{v}}{4\sigma} \tag{2.28}$$

On comparing Eqs. (2.24a) and (2.28), you will note that $\frac{\eta_F}{\eta_{Zero}} = \frac{2}{3} = 0.67$. It means that in the first order approximation, the multiplication coefficient changes by 0.667.

On substituting the expression for λ corresponding to Maxwellian distribution of velocities in Eq. (2.28) from Eq. (2.11), we get

$$\eta_{Max} = \frac{1}{3\sqrt{2}} \frac{m\overline{v}}{\sigma} = \frac{m\overline{v}}{3\sqrt{2}\pi d^2} \tag{2.29}$$

This result provides us a way to estimate a microscopic property such as molecular diameter in terms of a macroscopic property such as viscosity η, which is directly measureable. It therefore afforded a remarkable opportunity for proponents of kinetic theory to put it on a more sound pedastal.

Let us now discuss temperature and pressure dependence of the coefficient of viscosity.

Effect of Temperature and Pressure on Viscosity From Eq. (2.29), we note that the viscosity of a gas is independent of number density (pressure) but directly proportional to $\bar{v}$ and hence the square root of absolute temperature. Both these conclusions are well supported by experiments. For instance, for pressures from a few mm of nercury up to several atmosphere, Meyer showed that the coefficient of viscosity of a gas is independent of pressure. However, at very low or high pressures, this relation is not valid. This is essentially because at low pressures, the intermolecular collisions are rare and the mean free path attains a constant value; comparable with the dimensions of the apparatus. However, the number density will decrease gradually as pressure is lowered further. Consequently, the coefficient of viscosity decreases as pressure decreases. This aspect was verified experimentally by Crooks.

Warburg and von Babo showed that at very high pressures, the coefficient of viscosity increases as pressure increases. This is due to the fact that at such high pressures, the mean free path becomes comparable with the molecular diameter and the transport of momentum takes place through a distance $\lambda + d \cos \theta$ where θ is the angle which the projection of the freepath makes with the axis normal to the direction of mass motion.

In Fig. 2.8, we have plotted experimental values of viscosity of helium, argon and neon versus $T^{1/2}$ As may be noted, the curves are almost straight lines but suggest that viscosity of a gas increses with temperature at a slightly faster rate than that predicted by theory. The departure from the value 0.5 implies inadequacy of our assumption that gas molecules behave as elastic hard spheres devoid of intermolecular interactions. In fact, it provides us with an opportunity to get an insight into the nature of intermolecular forces. We know that σ is independent of T only for rigid hard spheres. The long range attractive part of force increases the scattering probability of a molecule. At lower temperatures, the molecules have lower velocities and can easily be scattered. If we denote the repulsive part of the potential by $V(r) = \dfrac{A}{r^s}$, it turns out that $\eta \; \alpha \; T^n$ with $n = \dfrac{1}{2} + \dfrac{2}{s-1}$. For rigid hard spheres, $s = \infty$, so that $n = 0.5$. From experimental observations made on η at different temperatures, we can deduce the value of s. Some representative values are given below:

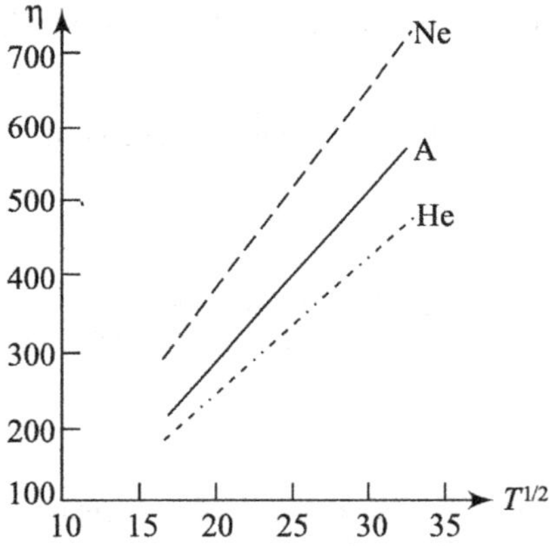

Fig. 2.8 Plot of measured values of η as a function of $T^{1/2}$ for helium, argon and neon.

Gas	s
H_2	11.3
He	14.6
Cl_2	5.0
HCl	4.97

You should now go through the following examples carefully. We have illustrated how Eq. (2.29) can be used to estimate molecular diameter.

Example 2.7 The coefficient of viscosity of helium at 27°C is 2×10^{-4} poise. Calculate the diameter of a helium molecule using the given data: $N_A = 6 \times 10^{23}$ mol^{-1}, $k_B = 1.38 \times 10^{-23}$ JK^{-1} and atomic weight of helium is 4 g.

Solution: From Eq. (2.29), we recall that

$$\eta = \frac{m\bar{v}}{3\sqrt{2}\pi d^2}$$

By rearranging terms, we can write

$$d^2 = \frac{m\bar{v}}{3\sqrt{2}\pi\eta}$$

Here $\eta = 2 \times 10^{-4}$ poise, $T = (273 + 27)$ K = 300 K

$$m = \frac{4\text{ g mol}^{-1}}{6 \times 10^{23}\text{ mol}^{-1}} = 6.67 \times 10^{-24}\text{ g} = 6.67 \times 10^{-27}\text{ kg}$$

and

$$\bar{v} = \left(\frac{2.55\, k_B T}{m}\right)^{1/2} = \left[\frac{2.55 \times (1.38 \times 10^{-23}\text{ JK}^{-1}) \times 300\text{ K.}}{6.67 \times 10^{-27}\text{ kg}}\right]^{1/2} = 1.26 \times 10^3\text{ ms}^{-1}$$

Hence, using the relation 1 poise = 10^{-1} kg m^{-1}s^{-1}, we get

$$d^2 = \frac{(6.67 \times 10^{-27}\text{ kg}) \times (1.26 \times 10^3\text{ ms}^{-1})}{3\sqrt{2} \times 3.14 \times (2 \times 10^{-5}\text{ kg m}^{-1}\text{s}^{-1})} = 3.15 \times 10^{-20}\text{ m}^2$$

$\therefore$

$$d = 1.77 \times 10^{-10}\text{ m}$$

Problem 2.4 The coefficient of viscosity of oxygen at 15°C is 196μ poise. Calculate the molecular diameter using the given data: R = 8.4 Jmol^{-1}K^{-1} Molecular weight of oxygen is 32 g.

Ans: 2.99×10^{25} m

Example 2.8 The molecules of a gas move with an average speed of 450 ms^{-1}. If $\eta = 16.6 \times 10^{-6}$ Nsm^{-2}, $\rho = 1.25$ kgm^{-3} and $n = 2.7 \times 10^{25}$ m^{-3}, calculate the mean free path and diameter of the gas molecules.

Solution: We can rewrite Eq. (2.28) as

$$\lambda = \frac{3\eta}{mn\bar{v}} = \frac{3\eta}{\rho\bar{v}}$$

On inserting the given numerical values, we get

$$\lambda = \frac{3 \times (16.6 \times 10^{-6}\text{ Nsm}^{-2})}{(1.25\text{ kg m}^{-3} \times (450\text{ ms}^{-1})} = 8.85 \times 10^{-8}\text{ m}$$

From Eq. (2.11), we recall that

$$\lambda = \frac{1}{\sqrt{2}\pi d^2 n}$$

We can invert this relation to express d in terms of λ:

$$d = \left(\frac{1}{\sqrt{2}\pi n\lambda}\right)^{1/2} = \left(\frac{1}{1.414 \times 3.14 \times (2.7 \times 10^{25}\ \text{m}^{-3}) \times (8.85 \times 10^{-8}\ \text{m})}\right)^{1/2}$$

$$= 3.42 \times 10^{-10}\ \text{m}$$

Example 2.9 Calculate the coefficient of viscosity of hydrogen gas at 27°C and one atmosphere pressure. Take molecular weight of hydrogen as 2.016 u and diameter of hydrogen molecule as 2.92×10^{-10} m. 1u = 1.66×10^{-27} kg. Assume that hydrogen obeys Maxwellian velocity distribution.

Solution: From Eq. (2.29), we recall that

$$\eta_{\text{Max}} = \frac{1}{3\sqrt{2}} \frac{m\overline{v}}{\sigma} = \frac{m\overline{v}}{3\sqrt{2}\pi d^2}$$

Here $m = 2.016 \times 1.66 \times 10^{-27}$ kg, $d = 2.92 \times 10^{-10}$ m, and average velocity

$$\overline{v} = \sqrt{\frac{8k_BT}{\pi m}} = \sqrt{\frac{8 \times (1.38 \times 10^{-23}\ \text{JK}^{-1}) \times (300\ \text{K})}{3.14 \times (3.35 \times 10^{-27}\ \text{kg})}}$$

$$= \sqrt{\frac{33.12}{10.51}} \times 10^3\ \text{ms}^{-1}$$

$$= 1.775 \times 10^3\ \text{ms}^{-1}$$

Hence, the coefficient of viscosity of hydrogen is

$$\eta = \frac{(3.35 \times 10^{-27}\ \text{kg}) \times (1.775 \times 10^3\ \text{ms}^{-1})}{3 \times 1.414 \times 3.14 \times (2.92 \times 10^{-10}\ \text{m})^2}$$

$$= \frac{5.95 \times 10^{-4}}{113.6}\ \text{kg m}^{-1}\text{s}^{-1} = 5.24 \times 10^{-6}\ \text{Nm}^{-1}\text{s}^2$$

You may now like to solve a few practise problem to assess your understanding.

Problem 2.5 Hydrogen and nitrogen are maintained under identical conditions of temperature and pressure. Calculate the ratio of their coefficients of viscosity if the diameters of these molecules are 2.5×10^{-10} m and 3.5×10^{-10} m.

Ans: 1.91

Problem 2.6 Calculate the coefficient of viscosity of hydrogen at STP using the following data: $\rho = 8.96 \times 10^{-2}$ kgm^{-3}; $\lambda = 1.69 \times 10^{-7}$ m, and $k_B = 1.38 \times 10^{-23}$ JK^{-1}.

Ans: 2.43×10^{-5} kg m^{-1}s^{-1}

2.3.2 Thermal Conductivity: Transport of Energy

The thermal conductivity of a gas is treated in the same way as its viscosity. We therefore again make reference to Fig. 2.6 and consider flow of thermal energy across the layers of the gas lying between parallel surfaces *aa′* and *bb′*. We assume that these are at rest but at different temperatures. That is, there is a temperature gradient rather than velocity gradient between different layers of the gas. Let us denote it as d*T*/d*y*. As before, we take the temperature gradient as positive along positve *y*-axis and normal to the surface *xx′*. The thermal energy flowing per unit area per second in the upward direction is characterised in terms of thermal condctivity. It is usually denoted by the symbol *K* and we can write

$$Q = -K\frac{\mathrm{d}T}{\mathrm{d}y} \tag{2.30}$$

where d*T*/d*y* is temperature gradient and temperature of the gas at the imaginary surface *xx′* within the gas is *T*. (You should not confuse the symbol for thermal conductivity with the symbol used for kelvin, the unit of temperature.)

To obtain an expression for thermal conductivity, we need to know the energy transported by the molecules across an imaginary surface/plane within the gas. As before, it requires information of (i) the average number of molecules crossing this surface and (ii) the average height at which a molecule makes its last collision before crossing it. For these, we can use the results obtained in the preceding section. That is, we use first order approximation and assume that each molecule crossing the given surface made its last collision at a distance (2/3)λ above and below it (Eq. (2.26)) and the energy carried by a molecule is characteristic of that distance.

The average energy of a gas molecule at temperature *T* is $\frac{f}{2}k_B T$, where *f* denotes the number of degrees of freedom. Hence, the energy of a molecule at a distance (2/3)λ above or below the surface *xx′* wll be $\frac{f}{2}k_B\left(T \pm \frac{2}{3}\lambda\frac{\mathrm{d}T}{\mathrm{d}y}\right)$. If the number of particles crossing the surface per unit area per second is $n\bar{v}/4$, the energy transported by molecules moving from above in the downward direction per unit area per second is given by

$$Q^+ = \frac{1}{4}n\bar{v}\frac{f}{2}k_B\left(T + \frac{2}{3}\lambda\frac{\mathrm{d}T}{\mathrm{d}y}\right)$$

Similarly, the energy transported by molecules moving from below in the upward direction per unit area per second is given by

$$Q^- = \frac{1}{4}n\bar{v}\frac{f}{2}k_B\left(T - \frac{2}{3}\lambda\frac{\mathrm{d}T}{\mathrm{d}y}\right)$$

Hence, the net rate of transport of energy in the upward direction per unit area will be equal to *Q* and is obtained by noting that

$$Q = Q^- - Q^+ = -\frac{1}{6}n\lambda\bar{v}fk_B\frac{\mathrm{d}T}{\mathrm{d}y} \tag{2.31}$$

On comparing Eqs. (2.30) and (2.31), we obtain

$$K = \frac{f}{6} n \bar{v} \lambda k_B = \frac{C_V}{3} n \bar{v} \lambda \tag{2.32}$$

where $C_V = \frac{f}{2} k_B T$ is molar heat capacity.

We thus find that under normal conditions, thermal conductivity, like viscosity, is independent of pressure. This is in good agreement with experimental results down to very low pressures. However, at extremely low pressures, thermal conductivity tends to decrease. From Eq. (2.32), we also note that thermal conductivity is directly proportional to average molecular speed. Therefore, theory predicts that thermal conductivity is directly proportional to half-power of temperature. However, in actual practise, K increases somewhat more rapidly suggesting that intermolecular forces come into play and begin to influence energy transport.

Relation between η and K When we closely examine the expressions for η and K, we expect some connection between them. It is instructive to ascertain this relationship as both these quantities are physically measureable. Therefore, we divide Eq. (2.32) by Eq. (2.28) and obtain

$$\frac{K}{\eta} = \frac{f}{2} \frac{k_B}{m} = \frac{f}{2} \frac{k_B N_A}{M} = \frac{f}{2} \frac{R}{M}$$

where $m = \frac{M}{N_A}$ and M is molecular weight of the substance.

From Sec. 1.3, we recall that molar heat capacity at constant volume is given by

$$C_V = \frac{f}{2} R$$

Using this result, we can rewrite

$$\frac{K}{\eta} = \frac{C_V}{M}$$

or

$$\frac{KM}{\eta C_V} = 1 \tag{2.33}$$

Equation (2.33) predicts that the ratio $\frac{KM}{\eta C_V}$ is constant, equal to one and same for all gases. But experiments show that this ratio is greater than one. A more rigourous calculation due to Chapman and Enskog showed that this ratio varies between 1.5 and 2.5 and is different for different gases; decreasing with increasing atomicity. There can be several reasons for this lack of agreement. But the most obvious reason seems to be non-inclusion of the Maxwellian distribution of velocities while calculating the energy transported across the imaginary plane in the body of the gas. Obviously, the faster molecules will cross the given plane more frequently and result in increased energy transport, effecting an increase in the value of the ratio $\frac{KM}{\eta C_V}$. This also suggests that probably the hard sphere model of a molecule is not so very adequate. More sophisticated theories do account for this disagreement. However, we will not consider these here.

You should now go through the following example carefully.

Example 2.10 Calculate the thermal conductivity of a gas using the following data: $\rho = 1.6$ kg m^{-3}; $\bar{v} = 480$ ms^{-1}, $\lambda = 8 \times 10^{-8}$ m, $M = 32$, $\gamma = 1.4$ and $R = 8.31$ kJ kmol^{-1}K^{-1}.

Solution: From Eq. (2.32), we recall that

$$K = \frac{C_V}{3} n\bar{v}\lambda = \frac{C_V}{3M} \rho\lambda\bar{v}$$

Since $\gamma = 1.4$, the gas is diatomic. So we can take $C_V = \frac{5}{2}R$. Since $R = 8.31$ kJ kmol^{-1}K^{-1}, $C_V = 2.5 \times 8.31$ kJ kmol^{-1}K^{-1} $= 20.8 \times 10^3$ J kmol^{-1}K^{-1}, using the given values of various quantities, we get

$$K = \frac{(1.6 \text{ kg m}^{-3}) \times (480 \text{ ms}^{-1}) \times (20.8 \times 10^3 \text{ J kmol}^{-1} \text{ K}^{-1}) \times (8 \times 10^{-8} \text{ m})}{3 \times (32 \text{ kg kmol}^{-1})}$$

$$= 13.31 \times 10^{-3} \text{ Jm}^{-1}\text{s}^{-1}\text{K}^{-1}$$

We will now like you to work out a few practise problems.

Problem 2.7 The coefficient of viscosity of helium is 18.6×10^{-6} Nsm^{-2}, $M = 4$ kg kmol^{-1} and $C_V = 12.5 \times 10^3$ Jkmol^{-1}K^{-1}. Calculate the thermal conductivity of helium at 0°C.

Ans: 5.8×10^{-2} Jm^{-1}s^{-1}K^{-1}

Problem 2.8 The collision diameters of hydrogen, nitrogen and chlorine are 25 nm, 35 nm and 40 nm, respectively. Compare their thermal conductivities.

Ans: $K(H_2) : K(N_2) : K(Cl_2)$ :: 15 : 2 : 1

Problem 2.9 The thermal conductivity of helium is 8.7 times the thermal conductivity of argon under STP. Calculate the ratio of their diameters. Take $M_{He} = 4$ kg kmol^{-1} and $M(\text{Ar}) = 40$ kg kmol^{-1}

Ans: $d_{\text{He}} : d_{\text{Ar}}$ :: 1 : 1.78

2.3.3 Diffusion: Transport of Matter

Consider two gases, say hydrogen and oxygen, contained in two different glass jars at the same temperature and pressure. Let us invert the jar containing hydrogen over the jar containing oxygen and remove the lids. As such, no large scale movement of these gases will be visible. But after some time, the two gases get mixed with one another. (Oxygen molecules will have moved against gravity!) This phenomenon as a result of which each gas gradually permeates the other is known as *diffusion*. This phenomenon has become possible due to random thermal motion and continues as long as inequalities in concentration exist. (After infinite time, the concentration would be same everywhere in the body of the gas.) Molecules diffuse from regions of higher concentration towards regions of lower concentration. Figure 2.9(a) schematically depicts the diffusion of like molecules—self-diffusion. Diffusion of unlike molecules is shown in Fig. 2.9(b). Diffusion is responsible for the smell of flowers/scent reaching us. This is also why while sitting in your study, you can know as to what is being cooked in the kitchen.

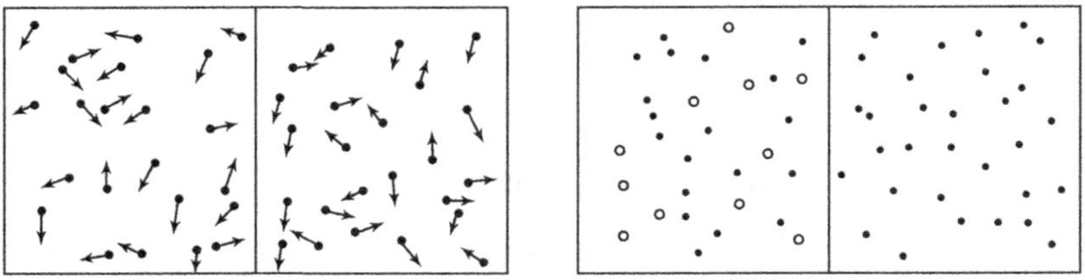

Fig. 2.9 (a) Self-diffusion across a barrier and (b) Diffusion of unlike molecules.

The intermixing of gases is described in terms of the diffusion coefficient, D. Let n be the molecular concentration along a horizontal plane xx'. Suppose that there is positive concentration gradient dn/dy in the vertical plane. Then the number of particles crossing the given surface per unit area per unit time is given by

$$\Gamma = -D\frac{dn}{dy} \tag{2.34}$$

As before, the negative sign is included because gas molecules move from regions of higher concentration to regions of lower concentration but we will calculate the number of molecules crossing in the opposite direction.

When the intermixing gases are different, the calculation of the diffusion coefficient is quite complicated because the rates of diffusion of two gases may not be the same. However, to simplify this problem and still bring out the essential ideas, we consider the diffusion of like molecules, i.e., *self-diffusion*. The diffusion of the isotopes of the same element (say ^{235}U and ^{238}U) is an excellent example of self-diffusion.

Self-diffusion finds very important application in uranium enrichment required for nuclear power generation. (The concentration of fissile isotope ^{235}U in natural uranium is 0.71%. For attaining criticality and generate power in a nuclear power plant, it is desirable to increase its concentration to 2 – 3%, which is above the natural value. Then we say that enriched uranium has been produced.) For this, we convert uranium into uranium hexafluoride (UF_6) gas and make it to diffuse through a porous barrier, which is a ceramic material and consists of fine capillary pores. Since ^{235}U has slightly smaller mass than the more abundant isotope ^{238}U, it diffuses faster. As a result, the gas coming out of the barrier is richer in ^{235}U content. Repetition of this process results in the desired level of enrichment.

To proceed with the calculation of diffusion coefficient D, as before, we assume that each molecule makes its last collision before crossing the reference plane at an average distance of $(2/3)\lambda$. If concentration of gas molecules at the reference plane is n_0, the number density at $(2/3)\lambda$ above or below xx' will be

$$n^{\pm} = n_0 \pm \frac{2}{3}\lambda\frac{dn}{dy}$$

As before, we assume that the number of particles crossing the surface per unit area per second is $n\bar{v}/4$. Hence, the number of molecules crossing the given surface per unit area per unit time from below is

$$\Gamma^- = \frac{1}{4}\bar{v}\left(n_0 - \frac{2}{3}\lambda\frac{dn}{dy}\right)$$

Similarly, the number of molecules transferred upward across the given surface per unit area per unit time from above is

$$\Gamma^+ = \frac{1}{4}\bar{v}\left(n_0 + \frac{2}{3}\lambda\frac{dn}{dy}\right)$$

Hence, the net number of molecules transferred upward across the given surface per unit area per second is given by

$$\Gamma = -\frac{1}{3}\bar{v}\lambda\frac{dn}{dy} \tag{2.35}$$

On comparing Eqs. (2.34) and (2.35), we get a compact expression for diffusion coefficient:

$$D = \frac{1}{3}\bar{v}\,\lambda \tag{2.36}$$

For air at STP, $\lambda = 100$ nm, and $\bar{v} = 450$ ms^{-1}. Therefore, the diffusion coefficient for air is of the order of 10^{-5} m^2 s^{-1}. Equation (2.36) has some important implications. Since $\lambda \propto (T/p)$ and $\bar{v} \propto T^{1/2}$, Eq. (2.36) implies that diffusion coefficient will vary inversely with pressure and directly with temperature as $T^{3/2}$. It is observed that the predicted variation with pressure is in agreement with the experimental results but the power of T lies between 1.75 and 2. The reason for the more rapid increase with temperature is attributed to the presence of intermolecular forces in real gases. To be precise, it is due to the attractive part of the intermolecular potential. Obviously, we do not get it as we have ignored these in our discussion.

Relation between D and η If we combine Eqs. (2.28) and (2.36), we can write

$$\frac{D\rho}{\eta_F} = 1 \tag{2.37}$$

From this result, we note that according to simple kinetic theory, the ratio $\frac{D\rho}{\eta_F}$ is constant for all gases and equal to one. However, its observed value lies between 1.3 and 1.5. A rigorous theoretical analysis by Chapman and Enskog led to a value1.2 for hard spheres and 1.543 when intermolecular potential was modelled as $V(r) = \frac{A}{r^5}$. These findings provided indirect evidence in support of kinetic theory and confidence to early workers. However, they continued their search for direct experimental evidence, which came from the work of a botanist Robert Brown. This coincidence puts faith in the integrated nature of scientific investigations.

So far we have been discussing self-diffusion. But in actual practise, the diffusing molecules are invariably dissimilar, which are hard to distinguish physically. Therefore, we have to look out ways which facilitate us to identify them correctly and follow their trajectories. In other words, molecules must be labelled or tagged somehow. One possible way forward is to induce radioactivity on the nuclei of some molecules and follow their diffusion by a Geiger counter. However, the sizes of the radioactive and the non-radioactive molecules may not be identical.

You should now go through the following examples carefully.

Example 2.11 The coefficients of viscosity and diffusion for oxygen are 1.95×10^{-5} Nsm^{-2} and 1.22×10^{-5} m^2 s^{-1}, respectively. If the average molecular speed is 440 ms^{-1}, calculate the density and mean free path.

Solution: From Eq. (2.37), we can write

$$\rho = \frac{\eta}{D}$$

On substituting the given values, we get

$$\rho = \frac{1.95 \times 10^{-5}\ \text{Nsm}^{-2}}{1.22 \times 10^{-5}\ \text{m}^2\ \text{s}^{-1}} = 1.6\ \text{kg m}^{-3}$$

We can rearrange Eq. (2.36) as

$$\lambda = \frac{3D}{\bar{v}} = \frac{3 \times (1.22 \times 10^{-5}\ \text{m}^2\ \text{s}^{-1})}{440\ \text{ms}^{-1}} = 8.32 \times 10^{-8}\ \text{m}$$

Example 2.12 Calculate the mass of nitrogen diffusing through an area 10^{-2} m^2 in 10 s, if the concentration gradient is 1.26 kg m^{-4}, $\lambda = 10^{-7}$ m and $\bar{v} = 480$ ms^{-1}.

Solution: The mass diffusing per unit time will be obtained by multiplying the expression for Γ with the mass of gas molecules. Hence, we can write

$$|M| = -\frac{1}{3}\bar{v}\lambda m\frac{dn}{dy} = -\frac{1}{3}\bar{v}\lambda\frac{d\rho}{dy}$$

On inserting the given data, we get

$$|M| = \frac{1}{3}(480\ \text{ms}^{-1}) \times (10^{-7}\ \text{m}) \times (1.26\ \text{kg m}^{-4}) = 2.02 \times 10^{-5}\ \text{kg m}^{-2}\,\text{s}^{-1}$$

For diffusion of two different gases (labelled 1 and 2, having concentrations n_1 and n_2, average molecular speeds $\bar{v}_1$ and $\bar{v}_2$ and mean free paths λ_1 and λ_2), the expression for diffusion coefficient modifies to

$$D = \frac{1}{3}\frac{\lambda_1 \bar{v}_1 n_1 + \lambda_2 \bar{v}_2 n_2}{n_1 + n_2} \tag{2.38}$$

A study of the variation of diffusion coefficient with the composition of a mixture is quite interesting. We denote D as D_{12} when $n_1 << n_2$ and as D_{21} when $n_2 << n_1$. Note that D_{12} will be proportional to $\sqrt{m_2/m_1}$ and D_{21} will be proportional to $\sqrt{m_1/m_2}$. So we can write $\frac{D_{12}}{D_{21}} = \frac{m_2}{m_1}$. For hydrogen diffusing into CO_2, $\frac{m_2}{m_1} = 22$. Experimental investigations on this problem show variations amongst the observed values as well as departures from theoretical predictions. This brings out the limitations of elementary kinetic theory model, in spite of its aesthetic appeal. A rigorous analysis, however, shows a fairly good agreement but this discussion is beyond the scope of this book.

You should now go through the following example.

Example 2.13 Calculate the coefficient of viscosity, thermal conductivity and diffusion coefficient for air. Given $\rho = 1.29$ kg m^{-3}, $\bar{v} = 460$ ms^{-1} and $\lambda = 6.4 \times 10^{-8}$ m. Take molecular weight of air as 29 kg kmol^{-1} and $R = 8.31$ kJ K^{-1} kmol^{-1}.

Solution: We know that the coefficient of viscosity η is given by

$$\eta = \frac{1}{3}\rho\bar{v}\lambda$$

On inserting the given values, we get

$$\eta = \frac{1}{3} \times (1.29 \text{ kg kmol}^{-1}) \times (460 \text{ ms}^{-1}) \times (6.4 \times 10^{-8} \text{ m}) = 1.26 \times 10^{-5} \text{ Ns m}^{-2}$$

Recall that the coefficient of thermal conductivity K is given by

$$K = \frac{1}{3}\rho\bar{v}C_V^*\lambda = \eta C_V^*$$

Here $C_V^* = \dfrac{C_V}{M}$ is specific heat capacity per unit mass, and $M = 29$ kg kmol^{-1}. For a diatomic gas, $C_V = \dfrac{5}{2}R = \dfrac{5}{2} \times 8.31 \text{ kJ K}^{-1} \text{kmol}^{-1} = 20.775 \times 10^3 \text{ JK}^{-1} \text{ kmol}^{-1}$. Hence,

$$K = \frac{(1.26 \times 10^{-5} \text{ Nsm}^{-2}) \times (20.775 \times 10^3 \text{ JK}^{-1} \text{ kmol}^{-1})}{29 \text{ kg kmol}^{-1}}$$

$$= 9.07 \times 10^{-3} \text{ J m}^{-1} \text{ s}^{-1} \text{ K}^{-1}.$$

The coefficient of diffusion D is given as

$$D = \frac{1}{3}\bar{v}\lambda = \frac{1}{3} \times (460 \text{ ms}^{-1}) \times (6.4 \times 10^{-8} \text{ m})$$

$$= 9.8 \times 10^{-6} \text{ m}^2 \text{ s}^{-1}.$$

2.4 Brownian Motion

We now know that elementary kinetic theory successfully explains many observed properties of gases. But the first experimental evidence for the existence of molecules and their continuous chaotic motion was provided by Robert Brown. He observed the motion of very small pollen grains suspended in water using a high power microscope. The suspended particles were seen to move completely haphazardly and execute perpetual movement. This irregular motion of the particles is termed *Brownian motion*. It is depicted in Fig. 2.10.

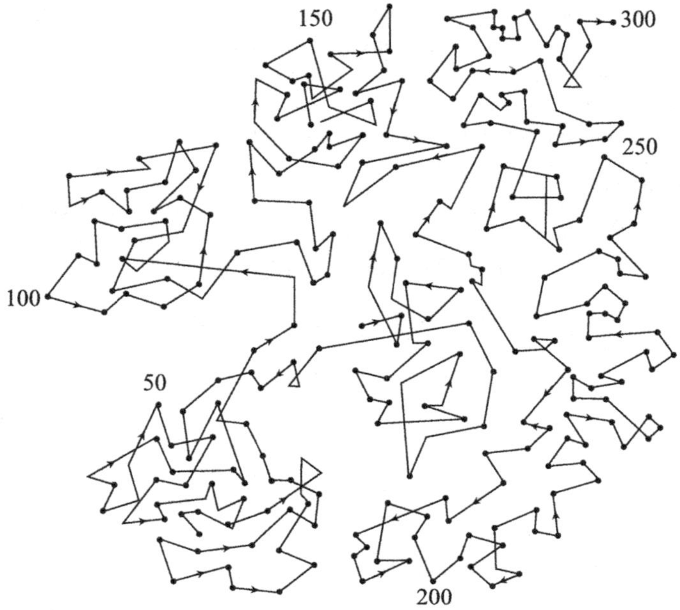

Fig. 2.10 Brownian motion of a fine particle suspended in an aqueous solution.

The observed characteristics of Brownian motion are:

1. The motion is continuous, completely random and irregular.
2. No two particles execute the same motion.
3. Smaller particles execute faster and hence more noticeable motion.
4. The motion becomes more vigorous and lively when temperature is increased or a less viscous liquid is taken. (It is perceptible in glycerine and very active in a gas.)
5. The movement is about the same in all directions.
6. The motion is independent of external influences.

The suspended particles can be likened to a fleet of buoys charting their course through a turbulent sea of molecules. The course of a suspended particle is similar to the groggy steps of a drunken man who starts walking from a lamp post on a city square.

The discovery of Brownian motion puzzled scientists for a long time. At first, it was regarded as a property of organic matter. The first dynamical theory of Brownian motion proposed that the particles were alive. It was argued that vitality is retained by the molecules of a plant for long after its death. However experiments by Brown and others proved conclusively that the motion is not due to any biological or chemical factor. In fact, Brown made observations on a drop of water trapped in a chunk of igneous rock as the rock cooled from its melt. By focusing a microscope on the drop, he saw scores of tiny particles suspended in the drop executing a random dance. Thus, it was established that small particles of inorganic matter suspended in both liquids and gases also executed Brownian motion. Thereafter, it was attributed to (i) surface tension, (ii) non-homogeneity in temperature, (iii) chemical or electrical action, etc, among others. However, none of these explanations were found adequate. After a systematic study, Wiener and Guy proposed that Brownian motion is perhaps due to the bombardment of suspended particles by the molecules of the surrounding fluid. For bigger particles, the forces due to molecular impacts almost completely balance. However, for smaller particles, molecular impacts do not exactly balance and the unbalanced force makes them move in a random manner.

Brownian motion provides a very useful picture of the gaseous state. One can suppose that like Brownian particles, gas molecules are in random motion and frequently collide with each other. In fact, Brownian motion is readily observed in gases because intermolecular forces are negligibly small. In a sense, this phenomenon provides us a way of visualising a microstate. It is for this reason that the work of a botanist gained fundamental importance as a subject matter of importance in physics.

Brownian motion has been holding unending charm for mathematicians as well as for physicists ever since its discovery. But it required the genius of Einstein to work out a systematic mathematical theory of Brownian motion. His arguments were based on physical processes that take place inside the medium, i.e., in terms of the collisions between fluid molecules and the suspended particles. He argued that although each impact is very small, the net result of a large number of random collisions gives rise to 'drunken man's walk'. Einstein quantified this problem by relating the diffusion of particles to the properties of the molecules responsible for the collisions. That is, Einstein related the molecular theory of gases to the observed diffusion of particles. Einstein's predictions were found to be precisely correct by the beautiful experiments of Perrin. This also paved the way for accurate determination of molecular masses. This work convinced everyone about the reality of the molecular nature of matter and launched the subject to higher horizons. In 1908, Langevin rederived Einstein's formula for mean square displacement by considering the equation of motion of suspended particles. We will now discuss Einstein's theory.

2.4.1 Einstein's Theory

Einstein gave an exact description of Brownian motion in terms of the effects of random collisions between the molecules of the liquid and the suspended particles. He argued that although each impact is very small, the net result of a large number of random collisions—nearly 10^{21} per second under normal conditions—gives rise to 'drunken man's walk'. To quantify this problem, Einstein related the diffusion of particles to the properties of the molecules responsible for the collisions. That is, *he calculated the diffusion coefficient from the erratic motion of particles arising from molecular bombardment.*

We also know that molecules of the solute dissolved in a dilute solution exert pressure, called *osmotic pressure.* (This is known as van't Hoff's law and states that the osmotic pressure is numerically equal to the pressure which the dissolved substance would exert if it were assumed to behave like a gas having the same volume and temperature as the given solution.) If a concentration gradient exists between different parts of the solution, *suspended particles will diffuse under the osmotic pressure difference.* This can also be used to calculate the diffusion coefficient.

Einstein calculated the diffusion coefficient from the random motion of the suspended particles as well as the osmotic pressure difference between different parts caused by difference in concentration of suspended particles. He then equated these expressions to calculate mean square displacement of a Brownian particle. We now give a simple derivation of D based on random molecular motion.

Calculation of D from random molecular motion We know that random molecular motion causes Brownian particles to diffuse and their motion is totally erratic. For simplicity, we confine ourselves to one-dimensional Brownian motion and assume that, on an average, each particle is displaced through a distance s in time τ. Within the medium,

Take an open glass tube and cover its one end with a 'semi-permeable membrane', which is permeable to water but not to sugar in solution. Fill this tube with a dilute sugar solution and dip it into a beaker of pure water. You will observe that the solution rises above the water level in the beaker. This means that the solution has a pressure ρgh higher than that of pure water at the same temperature. This pressure, exerted by the sugar dissolved in solution, is called *osmotic pressure*. For dilute solutions, van't Hoff proved that

$$p_{\text{osmotic}} = nk_B T$$

where n denotes the concentration of solution. So we can say that for dilute solutions, the osmotic pressure is equal to the pressure which the solute would exert if it were assumed to behave like an ideal gas having the volume and temperature of the solution.

let us imagine a cylinder of cross-sectional area A and length s with its axis parallel to the x-axis. Its end faces are denoted by P and Q in Fig. 2.11.

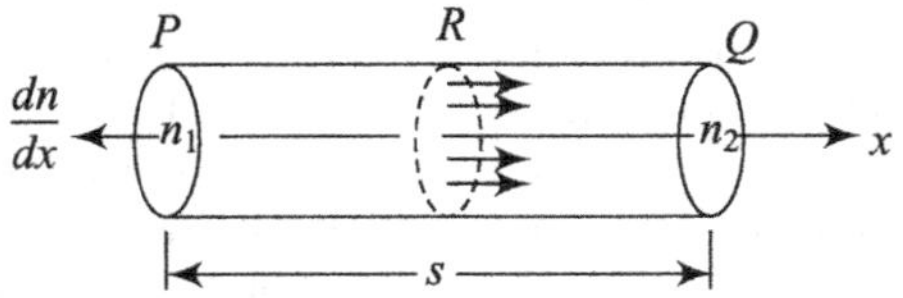

Fig. 2.11 Calculation of diffusion coefficient for one-dimensional Brownian motion.

Let the molecular concentration of Brownian particles at P be n_1 and that at Q be n_2 such that $n_1 > n_2$. That is, along the cylinder, there exists a molecular concentartion gradient $\frac{dn}{dx}$, where n is mean concentration,. This makes the suspended particles to diffuse. Then the number of particles crossing P to the right in time τ is

$$\vec{\Gamma} = \frac{n_1}{2} sA$$

Note that the factor (1/2) has been introduced because only half of the particles contained in a cylinder of volume sA situated to the left of P will enter the cylinder PQ at P in this time. Similarly, the number of particles entering Q and moving in the negative x-direction in time t is

$$\overleftarrow{\Gamma} = \frac{n_2}{2} sA$$

So the excess number of particles crossing the vertical plane R at the centre of the cylinder in the positive x-direction in time τ is given by

$$\Gamma = \vec{\Gamma} - \overleftarrow{\Gamma} = (n_1 - n_2)\frac{sA}{2}$$

Hence, the number of Brownian particles diffusing across R to the right per unit time is

$$\frac{\Gamma}{A\tau} = (n_1 - n_2)\frac{s}{2\tau}$$

$$= -\frac{s^2}{2\tau}\frac{dn}{dx} \tag{2.39}$$

since $n_1 - n_2 = -s\dfrac{dn}{dx}$.

From the definition of diffusion coefficient, it readily follows from Eq. (2.39) that

$$D = \frac{s^2}{2\tau} \tag{2.40}$$

Calculation of D from osmotic pressure difference Let us now calculate D from the osmotic pressure difference. Einstein argued that van't Hoff's law should hold for dilute solutions as well as for dilute suspensions. Let the osmotic pressures exerted by Brownian particles on the faces P and Q of the cylinder shown in Fig. 2.11 be p_1 and p_2, respectively. If we treat Brownian particles like the molecules of a gas, we can write

$$p_1 = n_1 k_B T$$

and
$$p_2 = n_2 k_B T$$

where k_B is Boltzmann constant and T is temperature of the solution. Since $n_1 > n_2$, p_1 will be greater than p_2 and the osmotic pressure difference $(p_1 - p_2)$ will give rise to a force

$$F = (p_1 - p_2)A = (n_1 - n_2)k_B TA$$

This force tends to push the molecules in the cylinder toward the right and is experienced by all (nAs) particles contained therein. Hence, the magnitude of force experience by any one of the Brownian particles is given by

$$f = \frac{(n_1 - n_2)k_B TA}{nAs} = \frac{(n_1 - n_2)}{n}\frac{k_B T}{s} = -\frac{k_B T}{n}\frac{dn}{dx}$$

As a particle moves under the influence of this force, it will experience viscous drag. If we assume that all Brownian particles are spherical and have a radius r, the viscous force is given by the Stokes' law ($f = 6\pi\eta rv$). Hence, we can write

$$f = 6\pi\eta r v = -\frac{k_B T}{n}\frac{dn}{dx}$$

or
$$nv = -\frac{k_B T}{6\pi\eta r}\frac{dn}{dx}$$

The product nv defines the number of particles moving to the right per second per unit area and is, by definition, equal to $-D\dfrac{dn}{dx}$. It, therefore, readily follows that

$$D = \frac{k_B T}{6\pi\eta r}. \tag{2.41}$$

On comparing Eqs. (2.40) and (2.41), we get

$$\frac{s^2}{2\tau} = \frac{k_B T}{6\pi\eta r}$$

or

$$s^2 = \frac{k_B T}{3\pi\eta r}\tau = \frac{RT}{N_A}\frac{1}{3\pi\eta r}\tau \tag{2.42}$$

This result is known as *Einstein's equation* for mean square displacement of a Brownian particle. Probabilistically speaking, it is a fluctuation-dissipation relation where mean square displacement (fluctuation) is connected to a dissipative mechanism (phenomenologically signified by η).

It is instructive to note that Brownian mean square displacement is independent of the mass of particles. Through a brilliant series of experiments, Perrin confirmed this prediction by varying m through a factor of 15,000. We further note that diffusion of particles is related to molecular motion. Moreover, since s^2, τ, η and r are measureable quantities, Eq. (2.42) provided a ready tool for the determination of Avogadro's number. In fact, this equation was verified by Perrin, and was found to be precisely correct. His experiments established beyond doubt the existence of molecules and provided kinetic theory the general acceptability. We shall return to these details in a later section.

ALTERNATIVE DERIVATION OF EINSTEIN'S EQUATION FOR BROWNIAN MOTION

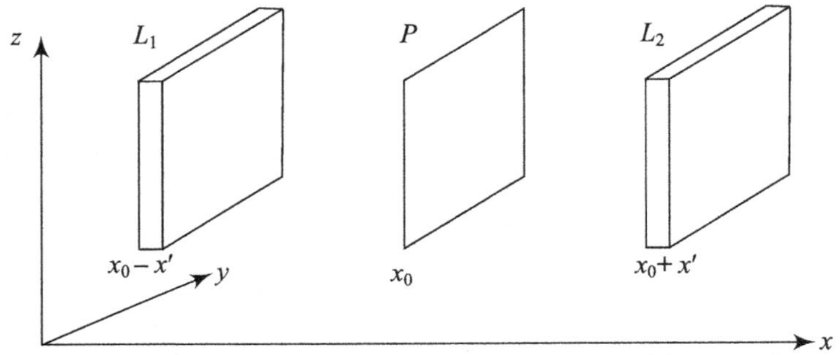

Fig. B2.2 Reference plane and symmetrically situated planes of thickness dx.

Let P be a plane perpendicular to the x-axis at the position $x = x_0$. We consider two layers L_1 and L_2, each of width dx and area A situated symmetrically with respect to P at $x_0 - x'$ and $x_0 + x'$ respectively. Suppose that the concentration at x_0 is n_0. If average concentration is n and there exists a concentration gradient $\frac{\mathrm{d}n}{\mathrm{d}x}$, the concentrations at L_1 and L_2 can be expressed as $n_0 - \frac{\mathrm{d}n}{\mathrm{d}x}x'$ and $n_0 + \frac{\mathrm{d}n}{\mathrm{d}x}x'$. Owing to Brownian motion, the particles will undergo random motion. If the probability of finding a particle between $x = b$ and $x = b + db$ in time τ is $f(b)db$, we can write

$$\int_{-\infty}^{\infty} f(b)db = 1$$

and arising out of symmetry, $f(b)$ must be an even function. Thus $f(b) = f(-b)$.

Let us now calculate the number of particles leaving L_1 and crossing P in time τ. This is given by

$$\Gamma^+ = A\,\mathrm{d}x'\left(n_0 - \frac{\mathrm{d}n}{\mathrm{d}x}x'\right)\int_{x'}^{\infty} f(b)\mathrm{d}b \tag{B.1}$$

Similarly, the number of particles leaving L_2 and crossing P in time τ is

$$\Gamma^- = A\,\mathrm{d}x'\left(n_0 + \frac{\mathrm{d}n}{\mathrm{d}x}x'\right)\int_{-\infty}^{-x'} f(b)\mathrm{d}b \tag{B.2}$$

From Eqs. (B.1) and (B.2), we note that the total number of particles crossing P in time τ is given by

$$N = \Gamma^- - \Gamma^+ = A\,\mathrm{d}x'\left[\left(n_0 + \frac{\mathrm{d}n}{\mathrm{d}x}x'\right)\int_{-\infty}^{-x'} f(b)\mathrm{d}b - \left(n_0 - \frac{\mathrm{d}n}{\mathrm{d}x}x'\right)\int_{x'}^{\infty} f(b)\mathrm{d}b\right]$$

Since $f(b)$ is symmetric, i.e. $\int_{-\infty}^{-x'} f(b)\mathrm{d}b = \int_{x'}^{\infty} f(b)\mathrm{d}b$, the expression for N can be rewritten as

$$N = \int_{x'=0}^{\infty} 2A\,dx'\frac{\mathrm{d}n}{\mathrm{d}x}x'\int_{x'}^{\infty} f(b)\mathrm{d}b$$

$$= 2A\frac{\mathrm{d}n}{\mathrm{d}x}\int_{b=0}^{\infty} f(b)\mathrm{d}b\int_{x'=0}^{b} x'\,\mathrm{d}x' = A\frac{\mathrm{d}n}{\mathrm{d}x}\int_0^{\infty} b^2 f(b)\mathrm{d}b = \frac{1}{2}A\frac{\mathrm{d}n}{\mathrm{d}x}\overline{b^2}$$

where $\overline{b^2} = \int_{-\infty} b^2 f(b)\,\mathrm{d}b$. Note that while interchanging the integration over b and x we have changed the limits of x'; now it runs from 0 to b instead of 0 to ∞. As such, this is mathematically consistent since b runs from 0 to ∞.

In terms of the diffusion coefficient and the concentration gradient, we note that $N = DA\tau\frac{\mathrm{d}n}{\mathrm{d}x}$ so that we can write

$$DA\tau\frac{\mathrm{d}n}{\mathrm{d}x} = \frac{1}{2}A\frac{\mathrm{d}n}{\mathrm{d}x}\overline{b^2}$$

so that

$$D = \frac{\overline{b^2}}{2\tau} \tag{B.3}$$

We know that in the presence of concentration gradient between different parts of the solution, Brownian particles diffuse under the osmotic pressure difference. To calculate the diffusion coefficient, we note that if Brownian particles can be treated like the molecules of a gas, we can write

$$p = n\,k_B T$$

so that

$$\frac{\mathrm{d}p}{\mathrm{d}x} = k_B T\frac{\mathrm{d}n}{\mathrm{d}x}$$

This result shows that the x-component of force exerted on each particle is $\frac{k_B T}{n}\frac{\mathrm{d}n}{\mathrm{d}x}$. Particles moving under the influence of this force experience viscous drag. For spherical Brownian particles of radius r, this is given by Stoke's law. Therefore, we can write

$$\frac{k_B T}{n}\frac{\mathrm{d}n}{\mathrm{d}x} = 6\pi\eta r v \tag{B.4}$$

where v is terminal velocity and η is coefficient of viscosity.

We further note that in unit time nv number of particles cross unit cross section in the yz-plane. Therefore, we can write

$$nv = D\frac{\mathrm{d}n}{\mathrm{d}x}$$

On rearrangement, we get an expression for diffusion coefficient:

$$D = \frac{nv}{\dfrac{\mathrm{d}n}{\mathrm{d}x}} \tag{B.5}$$

On substituting for *dn/dx* from Eq. (B.4) in (B.5), we get

$$D = n\frac{v}{6\pi\eta r v\cdot\dfrac{n}{k_B T}} = \frac{k_B T}{6\pi\eta r}$$

$$= \frac{RT}{6\pi r N_A} \tag{B.6}$$

On comparing Eqs. (B.3) and (B.6), we get

$$\overline{b^2} = \frac{RT}{3\pi r\,\eta\,N_A}\tau \tag{B.7}$$

which is analogous to Eq. (2.42).

You must have realised that in Einstein's derivation, emphasis is on relating Brownian motion to physical processes. A somewhat more elegant theory was given by Langevin. We now discuss it.

2.4.2 Langevin's Theory

Langevin argued that each Brownian particle suffers about 10^{21} collisions per second with the molecules of the liquid and its velocity changes (in magnitude as well as direction) about 10^7 times in one second. So it is not advisable to think in terms of individual collisions. That is, it is impossible to trace the path of a particle and predict its exact position at a given time. (Mathematically speaking, the trajectory of a particle is nowehere a differentiable function.) Following Langevin, we asume that the average force acting on a suspended particle due to molecular bombardment is made up of a frictional and a fluctuating component. The equation of motion of a free Brownian particle along any arbitrarily chosen x-axis is

$$m\ddot{x} = F + F_x \tag{2.43}$$

where m is mass of the suspended particle, double dot over x denotes its second order derivative with respect to time, F is the frictional force and F_x is x-component of the fluctuating force due to molecular bombardment. Phenomenologically, this equation combines the microscopic and the macroscopic viewpoints as signified by the frictional force and the fluctuating force, respectively.

Since a suspended particle suffers, on an average, one collision in about 10^{-21} s with the molecules of the liquid, the mean free path of the fluid molecules is small compared with the size of the suspended particles. Therefore, the surrounding medium can be considered as continuous. Langevin further assumed that suspended particles are spherical in shape and the frictional force is given by Stokes' law*:

$$F = -6\pi\eta r\, \dot{x} = -C\, \dot{x} \tag{2.44}$$

where $C = 6\pi\eta r$; r being the radius of the suspended particle. On combining Eqs. (2.43) and (2.44), we can write

$$m\ddot{x} = -C\,\dot{x} + F_x \tag{2.45}$$

It is important to note here that the direction of motion of each suspended particle changes at each collision. As a result, F_x will be quite irregular in its value; as often positive as negative. Therefore, if we follow the motion of suspended particles over a time $t >> 10^{21}$ s, the expected displacement will be zero. This is because the probabilities of positive and negative displacements are equal. To overcome this problem, we evaluate mean square displacement and work with x^2 rather than x. It means that the equation of motion should be written in terms of x^2. To do so, we multiply Eq. (2.45) throughout by x. This gives us

$$m\,x\ddot{x} = -Cx\,\dot{x} + xF_x \tag{2.46}$$

Proceeding further, we note that

$$\frac{d}{dt}(x^2) = 2x\,\dot{x}$$

and

$$\frac{d^2}{dt^2}(x^2) = 2x\ddot{x} + 2(\dot{x})^2$$

so that use of these relations in Eq. (2.46) leads to

$$\frac{m}{2}\frac{d^2}{dt^2}(x^2) - m(\dot{x})^2 = -\frac{C}{2}\frac{d}{dt}(x^2) + xF_x \tag{2.47}$$

This equation is valid for each suspended particle. If we average over a large number of particles, we get

$$\frac{m}{2}\overline{\frac{d^2}{dt^2}(x^2)} - m\overline{(\dot{x})^2} = -\frac{C}{2}\overline{\frac{d}{dt}(x^2)} + \overline{xF_x} \tag{2.48}$$

The bar denotes the average over all particles.

Since both x and F_x vary randomly, the term $\overline{xF_x}$ in Eq. (2.48) will become zero. Further, Brownian particles are in thermal equilibrium with fluid molecules and their mean kinetic energy (associated with one degree of freedom) is $k_BT/2$. That is,

$$m\overline{(\dot{x})^2} = k_B T$$

* Note that in the case of rarefield gases, the mean free path of the molecules becomes large as compared to the size of the suspended particles and the frictional force will be given by Doppler's law.

Using these results in Eq. (2.48), we obtain

$$\frac{m}{2}\overline{\frac{d^2}{dt^2}(x^2)}+\frac{C}{2}\overline{\frac{d}{dt}(x^2)}=k_B T$$

or

$$\overline{\frac{d^2}{dt^2}(x^2)}+\omega\overline{\frac{d}{dt}(x^2)}=\frac{2k_B T}{m} \tag{2.49}$$

where $\omega = C/m$. Note that ω has dimension of inverse time and signifies the frequency of collisions.

To solve this equation, we assume that averaging over space and differentiation with respect to time are commutative:

$$\overline{\frac{d^2}{dt^2}(x^2)}=\frac{d}{dt}\overline{(x^2)}$$

and introduce a change of variable by defining

$$\frac{d}{dt}\overline{x^2}=u \tag{2.50}$$

Then Eq. (2.49) takes the form

$$\dot{u}+\omega u=\frac{2k_B T}{m} \tag{2.51}$$

This is a first order inhomogeneous differential equation. Its most general solution is*

$$u=\frac{2k_B T}{C}+A\exp(-\omega t) \tag{2.52}$$

Since ω is very large (m being small), the exponential term decays very rapidly and may be safely neglected. Then, Eq. (2.52) takes a very compact form:

$$\frac{d}{dt}\overline{x^2}=u=\frac{2k_B T}{C} \tag{2.53}$$

This equation can be readily integrated, say from $t = 0$ to $t = \tau$:

$$\int_0^\tau \frac{d}{dt}\overline{(x^2)}\,dt=\frac{2k_B T}{C}\int_0^\tau dt$$

---

* This consists of the particular integral and complementary function. The particular integral is obtained by noting that

$$u=\left(\frac{1}{\frac{d}{dt}+\omega}\right)\left(\frac{2k_B T}{m}\right)=\frac{1}{\left(\frac{d}{dt}+\omega\right)\frac{m}{2k_B T}}=\frac{2k_B T}{m\omega}=\frac{2k_B T}{C}$$

On the other hand, the complementary function is obtained by equating the RHS of Eq. (2.51) equal to zero. The resulting equation can readily be solved to get

$$u=A\exp(-\omega t)$$

where A is a constant of integration.

The desired result is

$$\overline{x_\tau^2} - \overline{x_0^2} = \overline{\Delta(x^2)} = \frac{2k_B T}{C}\tau \tag{2.54}$$

On substituting for C, we get

$$\overline{\Delta(x^2)} = \frac{k_B T}{3\pi\eta r}\tau \tag{2.55}$$

This result is same as in Eq. (2.42). Note that $\overline{\Delta(x^2)}$ is in no sense the actual displacement of Brownian particles. It is the mean of the squares of the projections of actual displacements on the x-axis. That is, if we take a snapshot of the suspesion at t = 0 and $t = \tau$, we should measure the component of displacement along any arbitrarily chosen direction, say x-axis and determine $\delta(x^2)$ for each particle. Then we should add all values of $\delta(x^2)$ and divide by the number of particles to obtain $\overline{\Delta(x^2)}$. In his experiments, Perrin worked with 100 different particles of known size. And, if we closely re-examine Fig. 2.10, we note that the motion is so complex that an experimentalist may find it quite inconvenient to work with such a large number of particles. Therefore, motion of one particle is followed for N succesive intervals of time (when N is a large number), the motion is almost equivalent to the motion of N particles during a single time interval. (This corresponds to the assumption that differentiation and averaging are commutative.)

Perrin used Eq. (2.55) to measure Avogadro's number⁺, N_A and established beyond doubt the existence of molecules as well as intermolecular collisions. In a way, these studies put kinetic theory on a sound basis.

2.4.3 Examples of Brownian Motion

We have just now seen that colloidal suspensions in a fluid exhibit Brownian motion. We come across many other intersting examples of Brownian motion. These include sedimentation, diffusion of pollutants in our atmosphere or smoke particles in air, motion of a galavanometer mirror and Johnson noise in amplifiers (electrical appliances). These are discussed below.

Sedimentation In sedimentation, the distribution of particles is determined by the influence of gravity and diffusion. Whereas gravity tends to make them settle down, diffusion caused by molecular bombardment tends to homogenise them. (The same is true of pollutants in our atmosphere.) To calculate the number of particles at a given height, we consider a layer of particles bound by surfaces B and A at heights z and $z + \Delta z$, respectively. Let p be the pressure on the lower face and $p + \Delta p$ be the pressure at the upper face, as shown in Fig. 2.12.

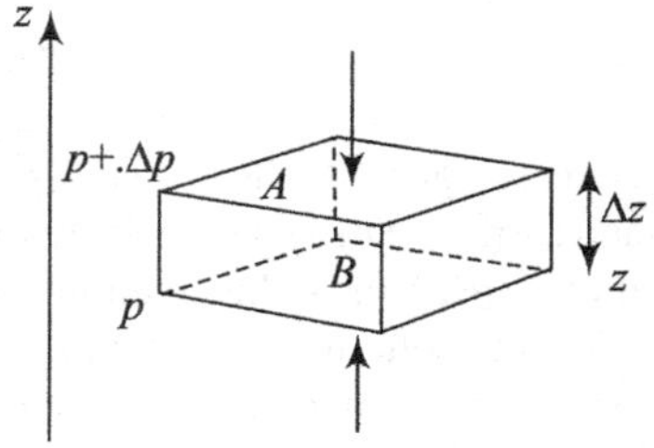

Fig. 2.12 A shallow box of depth Δz and cross sectional area A. The pressure on the upper and lower faces are $p + \Delta p$ and p, respectively.

Let us consider the equilibrium of unit area of the layer. If ρ and g denote the density of particles and acceleration due to gravity, respectively, then for equilibrium we must have

⁺ He was awarded the Nobel Prize in 1926 for this work.

$$\Delta p = -g\rho\Delta z \tag{2.56}$$

The negative sign signifies that pressure decreases as height increases.
If m is the mass of a single particle and n is number density, then $\rho = mn$ and Eq. (2.56) takes the form

$$\Delta p = -mng\Delta z = -\frac{mgN}{V}\Delta z$$

If we assume that Brownian particles obey gas laws, we can replace V by $\mu RT/p$. Further, on replacing Δp and Δz by dp and dz, respectively, the above equation can be rewritten as

$$\frac{dp}{p} = -\frac{mgN}{\mu RT}dz$$

This can readily be integrated to obtain

$$p = p_0 \exp\left(-\frac{mgN}{\mu RT}z\right) \tag{2.57}$$

where $p = p_0$ at $z = 0$.

Since pressure is directly proportional to number density $\left(p = \frac{1}{3}mn\overline{v^2}\right)$, we can rewrite Eq. (2.57) as

$$n(z) = n_0 \exp\left(-\frac{mgN}{\mu RT}z\right) \tag{2.58}$$

In a colloidal suspension, the suspended particles are buoyed up by the liquid, say of density ρ'. Then, their effective mass m_{eff} will be given by

$$m_{\text{eff}} = \frac{4\pi}{3}(\rho - \rho')r^3$$

where r is radius of the suspended particles.
Using this expression in Eq. (2.58) we get

$$n(z) = n_0 \exp\left(-\frac{4\pi}{3}(\rho - \rho')\frac{r^3 gN}{\mu RT}z\right) \tag{2.59}$$

This equation tells us that if $\rho = \rho'$, the effective gravitational field is greatly reduced and there will be an appreciable variation in particle concentration with height. However, if ρ' is small, the sedimentation due to gravity will be so rapid that all suspended particles reach the bottom of the container quickly and lie in a thin layer. The observed brownish skyline of cosmopolitan cities and sedimentation of hydrocarbons are vivid examples of these predictions.

Since $\frac{N}{\mu} = N_A$, we can readily rewrite Eq. (2.59) as

$$N_A = \frac{3RT}{4\pi r^3(\rho - \rho')gz}\ln\left(\frac{n_0}{n}\right) \tag{2.60}$$

This result shows that we can study the variation of number density with height for a fine suspension and conveniently determine Avogadro's number. Perrin worked with emulsions of gamboge and mastic and obtained a value of 6×10^{26} molecules per kilo-mole. (This value is virtually identical to the value obtained by other sophisticated methods.) This suggests that fine particles in thermal equilibrium behave like gas molecules.

Example 2.14 In his experiment on water suspension of gamboge at 20°C, Perrin observed an average of 49 particles per cm^2 in a layer at one level and 14 particles per cm^2 in a layer 60 microns higher (1 micron = 10^{-6} m). If the density of gamboge is 1.194 g cm^{-3} and radius of each particle is 0.212 micron, calculate Avogadro's number.

Solution: From Eq. (2.60), we recall that

$$N_A = \frac{3RT}{4\pi r^3(\rho - \rho')gz} \ln\left(\frac{n_0}{n}\right)$$

On substituting the values of various quantities, we get

$$N_A = \frac{3\times(8.31\,\text{J mol}^{-1}\,\text{K}^{-1})\times(293\,\text{K})\ln\left(\frac{49}{14}\right)}{4\times3.14\times(0.212\times10^{-6}\,\text{m})^3\times((1.194-1)\times10^3\,\text{kg m}^{-3})\times(9.8\,\text{ms}^{-2})\times(60\times10^{-6}\,\text{m})}$$

$$= \frac{3\times8.31\times293\times1.25}{4\times3.14\times(0.212)^3\times0.194\times9.8\times60}\times10^{21}\,\text{mol}^{-1}$$

$$= 6.7\times10^{23}\,\text{mol}^{-1}$$

Galvanometer Mirror In your physics laboratory, you may have noted that a delicately suspended galvanometer mirror undergoes small random oscillations which are due to its thermal energy. The net result is that the galvanometer reading fluctuates, i.e., the system shows an unsteady zero position. This provides us another example of Brownian motion in physics.

The Brownian fluctuations of the galvanometer mirror are expressed in terms of the root mean square angular deflection, θ_{rms}. To calculate this, we note that the mirror has a single degree of oscillation about its axis so that the thermal energy associated with mean square angular deflection $\overline{\theta^2}$ is $k_BT/2$, i.e.,

$$\frac{1}{2}C\overline{\theta^2} = \frac{k_B T}{2}$$

so that

$$\theta_{rms} = \sqrt{\frac{k_B T}{C}} \tag{2.61}$$

where C is torsional rigidity of the galvanometer suspension. This implies that smaller the value of C, larger will be θ_{rms}. That is, fluctuations of zero point will be more in a more sensitive galvanometer.

For a fine quartz fibre, $C = 10^{-13}$ Nm rad^{-1} so that at 300 K, the amount by which a galvanometer reading will fluctuate as a result of thermal motion is given by

$$\theta_{rms} = 2\times10^{-4}\ \text{rad}$$

If the lamp and scale arrangement is at a distance of 1 m from the galvanometer mirror, then a fluctuation of 2×10^{-4} rad corresponds to an rms fluctuation of the light spot of 4 ×

10^{-4} m (0.4 mm). This is depicted in Fig. 2.13. Note that these fluctuations do not depend on the presence of air surrounding the mirror. In fact, the gas molecules may also contribute to the oscillation but these also contribute to damping so that the net result is unchanged. The mirror behaves like a big molecule with a single mode of torsional oscillation. In a perfect vacuum, the oscillations can be considered as arising from random absorption and radiation of electromagnetic energy associated with the temperature T.

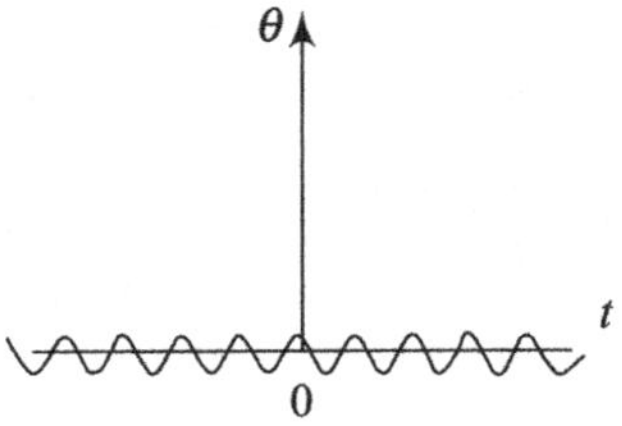

Fig. 2.13 Brownian motion of a galvanometer mirror.

Suppose that the temperature and pressure of air are reduced below normal. Can we eliminate random fluctuations in θ completely? Think about it and read the following example carefully.

Example 2.15 In his experiment, Kapler found that $\overline{\theta^2} = 4.178 \times 10^{-6}$ rad^2. If the suspended system has torsion constant 9.428×10^{-16} Nm rad^{-1} and $T = 287.1$ K, calculate the Boltzmann constant.

Solution: From Eq. (2.61), we can write

$$k_B = \frac{C\,\overline{\theta^2}}{T}$$

On substituting the value of various quantities, we get

$$k_B = \frac{(9.428 \times 10^{-16}\ \text{Nm rad}^{-1}) \times (4.178 \times 10^{-6}\ \text{rad}^2)}{287.1\ \text{K}}$$

$$= 1.4 \times 10^{-23}\ \text{JK}^{-1}$$

Note that radian is dimensionless.

Johnson Noise The conduction electrons in a metal may be regarded as a gas with random velocities. One expects that these may give rise to fluctuating voltage across a resistor. This indeed was observed by Johnson and is known as *Johnson noise* or *electrical noise*. The rms potential fluctuation across a resistor of resistance R in thermal equilibrium at temperature T is given by

$$V_{rms} = [4\,Rk_BT\,(f_2 - f_1)]^{1/2} \tag{2.62}$$

where $(f_2 - f_1)$ is the frequency bandwidth over which the measurements are made. Equation (2.62) shows that there is a limit to useful amplification. If the original signal is too feeble compared to the random electrical noise, reception will not improve with amplification.

Similarly, in the thermionic emission of current I. there is a fluctuation in the emitted current, which is given by

$$\left|\overline{i^2}\right|^{1/2} = \left(\frac{Ie}{\varsigma}\right)^{1/2}$$

where e is electron charge and ς is the time constant of the measuring system. This effect has indeed been used to determine the value of e and the result agrees with other methods to within one per cent.

In the preceding sections, we have discussed theoretical explanations and examples of Brownian motion in physics. One of the most fundamental constants whose determination was undertaken in very elaborate experiments by Perrin was Avogadro's number, which denotes the number of atoms/molecules in one mole of the substance. We will therefore now discuss Perrin's experiments on colloidal suspensions for determination of Avogadro's number. As such, this work signifies a great triumph of molecular theory.

2.4.4 Perrin's Experiment: Determination of Avogadro's Number

To determine Avogadro's number, we have to measure $\overline{x^2}$, the mean square displacement of a Brownian particle. Perrin observed the motion of a single gamboge grain suspended in water at intervals of thirty seconds with the help of a microscope using the camera Lucida. To locate the particles, the microscope had in its field of view a series of mutually perpendicular lines as shown on a graph paper in Fig. 2.14, having 16 divisions being equal to 5×10^{-3} cm. The projections of the successive displacements along the x-axis give a set of values of x, which were used to calculate $\overline{x^2}$. You may now like to ask: How could Perrin make such wonderful observations with a simple arrangement? Perrin derived his argument from the fact that $v_{rms} = 2 \times 10^{-2}$ ms^{-1} at 300 K for a grain of gamboges of radius 2×10^{-7} m and mass about 3×10^{-17} kg (which is 10^9 times the mass of the H_2O molecule). This combination of slow speeds and large size was harnessed by Perrin to observe the motion of suspended particles. It justifies the popular belief that Nature likes logic and simplicity! And most natural laws have been unfolded using very simple arguments. Our own Sir. C.V. Raman explained the blue colour of ocean using a fundamental logic to discard Rayleigh's explanation that the colour of the ocean was just a reflection of the colour of the sky. He conjectured that the colour of ocean had genesis in the scattering of solar light by water molecules and proved it using a very modest apparatus.

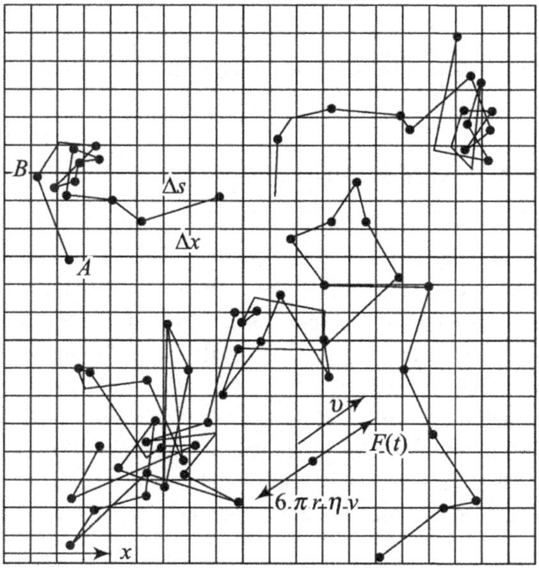

Fig. 2.14 Calculation of $\overline{x^2}$ for a Brownian particle.

Refer to Fig. 2.14. Note that the straight line segments shown here are in no way a representation of the actual path of the particle. The particle is hit millions of times in a second, and hence, its trajectory has a jagged and irregular structure. For example, if we magnify the part *AB* of the trajectory say 100 times, it will appear as shown in Fig. 2.15.

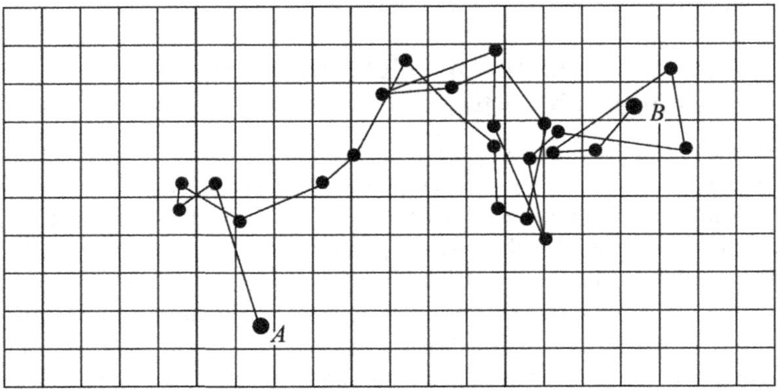

Fig. 2.15 The path *AB* after magnification.

Working with the colloidal suspensions of gum mastic and gum gamboge, Perrin measured all the quantities appearing on the RHS of Eq. (2.60) and obtained the value of Avogadro's number as 6.85×10^{26} molecules kmol^{-1}. (Subsequently, Westgren obtained the value 6.05×10^{26} molecules kmol^{-1}, which differs from the presently accepted value by just 1%.) From his measured value of Avogadro's number, Perrin estimated the mass of a molecule. For example, one kmol of nitrogen gas has a mass of 14 kg. Hence, mass of a nitrogen molecule

$$m_{N_2} = \frac{14 \text{ kg}}{6.85 \times 10^{26}} = 2.04 \times 10^{-26} \text{ kg}$$

Perrin is, therefore credited to be the first person to have weighed the atom with kinetic theory as the tool.

2.5 Random Walk Problem

In physics, we encounter many situations where a system is developing in time or space through individual discrete steps. If each step is random in direction, independent of the preceding or succeeding ones, the study of the net motion is referred to as the *random walk problem*. A simple random walk problem can be realised if we perform the experiment of coin tossing. A coin is tossed at regular intervals, say τ. We move a step to the right if it heads and to the left if it tails. Then our position $x(t)$ at time $t = N\tau$ will depend on the sequence of heads and tails. Note that in this case both time and space parameters are discrete. The net motion of a drunk person, who begins to stroll from a single light post, but is so intoxicated that each step may be in any random direction and a range of different lengths may be visualised as another common example of random walk (Fig. 2.16).

Random walks provide models for numerous applications such as (i) Brownian motion, (ii) turbulence in fluids, (iii) neutron diffusion, (iv) motion of electrons through a metal and of a hole in a semiconductor, (v) motion of defects in crystals, among others. It is therefore important to gain a good understanding of this problem. However, note that since motion in more than one dimension can be broken up into its individual components, it is sufficient to consider random walk in only one dimension.

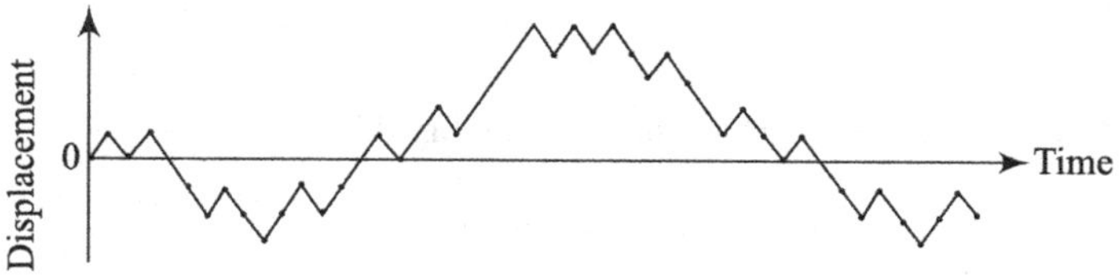

Fig. 2.16 Schematics of one-dimensional random walk. The position of the person is marked after each step.

2.5.1 Random Walk in One Dimension

A particle performing a random walk on a one-dimensional lattice moves to an adjoining position at every step. Let the probability of moving to the right be p (a constant) and to the left be q such that $p + q = 1$. When $p = q = 1/2$, the random walk is said to be *symmetric*. This corresponds to the motion of a free particle. When $p > 1/2$, there is shift to the right and vice versa, the walk is *asymmetric*. This happens in case a force is acting on a particle. If p and q are constant and successive steps are independent, the random walk is termed *Bernoullian*. The random walk where the movement of a particle depends on the direction of the preceding step is known as *correlated random walk*. When there is no limit to the extent to which a particle can move, we say that the walk is *unrestricted*. Otherwise, the walk is restricted to within barriers. When the probability of movement depends on the position of the particle, the random walk is said to be *biased*. When both the state and time parameters become continuous, the random walk is called *diffusion*.

Consider an assembly of particles moving randomly. The basic questions to be answered in this problem are:

1. After each particle has taken N steps, what will be their relative positions vis-à-vis the starting point?
2. How spread out will they be? That is, what will be the standard deviation of their positions relative to their average position?

For simplicity, we consider the motion of a free particle along a line, say x-axis. At the end of each successive interval τ, it may change its direction of motion or not but the two alternatives are equally likely, i.e., it has a probability (1/2) of moving in either direction and distributed at random. We are interested in determining the probability $W(m, N)$ that the particle is at the point m after N steps, if it started from the origin. If the particle makes n_R jumps to the right and n_L jumps to the left, then $n_R + n_L = N$ and $n_R - n_L = m$ so that

$$n_R = \frac{N+m}{2}$$

and

$$n_L = \frac{N-m}{2} \tag{2.63}$$

The number of sequences which contain n_R positive jumps and n_L negative jumps is given by the binomial coefficient and is equal to $N!/n_R!n_L!$* The probability of each specified sequence is $(1/2)^N$. Therefore,

$$W(m, N) = \frac{N!}{n_R!\, n_L!}(1/2)^N$$

$$= \frac{N!}{\frac{1}{2}(N+m)!\,\frac{1}{2}(N-m)!}(1/2)^N \tag{2.64}$$

For large N, we can use Stirling's formula to simplify this expression:

$$\ln n! = \left(n + \frac{1}{2}\right)\ln n - n + \frac{1}{2}\ln 2\pi$$

On substituting this in Eq. (2.64) and simplifying the resultant expression, we get

$$\ln W(m, N) = \left(N + \frac{1}{2}\right)\ln N - N\ln 2 - \frac{1}{2}\ln 2\pi$$

$$-\frac{1}{2}(N+m+1)\ln\left(\frac{N}{2}\left(1+\frac{m}{N}\right)\right) - \frac{1}{2}(N-m+1)\times\ln\left(\frac{N}{2}\left(1-\frac{m}{N}\right)\right) \tag{2.65}$$

For $m << N$, we can write

$$\ln\left(1 \pm \frac{m}{N}\right) = \pm\frac{m}{N} - \frac{1}{2}\frac{m^2}{N^2} \pm \ldots$$

where we have retained terms of order up to N^{-2} and neglected smaller terms. Using this expansion in Eq. (2.65), we get

$$\ln W(m, N) = \left(N + \frac{1}{2}\right)\ln N - N\ln 2 - \frac{1}{2}\ln 2\pi$$

$$-\frac{1}{2}(N+m+1)\left(\ln N - \ln 2 + \frac{m}{N} - \frac{m^2}{2N^2} + \ldots\right)$$

$$-\frac{1}{2}(N-m+1)\left(\ln N - \ln 2 - \frac{m}{N} - \frac{m^2}{2N^2} + \ldots\right)$$

This may be simplified to obtain a compact form:

$$\ln W(m, N) = \ln 2 - \frac{1}{2}\ln 2\pi - \frac{1}{2}\ln N - \frac{m^2}{2N} = \frac{1}{2}\ln\left(\frac{2}{\pi N}\right) - \frac{m^2}{2N} \tag{2.66}$$

On taking the antilog, we get the desired result:

$$W(m, N) = \sqrt{\frac{2}{\pi N}}\exp\left(-\frac{m^2}{2N}\right) \tag{2.67}$$

---

* If we have some criterion for the behaviour of a single element of a system, and if p is the probability that the criterion is satisfied and q is the probability that it is not satisfied, then for a system of N elements, the probability of this system being in a state where n elements satisfy the criterion and the remaining $(N - n)$ elements do not satisfy it is given by

$$P_N(n) = \frac{N!}{N!(N-n)!}p^n q^{N-n}$$

From this we note that although $W(m.N)$ is a discontinuous function, the terminal points of a random walk form the outline of a smooth bell-shaped curve.

Now let us suppose that the length of each step r is very small and $x = mr$. One can then approximate $W(m.N)$ by a continuous probability density function $W_c(x, N)$ such that

$$W(m, N) = KW_c(x, N) \tag{2.68}$$

where K is constant of proportionality. As the probability of finding the random walker is a certainty, K can be calculated by setting

$$\int_{-\infty}^{\infty} \mathrm{d}x\, W_c(x, N) = 1 \tag{2.69}$$

This gives $K = 2r$, so that

$$W_c(x, N) = \frac{1}{\sqrt{2\pi N r^2}} \exp\left(-\frac{x^2}{2Nr^2}\right) \tag{2.70}$$

Problem 2.10 Using Eq. (2.69) prove that $K = 2r$.

It is important to point out here that the constant $K = 2r$ can also be chosen from the physical consideration that in two consecutive non-zero values of $W(m, N)$, m increases by 2 so that x increases by $2r$.

Equation (2.70) can be rewritten in a slightly different form by noting that if the time taken in N jumps is t and $\nu = \tau^{-1}$ denotes the number of jumps per unit time, then $N = \nu t = t/\tau$. This gives

$$W_c(x, t) = \frac{1}{\sqrt{4\pi Dt}} \exp\left(-\frac{x^2}{4\,Dt}\right) \tag{2.71}$$

where $D = r^2/2\tau$ is known as *diffusion coefficient*.

Equation (2.71) gives the desired probability that the particle is at a distance x at time t. It represents a Gaussian distribution curve.

The mean square displacement of the particle is defined as

$$\overline{x^2} = \int_{-\infty}^{\infty} x^2 W_c(x, t)\,\mathrm{d}x$$

We now substitute for $W_c(x, t)$ from Eq. (2.71) to obtain

$$\overline{x^2} = \frac{1}{2\sqrt{\pi Dt}} \int_{-\infty}^{\infty} x^2 \exp\left(-\frac{x^2}{4Dt}\right) \mathrm{d}x$$

$$= \frac{1}{\sqrt{\pi Dt}} \int_{0}^{\infty} x^2 \exp\left(-\frac{x^2}{4Dt}\right) \mathrm{d}x$$

since the integrand is symmetric in x. Now, we introduce a new variable by defining $\frac{x^2}{4Dt} = \alpha$ so that $x\,dx = 2Dt\,d\alpha$ and $x^2 dx = 4(Dt)^{3/2}\alpha^{1/2} d\alpha$. The expression for $\overline{x^2}$ takes a compact form:

$$\overline{x^2} = \frac{4Dt}{\sqrt{\pi}} \int_0^\infty \alpha^{1/2} \exp(-\alpha)\, d\alpha$$

$$= \frac{4Dt}{\sqrt{\pi}} \Gamma(3/2)$$

Since $\Gamma(3/2) = \frac{\sqrt{\pi}}{2}$, we obtain

$$\overline{x^2} = 2Dt \tag{2.72}$$

Do you recognise this result? It is same as that obtained by Einstein for mean square displacement by considering the molecular concentration gradient.

Additional Examples

Example 2.16 Calculate the mean time between collisions for a nitrogen molecule in air at 27°C and a pressure of 1 atm. Take $k_B = 1.38 \times 10^{-23}$ JK^{-1} and $d_{N_2} \approx 10^{-10}$ m. Assume Maxwellian distribution of speeds.

Solution: We know that mean free path of gas molecules obeying Maxwellian distribution is given by

$$\lambda = \frac{1}{\sqrt{2}\pi d^2 n} = \frac{1}{\sqrt{2}\pi d^2} \cdot \frac{k_B T}{p}$$

On substituting the given values, we get

$$\lambda = \frac{1}{\sqrt{2}\pi (10^{-20}\ \text{m}^2)} \times \frac{(1.38 \times 10^{-23}\ \text{JK}^{-1}) \times (300\ \text{K})}{(1.013 \times 10^5\ \text{Nm}^{-2})}$$

$$= 0.92 \times 10^{-6}\ \text{m}$$

The average speed of a Maxwellian gas molecule is given by

$$\bar{v} = \sqrt{\frac{2.55 k_B T}{m}}$$

The mass of nitrogen molecule

$$m = \frac{28}{6.023 \times 10^{26}}\ \text{kg}$$

$$= 4.65 \times 10^{-26}\ \text{kg}$$

Hence,

$$\bar{v} = \sqrt{\frac{2.55 \times (1.38 \times 10^{-23}\ \text{JK}^{-1}) \times (300\ \text{K})}{4.65 \times 10^{-26}\ \text{kg}}}$$

$$= \sqrt{2.27 \times 10^{5}}\ \text{ms}^{-1}$$

$$= 476\ \text{ms}^{-1}$$

Hence, mean time between collisions

$$t = \frac{\lambda}{\bar{v}} = \frac{0.92 \times 10^{-6}\ \text{m}}{476\ \text{ms}^{-1}} = 1.93 \times 10^{-9}\ \text{s}$$

Example 2.17 Calculate the mean free path of the molecules of a gas of diameter 0.2 nm in a closed chamber maintained at 10^{-6} mm of Hg and 273 K. One gram molecule of the gas occupies 22.4 litre at STP.

Solution: We know that at 0.76 m of Hg pressure and 273 K, the number of molecules in 22.4×10^{-3} m^3 of a gas is 6.023×10^{23}.

$\therefore$ No. of molecules per m^3 at 0.76 m pressure $= \dfrac{6.023 \times 10^{23}}{22.4 \times 10^{-3}\ \text{m}^3}$

and

No. of molecules per m^3 at 10^{-6} mm pressure $= \dfrac{6.023 \times 10^{23} \times 10^{-9}\ \text{m}}{22.4 \times 10^{-3}\ \text{m}^3 \times 0.76\ \text{m}}$

$$= 0.354 \times 10^{17}\ \text{m}^{-3}$$

We know that $d = 2 \times 10^{-10}$ m. Hence, if the gas obeys Maxwell's distribution law of speeds, the expression for mean free path is given by

$$\lambda = \frac{1}{\sqrt{2}\,\pi\,d^2 n} = \frac{1}{4\sqrt{2}\,\pi\,(10^{-20}\ \text{m}^2) \times (0.354 \times 10^{17}\ \text{m}^{-3})}$$

$$= \frac{1}{6.29 \times 10^{-3}}\ \text{m}$$

$$= 1.589 \times 10^{2}\ \text{m}$$

$$= 159\ \text{m}$$

Note that the mean free path has a very large value, which may be greater than the size of a typical chamber. This indicates the possibility of obtaining unhindered movement of molecules in a chamber.

Example 2.18 Calculate the frequency of sound at which the wavelength of sound waves becomes equal to the mean free path in oxygen at 273 K and 1 atm. Take diameter of oxygen molecule as 3×10^{-10} m and velocity of sound waves as 330 ms^{-1}.

Solution: We know that the mean free path for a gas obeying Maxwell's distribution law is given by

$$\lambda = \frac{1}{\sqrt{2}\pi d^2 n}$$

At 273 K and 1 atm pressure,

$$n = \frac{6.023 \times 10^{23}}{22.4 \times 10^{-3}\ \text{m}^3} = 2.689 \times 10^{25}\ \text{m}^{-3}$$

Since $d = 3 \times 10^{-10}$ m, the mean free path is

$$\lambda = \frac{1}{1.414 \times 3.1417 \times (3 \times 10^{-10}\ \text{m})^2 \times 2.689 \times 10^{25}\ \text{m}^{-3}}$$

$$= \frac{1}{107.51 \times 10^5}\ \text{m}$$

$$= 9.3 \times 10^{-8}\ \text{m}$$

According to the given condition, this distance is equal to the wavelength of sound waves. Therefore, the frequency corresponding to this wavelength is given by

$$f = \frac{v}{\lambda} = \frac{330\ \text{ms}^{-1}}{9.3 \times 10^{-8}\ \text{m}}$$

$$= 3.5 \times 10^9\ \text{s}^{-1}$$

Example 2.19 Calculate the mean free path for the molecules of a gas at STP. Assume that gas molecules behaves like spheres of radius 3×10^{-10} m.

Solution: We know that at STP 1 kmol of a substance occupies 22.4 m^3 and has 6.023×10^{26} molecules. Hence,

$$\lambda = \frac{1}{\sqrt{2}\pi d^2 n} = \frac{1}{\sqrt{2} \times 3.1417 \times (6 \times 10^{-10}\ \text{m})^2} \times \left(\frac{22.4\ \text{m}^3}{6.02 \times 10^{26}} \right)$$

$$= 2.4 \times 10^{-8}\ \text{m}$$

Example 2.20 The viscosity of a gas at STP was measured to be 1.66×10^{-5} Nm^{-2} per unit velocity gradient. The average speed of the molecules is 450 ms^{-1} and the density of the gas is 1.25 kg m^{-3}. Calculate **(a)** mean free path of the gas, **(b)** collision frequency and **(c)** molecular diameter of the gas molecules.

Solution:

(a) From Eq. (2.28), we recall that

$$\eta = \frac{1}{3} \rho \bar{v} \lambda$$

$$\Rightarrow \qquad \lambda = 3\eta / \rho\bar{v} \qquad \text{(i)}$$

Here $\rho = 1.25$ kg m^{-3}, $\bar{v} = 450$ ms^{-1} and $\eta = 1.66 \times 10^{-5}$ Nm^{-2}s. Hence,

$$\lambda = \frac{3 \times (1.66 \times 10^{-5}\ \text{Nm}^{-2}\,\text{s})}{(1.25\ \text{kg m}^{-3}) \times (450\ \text{ms}^{-1})}$$

$$= \frac{4.98 \times 10^{-5}\ \text{kg m}^{-1}\text{s}^{-1}}{562.5\ \text{kg m}^{-2}\text{s}^{-1}}$$

$$= 8.85 \times 10^{-8}\ \text{m}$$

(b) Frequency of collision $= \frac{\bar{v}}{\lambda} = \frac{450\ \text{ms}^{-1}}{8.85 \times 10^{-8}\ \text{m}} = 5.08 \times 10^9\ \text{s}^{-1}$

(c) We know that for a gas obeying Maxwellian distribution law

$$\lambda = \frac{1}{\sqrt{2}\pi d^2 n} \tag{ii}$$

$$\Rightarrow \quad d^2 = \frac{1}{\sqrt{2}\,\pi n \lambda} \tag{iii}$$

To calculate n, we recall that 22.4 litre of every gas contains 6.023×10^{23} molecules. Hence,

$$\text{Number density (per cm}^3\text{)} = \frac{6.023 \times 10^{23}}{22400} = 0.269 \times 10^{20}$$

$$\Rightarrow \quad \text{Number density (per m}^3\text{)} = 2.69 \times 10^{25}.$$

On substituting the given values in Eq. (iii), we get

$$d^2 = \frac{1}{\sqrt{2} \times 3.1417 \times (2.69 \times 10^{25}\ \text{m}^{-3}) \times (8.85 \times 10^{-8}\ \text{m})}$$

$$= 9.456 \times 10^{-20}\ \text{m}^2$$

$$\therefore \quad d = 3.08 \times 10^{-10}\ \text{m}$$

Example 2.21 The molecular cross section is defined as $\sigma = \pi d^2$. For slow neutrons in hydrogen, $\sigma = 80 \times 10^{-28}\ \text{m}^2$. Assume that neutrons obey Maxwell's distribution law. Calculate their mean free path at $T = 273$ K and $p = 1$ atm. Take $k_B = 1.38 \times 10^{-23}\ \text{JK}^{-1}$.

Solution: We know that mean free path is given by

$$\lambda = \frac{1}{\sqrt{2}\pi d^2 n} = \frac{1}{\sqrt{2}\, n\sigma} = \frac{k_B T}{\sqrt{2}\, p\sigma}$$

On substituting the given values, we get

$$\lambda = \frac{1}{1.414} \times \frac{(1.38 \times 10^{-23}\ \text{JK}^{-1}) \times (300\ \text{K})}{(1.013 \times 10^5\ \text{Nm}^{-2}) \times (80 \times 10^{-28}\ \text{m}^2)}$$

$$= \frac{4.18 \times 10^{-21}}{1.414 \times 1.013 \times 80 \times 10^{-23}}\ \text{m}$$

$$= 3.65\ \text{m}$$

Example 2.22 The viscosity of oxygen at 16°C is 169×10^{-6} poise. Calculate the diameter of gas molecules. Take $N_A = 6.023 \times 10^{23}\ \text{mol}^{-1}$, molecular weight of oxygen is 32 g mol^{-1} and $k_B = 1.38 \times 10^{-23}\ \text{JK}^{-1}$.

Solution: From Eq. (2.29), we recall that

$$\eta = \frac{m\bar{v}}{3\sqrt{2}\,\pi d^2}$$

By rearranging terms, we can write

$$d^2 = \frac{m\bar{v}}{3\sqrt{2}\,\pi\eta}$$

Here $\eta = 169 \times 10^{-6}$ poise $= 169 \times 10^{-7}$ kg m^{-1}s^{-1}, and $T = (273 + 16)$ K $= 289$ K,

$$m = \frac{32 \text{ g mol}^{-1}}{6.023 \times 10^{23} \text{ mol}^{-1}} = 5.313 \times 10^{-23} \text{ g} = 5.313 \times 10^{-26} \text{ kg}$$

and

$$\bar{v} = \sqrt{\frac{2.55\, k_B T}{m}} = \sqrt{\frac{2.55 \times (1.38 \times 10^{-23} \text{ JK}^{-1}) \times (289 \text{ K})}{5.313 \times 10^{-26} \text{ kg}}}$$

$$= \sqrt{191.4 \times 10^3} \text{ ms}^{-1}$$

$$= 4.37 \times 10^2 \text{ ms}^{-1}$$

Hence,

$$d^2 = \frac{(5.313 \times 10^{-26} \text{ kg}) \times (437 \text{ ms}^{-1})}{3 \times 1.414 \times 3.1417 \times (169 \times 10^{-7} \text{ kg m}^{-1}\text{s}^{-1})}$$

$$= 1.031 \times 10^{-19} \text{ m}^2$$

$\therefore \qquad d = 3.21 \times 10^{-10}$ m.

Example 2.23 Calculate the coefficient of viscosity of hydrogen at STP. Given $\rho = 8.90 \times 10^{-2}$ kg m^{-3}, $\lambda = 2 \times 10^{-7}$ m and $k_B = 1.38 \times 10^{-23}$ JK^{-1}.

Solution: From Eq. (2.28), we recall that

$$\eta = \frac{1}{3}\rho\bar{v}\lambda$$

Here $\qquad \rho = 8.90 \times 10^{-2}$ kg m^{-3}, $\lambda = 2 \times 10^{-7}$ m and $\bar{v} = \sqrt{\dfrac{2.55 k_B T}{m}}$.

We know that

$$m = \frac{2 \text{ g mol}^{-1}}{6.023 \times 10^{23} \text{ mol}^{-1}} = 3.32 \times 10^{-24} \text{ g} = 3.32 \times 10^{-27} \text{ kg}$$

Hence,

$$\bar{v} = \sqrt{\frac{2.55 \times (1.38 \times 10^{-23} \text{ JK}^{-1}) \times (273 \text{ K})}{3.32 \times 10^{-27} \text{ kg}}}$$

$$= \sqrt{289 \times 10^4} \text{ ms}^{-1}$$

$$= 1700 \text{ ms}^{-1}$$

Hence,

$$\eta = \frac{1}{3} \times (8.9 \times 10^{-2}\ \text{kg m}^{-3}) \times (17 \times 10^{2}\ \text{ms}^{-1}) \times (2 \times 10^{-7}\ \text{m})$$
$$= 1.01 \times 10^{-5}\ \text{kg m}^{-1}\text{s}^{-1}$$

Example 2.24 In a vacuum flask, the gap between two concentric glass cylinders is 4 mm. Suppose that the pressure is lowed to p. Calculate the pressure at which thermal conductivity between the walls of the flask drops below its value for $p = 1$ atm. At what value of p will thermal conductivity be 10^{-3} times the value at 1 atm? Take 1 atm = 10^5 Nm^{-2} and $\lambda = 100$ nm.

Solution: From Eq. (2.32), we know that

$$K = \frac{C_V}{3} n \lambda \overline{v}$$

Also,

$$n = \frac{p}{k_B T} \quad \text{and} \quad \lambda = \frac{1}{\sqrt{2}\, n\sigma} = \frac{1}{\sqrt{2}} \frac{k_B T}{p\sigma}$$

When the pressure is reduced below the value at which the mean free path becomes 4 mm, the value of λ will remain constant. Further since $\lambda \propto \frac{1}{p}$ at constant T, this will occur when p is reduced below

$$\frac{100 \times 10^{-9}\ \text{m}}{4 \times 10^{-3}\ \text{m}} \times 10^5\ \text{Pa} = 2.5\ \text{Pa}$$

Below this pressure, K will be universally proportional to pressure, so it will be reduced to 10^{-3} of its value at 1 atm when pressure is reduced to 2.5×10^{-3} Pa.

Let us now sum up what you have learnt in this chapter.

SUMMARY

- The mean free path is the average distance travelled by a molecule between two successive collisions. If we assume that all molecules have average speed $\overline{v}$, the expressions for λ is given by

$$\lambda_{Cl} = \frac{0.75}{n\sigma}$$

  where n is the molecular number density and σ is the collision cross section. For a sphere of diameter d, $\sigma = \pi d^2$.
- For a sample of N_0 molecules, the number of molecules which travel a distance x without making any collision is given by survival equation:

$$N = N_0 \exp(-x/\lambda)$$

- When a gas is endowed with mass motion, random molecular motion can lead to transport of energy, momentum or mass, depending on the physical conditions, which lead to the phenomena of thermal conduction, viscosity and diffusion, respectively.

- The coefficient of viscosity for a gas having velocity gradient in a direction normal to the direction of mass motion is given by

$$\eta = \frac{1}{3} m\, n \bar{v}\, \lambda$$

where m is the molecular mass and $\bar{v}$ is the average speed of a molecule.
At a given temperature, η is independent of pressure.
- The thermal conductivity, K is given by

$$K = \frac{f}{3} n \bar{v}\, \lambda k_B$$

where f denotes the number of degrees of freedom.
- K and η are connected by the relation

$$\frac{KM}{\eta C_V} = 1$$

- The coefficient of diffusion D is given by

$$D = \frac{\bar{v}}{3} \lambda = \frac{0.376}{\sigma} \frac{(k_B T)^{3/2}}{p\sqrt{m}}$$

- Brownian motion is perpetual, irregular motion of the particles immersed in fluid caused by their continuous bombardment by the molecules of surrounding medium.
- The mean square displacement of a Brownian particle in terms of the diffusion coefficient D, is

$$\overline{x^2} = 2\,Dt$$

- The Einstein's relation for mean square displacement of a Brownian particle is

$$\overline{x^2} = \frac{RT}{N_A} \frac{1}{3\pi\eta r} \tau$$

- The variation of particle concentration with height, z, in sedimentation is given by

$$n = n_0 \exp\left[-\frac{mgN}{\mu RT} z \right]$$

- The probability of finding a particle at $x = m$ after N steps in a random walk is given by

$$W(m, N) = \frac{N!}{\left(\frac{N+m}{2}\right)! \left(\frac{N-m}{2}\right)!} \left(\frac{1}{2}\right)^N$$

For large N,

$$W(m, N) = \sqrt{\frac{2}{\pi N}} \exp\left(-\frac{m^2}{2N} \right)$$

- The probability that after N steps, the particle is between x and $x + dx$ is given by

$$W_c(x, N) = \frac{1}{\sqrt{2\pi\sigma}} \exp\left(-\frac{x^2}{2\sigma^2} \right)$$

with $\sigma^2 = Nr^2$.

EXERCISES

2.1 A billiard table measures 1.2 m × 2.4 m. The diameter of a billiard ball is 6 cm. Estimate the mean free path for collisions between balls when there are only four balls moving randomly on the table. (*Ans:* 3.8 m)

2.2 A one litre bulb at room temperature contains hydrogen gas at a pressure of 10^{-4} torr. At $t = 0$, a filament of area 0.2 cm^2 is suddenly heated to incandescence. Under these conditions hydrogen molecules striking the filament are dissociated. Neutral hydrogen atoms that are produced stick to the walls of the bulb on striking. How long is the mean free path for hydrogen molecules at the initial pressure? (*Ans:* $\lambda = 2.3$ m)

2.3 Calculate the mean free path and collision frequency of air molecules under standard conditions. Take radius of air molecules as 10^{-10} m, number density $n = 10^{25}$ m^{-3} and mean velocity $\overline{v} = 500$ ms^{-1}. (*Ans:* $\lambda = 5.63 \times 10^{-7}$ m; $z = 8.88 \times 10^8$ s^{-1})

2.4 The diameter of the molecules of a gas is 3×10^{-10} m. Calculate the mean free path at STP. Take $k_B = 1.38 \times 10^{-23}$ JK^{-1}. (*Ans:* 9.31×10^{-8} m)

2.5 An ion of mass m and charge e is moving in a dilute gas of molecules with which it collides. The mean time between collisions suffered by the ion is τ. Suppose that a uniform electric field **E** is applied in the x-direction. Calculate the mean distance $\overline{x}$ (in the direction of **E**) which the ion travels between collisions if it starts out with zero x-component of velocity after each collision. (*Ans:* $(eE/m)\tau^2$)

2.6 Obtain an expression for pressure exerted by gas molecules on the walls of the container using the survival equation.

2.7 The mean free path of the molecules of a gas at 25°C is 2.63×10^{-5} m. If the radius of the molecule is 2.56×10^{-10} m, calculate the pressure of the gas. (*Ans:* 134 Nm^{-2})

2.8 Calculate the frequency of sound at which the wavelength of sound waves will be equal to the mean free path of nitrogen molecules. Take diameter of nitrogen molecules $d = 0.3 \times 10^{-9}$ m and number density $n = 3 \times 10^{25}$ molecules m^{-3}. (*Ans:* 1.33×10^9 Hz)

2.9 Assuming $K/\eta C_V = 5/2$ for translational modes and $K/\eta C_V = 1$ for other internal modes, show that for a polyatomic gas $K/\eta C_V = \dfrac{9\gamma - 5}{4}$, where $\gamma = \dfrac{C_p}{C_V}$.

2.10 The thermal conductivity and specific heat capacity of a gas at 273 K are 24×10^{-3} Jm^{-1} s^{-1} K^{-1} and 20.9 kJ kmol^{-1} K^{-1}, respectively. Calculate the diameter of gas molecules. Take mass of gas molecule as 5.31×10^{-26} kg and $k_B = 1.38 \times 10^{-23}$ JK^{-1}. (*Ans:* 3.04×10^{-11} m)

3 REAL GASES: van der WAALS' EQUATION OF STATE

Learning Objectives

In this chapter, you will learn how to

- discuss experimental findings of Regnault, Andrews and Amagat for real gases;
- establish van der Waals' equation of state and explain how it helps to understand the behaviour of a real gas at different temperatures;
- discuss limitations of van der Waals' equation and how far these have been overcome by other equations of state;
- obtain reduced equation of state and discuss the law of corresponding states;
- describe Joule–Thomson experiment and obtain an expression for Joule–Thomson coefficient; and
- define inversion temperature and discuss its importance for liquefaction of gases.

3.1 INTRODUCTION

In the preceding two chapters, we discussed the behaviour of ideal gases and studied how it helped in the development of kinetic theory of gases in its formative stages. The modifications such as finite size hinted at the possibility that a real gas may show considerable deviation from the ideal equation of state for a perfect gas, $pV = \mu RT$, even at room temperature. As we now know, noble gases show marked deviation at high pressures and low temperatures. To describe the behaviour of such gases, many equations have been proposed. Some of these are purely empirical and a few phenomenological. van der Waals derived an equation by making allowance for the finite size of gas molecules and for the existence of intermolecular forces. Though van der Waals equation has been used with fair degree of success to understand the behaviour of real gases with particular reference to their liquefaction, subsequent experiments brought forth some of its inherent limitations. These essentially arise because no account is made for repulsive forces between molecules.

In Sec 3.2, we have discussed results of various classical experiments conducted to understand the behaviour of real gases. These hinted at the possibility of liquefaction of gases and their use for production of low temperatures, leading to a new and fertile field of study. In fact, it opened the gateway to some very exciting discoveries, which helped improve our understanding of different states of matter. This is followed by the discussion of Onnes empirical equation of state in Sec. 3.3. A detailed discussion of van der Waals equation forms the subject matter of Sec. 3.4. You will learn that it explains almost all experimental results. The reduced equation of state and other equations of state are considered in Sec. 3.5 and 3.6. The evidence for the existence of intermolecular forces of attraction, assumed by van der Waals in arriving at his equation, proved elusive for quite some time. And it needed the genius of Joule and Thomson, who devised an innovative experiment to show that temperature of a gas drops when it is made to expand adiabatically at constant pressure. This is known as porous-plug experiment and is discussed in Sec. 3.7 in detail.

3.2 Deviations from Perfect Gas Behaviour

The equation of state for a perfect gas implies that at a given temperature, the product pV will be constant. So if we plot pV as a function of p, the curve should be a straight line parallel to the pressure axis. Similarly, if pressure variation is studied versus volume, the curve, called an *isotherm*, will be part of a rectangular hyperbola. For a set of temperatures, these isotherms should be parallel. But experimental results reveal to the contrary, some of which are summarised below.

3.2.1 Regnault's Experiments

Regnault carried out a series of classical experiments. He applied pressures up to about 30 atmospheres and temperature was varied in the range 0–100°C. His main results for hydrogen, oxygen, nitrogen and carbon dioxide are depicted in Fig. 3.1, where we have plotted pV versus p. We note that

1. When the pressure is in the range 0 – 1 atm, the curves are straight lines inclined to the pressure axis (Fig. 3.1a). However, the product pV increases with pressure for hydrogen and decreases in case of nitrogen, oxygen and CO_2.
2. At high pressures, the product pV continues to rise with pressure for hydrogen but in case of nitrogen, oxygen and CO_2, it first decreases and then increases (Fig. 3.1b).

These observations point to the imperfect nature of real gases.

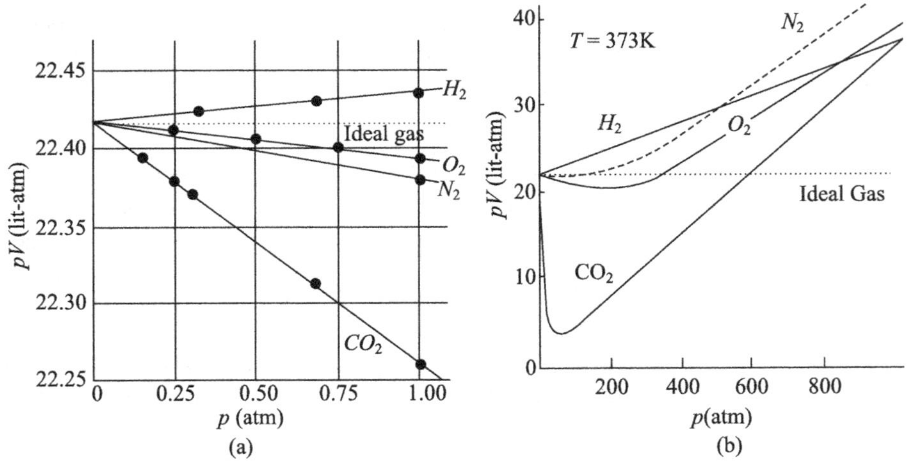

Fig. 3.1 Variation of pV versus p for hydrogen, oxygen, nitrogen and carbon dioxide at (a) low pressures (0–1 atm) and (b) high pressures (0–1000 atm). The dotted horizontal line is for an ideal gas.

3.2.2 Andrews' Experiments on Carbon Dioxide

Andrews carried out very exhaustive experiments on the compressibility of gases while attempting to liquefy them. These experiments threw considerable light on the actual behaviour of gases. He investigated the behaviour of CO_2 in great detail and his results are depicted in Fig. 3.2. The main conclusions were:

1. Above a temperature of about 48°C, the behaviour of CO_2 resembles that of a perfect gas.
2. As the temperature is lowered, isotherms exhibit distortion which gradually increases. This signifies deviation from perfect gas character.
3. At 31.4°C, a kink is observed, which suggests that gas can be liquefied under compression. Above this temperature, liquefaction cannot be produced how high the pressure may be.

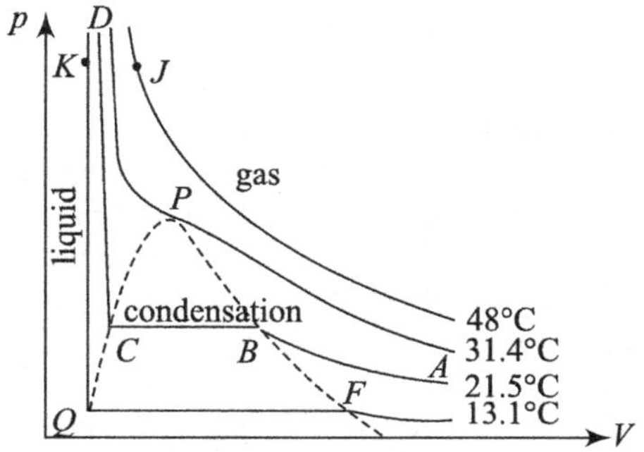

Fig. 3.2 Variation of pressure with volume for CO_2 at different temperatures: Andrews' curves.

4. As temperature is lowered further, say to 21.5°C, the kink spreads into a horizontal line, i.e., compression produces liquefaction.

   From *A* to *B*, CO_2 behaves as a gas. At point *B*, the liquefaction of the gas just starts. The gas condenses at a constant pressure from *B* to *C* so that liquid and vapour phases co-exist. At *C*, the gas is completely in the liquid phase. From *C* to *D* the slope is very steep since a liquid is almost incompressible.
5. These changes proceed in the same direction at still lower temperatures and the isotherm at 13.1°C is of the same general form as the one at 21.5°C, except that vapour will co-exist over a larger range. As can be noted, the volume of the vapour at the time condensation sets in, is greater than its volume at 21.5°C but the volume of the liquid after condensation is smaller. Also, the gas can be liquefied at a comparatively lower pressure.

The temperature at which it just becomes possible to liquefy a gas under compression is known as the *critical temperature*. We denote it by T_c. At this temperature, the properties of a liquid and its saturated vapour are identical. The pressure required to liquefy a gas at the critical temperature is called the *critical pressure* and the corresponding volume (for one mole of the gas) is called the *critical volume*. We denote these by p_c and V_c, respectively. T_c, p_c and V_c are collectively known as *critical constants*. For some typical gases, values of critical constants are given in Table 3.1.

Table 3.1 *Observed values of critical constants*

Gas	T_c (°C)	p_c (atm)	V_c (c.c)	$\frac{RT_c}{p_c V_c}$
He	– 268	2.25	15.4	3.13
H_2	– 240	12.8	32.2	3.28
O_2	– 119	49.7	2.32	3.42
CO_2	31.0	72.8	2.17	3.48
Ether	194	35.6	3.85	3.81

From the preceding discussion, it is clear that Andrews' experiments emphasise the following two aspects:

1. *A gas can be liquefied only if it is cooled up to or below its characteristic critical temperature*. (The observed upward rise in pV for hydrogen in Regnault's experiments depicted in Fig. 3.1 arises due to the fact that at ordinary temperature, the gas is much above its critical temperature.)
2. *There exists a continuity of liquid and gaseous states*, i.e., they are two distinct stages of a continuous physical phenomenon.

You can understand this by considering two isotherms defined by $T_1 < T_c < T_2$, i.e., one isotherm above the critical isotherm and the other below the critical isotherm. For $p > p_c$, imagine points J and K on them as shown in Fig. 3.2. At J, the substance is a gas ($T > T_c$) and at K, it is entirely a liquid ($T < T_c$). This means that if we gradually reduce the temperature (at constant pressure), the substance passes from the gaseous to the liquid state without any abrupt change in its characteristic properties.

3.2.3 Amagat's Experiments

Amagat investigated the behaviour of several gases at various temperatures and up to very high pressures (3000 atmosphere). This work lent support to the findings of Regnault and Andrews. His results for CO_2 are shown in Fig. 3.3. Following are the main conclusions:

1. For $T < T_c$, isotherms have a straight line portion which is parallel to the pV-axis. That is, below the critical temperature, volume decreases even though pressure remains constant. This corresponds to the condensation of the gas (as in the horizontal part in Fig. 3.2).
2. The curvature of the isotherms decreases as temperature increases. At the critical temperature, the straight line part diminishes to a point.
3. Each isotherm has a minimum which gradually shifts away from the origin towards the right as temperature increases. However, after a particular temperature, it begins to shift towards the left. The locus of the minima, shown by the dashed curve, is almost parabolic.
4. The parabola cuts the $p = 0$ axis at some higher temperature. This temperature is called the *Boyle* temperature*, T_B.
5. For $T > T_B$, the value of pV steadily increases.

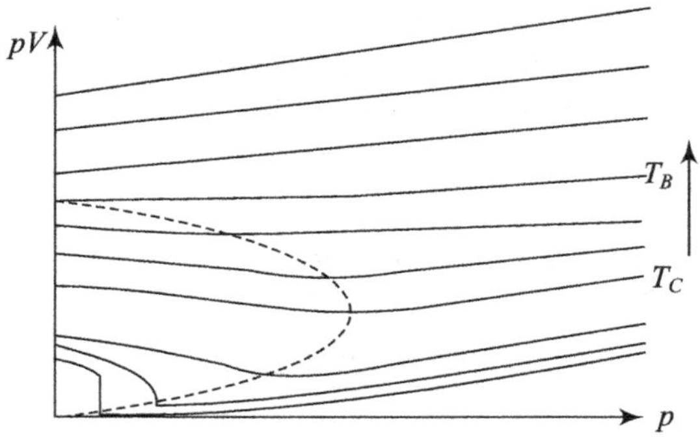

Fig. 3.3 pV isotherms for CO_2 obtained by Amagat.

*This nomenclature stems from the observation that at this temperature, all gases obey Boyle's law up to fairly high pressures.

A proper explanation of these experimental results posed a challenge for theoretical physicists. Numerous equations of state were proposed but none of them was found suitable over the entire range of temperature and pressure. Kammerlingh Onnes proposed an empirical, yet useful, equation and we will now discuss it in some detail.

3.3 ONNES' EQUATION OF STATE

Onnes proposed an empirical equation of the form*

$$pV = A + Bp + Cp^2 + \dots \tag{3.1}$$

where the constants A, B, C, ... are characteristic of a gas at a fixed temperature. (They vary with temperature in a complex manner.) These constants are called *virial coefficients.*

Since Eq. (3.1) is expected to reproduce the equation for a perfect gas as $p \to 0$, the first virial coefficient will be equal to RT for one mole of the gas. That is,

$$A = RT \tag{3.2}$$

The second virial coefficient is of special interest. It varies in a similar way for all gases. At very low temperatures, it has a negative value. As temperature increases, it shows a gradual rise through zero and becomes positive. At room temperature, $B < 0$ for oxygen, nitrogen and carbon dioxide, whereas $B > 0$ for hydrogen and helium. For all gases, $B = 0$ at the Boyle temperature. If one is not working at very high pressures, Eq. (3.1) implies that at $T = T_B$

$$pV = A$$

and

$$\frac{\partial(pV)}{\partial p} = B \tag{3.3}$$

The constant C is always positive but very small.

In his studies, Onnes had to use twenty five constants for best representation of experimental results. Later, Holborn and Otto investigated the behaviour of several gases up to 100 atmospheres in the temperature range – 183°C to 400°C and found that it is sufficient to retain only four virial coefficients.

It is important to mention here that the empirical equation of Onnes lacked physical basis. To explain the observed temperature dependent behaviour of real gases, van der Waals proposed another equation of state by modifying some of the assumptions of elementary kinetic theory. This equation is known by his name. We will now discuss it in some detail.

3.4 van der WAALS' EQUATION OF STATE

In deriving van der Waals' equation, some simplifying assumptions are made. We first state these.

1. Gas molecules have finite size and cannot be regarded as point masses**.

---

*Another useful form of Eq. (3.1) is

$$pV = \alpha + \frac{\beta}{V} + \frac{\gamma}{V^2} + \dots \tag{3.1a}$$

**This was considered by Clausius for the first time when he introduced the concept of mean free path (Chapter 2).

2. Molecules attract one another with a weak force+, which depends only upon the distance between them. (This implies that gas molecules have both kinetic and potential energies. However, only nearest neighbour interactions are important.)
3. The number of collisions with the walls of the container are exactly the same for point and finite size molecules.
4. The molecular density is small.

Some of these assumptions may not be justified under all conditions but we will discuss that when the occasion arises.

Derivation

1. ***Correction for finite size*** Consider one mole of a gas enclosed in a container of volume V. If this gas were composed of point masses, all this space would be available to them for free motion. When molecular size is taken into account, the volume available to a single molecule for free movement will be somewhat less than V. Let us denote this reduction by b. (This is called the *co-volume* and depends on the nature of the gas.) Therefore, van der Waals argued that the factor V occurring in the equation for perfect gas should be replaced by $(V - b)$. For one mole of the gas, we obtain**

$$p\,(V - b) = RT \tag{3.4}$$

The magnitude of b is equal to four times the total molecular volume for one mole of a van der Waals' gas. There are several ways of arriving at this result but we will give a relatively simple derivation.

Let us suppose that the radius of each molecule is r. At the instant of collision, the centre-to-centre distance of the two colliding molecules will be $d = 2r$ as shown in Fig. 3.4a. This implies that around any molecule, a spherical volume V_s $(= 4\pi(2r)^3/3)$ will be denied to every other molecule. (This volume is called the *sphere of exclusion* and is eight times the volume V_m $(= 4\pi/3)r^3$ of a molecule, i.e., $V_s = 8\,V_m$.)

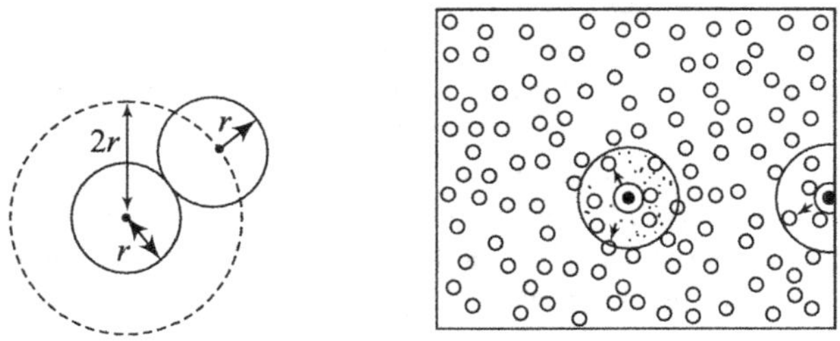

(a) (b)

Fig.3.4 (a) Collision of two molecules of radii r, and (b) Intermolecular attraction decreases with pressure : a physical picture.

Proceeding further, let us imagine filling the container with N molecules, one by one. For that

the volume available to the first molecule = V
the volume available to the second molecule = $V - (2\text{-}1)V_s$
the volume available to the third molecule = $V - (3\text{-}1)\,V_s$

+ This concept was proposed by Hirn in 1864.
**The equation is due to Clausius.

and by induction

the volume available to the Nth molecule = $V - (N - 1)\,V_s$

Hence, the average volume available to each molecule, obtained by taking arithmetic mean is

$$\bar{V} = \frac{1}{N}\sum_{i=1}^{N}(V - (i-1)V_s)$$

$$= V - \frac{(N-1)}{2}V_s \qquad \because \left(\sum_{i=1}^{N}(i-1) = \frac{N}{2}(N-1)\right)$$

For large N, one can neglect 1 in the second term on the right-hand side of the above equation and obtain

$$\bar{V} = V - \frac{N}{2}V_s$$

$$= V - 4N\,V_m$$

$$= V - b \qquad (3.5a)$$

where we have put $b = 4\,NV_m$.

2. ***Correction for intermolecular attraction*** To account for intermolecular attractions, we invoke Assumption 2. As will be noted from Fig. 3.4(b), a molecule in the interior of the gas is, on an average, attracted equally in all directions so that there is no resultant force on it. (It will, therefore, behave as if there were no intermolecular interactions in the gas and the effective pressure is the same as that for an ideal gas.) However, this is not true for a molecule in the outermost layer closest to the walls of the container. Since the molecular distribution is only on one side, there will be a net inward force. So whenever a molecule strikes the walls of the container, it has to overcome this molecular attraction. In this process, some kinetic energy is lost and molecular velocity decreases. Therefore, the momentum communicated to the wall on impact and rebound will be less than that for an ideal gas (Sec. 1.3). It means that intermolecular (cohesive) forces cause decrease in pressure. This drop is known as *cohesive pressure*. Let us denote it by Δp. If we write the ideal gas equation, modified to allow for these effects, we have

$$p_{\text{ideal}}V_{\text{ideal}} = RT$$

or

$$(p + \Delta p)(V - b) = RT$$

Note that the cohesive pressure will be proportional to the number of molecules per unit volume in the surface layer (on which the inward forces act) and the number of molecules per unit volume in the layer just below the surface layer (which are pulling the striking molecules due to attraction).

Thus,

$$\Delta p \;\alpha \left(\frac{N}{V}\right)^2$$

or

$$\Delta p = \frac{a'N^2}{V^2}$$

where a' is constant of proportionality. If we put $a' N^2 = a$, we can write

$$\Delta p = \frac{a}{V^2} \tag{3.5b}$$

Let us pause for a while and reflect as to what we have achieved so far. Equations (3.5a) and (3.5b) suggest that finite size and intermolecular attractions respectively modify the volume available to the molecules and the pressure communicated to the walls of the container. That is, in the ideal gas equation, we must replace p by the sum of the observed pressure for a real gas and the drop caused by intermolecular attractions, i.e., $\left(p + \frac{a}{V^2}\right)$ and the volume V by $(V\text{-}b)$. Using these results in Eq. (3.4), we find that the equation of state for an ideal gas modifies to*

$$\left(p + \frac{a}{V^2}\right)(V - b) = RT \tag{3.6}$$

This is known as *van der Waals' equation of state*. The quantities a and b are known as *van der Waals' constants*. These are assumed to be same for a gas at all temperatures but different for different gases.

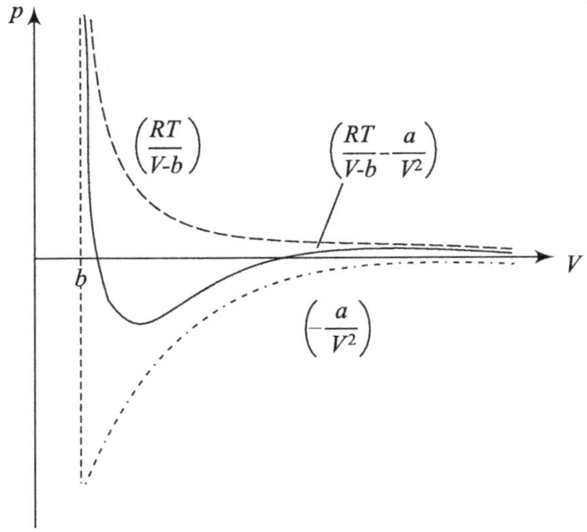

Fig. 3.5 A plot of van der Waals' equation for a real gas. Plots of $\left(\frac{RT}{V-b}\right)$ and $\left(-\frac{a}{V^2}\right)$ as a function of V are shown by dashed and dash dot curves, respectively.

Note that for large V, $p \gg a/V^2$ and b becomes negligible in comparison with V. The van der Waals' equation then reduces to the equation of state for an ideal gas. A plot of Eq. (3.6) on p–V diagram is shown in Fig. 3.5 for a given temperature. The contributions of thermal pressure and self-attracting terms to van der Waals' equation are shown by dashed and dash dot curves.

To appreciate these concepts and get an idea about the figures, go through the following example carefully.

*If we consider n moles of the gas, the constants a and b must be multiplied by n^2 and n, respectively.

Example 3.1 One mole of a gas occupies a volume of 0.55 litres at 0°C. Calculate the pressure it will exert if it behaves as (a) an ideal gas and (b) as a van der Waals' gas. Given $a = 0.37$ Nm^4 mol^{-2}, $b = 43 \times 10^{-6}$ m^3 mol^{-1} and $R = 8.31$ J mol^{-1} K^{-1}.

Solution: Here $V = 0.55$ litre mol^{-1} = 550 cm^3 mol^{-1} = 550×10^{-6} m^3 mol^{-1}

(a) For one mole of an ideal gas,

$$p = \frac{RT}{V}$$

On substituting the values of various quantities, we get

$$p = \frac{(8.31\ \text{J mol}^{-1}\text{K}^{-1}) \times (273\ \text{K})}{(550 \times 10^{-6}\ \text{m}^3\ \text{mol}^{-1})}$$

$$= 4.12 \times 10^6\ \text{Nm}^{-2}.$$

(b) For one mole of van der Waals' gas,

$$p = \frac{RT}{V-b} - \frac{a}{V^2}$$

$$= \frac{(8.31\ \text{J mol}^{-1}\text{K}^{-1}) \times (273\ \text{K})}{(550-43) \times 10^{-6}\ \text{m}^3\ \text{mol}^{-1}} - \frac{0.37\ \text{Nm}^4\ \text{mol}^{-2}}{(550 \times 10^{-6})^2\ \text{m}^6\ \text{mol}^{-2}}$$

$$= \left(\frac{8.31 \times 273}{507 \times 10^{-6}} - \frac{0.37}{(550 \times 10^{-6})^2}\right) \text{Nm}^{-2}$$

$$= (4.48 \times 10^6 - 1.22 \times 10^6)\ \text{Nm}^{-2}$$
$$= 3.26 \times 10^6\ \text{Nm}^{-2}.$$

As expected, the pressure exerted by a van der Waals' gas is less than that exerted by an ideal gas.

Example 3.2 For a gas obeying van der Waals' equation, the constants are $a = 1.32$ litre^2 atm. mol^{-2} and $b = 3.12 \times 10^{-2}$ litre mol^{-1}. Calculate the temperature at which 5 moles of the gas at 5 atmospheric pressures will occupy a volume of 20 litre. Given $R = 8.31 \times 10^7$ erg mol^{-1}K^{-1}

Solution: The van der Waals' equation for n moles of the gas is

$$\left(P + \frac{n^2 a}{V^2}\right)(V - nb) = nRT$$

By inverting it, we can write

$$T = \frac{\left(P + \frac{n^2 a}{V^2}\right)(V - nb)}{nR}$$

Here,

$$a = 1.32\ \text{litre}^2\ \text{atm. mol}^{-2}$$

$$= 1.32 \times (10^3)^2\ \text{cm}^6 \times (1.013 \times 10^6\ \text{dyne. cm}^{-2})\ \text{mol}^{-2}$$

$$= 1.34 \times 10^{12}\ \text{dyne cm}^4\ \text{mol}^{-2}$$

$$b = 3.12 \times 10^{-2}\ \text{litre mol}^{-1}$$

$$= 3.12 \times 10^{-2} \times 10^3 \text{ cm}^3 \text{ mol}^{-1}$$

$$= 31.2 \text{ cm}^3 \text{ mol}^{-1}$$

$$p = 5 \text{ atm}$$
$$= 5 \times 1.013 \times 10^6 \text{ dyne cm}^{-2}$$

$$= 5.065 \times 10^6 \text{ dyne cm}^{-2}$$

$$V = 20 \text{ litre} = 20 \times 10^3 \text{ cm}^3$$
$$n = 5,\ R = 8.31 \times 10^7 \text{ erg mol}^{-1}\text{K}^{-1}$$

$$\therefore \quad T = \frac{\left[(5.065 \times 10^6 \text{ dyne cm}^{-2}) + \dfrac{25 \times (1.34 \times 10^{12} \text{ dyne cm}^4 \text{ mol}^{-2})}{(20 \times 10^3 \text{ cm}^3)^2}\right] \times [(20 \times 10^3 \text{ cm}^3 \text{ mol}^{-1}) - (5 \times 31.2 \text{ cm}^3 \text{ mol}^{-1})]}{5 \times 8.31 \times 10^7 \text{ erg mol}^{-1} \text{ K}^{-1}}$$

$$= 245.9 \text{ K}$$

You may now like to answer a practise problem.

Problem 3.1 One mole of a gas, stored at 300 K, occupies a volume 1.2×10^{-3} m^3 mol^{-1}. Compare the pressures calculated on the basis of van der Waals' equation of state and ideal equation of state. Given a = 1.32×10^{-6} atm m^3 mol^{-2} and $b = 3.12 \times 10^{-5}$ m^3 mol^{-1}.

Ans: 1 : 1.018

3.4.1 Discussion of van der Waals' Equation: Comparison with Experimental Results

We will now discuss the van der Waals' equation in detail and see how far it explains the experimental results. Refer to Fig. 3.6, where we have depicted van der Waals' equation at different temperatures above and below the critical temperature. Note that these curves qualitatively reproduce several features of Andrew's experiments.

To facilitate rigorous analysis, we first rewrite Eq. (3.6) as

$$p = \frac{RT}{V - b} - \frac{a}{V^2} \tag{3.7a}$$

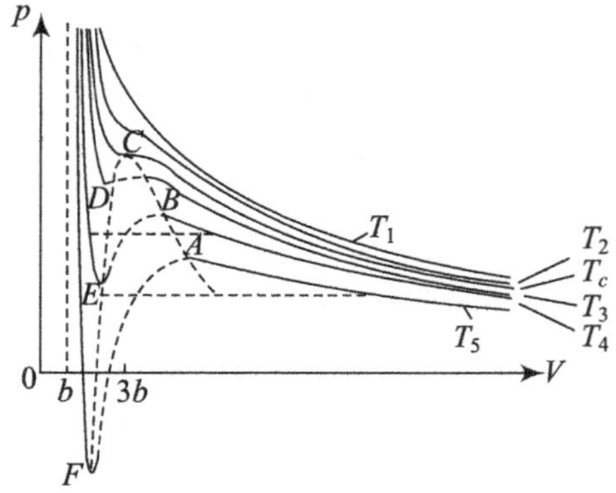

Fig. 3.6 A plot of Eq. (3.6) at different temperatures.

When expanded in powers of V, this equation takes the form

$$pV^3 - (pb + RT)V^2 + a(V - b) = 0 \tag{3.7b}$$

From this, we note the following points:

1. van der Waals' equation is cubic in V. This means that for each pair of values of p and T, V will have three values; either all of them are real or one is real and two are imaginary. For low temperatures, three positive real roots exist over a range of values of p. As temperature increases, these roots approach one another. They become equal at $T = T_c$. Above T_c, only one real root exists for all values of p implying that there can be no condensation in this region.
2. For large values of V, p would be small and in the limit $V \to \infty$, $p \to 0$. On the other hand, as $V \to b$, $p \to \infty$. This implies that in the p versus V plot, the curve will have a concavity upwards. Further, V cannot be less than b as this will lead to negative values of p, which is not physically acceptable.
3. For CO_2 at various temperatures, the general form of theoretical isotherms based on van der Waals' equation is shown in Fig. 3.7. As will be noted, the qualitative shape of the curve resembles the experimental isotherms of Andrews (Fig. 3.2). However, there are differences in details, particularly at lower temperatures. For instance, van der Waals' equation predicts maxima and minima in the straight line region of Andrews' curves. To explain this, let us consider van der Waals' isotherm at 13.1 °C shown in Fig. 3.7. The portions AB and FK represent gaseous and liquid states respectively. However, the portion BC represents *supersaturated vapour* and the portion EF represents a *superheated liquid*. Both of these correspond to metastable state and do not occur in Andrews' curves (which represent state of stable equilibrium). The portion CDE of the theoretical curve indicates that the volume should decrease as pressure decreases. This corresponds to a collapsible state and can never be realised in practice. We thus find that a proper interpretation of van der Waals' curves helps to resolve an apparent discrepancy with observed results.

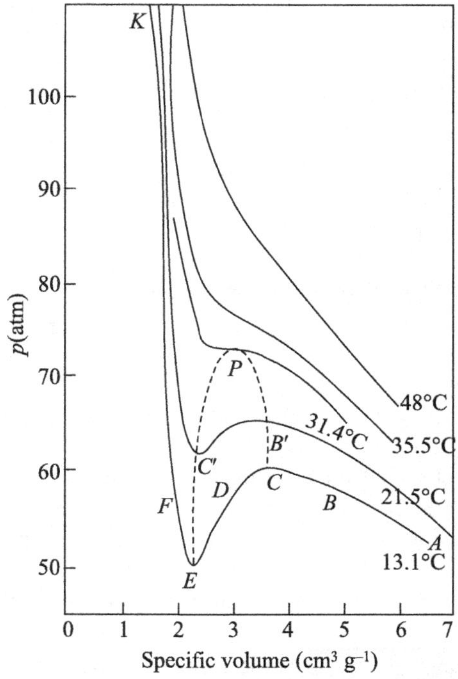

Fig. 3.7 Theoretical curves for CO_2 based on van der Waals' equation.

The equation of the locus of the maxima and minima, shown by the dashed curve in Fig. 3.7 can be obtained by putting $(\partial p/\partial V)_T$ equal to zero. Hence, on differentiating Eq. (3.7a) with respect to V, we get

$$\left(\frac{\partial p}{\partial V}\right)_T = -\frac{RT}{(V-b)^2} + \frac{2a}{V^3} = 0$$

or

$$T = \frac{2a(V-b)^2}{RV^3} \tag{3.8}$$

As will be noted, this equation is also cubic in V, so that for every isotherm there are three real or one real and two imaginary points of maxima and minima.

Eliminating T between Eqs. (3.7a) and (3.8), we get the required equation for the curve passing through the maxima and minima points:

$$p = \frac{a(V-2b)}{V^3} \tag{3.9}$$

From Fig. 3.2, we recall that all isotherms below P cut the dashed curve *FBPCQ* at two points (characterised by equal pressures) and a gas can be liquefied by changing pressure alone. For the critical isotherm these two points coalesce into one. Similarly, from Fig. 3.7, we note that below P all isotherms have maxima and minima but above P no isotherm exhibits these characteristics. Therefore, the point P must be identified with the *critical point* and the isotherm passing through it with the *critical isotherm.*

For P to be the point of maximum on the dashed curve, we differentiate Eq. (3.9) with respect to V and equate the resultant expression to zero. This gives

$$\frac{a}{V^3} - \frac{3a(V-2b)}{V^4} = 0$$

or

$$V_c = 3\,b \tag{3.10a}$$

Hence, from Eqs. (3.8) and (3.9), we have

$$T_c = \frac{8a}{27Rb} \tag{3.10b}$$

and

$$p_c = \frac{a}{27b^2} \tag{3.10c}$$

On combining these results, we get

$$\frac{RT_c}{p_c V_c} = C_c = \frac{8}{3} \tag{3.11}$$

That is, van der Waals' equation predicts that the critical coefficient is equal to 8/3 and same for all gases, irrespective of their nature. (For an ideal gas, C_c equals unity.) The observed values of the critical coefficient for some gases are also given in Table 3.1. As may be noted, van der Waals' equation predicts a lower value of the critical coefficient. Moreover, it varies from gas to gas and appears to depend on the molecular structure of a gas. To that extent, van der Waals equation is expected to have limited utility.

You may now like to answer a practise problem.

Problem 3.2 For the critical isotherm, the point P is a maximum as well as a point of inflexion since the isotherm is concave upward at the left of this point and concave downward at the right. Mathematically, it is defined by $(\partial p/\partial V)_T = 0$ and $(\partial^2 p/\partial V^2)_T = 0$, respectively. Using these definitions, verify expressions for critical constants.

Example 3.3 Calculate the critical temperature of helium if the critical pressure is 2.26 atm and critical density is 69 kg m^{-3}.

Solution: From Eq. (3.11) we note that the critical coefficient is given by

$$\frac{RT_c}{p_c V_c} = \frac{8}{3}$$

so that

$$T_c = \frac{8}{3} \frac{p_c V_c}{R}$$

Here, $p_c = 2.26$ atm $= 2.26 \times 1.013 \times 10^5$ Nm^{-2} and $\rho_c = 69$ kg m^{-3} so that the critical volume for 1 kg of the gas is (1/69)m^3. Hence, the critical volume per kmol of the gas is (4/69)m^3, i.e., $V_c = (4/69)$m^3 kmol^{-1}.

On substituting the values of various quantities, we get

$$T_c = \frac{8}{3} \frac{\times (2.26 \times 1.013 \times 10^5 \text{ Nm}^{-2}) \times (4/69) \text{ m}^3 \text{ kmol}^{-1}}{8.31 \times 10^3 \text{ Jkmol}^{-1} \text{K}^{-1}}$$

$$= 4.26 \text{ K}$$

We would now like you to answer a few practise problems.

Problem 3.3 Calculate the critical temperature for a gas obeying van der Waals' equation of state. Given $a = 0.00874$ atm cm^6 and $b = 0.0023$ cm^3 for 1 cm^3 of the gas at S.T.P.

Ans: $T_c = 307$ K

Hint: First calculate the values of a and b for 1 mole of the gas.

Problem 3.4 The van der Waals' constants for one gm molecule of hydrogen are $a = 0.245 \times 10^{-6}$ atm m^3 mol^{-2}, $b = 2.67 \times 10^{-5}$ m^3 mol^{-1} and $R = 8.31$ J mol^{-1} K^{-1}. Calculate critical temperature of the gas.

Ans: $T_c = 33.1$ K

3.4.2 Determination of van der Waals' Constants

Some of the important methods for determining the values of van der Waals' constants are given below:

From the critical data In the preceding section we observed that the critical constants for one mole of a van der Waals' gas are defined in terms of a and b. One can readily invert these relations (Eq. (3.10)) to obtain

$$a = \frac{27 R^2 T_c^2}{64 p_c} \tag{3.12a}$$

and

$$b = \frac{RT_c}{8 p_c} = \frac{V_c}{3} \tag{3.12b}$$

Thus, if T_c, p_c and V_c are known from experiments, we can calculate a and b using these relations. If we have n moles of the gas, the value of a has to be multiplied by n^2 and that of b by n.

From constant volume method A very simple and reliable method is to observe the rate of change of pressure with temperature at constant volume. To illustrate this, we note that from Eq. (3.7a), we can write

$$\left(\frac{\partial p}{\partial T}\right)_V = \frac{R}{V-b} \tag{3.13a}$$

so that

$$b = V - \frac{R}{(\partial p/\partial T)_V} \tag{3.13b}$$

We can also rewrite Eq. (3.13a) as

$$\left(\frac{\partial p}{\partial T}\right)_V = \frac{1}{T}\left(p + \frac{a}{V^2}\right)$$

Hence,

$$a = V^2\left(T\left(\frac{\partial p}{\partial T}\right)_V - p\right)$$

$$= V^2 Tp\left(\frac{1}{p}(\partial p/\partial T)_V - \frac{1}{T}\right) \tag{3.13c}$$

Equations (3.13b) and (3.13c) show that once we know $(\partial p/\partial T)_V$, it is possible to obtain both the constants.

It is instructive to note that apart from a, Eq. (3.13c) also gives us information about the magnitude of the cohesive pressure at any temperature. As will be noted, (a/V^2) is proportional to pressure (and hence density) of the gas.

From pressure and volume expansion coefficients The pressure coefficient of expansion, β, is defined as

$$\beta = \frac{1}{p}(\partial p/\partial T)_V \tag{3.14}$$

For a van der Waals' gas, we have from Eqs. (3.13c) and (3.14)

$$a = V^2 Tp\left(\beta - \frac{1}{T}\right) \tag{3.15}$$

Equation (3.15) predicts that for a perfect gas, $\beta = 1/T$ so that $a = 0$.

The volume expansion coefficient, α, is defined as

$$\alpha = \frac{1}{V}\left(\frac{\partial V}{\partial T}\right)_p \tag{3.16}$$

For a van der Waals' gas, defined by the equation

$$\left(p + \frac{a}{V^2}\right)(V - b) = RT$$

we can write

$$\left(\frac{\partial V}{\partial T}\right)_p = \frac{R}{\left(p + \dfrac{a}{V^2}\right) - \dfrac{2a(V-b)}{V^3}}$$

On multiplying the numerator as well as the denominator by $(V - b)$ and using Eq. (3.6), we obtain

$$\left(\frac{\partial V}{\partial T}\right)_p = \frac{R(V-b)}{RT - \dfrac{2a(V-b)^2}{V^3}} = \frac{R(V-b)}{RT\left(1 - \dfrac{2a}{RTV}\right)} = \frac{V}{T}\left(1 - \frac{b}{V}\right)\left(1 + \frac{2a}{RTV}\right)$$

If $b << V$ we can ignore the second and higher powers of b as well as the product of van der Waals' constants. In this approximation, the right-hand side of the above equation takes a very compact form:

$$\left(\frac{\partial V}{\partial T}\right)_p = \frac{V}{T}\left(1 + \frac{1}{V}\left(\frac{2a}{RT} - b\right)\right)$$

so that

$$\alpha = \frac{1}{V}\left(\frac{\partial V}{\partial T}\right)_p = \frac{1}{T} + \frac{1}{TV}\left(\frac{2a}{RT} - b\right)$$

or

$$\frac{2a}{RT} - b = V(\alpha T - 1) \tag{3.17}$$

This equation may be used to calculate b, once a and α are known.

Example 3.4 Calculate van der Waals' constants a and b and diameter of a molecule of helium. Given $T_c = 5.3$K, $p_c = 2.25$ atm and $R = 8.31$ J mol^{-1} K^{-1}. Note that 1 atm = 1.013×10^5 Nm^{-2}.

Solution: From Eq. (3.12a), we recall that

$$a = \frac{27R^2}{64}\frac{T_c^2}{p_c}$$

On substituting the values of various quantities, we get

$$a = \frac{27 \times (8.31\text{J mol}^{-1}\text{K}^{-1})^2 \times (5.3\text{ K})^2}{64 \times (2.25 \times 1.013 \times 10^5\text{ Nm}^{-2})}$$

$$= 3.59 \times 10^{-3}\text{ Nm}^4\text{mol}^{-2}$$

To calculate b, we note that $b = V_c/3$ so that the expression for critical coefficient can be rewritten as

$$b = \frac{RT_c}{8p_c}$$

Hence, on inserting the numerical values, we get

$$b = \frac{(8.31\text{J mol}^{-1}\text{K}^{-1}) \times (5.3\text{ K})}{8 \times (2.25 \times 1.013 \times 10^5\text{ Nm}^{-2})}$$

$$= 2.42 \times 10^{-5}\text{ m}^3\text{ mol}^{-1}$$

Since b, known as co-volume, is four times the volume occupied by all the molecules of 1 mole of a gas, we can write

$$b = 4 \times \frac{4}{3}\pi\left(\frac{d}{2}\right)^3 \times N_A$$

where d is molecular diameter and N_A is Avogadro's number.

On inverting this relation, we get

$$d^3 = \frac{3}{2}\frac{b}{\pi N_A}$$

$$= \frac{3}{2} \times \frac{(2.42 \times 10^{-5}\ m^3 mol^{-1})}{3.14 \times (6.02 \times 10^{23}\ mol^{-1})}$$

$$= 1.92 \times 10^{-29} m^3$$

Hence, $$d = (19.2 \times 10^{-30})^{1/3} m = 2.68 \times 10^{-10}\ m$$

You may now like to answer a few practise problems.

Problem 3.5 Calculate van der Waals' constants for dry air using the following data:

$$T_c = 132K,\ p_c = 37.2\ \text{atm and } R\ \text{(per mol)} = 82.07\ cm^3\ atm\ K^{-1}$$

Ans: $a = 13.3 \times 10^5$ atm cm^6 mol^{-2}; $b = 36.4$ cm^3 mol^{-1}

Problem 3.6 The critical temperature of CO_2 is 31 °C and the critical pressure is 73 atm. By assuming that CO_2 obeys van der Waals' equation, estimate the diameter of its molecule.

Ans: $d = 3.23$ Å

3.4.3 van der Waals' Equation and Virial Coefficients: Boyle Temperature

To put van der Waals' equation in virial form, we first rewrite it as

$$\left(p + \frac{a}{V^2}\right)V\left(1 - \frac{b}{V}\right) = RT$$

or $$pV = RT\left(1 - \frac{b}{V}\right)^{-1} - \frac{a}{V}$$

Using binomial theorem, we can write

$$\left(1 - \frac{b}{V}\right)^{-1} = 1 + \frac{b}{V} + \frac{b^2}{V^2} + \cdots$$

Using this result in the above expression, we get

$$pV = RT + \frac{RTb - a}{V} + \frac{RTb^2}{V^2} + \cdots$$

As will be noted, van der Waals' equation has only three virial coefficients and a comparison with Eq. (3.1a) yields $\alpha = RT$, $\beta = RTb - a$ and $\gamma = RT\,b^2$.

At the Boyle temperature, the second virial coefficient is zero. Hence,

$$RT_B b - a = 0$$

or

$$T_B = \frac{a}{Rb} \tag{3.18}$$

From the preceding section, you may recall that the critical temperature of a gas obeying van der Waals' equation of state is

$$T_c = \frac{8\,a}{27\,Rb}$$

On comparing these expressions, we get

$$T_B = \frac{27}{8} T_c = 3.375\, T_c \tag{3.19}$$

That is, the Boyle temperature, on the basis of van der Waals' equation, is 3.375 times the critical temperature.

3.4.4 Limitations of van der Waals' Equation

From the preceding discussion it is clear that van der Waals' equation satisfactorily explains the general features of transition from liquid to gaseous state. However, it fails to explain the quantitative details. Some of the points which bring out its limitations are listed below:

Values of a and b van der Waals assumed that for any gas, *a* and *b* are constant, independent of temperature. However, experiments do not support this assertion. This is due to the fact that one can make any isotherm agree closely with the equation by suitably choosing the values of *a* and *b*. However, for the same values, the agreement with other isotherms will not be so good, suggesting that they vary with temperature.

Critical volume From Eq. (3.10a), we note that the critical volume $V_c = 3b$ for all gases, independent of their nature. However, experiments show that V_c varies from 1.4*b* for Argon to 2.8*b* for hydrogen. This implies that van der Waals' equation is not followed exactly by any gas near its critical point.

Critical coefficient Theory predicts that the critical coefficient $RT_c/p_cV_c = 8/3$ for all gases. But from Table 3.1, you would note that it is different for different gases with values ranging from 3.28 for hydrogen to 3.48 for CO_2.

Relation between T_B and T_C From Eq. (3.19), we note that $T_B = 3.375\, T_c$. However, the observed values of Boyle temperature are found to lie between 2.5 T_c and 3.7 T_c.

Physically, these limitations have genesis in the fact that we have ignored the repulsive forces between molecules near fluid phase boundaries. However, in spite of these limitations, van der Waals' equation is commonly used because of its inherent simplicity.

3.5 Reduced Equation of State

We now know that the constants *a* and *b* depend on temperature as well as the nature of a gas. It would therefore be desirable to eliminate these and put van der Waals' equation in a form that is applicable to all gases. To do so, we define *reduced pressure, reduced volume* and *reduced temperature* through the relations

$$p_r = \frac{p}{p_c}, \quad V_r = \frac{V}{V_c} \quad \text{and} \quad T_r = \frac{T}{T_c} \tag{3.20}$$

where p, V and T are the actual pressure, volume and temperature of a gas and p_c, V_c and T_c, respectively denote their critical values. As may be noted, p_r, V_r and T_r are dimensionless numbers. These are known as *reduced variables* of state. On combining Eqs. (3.6) and (3.20), we can write

$$\left(p_r p_c + \frac{a}{V_r^2 V_c^2}\right)(V_r V_c - b) = R\,T_r T_c \tag{3.21}$$

On substituting for p_c, V_c and T_c from Eq. (3.10) and simplifying the resultant expression, we get

$$\frac{a}{9b}\left(p_r + \frac{3}{V_r^2}\right)\left(V_r - \frac{1}{3}\right) = \frac{8aT_r}{27b}$$

On dividing both sides by $\frac{a}{9b}$, van der Waals' equation takes a compact form

$$\left(p_r + \frac{3}{V_r^2}\right)\left(V_r - \frac{1}{3}\right) = \frac{8}{3}T_r \tag{3.22}$$

This equation is known as the *reduced equation of state* or *the law of corresponding states*, in that if any two of the variables p_r, V_r and T_r are the same for two different substances, the third will also be the same for them (and the states of gases will correspond to each other).

3.6 Other Equations of State

To overcome the shortcomings of van der Waals' equation, a number of other equations have been proposed by later researchers. Some of these equations are based on rigourous theoretical considerations while others consist of adding more empirical constants which can be suitably chosen. We will first mention some of the latter type.

1. ***Clausius' equation*** Clausius proposed the equation,

$$\left(p + \frac{a'}{T\,(V + c)}\right)(V - b) = RT \tag{3.23}$$

where van der Waals' constant a has been modified to a'/T (to account for its temperature variation) and an additional constant c has been introduced. Although this equation explains Andrews' results for CO_2 better than van der Waals' equation, it fails for other gases. If anything, this has no substantial advantage over van der Waals' equation.

2. ***Berthelot's equation*** Berthelot proposed an empirical equation which differs from Clausius' equation in respect of the constant c as well as the power of T:

$$\left(p + \frac{a'}{TV^2}\right)(V - b) = RT \tag{3.24}$$

It gives better agreement with experiments than van der Waals' equation over a wider range of pressures. However, near the critical point, it fails as badly as van der Waals' equation.

We would like you to now do some mathematical calculations so as to get practise.

Problem 3.7 For Berthelot's equation, calculate the expression for critical coefficient.

Ans: 8/3

3. ***Dieterici's equation*** By assuming that the number density near the walls of the container is smaller than that in the interior, Dieterici deduced the following equation for pressure exerted by a gas:

$$p = \frac{RT}{V-b} \exp\left(-\frac{a}{RTV}\right) \tag{3.25}$$

It is instructive to note that at low pressures, Eq. (3.25) reduces to van der Waals' equation, if $b << V$.

In general, this equation is more accurate than van der Waals' equation. It gives $V_c = 2b$ and $RT_c/p_cV_c = 3.695$ (Problem 3.8).

You may now like to answer a few practise problems.

Problem 3.8 By differentiating Eq. (3.25) twice with respect to V, keeping T constant, and equating the results to zero, show that $V_c = 2b$, $T_c = \frac{a}{4Rb}$, $p_c = \frac{a}{4e^2 b^2}$ and $C_c = 3.695$. Show that for large volumes, Dieterici's equation reduces to van der Waals' equation.

Problem 3.9 Obtain reduced equation for Dieterici's equation.

4. ***Saha and Bose's equation*** From thermodynamic considerations and the theory of probability, Saha and Bose derived the following equation of state:

$$p = -\frac{RT}{2b} \exp\left(-\frac{a}{RTV}\right) \ln\left(\frac{V-2b}{V}\right) \tag{3.26}$$

This equation predicts that the critical coefficient $RT_c/p_c V_c = 3.53$, which gives a better agreement with observed values for simpler gases.

5. ***Redlich-Kwong equation*** Yet another empirical equation of state of interest, particularly in engineering, is due to Redlich and Kwong:

$$p = \frac{RT}{V-b} - \frac{a}{V(V-b)T^{1/2}} \tag{3.27}$$

Note that like van der Waals' equation, Redlich-Kwong equation also contains two constants. It gives particularly good results at high pressures and appears fairly reliable for $T > T_c$. The constants a and b are given by

$$a = 0.4275 \frac{R^2 T_c^{2.5}}{p_c}$$

and

$$b = \frac{0.0867\, RT_c}{p_c} \tag{3.27a}$$

It may be mentioned here that a common inherent limitation of two constant equations of state is the lack of accuracy in predicting the behaviour of a given substance over a wide range of p–V–T conditions. To circumvent this problem, Benedict-Webb-Rubbin proposed an equation with eight adjustable constants. This equation is particularly suited for hydrocarbons. However, various other equations have been proposed but none of these explains the behaviour of all gases under different conditions of temperature and pressure. For this reason, one generally evaluates the virial coefficients from theoretical considerations and compares them with the observed values.

3.7 Molecular Attraction: Existence and Implications

In Sec. 3.2.3, we discussed Amagat's experiments with several real gases. It was observed that below the Boyle's temperature, the product pV first decreases and then increases with pressure. This suggests that a real gas is more compressible than a perfect gas. Theoretically, van der Waals anticipated, and rightly so, that the intermolecular attraction forces the molecules to come closer together leading to greater decrease in volume than expected for a perfect gas on increasing pressure. Similarly, once the molecules have come closer, he argued that due to finite size, molecules repel each other and oppose the decrease in pressure caused by intermolecular forces. That is, two competing mechanisms—due to intermolecular forces and finite size—operate in a gas at any temperature.

To confirm van der Waals assertions about the existence of intermolecular forces, Joule performed a simple experiment. He argued that if intermolecular forces do exist, some work will be done against these when a gas is allowed to expand. And since this work can only be done at the expense of the internal energy of the gas, its temperature should drop producing a cooling effect.

Joule Experiment Figure 3.8 depicts two configurations of Joule's experiment. In configuration (a), the gas of interest was filled in vessel A at a high pressure ($\approx$ 22 atm) and connected to an evacuated vessel B through a tube having a stopcock. Both the vessels were placed in a water bath and a sensitive thermometer was kept to measure temperature. When this configuration attained thermal equilibrium, the thermometer reading was noted. Next, the stopcock was opened and the gas was allowed to expand into the evacuated vessel. The system reached a new equilibrium state in which pressure was same in both the vessels but no change in temperature was detected. On the basis of this observation, it was concluded that *a perfect gas showed no change in temperature on expansion.*

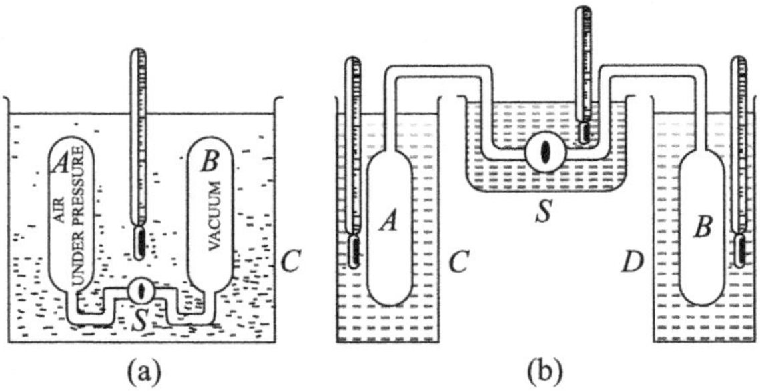

Fig. 3.8 Joule's experiment: Free expansion.

Next the vessels A and B were put in two separate water baths and the connecting tubes and stopcock were placed in the third water bath. When the gas in vessel A was made to expand, a slight drop in temperature of the water bath containing vessel A was detected and an equal rise of temperature in the water bath containing vessel B was observed. The drop in temperature was attributed to the heat absorbed by the air in vessel A while expanding against the rising pressure in B and the rise in temperature of B was interpreted as being due to the heat given out in doing work in compressing the air in B against the rising pressure. However, on the whole the total quantity of air in vessels A and B neither gained nor lost any heat during the *free expansion*. It was then argued that if intermolecular forces were present, the total internal energy of the gas would be made up of kinetic and potential energies and show volume dependence. Moreover, finite change in temperature was expected, i. e., $\left(\frac{\partial T}{\partial p}\right)_U > 0$. But neither was observed. This negative result was interpreted as: *Conservation of internal energy implied non-existence of intermolecular forces*. This is referred to as *Joule's law*. It states that *internal energy of a perfect gas is independent of volume at constant temperature or pressure*. Mathematically, if U denotes internal energy of a perfect gas, we write

$$\left(\frac{\partial U}{\partial V}\right)_T = \left(\frac{\partial U}{\partial V}\right)_p = 0$$

Soon after the publication of this result, Joule carried out a series of experiments in collaboration with Thomson (later Lord Kelvin) under more carefully controlled conditions than his earlier experiments and they showed conclusively that a real gas, if allowed to expand adiabatically, does show change in temperature. In fact, this so-called *porous plug experiment* conclusively proved that under certain conditions, real gases show cooling effect, which can be attributed to the existence of intermolecular attraction. (This is also known as *Joule–Thomson expansion*.) We will now discuss this experiment in detail.

3.7.1 The Porous Plug Experiment

In this experiment, a highly compressed gas is continuously forced at constant pressure to expand adiabatically and irreversibly through a porous plug kept in a non-conducting cylinder (Fig. 3.9). The plug consists of a porous material, say cotton wool, silk, etc.,

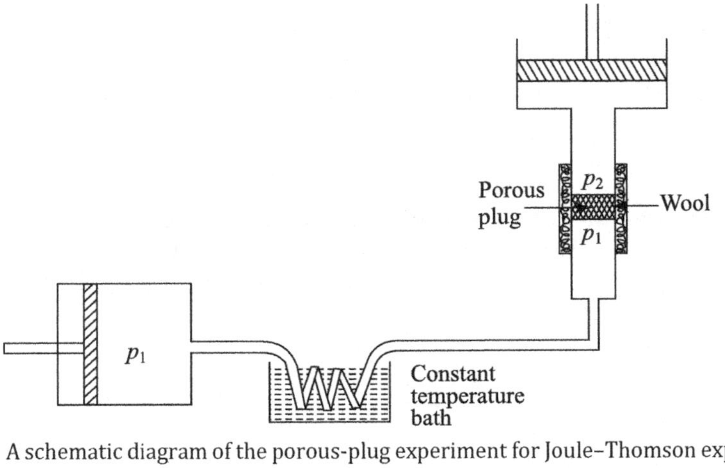

Fig. 3.9 A schematic diagram of the porous-plug experiment for Joule–Thomson expansion.

and has a number of fine holes. Thus, a porous plug is equivalent to a number of narrow orifices held in parallel. To avoid loss or gain of heat in the neighbourhood of the plug, it is surrounded by an insulation of cotton wool packed in a metallic vessel.

Before allowing the gas to pass through the porous plug, it is made to flow through a copper spiral immersed in a constant temperature bath so that it acquires a particular initial temperature. The pressures on the opposite sides of the porous plug are kept constant by moving pistons at an appropriate rate. Suppose the expansion takes place from a constant high pressure p_1 to a constant low pressure $p_2(< p_1)$. Let the temperatures on the two sides of the porous plug be T_1 and T_2. Due to decrease in pressure, the gas expands while it passes through a large number of long and narrow orifices in the plug. In this process, the molecules of the gas are gradually pulled apart. This process is known as *throttling process*. The change in temperature, if any, is measured by thermocouples.

The findings of Joule-Thomson experiment can be summarised as follows:

1. All gases show a change of temperature after passing through the porous plug. It means that no gas behaves perfectly. (In fact, the very purpose of this experiment was to establish the dependence of internal energy on volume and thereby verify the existence of intermolecular forces between gas molecules.)
2. At ordinary temperatures, all gases, except hydrogen and helium, show a cooling effect. In fact, these gases show slight heating. This was totally unexpected. (This paradox was subsequently resolved when it was recognised that to obtain any cooling, the initial temperature of the gas must be less than its inversion temperature.)
3. At low enough temperatures, all gases show cooling effect.
4. The fall in temperature is directly proportional to the pressure difference on the two sides of the porous plug. However, for a given difference of pressure, the drop in temperature is more if the initial temperature of the gas is less.
5. For every gas, there exists a characteristic temperature, called *temperature of inversion*, denoted as T_i, at which no change in temperature occurs. If the initial temperature of the gas is less than T_i, it cools after passing through the porous plug and vice versa.

Joule–Thomson Coefficient for a van der Waals' gas To understand these results theoretically, let us assume that one mole of the gas is made to expand through the porous plug from a pressure p_1 and volume V_1 to a pressure p_2 and volume V_1 and its temperature changes from T_1 to T_2. In this process, the work is done on compressing the gas, work is done by the gas as it expands as well as in overcoming internal forces due to intermolecular attraction. From your earlier physics classes, you will recall that

work done on the gas in compressing it at constant pressure p_1 will be p_1V_1 and
work done by the gas in expanding against constant pressure p_2 will be p_2V_2.

Hence, the net work done by the gas during expansion,

$$W_{ext} = p_2V_2 - p_1V_1 \tag{3.28a}$$

To obtain an expression for the work done against intermolecular forces, we assume that the gas obeys van der Waals' equation of state so that the attraction between the molecules is equivalent to an internal pressure $\frac{a}{V^2}$. Therefore, the work done against the molecular attraction is given by

$$W_{int} = \int_{V_1}^{V_2} \frac{a}{V^2} dV = a\left(\frac{1}{V_1} - \frac{1}{V_2}\right) \tag{3.28b}$$

Hence, the total work done by one mole of the gas as it flows through the porous plug is obtained by adding Eqs. (3.28a) and (3.28b):

$$W = p_2 V_2 - p_1 V_1 + \frac{a}{V_1} - \frac{a}{V_2} = \left(p_2 V_2 - \frac{a}{V_2}\right) - \left(p_1 V_1 - \frac{a}{V_1}\right) \tag{3.29}$$

The work done by the gas may cause fall in temperature of the gas. But if there is to be no cooling of the gas, heat equivalent to the work W given by Eq. (3.29) will have to be supplied. Let us calculate the amount of heat involved for a van der Waals gas. We start with the equation,

$$\left(p + \frac{a}{V^2}\right)(V - b) = RT$$

On expanding this and rearranging terms, we can write

$$pV = RT - \frac{a}{V} + bp + \frac{ab}{V^2}$$

The last term on the right-hand side of this equation involves the product of van der Waals constants, which are very small, and may therefore be ignored in comparison with other terms. Then on subtracting (a/V) from both sides, we can write

$$pV - \frac{a}{V} = RT - \frac{2a}{V} + bp \tag{3.30}$$

We now make another approximation. Since a and b are very small, we replace V by RT/p in the second term on the right hand side. (It essentially means that the gas is not very far from ideal gas behaviour.) Then Eq. (3.30) takes the form,

$$pV - \frac{a}{V} = RT - \frac{2a}{RT}p + bp = RT + \left(b - \frac{2a}{RT}\right)p \tag{3.31}$$

If T is to be maintained constant in Joule–Thomson expansion, Eq. (3.31) implies that we must have

$$\left(p_2 V_2 - \frac{a}{V_2}\right) - \left(p_1 V_1 - \frac{a}{V_1}\right) = (p_1 - p_2)\left(\frac{2a}{RT} - b\right) \tag{3.32}$$

Since $p_1 > p_2$, the heat supplied to maintain T constant will be positive if $\left(\frac{2a}{RT} - b\right) > 0$ or $\left(\frac{2a}{RT}\right) > b$. It means that if no heat is supplied, the gas will cool. Experiments show that except hydrogen and helium, all gases show cooling at room temperature after a Joule–Thomson expansion. In fact, even hydrogen and helium show cooling effect, if these are cooled below their respective critical temperatures before allowing them to expand through the porous plug.

Since the system is thermally isolated, we can obtain an expression for the temperature difference produced during a Joule–Thomson expansion by rewriting Eq. (3.31) as

$$\left(p_2 V_2 - \frac{a}{V_2}\right) - \left(p_1 V_1 - \frac{a}{V_1}\right) = R(T_2 - T_1) + (p_1 - p_2)\left(\frac{2a}{RT} - b\right)$$

On combining this result with Eq. (3.29), we get

$$W = \left(\frac{2a}{RT} - b\right)\Delta p - R\Delta T \tag{3.33}$$

where $\Delta T = (T_1 - T_2)$ signifies drop in temperature, and Δp $(= p_1 - p_2)$ denotes fall in pressure.

We cannot complete this derivation rigourously without reference to the first law of thermodynamics. Since gas exchanges no heat with the surroundings, the work is done by the gas at the expense of its internal energy, which causes fall in temperature of the gas. It turns out that the net work done by the gas is equal to the heat absorbed and can also be expressed as $C_V \Delta T$, where C_V is heat capacity at constant volume (Chapter 5). Hence, we can rewrite Eq. (3.33) as

$$C_V \Delta T = \left(\frac{2a}{RT} - b\right)\Delta p - R\Delta T$$

or

$$(C_V + R)\Delta T = C_p \Delta T = \left(\frac{2a}{RT} - b\right)\Delta p$$

Using this result, we can express the fall in temperature of a van der Waals gas in Joule–Thomson expansion as

$$\Delta T = \frac{1}{C_p}\left(\frac{2a}{RT} - b\right)\Delta p \tag{3.34}$$

so that Joule–Thomson coefficient is given by

$$\mu = \frac{\Delta T}{\Delta p} = \frac{1}{C_p}\left(\frac{2a}{RT} - b\right) \tag{3.35}$$

Since Δp is always greater than zero, Eq. (3.34) or Eq. (3.35) suggests that cooling or heating in Joule–Thomson expansion is determined by the competition between terms characterising intermolecular forces and finite size of gas molecules. If intermolecular forces are strong so that $a >> b$, $\left(\frac{2a}{RT} - b\right) > 0$. Hence, $\Delta T > 0$ and the gas will cool down. That is, $\mu > 0$ signifies cooling. On the other hand, when intermolecular forces are weak and volume correction term dominates in Eq. (3.34), i.e., $\left(\frac{2a}{RT} - b\right) < 0$, $\Delta T < 0$. Hence, a gas with weak intermolecular forces will warm upon J–T expansion. However, there will be neither cooling nor heating if

$$\frac{2a}{RT_i} = b \tag{3.36}$$

Here T_i signifies the temperature of the gas below which it will cool down when subject to Joule–Thomson expansion and above which it will warm up. It is referred to as *inversion temperature*. This is a significant result and we can say that if the initial temperature of a gas is below its inversion temperature, it will cool down and vice versa.

Table 3.2 *Inversion temperatures of some gases*

Gas	He	H_2	N_2	A	O_2	CO_2	Air
T_i(K)	23.6	195	621	723	893	1500	603

The inversion temperatures for some gases of interest are given in Table 3.2. Note that but for H_2 and He, T_i is well above the room temperature. This explains why He and H_2 could not be liquefied easily. As we now know, these gases ought to be pre-cooled below their inversion temperatures using other liquids. For example, we pre-cool hydrogen with liquid nitrogen (boiling point = 77.3 K at one atmosphere pressure) and helium is precooled with liquid hydrogen (boiling point = 20.4 K).

Since ΔT is directly proportional to Δp, for greater cooling, a gas should be highly compressed before allowing it to expand through the porous plug. In general, the Joule–Kelvin coefficient is quite small; a few tenths of a degree per atmosphere. Note that J-T expansion is a steady-state process. So we can accumulate the total effect by using the cooled gas in a heat exchanger to pre-cool the incoming gas. For this reason, the Joule–Kelvin expansion is of great practical importance in the liquefaction of gases and in studying the properties of matter at low temperatures. You will learn these details in Chapter 10.

By combining Eqs. (3.18), (3.19) and (3.36), we can relate temperature of inversion with Boyle temperature and critical temperature as

$$T_i = 2T_B = \frac{27}{4}T_c \tag{3.37}$$

To get an idea about the typical drop in temperature, we would like you to go through the following examples.

Example 3.5 Oxygen gas is made to undergo adiabatic throttling and the pressure is reduced by 50 atm. If the initial temperature is 27°C, calculate the drop in temperature. Given $a = 13.2 \times 10^{-2}$ Nm^{4}mol^{-2}, $b = 31.2 \times 10^{-6}$ m^{3}mol^{-1}, $R = 8.3$ J mol^{-1}K^{-1} and $C_p = 3.4\,R$.

Solution: The fall of temperature is given by Eq. (3.34):

$$\Delta T = \frac{(p_1 - p_2)}{C_p}\left(\frac{2a}{RT} - b\right)$$

Here $p_1 - p_2 = 50$ atm $= 50 \times 1.013 \times 10^5$ Nm^{-2}, $T = 27°C = 300$ K,

$C_p = 3.5R = 3.5 \times 8.3$ J mol^{-1}K^{-1} $= 29.05$ J mol^{-1}K^{-1}

$a = 13.2 \times 10^{-2}$ Nm^{4}mol^{-2}, and $b = 31.2 \times 10^{-6}$ m^{3}mol^{-1}. Hence,

$$\therefore \Delta T = \frac{(50 \times 1.013 \times 10^5 \text{ Nm}^{-2})}{(29.05\,\text{J}\,\text{K}^{-1}\text{mol}^{-1})}\left(\frac{2 \times (13.2 \times 10^{-2}\text{ Nm}^4\text{mol}^{-2})}{(8.3\,\text{J}\,\text{mol}^{-1}\text{K}^{-1}) \times (300\text{ K})} - 31.2 \times 10^{-6}\text{ m}^3\text{ mol}^{-1}\right)$$

$$= \frac{50.65 \times 10^5 \times 74.8 \times 10^{-6}}{29.05}\text{K}$$

$$= 13.04 \text{ K}$$

Example 3.6 Calculate the change in temperature when helium is made to undergo Joule–Thomson expansion at –173°C and the pressure difference on the two sides of the plug is 20 atm. Does the gas show a heating effect or a cooling effect in this expansion? Given $R = 8.3$ JK^{-1}mol^{-1}; $a = 3.41 \times 10^{-3}$ Nm^{4}mol^{-2}; $b = 23.7 \times 10^{-6}$ m^{3}mol^{-1}; and $C_p = 2.5\,R$.

Solution: The fall of temperature is given by Eq. (3.34)

$$\Delta T = \frac{(p_1 - p_2)}{C_p}\left(\frac{2a}{RT} - b\right)$$

Here, $(p_1 - p_2) = 20$ atm $= 20 \times 1.013 \times 10^5$ Nm^{-2} $= 20.26 \times 10^5$ Nm^{-2}, $T = 100$ K, $C_p = 2.5\ R = 2.5 \times 8.3$ J K^{-1}mol^{-1} $= 20.75$ J K^{-1}mol^{-1}, $a = 3.41 \times 10^{-3}$ Nm^{4}mol^{-2}, and $b = 23.7 \times 10^{-6}$ m^{3}mol^{-1}. Hence,

$$\therefore \Delta T = \frac{(20.26 \times 10^5 \text{ Nm}^{-2})}{(20.75 \text{ J K}^{-1} \text{mol}^{-1})} \left(\frac{2 \times (3.41 \times 10^{-3} \text{ Nm}^4 \text{ mol}^{-2})}{(8.3 \text{ J K}^{-1} \text{mol}^{-1}) \times (100 \text{ K})} - (23.7 \times 10^{-6} \text{ m}^3 \text{ mol}^{-1}) \right)$$

$$= \frac{20.26 \times 10^5}{20.75} (8.217 \times 10^{-6} - 23.7 \times 10^{-6}) \text{ K}$$

$$= -1.51 \text{ K} \qquad \text{or} \qquad -1.51\text{°C}$$

Since $\Delta T < 0$, the temperature of the gas rises by 1.51°C. (This is an expected result since the gas is made to undergo J-T expansion at –173°C, which is above the temperature of inversion of helium, – 240°C.).

Example 3.7 The van der Waals' constants of hydrogen are $a = 2.47 \times 10^{-2}$ Nm^{4}mol^{-2} and $b = 26.5 \times 10^{-6}$ m^{3}mol^{-1}. Calculate (a) the temperature of inversion and (b) the Joule–Thomson cooling if pressure drops by 2 atm. The initial temperature of the gas is 100 K. Given $R = 8.3$ J K^{-1}mol^{-1}.

Solution: **(a)** The inversion temperature is given by Eq. (3.36):

$$T_i = \frac{2a}{Rb}$$

Here $a = 2.47 \times 10^{-2}$ Nm^{4}mol^{-2} and $b = 26.5 \times 10^{-6}$ m^{3}mol^{-1} and $R = 8.3$ J K^{-1}mol^{-1}.

$$\therefore \qquad T_i = \frac{2 \times (2.47 \times 10^{-2} \text{ Nm}^4 \text{ mol}^{-2})}{(8.3 \text{ J K}^{-1} \text{mol}^{-1}) \times (26.5 \times 10^{-6} \text{ m}^3 \text{ mol}^{-1})}$$

$$= 224.6 \text{ K}$$

(b) The Joule–Thomson cooling is given by Eq. (3.34):

$$\Delta T = \frac{(p_1 - p_2)}{C_p} \left(\frac{2a}{RT} - b \right)$$

Here, $(p_1 - p_2) = 2$ atm $= 2 \times 1.013 \times 10^5$ Nm^{-2} $= 2.026 \times 10^5$ Nm^{-2}, $T = 100$ K,

and $C_p = \frac{7}{2} R = \frac{7}{2} \times 8.3$ J K^{-1}mol^{-1}

$$\therefore \quad \Delta T = \frac{(2.026 \times 10^5 \text{ Nm}^{-2})}{7 \times (4.15 \text{ J K}^{-1} \text{mol}^{-1})} \left(\frac{2 \times (2.47 \times 10^{-2} \text{ Nm}^4 \text{ mol}^{-2})}{(8.3 \text{ J K}^{-1} \text{mol}^{-1}) \times (100 \text{ K})} - 26.5 \times 10^{-6} \text{ m}^3 \text{ mol}^{-1} \right)$$

$$= 0.23 \text{ K}$$

ADDITIONAL EXAMPLES

Example 3.8 Calculate the deviation of van der Waals equation from the perfect gas law at critical temperature.

Solution: To calculate the deviation of van der Waals' equation from the perfect gas equation, we express pV in terms of temperature in both cases for one mole of the gas. For van der Waals' gas, we can write

$$pV - bp + \frac{a}{V} - \frac{ab}{V^2} = RT \tag{i}$$

and for a perfect gas, we have

$$pV = RT \tag{ii}$$

At $T = T_c$, the deviation is obtained by subtracting Eq. (ii) from Eq. (i). This gives

$$-bp_c + \frac{a}{V_c} - \frac{ab}{V_c^2} = R(T_c - T)$$

so that

$$R\Delta T = -bp_c + \frac{a}{V_c} - \frac{ab}{V_c^2} \tag{iii}$$

We know that at critical temperature

$$p_c = \frac{a}{27b^2}$$

and

$$V_c = 3b$$

On inserting these values in Eq. (iii), we get

$$R\Delta T = -b \cdot \left(\frac{a}{27b^2}\right) + \frac{a}{3b} - \frac{ab}{9b^2}$$

$$= -\frac{a}{27b} + \frac{a}{3b} - \frac{a}{9b}$$

$$= \frac{5a}{27b} \tag{iv}$$

From Eqs (3.12a) and (3.12b) we note that $\frac{a}{b} = \frac{27RT_c}{8}$, so that we can rewrite (iv) as

$$R\Delta T = \frac{5RT_c}{8}$$

or

$$\frac{\Delta T}{T_c} = \frac{5}{8} = 0.625$$

This result shows that at the critical point, van der Waals equation deviates from the perfect gas equation by 62.5%. Moreover, it is independent of the nature of the gas.

Example 3.9 For helium gas, $T_c = 5.0$ K and $p_c = 2.3$ atm. Calculate the van der Waals' constants. Given $R = 8.314$ JK^{-1}mol^{-1} and 1 atm = 1.013×10^5 Nm^{-2}.

Solution: We know that critical constants for van der Waals gas are given by

$$T_c = \frac{8a}{27Rb} \tag{i}$$

and

$$p_c = \frac{a}{27b^2} \tag{ii}$$

On substituting the given values, we get

$$5.0\text{ K} = \frac{8a}{27b \times (8.314\text{ JK}^{-1}\text{mol}^{-1})} \Rightarrow \frac{a}{b} = 140.299\text{ J mol}^{-1} \tag{iii}$$

and

$$2.3 \times 1.013 \times 10^5\text{ Nm}^{-2} = \frac{a}{27b^2} \Rightarrow \frac{a}{b^2} = 62.907 \times 10^5\text{ Nm}^{-2} \tag{iv}$$

On dividing Eq. (iii) by Eq. (iv), we get

$$b = \frac{140.299\text{ J mol}^{-1}}{62.907 \times 10^5\text{ Nm}^{-2}}$$
$$= 2.23 \times 10^{-5}\text{ m}^3\text{ mol}^{-1}. \tag{v}$$

Using this value of b in Eq. (iii), we get

$$a = 140.299 \times 2.23 \times 10^{-5}\text{ Jmol}^{-2}\text{m}^3$$
$$= 312.90 \times 10^{-5}\text{ Jm}^3\text{mol}^{-2}$$
$$= 3.13 \times 10^{-3}\text{ Nm}^4\text{mol}^{-2}$$

Example 3.10 Calculate the change in temperature produced by adiabatic throttling process of one gm – mole of oxygen when pressure drops by 50 atm. The initial temperature of the gas is 27°C. Use $C_p = 7.03$ cal mol^{-1}K^{-1}, $a = 1.32$ litre^2 atm mol^{-2}, $b = 3.12 \times 10^{-2}$ litre mol^{-1}, $R = 8.314$ JK^{-1}mol^{-1} and 1 atm = 1.013×10^5 Nm^{-2}.

Solution: Here,

$$a = 1.32\text{ litre}^2\text{ atm mol}^{-2}$$
$$= 1.32 \times ((10^3)^2 \times 10^{-12}\text{ m}^6) \times (1.013 \times 10^5\text{ Nm}^{-2})\text{ mol}^{-2}$$
$$= 1.32 \times 10^{-1} \times 1.013\text{ Nm}^4\text{ mol}^{-2}$$
$$= 0.134\text{ Nm}^4\text{ mol}^{-2}$$

$$b = 3.12 \times 10^{-2}\text{ litre mol}^{-1}$$
$$= 3.12 \times 10^{-2} \times 10^3 \times 10^{-6}\text{m}^3\text{mol}^{-1}$$
$$= 3.12 \times 10^{-5}\text{ m}^3\text{ mol}^{-1}$$

$$R = 8.314\text{ JK}^{-1}\text{mol}^{-1},\ C_p = 7.03\text{ cal K}^{-1}\text{mol}^{-1} = 29.53\text{ JK}^{-1}\text{mol}^{-1}.$$

$$T = 300\text{ K and } \Delta p = 50\text{ atm} = 5.065 \times 10^6\text{ Nm}^{-2}$$

We know that change in temperature produced by adiabatic throttling process is given by

$$\Delta T = \frac{1}{C_p}\left(\frac{2a}{RT} - b\right)\Delta p$$

On substituting the given values, we get

$$\Delta T = \frac{1}{(29.53\text{J K}^{-1}\text{mol}^{-1})}\left[\frac{2\times 0.134\text{ Nm}^4\text{mol}^{-2}}{(8.314\text{ JK}^{-1}\text{mol}^{-1})\times 300\text{ K}} - 3.12\times 10^{-5}\text{ m}^3\text{mol}^{-1}\right] \times (5.065\times 10^6\text{ Nm}^{-2})$$

$$= \frac{1}{29.53\text{ JK}^{-1}\text{mol}^{-1}}[1.0745\times 10^{-4}\text{ m}^3\text{mol}^{-1} - 3.12\times 10^{-5}\text{ m}^3\text{mol}^{-1}] \times (5.065\times 10^6\text{ Nm}^{-2})$$

$$= \frac{5.065\times 10^6}{29.53}[7.625\times 10^{-5}]\text{ K}$$

$$= \frac{386.20}{29.53}\text{K} = 13.08\text{ K}$$

Since $\Delta T > 0$, adiabatic throttling will produce cooling.

Example 3.11 Helium gas is made to undergo Joule–Thomson expansion at 100 K. If the pressure difference on the two sides is 20 atm, calculate the change in temperature. Given $R = 8.3$ JK^{-1}mol^{-1}, $a = 0.0341$ litre^2 atm mol^{-2} and $b = 0.0237$ litre mol^{-1}.

Solution: Here,

$$a = 0.0341\text{ litre}^2\text{ atm mol}^{-2}$$
$$= 0.0341\times(10^{-6}\text{ m}^6)\times(1.013\times 10^5\text{ Nm}^{-2})\text{ mol}^{-2}$$
$$= 0.0341\times 1.013\times 10^{-1}\text{ Nm}^4\text{ mol}^{-2}$$

$$b = 0.0237\text{ litre mol}^{-1}$$
$$= 0.0237\times 10^{-3}\text{ m}^3\text{ mol}^{-1}$$
$$= 2.37\times 10^{-5}\text{ m}^3\text{ mol}^{-1}$$

$$R = 8.314\text{ JK}^{-1}\text{mol}^{-1},\ T = 100\text{ K and } \Delta p = 20\text{ atm} = 20.26\times 10^5\text{ Nm}^{-2}$$

Since helium is monatomic gas, $C_p = \frac{5}{2}R = 20.335$ JK^{-1}mol^{-1}. The change in temperature produced in adiabatic throttling process is given by

$$\Delta T = \frac{1}{C_p}\left(\frac{2a}{RT} - b\right)\Delta p$$

On substituting the given values, we get

$$\Delta T = \frac{1}{20.335\text{ JK}^{-1}\text{mol}^{-1}}\left[\frac{2\times 0.00345\text{ Nm}^4\text{mol}^{-2}}{(8.314\text{ JK}^{-1}\text{ mol}^{-1})\times(100\text{ K})} - (2.37\times 10^{-5}\text{ m}^3\text{mol}^{-1}\right] \times (20.26\times 10^5\text{ Nm}^{-2})$$

$$= \frac{1}{20.335\ \text{JK}^{-1}\text{mol}^{-1}}[0.8299 \times 10^{-5}\ \text{m}^3\text{mol}^{-1} - 2.37 \times 10^{-5}\ \text{m}^3\text{mol}^{-1}] \times (20.26 \times 10^5\ \text{Nm}^{-2})$$

$$= \frac{1}{20.335\ \text{JK}^{-1}\text{mol}^{-1}}[-1.5401 \times 10^{-5}\ \text{m}^3\text{mol}^{-1}] \times (20.26 \times 10^5\ \text{Nm}^{-2}) = -1.53\ \text{K}$$

Note that $\Delta T < 0$. It means that He will show heating effect implying that it has been throttled above its inversion temperature.

Example 3.12 A diatomic gas obeys van der Waals' equation of state with $a = 24.8 \times 10^{-3}\ \text{Nm}^4\text{mol}^{-2}$ and $b = 0.0266 \times 10^{-3}\ \text{m}^3\text{mol}^{-1}$. (a) Calculate Boyle temperature as well as inversion temperature. Use $R = 8.314\ \text{JK}^{-1}\text{mol}^{-1}$. (b) Change in temperature due to J–T expansion when pressure drops by 2 atm and the gas is held at 100 K.

Solution: **(a)** We know that Boyle temperature of a gas is given by

$$T_B = \frac{a}{bR}$$

On substituting the values of a, b and R, we get

$$T_B = \frac{24.8 \times 10^{-3}\ \text{Nm}^4\text{mol}^{-2}}{(0.0266 \times 10^{-3}\ \text{m}^3\text{mol}^{-1}) \times (8.314\ \text{JK}^{-1}\text{mol}^{-1})}$$

$$= 112\ \text{K}$$

The inversion temperature of the gas is given by

$$T_i = \frac{2a}{Rb} = 2T_B = 224\ \text{K}$$

(b) Since the diatomic gas is kept at 100 K, we note that $T < T_i$. So we expect that J–T expansion will lead to cooling. Recall that

$$\Delta T = \frac{1}{C_p}\left(\frac{2a}{RT} - b\right)\Delta p$$

For a diatomic gas, $C_p = \frac{7}{2}R = \frac{7}{2} \times 8.314\ \text{JK}^{-1}\text{mol}^{-1} = 29.099\ \text{JK}^{-1}\text{mol}^{-1}$.

On substituting the given values, we get

$$\Delta T = \frac{1}{(29.099\ \text{JK}^{-1}\text{mol}^{-1})}\left[\frac{2 \times 24.8 \times 10^{-3}\ \text{Nm}^4\text{mol}^{-2}}{(8.314\ \text{JK}^{-1}\text{mol}^{-1}) \times (100\ \text{K})} - 0.0266 \times 10^{-3}\ \text{m}^3\text{mol}^{-1}\right] \times (2 \times 1.013 \times 10^5\ \text{Nm}^{-2})$$

$$= \frac{(5.966 \times 10^{-5} - 2.66 \times 10^{-5})\text{m}^3\text{mol}^{-1} \times (2.026 \times 10^5\ \text{Nm}^{-2})}{29.099\ \text{JK}^{-1}\text{mol}^{-1}}$$

$$= 0.23\ \text{K}$$

Example 3.13 The critical volume of nitrogen gas is 81 cm^3. The van der Waals constant b is three times the actual volume of a mole of the gas. If N_2 molecule can be treated as a billiard ball, calculate its radius.

Solution: We know that

$$b = \frac{V_c}{3} = \frac{81}{3}\text{cm}^3 = 27\text{ cm}^3$$

$\therefore$ Volume of a mole of $N_2 = \frac{b}{3} = \frac{27}{3}\text{cm}^3 = 9\text{ cm}^3$

From this it readily follows that

$$\text{volume of one molecule of } N_2 = \frac{9\text{ cm}^3}{6.023 \times 10^{23}} = 1.494 \times 10^{-23}\text{ cm}^3$$

Since N_2 molecule behaves as a sphere of radius r, we can write

$$\frac{4\pi}{3}r^3 = 1.494 \times 10^{-23}\text{ cm}^3$$

$$\Rightarrow \quad r^3 = \frac{3 \times 1.494 \times 10^{-23}}{4\pi}\text{cm}^3$$

$$= 3.566 \times 10^{-24}\text{ cm}^3$$

Hence,

$$r = 1.528 \times 10^{-8}\text{ cm}$$

Example 3.14 One mole of a gas is contained in a closed vessel of volume 0.25 litre. It is assumed to obey van der Waals' equation of state and is maintained at a pressure of 90 atm. As the pressure is increased to 100 atm at 300 K, its temperature rises to 325 K. Calculate a and b.

Solution: Let us denote the pressures as p_1 and p_2. Then we can write

$$\left(p_1 + \frac{a}{V^2}\right)(V - b) = RT_1 \Rightarrow \frac{a}{V^2} = \frac{RT_1}{V - b} - p_1 \quad \text{(i)}$$

and

$$\left(p_2 + \frac{a}{V^2}\right)(V - b) = RT_2 \Rightarrow \frac{a}{V^2} = \frac{RT_2}{V - b} - p_2 \quad \text{(ii)}$$

On comparing Eqs. (i) and (ii) we can eliminate $\frac{a}{V^2}$ to obtain

$$\frac{RT_1}{V - b} - \frac{RT_2}{V - b} = p_1 - p_2$$

On simplification, we get

$$V - b = \frac{R(T_1 - T_2)}{p_1 - p_2}$$

so that

$$b = V - \frac{R(T_2 - T_1)}{p_2 - p_1} \tag{iii}$$

On substituting the value of $V - b$ in Eq. (i), we can write

$$\frac{a}{V^2} = \frac{RT_1\,(p_1 - p_2)}{R(T_1 - T_2)} - p_1$$

On simplification, we get

$$a = \frac{V^2(T_1 p_2 - T_2 p_1)}{T_2 - T_1} \tag{iv}$$

On substituting the given values, we get

$$a = \frac{(0.25 \text{ lit mol}^{-1})^2 \times (300 \text{ K} \times 100 \text{ atm} - 90 \text{ atm} \times 395 \text{ K})}{15 \text{ K}}$$

$$= \frac{(625 \times 10^{-4} \text{ lit}^2 \text{ mol}^{-2}) \times (30{,}000 - 29{,}250) \text{ atm K}}{25 \text{ K}}$$

$$= 1.875 \text{ atm lit}^2 \text{ mol}^{-2}$$

Similarly,

$$b = (0.25 \text{ lit mol}^{-1}) - \frac{(8.314 \text{ J mol}^{-1}\text{K}^{-1}) \times 25 \text{ K}}{10 \times 1.013 \times 10^5 \text{ Nm}^{-2}}$$

$$= (0.250 \times 10^{-3}\text{m}^3\text{mol}^{-1}) - (0.205 \times 10^{-3}\text{m}^3\text{mol}^{-1})$$
$$= 0.045 \times 10^{-3}\text{m}^3\text{mol}^{-1}$$
$$= 0.045 \text{ lit mol}^{-1}$$

Example 3.15 Obtain expression for Boyle temperature for a gas obeying (a) Berthelot equation of state and (b) Dieterici equation of state.

Solution: **(a)** From Eq. (3.24), we recall that Berthelot's equation of state for one mole of gas is

$$\left(p + \frac{a'}{TV^2}\right)(V - b) = RT$$

We rewrite it as

$$p = \frac{RT}{V - b} - \frac{a'}{TV^2} \tag{i}$$

To obtain expressions for Boyle temperature, we put

$$\frac{\partial(pV)}{\partial p} = 0 \quad \text{for} \quad p \to 0$$

Therefore, we multiply both sides of Eq. (i) by V and obtain

$$pV = \frac{RTV}{V - b} - \frac{a'}{TV}$$

On differentiating this equation with p at constant T, we get

$$\frac{\partial(pV)}{\partial p} = \left[\frac{RT}{V-b} - \frac{RTV}{(V-b)^2} + \frac{a'}{TV^2}\right]\left(\frac{\partial V}{\partial p}\right)_T$$

$$= \left[-\frac{RTb}{(V-b)^2} + \frac{a'}{TV^2}\right]\left(\frac{\partial V}{\partial p}\right)_T \qquad \text{(ii)}$$

At $T = T_B$, $\dfrac{\partial(pV)}{\partial p} = 0$ and $p \to 0$ or $V \to \infty$. Then we can write

$$\frac{a'}{T_B V^2} - \frac{RT_B b}{V^2} = 0$$

or

$$T_B = \sqrt{\frac{a'}{Rb}} \qquad \text{(iii)}$$

(b) For Dieterici equation of state, we recall from Eq. (3.25) that

$$p = \frac{RT}{V-b}\exp\left(-\frac{a}{RTV}\right)$$

As before, we multiply both sides by V to obtain

$$pV = \frac{RTV}{(V-b)}\exp\left(-\frac{a}{RTV}\right) \qquad \text{(iv)}$$

On differentiating both sides with respect to p, we get

$$\frac{\partial(pV)}{\partial p} = \left[\frac{RT}{V-b} - \frac{RTV}{(V-b)^2} + \frac{RTV}{V-b}\frac{a}{RTV^2}\right]\exp\left(-\frac{a}{RTV}\right)\left(\frac{\partial V}{\partial p}\right)_T$$

$$= \left[-\frac{RTb}{(V-b)^2} + \frac{a}{V(V-b)}\right]\exp\left(-\frac{a}{RTV}\right)\left(\frac{\partial V}{\partial p}\right)_T \qquad \text{(v)}$$

On equating the RHS of (v) to zero and taking the limit $V \to \infty$ for $T = T_B$, we get

$$-\frac{RT_B b}{V^2} + \frac{a}{V^2} = 0$$

or

$$T_B = \frac{a}{Rb} \qquad \text{(vi)}$$

Note that Boyle temperature for Dieterici equation is the same as that for a van der Waals gas.

Example 3.16 A monatomic van der Waals' gas is contained in a cylinder of molar volume 0.8 lit mol^{-1} at a pressure of 36 atm. If $a = 4.05$ atm lit^2 mol^{-2}, $b = 0.037$ lit mol^{-1} and $R = 0.082$ atm.lit K^{-1}mol^{-1}, calculate internal pressure and temperature of the gas.

Solution: The internal pressure

$$\frac{a}{V^2} = \frac{4.05 \text{ atm lit}^2 \text{ mol}^{-2}}{(0.8 \text{ lit mol}^{-1})^2}$$

$$= 6.33 \text{ atm}$$

The temperature of the gas is given by

$$T = \frac{1}{R}\left(p + \frac{a}{V^2}\right)(V - b)$$

$$= \frac{1}{(0.082 \text{ atm lit K}^{-1}\text{mol}^{-1})}[(36 \text{ atm} + 6.33 \text{ atm})(0.8 - 0.037) \text{ lit mol}^{-1}]$$

$$= \frac{42.33 \times 0.763}{0.082} \text{ K}$$

$$= 393.9 \text{ K.}$$

We now summarise what you have learnt in this chapter.

SUMMARY

- The experiments of Regnault, Andrews and Amagat showed that a real gas shows deviation from perfect gas behaviour at high pressures.
- The temperature above which a gas cannot be liquefied by applying pressure is known as critical temperature (T_c). The minimum pressure at which a gas begins to liquefy is called critical pressure and the volume corresponding to critical values of pressure and temperature is known as critical volume (V_c). The quantity $RT_c / p_c V_c$ is referred to as critical coefficient of the gas.
- Onnes equation is empirical and is heuristic rather than having any physical basis:

  $$pV = A + Bp + Cp^2 + \ldots,$$

  where A, B, C,... are known as virial constants.
- van der Waals' equation of state for one mole of a real gas is

  $$\left(p + \frac{a}{V^2}\right)(V - b) = RT$$

- The term (a/V^2) arises due to intermolecular interactions and is called cohesive pressure. b is known as co-volume and has its genesis in the finite size of the gas molecules.
- Critical constants of a van der Waals' gas are given by

  $$V_c = 3b, p_c = a / 27b^2 \text{ and } T_c = 8a / 27Rb$$

- The reduced equation of state is given by

  $$\left(p_r + \frac{3}{V_r^2}\right)\left(V_r - \frac{1}{3}\right) = \frac{8}{3} T_r$$

- Saha and Bose equation of state predicts a value 3.53 for the critical constant and gives better agreement with observed behaviour for simpler gases:

$$p = -\frac{RT}{2b} e^{\frac{a}{RTV}} \ln\left(\frac{V-2b}{V}\right)$$

- Redlich-Kwong equation of state is particularly good at high pressures. Like van der Waals' equation, it is also two-constant equation:

$$p = \frac{RT}{V-b} - \frac{a}{V(V-b)T^{1/2}}$$

- Joule–Thomson coefficient is defined as

$$\mu = \frac{\Delta T}{\Delta p} = \frac{1}{C_p}\left(\frac{2a}{RT} - b\right) = \frac{b}{C_p}\left(\frac{T_i}{T} - 1\right)$$

where $T_i = \frac{2a}{Rb}$ is temperature of inversion of the gas. Every gas undergoing Joule–Thomson expansion at a temperature below the inversion temperature shows cooling and vice versa.

EXERCISES

3.1 The critical temperature and pressure of argon are – 122°C and 48 atm, respectively. Calculate the radius of an argon atom. (*Ans:* 1.37 Å)

3.2 The second virial coefficient of an imperfect gas is 0.02 litre mol^{-1}. Calculate the volume of a gram mole of the gas at 27°C and 5 atm pressure. (*Ans:* 4.94 litre)

3.3 A gaseous hydrocarbon obeys Dieterici's equation. Its critical temperature and volume are 33°C and 0.142 litre mol^{-1}, respectively. If 10 moles of this gas have a volume of 4.86 litre at 27°C, calculate its pressure. (*Ans:* 32.65 atm)

3.4 Show that van der Waals' equation is consistent with the statement that all gases approach ideal gas behaviour at low pressures.

3.5 Using the reduced van der Waals' equation of state, find how many times the gas temperature exceeds its critical value if the volume of the gas is equal to half the critical volume. (*Ans:* 1.5)

3.6 Assuming that water vapour obeys van der Waals' equation, calculate (a) the maximum volume which 1 kg water can occupy in liquid state; and (b) the maximum pressure of the saturated water vapour. It is given that $a = 5.47$ atm $litre^2$ mol^{-2}, and $b = 0.03$ mol^{-1}. (*Ans:* $V_{max} = 5$ litre, $p_{max} = 230$ atm)

3.7 What fraction of the volume of a vessel must liquid ether occupy at room temperature in order to pass into critical state when critical temperature is reached? For ether $T_c = 467$K, $p_c = 35.5$ atm, density = 720 kg m^{-3} and its molecular weight is 74. (*Ans:* 0.25)

3.8 Calculate the drop in temperature when carbon dioxide gas suffers Joule-Thomson expansion at 27°C. The pressures on the two sides of the porous plug are 50 atm and 1 atm, respectively. The van der Waals' constants of the gas are $a = 36.56 \times 10^{-2}$ Nm^4mol^{-2}; $b = 4.28 \times 10^{-5}$ m^3 mol^{-1}. Take $C_p = 36.575$ J K^{-1} mol^{-1} and $R = 8.3$ J K^{-1} mol^{-1}. (*Ans:* 34.14 K)

3.9 Calculate the drop in temperature produced in adiabatic throttling of oxygen when the pressure is reduced by 49 atm. The initial temperature of the gas is 27°C. Given that $R = 8.31$ kJ (kg mol)$^{-1}$ K^{-1}, $a = 132 \times 10^3$ N – m^4 (kgmol)$^{-2}$, $b = 0.0318$m^3 (kgmol)$^{-1}$, and $C_p = 7 \times 10^3$ cal (kgmol)$^{-1}$ K^{-1}. (*Ans:* 12.56 K)

3.10 Calculate the Joule–Thomson coefficient for oxygen, if $a = 1.36 \times 10^6$ atm cm^{6}mol^{-2}, and $b = 32.0$ cm^{3}mol^{-1} at S.T.P. and $C_p = 7.03$ cal K^{-1}mol^{-1}. (*Ans:* 0.31K atm^{-1})

3.11 Helium, is assumed to behave as a van der Waals' gas. It is throttled at 20 K and 150 K. Calculate Joule–Thomson coefficient. Given $C_p = 5R/2$, $b = 0.0234$ m^3 (kmol)$^{-1}$ and $a = 3.44 \times 10^3$ Jm^3 (kmol)$^{-2}$. If 1 kmol of helium gas at 20 K occupies a volume 0.12 m^3, calculate the change in temperature if the final pressure is 1 atm in each case. (*Ans:* –0.088 K atm^{-1}, 0.087 K atm^{-1}, 1.12 K, –1.12 K)

3.12 At STP, argon obeys the equation

$$pV = RT + B(T)p$$

with $B(T) = -22$ cm^{-3} mol^{-1} and $\dfrac{dB}{dT} = 0.25$ cm^3 K^{-1} mol^{-1}. Calculate the value of μ. Take $C_p = 20.38 \times 10^7$ erg K^{-1} mol^{-1}. (*Ans:* 0.44 K atm^{-1})

4 BASIC CONCEPTS OF THERMODYNAMICS

Learning Objectives

In this chapter, you will learn how to

- identify thermodynamic systems with their surroundings and boundaries;
- define thermodynamic variables and characterise different systems with appropriate variables;
- distinguish between reversible and irreversible processes;
- explain the term thermodynamic equilibrium;
- discuss the zeroth law and explain how it helped to define temperature;
- apply the equation of state to physical systems and obtain expressions for volume expansivity and compressibility;
- discuss the principle of measurement of temperature with different devices; and
- describe thermodynamic scale of temperature.

4.1 INTRODUCTION

In the preceding three chapters, we discussed the behaviour of gases using kinetic theory. A complementary approach to kinetic theory is thermodynamics where one deals with (directly observable) macroscopic properties of matter without making any reference to their microscopic structure. For example, in kinetic theory, we interpret the temperature of a gas in terms of the random and chaotic motion of its molecules. But in thermodynamics, temperature is a property of the gas based only on direct observation of the system as a whole.

Thermodynamics deals with transformation of thermal energy into other forms of energy and vice versa. This subject evolved out of the search for an efficient engine during industrialisation era in Europe. But today, it finds wide applications in all facets of physical sciences and beyond, from combustion to power plants, refrigeration and rocketry. Even at home, we have several appliances/gadgets in which thermodynamic considerations play an important role. It helps in understanding the evolution of a natural process on one hand and the behaviour of radiation in a cavity and low temperature properties of materials on the other. It assumes great significance in the study of chemical reactions.

Like classical mechanics, thermodynamics is also based on empirical laws—there is no way to prove them—and is therefore phenomenological. Nevertheless, it is very exact and powerful. Each of these laws introduces a new concept (like temperature, internal energy, entropy) which gives a definite meaning to physically measurable quantities and provides useful correlation between observable properties of matter. For this, we need to develop a mathematical formulation of thermodynamics. The purpose of this chapter is to introduce the basic concepts that will help in this development.

In Sec. 4.2, we have introduced the basic concepts of a thermodynamic system, surroundings and boundaries. This is followed by a discussion of the state of a system and thermodynamic variables. Here you will learn about intensive and extensive variables. The concept of thermodynamic equilibrium, which entails thermal, mechanical and chemical equilibria, has been introduced in Sec. 4.4. Thermodynamic processes—quasistatic and non-quasistatic are discussed in Sec. 4.5. The zeroth law of thermodynamics, which helps us to introduce the concept of thermodynamic temperature and obtain the equation of state is discussed in Sec. 4.6. This is followed by the applications of the equation of state to basic thermodynamic processes. In Sec. 4.8, we have discussed the principle of measurement of temperature and scale of temperature. Finally, in Sec. 4.9 we have presented the International Practical Temperature Scale.

4.2 THERMODYNAMIC SYSTEM, SURROUNDINGS AND BOUNDARIES

A *thermodynamic system* refers to a certain portion of universe selected for investigation. (It is imagined to be distinct and set apart from everything else.) This means that a thermodynamic system is essentially macroscopic. In engineering, some important thermodynamic systems are (i) gases such as helium or air (ii) vapour such as steam (iii) mixtures such as air-gasoline vapour or phenol-water mixture, and (iv) vapour in contact with its liquid such as liquid ammonia and vaporised ammonia. In physics, in addition to the aforesaid systems, we consider systems such as stretched wires, electrostatic capacitors, thermocouples, magnetic materials, surface films, electric cells, etc. Obviously, a thermodynamic system will be perceptible to our senses.

A system may be simple or complex; it may be *homogeneous* or *heterogeneous* (where each *component* can exist in a different phase). It will however, always be finite. A gas in a cylinder (fitted with a frictionless leak-proof piston) constitutes a simple homogeneous system, whereas a phenol-water mixture constitutes a more complicated heterogeneous system.

Everything outside the system that can influence its behaviour constitutes the *surroundings* and the envelope which encloses a system and separates it from its surroundings is called the *boundary* of the system (Fig. 4.1). The boundary of a system may be real, as in the case of a gas enclosed in a cylinder, or imaginary, as for a gas flowing through a zig-zag pipe. However, their existence is important to visualise the system distinctly. The boundary of a system may or may not allow it to interact with the surroundings. A boundary which prevents *any* exchange of energy and matter between the system and the surroundings is termed an *isolating boundary*, and the system is then said to be an *isolated system*. Such systems are not

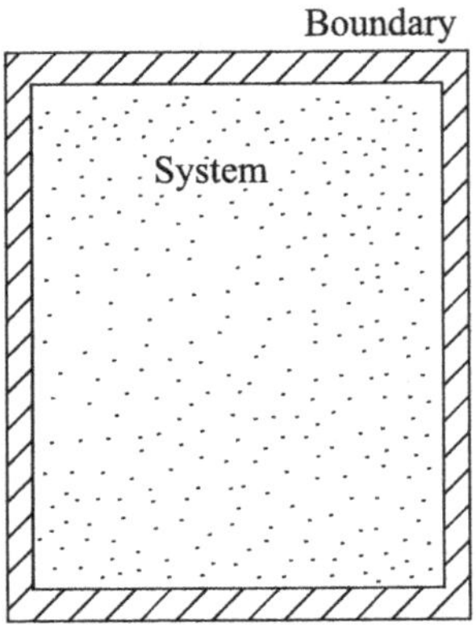

Fig. 4.1 The gas in the enclosure is a thermodynamic system.

affected by their surroundings and will not be of much interest from the thermodynamic viewpoint. If the surface of a system permits exchange of matter between the system and its surroundings, the system is said to be an *open system*. On the other hand, if exchange of energy only and no matter occurs, the system is said to be a *closed system*. (However, it is not an isolated system.) The exchange of energy can take place either thermally or *by doing work on a system*. For instance, heating a system belongs to the former category, whereas compression and magnetisation fall in the latter category. A surface which prevents thermal interaction is known as *adiabatic* and the system is said to be *thermally isolated*. On the other hand, if exchange of heat occurs through a boundary, it is called *diathermic*. Obviously, a system with diathermic boundary will be in *thermal contact* with its surroundings.

4.3 State of a System and Thermodynamic Variables

To describe the motion of a mechanical system, we first write down its equation of motion and then solve it. The solution gives us information about the time variation of the position vector associated with the system. That is, the instantaneous state of a mechanical system is represented by its trajectory. Similarly in thermodynamics, the *state* of a system at any instant represents its condition at that time. It is specified by a set of experimentally measurable quantities, called *variables of state* or *thermodynamic variables*, such as temperature, pressure, volume, electric field, surface area, etc. An equilibrium state is a particularly simple condition of a system in which variables characterising it do not change with time and are reproducible.

Thermodynamic variables may be divided into two categories;* *intensive* and *extensive*. Let y be a macroscopic parameter specifying the state of a homogeneous system. Suppose that the system is divided into two parts, say by introducing a partition. Let y_1 and y_2 be the values of this parameter for two subsystems. Then y is said to be extensive if $y = y_1 + y_2$ and intensive if $y = y_1 = y_2$. Thus, those variables of a system in a given state that are independent of its mass or the number of particles are called *intensive variables*. Some systems of interest are listed in Table 4.1. We know that pressure, surface tension, electric field and temperature are some of the intensive parameters. If f and g are arbitrary intensive variables, then fg, f/g, $\partial f/\partial g$ and $f + g$ will also be intensive variables.

The variables that are proportional to the mass and to the number of particles in the system belong to the category of extensive variables. The extensive variables corresponding to above mentioned intensive variables are also listed in Table 4.1. If F and G are two arbitrary extensive variables, then $F + G$ will be an extensive variable. But F/G and $\partial F/\partial G$ will be intensive. Similarly, we find that if f is an intensive variable and F is an extensive variable, then fF, F/f and $\partial F/\partial f$ will be extensive variables. For extensive variables, one often refers to their *specific values*, i.e., the value per unit mass of the system. They are then called *specific variables*. It is important to note that a specific variable corresponding to any extensive variable will be independent of mass, and hence be an intensive variable.

---

*Sometimes, thermodynamic variables are also categorised as *external* and *internal*. If a variable is determined by the position of external bodies, which are not a part of our system, it is said to be external. Familiar examples are volume and intensity of a field. On the other hand, when a variable is determined by the combined motion and distribution of particles of the system under consideration, it is said to be internal. Examples of internal variables include density, pressure, energy, polarisation, etc.

Table 4.1 *Intensive and extensive variables*

System	Intensive variable	Extensive variable
Hydrostatic system	Pressure, p	Volume⁺, V
Paramagnetic substance	Flux density, B	Magnetic dipole moment, m
Electric cell	Potential difference, V_d	Charge, Z
Surface film	Surface tension, σ	Area, A
Stretched wire	Tension, F	Length, L
Dielectric solid	Electric field, E	Electric dipole moment, p_e

Usually, extensive variables are designated by capital letters and their specific values by corresponding small letters. For instance, the total volume of a system is represented by V and the specific volume by v.

4.4 THERMODYNAMIC EQUILIBRIUM

The state of a system undergoes a change* when it involves work-like and/or thermal interactions with its surroundings. After a period of time, a state is reached when all observable changes in its measurable properties cease to occur as long as external conditions remain unaltered. This state is known as *equilibrium* state of the combined configuration and the system is said to be in *thermodynamic* equilibrium. Since a thermodynamic system may be hydrostatic, physical or chemical in general, we define *thermodynamic equilibrium as the state achieved by a system when it is in mechanical, thermal and chemical equilibria with its surroundings. We define mechanical equilibrium as that state of a system in which it experiences no pressure or elastic stresses and there is no unbalanced force between the system and its surroundings.* A gas in a cylinder fitted with a piston is said to be in mechanical equilibrium, if there is no unbalanced force on the piston. If pressure is not uniform throughout the system, turbulent motion, expansion and/or contraction will be produced. Similarly, if the forces between the system and the surroundings are not balanced, the system would experience mass drift. All these are non-equilibrium states and a lot of good physics is involved in their study. However, at this stage, these are of little interest to us.

To understand the concept of thermodynamic equilibrium further, let us consider two systems in thermal contact through a diathermic wall. One observes that changes occur in their properties, such as volume and pressure. Soon these changes cease and the two systems are said to be in *thermal equilibrium.* One can, therefore, define thermal equilibrium as the *state attained by two (or more) systems placed in thermal contact through a diathermic wall.* (As we show later, all systems in thermal equilibrium will be characterised by the same temperature.) If a hot body is dropped into a liquid in a calorimeter, after sometime the calorimeter and its contents attain an intermediate temperature. The system is then said to be in thermal equilibrium. You will note that in this case work-like interactions are not allowed.

⁺Note that the product of intensive and extensive variables has the dimensions of energy. For example, the dimensions of pV are $[MLT^{-2}L^{-2}L^{3}] = [ML^{2}T^{-2}]$.

*Here the term change of state does not signify a process such as transition from solid to liquid or liquid to vapour. For these, the term used in thermodynamics is *change of phase*.

If a system is such that it neither shows a tendency to undergo a spontaneous change in its internal structure (such as during a chemical reaction) *nor allows transfer of matter from one portion of it to another* (such as in diffusion), *it is said to be in chemical equilibrium.* For example, a mixture of caustic soda and hydrochloric acid will attain chemical equilibrium only after the reaction forming common salt and water is complete.

You will note that mechanical equilibrium refers to uniformity of pressure, thermal equilibrium to constancy of temperature and chemical equilibrium to uniformity of chemical composition. If any of these conditions is not satisfied, the system will not be in thermodynamic equilibrium.

4.5 THERMODYNAMIC PROCESSES

When the values of thermodynamic variables associated with a system change from one equilibrium state to another, the system is said to undergo a *thermodynamic process.*

In order to analyse a thermodynamic process, it is necessary to know the variation of one thermodynamic variable with respect to another. Such a plot is known as *indicator diagram* (Fig. 4.2). The state of a system is uniquely determined by a point on this diagram and a process just means a line joining a series of such points. For example, the initial and final states of a system undergoing expansion are defined by points (p_1, V_1) and (p_2, V_2), respectively. It may be emphasised here that the process of joining the initial and final points with a line has an important implication; the intermediate states also signify equilibrium states. Of course, this also imposes some restrictions on the evolution of a thermodynamic process. Let us think of the following situation. You walk from your home to your college and are asked to retrace your path back home. Then you will have to go by the same route. Similarly, a thermodynamic process can be made to retrace its original path from the final to the initial state, provided intermediate states are exactly defined, i.e., they are in equilibrium states. Such a process has to be executed very slowly so that we can retrace every possible state of the forward process (without affecting the system and the surroundings) and is said to be *reversible.* If we cannot retrace every possible state of the forward process, then it is said to be *irreversible.* All natural processes are irreversible.

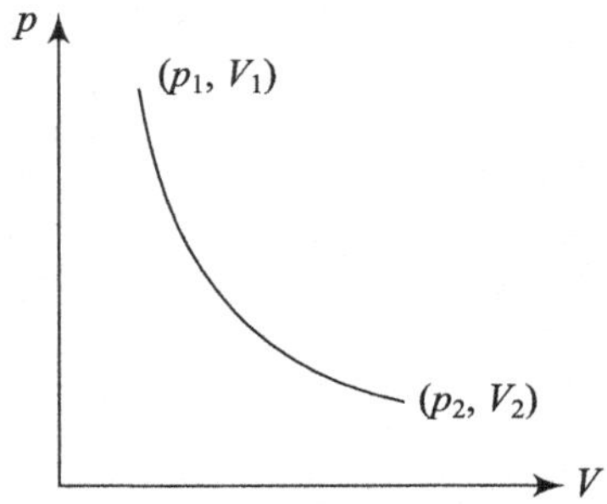

Fig. 4.2 An indicator diagram.

Example 4.1 A reversible process is an ideal process and can never be achieved in practise. Explain.

Solution: Consider a gas contained in a cylinder fitted with an air-tight frictionless piston (Fig. 4.3a). We wish to change the volume of the gas from V to $2V$. This can be done by a rapid movement of the piston from A to B. Then the process will be irreversible as all intermediate states are non-equilibrium states. An alternative way can be to have a piston which remains at A when a weight W is placed on its top, so that if W is removed, the piston moves to B (Fig. 4.3b). But, even then, the process will be irreversible. We can think of placing two weights (each equal to $W/2$) instead of one weight, as shown in Fig. 4.3c. On removing one weight we will get an intermediate equilibrium state, say C.

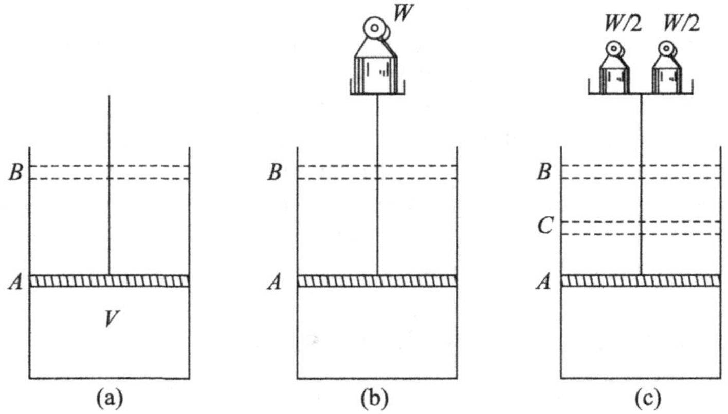

Fig. 4.3 Schematic representation of thought experiment to demonstrate that a reversible process is an idealised concept.

So if we use n weights (each equal to W/n), we can realise $(n-1)$ intermediate equilibrium states between A and B. Thus, if n is made very large (infinite), we can have equilibrium states which do not differ much from each other. This obviously is not practicable implying that the reversible process is an ideal process. But it provides the norm against which we can judge all other real processes.

The concept of reversible process facilitates mathematical formulation of thermodynamics. But the question arises: How can we analyse a natural process within the framework of such a formulation? The problem is overcome by defining what we call a *quasistatic process*. When a process is carried out extremely slowly so that every state through which the system passes departs only infinitesimally from equilibrium, it is said to be *quasistatic*. This means that a quasi-static process corresponds to *a succession of thermodynamic equilibrium states*. If a system undergoes finite departures from equilibrium, the process is said to be *non-quasistatic*. For this case, system variables do not define the states through which the system passes nor can they describe the processes it undergoes. Therefore, a non-quasi-static process cannot be represented by a line on an indicator diagram.

Many a time, the series of processes undergone by a system may revert it back to its initial state. Such a series is said to constitute a *cyclic process*. A cyclic process is represented by a closed path on the indicator diagram (Fig. 4.4).

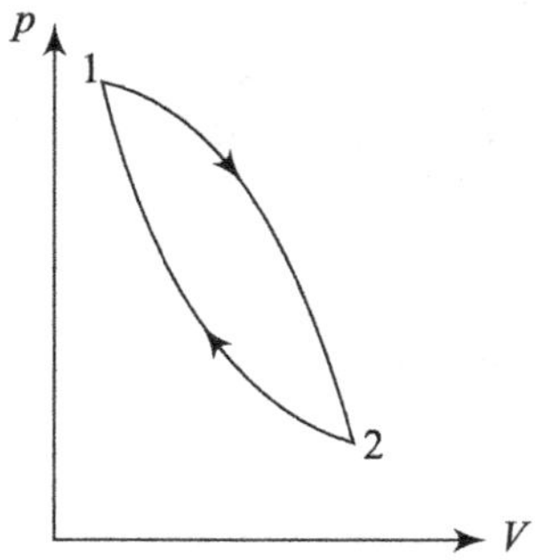

Fig. 4.4 A cyclic process on the indicator diagram.

Many processes are characterised by the fact that a thermodynamic co-ordinate of a system remains constant throughout. A process in which the volume of a system remains constant is called *isochoric*. On the indicator diagram this process will be represented by a straight line parallel to the *p-axis* (Fig. 4.5a). A process in which pressure remains constant

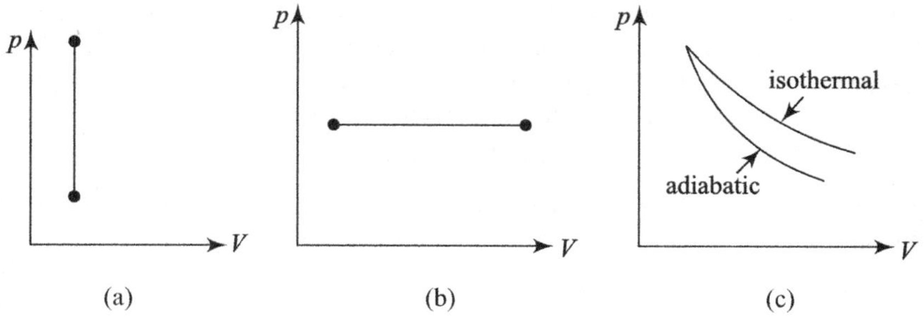

Fig. 4.5 Representation of a process on the indicator diagram: (a) isochoric process, (b) isobaric process and (c) isothermal and adiabatic processes.

is called *isobaric*. Such a process will be represented by a line parallel to the V-axis (Fig. 4.5b). Similarly, a process in which no thermal interaction takes places between a system and its surroundings is said to be *adiabatic*, whereas a process taking place at constant temperature is called *isothermal*. The adiabatic and isothermal processes are depicted in Fig. 4.5c on a p-V diagram.

4.6 THE ZEROTH LAW AND THE CONCEPT OF TEMPERATURE

The zeroth law of thermodynamics*, enunciated by Fowler, deals with systems in thermal equilibrium. This law enables us to give a precise meaning to *temperature*. It can be stated as follows:

If two systems are separately in thermal equilibrium with a third system, they will also be in thermal equilibrium with one another.

To illustrate the experimental basis of this law, we consider two systems 1 and 2, each of which consists of a fluid in a cylinder, separated by adiabatic walls and a mercury thermometer. If the thermometer is placed in thermal contact with them and both exhibit thermal equilibrium simultaneously or successively, then the zeroth law implies that there will be no change in 1 and / or 2 when they are placed in thermal contact, i.e., when the adiabatic wall between them is replaced by a diathermic wall. The state of these systems may be specified by associating a parameter which we call *temperature*. Thus, thermodynamically speaking, *temperature of a system is a property that determines whether or not a system is in thermal equilibrium with other systems.* Obviously, two bodies in thermal equilibrium will be at the same temperature.

We can demonstrate the existence of this concept in a more elegant fashion. To this end, as before, consider three isolated systems 1, 2 and 3 each of which is in equilibrium individually. Let pressure and volume be used as independent variables to describe their equilibrium states. We choose system 3, defined by p_3 and V_3, as the reference system. Then, the condition for thermal equilibrium between systems 1 and 3 may be expressed in the functional form as

$$f_{13}(p_1, V_1, p_3, V_3) = 0 \tag{4.1}$$

*This law was formulated after the first and the second laws had been enunciated. That explains the nomenclature zeroth law.

This equation implies that when two systems are in thermal equilibrium, there is a definite relationship between the four variables and once p_3, V_3 and p_1 are fixed, V_1 will be uniquely determined. That is, out of the four variables, only three are independent. Similarly, for thermal equilibrium between systems 2 and 3, we can write

$$f_{23}(p_2, V_2, p_3, V_3) = 0 \tag{4.2}$$

Equations (4.1) and (4.2) can, in principle, be solved for, say, p_3 and the solutions can be expressed as

$$p_3 = F_{13}(p_1, V_1, V_3) \tag{4.3}$$

and

$$p_3 = F_{23}(p_2, V_2, V_3) \tag{4.4}$$

On equating these values of p_3, we get

$$F_{13}(p_1, V_1, V_3) = F_{23}(p_2, V_2, V_3) \tag{4.5}$$

This may be solved for p_1 to give

$$p_1 = G\,(V_1, p_2, V_2, V_3) \tag{4.6}$$

But according to the Zeroth law, if systems 1 and 2 are separately in thermal equilibrium with system 3, they will also be in thermal equilibrium with one another. This requires

$$f_{12}(p_1, V_1, p_2, V_2) = 0 \tag{4.7}$$

Again, we can solve this for p_1 to obtain

$$p_1 = F_{12}\,(V_1, p_2, V_2) \tag{4.8}$$

This equation states that under given conditions, p_1 is uniquely determined by V_1, p_2 and V_2 so that V_3 must drop out in Eq. (4.6) or its earlier form, Eq. (4.5). For this to happen, the functions F_{13} and F_{23} must be of the general form

$$F_{13} = \phi_1\,(p_1, V_1)\,\eta\,(V_3) + \xi(V_3) \tag{4.9}$$

and

$$F_{23} = \phi_2\,(p_2, V_2)\,\eta\,(V_3) + \xi(V_3) \tag{4.10}$$

This implies that Eq. (4.5) should, in fact, be of the form

$$\phi_1\,(p_1, V_1) = \phi_2\,(p_2, V_2) \tag{4.11}$$

This is the condition for thermal equilibrium between systems 1 and 2. In general, we can write

$$\phi_1\,(p_1, V_1) = \phi_2\,(p_2, V_2) = \phi_3\,(p_3, V_3) \tag{4.12}$$

This equation shows that when two or more systems are in thermal equilibrium, for each system, state variables p and V define a single valued function which has a common numerical value for all of them. The common value of this function is known as *empirical temperature**. Thus, for thermal equilibrium between the system of interest and the reference system, we may write

$$\phi\,(p, V) = \theta \tag{4.13}$$

---

*You will come across another nomenclature like *absolute* or *thermodynamic temperature* which denotes temperature on absolute scale. As we show in Chapter 6, measurement of absolute temperature does not depend on thermometric property of any particular material.

This equation may be rearranged in the more general form as

$$\phi(p, V, \theta) = 0 \tag{4.14}$$

This equation is called the *equation of state*. Note that ϕ is a single valued function of pressure and volume at temperature θ.

Note that Eq. (4.14) is parametric*. Also, it shows that θ or the absolute temperature T is another thermodynamic variable to express the state of a system. Further, it also emphasises that out of the three variables p, V and T, only two are independent and any two of these completely specify the state of a system. The equations of state in parametric as well as their exact forms are given in Table 4.2 for some systems of interest.

Table 4.2 *Equations of state for some physical systems*

System	Equation of state	
	Parametric form	**Exact**
Hydrostatic	$f(p, V, T) = 0$	Ideal gas $pV = nRT$ real gas $\left(p + \frac{n^2 a}{V^2}\right)(V - nb) = nRT$
Paramagnetic substance	$f(M, B, T) = 0$	$M = \frac{kB}{T}$
Electric cell	$f(V_d, T, Z) = 0$	$V_d = c_0 + c_1 T + c_2 T^2 + c_3 T^3$
Surface film	$f(\sigma, T, A) = 0$	$\sigma = \sigma_0 \left(\frac{T_c - T}{T_c - T_0}\right)$
Stretched wire	$f(L, T, F) = 0$	$L = L_0 \left(1 + \frac{F}{YA} + \alpha(T - T_0)\right)$
Dielectric solid	$f(p, E, T) = 0$	$p = \left(a + \frac{b}{T}\right)E$

It may be pointed out here that we can feel the effect of temperature through our senses but these can not quantify it in absolute numbers. It means that to be able to assign a numerical value, we must make use of some thermometric property, i.e., use a substance which exhibits temperature dependence. For example, the clinical thermometer used to measure body temperature utilises the temperature dependence of volume of a liquid (mercury). However, you may be interested in measuring temperature in a wide range—from as low as 0.01 K to as high as 6000 K or beyond. The principle of measurement of temperature is discussed in Sec. 4.8.

4.7 Some Deductions from The Equation of State

As mentioned earlier, we can use the equation of state in the parametric form to study the behaviour of any substance under different conditions. To show this, we first consider a hydrostatic system and note that Eq. (4.14) may be solved for, say, p**

*In fact, it is not possible to obtain a closed mathematical form of the equation of state on the basis of thermodynamics. However, one can use this equation to study the behaviour in various processes.

**We have replaced θ by T in writing this expression.

$$p = p\,(V, T) \tag{4.15}$$

An infinitesimal change in p can be expressed in terms of infinitesimal changes in V and T as

$$\mathrm{d}\,p = \left(\frac{\partial p}{\partial V}\right)_T \mathrm{d}V + \left(\frac{\partial p}{\partial T}\right)_V \mathrm{d}T \tag{4.16}$$

where $(\partial p/\partial V)_T$ expresses the rate of change of pressure with volume when temperature is kept constant (an isothermal process) and $(\partial p/\partial T)_V$ expresses the rate of change of pressure with temperature when volume is kept constant (an isochoric process). For an isobaric process, $dp = 0$ and Eq. (4.16) takes the form

$$\left(\frac{\partial p}{\partial T}\right)_V = -\left(\frac{\partial p}{\partial V}\right)_T \left(\frac{\partial V}{\partial T}\right)_p \tag{4.17}$$

To give a physical meaning to Eq. (4.16), we note that the *coefficient of volume expansion* or *expansivity*, α, of a material, in an isobaric process is defined as

$$\alpha = \frac{1}{V}\left(\frac{\partial V}{\partial T}\right)_p \tag{4.18}$$

and isothermal elasticity E_T, which is inverse of isothermal compressibility, β_T, of a material is defined as

$$E_T = -\,V\,(\partial p\,/\,\partial V)_T \tag{4.19}$$

Since volume decreases as pressure increases at constant temperature, $(\partial p/\partial V)_T$ is always a negative quantity. Therefore, a negative sign has been included in the definition of E_T so as to keep this quantity positive and physically meaningful.

On combining Eqs. (4.17), (4.18) and (4.19), we get

$$\left(\frac{\partial p}{\partial T}\right)_V = \frac{\alpha V E_T}{V} = \alpha\, E_T \tag{4.20}$$

This equation implies that for a given change of temperature, the increase in pressure will be more when expansivity as well as elasticity are large (compressibility is small).

On using this result in Eq. (4.16), we can write

$$\mathrm{d}\,p = -\frac{E_T}{V}\mathrm{d}V + \alpha\, E_T\, \mathrm{d}T \tag{4.21}$$

This result shows that if a system is allowed to expand, change in pressure will be less than that for an isochoric process.

For an isochoric process, this equation reduces to

$$\mathrm{d}p = \alpha E_T\,\mathrm{d}T \tag{4.22}$$

Thus, for a finite change in temperature, the change in pressure is given by

$$\int_{p_1}^{p_2} \mathrm{d}\,p = p_2 - p_1 = \int_{T_1}^{T_2} E_T\, \alpha \mathrm{d}T \tag{4.23}$$

If α and E_T are assumed to be constant, independent of temperature in the small range considered here, we get

$$p_2 - p_1 = \alpha E_T (T_2 - T_1) \tag{4.24}$$

This equation shows that if α and E_T are known experimentally, then for an isochoric process we can calculate the final pressure for a given rise in temperature and vice versa. This is illustrated in the following examples. You should go through these carefully.

Example 4.2 A block of copper whose expansivity is 48.0×10^{-6} K^{-1} and isothermal elasticity is 1.30×10^{11} N m^{-2} is at atmospheric pressure and a temperature of 0°C. Its temperature is raised to 10°C. Calculate the final pressure when (a) volume is kept constant and (b) volume changes from 1000 c.c. to 1000.1 c.c.

Solution: **(a)** For an isochoric process, we have from Eq. (4.24)

$$p_2 - p_1 = \alpha E_T (T_2 - T_1)$$

On substituting the values of α, E_T and $(T_2 - T_1)$, we get

$$p_2 - p_1 = (48.0 \times 10^{-6}\ \text{K}^{-1}) \times (1.30 \times 10^{11}\ \text{N m}^{-2}) \times 10\text{K}$$

$$= 624 \times 10^5\ \text{Nm}^{-2}$$

$$= 624\ \text{atm}$$

so that final pressure p_2 is

$$p_2 = (624 + 1)\ \text{atm} = 625\ \text{atm}$$

That is, to keep the volume of the copper block constant when its temperature is raised from 0°C to 10°C, one must increase the pressure to 625 atm.

(b) To calculate the pressure when the volume of the block is allowed to increase, we use Eq. (4.21). This gives

$$dp = -\frac{E_T}{V} dV + \alpha E_T\, dT$$

As before, on substituting the values of various quantities, we get

$$dp = p_2 - p_1 = -\frac{(1.30 \times 10^{11}\ \text{Nm}^{-2}) \times (0.10 \times 10^{-6}\ \text{m}^3)}{(10^3 \times 10^{-6}\ \text{m}^3)} + (48.0 \times 10^{-6}\ \text{K}^{-1}) \times (1.30 \times 10^{11}\ \text{N m}^{-2}) \times 10\ \text{K}$$

$$= (-130 \times 10^5 + 624 \times 10^5)\ \text{N m}^{-2}$$

$$= 494 \times 10^5\ \text{N m}^{-2}$$

$$= 494\ \text{atm}$$

Hence,

$$p_2 = (494 + 1)\ \text{atm} = 495\ \text{atm}$$

As will be noted, when the specimen is allowed to expand, the final pressure is much less

Example 4.3 Mercury is heated at constant volume. Calculate the rise in temperature necessary to produce a pressure of 4000 atm. Take isothermal compressibility $\beta_T = 3.5 \times 10^{-12} \text{m}^2\text{N}^{-1}$ and coefficient of cubical expansion, $\alpha = 17.5 \times 10^{-5}\text{K}^{-1}$.

Solution: We know for an isochoric process $dV = 0$, so that we can write

$$p_2 - p_1 = \alpha E_T (T_2 - T_1)$$

Since isothermal elasticity

$$E_T = \frac{1}{\text{isothermal compressibility, } \beta_T}$$

$$= \frac{1}{35 \times 10^{-12} \text{ m}^2 \text{N}^{-1}} = 0.0286 \times 10^{12} \text{ Nm}^{-2} = 2.86 \times 10^{10} \text{ Nm}^{-2}$$

$$\approx 2.86 \times 10^5 \text{ atm} \qquad (\because 1 \text{ atm} \approx 10^5 \text{ Nm}^{-2})$$

Hence, change in temperature

$$T_2 - T_1 = \frac{p_2 - p_1}{\alpha E_T} = \frac{4000 \text{ atm}}{(17.5 \times 10^{-5} \text{ K}^{-1}) \times (2.86 \times 10^5 \text{ atm})}$$

$$= \frac{4 \times 10^3}{17.5 \times 2.86} \text{K}$$

$$= 79.92 \text{ K}$$

Since $T_2 - T_1 > 0$, we can say that temperature rises by 79.92 K.

Example 4.4 Calculate the final pressure if at constant volume, 50 g of mercury is heated at 1 atm and 0°C slowly to 24°C. Use the data given in Example 4.3.

Solution: For an isochoric process, we have

$$p_2 - p_1 = \alpha E_T (T_2 - T_1)$$

Here $\alpha = 17.5 \times 10^{-5}\text{K}^{-1}$, $T_2 - T_1 = 24 - 0 = 24°C = 24$ K. Hence,

$$p_2 - p_1 = (17.5 \times 10^{-5}\text{K}^{-1}) \times (0.286 \times 10^6 \text{ atm}) \times (24 \text{ K})$$

$$= 1194 \text{ atm}$$

$\therefore$ Final pressure $p_2 = 1194 + 1 = 1195$ atm.

You may now like to answer a few practise problems.

Problem 4.1 A given mass of mercury at atmospheric pressure and 0°C is heated to 10°C. Calculate the pressure coefficient and final pressure necessary to keep the volume of mercury constant. Given $\alpha = 1.81 \times 10^{-4}\ °\text{C}^{-1}$ and $\beta_T = 3.87 \times 10^{-6} \text{ atm}^{-1}$.

Ans: 46.8 K^{-1}, 469 atm.

Problem 4.2 Calculate the coefficient of volume expansion and isothermal compressibility of a van der Waals' gas.

Ans: $\alpha = \dfrac{RV^2(V-b)}{RTV^3 - 2a(V-b)^2}$; $\beta_T = \dfrac{V^2(V-b)^2}{RTV^3 - 2a(V-b)^2}$

In the above analysis, we studied how we can use the equation of state to calculate the change in pressure when temperature of a system is raised during an isochoric process. To do so, we considered volume and temperature as independent variables and pressure was uniquely fixed. One may as well take pressure and temperature as independent variables and solve Eq. (4.14) for V. Then we can express the functional dependence of V on p and T as

$$V = V(p, T) \tag{4.25}$$

As before, by partial differentiation we can relate an infinitesimal change in V with infinitesimal changes in p and T as

$$\mathrm{d}V = \left(\frac{\partial V}{\partial p}\right)_T \mathrm{d}p + \left(\frac{\partial V}{\partial T}\right)_p \mathrm{d}T \tag{4.26}$$

In terms of α and β_T, this equation can be expressed as

$$\frac{\mathrm{d}V}{V} = -\beta_T\,\mathrm{d}p + \alpha\,\mathrm{d}T \tag{4.27}$$

For an isobaric process, $\mathrm{d}p = 0$ and Eq. (4.27) reduces to

$$\mathrm{d}V = \alpha V\,\mathrm{d}T \tag{4.28}$$

For a finite change in temperature, we can write

$$V_2 - V_1 = \alpha V(T_2 - T_1) \tag{4.29}$$

Note that as before, α has been assumed to be independent of temperature.

Example 4.5 A block of metal whose volume expansivity is 5.0×10^{-5} C^{-1} and isothermal compressibility is 1.2×10^{-6} atm^{-1} has volume 5 litre at 1 atm and 20°C. On applying pressure, its temperature rises by 12°C and volume increases by 0.5 cm^3. Calculate the pressure applied.

Solution: An infinitesimal change in pressure dp can be expressed in terms of the volume expansivity, α and the isothermal compressibility β_T using Eq. (4.27) as

$$\mathrm{d}p = \frac{\alpha}{\beta_T}\mathrm{d}T - \frac{1}{\beta_T V}\mathrm{d}V$$

Here $\alpha = 5 \times 10^{-5}$ °C^{-1}, $\beta_T = 1.2 \times 10^{-6}$ atm^{-1}, dT = 12°C, V = 5 litre = 5×10^3 cm^3 and dV = 0.5 cm^3. Therefore,

$$\mathrm{d}p = \frac{(5\times10^{-5}\,{}^\circ\mathrm{C}^{-1})}{(1.2\times10^{-6}\,\mathrm{atm}^{-1})}\times 12 - \frac{1}{(1.2\times10^{-6}\,\mathrm{atm}^{-1})\times(5\times10^3\,\mathrm{cm}^3)}\times(0.5\,\mathrm{cm}^3)$$

$$= (500 - 83.3) \text{ atm}$$
$$= 416.7 \text{ atm}$$

$\therefore$ The pressure applied

$$p_2 = p_1 + dp = 1 + 416.7 = 417.7 = 418 \text{ atm}$$

From Eq. (4.27) we note that once β_T and α are known experimentally, one can calculate the change in volume for a given change in temperature and pressure. Alternatively, one may use this information to arrive at the equation of state for the system of interest. To illustrate this, we assume that isothermal compressibility $\beta_T = a/V$ and expansivity $\alpha = \frac{2bT}{V}$. Then, Eq. (4.27) implies that

$$\frac{dV}{V} = -\frac{a}{V}dp + \frac{2bT}{V}dT$$

or

$$dV + a\,dp - 2bT\,dT = 0$$

This can readily be integrated to obtain

$$V + ap - bT^2 = \text{const} \tag{4.30}$$

In order to give you practise in using the results derived in this section and test your understanding, we would like you to answer a few practise problems.

Problem 4.3 The isothermal compressibility and expansivity of a substance are respectively given by $\beta_T = \frac{aT^3}{p^2}$ and $\alpha = \frac{bT^2}{p}$, where a and b are constant. Derive the equation of state for the system under consideration.

Ans: $V = V_0 \exp(aT^3/p)$

Problem 4.4 The equation of state of a paramagnetic substance is $M = \frac{kB}{T}$. Show that

$$\left(\frac{\partial M}{\partial B}\right)_T \left(\frac{\partial B}{\partial T}\right)_M \left(\frac{\partial T}{\partial M}\right)_B = -1$$

To further illustrate the use of the equation of state, we consider a wire attached to a rigid support. The state of the wire will be defined by the tension, length and the temperature since p as well as V can be assumed to be constant. Thus, we can write

$$F = F(L, T) \tag{4.31}$$

On partial differentiation, we can write

$$dF = \left(\frac{\partial F}{\partial L}\right)_T dL + \left(\frac{\partial F}{\partial T}\right)_L dT \tag{4.32}$$

If tension remains constant, dF = 0 and Eq. (4.32) reduces to

$$\left(\frac{\partial F}{\partial L}\right)_T \left(\frac{\partial L}{\partial T}\right)_F = -\left(\frac{\partial F}{\partial T}\right)_L \tag{4.33}$$

We now introduce isothermal Youngs' modulus defined as

$$Y = \frac{L}{A}\left(\frac{\partial F}{\partial L}\right)_T \tag{4.34}$$

where A is the cross-sectional area of the wire. Similarly, the coefficient of linear expansion is defined as

$$\alpha_\ell = \frac{1}{L}\left(\frac{\partial L}{\partial T}\right)_F \tag{4.35}$$

On combining Eqs. (4.33), (4.34) and (4.35), we get

$$\left(\frac{\partial F}{\partial T}\right)_L = -AY\alpha_\ell \tag{4.36}$$

On using this result in Eq. (4.32), we find that when a wire undergoes an infinitesimal change, the change in tension is given by

$$\mathrm{d}F = \frac{AY}{L}\mathrm{d}L - AY\alpha_\ell\, \mathrm{d}T \tag{4.37}$$

If we consider a wire vibrating between two fixed rigid supports, dL = 0 and Eq. (4.37) reduces to

$$\mathrm{d}F = -A\,Y\,\alpha_\ell\,\mathrm{d}T \tag{4.38}$$

Note that this does not depend on the length of the wire. For a small change in temperature, we get

$$\int_{F_1}^{F_2} \mathrm{d}F = -A\int_{T_1}^{T_2} Y\alpha_\ell\, \mathrm{d}T$$

If Y and α_ℓ are independent of temperature, we can write

$$F_2 - F_1 = AY\,\alpha_\ell\,(T_1 - T_2) \tag{4.39}$$

That is, temperature of a wire vibrating between two fixed supports will drop if tension is increased and vice versa.

You should now go through the following example carefully.

Example 4.6 A student is working in a physics laboratory at a temperature of 20°C on a sonometer. The sonometer wire is of cross-sectional area 0.85×10^{-6} m^2 and a tension of 20 N is applied at the other end. It is stretched between two rigid supports 1.2 m apart. If the temperature is reduced to 8°C, calculate (a) the final tension and (b) frequency of vibration of the wire. The density of the material of the wire is 9.0×10^3 kg m^{-3}. The coefficient of linear expansion and isothermal Young's modulus may be assumed to be constant at the values 1.5×10^{-5} K^{-1} and 2.0×10^{11} Nm^{-2}.

Solution: **(a)** Since the wire is stretched between two rigid supports, $dL = 0$. Therefore, we can use Eq. (4.39):

$$F_2 - F_1 = A\alpha_\ell\, Y(T_1 - T_2)$$

On substituting the values of various quantities, we have

$$F_2 - F_1 = (0.85 \times 10^{-6}\ \text{m}^2) \times (1.5 \times 10^{-5}\ \text{K}^{-1}) \times (2.0 \times 10^{11}\ \text{Nm}^{-2}) \times (12\ \text{K})$$

$$= 31\ \text{N}$$

so that $F_2 = (31 + 20)\ \text{N} = 51\ \text{N}$

(b) The fundamental frequency of vibration of a sonometer wire is given by

$$\nu = \frac{1}{2L}\sqrt{\frac{F}{m}}$$

where m is mass per unit length ($= M / L$). If A is cross-sectional area and d is density of the wire, we can write $m = Ad$ so that on substituting for F, L, A and d, we get

$$\nu = \frac{1}{2 \times (1.2\text{m})}\sqrt{\frac{20\ \text{N}}{(85 \times 10^{-8}\ \text{m}^2) \times (9.0 \times 10^{3}\ \text{kgm}^{-3})}}$$

$$= 21\ \text{Hz}$$

You may now like to answer a practise problem.

Problem 4.5 An iron wire of cross-sectional area 0.01 cm^2 under a tension of 10 N and a temperature of 20°C is stretched between two rigid supports 1 m apart on a sonometer board. If the temperature falls to 0°C calculate (a) the final tension and (b) the frequency of vibration of the wire. The density of the wire is 7.8×10^3 kg m^{-3}. The coefficient of linear expansion and Young's modulus are respectively equal to 12×10^{-6} K^{-1} and 2.0×10^{11} N m^{-2}.

Ans: 58 N, 17.9 Hz

One can similarly use the equation of state to study the behaviour of other thermodynamical systems. However, we will not go into a detailed discussion of all these here.

4.8 Measurement of Temperature

Measuring temperature is a very common experience. The body temperature of a patient is determined by putting a clinical thermometer below the tongue. (For a child, it is inserted in the armpit.) The length of the column of mercury changes and the reading on a scale is an indicator of body temperature. As mentioned earlier, in this process, we make use of the temperature dependence of the volume of mercury. In fact, depending on the range of interest, we make use of some such physical property. We now discuss these.

4.8.1 Principle of Measurement of Temperature

The properties which exhibit smooth and preferably large variation with small change in temperature are length of a liquid column in a capillary, volume of a fluid and pressure of a gas kept in a container of fixed volume. Other less common but more versatile properties include electrical resistance of a metal or a semiconductor, thermo-emf generated in a closed circuit of two dissimilar metals whose junctions are kept at different temperatures, the emissive power of the radiation emitted by a black body, the magnetic susceptibility of a paramagnetic substance and the like. These properties are referred to as *thermometric properties*. The thermometric substance, thermometric property and some common thermometers are listed in Table 4.3.

Table 4.3 *Some common thermometers*

Thermometric Substance	Thermometric property	Thermometer	Principle
Gas	Pressure (p)	Constant volume gas thermometer	$T(p) = 273.16(p/p_{tp})$
Gas	Volume (V)	Constant pressure thermometer	$T(V) = 273.16(V/V_{tp})$
Liquid	Length (L)	Liquid thermometer	$T(L) = 273.16(L/L_{tp})$
Pure metal	Resistance (R)	Resistance thermometer	$T(R) = 273.16(R/R_{tp})$; $R = R_0(1 + \alpha T + \alpha T^2)$
Dissimilar metallic junctions	Thermo-emf (E)	Thermocouple	$E = at + bt^2$
Paramagnetic salt	Magnetic susceptibility	Susceptibility thermometer	$\chi = C/T$
Black body radiation	Emissive power (e_λ)	Radiation pyrometer	$e_\lambda = \sigma T^4$

In addition to expressing a thermometric property as an explicit function of temperature, we must assign a numerical value to temperature of some standard state of the chosen system to be able to construct a temperature scale. For example, in the clinical thermometer, the scale has graduations from 94°F to 108°F. On the other hand, a common laboratory thermometer has graduations from –10°C to 110°C. In both cases, the entire range is divided into a number of equal divisions. Note that selection of the range and its subsequent division are determined by the nature of thermometric substance. As such, these are vital for measurement of temperature. This is referred to as a scale of temperature. We will now discuss it in detail.

4.8.2 Scale of Temperature

We now know that to define a scale of temperature, we select a suitable substance and assign a set of numbers to the values of the thermometric property during its variation according to some well-defined rule. When the thermometer is brought in contact with a body/system and is allowed to attain thermal equilibrium, the thermometric property attains a specific value, which is read in terms of the number assigned to it. This gives a measure of the temperature of the body. If the thermometric property has a linear dependence on temperature, the scale will be relatively simpler.

Suppose we choose to denote the value of thermometric property by X_θ and the temperature corresponding to it by $\theta(X)$. For linear variation between X_θ and $\theta(X)$, we can write

$$X_\theta = a\,\theta(X) + b \tag{4.40}$$

where a and b are constants. To determine these constants, we need to know the values of X_θ at two different fixed points. (A fixed point is an arbitrarily chosen but easily reproducible state of a standard system.) Though this is somewhat arbitrary, the choice fell on water at ice point (0°C) and steam point (100°C). (The ice point and steam point are respectively the temperatures at which pure water freezes and boils at atmospheric pressure at the sea level and 45° latitude.) Then we can rewrite Eq. (4.40) as

$$X_0 = a \times 0 + b = b$$

and $$X_{100} = a \times 100 + b = 100a + X_0$$

This can be rewritten to obtain the value of a:

$$a = \frac{X_{100} - X_0}{100}$$

Hence, we can rewrite Eq. (4.40) as

$$\theta(X) = \frac{X_\theta - b}{a} = \frac{X_\theta - X_0}{X_{100} - X_0} \times 100 \tag{4.41}$$

The scale defined based on Eq. (4.41) is the familiar Celsius scale.

It is important to mention here that the ice point and the steam point are very sensitive to atmospheric pressure, presence of impurities in the liquid, nature of the glass used to hold the liquid and so on. So it invariably became difficult to reproduce identical values of fixed points. To overcome this problem, it was decided that the *triple point of water*—the temperature at which all the three phases, ice, water and water vapour co-exist in equilibrium—be taken as the fixed point. The temperature at the triple point was arbitrarily assigned the numerical value 273.16 and the unit is called kelvin (K). This is the fundamental fixed point of thermometry. On this scale, the ice point is 273.16 K and the steam point is 373.16 K. This scale is called the *absolute scale*.

Let us denote the value of the thermometric property at the triple point by X_{tp}. Then Eq. (4.40) with $b = 0$ reduces to

$$X_{tp} = 273.16a$$

and the temperature $\theta(X)$ can be expressed as

$$\theta(X) = 273.16(X_\theta / X_{tp}) \tag{4.42}$$

This result shows that there is one-to-one correspondence between temperature and a thermodynamic property; all that one has to do is to know the ratio of the observed value of the property to its value at the triple point and multiply it by 273.16. For the specific cases of a constant volume gas thermometer, we have

$$\theta(p) = 273.16\left(\frac{p}{p_{tp}}\right)_V . \tag{4.43}$$

whereas for a resistance thermometer, we can write

$$\theta(R) = 273.16\left(\frac{R}{R_{tp}}\right). \tag{4.44}$$

Similarly, for a liquid thermometer, we can write

$$\theta(L) = 273.16\left(\frac{L}{L_{tp}}\right) \tag{4.45}$$

It is important to realise that this situation is not so simple as it appears because the readings $\theta(R)$, $\theta(p)$, $\theta(L)$ ... of the same temperature given by different thermometers differ from each other. This has genesis in the assumption of linearity between $\theta(X)$ and X. That is, each type of thermometer has its own scale of temperature. In such a situation, it is advisable to choose one type of thermometer as the standard and others may be calibrated in terms of it. Experimental evidence suggests that the variation in X_θ/X_{tp} values is the smallest for a constant volume gas thermometer. It means that a gas can be chosen as the standard thermometric substance.

The *perfect gas temperature scale* is defined by the equation

$$\theta(p) = (273.16\text{ K}) \lim_{p_{tp}\to 0}\left(\frac{p}{p_{tp}}\right)_V \tag{4.46}$$

From Eq. (4.46) we note that $\theta \to 0$ as $p \to 0$. It means that zero on the perfect gas scale is the temperature at which the gas exerts no pressure. However, you will learn later that at temperatures close to zero kelvin, intermolecular interactions cannot be ignored as all gases liquefy. In fact, the absolute zero, as it is called, has neither been attained in practise nor can be defined in terms of ideal gas behaviour.

The dependence on the properties of a substance in defining temperature is an obvious limitation for any temperature scale. Lord Kelvin overcame this problem when he developed what came to be known as *the kelvin scale or the thermodynamic scale of temperature.* He defined the ratio of two temperatures based on the second law of thermodynamics. It is discussed in Chapter 6. You will note that the thermodynamic scale and the perfect gas scale are identical. With this understanding, it is customary to replace the ideal gas symbol θ by thermodynamic scale temperature, T. (The unit of temperature was named after Lord Kelvin.) On the Celsius temperature scale, ice point is chosen as 0°C and is expressed on the kelvin scale through the relation,

$$t(°\text{C}) = T(\text{K}) - 273.15 \tag{4.47}$$

Note that one unit of temperature on Celsius scale is equal to one kelvin.

To give you some idea about the use of different scales, we would like you to go through the following examples.

Example 4.7 The resistance of a platinum wire at θ°C on the ideal gas scale is given by $R_\theta = R_0(1 + \alpha\theta + \beta\theta^2)$, where R_0 denotes resistance at 0°C, $\alpha = 3.5 \times 10^{-3}$ °C^{-1} and $\beta = -3.0 \times 10^{-6}$ °C^{-2}. Calculate the temperature on the resistance scale when the temperature is 50°C.

Solution: If we assume that the ice point and the steam point are the same on both scales, the temperature on the resistance scale can be defined using Eq. (4.41) as

$$\theta_R = \frac{R_\theta - R_0}{R_{100} - R_0} \times 100 \qquad \text{(i)}$$

Using the given equation, we can write

$$R_\theta = R_0\,(1 + 50\,\alpha + 2500\,\beta) \;\Rightarrow\; R_\theta - R_0 = R_0\,(50\,\alpha + 2500\,\beta) \qquad \text{(ii)}$$

and $$R_{100} = R_0\,(1 + 100\,\alpha + 10^4\,\beta) \;\Rightarrow\; R_{100} - R_0 = R_0\,(100\,\alpha + 10^4\,\beta) \qquad \text{(iii)}$$

On substituting the values of $R_\theta - R_0$ and $R_{100} - R_0$ in Eq. (i), we get the temperature on the resistance scale:

$$\theta_R = \frac{(50\,°\text{C})\,\alpha + (2500\,°\text{C}^2)\,\beta}{(100\,°\text{C})\,\alpha + (10^4\,°\text{C}^2)\beta} \times (100\,°\text{C}) = \frac{50 \times 3.5 \times 10^{-3} - 2500 \times 3 \times 10^{-6}}{3.5 \times 10^{-3} - 300 \times 10^{-6}}\,°\text{C}$$

$$= \frac{0.175 - 0.0075}{3.2 \times 10^{-3}}\,°\text{C} = 52.3°\text{C}$$

Example 4.8 The length of the mercury column in a liquid thermometer is 6.00 cm when it is placed in water at its triple point (T = 273.16K). **(a)** Calculate the length of the column at the steam point and **(b)** At what temperature will the length of the column be 7.2 cm?

Solution

(a) By inverting Eq. (4.45), we can write

$$L = \frac{\theta(L)}{273.16} \times L_{tp}$$

At steam point, θ_{steam} = 373.15K. Hence, on substituting the given value of L_{tp}, the length at steam point is given by

$$L = \frac{373.15 \times 6.0}{273.16} = 8.20 \text{ cm}$$

(b) From Eq. (4.45), we recall that

$$\theta(L) = 273.16\left(\frac{L}{L_{tp}}\right)$$

Hence, $$\theta(7.2) = 273.16\left(\frac{7.2}{6.0}\right) = 327.8 \text{ K}$$

4.9 The International Practical Temperature Scale (IPTS)

The international practical temperature scale evolved out of concerns emanating from difficulties and inconveniences faced in calibration of other practical thermometers by an ideal gas thermometer. It was first adopted in 1927 and has been revised subsequently several times. The IPTS was so designed as to approximate thermodynamic temperature as closely as possible without the use of gas thermometry. The essential features of IPTS are:

1. A set of (eleven) fixed points with assigned values of thermodynamic temperature from the most accurate measurements made by a constant volume gas thermometer.
2. Prescriptions stating the practical thermometers to be used for interpolation between fixed points.
3. Interpolation procedures to be used.

The fixed reference points with their assigned temperatures, together with the specified thermometers and interpolation procedures establish an empirical scale—the IPTS— which is arranged to coincide as closely as possible with thermodynamic temperature. The latest revision/amendment was made in 1990 and the current practical temperature scale is called the *International Temperature scale 1990* (ITS-90).

The values of temperatures and the standard thermometers are given in Table 4.4.

Table 4.4 *International temperature scale – 1990*

	Temperature Range	Thermometer
1.	0.65 K to 5 K	Saturation vapour pressure thermometer using
		^{3}He from 0.65 K to 3.2 K
		^{4}He from 2.1768 K to 5K
2.	3.0 K to 24.5561 K	^{3}He or ^{4}He constant volume gas thermometer
3.	13.8033 K to 1234.93 K	Platinum resistance thermometer with specified interpolation procedures for eleven sub-intervals in this range
4.	> 1234.93 K	Optical pyrometer

You should now go through the following example carefully.

Example 4.9 A constant volume gas thermometer contains a gas, which obeys van der Waals' equation of state. Another thermometer of identical construction contains another gas which obeys the ideal equation of state. The thermometers are calibrated at the ice point and the steam point. Will they show identical values of temperature?

Solution: From Eq. (4.41), we recall that on a two point scale, the temperature is defined as

$$\theta(X) = \frac{X_\theta - X_0}{X_{100} - X_0} \times 100 \tag{i}$$

Since the gas in the first thermometer obeys van der Waals' equation of state, we can write

$$p = \frac{RT}{V - b} - \frac{a}{V^2}, \tag{ii}$$

where T is defined on the perfect gas scale and V is constant.

We use Eq. (ii) for the first gas and substitute for p_θ, p_0, and p_{100}. On simplification, we get a compact relation for temperature on the first thermometer for an arbitrary temperature T on the ideal gas scale:

$$\theta_1 = \frac{T_\theta - T_0}{T_{100} - T_0} \times 100 \qquad \text{(iii)}$$

Note that in arriving at Eq. (iii), we have assumed that van der Waals' constants do not vary with temperature and/or pressure.

Suppose that for the second thermometer, the gas obeys the ideal gas equation: $p = \frac{RT}{V}$. Then if the reading on the thermometer is θ_2, we get

$$\theta_2 = \frac{T_\theta - T_0}{T_{100} - T_0} \times 100 \qquad \text{(iv)}$$

From Eqs. (iii) and (iv), we note that both thermometers show identical values of temperature.

You may now like to answer a practise problem.

Problem 4.6 The length of the mercury column in a liquid thermometer is 5 cm when in contact with water at its triple point. Calculate (a) the empirical temperature when the length of the column is 7.0 cm and (b) length of the mercury column at the steam point.

Ans: (a) 382.4; (b) 6.83 cm

ADDITIONAL EXAMPLES

Example 4.10 The perfect gas temperature at the steam point is 373.15 K. Calculate the limiting value of the ratio of the pressures of a gas at the steam point and at the triple point of water. Assume that the gas is kept at constant volume.

Solution: From Eq. (4.43), we recall that

$$\theta(p) = 273.16\left(\frac{p}{p_{tr}}\right)_V$$

$$\therefore \quad \left(\frac{p}{p_{tr}}\right)_V = \frac{\theta(p)}{273.16} = \frac{373.15\text{ K}}{273.16\text{ K}} = 1.37$$

Example 4.11 The resistance of a platinum thermometer at the triple point is 60 Ω. Calculate the temperature of a bath in which resistance of the bulb becomes 75 Ω.

Solution: From Eq. (4.44), we recall that

$$\theta_R = 273.16\left(\frac{R}{R_{tr}}\right)$$

On substituting the given values, we get

$$\theta_R = (273.16\text{ K}) \times \left(\frac{75\ \Omega}{60\ \Omega}\right)$$

$$= 341.45\text{ K}$$

Example 4.12 The resistance of a platinum thermometer at the ice point is 5Ω and 5.9Ω at the steam point. In a constant volume gas thermometer, the pressure is 1 mHg at the ice point and 1.366 mHg at the steam point. Both thermometers are inserted in a bath. The resistance of platinum thermometer changes to 5.8Ω and the pressure of gas to 1.325 mHg. Calculate the temperature of the bath in °C (a) on the platinum scale and (b) on the gas scale.

Solution: On the resistance scale, the temperature is defined as

$$\theta_R = \frac{R_\theta - R_0}{R_{100} - R_0} \times 100$$

On substituting the given values, we get

$$\theta_R = \frac{5.8\ \Omega - 5\ \Omega}{5.9\ \Omega - 5\ \Omega} \times (100°\text{ C}) = \frac{0.8\ \Omega}{0.9\ \Omega} \times 100°\text{ C}$$

$$= 88.89\ °\text{C}$$

The temperature on the gas scale is defined as

$$\theta = \frac{p_\theta - p_0}{p_{100} - p_0} \times (100°\text{ C})$$

$$= \frac{1.325\text{ mHg} - 1.0\text{ mHg}}{1.366\text{ mHg} - 1.0\text{ mHg}} \times (100°\text{ C})$$

$$= 88.80°\text{C}$$

Example 4.13 Using a constant-volume gas thermometer, the ideal gas scale was defined as $p_\theta = p_{tr}(1 + \alpha\theta)$, where θ is in °C and $\alpha = 3.661 \times 10^{-3}\ °\text{C}^{-1}$. Calculate the temperature at which $p_\theta = 0$ and show that this definition is consistent with Eq. (4.46).

Solution: We are told that the ideal gas scale is defined by the relation

$$p_\theta = p_{tr}\ (1 + \alpha\theta) \tag{i}$$

We can rewrite it as

$$\frac{p_\theta}{p_{tr}} = 1 + \alpha\theta \tag{ii}$$

For $p_\theta = 0$, we get

$$\theta = -\frac{1}{\alpha} = -\frac{1}{3.661 \times 10^{-3}\ °\text{C}^{-1}} = -273.149°\text{C}$$

Using this result in Eq. (ii), we can write

$$\theta = \frac{1}{\alpha}\left(\frac{p_\theta}{p_{tr}} - 1\right)$$

or

$$\theta + \frac{1}{\alpha} = \frac{1}{\alpha}\frac{p_\theta}{p_{tr}}$$

$\Rightarrow$

$$(\theta + 273.149)°\text{C} = 273.149\frac{p_\theta}{p_{tr}}$$

or

$$T(°\text{C}) = 273.149(p_\theta / p_{tr})$$

This result is consistent with Eq. (4.46), except that the multiplier is 273.149 instead of 273.16.

Example 4.14 A stretch of railway track in Thar desert is laid without expansion joints. The days are hot and nights are cold with temperature difference of 20°C. Calculate the change in tension in the rails between the day and the night, if length of the track remains constant. Given area of cross section $A = 3.6 \times 10^{-3}$ m^2, Young's modulus of material $Y = 2 \times 10^{11}$ Nm^{-2} and coefficient of linear expansion $\alpha_l = 8 \times 10^{-6}$ °C^{-1}.

Solution: The state of the track will be defined by the tension, temperature and length of the track. Therefore, the parametric equation will be of the form

$$f(L, F, T) = 0$$

If we take F and T as independent variables, we can write

$$L = L(F, T) \tag{i}$$

On partial differentiation, we can write

$$\text{d}L = \left(\frac{\partial L}{\partial F}\right)_T \text{d}F + \left(\frac{\partial L}{\partial T}\right)_F \text{d}T \tag{ii}$$

At constant L, d$L = 0$ so that

$$\left(\frac{\partial L}{\partial F}\right)_T \left(\frac{\partial F}{\partial T}\right)_L + \left(\frac{\partial L}{\partial T}\right)_F = 0$$

or

$$\left(\frac{\partial L}{\partial F}\right)_T \left(\frac{\partial F}{\partial T}\right)_L \left(\frac{\partial T}{\partial L}\right)_F = -1 \tag{iii}$$

We now recall that the Young's modulus is defined as

$$Y = \frac{L}{A}\left(\frac{\partial F}{\partial L}\right)_T \tag{iv}$$

and the coefficient of linear expansion is defined as

$$\alpha_l = \frac{1}{L}\left(\frac{\partial L}{\partial T}\right)_F \tag{v}$$

On combining Eqs. (iii), (iv) and (v), we get

$$\frac{L}{YA}\left(\frac{\partial F}{\partial T}\right)_L \frac{1}{\alpha_l L} = -1$$

or

$$\left(\frac{\partial F}{\partial T}\right)_L = -YA\alpha_l \tag{vi}$$

Hence, change in tension corresponding to a change in temperature ΔT at constant L is given by

$$\Delta F = -YA\alpha_l \Delta T$$

Note that in this case, temperature drops as night descends. Therefore, ΔT is negative. On inserting the given values, we get

$$\Delta F = -(2 \times 10^{11}\ \text{Nm}^{-2}) \times (3.6 \times 10^{-3}\ \text{m}^2) \times (8 \times 10^{-6}\ {}^\circ\text{C}^{-1}) \times (-20\ {}^\circ\text{C})$$

$$= 1.15 \times 10^5\ \text{N}$$

Example 4.15 Explain

(a) Is temperature microscopic or macroscopic?
(b) A system composed of 100 molecules cannot be termed as a thermodynamic system.
(c) A bottle of soft drink at room temperature put inside a refrigerator so that it can cool is a closed system.
(d) A plant in a garden is an example of an open system.

Solution

(a) Temperature is a macroscopic quantity. Moreover, it is not additive and is independent of mass of the substance.
(b) A thermodynamic system should be so large that it can be treated as a continuous collection. Obviously, a system of 100 molecules does not satisfy this condition.
(c) The mass of soft drink bottle is constant. Therefore, it behaves as a closed system.
(d) A plant in a garden forms part of a system with other plants, soil and atmosphere as its surroundings. It absorbs solar energy and CO_2 (matter) from the atmosphere and gives out O_2. It also takes nutrients from the soil and grows, increasing its own matter content. It means that exchange of energy as well as matter takes place. Therefore, a plant in the garden behaves as an open system.

Example 4.16 For a platinum wire, the temperature variation of resistance is modelled by the equation,

$$R_t = R_0\ (1 + \alpha t + \beta t^2)$$

If $\alpha = 3.90 \times 10^{-3}$°C^{-1}, $\beta = -5.7 \times 10^{-7}$°C^{-2} and $R_0 = 20\Omega$, calculate the temperature of a heat bath in whose contact the resistance of the wire is found to be 26.24Ω.

Solution: The temperature variation of Platinum wire is given by

$$R_t = R_0\ (1 + \alpha t + \beta t^2)$$

On substituting the given values, we get

$$26.24\,\Omega = 20\,\Omega\,[1 + (3.90 \times 10^{-3}\,°\text{C}^{-1})\,t - (5.7 \times 10^{-7}\,°\text{C}^{-2})t^2]$$

or
$$1.312 = 1 + 3.90 \times 10^{-3}\ t - 5.7 \times 10^{-7}\ t^2$$

We can rewrite it as

$$5.7 \times 10^{-7}\ t^2 - 3.90 \times 10^{-3}\ t + 0.312 = 0.$$

This is a quadratic equation in t and its roots are given by

$$t = \frac{3.9 \times 10^{-3} \pm \sqrt{15.21 \times 10^{-6} - 4 \times 0.312 \times 5.7 \times 10^{-7}}}{2 \times 5.7 \times 10^{-7}}\ °\text{C}$$

$$= \frac{3.9 \times 10^{-3} \pm \sqrt{15.21 \times 10^{-6} - 0.711 \times 10^{-6}}}{11.4 \times 10^{-7}}\ °\text{C}$$

$$= \frac{3.9 \times 10^{-3} \pm 3.81 \times 10^{-3}}{11.4 \times 10^{-7}}\ °\text{C}$$

$$= \frac{0.09 \times 10^{-3}}{1.14 \times 10^{-6}}\ °\text{C},\ \frac{7.71 \times 10^{-3}}{1.14 \times 10^{-6}}\ °\text{C}$$

$$= 78.95°\text{C},\ \ 6.76 \times 10^3\,°\text{C}$$

The second of these roots is unrealistic and we ignore it. Hence, the temperature of heat bath is 79°C.

Example 4.17 The thermo-emf developed between the junctions of two dissimilar metals maintained at a temperature difference t is given by

$$E = C_1 t + C_2 t^2$$

where C_1 and C_2 are thermo-electric constants for the pair of metals or alloys used. For a particular thermocouple, $C_1 = 40.0 \times 10^{-6}$ V°C^{-1} and $C_2 = -0.01 \times 10^{-6}$ V°C^{-2}. If the thermo-emf between the junctions is $2.3 \times 10^{-2} V$ and the cold junction is kept at the ice point, calculate the temperature of hot junction.

Solution: Since the cold junction is kept at 0°C, t in the modelling equation refers to the temperature of the hot junction. Therefore, on using the given conditions, we can write

$$2.3\times 10^{-2}\ V = (40.0\times 10^{-6}\ V\,°C^{-1})\,t - (0.01\times 10^{-6}\ V\,°C^{-2})\,t^2$$

As before, this is a quadratic equation in t and to calculate its roots, we first rewrite it as

$$0.01\times 10^{-6}\ t^2 - 40.0\times 10^{-6}\ t + 2.3\times 10^{-2} = 0$$

Hence,

$$t = \frac{40\times 10^{-6} \pm \sqrt{1600\times 10^{-12} - 4\times 0.01\times 10^{-6}\times 2.3\times 10^{-2}}}{2\times 0.01\times 10^{-6}}\ °C$$

$$= \frac{40\times 10^{-6} \pm \sqrt{16\times 10^{-10} - 9.2\times 10^{-10}}}{0.02\times 10^{-6}}\ °C$$

$$= \frac{4\times 10^{-5} \pm 2.6\times 10^{-5}}{2\times 10^{-8}}\ °C$$

$$= 3.3\times 10^{3}\ °C,\quad 0.7\times 10^{3}\ °C$$

The first of these roots is unrealistic for a thermo-couple. Therefore, the temperature of the bath is 700 °C.

Example 4.18 The thermometric property X_θ of a material at temperature θ is given by the linear relation $X_\theta = a\theta^n + b$, where a and b are constants and n is a positive number. Show that the temperature $\theta = \left(\dfrac{X_\theta - X_0}{X_{100} - X_0}\right)^{1/n} \times 100.$

Solution: Since $X_\theta = a\theta^n + b$, we can determine a and b from the values of X_θ at two different temperatures. Suppose X_θ takes values X_0 and X_{100} at $\theta = 0$°C and $\theta = 100$°C. Then it readily follows that

$$b = X_0$$

and

$$X_{100} = 100^n\,a + X_0$$

so that

$$a = \frac{X_{100} - X_0}{100^n}$$

Hence,

$$X_\theta = \frac{X_{100} - X_0}{100^n}\theta^n + X_0$$

or

$$X_\theta - X_0 = \frac{X_{100} - X_0}{100^n}\,\theta^n$$

We can rewrite it as

$$\theta^n = \frac{X_\theta - X_0}{X_{100} - X_0} \times 100^n$$

or

$$\theta = \left(\frac{X_\theta - X_0}{X_{100} - X_0}\right)^{1/n} \times 100$$

Example 4.19 For a gas, the critical point is defined as $\left(\frac{\partial p}{\partial V}\right)_T = 0$. What happens to the values of isobaric volume expansivity and the isothermal compressibility at the critical point?

Solution We know that isobaric volume expansivity is defined as

$$\alpha = \frac{1}{V}\left(\frac{\partial V}{\partial T}\right)_p$$

Using Eq. (4.17), we can rewrite it as

$$\alpha = \frac{1}{V}\left(\frac{\partial V}{\partial p}\right)_T \left(\frac{\partial p}{\partial T}\right)_V = \frac{1}{V}\frac{\left(\frac{\partial p}{\partial T}\right)_p}{\left(\frac{\partial p}{\partial V}\right)_T}$$

Since $\left(\frac{\partial p}{\partial V}\right)_T = 0$, we note from the above expression for α that at the critical point

$$\alpha \to \infty$$

Similarly, isothermal compressibility is defined as

$$\beta_T = -\frac{1}{V}\left(\frac{\partial V}{\partial p}\right)_T$$

$$= -\frac{1}{V}\frac{1}{\left(\frac{\partial p}{\partial V}\right)_T}$$

As before, at the critical point

$$\beta_T \to \infty$$

That is, at the critical point, isobaric volume expansivity as well as isothermal compressibility attains infinite value.

We now sum up what you have learnt in this chapter.

SUMMARY

- A region of space under study constitutes the system and everything else around it which can influence its behaviour is the surroundings. An arbitrary surface separating the system from its surroundings is known as the boundary.
- The macroscopic quantities that determine the properties of the state of the system are called thermodynamic variables.
- Thermodynamic variables are characterised as extensive and intensive depending on whether these depend on the mass of the system or not.
- A system is said to be in thermodynamic equilibrium if its state does not change in any way with time.
- The zeroth law of thermodynamics states that if two systems are separately in thermal equilibrium with a third, they will also be in thermal equilibrium with one another.
- The equation of state of a thermodynamic system is represented as

$$f(p, V, \theta) = 0$$

where f is a single-valued function of pressure and volume at temperature θ.
- The absolute temperature scale is one-point (triple point of water) temperature scale. The temperature on this scale is obtained by multiplying the ratio of the observed value of a thermometric property and its value at the triple point of water by 273.16.
- The International Practical Temperature Scale (IPTS) uses a set of (eleven) fixed points with assigned values of thermodynamic temperature from the most accurate measurements made by a constant volume thermometer. Moreover, there are prescriptions stating the practical thermometers to be used for interpolation between fixed points along with well laid down interpolation procedures.

EXERCISES

4.1 Calculate the pressure of air in a vessel being evacuated, as a function of evacuation time t. The volume of vessel is V and the initial pressure is p_0. The process is assumed to be isothermal and the evacuation rate C is independent of pressure. (The evacuation rate is the gas volume being evacuated per second.) (*Ans:* $p = p_o e^{-Ct/V}$)

4.2 Using the result of Problem 4.2 compute the temperature at which β_T of a van der Waals' gas is greater than that of an ideal gas. Examine the case when the molar volume is much greater than b. (*Ans:* $T > a/b\text{R}$)

4.3 Using Eq. (4.27) and the fact that dV is an exact differential, prove that

$$\left(\frac{\partial \beta_T}{\partial T}\right)_p = -\left(\frac{\partial \alpha}{\partial p}\right)_T$$

4.4 Calculate the values of isobaric volume expansivity and isothermal compressibility for one mole of a perfect gas at STP.

(*Ans:* $\alpha = T^{-1} = 3.73 \times 10^{-3}\text{K}^{-1}$, $\beta_\text{T} = \text{p}^{-1} = 1\ \text{atm}^{-1}$)

4.5 The equation of state of an ideal elastic substance is $T = k\theta\left(\frac{L}{L_0} - \frac{L_0^2}{L^2}\right)$ where T is the tension and θ the temperature. k is a constant and L_0 (the value of L at zero tension) is a function of temperature only.

Show that the isothermal Young's modulus is given by $Y = \frac{k\theta}{A}\left(\frac{L}{L_0} + \frac{2L_0^2}{L^2}\right)$ and that its value at zero tension is given $Y_0 = \frac{3k\theta}{A}$, where A is the cross-sectional area of the wire.

4.6 Two thermally insulated vessels 1 and 2 are filled with air. They are connected by a short tube equipped with a valve. The volumes of the vessels and the pressure and temperature of air in them are (V_1, p_1, T_1) and (V_2, p_2, T_2). Calculate the air temperature and pressure established after the opening of the valve.

(*Ans:* $T = T_1T_2\,(p_1V_1 + p_2V_2)\,/\,(p_1V_1T_2 + p_2V_2T_1)$

$p = (p_1V_1 + p_2V_2)\,/\,(V_1 + V_2)$)

4.7 Calculate the minimum attainable pressure of an ideal gas during a process governed by the relation

$$T = T_0 + \alpha V^2$$

where T_0 and α are positive constants and V is the volume of one mole of the gas.

(*Ans.* $2R\,(\alpha T_0)^{1/2}$)

4.8 A thick surrounding cover of invar, whose expansivity and compressibility are almost negligible, is made to fit quite approximately a metal for which $\alpha = 4.5 \times 10^{-5}$ K^{-1} and $\beta_T = 1.0 \times 10^{-11}$ N^{-1} m^2. The initial temperature and pressure of the metal piece are 300K and 1atm, calculate (a) the final pressure if the temperature is raised to 308K, (b) rise in pressure if the temperature changes to 303 K, and (c) the temperature of the system if it can withstand a maximum pressure of 10^8 Nm^{-2}.

(*Ans:* (a) 361 atm, (b) 135 atm, (c) 322.2 K)

4.9 The resistance of a carbon resistor is related to temperature through the relation $a + b\log R = \left(\frac{\log R}{T}\right)^{1/2}$. If $a = -1.25$ and $b = 0.675$, calculate the temperature for $R = 10^3\Omega$.

(*Ans:* 3K)

5

THE FIRST LAW OF THERMODYNAMICS

arning Objectives

In this chapter, you will learn how to

explain the thermodynamic concept of work and obtain expressions for work done by different thermodynamical systems;

describe heat as energy in transit;

e internal energy and use it to formulate the first law of thermodynamics;

y the first law of thermodynamics to explain the behaviour of typical physical systems; and

e adiabatic lapse rate and explain the observed drop in temperature with height in our atmosphere.

TRODUCTION

In Chapter 4, we introduced the concept of temperature on the basis of the zeroth law of thermodynamics. In this chapter, we will begin our study with the internal energy of a system and show how it changes by thermal, mechanical or diffusive interactions. We will then establish the first law of thermodynamics and apply it to study the behaviour of different physical systems.

The first law is essentially a statement about the conservation of energy for thermodynamical systems and recognition of heat as a form of energy. It tells us that heat and other forms of energy are equivalent. Do you know that much of our world works because of this equivalence? The electrical energy that lights our houses, operates machines and runs trains originates in heat released in burning of coal, oil, gas or fissioning of uranium. This law has evolved as an expression of our experience of thermal behaviour of matter. It has been applied to a wide variety of phenomena and has proved correct without fail. To most undergraduate students, it is easy to understand today but the nineteenth century pioneers confronted many conceptual problems until it was established by Joule. In fact, the difficulties with the first law were associated with the understanding of the nature of heat. For brevity, in Sec. 5.2, we have captured the essence of developments leading to the formulation of the first law. You will learn that the caloric and dynamic theories came into conflict at the end of the 18th century and how the latter ultimately triumphed.

The efforts to develop analytical theory of heat based on dynamical viewpoint led to the introduction of the concept of *internal energy*, which is defined as the sum of energies of the individual constituents of the system. This was in essence recognition of the molecular nature of matter. We will discuss it in detail in Sec. 5.3. The first law of thermodynamics is discussed in Sec. 5.4. You will learn that this law is universal and explains various phenomena observed in nature as well as in a laboratory—from fall in temperature with height in outer atmosphere to pressure oscillations in a sound wave. In Sec. 5.5, we will discuss the various applications of the first law of thermodynamics. Starting with heat capacities of an ideal and a real gas, we have explained how it can be used to obtain the equation of state for an adiabatic process and adiabatic lapse rate.

RIGIN OF THE FIRST LAW

The researches which led to the recognition of heat as a form of motion are now well known. However, the earliest pioneers confronted certain conceptual problems and their resolution required the brilliance of Mayor, Joule, Thomson (later Lord Kelvin), Helmholtz and Clausius. Briefly, eighteenth century philosophers regarded heat to be an 'imponderable fluid'; an indestructible hypothetical fluid, called *caloric*, which permeated matter, had no mass, and was highly elastic and non-creatable. It was further supposed that addition of caloric to a body made it warmer while removal of this fluid made the body cooler. Also, if a body is at a higher temperature than another and they are brought into thermal contact, then the caloric was supposed to flow spontaneously from the hotter body to the colder body, until they attained the same temperature. However, there were problems with this theory. For example, when a warm body was brought in contact with ice, the caloric was supposed to flow from the warm body into ice. Though ice got converted into water, the temperature of the ice-water mixture remained the same. It had to be supposed that caloric combined with ice to form water. An alternative view—the kinetic theory of matter—associated heat with the motions/vibrations of the microscopic particles (atoms/molecules), which are building blocks of matter.

These theories soon came into conflict and doubts were raised about the caloric theory of heat towards the end of the eighteenth century. The correct idea of the nature of heat was, for the first time, given by Count Rumford in 1798. While drilling the barrel of a gun with a blunt tool, he observed that apparently inexhaustible heat is produced. In this experiment, there was no source of caloric. Moreover, the heat produced was proportional only to the amount of work done and was associated with the kinetic energy of the motions of microscopic constituents of the material. In a paper read before the Royal Society, he wrote

it appears to me to be extremely difficult, if not quite impossible, to form any distinct idea of anything capable of being excited and communicated in the manner the Heat was excited and communicated in these Experiments, except it be **MOTION**".

Through these words, he asserted that *heat is some form of motion*.

The inadequacy of the caloric theory was further exposed by Davy in 1799 when he showed that rubbing two pieces of ice together resulted in water. According to the caloric theory, rubbing squeezed caloric out of the solid so that the liquid produced by friction should contain *less* caloric than the solid. The liquid (water) produced in this experiment contained more caloric than ice and contradicted the caloric theory.

The work of Rumford and Davy was not given much credit and caloric theory continued to enjoy great support until Joule's work of 1840s. In fact, Joule put the *molecular motion* on a firm footing by demonstrating that there is a definite relation between work and heat. Through paddle wheel experiment, he conclusively established the equivalence of mechanical work and heat; the work done by the weights in driving the paddle wheel changed into heat through the frictional force between the water and the vanes of the paddle wheel. According to him, heat is energy in transit. (Just as it is wrong to say 'work' in a body, it is also wrong to say 'heat' in a body.) These findings greatly excited Thomson and he used these as the foundation of thermodynamics, a nomenclature given by him to this subject. Helmholtz and Clausius formulated the first law of thermodynamics in such a way as to incorporate mechanical and electrical phenomena, heat and work.

HE INTERNAL ENERGY

It is a common experience that ice melts on heating to give water at 0°C. On the basis of the kinetic theory of gases, we can say that if temperature remains constant, the kinetic energy does not change. Since energy has to be conserved, what can you say about the added energy? When ice is converted into water, its specific volume changes from $^{-1}$ to 1 cm^3 g^{-1}. To bring about this change, some work has to be done at the expense of added energy. However, the major share of the additional energy is used up in breaking the bonds of intermolecular attraction in ice and this energy is stored in water, which is released when water freezes into ice. The stored energy apparently shown by a system during a phase transition is termed *internal energy*. In thermodynamics, it is not necessary to know the source of this energy. But we can easily conceptualise it on the basis of molecular theory. From Chapter 1 we recall that matter is made up of a large number of atoms/molecules. At temperatures above absolute zero, they are in a state of constant motion and hence each one of them has kinetic energy. The sum total of kinetic energies of all molecules is known as *internal kinetic energy*. Also, due to inter-atomic/ intermolecular interactions, each atom/molecule making up the system possesses potential energy. The total energy stored in the system due to interactions is called *total internal potential energy. The sum of total internal kinetic and potential energies of all molecules constitutes the internal energy of the system.* We denote it by the symbol U. We are usually only in the changes in the value of internal energy which may be accomplished by the following three types of interactions:

Thermal interaction, i.e., by exchanging heat with a system.
Mechanical interactions, i.e., work being done on or by a system.
Diffusive interactions, i.e., exchanging matter with a system.

We now discuss these in turn.

5.3.1 Thermal Interaction

The internal energy of a system may be changed by adding or removing heat. For instance, we may add heat energy by holding a body over a flame or remove heat by putting it in ice cold water.

The thermal interaction between two systems is schematically depicted in Fig. 5.1, where a hotter system A is separated from a colder by a thin thermally conducting rigid boundary. These are together insulated from the outside by a rigid insulating surface. Since temperature is a measure of the average kinetic energy per particle, the particles making up the system A have higher kinetic energy than do those B. The particles making up these systems will collide along the surface of contact. The particles A will lose energy in collisions. When a large number of collisions have occurred, energy will be transferred from system A to system B. That is, heat flows from the hotter system to the cooler system.

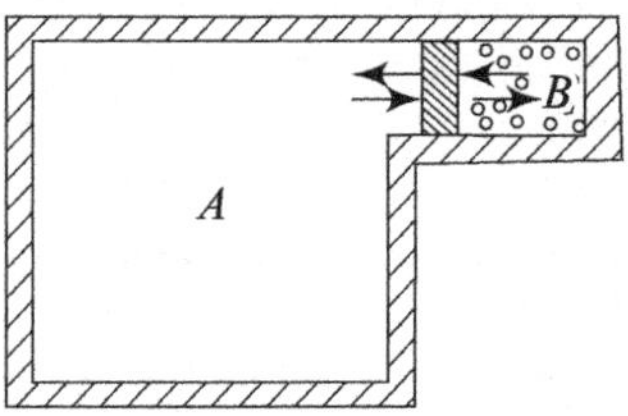

Fig. 5.1 Thermal interaction between two systems separated by a conducting wall.

Heat may also be transferred by electromagnetic radiation. More energetic particles tend to radiate greater energy than do less energetic particles so that radiative interactions also result in net transfer of energy from the hotter to the colder system.

We may conclude that when two systems are in thermal contact, heat is transferred from one to the other due to collisions of their constituent particles, along the surface separating them. In collisions, a more energetic particle loses energy so that heat flows from the hotter to the colder system.

5.3.2 Mechanical Interaction

The internal energy of a system may also change when work is done on or by the system. We know that work is the scalar product of force and displacement, i.e.,

$$W = \mathbf{F} \cdot \mathbf{s}$$

is force and **s** is displacement. There are a wide variety of forces that can do work on a system. Depending on the type of force, work may be mechanical, electrical or magnetic in nature (Table 5.1). The work done by a thermodynamic system is taken as positive and the work done on a system is taken as negative. It may be mentioned here that the nature of force is irrelevant when one considers the effect of work on changing the internal energy.

Expression for work done by different systems

System	Intensive variable (generalised force)	Extensive variable (generalised displacement)	Infinitesimal work
Hydrostatic	p	V	$p\,\mathrm{d}V$
Paramagnetic substance	B	m	$-B\,\mathrm{d}m$
Electric cell	V_d	Z	$V_\mathrm{d}\,\mathrm{d}Z$
Surface film	σ	A	$-\sigma\,\mathrm{d}A$
Stretched wire	T	l	$-T\mathrm{d}l$

To understand how work changes the internal energy of a system, we consider collisions between the particles and the walls of container. Since we are not interested in heat transfer to the walls of the container, we assume that collisions are elastic. That is, when particles collide against a stationary wall, their energy does not change. But they tend to gain energy if the system is being compressed and walls are moving inward. On the other hand, they tend to lose energy if the system expands and the walls move outward. The internal energy of a system under pressure will increase with further compression and decrease with expansion. It may be noted here that a mechanical force gives energy to particles at the boundary where it acts and this energy is communicated to other particles gradually. In the case of electric and magnetic forces, the energy is given directly and instantaneously to all particles.

Now let us apply this concept to a real-life situation. As a child grows, she/he is able to gradually sit on her/his own, stand on the feet and take a few steps forward in due course of time. A more energetic (healthy) child is more confident. Thermodynamically speaking, the internal energy of the child increases with months and years and enables greater movement/action. This energy is supplied in the form of milk fed by the lactating mother and/or nutritional food supplements. For normal growth, internal energy must increase. You should now go through the following example.

Example 5.1 A steel wire of length 2.5 m and area of cross-section 2.5×10^{-6} m^2 is suspended from torsion head. A 5 kg weight is suspended at its free end. Calculate the work done on the wire. Take $Y = 2 \times 10^{11}$ Nm^{-2}.

We know that

$$Y = \frac{\text{Stress}}{\text{Strain}}$$

By rearrangement, we can write

$$\text{Strain} = \frac{\Delta L}{L} = \frac{\text{Stress}}{Y}$$

$$\Delta L = \frac{L \times (Mg / A)}{Y} = \frac{LMg}{YA}$$

On substituting the given values, we get

$$\Delta L = \frac{(2.5\,\text{m}) \times (5\,\text{kg}) \times (9.8\,\text{ms}^{-2})}{(2 \times 10^{11}\,\text{Nm}^{-2}) \times (2.5 \times 10^{-6}\,\text{m}^2)} = 2.45 \times 10^{-4}\ \text{m}$$

Hence, work done on the wire is given by

$$W = F \times \Delta L = Mg\Delta L = (5\,\text{kg}) \times (9.8\text{ms}^{-2}) \times (2.45 \times 10^{-4}\,\text{m})$$
$$= 1.2 \times 10^{-2}\,\text{J}$$

Diffusive Interaction

When two systems exchange particles (atoms/molecules), they are said to be diffusively interacting. To understand how this exchange gives rise to a change in internal energy of a system, we note that the particles making up the system possess potential energy due to intermolecular interactions. From elementary physics curriculum, you may recall that one has to choose a zero energy level to measure potential energy. Usually, the bottom of the potential well in which a particle sits serves as the reference level and as long as the system is not in diffusive interaction, all energies are measured relative to it (Fig. 5.2). Usually, this level will be different for different systems. Therefore, for systems exchanging particles, one has to establish a *universal* reference level to measure the energies of all interacting systems. This universal

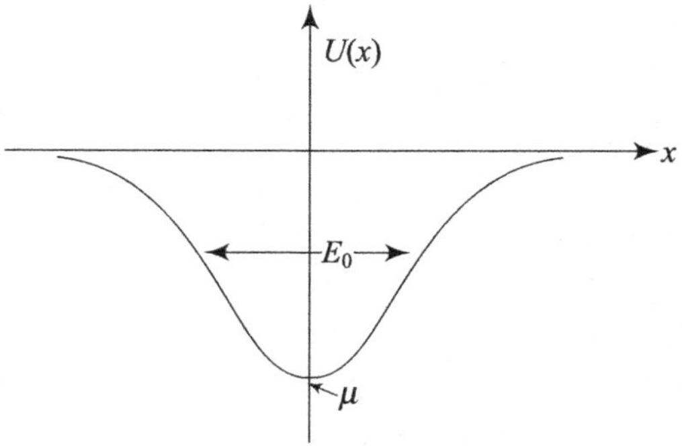

Fig. 5.2 Plot of potential energy of a particle as a function of its position.

zero energy reference level is taken as the potential energy of an isolated particle. The energy of the lowest point in a particle potential well measured with respect to the universal reference level is called the *chemical potential energy* and is denoted by the symbol . It may be mentioned here that when intermolecular interactions are predominantly attractive, the chemical potential is negative, and when the interactions are repulsive, the chemical potential is positive. Furthermore, interactions among neighbours, and therefore the chemical potential are influenced by the temperature, volume and number of particles in the system.

The energy of a particle trapped in a harmonic oscillator potential well, whose lowest point has energy μ, can be written as

$$\varepsilon = \varepsilon_{th} + \mu \tag{5.1}$$

$$\varepsilon_{th} = \frac{1}{2}kr^2 + \frac{1}{2}mv^2 \tag{5.2}$$

thermal energy and is equal to the sum of its kinetic and potential energies; the potential energy being measured relative to the lowest point in the potential well.

So, the total internal energy of a system of N particles may be written as

$$U = E_{th} + N\mu \tag{5.3}$$

$= N\,\varepsilon_{th.}$

When a particle goes from one system to another, interactions with new neighbours in the changed environment may change its chemical potential, and hence the internal ΔN particles are added to the system without adding any thermal energy at all, the change in internal energy is given by

$$\Delta U = \mu\,\Delta N \tag{5.4}$$

This is illustrated in Fig. 5.3. Note that the two systems are separated by a porous, rigid, thermally insulating wall.

It may be pointed out here that when systems are not exchanging particles, one will not take note of the changes in zero energy reference level and the internal energy will be equal to the thermal energy. These two systems can be envisaged as being separated by a flexible thermally conducting membrane (Fig. 5.4). This will be true for most systems to be discussed in this book.

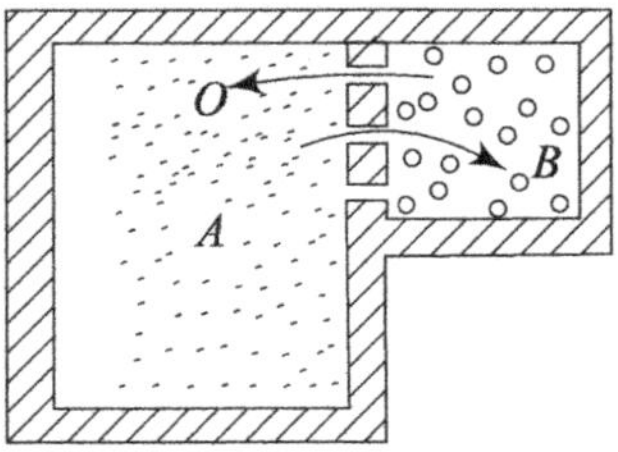

Diffusive interactions between two systems separated by a porous rigid insulating wall.

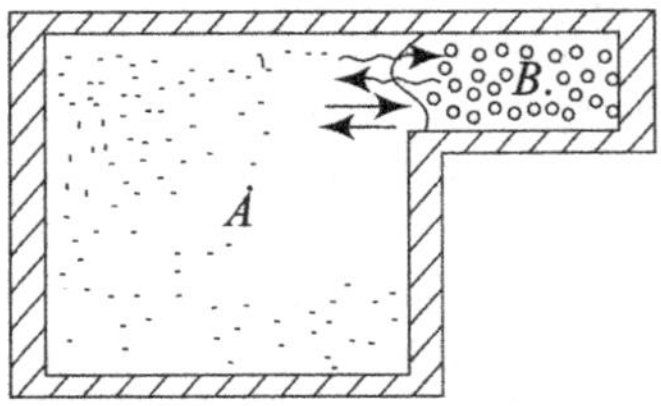

Fig. 5.4 Two systems in thermal and mechanical interactions.

HE FIRST LAW

The first law of thermodynamics states that *when a system is constrained to undergo a change by mechanical, diffusive or thermal interactions, its internal energy changes by an amount equal to the heat transferred to it, work done on it and matter exchanged.* The different ways in which internal energy of a system may be changed are illustrated in Fig. 5.5. Let dU be the change in internal energy of a system when δQ amount of heat is added to it, δW work is done by it and dN is the change in the number of particles. Then, the first law of thermodynamics can mathematically be written as

$$dU = \delta Q - \delta W + \mu \, dN \tag{5.5}$$

is chemical potential. By convention, the heat flow into the system is taken as positive, whereas work done by a system is considered positive.

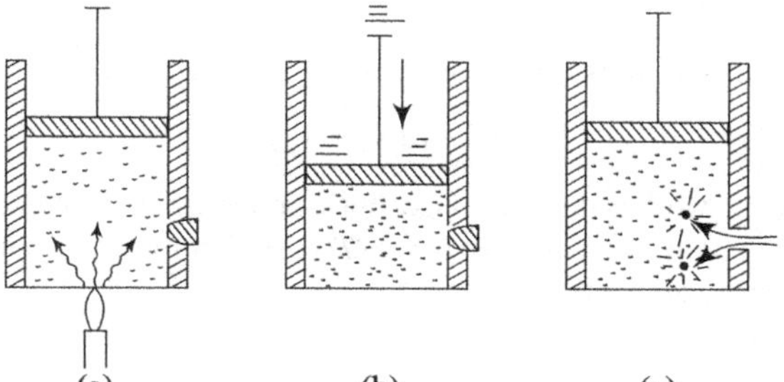

(a) (b) (c)

Illustration of processes by which internal energy of a system can change.

If more than one kind of work is done by the system, δW is replaced by $\Sigma_j \delta W_j$. Similarly, if there are several different kinds of particles in the system, the last term in Eq. (5.5) will $\mu_i \, dN_i$. On the other hand, a system in diffusive equilibrium may undergo thermal and mechanical interactions. Then, the first law takes the form

$$dU = \delta Q - \delta W \tag{5.6}$$

This is the *differential form* of the first law of thermodynamics for a non-diffusively interacting system.

Note that we have used 'd' to denote changes in internal energy and 'δ' to signify changes in heat added and work done. This is to show the difference that internal energy of a system is a function of state rather than the path followed. In mathematical language, we say that dU is an 'exact differential'. On the other hand, total heat added and the work done in going from initial to the final state depend on the path followed and cannot be determined from the end points alone. Mathematically, the differentials of such quantities are said to be *inexact*. (We have used different symbols to emphasise the difference between exact and inexact differentials; to express it mathematically, we have used d for exact and for inexact differential.)

To develop an appreciation for this concept, go through the following example carefully.

Example 5.2 Show that the work done by a system during a thermodynamic process depends on the path followed.

For a hydrostatic system, the work done during a thermodynamic process occurring between states 1 and 2 is given by

$$W = \int_1^2 p\,dV \qquad \text{(i)}$$

Geometrically speaking, W will be equal to the area under the curve along which we go from the initial to the final state. And from Fig. 5.6, you will observe that we can go from state 1 to state 2 along different paths. It means that the value of W will be different for different paths. This implies *work is a path function* and the work done in a cyclic process is non-

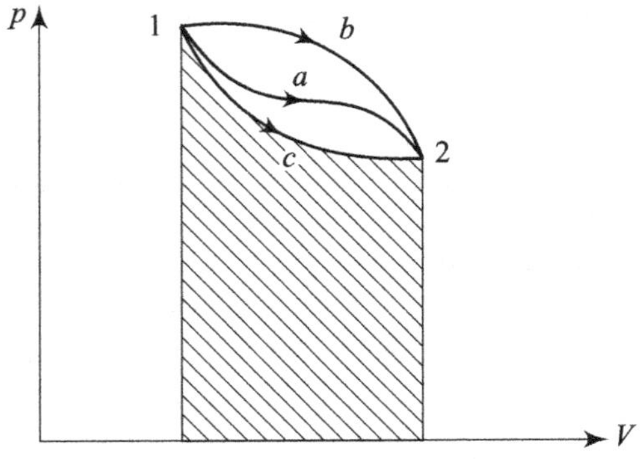

Fig. 5.6 Work is a path function.

In an isobaric process, pressure is constant and the work done is

$$W = p(V_f - V_i) \qquad \text{(ii)}$$

On the other hand, no work is done in an isochoric process, since volume does not change. Using ideal gas equation, we can write $p = nRT/V$, so that the work done by an ideal gas in an isothermal expansion is given by

$$W = nRT \int_{V_i}^{V_f} \frac{dV}{V} = nRT \ln\left(\frac{V_f}{V_i}\right) \qquad \text{(iii)}$$

Using Boyle's law $[p_1V_1 = p_2V_2]$, we can write the expression for work done by an ideal gas during an isothermal process as

$$W = nRT \ln\left(\frac{p_1}{p_2}\right) \qquad \text{(iv)}$$

signifies the number of moles of the gas. Note that $\ln\left(\frac{V_f}{V_i}\right)$ is positive in an expansion so that the work done is positive. On the other hand, for a compression, the ratio

is less than one and work done is negative, i. e., work is done on the gas.

Like work, heat is also path-dependent. That is, it depends on the type of process concerned. Consider the following situation: Suppose you want to raise the temperature of a highly viscous liquid by 10°C above room temperature. You can do this by (a) heating it, (b) stirring it vigourously and (c) by combining heating and stirring. The heat given to the system will be different in these three cases, though temperature is raised by the same amount. We may therefore conclude from the first law of thermodynamics that

No useful work can be obtained without expenditure of an equivalent amount of some other form of energy. It means that we can not devise a machine which can run perpetually.

Internal energy is a function of state, i.e., its differential is exact and can be expressed as the algebraic sum of two inexact differentials.

You may now like to answer some practise problems.

Problem 5.1 The work done by hydrostatic pressure is defined as $W = \int_{V_1}^{V_2} p\, dV$. moles of a real gas are made to undergo isothermal process. Show that the work done is

$$W_{real} = nRT \ln\left(\frac{V_2 - b}{V_1 - b}\right) + n^2 a\left(\frac{1}{V_2} - \frac{1}{V_1}\right)$$

Problem 5.2 A steel wire in a torsional pendulum is 1.5 m long and has area of cross section 1.5×10^{-6} m^2. It is loaded with a weight of 2 kg at the free end. Calculate the work done on the wire. Take $Y = 2 \times 10^{11}$ Nm^{-2} and $g = 9.8$ ms^{-2}.

-1.92×10^{-3} J

Now go through the following examples.

Example 5.3 n moles of a perfect gas is made to undergo cyclic change $ABCA$ consisting of the following processes:

: Isothermal expansion at temperature T so that the volume is doubled from V_1 to and pressure changes from p_1 to p_2.

: Isobaric compression at pressure p_2 to initial volume V_1.

: Isochoric change leading to change of pressure from p_2 to p_1.

(a) Obtain expression for the work done and change in internal energy for each part.

(b) By how much will the internal energy change for the complete cycle.

(a) Refer to Fig. 5.7. From (iii) in Example 5.2, you can express that the work done by the gas in isothermal expansion, i.e., in going from $A \rightarrow B$ as

$$W = nRT \ln\left(\frac{2V_1}{V_1}\right) = nRT \ln 2$$

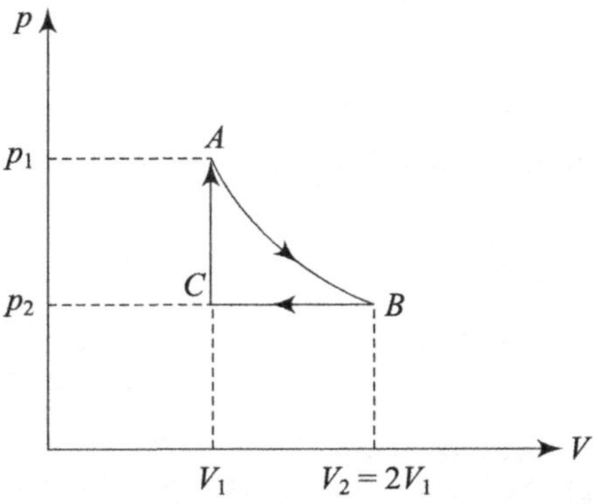

Fig. 5.7 A typical cyclic process.

Since the change takes place at constant temperature, there will be no change in internal energy in this part of the cycle.

In compressing the gas at constant pressure from $B \rightarrow C$, the work is done on the gas and is given by $W_{BC} = p_2(V_1 - 2V_1)$

From the ideal gas equation, we can write that at B, $p_2V_2 = 2p_2V_1 = nRT$

Similarly, at C, we can write $p_2 V_1 = \frac{nRT}{2}$. This shows that the temperature at C drops to half of the initial value. Hence, we can express the work done on the gas in going from

$$W_{AB} = -p_2 V_1 = -\frac{nRT}{2}$$

The change in internal energy is given by the first law of thermodynamics:

$$\Delta U_{BC} = Q_{BC} - W_{BC}$$

$$Q_{BC} = nC_p \Delta T = nC_p \left(\frac{T}{2} - T\right) = -nC_p \frac{T}{2}.$$

$$\Delta U_{BC} = -\left(nC_p \frac{T}{2} - nR\frac{T}{2}\right) = -nC_V \frac{T}{2}.$$

In going from $C \to A$, the gas is compressed at constant volume. Therefore, no work will $W_{CA} = 0$.

$$\Delta U_{CA} = Q_{CA} - W_{CA} = nC_V \left(T - \frac{T}{2}\right) = nC_V \frac{T}{2}$$

The total change in internal energy is given by

$$\Delta U = 0 - \frac{nC_V T}{2} + \frac{nC_V T}{2} = 0$$

You should have expected this result for a cyclic process since internal energy is a function of state rather than the path followed.

Example 5.4 The temperature of 5 g of air is raised by 1°C at constant volume. Calculate the increase in its internal energy. Given $C_V = 0.172$ cal g^{-1} °C^{-1} and $J = 4.18$ J cal^{-1}.

From Eq. (5.6), we have

$$\begin{aligned} dU &= \delta Q - \delta W \\ &= \delta Q - p dV \end{aligned}$$

Since the air is heated at constant volume, $dV = 0$. Hence,

$$\begin{aligned} dU &= \delta Q \\ &= (\text{mass of air}) \times c_v \times \text{change in temperature} \\ &= 5\text{g} \times (0.172 \text{ cal g}^{-1} {}^\circ\text{C}^{-1}) \times (1^\circ\text{C}) \\ &= 0.86 \text{ cal} \\ &= 0.86 \times 4.18 \text{ J} = 3.59 \text{ J} \end{aligned}$$

Example 5.5 1 kg of water is boiled under 2 atm at 120°C. If the volumes occupied by water and steam under given conditions are 10^{-3}m^3 and 0.825 m^3 respectively, calculate (a) the work done and (b) increase in internal energy. Given $L = 2.20 \times 10^6$ J kg^{-1}, and 1.013×10^5 Nm^{-2}.

The work done against pressure

$$\begin{aligned} \delta W &= p dV \\ &= (2 \times 1.013 \times 10^5 \text{ Nm}^{-2}) \times (0.824 \text{ m}^3) \\ &= 0.167 \times 10^6 \text{ J} \end{aligned}$$

From the first law of thermodynamics, we find that increase in internal energy is given by

$$\begin{aligned} dU &= \delta Q - \delta W \\ &= mL - \delta W \\ &= (2.20 \times 10^6 - 0.167 \times 10^6)\ \text{J} \\ &= 2.03 \times 10^6\ \text{J} \end{aligned}$$

Note that most of the latent heat (92% in this case) is used up in increasing the internal energy, which brings about change of state. For this reason, dU is also known as internal

To gain experience in problem solving, you should answer a practise problem.

Problem 5.3 1g of water at 100° C changes into steam occupying a volume of 1760 cc at atmospheric pressure. Calculate (a) the work done in joules, and (b) the increase in internal energy. Given, latent heat of steam = 540 cal g^{-1} and 1 atm = 1.01×10^5 N m^{-2}.

178 J, 2079 J

Significance of the first law Let us pause for a while and reflect on the significance of the first law. Among others, following points need particular mention.

It defines 'internal energy' as a function of state of the system.
It highlights that heat is energy in transit.
It rules out the possibility of constructing a machine which can work on its own, without any input. Such a machine is called the *perpetual motion machine of the first* . Existence of such a machine would have resolved all our energy problems; once started, such a machine would continue working.

Limitations of the first law The first law of thermodynamics has some serious limitations. These include the following:

It gives no information about the way thermodynamic systems evolve. For instance, it fails to acknowledge unidirectional processes occurring in nature; i.e., it does not rule out the possibility of flow of heat from a colder body to a hotter body, which for sure never occurs. Similarly, we know that a storage battery gets discharged through a resistor when it is operated in a circuit. But the battery can not be charged by reversal of this arrangement.
It does not rule out a machine which can take energy from the environment and converts it fully into work. Such a machine is called *perpetual motion machine of second kind* and is not physically permissible.

We will now consider some typical applications of the first law thermodynamics.

PPLICATIONS OF THE FIRST LAW OF HERMODYNAMICS

The first law of thermodynamics has been successfully applied to diverse physical and chemical processes. In a sense, this law is universal and explains large scale phenomena occurring in nature as well as the microscopic processes that can be observed in a laboratory. For instance, the drop in temperature as we move upward in the outer atmosphere is explained rather well using the first law. We can also explain the pressure oscillations in a sound wave. Its applications to flow processes and chemical reactions leads to very interesting results. We begin by considering the heat capacities, which give us information about variation of internal energy of a substance with volume.

Heat Capacities of a Gas

From Table 5.1, we recall that for a gaseous system $\delta W = p\mathrm{d}V$. Then, the first law can be expressed as

$$\delta Q = \mathrm{d}U + p\mathrm{d}V \tag{5.7}$$

We know that the state of a gas can be described in terms of only two coordinates out and T. Let us choose T and V as independent variables. Since U is a function of state, we can express it in terms of these variables, i.e., $U = U(T, V)$. A small change in U can then be expressed as

$$\mathrm{d}U = \left(\frac{\partial U}{\partial T}\right)_V \mathrm{d}T + \left(\frac{\partial U}{\partial V}\right)_T \mathrm{d}V$$

Substituting this expression in Eq. (5.7), we get

$$\delta Q = \left(\frac{\partial U}{\partial T}\right)_V \mathrm{d}T + \left\{p + \left(\frac{\partial U}{\partial V}\right)_T\right\}\mathrm{d}V \tag{5.8}$$

We now introduce the concept of *heat capacity*. At constant volume, the second term in Eq. (5.8) will vanish and we can write

$$C_V = \left(\frac{\delta Q}{\partial T}\right)_V = \left(\frac{\partial U}{\partial T}\right)_V \tag{5.9}$$

This is an obvious result. If there is no change in volume, no work will be done and change in internal energy will be equal to the heat entering the system. The corresponding *specific heat* is

$$c_V = \frac{1}{m}\left(\frac{\partial U}{\partial T}\right)_V = \frac{C_V}{m} \tag{5.10}$$

Similarly, using Eq. (5.8) we can write the expression for heat capacity at constant pressure

$$C_p = \left(\frac{\delta Q}{\partial T}\right)_p = \left(\frac{\partial U}{\partial T}\right)_V + \left(\left(\frac{\partial U}{\partial V}\right)_T + p\right)\left(\frac{\partial V}{\partial T}\right)_p$$

On combining this result with Eq. (5.9), we get

$$C_p = C_V + \left(\left(\frac{\partial U}{\partial V}\right)_T + p\right)\left(\frac{\partial V}{\partial T}\right)_p$$

$$C_p - C_V = \left(\left(\frac{\partial U}{\partial V}\right)_T + p\right)\left(\frac{\partial V}{\partial T}\right)_p \tag{5.11}$$

Let us interpret this result. The first term in the brackets on the right-hand side of this equation tells us how internal energy changes with volume at constant temperature. (It must be associated with work done against intermolecular forces in the gas.) The second term tells us how much work is done in pushing back the surroundings at constant pressure. It means that a knowledge of the difference of heat capacities at constant pressure and at constant volume will give us information about variation of internal energy of a substance with volume.

So far, our approach has been quite general. We now consider two specific cases:

Perfect Gas: The Joule Expansion One way of determining the relation between is to allow free expansion of the gas into a larger volume, i.e., perform Joule expansion. Since there is no heat inflow and no work is done in free expansion, we do not expect any change in internal energy of the gas. This helps us to define a *perfect gas* as being characterised by the following properties:

(a) There is no change in the internal energy of a gas in a Joule expansion*. That is, there are no intermolecular attractions and internal energy is wholly kinetic.
(b) The equation of state is the perfect gas law: $pV = nRT$.

Property (a) implies that $(\partial U/\partial V)_T = 0$, so that Eq. (5.11) reduces to

$$C_p - C_V = p\left(\frac{\partial V}{\partial T}\right)_p \tag{5.12}$$

This shows that the difference in the two heat capacities depends on how the volume of a system changes as its temperature increases at constant pressure. We expect this difference to be large because a small change in temperature gives rise to a large change in volume of a gas.

From property (b), we can write

$$p\left(\frac{\partial V}{\partial T}\right)_p = nR$$

$$C_p - C_V = nR \tag{5.13}$$

This result is known as *Mayer's formula*. It demonstrates that internal energy of a perfect gas is a function only of temperature. Recall that we started by expressing internal energy as a function of temperature and volume since it is a function of state. But the Joule expansion clearly demonstrates that internal energy is independent of volume. (It must also be independent of pressure because the pressure decreases in Joule expansion.)

Using Mayer's formula, we can rewrite the first law of thermodynamics in a different form. To illustrate this, we note that for a perfect gas U is a function of T only. Hence, from Eq. (5.9), we can say that

$$C_V = \frac{dU}{dT}$$

$$\mathrm{d}U = C_V\,\mathrm{d}T$$

Using this result in Eq. (5.7), we get

$$\delta Q = C_V \mathrm{d}T + p\mathrm{d}V \tag{5.6'}$$

*It is free expansion of a gas into a larger volume at constant temperature. You may now like to know: Is the internal energy of a perfect gas independent of pressure? Yes, it is because a change in pressure affects only the volume of the gas, which means relative separation between molecules changes. Since there are no intermolecular forces in a perfect gas, internal energy remains unchanged.

Again, using the equation of state for one mole of a perfect gas we can correlate infinitesimal changes in thermodynamic variables as

$$p\mathrm{d}V + V\mathrm{d}p = R\mathrm{d}T$$

$$p\mathrm{d}V = R\mathrm{d}T - V\mathrm{d}p$$

Using this result in Eq. (5. 6′), we can rewrite the first law of thermodynamics as

$$\begin{aligned}\delta Q &= C_V\mathrm{d}T + R\mathrm{d}T - V\mathrm{d}p \\ &= C_p\mathrm{d}T - V\mathrm{d}p \end{aligned} \tag{5.6″}$$

On the other hand, if we eliminate dT between Eqs. (5.6′) and (5.6″), we get

$$\delta Q = \frac{1}{R}(C_p p\mathrm{d}V + C_V V\mathrm{d}p) \tag{5.6‴}$$

Note that Eqs. (5.6), (5.6′), (5.6″) and (5.6‴) are equivalent forms of the first law of thermodynamics.

2. Real Gas For a real gas, the internal energy changes with volume because work has to be done against the intermolecular forces, e.g., van der Waals' forces. Also at high pressures, molecules experience a repulsive force—the hard core repulsion. Thus we may expect some change, howsoever small, in temperature in the expansion. For one mole of a real gas, we write van der Waals' equation as the equation of state as

$$\left(p + \frac{a}{V^2}\right)(V - b) = RT \tag{5.14}$$

As we show later, for a real gas

$$\left(\frac{\partial U}{\partial V}\right)_T = \frac{a}{V^2}$$

Using this result in Eq. (5.11) and combing the resultant expression with Eq. (5.14), we get

$$C_p - C_V = \frac{RT}{(V-b)}\left(\frac{\partial V}{\partial T}\right)_p \tag{5.15}$$

To evaluate $(\partial V/\partial T)_p$, we differentiate van der Waals' equation. This gives

$$\left(p + \frac{a}{V^2} - (V-b)\frac{2a}{V^3}\right)\left(\frac{\partial V}{\partial T}\right)_p = R$$

$$\left(\frac{\partial V}{\partial T}\right)_p = \frac{R}{\left(p + \dfrac{a}{V^2} - \dfrac{2a}{V^3}(V-b)\right)}$$

Next, we multiply the numerator as well as the denominator in this expression by $(V - b)$, and simply the resultant expression using Eq. (5.14). The result is

$$\left(\frac{\partial V}{\partial T}\right)_p = \frac{R(V-b)}{RT - \dfrac{2a}{V^3}(V-b)^2}$$

$$\frac{1}{(V-b)}\left(\frac{\partial V}{\partial T}\right)_p = \frac{1}{T\left(1-\dfrac{2a}{RTV^3}(V-b)^2\right)}$$

Using binomial expansion we can write for very small a

$$\frac{1}{(V-b)}\left(\frac{\partial V}{\partial T}\right)_p = \frac{1}{T}\left(1+\frac{2a}{RTV^3}(V-b)^2\right) \quad (5.16)$$

where we have retained terms only up to first order in a.

Substituting this in Eq. (5.15), we get

$$C_p - C_V = R\left(1+\frac{2a}{RTV^3}(V-b)^2\right) \quad (5.17)$$

On comparing Eqs. (5.13) and (5.17) we note that the difference between heat capacities at constant pressure and at constant volume is more for a real gas than that for a perfect gas. However, if the gas is not very densely packed, i.e., for large molecular separations, we can ignore the second term within the bracket of Eq. (5.17). Then we obtain the ideal gas result.

For a real gas, the variation of $C_p - C_V$ with V is shown in Fig. 5.8. Note that the ideal gas behaviour is attained at large values of V.

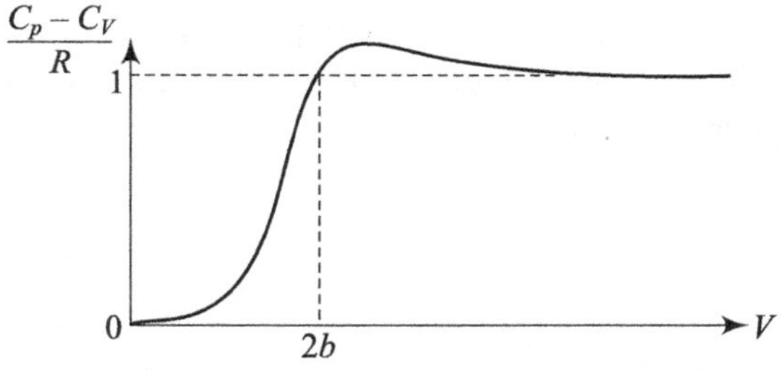

Plot of $C_p - C_V$ versus V for a van der Waals' gas.

You should now go through the following example.

Example 5.6 One mole of a gas, assumed to be perfect, at 0°C is heated at constant pressure till its volume is twice its initial value. Calculate the amount of heat absorbed. = 20.9 J mol^{-1} K^{-1} and R = 8.3 J mol^{-1} K^{-1}.

Since the gas is heated at constant pressure, we have from Charles' law

$$\frac{V_1}{V_2} = \frac{T_1}{T_2}$$

= 2 and T_1 = 273 K. Therefore,

$$T_2 = T_1\left(\frac{V_2}{V_1}\right)$$
$$= (273\text{ K}) \times 2$$
$$= 546\text{ K}$$

Hence, the amount of heat required to raise the temperature of n moles of gas by $(546 - 273)$ K = 273 K is

$$\begin{aligned}\delta Q &= nC_p\,\Delta T\\ &= n(C_V + R)\,\Delta T\end{aligned}$$

On substituting the given data, we get

$$\begin{aligned}\delta Q &= (1\text{ mol})\,(20.9 + 8.3)\text{ J mol}^{-1}\text{K}^{-1} \times (273\text{ K})\\ &= 29.2 \times 273\text{ J}\\ &= 79.7 \times 10^2\text{ J}\end{aligned}$$

To gain practise in working out new problems, answer the following problems.

Problem 5.4 For a pure paramagnetic substance, write down the differential form of the first law of thermodynamics and show that

$$C_B - C_M = \frac{kB^2}{T^2}$$

and C_M respectively are defined as $C_B = (\delta Q/\delta T)_B$ and $C_M = (\delta Q/\delta T)_M$.

Problem 5.5 A car weighing 1350 kg is going down a hill. When it is 60 m vertically above the bottom of the hill, the driver sees red light of traffic crossing at the bottom. His speed at the time brakes are applied is 20 ms^{-1}. How much energy will be dissipated by the brakes if wind and other frictional effects are neglected. Take $g = 9.80$ ms^{-2}.

1323.5 kJ

Problem 5.6 An ideal monatomic gas is made to undergo reversible isobaric $p = 1.0 \times 10^5$ Pa. If volume increases from 10 litre to 20 litre, calculate (i) the increase in internal energy, (ii) the work done by the gas and (iii) the heat absorbed

(i) 1.49 kJ; (ii) 1 kJ; (iii) 2.49 kJ

Adiabatic Transformation: Equation of State

In an adiabatic change, there is no thermal contact between the system and its surroundings so that no heat exchange takes place, i.e., $\delta Q = 0$. Expansion or compression of a gas within a perfectly insulated piston in a cylinder with non-conducting walls, sound propagation in air, drop in temperature with altitude, existence of deep ocean currents and the gravitational collapse of interstellar mass into a new star are some important examples of adiabatic processes. For an adiabatic process, the first law of thermodynamics takes the form

$$dU + \delta W = 0 \tag{5.18}$$

This shows that if a system, a gas say, enclosed within an adiabatic boundary is permitted to expand adiabatically, it does work at the cost of its internal energy. That is, the internal energy of a gas decreases, and hence its temperature falls during an adiabatic expansion. We may therefore conclude that adiabatic expansion produces cooling and adiabatic compression produces heating. For this reason, adiabatic demagnetisation of a paramagnetic substance is used to produce temperatures below 1 K. (We will discuss it in detail in Chapter 10.) Using the first law, one can easily study changes in any two of the three thermodynamic coordinates that occur during an adiabatic process. From Eq. (5.18), we note that for a gaseous system we can write

$$C_V\,\mathrm{d}T + p\mathrm{d}V = 0$$

$C_V\mathrm{d}T.$

During expansion, the gas passes through an infinite number of equilibrium states. If we assume that equation of state for the perfect gas holds for each state, we can write

$$C_V\,\mathrm{d}\,T + \frac{RT}{V}\mathrm{d}\,V = 0$$

Dividing throughout by $C_V T$ and integrating the resultant expression, we get

$$\ln T + \frac{R}{C_V}\ln V = K$$

is the constant of integration.

Changing from logarithms to numbers, we can write

$$TV^{R/C_V} = \text{constant}$$

$$TV^{\gamma-1} = \text{constant} \tag{5.19a}$$

C_p/C_V and we have made use of Eq. (5.13). Note that in this derivation we have and γ as independent of temperature. But from Sec. 1.4, we recall that they do T. However, this happens for very large changes in temperature (1000 K or

Since the equation of state holds at all stages in the expansion, this result can also be put in the following equivalent forms using the perfect gas equation:

$$pV^{\gamma} = \text{constant} \tag{5.19b}$$

$$T^{\gamma}p^{1-\gamma} = \text{constant} \tag{5.19c}$$

You must convince yourself about the authenticity of these results. Eq. (5.19c) predicts that if a gas is compressed adiabatically, the pressure increases which is accompanied by a corresponding increase in temperature. Alternatively, from Eq. (5.19b), we can say that in adiabatic compression, volume decreases and since no heat can leave the system, its temperature rises. On the other hand, the temperature drops during adiabatic expansion as the gas does work at the cost of its internal energy.

On differentiating Eq. (5.19b) with respect to V, we get

$$\gamma p V^{\gamma-1}\mathrm{d}V + V^{\gamma}\,\mathrm{d}p = 0$$

$$\left(\frac{\partial p}{\partial V}\right)_S = -\gamma\left(\frac{p}{V}\right) \tag{5.20}$$

For an isothermal process, $pV = RT$

$$pdV + Vdp = 0$$

$$\left(\frac{\partial p}{\partial V}\right)_T = -\left(\frac{p}{V}\right) \tag{5.21}$$

From Eqs. (5.20) and (5.21), we note that

An adiabat is steeper than an isotherm $\gamma > 1$. This is because the gas loses internal energy as it expands. This also implies that the relative change in the volume in an adiabatic process is less than that in an isothermal process. It may be mentioned here that on the $p - V$ diagram, the adiabats form a family like the isotherms. This is illustrated in Fig. 5.9.

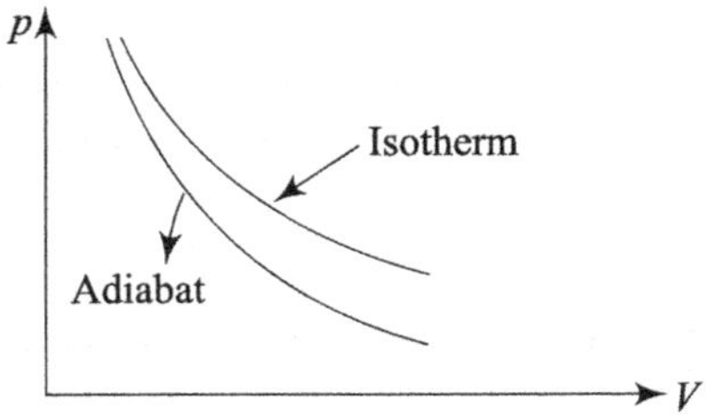

Fig. 5.9 Adiabates and isotherms of an ideal gas.

The ratio of adiabatic elasticity and isothermal elasticity is equal to γ:

$$\frac{E_S}{E_T} = \frac{\beta_T}{\beta_S} = \gamma$$

This result has also been obtained in Sec. 5.5.4 (Eq. (5.31)).

Problem 5.7 Derive Eqs. (5.19b) and (5.19c).

You should now go through the following examples.

Example 5.7 The nozzle of a bicycle is blocked. With no force on the handle, the pump contains a volume V of air at 300K and atmospheric pressure. The handle is pushed down with a constant force of magnitude F reducing the volume to half. No air escapes from the pump. Assume the change to be adiabatic. Taking γ for air to be 1.4, compute the final temperature of air in the pump.

Since the change is adiabatic, we can use Eq. (5.19b):

$$pV^\gamma = \text{constant}$$

$$p_i V^\gamma = p_f \left(\frac{V}{2}\right)^\gamma$$

is the pressure of the gas when the handle is in final equilibrium. On simplification, we can write

$$p_f = p_i\ 2^\gamma = 2^{1.4}\ p_i = 2.64\ p_i$$

Applying the perfect gas equation of state, we can write

$$\frac{p_i V}{T_i} = \frac{p_f \, (V/2)}{T_f}$$

$$T_f = \frac{p_f}{2 p_i} T_i$$

$$= \frac{2.64}{2} T_i$$

$$= 1.32 \times 300 \text{ K}$$

$$= 396 \text{ K}$$

This temperature is higher than the boiling point of water! So you are advised not to touch the nozzle of a blocked pump. Physically, this means that adiabatic compression produces

---

Example 5.8 An ideal gas is allowed to undergo adiabatic expansion so that its temperature drops from 320 K to 286 K and the pressure changes from 1.4×10^5 Pa to Pa. Determine its atomicity.

In this case, we use Eq. (5.19c):

$$T^{\gamma} p^{1-\gamma} = \text{constant}$$

and T_f respectively denote initial pressure, final pressure, initial temperature and final temperature, we can write

$$T_i^{\gamma} \, p_i^{1-\gamma} = T_f^{\gamma} \, p_f^{1-\gamma}$$

$$\left(\frac{p_i}{p_f} \right)^{\frac{1-\gamma}{\gamma}} = \frac{T_f}{T_i}$$

Taking natural logarithm on both sides, we can write

$$\frac{1}{\gamma} - 1 = \frac{\ln(T_f / T_i)}{\ln(p_i / p_f)} = \frac{\ln(286/320)}{\ln(1.4)} = -\frac{0.048784}{0.146128} = -0.3338$$

$$\frac{1}{\gamma} = 0.6662 \Rightarrow \gamma = 1.50$$

Since this value of γ is closer to diatomic gas (γ = 1.40) than a monatomic gas 1.67), the gas is diatomic.

Example 5.9 Obtain the equivalent of Eq. (5.19a) for a van der Waals' gas.

For an adiabatic change, the first law of thermodynamics can be expressed as

$$C_V dT + pdV = 0 \tag{i}$$

For a van der Waals' gas

$$p = \frac{RT}{V-b} - \frac{a}{V^2}$$

Using this in (i), we can write

$$C_V\, dT + \left[\frac{RT}{V-b} - \frac{a}{V^2}\right] dV = 0$$

On dividing throughout by C_V, we get

$$dT + \left[\frac{RT}{C_V\,(V-b)} - \frac{a}{C_V\, V^2}\right] dV = 0 \tag{ii}$$

The second term within the parentheses is very small. So if we ignore it, Eq. (ii) simplifies

$$\frac{R}{C_V\,(V-b)} dV = -\frac{dT}{T}$$

This expression can be easily integrated to obtain

$$\frac{R}{C_V} \ln\,(V-b) = -\ln T + C \tag{iii}$$

is the constant of integration.

Expressing Eq. (iii) in terms of numbers by taking antilog, we get the required result:

$$T(V-b)^{\gamma-1} = \text{Constant.}$$

Example 5.10 One mole of a perfect gas undergoes adiabatic change and its volume changes from V_1 to V_2 and pressure from p_1 to p_1. Show that the work done by it is given

$$W_{\text{adiabatic}} = \frac{p_2 V_2 - p_1 V_1}{1-\gamma}$$

We know that during an adiabatic process, a system is thermally insulated from its surroundings. When the gas expands from volume V_1 to V_2, the work done by it

$$W_{\text{adiabatic}} = \int_{V_1}^{V_2} p dV \tag{i}$$

We know that an adiabatic change is defined by

$$pV^{\gamma} = K \tag{ii}$$

On substituting the value for pressure in the expression for work done, we obtain

$$W_{\text{adiabatic}} = \int_{V_1}^{V_2} \frac{K}{V^{\gamma}} dV = \frac{K}{1-\gamma}\left(\frac{1}{V_2^{\gamma-1}} - \frac{1}{V_1^{\gamma-1}}\right) = \frac{1}{1-\gamma}[p_2 V_2 - p_1 V_1] \qquad \text{(iii)}$$

Note that in writing the last expression, we have used the result contained in Eq. (ii) above for an adiabatic change.

Example 5.11 One mole of an ideal gas defined by (p_1, T_1) is allowed to expand reversibly and isothermally till its pressure is reduced to one-half of the original pressure. This is followed by a constant volume cooling till its pressure is reduced to one-fourth of the initial value. Then it is restored to its initial state by a reversible adiabatic compression. Show that the net work done by the gas is equal to $RT\left[\ln 2 - \frac{1}{2(\gamma - 1)}\right]$.

The given series of processes is depicted on the indicator diagram in Fig. 5.10. $A \rightarrow B$ represents isothermal expansion where pressure reduces to one-half of the original value.

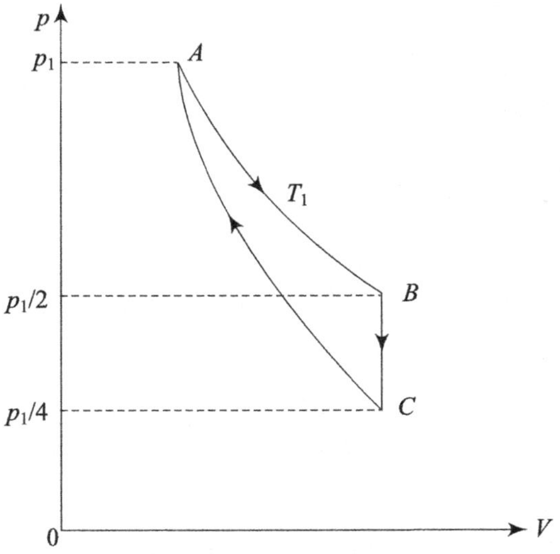

Indicator diagram for an isothermal expansion, isochoric process and adiabatic compression.

Hence, using the result contained in Eq. (iv) in Example 5.2, we can write the expression for work done by the gas as

$$W_{A \rightarrow B} = RT \ln 2 \qquad \text{(i)}$$

Note that Boyle's law implies that the volume of the gas will be doubled, since its pressure

The process $B \rightarrow C$ is isochoric. Hence, no work will be done by the gas but its temperature will drop. So we can write

$$W \quad = 0 \qquad \text{(ii)}$$

In going from $C \to A$, the gas is compressed adiabatically. Therefore, work will be done on the gas. Using the result obtained in Eq. (iii) in Example 5.10, we can write

$$W_{C\to A} = \frac{1}{1-\gamma}\left(p_1V_1 - \frac{p_1}{4}2V_1\right) = \frac{p_1V_1}{2(1-\gamma)} = -\frac{RT}{2(\gamma-1)} \quad \text{(iii)}$$

On combining Eqs. (i), (ii), (iii), we get

$$W = RT\left(\ln 2 - \frac{1}{2(\gamma-1)}\right) \quad \text{(iv)}$$

Example 5.12 One mole of air at STP is first compressed isothermally and reversibly to one-half of the original volume, and then expanded adiabatic and reversibly to its original pressure. Calculate the (a) net work done by the gas, (b) net heat flowing into the gas, (c) change in internal energy, and (d) final temperature. Take $R = 8.31$ $\text{Jmol}^{-1}\text{K}^{-1}$.

(a) The work done on one mole of the gas during the isothermal compression

$$W_{\text{isothermal}} = RT\log_e\frac{V_2}{V_1}$$

$$= (8.31\,\text{Jmol}^{-1}\text{K}^{-1})\times(273\,\text{K})\times\log_e\left(\frac{1}{2}\right)$$

$$= -1572.49\ \text{J}$$

For adiabatic expansion, we use Eq. (5.19c) to write

$$\frac{p_2^{\gamma-1}}{T_2^{\gamma}} = \frac{p_3^{\gamma-1}}{T_3^{\gamma}}$$

$T_2 = T_1 = 273$ K, $\quad p_2 = 2p_1;\ p_3 = p_1$ and $\gamma = 1.4$. Hence,

$$\frac{(2p_1)^{\gamma-1}}{273^{\gamma}} = \frac{p_1^{\gamma-1}}{T_3^{\gamma}}$$

$$T_3^{\gamma} = 273^{\gamma}\,2^{-(\gamma-1)}$$

$$T_3^{1.4} = 273^{1.4}\times 2^{-0.4}$$

On taking log of both the sides, we get

$$1.4\log_{10}T_3 = 1.4\log_{10}273 - 0.4\log_{10}2$$

$$= 3.4106 - 0.1204$$

$$\log_{10}T_3 = \frac{3.2902}{1.4} = 2.3501$$

Hence, the final temperature is

$$T_3 = 224\ \text{K}$$

The work done by one mole of the gas during adiabatic expansion

$$W_{\text{adiabatic}} = \frac{R}{\gamma - 1}(T_1 - T_3)$$

224 K and $T_1 = 273$ K. Hence,

$$W_{\text{adiabatic}} = \frac{(8.31\ \text{Jmol}^{-1}\text{K}^{-1})}{0.4} \times (49\ \text{K}) = 1017.98\ \text{J}$$

Net work done by the gas = (–1572.49 + 1017.98) J = –554.51 J

The net heat flowing into the gas:
Since no heat is exchanged in a reversible adiabatic process, the net heat flowing into the gas will equal to the work done during an isothermal process, which is –1572.49J.

The change in internal energy:
Since the change in internal energy of the system during a reversible isothermal process is zero, the change in internal energy (due to the fall in temperature) is equal to the work done during the adiabatic expansion which is 1017.98 J.

The final temperature is 224 K.

Example 5.13 One mole of oxygen at STP is adiabatically compressed to 5 atm. Calculate the new temperature. How much work is done on the gas? Take γ = 1.4 and 8.31 J mol^{-1} K^{-1}.

Here $T_1 = 0°\text{C} = 273$ K, $p_1 = 1$ atm, and $p_2 = 5$ atm. Hence, using Eq. (5.19c), we can write

$$T_2^\gamma = T_1^\gamma \left(\frac{p_2}{p_1}\right)^{\gamma-1} = (273\ \text{K})^{1.4} \times (5)^{0.4} = 2574.11 \times 1.9036\ \text{K} = 4900.2\ \text{K}$$

On taking logarithm of both sides, we get

$$\gamma \log T_2 = \log(4900.2) = 3.6902$$

$$\log T_2 = \frac{3.6902}{1.4} = 2.6359$$

$$T_2 = 432\ \text{K} = 159\ °\text{C}$$

The adiabatic work in compressing 1 mole gas is

$$W = \frac{R(T_1 - T_2)}{\gamma - 1}$$

$$= \frac{(8.31\ \text{J mol}^{-1}\text{K}^{-1}) \times (273 - 432)\ \text{K}}{1.4 - 1}$$

$$= -3303.2\ \text{J}$$

The –ve sign shows that the work is done on the gas.

You may now like to answer a few practise problems.

Problem 5.8 Use the results obtained in Example 5.10 to calculate the total work done in a four stage cyclic process comprising isothermal expansion, adiabatic expansion, isothermal compression and adiabatic compression (Fig. 5.11).

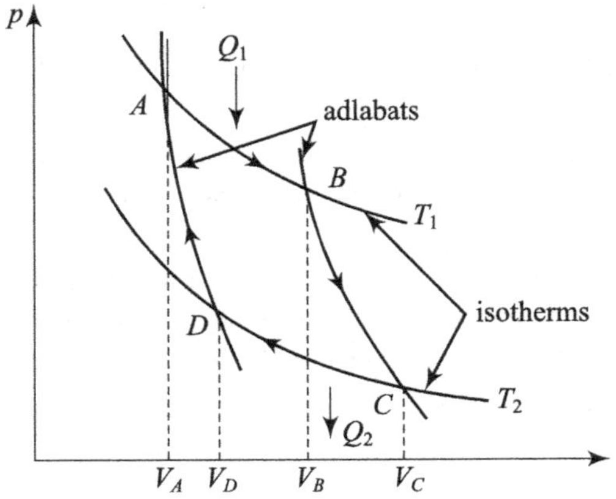

A four stage cyclic process. This cycle was used by Carnot for conversion of heat into work.

$$W = RT_1 \ln\left(\frac{V_2}{V_1}\right) - RT_2 \ln\left(\frac{V_3}{V_4}\right)$$

Problem 5.9 One mole of an ideal gas is heated isochorically till its temperature is doubled. Then it is expanded isothermally till it reaches the original pressure. Finally it is cooled by an isobaric process and restored to the original state. By assuming all the processes to be reversible, show that the resultant work done is RT [2 log 2 – 1].

Now we consider a large scale macroscopic system such as earth's atmosphere and obtain expression for adiabatic lapse rate.

Adiabatic Lapse Rate An interesting application of the adiabatic expansion of a gas is the calculation of the *adiabatic lapse rate*, i.e., variation of temperature in the atmosphere with height above sea level. This arises mainly due to convection currents in the troposphere—the lowest (and most dense) region of earth's atmosphere. The surface of the earth and hence air in its immediate contact gets heated due to heat radiations from the sun. As a result, a vertical density gradient is established. This gives rise to convection currents which continuously transport hot air from the lower regions to the higher ones and vice versa. When air from the sea level rises to the upper regions of lower pressure, it expands. Since air is a poor conductor of heat, very little heat is exchanged by the expanding air and we may consider the expansion as adiabatic.

To calculate the change in temperature with height, we note that in the logarithmic form, Eq. (5.19c) can be written as

$$\ln T = \frac{\gamma - 1}{\gamma} \ln p = \ln K$$

On differentiating this equation, we get

$$\frac{\mathrm{d}T}{T} - \frac{\gamma - 1}{\gamma}\frac{\mathrm{d}p}{p} = 0$$

$$\frac{\mathrm{d}p}{p} = \frac{\gamma}{\gamma - 1}\frac{\mathrm{d}T}{T} \tag{5.22}$$

If we assume that pressure decreases by dp as we go up a distance dh in the atmosphere, we can write

$$\mathrm{d}p = -\rho g \, \mathrm{d}h$$

is the density of air and g is acceleration due to gravity.
On dividing both sides by p and using the equation of state, we obtain

$$\frac{\mathrm{d}p}{p} = -\frac{gM}{RT}\mathrm{d}h \tag{5.23}$$

= 28.88u) is the average molecular weight of air.
On comparing Eqs. (5.22) and (5.23), we get

$$\frac{\mathrm{d}T}{\mathrm{d}h} = -\left(\frac{\gamma - 1}{\gamma}\right)\frac{Mg}{R} \tag{5.24}$$

This is the required expression of *adiabatic lapse rate*. The negative sign implies that temperature decreases with height.

Note that this result holds good for dry air. But there is a lot of moisture in the air. You may therefore logically ask: How will the presence of water vapour in air affect it? The lapse rate decreases in the presence of vapours. The decrease in temperature with height under static conditions as well as early morning air is depicted in Fig. 5.12(a) and (b) respectively.

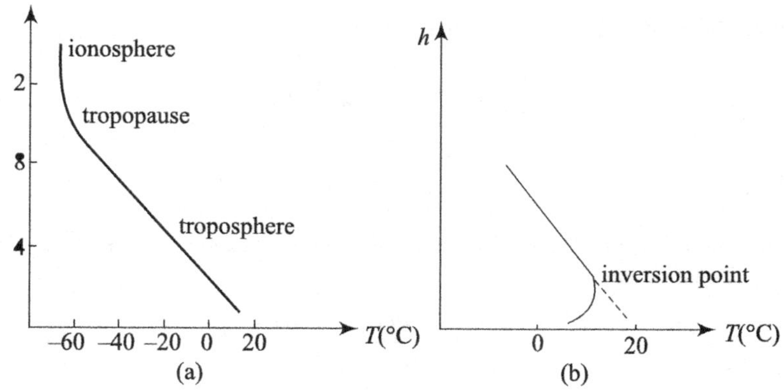

Adiabatic lapse rate under (a) steady conditions, and (b) early morning.

If we take $\gamma = 1.4$, $g = 9.81\ \text{ms}^{-2}$, $R = 8.31\ \text{J mol}^{-1}\text{K}^{-1}$ and $M = 0.029\ \text{kg mol}^{-1}$, the adiabatic lapse rate is given by

$$\frac{\mathrm{d}T}{\mathrm{d}h} = -\frac{0.4}{1.4} \times \frac{(0.029\ \text{kg mol}^{-1}) \times 9.81\ \text{ms}^{-2}}{8.31\ \text{J mol}^{-1}\text{K}^{-1}}$$

$$= -9.78 \times 10^{-3}\ \text{K m}^{-1}$$

This shows that temperature drop over 1 km will be about 10°C. This value is somewhat higher than the observed lapse rate of 5 – 6.5°C per kilometre. The difference may be attributed to the effect of water vapours in the expanding mass of air, which we have not taken into account. As water vapours rise with the convection currents, they cool and eventually condense into the liquid form. The latent heat of vaporisation released in this process tends to reduce the rate of cooling in the troposphere. Of course, the exact value of the adiabatic lapse rate will be determined by the amount of water vapour in the atmosphere at those heights. Some modifications to the vertical temperature gradient are also caused by mountains, glaciers, volcanoes and industrial/vehicular exhaust.

The Enthalpy

You may recall that in arriving at Eq. (5.11), we expressed the internal energy in terms , and discovered that C_V is the differential coefficient of U, a function of state, with respect to T. You could jolly well ask: Can we define a function of state which corresponds to C_p? To answer this question, we write U as a function of T and p. We can do so because we need two coordinates to specify the state of a gas. Then, an infinitesimal change in temperature and pressure induces change in internal energy. Mathematically, we can express this as

$$\mathrm{d}U = \left(\frac{\partial U}{\partial T}\right)_p \mathrm{d}T + \left(\frac{\partial U}{\partial p}\right)_T \mathrm{d}p$$

so that the first law of thermodynamics takes the form

$$\delta Q = \left(\frac{\partial U}{\partial T}\right)_p \mathrm{d}T + \left(\frac{\partial U}{\partial p}\right)_T \mathrm{d}p + p\mathrm{d}V$$

Hence, on dividing by dT at constant p, we can write

$$\left(\frac{\delta Q}{\partial T}\right)_p = C_p = \left(\frac{\partial U}{\partial T}\right)_p + p\left(\frac{\partial V}{\partial T}\right)_p$$

$$= \left(\frac{\partial}{\partial T}(U + pV)\right)_p \tag{5.25}$$

The sum $U + pV$ is denoted by a new function of state, known as *enthalpy*. We denote . Thus, we can write

$$C_p = \left(\frac{\partial Q}{\partial T}\right)_p = \left(\frac{\partial H}{\partial T}\right)_p$$

$$H = U + pV \tag{5.26}$$

The enthalpy often appears in flow processes. Joule–Thomson expansion is one of the most important examples where enthalpy is conserved. From Chapter 3, you may recall that in this experiment, a gas is allowed to expand from a high constant pressure against a lower constant pressure through a porous plug. The entire system is thermally isolated. We will discuss it again in Chapter 10.

We can use enthalpy to write another form of the first law of thermodynamics. To do so, as a function of T and p and write

$$H = H(T, p)$$

An infinitesimal change in temperature and/or pressure will induce infinitesimal change in enthalpy:

$$\mathrm{d}H = \left(\frac{\partial H}{\partial T}\right)_p \mathrm{d}T + \left(\frac{\partial H}{\partial p}\right)_T \mathrm{d}p$$

From the definition of enthalpy (Eq. (5.26)), we can write

$$\mathrm{d}H = \mathrm{d}U + p\mathrm{d}V + V\mathrm{d}p$$

On combining this result with the first law of thermodynamics, we can write

$$\delta Q = \mathrm{d}U + p\mathrm{d}V = \mathrm{d}H - V\mathrm{d}p$$

On substituting for d H, we get

$$\delta Q = \left(\frac{\partial H}{\partial T}\right)_p \mathrm{d}T + \left[\left(\frac{\partial H}{\partial p}\right)_T - V\right]\mathrm{d}p = C_p\,\mathrm{d}T + \left[\left(\frac{\partial H}{\partial p}\right)_T - V\right]\mathrm{d}p \tag{5.6a}$$

This is another form of the first law of thermodynamics.

Adiabatic and Isothermal Elasticities

The first law of thermodynamics can also be used to show that for a gaseous system, the ratio of adiabatic elasticity to isothermal elasticity is equal to the ratio of two heat capacities, i.e.,

$$\frac{E_S}{E_T} = \frac{C_p}{C_V} = \gamma$$

To show this we note that by definition, elasticity E is given by

$$E = \frac{\text{stress}}{\text{strain}} = -V(\Delta p/\Delta V)$$

Hence, adiabatic elasticity is given by

$$E_S = -V(\partial p/\partial V)_S$$

and isothermal elasticity is defined as

$$E_T = -V(\partial p/\partial V)_T$$

$$\frac{E_S}{E_T} = \frac{(\partial p/\partial V)_S}{(\partial p/\partial V)_T} \tag{5.27}$$

Let us take T as dependent variable and write

$$T = T(V, p)$$

Then an infinitesimal change in temperature can he expressed as

$$\mathrm{d}T = \left(\frac{\partial T}{\partial V}\right)_p \mathrm{d}V + \left(\frac{\partial T}{\partial p}\right)_V \mathrm{d}p \tag{5.28}$$

In an isothermal process, temperature remains constant, i.e., $\mathrm{d}T = 0$. Therefore, the above equation can be rewritten as

$$\left(\frac{\partial p}{\partial V}\right)_T = -\frac{\left(\frac{\partial T}{\partial V}\right)_p}{\left(\frac{\partial T}{\partial p}\right)_V} \tag{5.29}$$

From Eqs. (5.8) and (5.9), we recall that

$$\delta Q = C_V\, \mathrm{d}T + \left(p + \left(\frac{\partial U}{\partial V}\right)_T\right) \mathrm{d}V$$

Combining this result with Eq. (5.11), we can write

$$\delta Q = C_V\, \mathrm{d}T + \frac{(C_p - C_V)}{(\partial V/\partial T)_p} \mathrm{d}V$$

$$= C_V \mathrm{d}T + (C_p - C_V)\, (\partial T/\partial V)_p\, \mathrm{d}V$$

On dividing throughout by C_V, we can write

$$\frac{\delta Q}{C_V} = \mathrm{d}T + (\gamma - 1)\,(\partial T/\partial V)_p\, \mathrm{d}V$$

C_p/C_V is ratio of two heat capacities.

Combining this result with Eq. (5.28) and simplifying the resultant expression, we

$$\frac{\delta Q}{C_V} = \left(\frac{\partial T}{\partial p}\right)_V \mathrm{d}p + \gamma\,(\partial T/\partial V)_p\, \mathrm{d}V$$

For an adiabatic process, $\delta Q = 0$ and this equation gives

$$(\partial p/\partial V)_S = -\frac{\gamma\,(\partial T/\partial V)_p}{(\partial T/\partial p)_V} \tag{5.30}$$

On substituting for $(\partial p/\partial V)_S$ and $(\partial p/\partial V)_T$ from Eqs. (5.30) and (5.29) in Eq. (5.27), we

$$\frac{E_S}{E_T} = \frac{C_p}{C_V} = \gamma \tag{5.31}$$

You should now answer the following practise problems.

Problem 5.10 Propagation of sound in air involves adiabatic changes and the speed of sound is given by

$$v = \sqrt{\frac{E_S}{\rho}}$$

is density of the medium. Calculate the velocity of sound in ^{4}He at 6 K and 19.7 atm pressure. The isothermal compressibility and density of ^{4}He under given conditions are respectively $9.42 \times 10^{-8} m^2N^{-1}$ and 162 kg m^{-3}. Take $\gamma = 1.48$.

$\times 10^2$ ms^{-1}

Problem 5.11 Refer to Fig. 5.13. It shows the changes in the state of a gas occupying 10^{-3} m^3 at 3 K and pressure 10^3 Pa. The gas is initially compressed at constant volume so that its temperature increases to 300 K. It is then allowed to expand adiabatically till the original temperature is attained. This is followed by isothermal compression which restores the original volume. Assuming each process as reversible, calculate the total work done and the heat transferred.

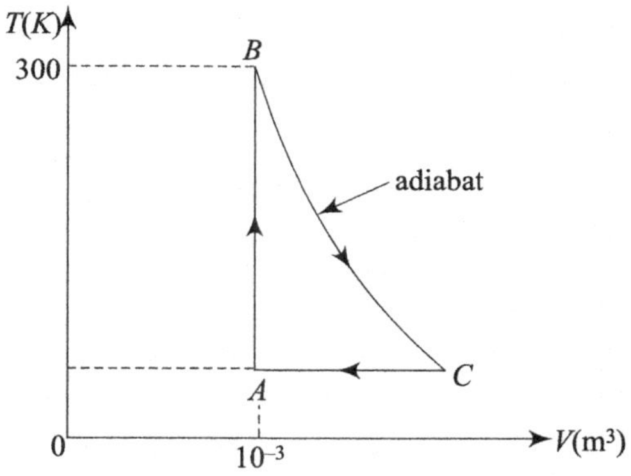

Fig. 5.13 Changes in the state of a gas.

141.6 J; 148.5 J

DDITIONAL EXAMPLES

Example 5.14 A 50 kg weight is made to fall through a height of 6 m on a paddle wheel which stirs 6 kg water. Calculate the rise in temperature, if no heat is lost in this process. Take $g = 9.8$ ms^{-2} and specific heat capacity of water as 332 J kg^{-1} K^{-1}.

The potential energy of falling weight $= mgh$

$$= 4(50 \text{ kg}) \times (9.8 \text{ ms}^{-2}) \times (6 \text{ m})$$
$$= 294 \text{ J}.$$

This energy is used up in raising the temperature of water. So we can write

$$2940 \text{ J} = (6 \text{ kg}) \times (332 \text{ J kg}^{-1}\text{K}^{-1}) \times \Delta T$$

$$\therefore \qquad \Delta T = \frac{2940 \text{ J}}{6 \times 332 \text{ JK}^{-1}} = 1.48 \text{ K}.$$

Example 5.15 A perfect gas at 300 K occupies a volume 0.2 m^3 at a pressure 10^6 Nm^{-2}. It is allowed to expand isothermally to a volume 0.5 m^3. Then the gas is compressed isobarically to its original volume. Finally, the pressure is increased isochorically and returns to its original state. Calculate the work done during this cycle. $= 8.3\ Jmol^{-1}K^{-1}$.

The cycle of changes described in the question is depicted in Fig. 5.14. To calculate the work done in this reversible process, we first determine the number of moles of the perfect gas using the equation of state:

$$n = \frac{pV}{RT}$$

On substituting the given values, we get

$$n = \frac{(5 \times 10^6\ Nm^{-2}) \times (0.2\ m^3)}{(8.3\ Jmol^{-1}K^{-1}) \times (300\ K)} = 400\ mol$$

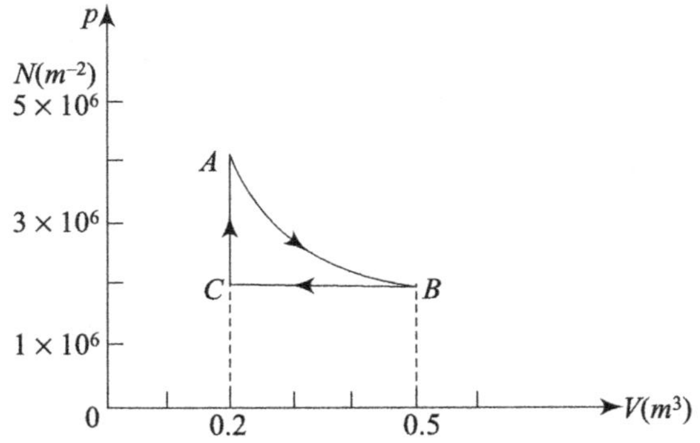

(a) The final pressure at the end of the isothermal expansion is given by

$$p_f = \frac{p_i V_i}{V_f} = \frac{(5 \times 10^6\ Nm^{-2}) \times (0.2 m^3)}{(0.5 m^3)} = 2 \times 10^6\ Nm^{-2}$$

The work done by the gas in this process is

$$\begin{aligned} W_{A \to B} &= nRT \ln(V_f/V_i) \\ &= (4 \times 10^2\ mol) \times (8.3\ Jmol^{-1}K^{-1}) \times (300\ K) \times \ln(0.5/0.2) \\ &= 9.1 \times 10^5\ J \end{aligned}$$

(b) The work done on the gas in going from $B \to C$ isobarically is given by

$$W_{B \to C} = p\Delta V = (2 \times 10^6\ Nm^{-2}) \times (0.3\ m^3) = 6 \times 10^5\ J$$

(c) Since the change from $C \to A$ is isochoric, no work is done.

Hence, total work done in the cycle by the gas is

$$W = W_{A\to B} - W_{B\to C} = (9.1 - 6) \times 10^5 \text{ J} = 3.1 \times 10^5 \text{ J}$$

Example 5.16 32 kg of oxygen gas kept at a pressure of 5 atm and 27°C is made to under a cycle of changes. It is first heated at constant volume till such time that its pressure increases by a factor of 3. In the second step, it is allowed to expand isothermally till it regains the initial pressure. Finally it is cooled isobarically to the original temperature. Calculate the change in its internal energy and work done. Use atomic weight of oxygen atom as 16 and $R = 8314 \text{ Jkmol}^{-1}\text{K}^{-1}$. You can assume that given mass of oxygen behaves like an ideal gas.

Refer to Fig. (5.15). Note that b, the gas is heated at constant volume; from $b \to c$, it undergoes isothermal expansion and from $c \to a$, the gas is cooled isobarically. Here $p_i = 5$ atm and $T_i = 300$ K. In the process $a \to b$, $V =$ constant so that 0 and the first law of thermodynamics implies that

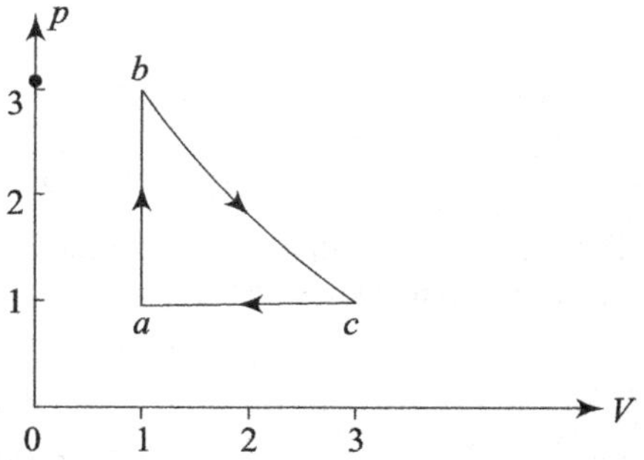

Fig. 5.15

$$\delta Q = \mathrm{d}U$$

Since pressure becomes three times, the final temperature T_b is given by

$$T_b = T_a\left(\frac{p_f}{p_i}\right)$$

$$= (300 \text{ K}) \times 3 = 900 \text{ K}$$

$\therefore$

$$\Delta T = T_b - T_a = 900 \text{ K} - 300 \text{ K} = 600 \text{ K}$$

For a diatomic gas, $C_V = \frac{5}{2}R = \frac{5}{2} \times 8314 \text{ J kmol}^{-1}\text{K}^{-1}$. Hence, change in internal energy

$$\mathrm{d}U = \mu C_V \Delta T$$

$$= \left(\frac{5}{2} \times 8314 \text{ J kmol}^{-1}\text{K}^{-1}\right) \times (600 \text{ K}) \times (1 \text{ kmol})$$

$$= 1.247 \times 10^7 \text{ J} \qquad \text{(i)}$$

During the process $b \to c$, the pressure decreases from 3 atm to 1 atm at constant temperature so that Boyle's law applies:

$$pV = \text{constant}$$

$$\frac{V_f}{V_i} = \frac{p_i}{p_f} = 3$$

The work done by an ideal gas during the isothermal expansion is given by

$$W = 2.3\ \mu RT \log_{10}\left(\frac{V_f}{V_i}\right)$$

$$= 2.3 \times (8314\ \text{J kmol}^{-1}\text{K}^{-1}) \times (900\ \text{K}) \times \log_{10}(3) \times (1\ \text{kmol})$$

$$= 0.822 \times 10^7\ \text{J}$$

Since temperature remains constant during isothermal expansion, the internal energy remains unchanged, i.e., $dU = 0$.
During the process $c \rightarrow a$, p remains constant but temperature drops from 900 K to 300 K. Hence, change in internal energy

$$dU = +\frac{5}{2}\mu R\Delta T$$

$$= +\frac{5}{2} \times (8314\ \text{Jkmol}^{-1}\text{K}^{-1}) \times (-600\ \text{K}) \times (1.0\ \text{kmol})$$

$$= -1.247 \times 10^7\ \text{J}$$

The negative sign implies that internal energy of the gas decreases.
The heat given out at constant pressure

$$\delta Q = \mu C_p \Delta T$$

$$= (C_V + R)\Delta T$$

$$= \frac{7}{2}\mu R\Delta T$$

$$= \frac{7}{2} \times (8314\ \text{Jmol}^{-1}\text{K}^{-1}) \times (-600\ \text{K}) \times (1.0\ \text{kmol})$$

$$= -1.746 \times 10^7\ \text{J}$$

Hence work done on the gas is given by

$$\delta W = \delta Q - dU$$

$$= (-1.746 \times 10^7\ \text{J}) + (1.247 \times 10^7\ \text{J})$$

$$= -0.387 \times 10^7\ \text{J}$$

Note that the total change in internal energy in a cyclic process is zero. But the network done by the gas is given by

$$\Delta W = 0.822 \times 10^7\ \text{J} - 0.387 \times 10^7\ \text{J}$$

$$= 0.435 \times 10^7\ \text{J}$$

Geometrically, this is equal to the area under the curve *abc*.

Example 5.17 A certain mass of a diatomic gas at STP is expanded to three times its volume adiabatically. Calculate the change in temperature and pressure.

(i) To calculate the final temperature when an adiabatic transformation is accompanied by change in volume, we use the relation

$$TV^{\gamma-1} = \text{constant}$$

$$T\ V^{\gamma-1} = T\ V^{\gamma-1}$$

$3V_1$ and $T_1 = 273$ K. Hence,

$$T_2 = T_1 \left(\frac{V_1}{V_2}\right)^{\gamma-1}$$

For a diatomic gas $\gamma = 1.4$ so that $\gamma - 1 = 0.4$. Hence,

$$T_2 = (273 \text{ K}) \times \left(\frac{1}{3}\right)^{0.4}$$
$$= 175.9 \text{ K}$$

Hence change in temperature $\Delta T = (175.9 - 273)$ K $= -97.1$ K.

To calculate the change in pressure, we use the relation

$$pV^{\gamma} = \text{constant}$$

$$p_1 V_1^{\gamma} = p_2 V_2^{\gamma}$$

$$p_2 = p_1 \left(\frac{V_1}{V_2}\right)^{\gamma}$$

$$= (1 \text{ atm}) \times \left(\frac{1}{3}\right)^{1.4}$$

$$= 0.215 \text{ atm}$$

Hence, change in pressure $= -0.785$ atm.

Example 5.18 The front types of a sedan can have a pressure of 2 atm at 300 K. If one of the tyres suddenly bursts, calculate the final temperature. Take $\gamma = 1.40$.

When a tyre bursts suddenly, the process is so fast that no transfer of heat takes place. This exemplifies an adiabatic change. To calculate temperature when pressure changes adiabatically, we use the relation

$$\frac{p^{\gamma-1}}{T^{\gamma}} = \text{constant}$$

$$\frac{p_1^{\gamma-1}}{T_1^{\gamma}} = \frac{p_2^{\gamma-1}}{T_2^{\gamma}}$$

$$T_2^{\gamma} = \left(\frac{p_2}{p_1}\right)^{\gamma-1} T_1^{\gamma}$$

$$T_2 = \left(\frac{p_2}{p_1}\right)^{\frac{\gamma-1}{\gamma}} T_1$$

$$p_1 = 2 \text{ atm}, \quad p_2 = 1 \text{ atm} \quad \text{and} \quad T_1 = 300 \text{ K}$$

$$T_2 = (300\text{ K})\left(\frac{1\text{ atm}}{2\text{ atm}}\right)^{\frac{0.4}{1.4}}$$

$$= (300\text{ K})\left(\frac{1}{2}\right)^{\frac{0.4}{1.4}}$$

$$= (300\text{ K}) \times 0.82$$
$$= 246.1\text{ K}$$
$$= -26.9°\text{C}$$

Example 5.19 A monatomic ideal gas at 27°C is suddenly compressed to one-tenth of its original volume. Calculate the temperature after compression. How will this result be different for a diatomic gas?

Since the gas is compressed suddenly, the process is adiabatic. It is described by the relation

$$TV^{\gamma-1} = \text{constant}$$

$$T_1 V_1^{\gamma-1} = T_2 V_2^{\gamma-1}$$

$$T_2 = T_1 \left(\frac{V_1}{V_2}\right)^{\gamma-1}$$

For a monatomic gas, $\gamma = 1.67$. Further, we are told that $\frac{V_1}{V_2} = 10$ and $T_1 = 300$ K. Using these values, we get

$$T_2 = (300\text{ K}) \times (10)^{1.67-1}$$
$$= (300\text{ K}) \times (10)^{0.67}$$
$$= 1403\text{ K}$$
$$= 1130°\text{C}$$

For a diatomic gas, $\gamma = 1.4$. Hence, the final temperature is given by

$$T_2 = (300\text{ K}) \times (10)^{0.4}$$
$$= 753.6\text{ K}$$
$$= 480.6°\text{C}$$

Example 5.20 Express the work done by a gas in terms of isothermal compressibility and the coefficient of thermal expansion, α.

The work done by a gas is given by

$$\delta W = p dV \tag{i}$$

V as a function of T and p, we can write

$$V = V(T, p)$$

Therefore, an infinitesimal change in volume is given by

$$dV = \left(\frac{\partial V}{\partial T}\right)_p dT + \left(\frac{\partial V}{\partial p}\right)_T dp$$

$$\beta = -\frac{1}{V}\left(\frac{\partial V}{\partial p}\right)_T \quad \text{and} \quad \alpha = -\frac{1}{V}\left(\frac{\partial V}{\partial T}\right)_p, \text{ we can write}$$

$$dV = \alpha V\, dT - V\beta_T\, dp \qquad \text{(ii)}$$

On using this result in Eq. (i), we get the required expression for the work done in terms α:

$$\delta W = p\alpha V\, dT - pV\beta_T\, dp$$

Example 5.21 Calculate the work done when one litre of a monatomic gas at STP is compressed adiabatically to half its volume.

The work done by a gas in adiabatic compression is given by

$$W = \left(\frac{1}{1-\gamma}\right)\left[\frac{1}{V_2^{\gamma-1} - V_1^{\gamma-1}}\right] \qquad \text{(i)}$$

10^3 cm^3 = 10^{-3} m^3, $V_2 = \frac{V_1}{2} = 5 \times 10^{-4}$ m^3 and for a monatomic gas $\gamma = 1.67$.

On inserting these values in Eq. (i), we get

$$W = \left(\frac{1}{1-1.67}\right)\left[\frac{1}{(5\times10^{-4}\ \text{m}^3)^{0.67}} - \frac{1}{(10^{-3}\ \text{m}^3)^{0.67}}\right]$$

$$= -\frac{1}{0.67}[(2\times10^3\ \text{m}^3)^{0.67} - (10^3\ \text{m}^3)^{0.67}]$$

$$= -\frac{1}{0.67}[162.81 - 102.33] = -90.3\ \text{J}$$

The negative sign signifies that work is done on the gas.

Example 5.22 A gas at 27°C is compressed suddenly to eight times its initial pressure. Calculate the rise in temperature. Take $\gamma = 1.5$.

An adiabatic process in which pressure and temperature vary is described by the relation

$$\frac{p^{\gamma-1}}{T^{\gamma}} = \text{constant}$$

So we can write

$$\frac{p_1^{\gamma-1}}{T_1^{\gamma}} = \frac{p_2^{\gamma-1}}{T_2^{\gamma}}$$

$$T_2 = T_1\left(\frac{p_2}{\ }\right)^{\frac{\gamma-1}{\gamma}}$$

$$\frac{p_2}{p_1} = 8 \text{ and } T_1 = 300 \text{ K. Hence,}$$

$$T_2 = (300 \text{ K}) \times (8)^{\frac{0.5}{1.5}}$$

$$= (300 \text{ K}) \times (8)^{\frac{1}{3}}$$

$$= (300 \text{ K}) \times 1.9999$$

$$= 599.96 \text{ K}$$

Hence, rise in temperature = (599.96 – 300) K = 299.96 K = 299.96°C.

Example 5.23 Air at STP is compressed (i) slowly and (ii) suddenly to half its volume. Calculate the change in its temperature.

(i) When the gas is compressed slowly, the process is isothermal and there will be no change in its temperature.

When the gas is compressed suddenly, the process is adiabatic. To calculate the change in temperature, we use the relation

$$TV^{\gamma-1} = \text{constant}$$

So we can write

$$T_1 V_1^{\gamma-1} = T_2 V_2^{\gamma-1}$$

$$T_2 = T_1 \left(\frac{V_1}{V_2} \right)^{\gamma-1}$$

Since air is predominantly oxygen, which is diatomic, we have $\gamma = 1.4$. Also $\frac{V_1}{V_2} = 2$ and 273 K. Hence,

$$T_2 = (273 \text{ K}) \times (2)^{0.4}$$

$$= 360.2 \text{ K}$$

Hence, change in temperature

$$\Delta T = T_2 - T_1 = 87.2 \text{ K.}$$

Example 5.24 A certain mass of an ideal gas at 27°C and 8 atm pressure is expanded suddenly to four times its initial volume. Calculate final pressure and temperature of the $\gamma = 1.5$.

To calculate final pressure when adiabatic transformation is accompanied by change in volume, we use the relation

$$pV^{\gamma} = \text{constant}$$

$$p_1 V_1^{\gamma} = p_2 V_2^{\gamma}$$

$$p_2 = p_1 \left(\frac{V_1}{V_2} \right)^{\gamma}$$

8 atm and $V_2 = 4V_1$ so that $\frac{V_1}{V_2} = \frac{1}{4}$. Using these values in Eq. (i), we get

$$p_2 = (8\text{ atm}) \times \left(\frac{1}{4}\right)^{1.5}$$
$$= 1\text{ atm}$$

To calculate final temperature, we use the relation

$$TV^{\gamma} = \text{constant}$$

$$T_1 V_1^{\gamma-1} = T_2 V_2^{\gamma-1}$$

$$T_2 = T_1 \left(\frac{V_1}{V_2}\right)^{\gamma-1}$$

300 K and $\gamma - 1 = 0.5$. Hence,

$$T_2 = (300\text{ K}) \times \left(\frac{1}{4}\right)^{0.5}$$
$$= 150\text{ K}.$$

Example 5.25 The temperature of 5 kg of air is raised by 1°C at constant volume. Calculate the increase in its internal energy. Given $C_p = 999\text{ J kg}^{-1}\text{K}^{-1}$ and $\gamma = \frac{5}{3}$.

We know that in an isochoric process, no work is done and internal energy equals heat taken. Since

$$dU = mC_V\, dT = \frac{mC_p}{\gamma} dT$$

$$dU = (5\text{ kg}) \times \frac{(999\text{ J kg}^{-1}\text{K}^{-1}) \times 3}{5} \times (1\text{ K})$$
$$= 2997\text{ J}.$$

Example 5.26 Ten litre of air at 17°C and 76 cm of Hg pressure is compressed adiabatically to a volume of 0.5 litre. Calculate final temperature and pressure. Take

To calculate the final temperature, we use the relation

$$TV^{\gamma-1} = \text{constant}.$$

$$T_2 = T_1 \left(\frac{V_1}{V_2}\right)^{\gamma-1}$$

17°C = 290 K, $V_1 = 10\ell$, $V_2 = 0.5\ell$ and $\gamma = 1.4$. Hence,

$$T_2 = (290\text{ K}) \left(\frac{10}{0.5}\right)^{1.4-1}$$

$$= 290 \times (20)^{0.4}\ \text{K}$$
$$= 961.19\ \text{K}$$
$$= 688.19°\text{C}$$

To calculate the final pressure, we use the relation

$$PV^{\gamma} = \text{constant}$$

$$p_2 = p_1 \left(\frac{V_1}{V_2}\right)^{\gamma}$$
$$= (76\ \text{cm}) \times (20)^{1.4}$$
$$= 5037.97\ \text{cm of Hg}$$

Since 76 cm of Hg = 1 atm, 5037.97 cm of Hg = 66.29 atm.

Example 5.27 For a gas obeying van der Waals' equation, the change in internal energy with volume at constant temperature is given by

$$\left(\frac{\partial U}{\partial V}\right)_T = \frac{a}{V^2}$$

Using this result, show that $\left(\frac{\partial T}{\partial V}\right)_U = -\frac{1}{C_V}\frac{a}{V^2}$, if this is made to undergo free expansion.

For a real gas, internal energy is a function of temperature as well as volume. Therefore, we can write

$$U = U(T, V)$$

We can express a small change in internal energy in terms of changes in T and V as

$$dU = \left(\frac{\partial U}{\partial T}\right)_V dT + \left(\frac{\partial U}{\partial V}\right)_T dV$$

If we assume that internal energy remains constant during free expansion, we can write

$$\left(\frac{\partial U}{\partial T}\right)_V \left(\frac{\partial T}{\partial V}\right)_U + \left(\frac{\partial U}{\partial V}\right)_T = 0$$

$$\left(\frac{\partial T}{\partial V}\right)_U = -\frac{\left(\frac{\partial U}{\partial V}\right)_T}{\left(\frac{\partial U}{\partial T}\right)_V}$$
$$= -\frac{1}{C_V}\frac{a}{V^2}$$

Since all the quantities on the RHS of this expression are positive definite, the temperature will fall when a van der Waals' gas is made to undergo free expansion.

Example 5.28 Using the result derived in the above example, calculate drop in temperature if $a = 0.0245\ \text{Nm}^4\text{mol}^{-2}$, $C_V = 21\ \text{J mol}^{-1}\text{K}^{-1}$ and the gas expands from a volume of 2 litre to triple its volume.

We know that change in temperature when a van der Waals gas undergoes free expansion is given by

$$dT = -\frac{1}{C_V}\left(\frac{a}{V^2}\right)dV$$

0.0245 Nm^4mol^{-2}, $C_V = 21\ JK^{-1}\ mol^{-1}$, $V = 2$ litre $= 2 \times 10^3\ cm^3 = 2 \times 10^{-3}m^3$ and change in volume. $dV = (2 \times 3 - 2)$ litre $= 4$ litre $= 4 \times 10^{-3}m^3$. On substituting these values in the expression for change in temperature, we get

$$-\frac{1}{(21\ Jmol^{-1}K^{-1})} \times \frac{0.0245\ Nm^4\ mol^{-2}}{(2\times10^{-3}\ m^3)^2} \times (4\times10^{-3}\ m^3)$$

$$-\frac{0.0245\times10^{-3}}{21\times10^{-6}}\ K$$

-1.2 K.

Example 5.29 One gram molecule of monatomic $\left(\gamma = \frac{5}{3}\right)$ perfect gas at 27°C is adiabatically compressed in a reversible process from an initial pressure of 1 atmosphere to a final pressure of 50 atmospheres. Calculate the resulting difference in temperature.

In this case we use the relation connecting p and T for an adiabatic change:

$$p^{\gamma-1}/T^{\gamma} = \text{constant}$$

$$\left(\frac{p_2}{p_1}\right)^{\gamma-1} = \left(\frac{T_2}{T_1}\right)^{\gamma}$$

$= 50$ atm, $p_1 = 1$ atm, $T_1 = 27°C = 300$ K and $\gamma = \frac{5}{3}$. On substituting the given values, we get

$$(50)^{\frac{2}{3}} = \left(\frac{T_2}{300}\right)^{5/3}$$

$$T_2 = 300(50)^{2/5}$$

$$= 1434.5\ K = 1161.5°C$$

Hence change in temperature $\Delta T = T_2 - T_1 = 1134.5°C$.

Example 5.30 Calculate the change in internal energy and state whether temperature will fall or rise when

a system absorbs 600 cal of heat and performs 420 J work.
No heat is absorbed by the system but 210 J work is done on it.
250 cal heat is evolved by the system and 350 J work is done on the system.

From the first law of thermodynamics, we know that

$$\delta Q = dU + \delta W = dU + pdV$$

$$dU = \delta Q - \delta W$$

$$\delta Q = 600 \text{ cal}, \ \delta W = \frac{420 \text{ J}}{4.2 \text{ J cal}^{-1}} = 100 \text{ cal}$$

$$dU = (600 - 100) \text{ cal}$$
$$= 500 \text{ cal}$$

> 0, temperature will rise.

$$\delta Q = 0, \ \delta W = -\frac{210 \text{ J}}{4.2 \text{ J cal}^{-1}} = -50 \text{ cal}$$

$$dU = 0 - (-50 \text{ cal})$$
$$= 50 \text{ cal}$$

As before, $dU > 0$ and temperature will rise.

$$\delta Q = -250 \text{ cal}, \ \delta W = -\frac{350 \text{ J}}{4.2 \text{ J cal}^{-1}} = -83.3 \text{ cal}$$

$$dU = [-250 - (-83.3)] \text{ cal}$$
$$= -166.7 \text{ cal}$$

< 0, temperature will fall.

Example 5.31 A gas is confined in a cylinder and 500 cal of heat is added to it. As a result, it expands against the atmospheric pressure. If the change in internal energy of the gas is 420 J, calculate the increase in volume. Take 1 atm = $1.013 \times 10^5 \text{Nm}^{-2}$.

From the first law of thermodynamics, we know that

$$\delta W = p dV = \delta Q - dU$$

$$500 \text{ cal}, \ dU = 420 \text{ J} = \frac{420 \text{ J}}{4.2 \text{ J cal}^{-1}} = 100 \text{ cal}$$

$$\delta W = 400 \text{ cal} = 1680 \text{ J}$$

Since the gas expands against constant pressure,

$$dV = \frac{1680 \text{ J}}{1.013 \times 10^5 \text{ Nm}^{-2}}$$

$$= 1.66 \times 10^{-2} \text{m}^3$$

$$= 1.66 \times 10^4 \text{ cm}^3$$

Example 5.32 The molar energy of a real gas is given by

$$U = \frac{3}{2} RT - \frac{a}{V}$$

is molar volume at temperature T and a is a constant. Suppose that one mole of the gas occupies volume V at temperature T. The gas is allowed to expand adiabatically into vacuum so that its volume changes to V_2. Obtain the expression for final temperature. Use your result to calculate the change in temperature of N_2 if $a = 1.39$ litre^{2}mol^{-2}atm, 0.082 litre atm mol^{-1}K^{-1}, $V_1 = 22.4$ litre and $V_2 = 2V_1$.

Since the gas expands adiabatically into vacuum, $\delta Q = 0$ and $p = 0$ so that 0. Hence, $dU = 0$. That is U remains constant. Hence we can write

$$\frac{3}{2}RT_1 - \frac{a}{V_1} = \frac{3}{2}RT_2 - \frac{a}{V_2}$$

$$\frac{3}{2}RT_2 = \frac{3}{2}RT_1 + a\left(\frac{1}{V_2} - \frac{1}{V_1}\right)$$

$$T_2 = T_1 + \frac{2a}{3R}\left(\frac{1}{V_2} - \frac{1}{V_1}\right)$$

On substituting the given values, we get

$$T_2 - T_1 = \frac{2 \times (1.39 \text{ litre}^2 \text{ mol}^{-2} \text{ atm})}{3 \times (0.082 \text{ litre atm mol}^{-1} \text{K}^{-1})}\left(\frac{1}{44.8} - \frac{1}{22.4}\right) \text{litre}^{-1} \text{ mol}$$

$$= -0.25 \text{ K}$$

will cool slightly due to Joule effect.

Example 5.33 The specific heat capacity of hydrogen at constant pressure is 3.456 . At STP, one litre of hydrogen weighs 0.0896 g. Calculate its specific heat capacity at constant volume. Take $J = 4.2$ J cal^{-1} and 1 atmospheric pressure = 10^6 dynes

We know that

$pV = nRT$, where n denotes the number of moles of the gas

$p = 10^6$ dynes cm^{-2}, specific volume $v = \dfrac{1000}{0.08968}$ cm^3 g^{-1} and $T = 273$ K.

is molecular weight of hydrogen, then for 1 g of the gas, $n = \dfrac{1}{M}$ mole

$$\frac{pv}{T} = nR = \frac{R}{M}$$

$$c_P - c_V = nR = \frac{R}{M} = \frac{pv}{T} = \frac{(10^6 \text{ dynes cm}^{-2}) \times (1000 \text{ cm}^3)}{(273 \text{ K}) \times (0.0896 \text{ g})} = \frac{10^9 \text{ dynes cm g}^{-1}}{273 \times 0.0896 \text{ K}}$$

$$= \frac{10^9 \text{ ergs g}^{-1}}{(273 \text{ K}) \times 0.0896 R} = \frac{10^2 \text{ Jg}^{-1}}{(273 \text{ K}) \times 0.0896}$$

$$= \frac{10^2 \text{ Jg}^{-1}}{(273 \text{ K}) \times 0.0896 \times (4.2 \text{ J cal}^{-1})}$$

$$= 0.973 \text{ cal g}^{-1}\text{K}^{-1}$$

$$c_V = (c_P - 0.973) \text{ cal g}^{-1}\text{K}^{-1}$$

$$= (3.456 - 0.973) \text{ cal g}^{-1}\text{K}^{-1}$$

$$= 2.483 \text{ cal g}^{-1}\text{K}^{-1}.$$

Example 5.34 Calculate the quantity of heat absorbed, the increase in internal energy and the work done by the gas when the temperature of 30 g of oxygen is raised from 30°C to 100°C **(a)** at constant pressure and **(b)** at constant volume. It is given that 2 cal mol^{-1}K^{-1}, c_V for oxygen = 0.156 cal g^{-1}K^{-1} and molecular weight of oxygen = 32.

We know that

$$c_P = c_V + \frac{R}{M}$$

$$= \left(0.156 + \frac{2}{32}\right) \text{cal g}^{-1}\text{K}^{-1}$$

$$= 0.219 \text{ cal g}^{-1}\text{K}^{-1}$$

Mass of the gas, $m = 30$ g and increase in temperature = (100 – 30)°C = 70°C = 70°K.

For constant volume

$$dU = C_V dT = mc_V dT$$

$$= (30 \text{ g}) \times (0.156 \text{ cal g}^{-1}\text{K}^{-1}) \times (70 \text{ K})$$

$$= 327.6 \text{ cal}$$

$$\delta W = pdV = 0 \qquad (\because \ dV = 0)$$

$$\delta Q = dU = \delta W = dU = 327.6 \text{ cab.}$$

For constant pressure

$$\delta Q = m\, c_p dT = (30\text{g}) \times (0.219 \text{ cal g}^{-1}\text{K}^{-1}) \times (70 \text{ K})$$

$$= 459.9 \text{ cal}$$

$$dU = m\, c_V\, dT = 327.6 \text{ cal}$$

$$\delta W = \delta Q - dU = (459.9 - 327.6) \text{ cal} = 132.3 \text{ cal} = 555.7 \text{ J.}$$

Example 5.35 Two moles of an ideal monatomic gas at 27°C and one atm pressure is expanded adiabatically to twice its volume. Calculate the **(a)** change in internal energy work done by the gas during the process. Take $R = 8.3$ J mol^{-1}K^{-1}.

For a monatomic gas,

$$\gamma = 1 + \frac{2}{f} = 1 + \frac{2}{3} = \frac{5}{3}$$

To calculate change in internal energy, we have to determine final temperature. An adiabatic change accompanied by changes in temperature and volume is defined by the

$$TV^{\gamma-1} = \text{constant}$$

$$T_1 V_1^{\gamma-1} = T_2 V_2^{\gamma-1}$$

$$T_2 = T_1 \left(\frac{V_1}{V_2} \right)^{\gamma-1}$$

300 K and $\frac{V_1}{V_2} = \frac{1}{2}$. Hence,

$$T_2 = (300\ K) \left(\frac{1}{2} \right)^{2/3} = 189\ \text{K}$$

$$T_2 = (189 - 273)\,^\circ\text{C} = -84^\circ\text{C}.$$

The change in internal energy,

$$\text{d}U = n\ C_V\ (T_2 - T_1)$$

$$= (2\text{mol}) \cdot \frac{3R}{2} (189\ \text{K} - 300\ \text{K})$$

$$= 3 \times (8.3\ \text{J mol}^{-1}\text{K}^{-1}) \times (-111\ \text{K})\ \text{mol} = -2764\ \text{J}.$$

For an adiabatic process,

$$\delta Q = 0$$

$$\text{d}U + \delta W = 0$$

$$\delta W = -\text{d}U = 2764\ \text{J}.$$

Let us now summarise what you have learnt in this chapter.

MMARY

The sum of total internal kinetic and potential energies of all molecules constitutes the internal energy of the system.

The work done by a thermodynamic system is taken as positive and that done on it as negative.

The changes in the value of internal energy which may be accomplished by exchanging heat with a system, by work done on or by a system and exchanging matter with a system.

The internal energy is a function of state whereas the work and heat are path functions.

The first law of thermodynamics states that when a system is constrained to undergo a change by mechanical, diffusive or thermal interactions, its internal energy changes by an amount equal to the heat transferred to it, work done on it and matter exchanged. In differential form, the first law of thermodynamics can be expressed

$$\text{d}U = \delta Q - \delta W + \mu\ \text{d}N$$

The work done by a system is by convention positive and for different systems can be expressed as

Hydrostatic	$\int_{V_i}^{V_f} p\mathrm{d}V$
Surface film	$-\int_{A_i}^{A_f} \sigma \mathrm{d}A$
Stretched wire	$-\int_{L_i}^{L_f} T\mathrm{d}l$
Paramagnetic substance	$-\int_{M_i}^{M_f} B\mathrm{d}m$
Electric cell	$-\int_{t_i}^{t_f} Ei\mathrm{d}t$

For one mole of a gas, the difference between heat capacities at constant pressure and at constant volume is

$$C_p - C_V = R \qquad \text{Perfect gas}$$

$$= R\left(1 + \frac{2a}{RTV^3}(V-b)^2\right) \qquad \text{Real gas}$$

The equation of state for an adiabatic process can be expressed as

$$pV^\gamma = \text{constant}; \quad TV^{\gamma-1} = \text{constant}; \quad T^\gamma p^{1-\gamma} = \text{constant}$$

The adiabatic lapse rate in the stratosphere is caused by convective currents. It is given by

$$\frac{\mathrm{d}T}{\mathrm{d}h} = -\frac{\gamma - 1}{\gamma}\frac{Mg}{R}$$

The theoretical value of adiabatic lapse rate is –9.8°C over 1 km.

The sum $H = U + pV$ is called enthalpy. Like internal energy, enthalpy is also a state function.

ERCISES

Refer to Fig. 5.16. It is given that $\Delta Q_{prq} = 80$ J, $_{prq} = 30$ J. Calculate

ΔQ_{psq} if $\Delta W_{psq} = 10$ J

When the system undergoes a change along the path *qtp* the work done on it is 20 J. How much does the system absorb or librate heat?

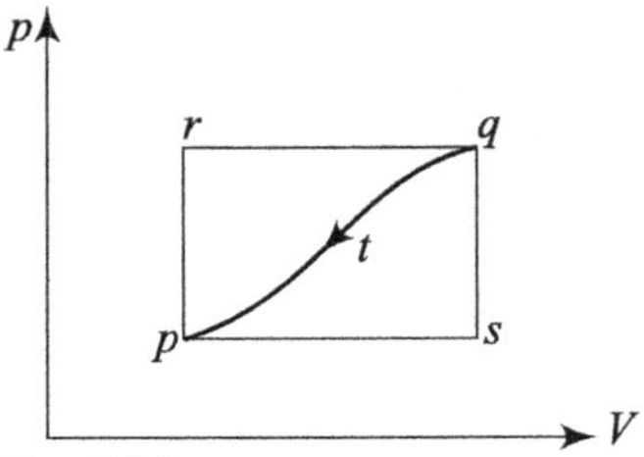

Fig. 5.16

If $U_p = 0$ and $U_s = 40$ J, calculate the heat absorbed during the processes $p \to s$ and $s \to q$.

(*Ans:* (a) 60J, (b) Librates, 70J (c) 50J, 10J)

Two moles of a perfect monatomic gas, initially kept in a cylinder at a pressure $\times 10^6$ Nm^{-2} and temperature 27°C, are made to expand until its volume is doubled.

How much work is done if the expansion is

(i) adiabatic (ii) isobaric (iii) isothermal?

Show each process on a p–V diagram.

Name the process in which heat transfer is maximum and minimum.

In which case is the change in internal energy the greatest and the least?

(*Ans:* (a) (i) 2.8×10^3 J (ii) 5.0×10^3 J (iii) 3.5×10^3 J, (c) Isobaric, adiabatic (d) Isobaric, isothermal)

One mole of oxygen, initially kept at 17°C, is adiabatically compressed so that its pressure becomes 10 times. Calculate (a) its temperature after the compression, and (b) the work done on the gas. (*Ans:* 5.6×10^2 K, 5.6×10^3 J)

One mole of an ideal gas is contained under a light frictionless piston of a vertical cylinder at a temperature T. The space above the piston opens into the atmosphere. Calculate the work done in increasing the volume of the gas to n times the initial value by isothermally raising the piston in a very slow and controlled manner.

(*Ans:* $RT\,(n - 1 - \ln n)$)

An ideal gas undergoes a process in which its internal energy is related to its volume $= aV^b$, where a and b are constants. Calculate the work done by the gas and the amount of heat transferred to this gas to increase its internal energy by ΔU.

(*Ans:* $\Delta U(\gamma - 1)$, ΔU_γ)

A fixed mass of air at one atm pressure is compressed adiabatically to 5 atm. It is then allowed to expand isothermally to its original volume. Calculate the pressure exerted by air at the end of the isothermal process. In which process is more work done? Take $\gamma = 1.4$.

(*Ans:* 1.58 atm; Isothermal)

The equation of state for NO is given by $p\,(V - nb) = nRT$, where $b = 40$ cm^3 mol^{-1}. Calculate the maximum work when 7.5 g of the gas expands isothermally at 47°C from 5 litre to 25 litre. (*Ans:* 10.5 litre atm)

The equation of state of an ideal elastic substance is

$$T = k\theta\left(\frac{L}{L_0} - \frac{L_0^2}{L^2}\right)$$

where k is a constant and L_0 (the value of L at zero tension) is a function of temperature only. Derive an expression for the work required to change the length from L_0 to $L_0/3$ quasistatistically and isothermally. $\left(Ans\text{:} \dfrac{14}{9}k\theta L_0\right)$

Express U as a function of p and V and prove that

$$\left(\frac{\partial U}{\partial p}\right)_V = \frac{C_V\,\beta_T}{\alpha}$$

$$\left(\frac{\partial U}{\partial V}\right)_p = \frac{C_p}{V\alpha} - p$$

One litre of a monatomic gas at STP is compressed adiabatically to half its volume. Calculate the work done on the gas. Take $R = 8.31$ Jmol^{-1} K^{-1}. (*Ans:* 89.32J)

Calculate the work done when a gram molecule of an ideal gas expands isothermally at 27°C to double of its original volume. Take $R = 8.3$ Jmol^{-1} K^{-1}. (*Ans:* 1725.8J)

A fixed quantity of air at 27°C is compressed (i) slowly and (ii) suddenly to one-third of its volume. Calculate the change in temperature in each case.

(*Ans:* (i) Zero for isothermal change; (ii) 165.5K, adiabatic change)

A certain amount of oxygen gas at STP is expanded to 3 times its volume under adiabatic conditions. Calculate the final temperature and pressure. Take $\gamma = 1.4$.

(*Ans:* 176 K; 0.215 atm)

Two ideal gases at initial pressure p_1 and volume V_1 are made to expand adiabatically to a final volume equal to four times the original volume. If the values for these gases are 1.67 and 1.40, respectively, compare their final pressures.

(*Ans:* 1:0.6875)

An ideal gas occupies 1 litre at 8 atm. It is made to undergo adiabatic expansion until pressure drops to 1 atm. Calculate the final volume and the work done in Joules during the expansion. Given $\gamma = 1.5$ and 1 atm $= 1.013 \times 10^5$ Nm^{-2}.

(*Ans:* 4 litre; 810.4 J)

Two moles of a monatomic gas at T°C are made to expand adiabatically to twice its initial volume so that the final temperature is $-84°C$. Calculate the initial temperature. Also calculate the change in internal energy and the work done in this process.

(*Ans:* 27°C, $dU = -2767$ J, $\delta W = 2767$ J)

A fixed mass of an ideal gas defined by thermodynamic coordinates (p, V, T) undergoes isothermal expansion followed by adiabatic expansion so that in each case, the final volume is twice the initial value and the final pressures attained are given by $\frac{p_{\text{isothermal}}}{p_{\text{adiabatic}}} = 2^{2/3}$. Calculate the value of γ. (*Ans:* 1.67)

A person consumes a diet of 10^4 J every day and spends total energy of 1.2×10^4 J. Calculate the change in internal energy. If the net energy spent comes from sucrose at the rate of 1.6×10^4 J kg^{-1}, calculate the days in which his weight will reduce by 5 kg. (*Ans:* 40 days)

6 THE SECOND LAW OF THERMODYNAMICS

arning Objectives

In this chapter, you will learn how to

e expression for the efficiency of a Carnot engine;

state the second law of thermodynamics and establish equivalence of n–Planck and Clausius statements;

e Carnot theorem of maximum efficiency;

e thermodynamic temperature and establish thermodynamic scale;

blish the connection between irreversibility and unavailable energy; and

n expressions for the efficiency of a Rankine cycle, Otto cycle, Diesel cycle and vapour compression cycle.

TRODUCTION

We now know that the first law of thermodynamics is a statement about conservation of energy for thermodynamic processes. However, you will agree that it gives us no information about the way thermodynamic systems evolve as well as about the direction of flow of natural processes. For instance, it is a common experience that heat can not flow by itself from a colder body to a hotter body. But the first law does not rule out this possibility. Similarly, one can achieve complete conversion of work into heat via dissipative processes. But the first law tells us of no definite limitations to the contrary. That is, it places no restriction on the extent of convertibility of heat into work. Can we achieve 100% efficiency? Engineering experience refrains this; if this were true, one could transform the heat of the environment into work. And since this supply is virtually unlimited, we would require no fuel to produce energy! Then energy crisis would not have been an issue for world economy. We can similarly consider many other natural processes in which energy is conserved but they never happen. This suggests that besides the first law, we must have some other fundamental principle which satisfactorily explains these facts of experience. This principle is known as the *second law of thermodynamics*. In fact, the second law goes far beyond the efficiency of heat engines. It also serves to define thermodynamic temperature and resolves the problem of how natural processes proceed or systems evolve thermodynamically.

In Sec. 6.2, we begin the discussion of second law of thermodynamics by tracing its origin and the contributions of Carnot, which facilitated industrial revolution in Europe. The convertibility of heat into work using a heat engine is discussed in Sec. 6.3. For simplicity, we confine ourselves to the framework of reversible Carnot cycle in Sec. 6.4. As we now know, the work of Carnot led Clausius, Thomson (later Lord Kelvin) and Planck, among others, to study convertibility of heat into work and vice versa. These led them to sum up generalisations of experiences in different statements of the second law. However, the two most well-known statements of the second law are the Kelvin–Planck statement and the Clausius statement. These are discussed in detail and their equivalence has also been established in Sec. 6.5. We show that if one statement is not obeyed, the other one is also violated. In Sec. 6.6, we have established Carnot theorem. This is followed by a discussion of thermodynamic temperature in Sec. 6.7. Here we also show that it is not possible to attain absolute zero in any way whatsoever. A brief discussion of irreversibility and unavailable energy is presented in Sec. 6.8.

We now know that Carnot engine has maximum efficiency but it is a theoretical idealisation. Yet it is of great value in understanding the operation of practical heat engines as well as refrigerators. The physical concepts developed in Secs. 6.3 and 6.4 have been applied to discuss power cycles with particular reference to Rankine cycle, Otto cycle and Diesel cycle in Sec. 6.9 and refrigeration cycle in Sec. 6.10. A comparison with Carnot engine has been made to give you an idea about the limits on actual

RIGIN OF THE SECOND LAW

Historically, the second law of thermodynamics has genesis in the work of Carnot. Prior to him, researchers working to improve efficiency of heat engines confined to empirical studies such as comparing fuel input with the work output. Carnot, who was proponent of caloric theory, assumed that caloric was conserved in the operation of heat engines. He postulated that work is done by a steam engine due to flow of caloric from the hotter body to the colder body. He envisaged the flow of caloric as analogous to flow of fluid, which produces work when falls on a potential gradient.

Carnot' genius lay in imaginative insights into the working of heat engines. He recognised that a heat engine is the most efficient machine when it is operated in a cycle and the amount of work done by an engine depends on the temperature difference between the source of heat and the sink into which the caloric flows. He devised a reversible cycle of operations which we now call *Carnot cycle* as an ideal for the operation of any heat engine. Moreover, it was theoretically possible to operate such an engine so that by reversing the sequence of operations, work can be done on the system and caloric transferred from the colder body to the hotter body. By joining together an arbitrary heat engine and a reversible Carnot engine, he demonstrated that the Carnot engine produces maximum work. He argued that if this were not true, one could very well join two heat engines and transfer heat from the colder body to the hotter body without doing any work or produce a net amount of work without any net transfer of heat, which contradicts common experience.

The profound significance of this work was recognised by Clapeyron, who reformulated Carnot's arguments analytically and related the ideal Carnot engine to the $p - V$ indicator diagram. A few years later, Lord Kelvin studied this problem all over again but he could not reconcile the conflicting view points of caloric conservation in the work of Carnot and interconvertibility of heat and work in Joules' work. The contradiction was however resolved by Clausius when he reformulated the problem wherein he justified Carnot's theorem of maximum efficiency but showed that the assumption of no heat loss was incorrect. He proved that heat is actually converted into work. In a way, this work led to the formulation of the second law of thermodynamics.

EAT ENGINES

We know that heat flows spontaneously from a hotter to a colder body. If we intercept this flow with a machine, some of it can be converted into work. *A machine that can convert heat into work is said to be a heat engine.* To be a useful device, a heat engine must operate continuously; absorb heat at a higher temperature and reject it at a lower temperature. *a heat engine has to operate between two heat reservoirs* (Fig. 6.1). Moreover, the processes which take place inside an engine must not cause permanent changes. This means *an engine has to operate in a cycle.*

The material used in the operation of an engine is called the *working substance.* The working substance can be a solid, liquid, gas, or a paramagnetic substance. In a steam engine, working substance is steam (water). Other familiar working substances for automobile engines are petrol, diesel and CNG. In a refrigerator, the most widely used working substance used to be chlorofluorocarbon compounds. But these are now being phased out to control ozone layer depletion in the stratosphere.

The information about the extent of convertibility of heat into useful work is contained in the physical parameter called efficiency. We now discuss it in some detail.

Efficiency The efficiency of an engine is defined as the ratio of the net work done by it to the heat absorbed during one complete cycle. It is usually denoted by the symbol η:

$$\eta = \frac{\text{Useful work done}}{\text{Heat absorbed}} \tag{6.1}$$

After one complete cycle, the engine returns to its original state. Therefore, there will be no change in its internal energy, i.e., $\Delta U = 0$. From the first law, we have

$$\Delta U = Q_1 - Q_2 - W = 0$$

$$W = Q_1 - Q_2 \tag{6.2}$$

is the heat absorbed, Q_2 is the heat rejected and W is the work done during one cycle (Fig. 6.1). Note that in a real engine, Q_2 contributes to *environmental* (*thermal*) , as it is in the form of hot exhaust gases or steam.

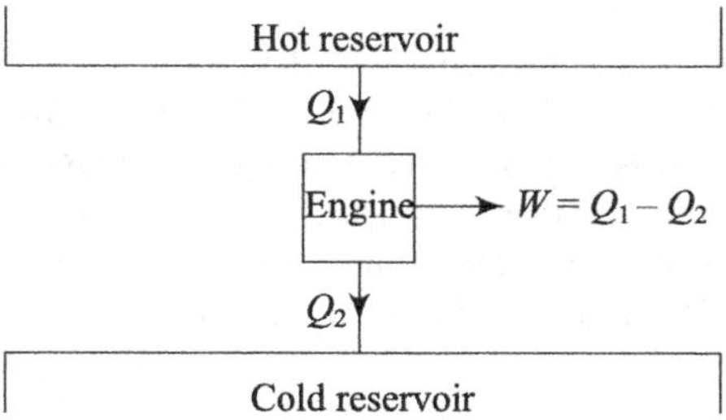

Schematics of operation of a heat engine.

From Eqs. (6.1) and (6.2), we can write

$$\eta = \frac{W}{Q_1} = \frac{Q_1 - Q_2}{Q_1} = 1 - \frac{Q_2}{Q_1} \tag{6.3}$$

This result shows that η will always be less than one. Can you think of some possible explanation for this? Is it due to irreversibility of thermal processes? Nevertheless it is desirable to design an engine with maximum efficiency. To know how high η can be, let us consider the following questions:

Under what conditions should an engine operate?
Does the choice of a working substance play some role?

Carnot was the first researcher who recognised that for maximum efficiency, a heat engine should be (thermodynamically) reversible. That is, all stages of operation should be carried out infinitely slowly so that there are no dissipative losses, such as friction or turbulence, to waste energy. (In practise, however, there are always some losses.) The simplest reversible cycle is due to Carnot. In *Carnot engine*, the working substance is made to exchange heat with heat reservoirs. We will discuss it in some detail now. However, it may be mentioned here that any heat engine operating in a Carnot cycle is called a *Carnot engine*.

HE CARNOT CYCLE

The Carnot cycle consists of four stages. These are schematically depicted in Fig. 6.2. T_2 be the temperatures of the heat reservoirs such that $T_1 > T_2$. The working substance, say a gas, is contained within a cylinder fitted with a frictionless piston. To simulate the working of a real engine, we consider the following reversible sequence:

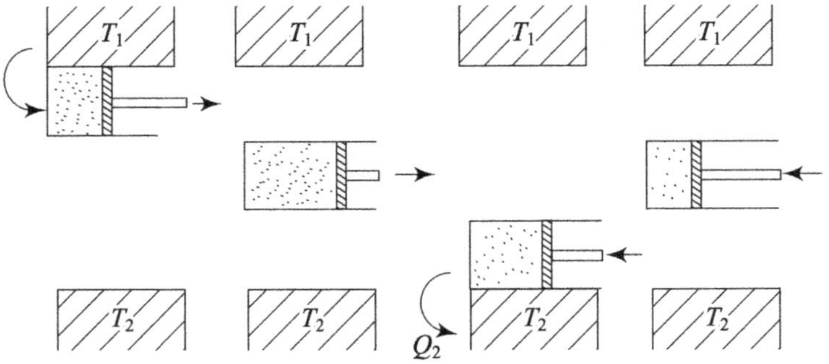

Four stages in the Carnot cycle.

We place the cylinder in thermal contact with the hot reservoir and let the gas undergo *reversible isothermal expansion.* Suppose Q_1 heat flows from the reservoir into the gas in the process. We have indicated this change as A to B on the p-V diagram in Fig. 6.3a. On the $T - V$ diagram (Fig. 6.3b), this is represented by a straight line. Note that the process is reversible so that the temperature of the working substance is equal to the temperature of the reservoir during heat transfer.

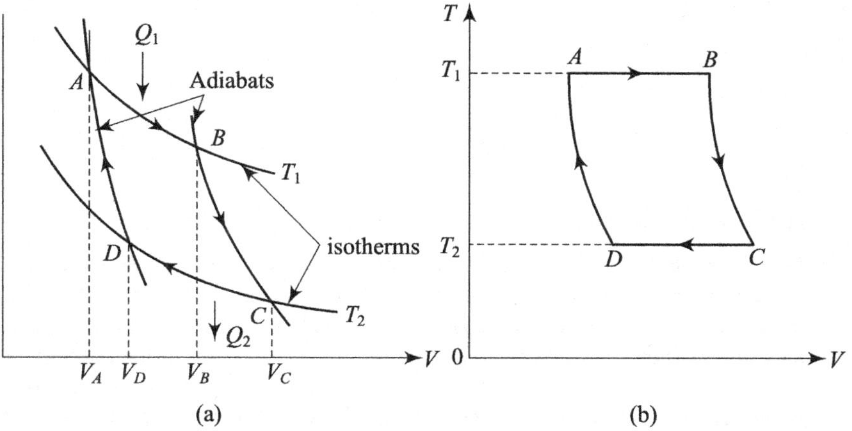

Carnot cycle on (a) p-V diagram, and (b) $T - V$ diagram.

Note that the work is done by the gas on the surroundings (via piston) and is given by

$$W_1 = \int_A^B p\,\mathrm{d}\,V$$

From the first law of thermodynamics, we know that for an isothermal process, Q_1 since $\Delta U = 0$. Hence, using the equation of state for n moles of an ideal gas, we can write

$$W_1 = Q_1 = nRT_1 \int_{V_A}^{V_B} \frac{\mathrm{d}V}{V} = nRT_1 \ln (V_B / V_A) \tag{6.4}$$

where V_B and V_A are volumes of the gas in states B and A, respectively.

Next, we thermally isolate the gas and let it undergo *reversible adiabatic expansion*. The temperature of the gas (working substance) falls from T_1 to T_2, the temperature of the cold reservoir. This change is indicated as B to C on the p - V diagram (Fig. 6.3a). In this process, work is done by the gas on the surroundings and is given by

$$W_2 = \int_B^C p\,\mathrm{d}V$$

Since the states B and C lie on the same adiabat, we can use the relation $pV^\gamma = K$ [Eq. (5.19b)]. Then the expression for W_2 takes the form

$$W_2 = K \int_{V_B}^{V_C} \frac{\mathrm{d}V}{V^\gamma}$$

$$= \frac{K}{1-\gamma}\left(\frac{1}{V_C^{\gamma-1}} - \frac{1}{V_B^{\gamma-1}}\right)$$

Using the relation $pV^\gamma = K$ again, we can write

$$W_2 = \frac{1}{1-\gamma}(p_C V_C - p_B V_B)$$

$$= nR\frac{(T_1 - T_2)}{(\gamma - 1)} \tag{6.5}$$

where we have used the relation $p_C V_C^\gamma = K = p_B V_B^\gamma$.

As you may note, the gas (working substance) is now at relatively low pressure. To be used in a cyclic order, it has to be restored to its initial state. To this effect, the gas is compressed in two stages: First isothermally and then adiabatically. Therefore, we place the cylinder in thermal contact with the cold reservoir at lower temperature and compress the gas isothermally and reversibly. Suppose Q_2 heat is given up by the gas to the cold reservoir. This change is indicated as C to D. The work done during this process is

$$W_3 = Q_2 = \int_C^D p\mathrm{d}V = nRT_2 \ln(V_D/V_C) \tag{6.6}$$

where V_D and V_C are the volume of the gas in states D and C respectively. Since V_C, Eq. (6.6) signifies that W_3 is negative. That is, *work is done on the gas.* The gas is again thermally isolated and compressed under reversible adiabatic conditions till its original state is restored. Obviously, the work is done on the gas in this process and is equal to

$$W_4 = \int_D^A p\mathrm{d}V = \frac{nR}{\gamma - 1}(T_2 - T_1) = -W_2 \tag{6.7}$$

$T_2 < T_1$, W_4 is a negative quantity (indicating that work is done *on* the gas during adiabatic compression).

Note that Carnot cycle is a curvilinear quadrilateral on the p-V diagram.

The net effect of this cycle is that we have extracted heat Q_1 (from the source) at temperature T_1 and put back heat Q_2 (in the sink) at temperature T_2. In the process, the total work done by the engine is

$$W = \oint p\mathrm{d}V = W_1 + W_2 - W_3 - W_4$$

where the symbol $\oint$ means that the integral is taken around the cycle. From Fig. 6.4 it is clear that the net work done by the engine is numerically equal to the area of closed curve in the p-V diagram (or T– V diagram). Since, $W_2 = -W_4$, the expression of work done by the engine simplifies to

$$W = W_1 - W_3 = nRT_1 \ln(V_B/V_A) - nRT_2 \ln(V_D/V_C) \tag{6.8}$$

To simplify this expression, we note that states B and C lie on the same adiabatic curve. Then, using Eq. (5.19a), we can write

$$T_1 V_B^{\gamma-1} = T_2 V_C^{\gamma-1}$$

$$\frac{T_1}{T} = \left(\frac{V_C}{V}\right)^{\gamma-1} \tag{6.9}$$

Similarly, for states D and A we can write

$$T_2 V_D^{\gamma-1} = T_1 V_A^{\gamma-1}$$

$$\frac{T_1}{T_2} = \left(\frac{V_D}{V_A}\right)^{\gamma-1} \tag{6.10}$$

On comparing Eqs. (6.9) and (6.10), we get

$$\left(\frac{V_C}{V_B}\right)^{\gamma-1} = \left(\frac{V_D}{V_A}\right)^{\gamma-1} \tag{6.11}$$

We can drop the powers in the exponents from both sides. Then cross multiplication leads to the result

$$\frac{V_B}{V_A} = \frac{V_C}{V_D}$$

Using this result in Eq. (6.8), we can write

$$W = nR(T_1 - T_2)\ln(V_B/V_A) \tag{6.12}$$

On combining Eqs. (6.4) and (6.12), we can express the efficiency of a Carnot engine in terms of the temperatures within which it operates:

$$\eta = \frac{W}{Q_1} = \frac{nR}{nRT_1}\frac{(T_1 - T_2)\ln(V_B/V_A)}{\ln(V_B/V_A)}$$

$$= \frac{(T_1 - T_2)}{T_1} = 1 - \frac{T_2}{T_1} \tag{6.13}$$

This result shows that the efficiency of a Carnot engine depends only on the temperatures of heat reservoirs; the working substance plays no role at all in determining η. In fact, it is the temperature difference $(T_1 - T_2)$ that really matters. If two or more Carnot engines are working between the same high temperature reservoirs and different low temperature reservoirs, the engine working over the greater temperature difference will have higher efficiency. Usually, T_2 is limited to the temperature of the surroundings of the engine (~ 300 K) and the efficiency depends on the temperature of the heat source. In a steam engine, T_1 is raised by working with saturated steam at high pressures. The efficiency of a typical engine is around 45%. It means that more fuel is wasted than is utilised. Moreover, it causes thermal as well as environmental pollution. This also means that energy has *quality* as well as quantity.

On comparing Eqs. (6.3) and (6.13), we further note that

$$\frac{T_1}{T_2} = \frac{Q_1}{Q_2} \tag{6.14}$$

That is, for an ideal gas, the ratio Q_1/Q_2 also depends on temperatures T_1 and T_2.

To test your understanding, we would like you to answer the following questions.

Problem 6.1 What is the more effective way to increase η of a Carnot engine: To or lower T_2? Comment.

Problem 6.2 A Carnot engine is independent of the state and nature of the working substance. We expect that real engines should behave similarly. Why then are we so concerned about fuels for energy generation?

We would now like you to go through the following examples to get an idea about the efficiency of a Carnot engine.

Example 6.1 A Carnot engine is made to work between 0°C and –200°C. Calculate its efficiency.

The efficiency of a Carnot engine is given by Eq. (6.13)

$$\eta = \frac{T_1 - T_2}{T_1}$$

= 0°C = 273 K and T_2 = – 200°C = 73 K. Hence,

$$\eta = \frac{273 - 73}{273}$$

$$= \frac{200}{273} = 0.73$$

Do we obtain this much efficiency with real engines? In fact, if we can realise such a figure, the pressure on petrol and diesel will be eased to a great extent.

Example 6.2 A Carnot engine has an efficiency of 30%. Its efficiency is to be raised to 50%. By how much must the temperature of the source be increased if the sink

From Eq. (6.13), we recall that

$$\eta = 1 - \frac{T_2}{T_1}.$$

= (273 + 27) K = 300 K and η = 0.30. Using this data, we can easily calculate the temperature of the source:

$$0.30 = 1 - \frac{300\text{ K}}{T_1}$$

$$T_1 = \frac{300\text{ K}}{0.70} = 428.6\text{ K} \qquad \text{(i)}$$

For the increased efficiency, we can write

$$T_1' = \frac{300\text{ K}}{0.5} = 600\text{ K} \qquad \text{(ii)}$$

Hence the temperature of the source should be raised by

$$\Delta T = T_1' - T_1 = (600 - 428.6)\text{ K} = 171.4\text{ K}.$$

Example 6.3 A Carnot engine is operated between the ice-point and steam-point. (a) If the engine receives 746.0 calories from the hot reservoir in each cycle, how many calories does it reject to the cold reservoir? (b) How much work is done by the engine in each cycle?

(a) From Eq. (6.14), we have

$$\frac{T_1}{T_2} = \frac{Q_1}{Q_2}$$

= 100°C = 373K, T_2 = 273K, and Q_1 = 746.0 cal. Hence, heat rejected to the cold reservoir is

$$Q_2 = \frac{T_2}{T_1} Q_1 = \frac{273\text{ K}}{373\text{ K}} \times (746.0\text{ cal}) = 546.0\text{ cal}$$

(b) Work done in each cycle is given by

$$W = Q_1 - Q_2 = (746.0 - 546.0)\text{ cal} = 200.0\text{ cal}$$

Example 6.4 A reversible heat engine converts one sixth of the heat input into work. When the temperature of the sink is reduced by 62°C, its efficiency is doubled. Calculate the temperatures of the source as well as the sink.

We know that efficiency of a heat engine is given by

$$\eta = \frac{\text{Heat converted into work}}{\text{Heat input}}$$

$$= \frac{1/6}{1} = \frac{1}{6}$$

In terms of the temperatures of the source and the sink, we can express the efficiency of the heat engine as

$$\eta = 1 - \frac{T_2}{T_1} = \frac{1}{6}$$

$$T_1 = \frac{6}{5} T_2 \qquad \text{(i)}$$

Since the efficiency of the engine is doubled when the temperature of the sink is reduced by 62°C or by 62 K, we can write

$$2 \times \frac{1}{6} = \frac{T_1 - (T_2 - 62\text{ K})}{T_1}$$

$$3T_2 - 2T_1 = 186\text{ K} \qquad \text{(ii)}$$

On substituting the value of T_1 from Eq. (i) in Eq. (ii), we get

$$3T_2 - 2 \cdot \frac{6}{5} T_2 = 186\text{ K}$$

$$T_2 = 310\ \text{K}$$

$$T_1 = \frac{6}{5}T_2 = \frac{6}{5}\times 310\ \text{K} = 372\ \text{K}.$$

Example 6.5 Two Carnot engines A and B are operated in series. Engine A receives heat at 900 K and rejects to a reservoir at temperature T K. The second engine B receives heat rejected by the first engine and in turn rejects to a heat reservoir at 400 K. Calculate the temperature T when (a) the work outputs of both the engines are equal and (b) the efficiencies of the two engines are equal.

(a) Suppose that engine A receives heat Q_1 at temperature T_1 and rejects heat at temperature T. Similarly, suppose that the engine B receives heat Q at temperature and rejects heat Q_2 at temperature T_2. Then the work performed by these engines are respectively given by

$$W_A = Q_1 - Q$$

$$W_B = Q - Q_2$$

= W_B, we can write

$$Q_1 - Q = Q - Q_2$$

$$\frac{Q_1 + Q_2}{Q} = 2 \qquad \text{(i)}$$

As both engines are Carnot engines, we have

$$\frac{Q_1}{Q} = \frac{T_1}{T} \quad \text{and} \quad \frac{Q_2}{Q} = \frac{T_2}{T}$$

$$\frac{Q_1 + Q_2}{Q} = \frac{T_1 + T_2}{T} \qquad \text{(ii)}$$

From Eqs. (i) and (ii), we get

$$\frac{T_1 + T_2}{T} = 2$$

= 900 K and T_2 = 400 K. Hence,

$$T = \frac{900\ \text{K} + 400\ \text{K}}{2} = 650\ \text{K}$$

(b) When the efficiencies are equal, we can write

$$\eta = 1 - \frac{T}{T_1} = 1 - \frac{T_2}{T}$$

$$\frac{T}{T_1} = \frac{T_2}{T}$$

$$T = (T_1 \times T_2)^{1/2}$$
$$= (900\ \text{K} \times 400\ \text{K})^{1/2} = 600\ \text{K}.$$

Example 6.6 It is claimed that an engine working on Carnot cycle between temperatures 1400°C and 30°C receives 4.2 kJs^{-1} of heat and develops a power of 3.675 kW. (a) Discuss its feasibility and (b) What change in condition(s) could validate the claim?

(a) We know that

$$\frac{Q_1}{Q_2} = \frac{T_1}{T_2}$$

= 1400°C = (1400 + 273) K = 1673 K, T_2 = 30°C = (30 + 273) K = 303 K and = 4.2 kJs^{-1}.

$$Q_2 = Q_1 \times \frac{T_2}{T_1} = (4.2\ \text{kJs}^{-1}) \times \frac{303\text{K}}{1673\text{K}} = \frac{1272.6}{1673}\text{kJs}^{-1} = 0.76\ \text{kJs}^{-1}$$

The work obtained from the engine per second is

$$W = Q_1 - Q_2$$
$$= (4.20 - 0.76)\ \text{kJs}^{-1}$$
$$= 3.44\ \text{kW}$$

We are told that the power developed by the engine = 3.675 kW, which is more than the actual power of 3.44 kW. So the claim is not valid.
(b) To validate the claim either T_1 should be raised or T_2 should be lowered.

Example 6.7 A Carnot engine whose low temperature reservoir is at 7°C has an efficiency of 50%. By how much degrees should the temperature of the hot reservoir be increased so that efficiency increases to 70%.

In the first case, η = 50% = 0.5, T_2 = 7°C = 280 K. We recall that

$$\eta = 1 - \frac{T_2}{T_1}$$

On substituting the given values, we get

$$0.5 = 1 - \frac{280\ \text{K}}{T_1}$$
$$T_1 = 560\ \text{K}$$

In the second case, η' = 70% = 0.7, T_2 = 280 K and we have to determine the temperature of the source. Let us denote it by T_1'. Hence, we can write

$$\eta' = 1 - \frac{T_2}{T_1'}$$

$$0.7 = 1 - \frac{280\,\text{K}}{T_1'}$$

$$T_1' = \frac{280\,\text{K}}{0.3} = \frac{2800\,\text{K}}{3} = 933.3\,\text{K}$$

Therefore, the temperature of the source should be increased by

$$\Delta T = 933.3\text{ K} - 560\text{ K} = 373.3\text{ K}.$$

Note that we always tend to change the temperature of the source to improve efficiency. This is both convenient and more efficient way because temperature of the sink is determined by the surroundings. Moreover, the quality of heat is better at a higher temperature.

You should now answer the following practise problems.

Problem 6.3 A Carnot engine absorbs 300 J heat from a hot reservoir at the normal boiling point of water (373.15 K) and rejects heat to a sink at the triple point of water (273.16 K). Calculate the quantity of heat rejected to the sink, amount of work done by the engine and its thermal efficiency.

219.6 J; 80.4 J; 0.268

Problem 6.4 A reversible engine takes in heat from a reservoir at 527°C and gives out heat to a sink at 127°C. How many calories per second must it take from the reservoir to produce mechanical work at the rate of 750 W?

358.8 cal s^{-1}

Problem 6.5 A Carnot engine is operated between two reservoirs at temperatures 450K and 350 K. If the engine receives 1,000 cal of heat from the source in each cycle, calculate the amount of heat rejected to the sink in each cycle. Also calculate the efficiency of the engine and the work done by it in each cycle.

777.8 cal, 22.2%, 933.3 cal

Carnot Cycle as Refrigerator

You must have realised that the most beautiful aspect of a Carnot engine is that we can run the whole system backward so that the sequence of events and their functions are reversed. Thus, the cycle working in the reverse direction will act as an ideal *refrigerator*, in which heat is extracted from the reservoir at lower temperature and transferred to the reservoir at higher temperature. Therefore, in a sense, a refrigerator is also a heat engine.

Let us re-examine Fig. 6.3. If the directions of the arrows are reversed, the cycle *ABCDA* *ADCBA*. Since each process is reversible, the cycles are also reversible. Therefore, the magnitudes of heat taken, heat rejected and the work done remain the same, except that their signs are reversed. It means that the heat Q_2 is absorbed by the working substance from the lower temperature reservoir and heat Q_1 is rejected to the reservoir at higher temperature. The work W represents the work done on the system (Fig. 6.4). In a domestic

ENVIRONMENTAL DEGRADATION BY THERMAL POLLUTION

We now know that the efficiency of a heat engine is less than one. This means that to get some useful work done, some energy is always lost to the environment. This adds to thermal pollution in our surroundings. For instance, an automobile or a locomotive, particularly air-conditioned, discharges a lot of hot gases in air. It is now well known that we have developed a taste for comfort. The way affluent groups are developing, particularly in privileged societies/nations, increased energy usage in household (to run washing machines, room coolers/air-conditioners, vacuum cleaners), transport (automobiles, aeroplanes) industry, power generation, space explorations (rocket propulsion), among other human activities are thermally polluting our environment at an alarming rate and degrading its quality.

It is common experience that no person can remain in a swimming pool, if its temperature is above her/his body temperature. How can then we expect fauna and flora or aquatic life to remain unaffected by thermal pollution? In fact, our natural resources, which are so essential to support various life forms on this planet, are getting degraded continuously. If present trends go unchecked, the delicate balance existing in nature may be threatened; it may bring about permanent changes in weather patterns and climatic conditions. (In fact serious debate on climate change is raging international fora and has polarised the world into several groups.) Once this sets in, it would mark the beginning of the end of life. The natural disasters will occur more frequently and with increasing intensity.

Scientists believe that continuous discharge of heat bearing gases such as CO_2, CO, $_2$, N_2O in our environment has led *to global warming*, i.e., increase in the atmospheric temperature by a fraction of a degree, by trapping solar radiations reflected by the earth's surface. (The other factors include oil spills, gradually diminishing forest cover and increasing concrete jungles due to growing population and urbanisation.) If the temperature goes up by a degree or so, polar ice and glaciers may begin to melt at a faster pace creating new water bodies. This will have disastrous consequences: sea level will rise, which can change the existing geographical as well as coastal boundaries globally by submerging cities and low lying islands, bring in fury of floods, which will erode the fertile layer of the soil and pose serious threat to human lives and property, in addition to raising the alarm bells for food production. The coral and with it other species will be wiped out. Doomsayers have put the time limit of about 50 years for such drastic changes, if corrective steps are not initiated immediately. The ultimate danger is climate change and with it weather patterns.

What measures would you take to avoid such a situation because environment belongs to and affects everyone? Without bringing our economic activities to a stand still, we have to strive for sustainable green development, live in harmony with nature rather than exploit it for our insatiable appetite for luxurious living and design more fuel-efficient devices to minimise heat rejection to the environment. As a proactive

refrigerator, heat is pumped out of its interior, which is at a temperature lower than the surroundings and work is done by the motor driving the refrigerator. Thermodynamically, a refrigerator makes heat flow against the direction of its spontaneous flow, i.e., from a lower temperature to a higher temperature.

The ability of a refrigerator is rated in terms of *coefficient of performance*, which is defined as

$$\frac{\text{Heat extracted at lower temperature from the object to be refrigerated}}{\text{Work input}}$$

$$= \frac{Q_2}{W} = \frac{Q_2}{Q_1 - Q_2} = \frac{T_2}{T_1 - T_2} \qquad (6.15)$$

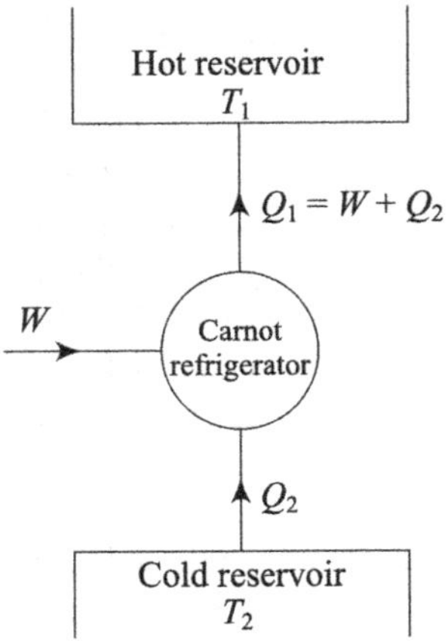

Fig. 6.4 A Carnot refrigerator.

Note that unlike the efficiency of a heat engine, the coefficient of performance of a refrigerator can be *greater than unity*. That is, the amount of heat removed from the refrigerated space can be greater than the work input. (In fact, one of the reasons for expressing the efficiency of a refrigerator by another nomenclature—the coefficient of performance—is the intention to avoid confusion and oddity of having thermal efficiencies greater than unity.) To give you an idea about the data, let us consider that freezer in a refrigerator or cold storage is maintained at –10°C and the room temperature is 30°C. You can readily calculate the maximum value of coefficient of performance in this case as $\omega = \frac{263}{40} = 6.58$.

You must have seen a household refrigerator. In this, the work is done by an electric motor and the working substance is Freon-12, which has a boiling point of –29°C. So Freon-12 is a gas at room temperature. More specifically, Freon gas in a tube is made to expand suddenly from a high pressure to a low pressure. In this process, it cools and a vapour-liquid mixture is obtained. This cold fluid, circulated through expansion coil around the region to be cooled, absorbs heat from the eatables kept inside the refrigerator and the entire liquid in the mixture changes into vapour. The vapour is compressed and work is done by the compressor on the vapour. The temperature as well as pressure of the vapour rises. The compressed vapour rejects heat to the surrounding medium such as the kitchen air and condenses through a set of tubes (called condenser and located at the back of the refrigerator).

It has been observed that CFCs adversely affect the life protecting layer of ozone in our atmosphere. So, there is now growing emphasis on phasing out CFCs. In India, non-CFC refrigerators are available now.

An air-conditioner is basically a refrigerator and the refrigerated space is a room or a building instead of the food compartment. A window air-conditioning unit produces cooling by absorbing heat from the room air and discharging it outside. (When you travel long distances by an aeroplane, sit in your air-conditioned study for long hours or work in a centrally air-conditioned building, it is advisable to drink water periodically to avoid dehydration.) The same unit can also be used as heat pump in winter by running it backward. Now-a-days, systems are fitted with controls so as to operate them as air-conditioners in summer and as heat pump in winter.

Now go through the following examples carefully.

Example 6.8 An air conditioner removes 1.5 kcal of heat per second from a large room. The power required to run it is 2 kW. If the typical value of Q_2/W is one-tenth that of a Carnot refrigerator, and the temperature outside is 40°C, calculate the temperature inside the room

From the given data and Eq. (6.15), we can write

$$\frac{Q_2}{W} = 0.1$$

$$\left.\frac{Q_2}{W}\right|_{\text{Carnot}} = 0.1\frac{T_2}{T_1 - T_2}$$

On rearrangement, we get

$$T_2 = \frac{10(Q_2/W) \times T_1}{1 + 10(Q_2/W)} = \frac{10 Q_2 T_1}{W + 10 Q_2}$$

= 1.5 kcal, T_1 = (273 + 40) K = 313 K and

$$W = P \times t = (2000\ \text{Js}^{-1}) \times 1\ \text{s} = 2000\ \text{J} = \frac{2000\ \text{J}}{4184\ \text{J kcal}^{-1}} = 0.478\ \text{kcal}$$

is the power of the air conditioner. Hence,

$$T_2 = \frac{10 \times (1.5\ \text{kcal}) \times (313\ \text{K})}{(0.478\ \text{kcal}) + 10 \times (1.5\ \text{kcal})} = \frac{4695\ \text{kcal K}}{15.478\ \text{kcal}} = 303\ \text{K} = 30\,°\text{C}$$

Example 6.9 A commercial refrigerator is driven by a 1000 W electric motor, which operates at an efficiency of 60%. If the refrigerator can be treated as a reversible heat engine operating between 0°C and 20°C, calculate the time required by it to freeze 100 kg of water which is at 0°C. Heat losses may be neglected. Take latent heat of fusion of ice as 332 kJ kg^{-1}.

We know that work done by a refrigerator is given by

$$W = Q_2\left(\frac{T_1}{T_2} - 1\right)$$

= 20°C = 293 K, and T_2 = 0°C = 273 K. Since the refrigerator is being driven by a motor of 1000 W with 60% efficiency, we can write

$$\text{Energy input } W = (1000\ \text{W}) \times \frac{60}{100} = 600\ \text{W} = 600 \text{Js}^{-1}$$

$$Q_2 = (600\ \text{Js}^{-1}) \times \frac{273\,\text{K}}{293\,\text{K} - 273\,\text{K}} = 8190\ \text{Js}^{-1}$$

But the heat required to freeze 100 kg of water = mL = 100 × 332 kJ

Hence, the time required to extract 332×10^5 J of heat $= \dfrac{332 \times 10^5\ \text{J}}{8190\,\text{Js}^{-1}} = 4054\,\text{s} = 67.6\ \text{min}$.

Example 6.10 Ice in a cold storage melts at the rate of 2 kg per hour when the external temperature is 20°C. Calculate the power of the motor required to drive the refrigerator so as to just prevent the ice from melting. Take latent heat of fusion of ice as J kg^{-1}.

We know that work done by a refrigerator is given by

$$W = Q_2 \left(\frac{T_1}{T_2} - 1 \right)$$

Now the ice melts at the rate of 2 kg hr^{-1}, i.e., (2/3600) kg s^{-1}. Therefore, the heat that must be removed to prevent ice from melting is given by

$$Q_2 = \frac{2}{3600} \text{kg s}^{-1} \times (332 \times 10^3 \text{ J kg}^{-1}) = 184.44 \text{ Js}^{-1}.$$

= (20 + 273) K = 293 K, and T_2 = (0 + 273) K = 273 K. Hence,

$$W = 184.44 \text{ Js}^{-1} \times \left(\frac{293 \text{ K}}{273 \text{ K}} - 1 \right)$$

$$= 13.51 \text{ W}.$$

We would now like you to answer the following problems to test your understanding.

Problem 6.6 Can a kitchen be cooled by leaving the door of an electric refrigerator open? Explain.

Problem 6.7 A domestic refrigerator is regarded as a reversible engine working between the temperature of melting ice and that of atmosphere at 17°C. Calculate the energy required to freeze 1 kg of water at 0°C.

20.93 kJ

Problem 6.8 A freezer is to be maintained at a temperature of 238 K on a summer day when the ambient temperature is 306 K. In order to maintain the freezer temperature, it is necessary to remove heat from it at the rate of 1230 Js^{-1}. Calculate the maximum possible coefficient of performance of the freezer, and the heat discarded to the surroundings.

3.5; 1581.4 W

HE KELVIN–PLANCK AND CLAUSIUS STATEMENTS

The Kelvin–Planck Statement *No process is possible whose sole result is complete conversion of heat into work.*

This statement implies that one cannot devise a machine which just absorbs heat and produces 100% work. However, let us consider isothermal expansion of an ideal gas. When the gas expands, it does work on the piston. For an ideal gas, the first law tells us that during an isothermal process, internal energy remains constant. That is,

OZONE LAYER DEPLETION: HOLE IN THE SKY

We all know that ozone layer is present in the stratosphere at a height of about 25 km above the sea level. At any time, the quantity of ozone in our atmosphere is about 10^6 ton and if it is compressed, it will make a layer of only about 3 mm. (The quantity of nitrogen in the atmosphere is 3.865×10^{12} ton.) But it is very vital to support various life forms on this planet; it absorbs ultraviolet radiations coming from the outer space and protects us from their ill-effects. But increased industrial and agricultural activities, transportation requirements and power generation have led to release of compounds of chlorine, nitrogen and sulphur, which destroy ozone layer by breaking it into oxygen. Scientific observations showed that the holes created in the ozone layer over Antarctica near the south pole and over Norway in the arctic region in the early 1990s have been ever increasing.

It is estimated that one per cent reduction in ozone layer causes up to three per cent increase in the amount of ultraviolet radiation reaching the surface of the earth. This will increase the incidence of skin cancer significantly. According to US National Academy of Sciences, if ozone layer reduces by about five per cent, there will be about 40,000 more cases of skin cancer each year in US alone. Similarly, ultraviolet radiations can destroy food crops and create food scarcity; it will then be impossible to feed the seven billion souls, cohabitating on the earth.

The most serious threat to ozone layer comes from *freons* or *chlorofluorocarbons* (CFCs), which are used as coolants in air conditioners and refrigerators, insecticides, deodorants, hair sprays, shaving creams and as fire extinguisher. Freons are extremely inert and once released into the atmosphere, they remain there for long. However, ultraviolet light causes their photo-dissociation and liberates chlorine atoms, which catalyse decomposition of ozone.

In the 1990s, nearly 95% CFCs were released by the industrialised countries in Europe, the United States, the Soviet Union and Japan. But the ever growing appetite for luxurious life style of civil societies in fast emerging economies such as Brazil, China, India, South Africa and the highly populated Third World countries such as Bangladesh, Malaysia, Nigeria, Pakistan, among others are also contributing a lot of ozone affecting pollutants now. CFCs also increase the level of CO_2, which is responsible for *greenhouse effect*. The ozone shield is also put to threat by oxides of nitrogen—nitrogen oxide and nitrous oxide—which emanate from nitrogen fertilisers, power automobiles and aeroplanes operating at high altitudes.

Atmosphere belongs to and affects everyone but no one seems to care to keep it clean. Though steps have been initiated to eliminate ozone depleting chemicals through mutual consultations among the scientists. Ministers responsible for environmental matters and even Heads of States of different nations facilitated by the United Nations, the politics of development has hindered these efforts. Nevertheless, we can initiate steps at individual and community levels by raising environmental awareness among

$$\Delta U = \delta Q - \delta W = 0$$

$$\delta W = \delta Q \tag{6.16}$$

That is, we have complete conversion of heat into work! Does this violate Kelvin–Planck statement? No, it does not because the state of the gas at the end of the process is different from that at the beginning (due to change in volume). So, conversion of heat into work is

There are other processes in which energy is conserved but they do not occur. For example, heat does not flow on its own from a body at a lower temperature to a body at higher temperature. That is, spontaneous heat flow is unidirectional. This fact of experience is contained in the Clausius statement of the second law of thermodynamics.

The Clausius Statement *No process is possible whose sole result is the transfer of heat from a body at a lower temperature to a body at a higher temperature.*

Note that this is a 'fact of experience' and needs no proof. However, is heat flow in a refrigerator from food items kept inside (a colder body) to the surroundings (a hotter body) in violation of Clausius statement? Definitely not because some work has to be done by an external source on the working substance. Therefore, an important implication of this law is that it is not possible to transfer heat from a cold body to a hot body without some change somewhere, including the working substance/surroundings of the system.

Note that the two statements of the second law apparently seem different or unconnected but this is not true; they are equivalent and so are their consequences. In fact, each statement implies the other and the two are equivalent. If one statement is untrue, will the other necessarily be not true? Indeed it is so and the truth of either form is both a necessary and sufficient condition for the truth of the other.

Equivalence of the Kelvin–Planck and the Clausius tatements

The equivalence of these statements implies that if one statement is untrue, the other statement is necessarily untrue.

Let us suppose that the Clausius statement of the second law is violated by a hypothetical refrigerator A, which transfers Q_2 units of heat in each cycle from a cold reservoir at temperature T_2 to a hot reservoir at temperature T_1 without expenditure of any work (Fig. 6.5a).

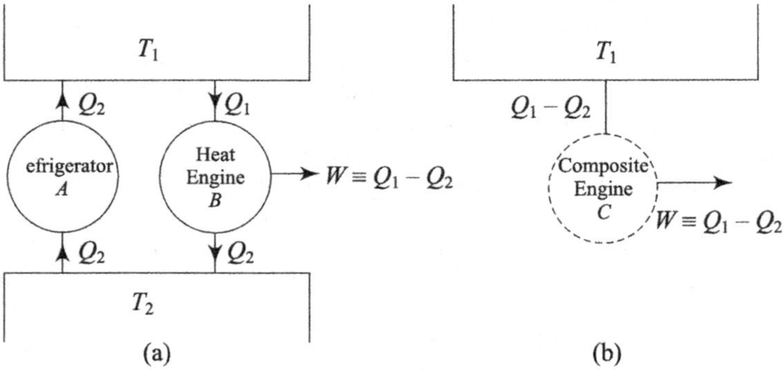

(a) Refrigerator A violates the Clausius statement of the second law, whereas B does not violate the law and (b) the composite engine violates the Kelvin–Planck statement.

Let us now assume that a heat engine working in a cycle between the same heat reservoirs draws an amount of heat Q_1 from the hot reservoir and rejects heat Q_2 to the low temperature reservoir and performs work $W = Q_1 - Q_2$ in one cycle. Further suppose that the heat engine operates at such a rate that it completes one cycle in the same period as does the refrigerator. Then, a composite engine is formed by draws heat $Q_1 - Q_2$ from the hot reservoir and converts it completely into work. Such an engine obviously violates Kelvin–Planck statement.

To prove that if Kelvin–Planck statement is violated, the Clausius statement is also violated, let us consider a hypothetical heat engine which extracts heat Q_1 from the hot reservoir, converts it completely into work and rejects no heat to the low temperature reservoir (Fig. 6.6a). Suppose that the work performed by the heat engine is used to drive a refrigerator operating between reservoirs at temperatures T_1 and T_2. Further, suppose that the refrigerator absorbs heat Q_2 from the low temperature reservoir and rejects heat $W + Q_2$ to the hot reservoir per cycle. As before, we assume that the refrigerator completes one cycle in the same period as does the heat engine. You will agree that the refrigerator by itself does not violate any law but when it is made to form a composite engine with a heat engine (Fig. 6.6b), we note that the net result of operation of the composite engine working as refrigerator will be to transfer heat Q_2 from the low temperature reservoir to the higher temperature reservoir without any external work. This obviously violates Clausius statement of the second law.

We can now conclude that both the statements of the second law are equivalent.

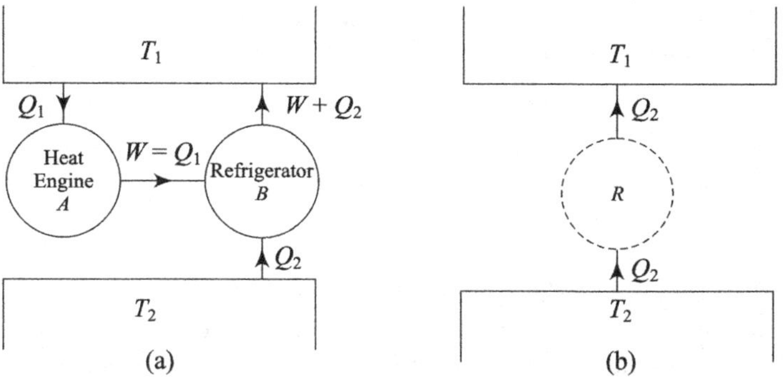

(a) *A* violates the Kelvin–Planck statement of the second law, whereas *B* does not violate the law, and (b) the refrigerator *R* violates the Clausius statement.

We can now conclude that the second law of thermodynamics completely rules out the so-called perpetual motion of the second kind*: A machine cannot run continuously at the expense of the internal energy of a body, say a heat reservoir (Fig. 6.7). Can you imagine perpetual motion of the third kind?** The impossibility of designing a perpetual motion machine of the second kind implies that a real machine cannot make use of the total energy it absorbs. So we must have a stage beyond its working part where unutilised heat can be rejected. It is this engineering limitation because of which work producing devices such as automobiles as well as energy generating systems reject enormous amount of heat

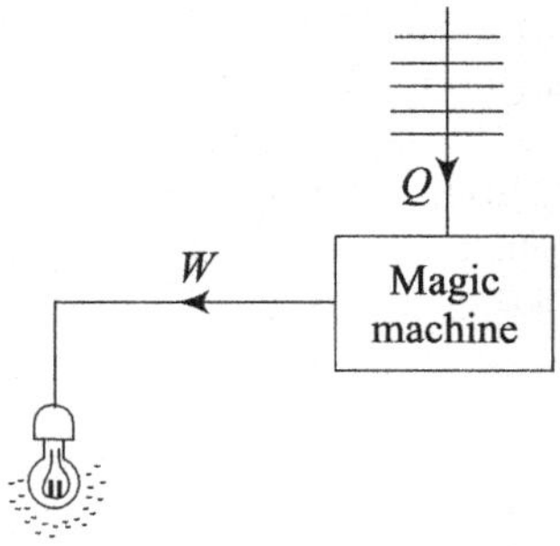

Fig. 6.7 Perpetual motion machine of the second kind.

---

* Recall that the first law forbids perpetual motion of the first kind: A machine cannot operate continuously by creating its own energy.

** If all dissipative effects such as friction, viscosity or electrical resistance are removed, the motion once started in a device would continue indefinitely.

(energy) into our environment. Is the ever increasing thermal pollution essentially a direct manifestation of the second law of thermodynamics? Indeed, it is so.

Before proceeding further, you may like to answer a conceptual question.

Problem 6.9 Use Kelvin–Planck statement of the second law of thermodynamics to show that only one adiabat can pass through one point.

Having analysed the operation of Carnot cycle, we can do two things: (a) show that no real engine can be more efficient than the Carnot engine, i.e., prove *Carnot theorem*, and (b) introduce the concept of *thermodynamic temperature*. We now discuss these in some detail.

ARNOT THEOREM

Carnot theorem states that *of all heat engines working between the same (constant) temperatures, the reversible Carnot engine has the maximum efficiency.* Let us consider an irreversible engine (I) and a reversible engine (R) operating between the same reservoirs which are at temperatures T_1 and T_2. Suppose that the irreversible engine is more efficient than a reversible engine. Then we can write

$$\eta_I > \eta_R$$

So we must have

$$\frac{W}{Q_1} > \frac{W}{Q_1'} \tag{6.17}$$

$W = Q_1 - Q_2 = Q_1' - Q_2'$. This implies that $Q_1' > Q_1$.

We now couple them and regard the system of combined engines to be a single device. And suppose that the work produced by the irreversible engine is used to drive the reversible engine backwards so that it acts as a (Carnot) refrigerator as shown in Fig 6.8a. Thus, the overall effect of the combined engine is to transfer a net amount of $(Q_1 - W) - (Q_1' - W) = Q_1 - Q_1'$ from the cold reservoir to the hot reservoir on its own. That is, the combined engine acts as a self-acting device, which requires no external input. But this is forbidden by the Clausius statement of the second law. Therefore, the assumption that $\eta_I > \eta_R$ is not valid, i.e., an irreversible engine cannot have efficiency greater than that of the reversible engine. From this we can conclude that the efficiency of a Carnot engine is maximum, i.e., $\eta_I < \eta_R$. We expect that the same will hold true for a Carnot refrigerator as well.

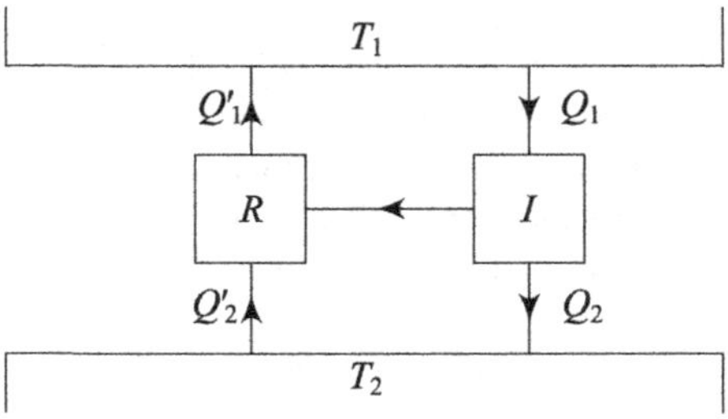

Illustrating the proof of Carnot's theorem.

A corollary of Carnot theorem is: *All reversible engines operating between the same temperature limits have the same efficiency.*

It may be remarked here that Carnot engine is an ideal device because all losses due to conduction, radiation or friction are assumed to be absent. However, in real heat engines, some useful energy is always dissipated. Yet a study of this idealised engine helps us to understand the working of a real engine.

You should now go through the following example.

Example 6.11 For a reversible engine, show that $\sum \frac{Q}{T} = 0.$

According to Carnot theorem, the efficiency of a Carnot engine is greater than or equal to the efficiency of any other engine operating between the same temperature

$$\eta \leq \eta_R$$

Note that the equality sign holds for a reversible cycle, whereas the inequality sign holds for an irreversible cycle.

In terms of heats exchanged, we can write

$$1 - \frac{Q_2'}{Q_1'} \leq 1 - \frac{Q_2}{Q_1}$$

$$\frac{Q_2'}{Q_1'} \geq \frac{Q_2}{Q_1} = \frac{T_2}{T_1}$$

$$\frac{Q_1'}{T_1} - \frac{Q_2'}{T_2} \leq 0$$

Therefore, for any cycle in which heat exchange takes place with two reservoirs only, the algebraic sum

$$\sum \frac{Q}{T} \leq 0.$$

In Chapter 4, we derived the parametric form of the equation of state with empirical temperature θ. We also defined a temperature scale in terms of the thermometric property of the fluid used in the thermometer. We will now define thermodynamic temperature based on Carnot cycle, i.e., on the fundamental laws of thermodynamics and see that it is independent of the properties of the working substance whatsoever. For this reason, it is *absolute temperature scale.*

HERMODYNAMIC TEMPERATURE

From Chapter 4, you may recall that we had asked you to take numerical equivalence of gas scale and thermodynamic scales for granted. However, we will prove this now. Let us therefore revert back to the original symbol θ designating empirical temperature as defined by an arbitrary thermometer using thermometric property denoted as X in Eq. (4.40).

Refer to Fig. 6.9a, which shows two Carnot cycles. In cycle C_1 heat Q_1 is absorbed at temperature θ_1 and heat Q_2 is rejected at temperature θ_2. We know that efficiency of a

$$\eta = 1 - \frac{Q_1}{Q_2} = \phi(\theta_1, \theta_2) \tag{6.18a}$$

$$\frac{Q_1}{Q_2} = f(\theta_1, \theta_2) \tag{6.18b}$$

where both ϕ and f are universal functions of θ_1 and θ_2.

Let another Carnot engine operate between temperatures θ_2 and θ_3 such that θ_3. Suppose that the working substance absorbs heat Q_2 at θ_2 and rejects heat . Then we can write

$$\frac{Q_2}{Q_3} = f(\theta_2, \theta_3) \tag{6.19}$$

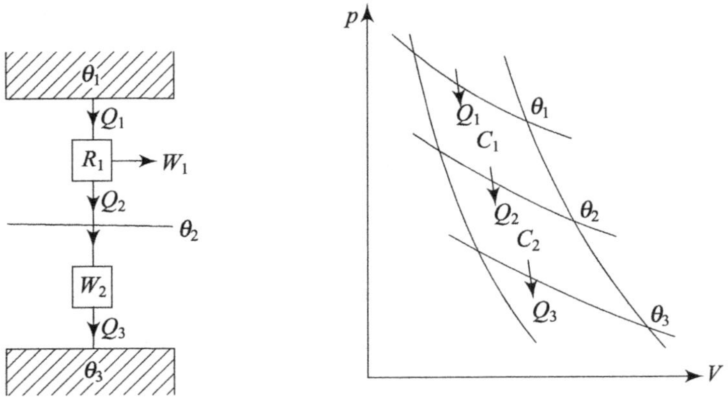

(a) (b)

Defining thermodynamic temperature. (a) Two reversible engines operated in series and (b) Representation on indicator diagram using Carnot cycles.

Now let us couple two heat engines together in such a way that no net heat is exchanged by the reservoir at θ_2. It means that essentially we have a single Carnot engine operating between temperatures θ_1 and θ_3. Therefore, we can write

$$\frac{Q_1}{Q_3} = f(\theta_1, \theta_3) \tag{6.20}$$

$$\frac{Q_1}{Q_3} = \frac{Q_1}{Q_2} \times \frac{Q_2}{Q_3}$$

it readily follows that

$$f(\theta_1, \theta_3) = f(\theta_1, \theta_2)\, f(\theta_2, \theta_3) \tag{6.21}$$

Note that the left-hand side is independent of θ_2. Therefore, for this equality to hold, the function f must be equal to the ratio of two functions so that the product on the right-hand side does not contain θ_2. Let us express this as

$$f(\theta_1, \theta_2) = \frac{T(\theta_1)}{T(\theta_2)}$$

$$f(\theta_2, \theta_3) = \frac{T(\theta_2)}{T(\theta_3)}$$

$$f(\theta_1, \theta_3) = \frac{T(\theta_1)}{T(\theta_3)} \tag{6.22}$$

) is a universal function of the empirical temperature. Note that the ratio $\frac{T(\theta_1)}{T(\theta_2)}$ is independent of the nature and state of the working substance. Obviously, several functional relations may satisfy Eq. (6.22). However, Kelvin argued that one can define a thermodynamic temperature corresponding to an empirical temperature by the relation

$$T_1 = T(\theta_1),$$

$$T_2 = T(\theta_2)$$

$$T_3 = T(\theta_3) \tag{6.23}$$

Hence, we can write

$$\frac{Q_1}{Q_2} = \frac{T_1}{T_2} \tag{6.24}$$

That is, ratio of two thermodynamic temperatures is equal to the ratio of heats exchanged in a Carnot cycle operating between these temperatures.

It is important to note that Eq. (6.24) does not give us any information about the magnitude of a degree on the absolute scale. To this end, let us perform a thought experiment wherein we operate a Carnot engine by keeping the cold reservoir at the triple point of $= T_{tp} = 273.16$ K) and hot reservoir at temperature T. Then we obtain the kelvin scale on which the temperature T is defined as

$$T = 273.16 \frac{Q}{Q_{tp}} \tag{6.25}$$

signifies the absolute value of heat exchanged by a Carnot engine with a reservoir at thermodynamic temperature T. Since the Q- values are necessarily positive numbers, it readily follows that the thermodynamic temperature will be positive definite. Further, is directly proportional to thermodynamic temperature T_2, the zero of the scale is that temperature at which no heat is exchanged, i.e., $Q_2 = 0$.

Proceeding further, in the thought experiment let us choose Q as the heat exchanged at the boiling point of water. Then, we can write

$$\frac{Q_{\text{steam}}}{Q_{tp}} = \frac{\theta_{\text{steam}}}{273.16} = \frac{T_{\text{steam}}}{273.16} \tag{6.26}$$

This implies that

$$\theta_{\text{steam}} = T_{\text{steam}} \tag{6.27}$$

If we divide the interval between T_{steam} and 273.16 K into 100 equal parts, we obtain the size of the one degree on absolute scale to be equal to that on the Celsius scale.

You should now answer the following practise problems to test your conceptual understanding.

Problem 6.10 Can we operate a Carnot engine between $T = 0$ and $T = T_{steam}$? What will be its efficiency?

Problem 6.11 Why do we refer to thermodynamic scale of temperature as absolute

Unattainability of Absolute Zero

From Eq. (6.24) we note that Q_2 is directly proportional to T_2. So if $T_2 \rightarrow 0$, $Q_2 \rightarrow 0$, i.e., no heat will be exchanged with the cold reservoir. Then Eq. (6.3) implies that $\eta = 1$ and complete conversion of heat into work is possible. However, this contradicts the second law of thermodynamics. Therefore, we interpret this result to mean that thermodynamic temperature of zero degrees absolute cannot be achieved by any mechanical means. That is, the second law of thermodynamics implies unattainability of absolute zero.

RREVERSIBILITY AND UNAVAILABLE ENERGY

We know that the efficiency of a heat engine depends on the temperatures within which it operates. It can be increased by raising the source temperature (T_1) or by lowering the sink temperature (T_2). However, T_2 is limited to the temperature of the surroundings (T_S); to below T_S, one would require a refrigeration device. For this reason, we say that heat is more useful when it is supplied at higher temperature. It means that the quality of *degraded* when it is made to flow from a higher temperature to a lower temperature and used there in the sense that less of it can be changed into useful work. To illustrate this, let us consider heat conduction along a bar having a temperature gradient. (Note that this is an irreversible process.) Suppose Q amount of heat is conducted from T_1 to T_2 and is the lowest available temperature to us. (This may be the temperature of surrounding air, river or lake.) If an engine is operated between T_1 and T_0, the *maximum work output* is

$$W_1 = Q\left(1 - \frac{T_0}{T_1}\right) \tag{6.28}$$

After conduction, this heat is available at the lower temperature T_2. If we now operate the engine between T_2 and T_0, the maximum work output will be

$$W_2 = Q\left(1 - \frac{T_0}{T_2}\right) \tag{6.29}$$

T_1, $W_2 < W_1$. That is, we get less work output. This may be viewed as a loss for conversion of chaos (heat) into order (motion). (From humanistic viewpoint, we can say *we have to discipline ourselves to inculcate virtues; degeneration is spontaneous.*) The lost energy is

$$LE = W_1 - W_2 = QT_0\left(\frac{1}{T_2} - \frac{1}{T_1}\right) \tag{6.30}$$

Note that this loss is associated with irreversibility of heat conduction from T_1 to T_2. Since natural processes are irreversible, available energy is continuously running downhill. Will it disappear ultimately in violation of the first law of thermodynamics? To answer this question we should learn to measure irreversibility of a process in terms of a physical quantity. In the next chapter, you will learn to characterise irreversibility while analysing evolution of a process in time. For now, we wish to continue discussing convertibility of heat into work in real engines.

OWER CYCLES

You are now familiar with the working principle of heat engines and refrigerators. In a heat engine, the working substance absorbs heat from a reservoir at a higher temperature, performs some work and rejects heat to a reservoir at lower temperature. The second law of thermodynamics precludes the possibility of complete conversion of heat into work; to increase efficiency; the temperature of the source should be as high as possible because temperature of the sink is dictated by the surrounding environment. Though efficiency of a Carnot engine is independent of the nature of the working substance, yet for a practical engine, one chooses a working substance that is readily available and that too in abundance, less expensive so as to keep cost of production of power to the minimum and thermally efficient. Two substances that attracted attention of early workers as viable alternatives were air and water. Of these, water was initially preferred because of its larger latent heat, whereas air had to be heated to a very high temperature. We now discuss these.

Steam Power Plant: Rankine Cycle

Refer to Fig. 6.10. Water at normal atmospheric pressure is pumped into the boiler kept at a constant high pressure. Water absorbs heat, boils and is superheated. High pressure superheated steam enters the turbine where it is made to expand adiabatically and drives the turbine to generator power. In this process, the temperature and pressure of the steam decrease and low pressure steam at a reduced temperature leaves the turbine to enter the condenser. In the condenser, steam is allowed to condense by rejecting heat to the sink, which is usually a stream of cold water. (Such an engine is also known as *external combustion engine*.)

It is important to point out here that an exact analysis of steam cycle is very complex because of the presence of energy dissipative processes such as friction and heat conduction. Therefore, for simplicity one resorts to the Rankine cycle, which is the idealised version of a simple steam cycle.

In the Rankine cycle, all processes are assumed to be quasistatic and devoid of all energy dissipative effects. It is illustrated in Fig. 6.11 on $p - V$ diagram as well as $T - S^*$ diagram and is defined by the following processes:

Reversible adiabatic compression of saturated water coming out of the condenser at pressure p_1 to pressure of the boiler p_2. In this cycle, the temperature may change slightly but the volume remains essentially constant.

Reversible isobaric heating of water to its boiling point at pressure p_2.

S is used to denote entropy of a system. It is a measure of chaos/disorder in the system and is connected to heat through the relation $dS = \frac{\delta Q}{T}$. For an adiabatic process, $\delta Q = 0$ and entropy is constant. It is discussed in detail in Chapter 7.

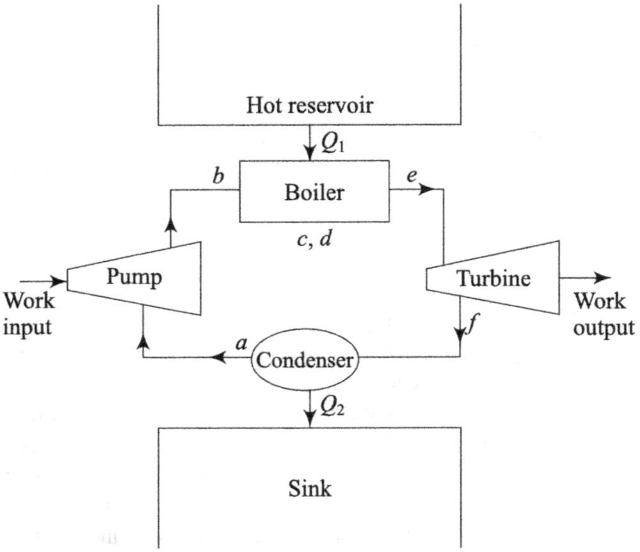

Schematic diagram of a steam power plant.

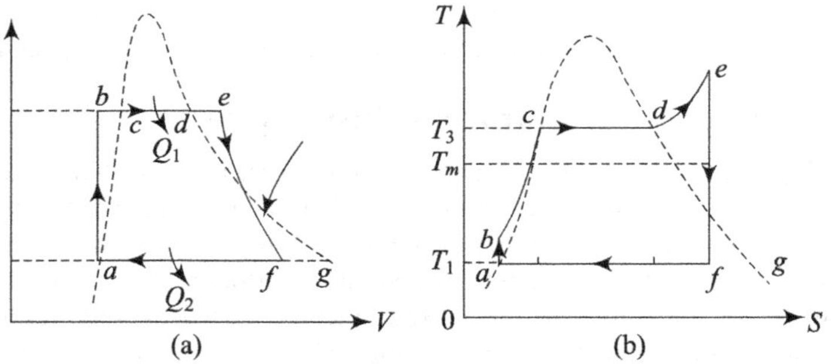

(a) $p - V$ diagram and (b) $T - S$ diagram of Rankine cycle. The dashed curves represent saturated liquid and saturated vapour.

Reversible isobaric-isothermal vaporisation of water into saturated steam.

Reversible isobaric superheating of saturated steam into superheated steam. It leads to further increase in temperature.

Reversible adiabatic expansion of superheated steam to wet steam at initial pressure p_1.

Reversible isobaric-isothermal condensation of wet steam into supersaturated water.

To calculate the efficiency of a Rankine cycle, let the heat absorbed by a unit mass of the working substance in going from stage $b \to e$ be Q_1 and the heat rejected per unit mass in the condensation process $f \to a$ be Q_2. Then we can write

$$\eta_{\text{Rankine}} = 1 - \frac{Q_2}{Q_1}.$$

If we assume that the mean temperature at which heat is taken by the working substance is , then you can easily convince yourself that we can rewrite the expression for efficiency of a Rankine cycle as

$$\eta_{\text{Rankine}} = 1 - \frac{T_1}{T_m} \tag{6.31}$$

$< T_3$, the efficiency of the Rankine cycle is less than that of a Carnot engine. Therefore, to increase the efficiency of the Rankine cycle with T_1 fixed, we need to raise . This is possible only by superheating the steam at high pressure. This however presents intricate engineering problems.

Air Standard Cycles

Heat engines using gas as a working substance are one of the most common work producing devices. The gas is taken as a mixture of air and a fuel such as petrol, or diesel. In metropolitan cities, use of compressed natural gas (CNG) based buses and three wheelers is being promoted now-a-days to prevent environmental pollution. The petrol and diesel engines essentially differ in their ignition systems. While petrol engine is a spark ignition engine, the diesel engine is compression ignition engine. (You will learn about the basic difference between them soon.) The input supply of heat is obtained by combustion (burning) of petrol/diesel/CNG and air inside a cylinder. (For this reason, these engines are also known as *internal combustion engines*. The engines used in buses, trucks, cars, scooters, auto-rickshaws, motor cycles belong to this category.) Note that unlike the steam engine, the working substance in this case suffers permanent change due to chemical reaction during the combustion and can not return to its initial state. The successive mechanical cycles start with a fresh mixture after discarding the burnt mixture. As such, thermodynamic analysis of internal combustion engines is fairly complex because of irreversible chemical reaction, presence of dissipative effects and intermittent nature of the processes. Therefore, it is customary to approximate the actual (open) cycles by (closed) isolated thermodynamic cycles. In addition, we make some simplifying assumptions:

Assumptions

The working substance is a fixed mass of air and behaves like an ideal gas with constant heat capacities.

The reservoirs with which heat is exchanged at high as well as low temperatures are external to the system.

All processes are quasistatic and there is no loss due to dissipative effects.

Based on these assumptions, we will first discuss the Otto cycle.

(a) The air standard Otto cycle The air standard Otto cycle is an ideal cycle to represent the spark-ignition internal combustion engine. You may have learnt about it in your school physics curriculum but we briefly outline the details for brevity and completeness. An internal combustion engine consists of a cylinder fitted with a piston, which is connected to the wheels through a crankshaft. The cylinder has two valves: an *intake valve* through which the fuel (petrol vapour + air) enters and an *exhaust valve* through which spent gases are expelled. A spark plug causes the fuel to ignite via an electric spark. The engine has four strokes, which operate cyclically:

Intake stroke: A stream of air is forced past the nozzle of the carburettor and petrol vapour from the carburettor is sprayed into the air stream. This explosive mixture goes through the intake valve in the cylinder by the suction stroke of the piston.

Compression stroke: The mixture is compressed adiabatically and leads to significant rise in temperature and pressure. In high compression engines, which have greater efficiency, the volume decreases almost eight-fold.

Power stroke: Adiabatic expansion of the products of combustion pushes the piston outward. An electric spark in the cylinder ignites the fuel-air mixture causing instantaneous combustion. The engine performs work in this stroke.

Exhaust stroke: At the end of the power stroke, the gases escape into the surroundings through the exhaust valve and the cylinder is available for fresh cycle.

These changes are depicted in Fig. 6.12 on pressure volume ($p - V$) diagram as well as temperature entropy ($T - S$) diagram of the working substance. It is defined by the following processes:

Intake stroke

Reversible adiabatic compression of n moles of air is accompanied by increase in temperature from T_1 to T_2. Mathematically, states a and b are connected through the relation

$$T_1 V_1^{\gamma-1} = T_2 V_2^{\gamma-1} \tag{6.32a}$$

Heat Q_1 is taken reversibly from the source reservoir at constant volume. This process stimulates combustion in the spark ignition engine.

Reversible adiabatic expansion with drop in temperature from T_3 to T_4. Mathematically, we can relate the states c and d as

$$T_3 V_2^{\gamma-1} = T_4 V_1^{\gamma-1} \tag{6.32b}$$

Heat Q_2 is rejected by air reversibly at constant volume to a series of reservoirs whose temperatures vary from T_4 to T_1. This process represents opening of exhaust valve.

Exhaust stroke

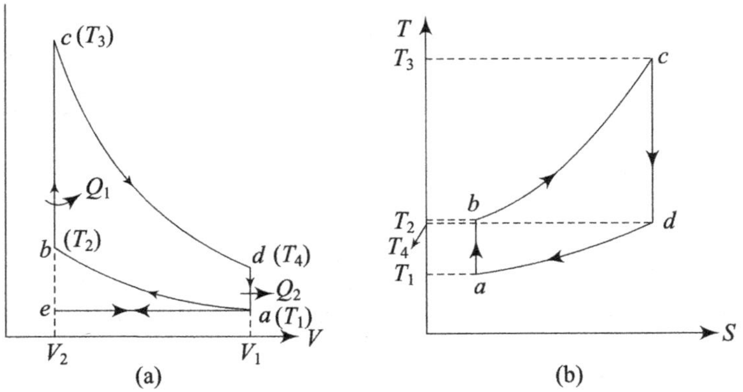

Air standard Otto cycle. (a) Pressure volume and (b) Temperature entropy diagrams.

As Otto cycle is a closed system, the intake and exhaust strokes are ignored. Therefore, in terms of energy, we can say that heat Q_1 is added during the ignition of the fuel and exhaust gases take away heat Q_2, which is rejected to the environment at constant volume. So, the efficiency of the Otto cycle is given by

$$\eta = 1 - \frac{Q_2}{Q_1}$$

$$= 1 - \frac{nC_V(T_4 - T_1)}{nC_V(T_3 - T_2)}$$

$$= 1 - \frac{T_4 - T_1}{T_3 - T_2} \tag{6.33}$$

On subtracting Eq. (6.32a) from Eq. (6.32b), we can write

$$(T_4 - T_1)V_1^{\gamma-1} = (T_3 - T_2)V_2^{\gamma-1}$$

$$\frac{T_4 - T_1}{T_3 - T_2} = \left(\frac{V_2}{V_1}\right)^{\gamma-1}$$

Hence, the expression for efficiency takes the form

$$\eta = 1 - \left(\frac{V_2}{V_1}\right)^{\gamma-1}$$

We can rewrite it as

$$\eta = 1 - \left(\frac{1}{V_1/V_2}\right)^{\gamma-1}$$

$$= 1 - \left(\frac{1}{\rho_c}\right)^{\gamma-1} \tag{6.34}$$

$= \frac{V_1}{V_2}$ is called the *adiabatic compression ratio*. This result shows that efficiency of an Otto engine depends only on the compression ratio; it increases with compression ratio, which is same as expansion ratio in this case. However, you should not think that we can make it approach unity; mechanical constraints put an upper limit of 10 on this $\rho_c = 8$ and $\gamma = 1.40$, we obtain $\eta = 1 - \frac{1}{8^{0.4}} = 1 - \frac{1}{2.2974} = 1 - 0.4353 = 0.5647$.

It means that efficiency of an Otto engine $\eta \approx 56\%$. This shows that the Otto engine is fairly efficient. However, it is interesting to point out here that only a small fraction (5 – 10%) of the fuel power is utilised by the automobile. According to rough estimates, nearly 36% fuel power is taken away by the exhaust gases and an almost the same amount is absorbed by the coolant. Similarly, about 15 – 20% is lost in overcoming friction (air, engine, transmission, tyres).

You must have noted that a heat engine does useful work only during a part of the working cycle. It means that for its smooth running in a cycle, a machine should have several engines connected so that they have their power strokes one after the other. You may have enjoyed long car drives under diverse road conditions. And more often than not, these must have been rather smooth. This is because a car engine has four or six pistons. The same is however not true of a scooter or a motorcycle.

You should now go through the following example carefully to get an idea about typical

Example 6.12 For what value of compression ratio in an air standard Otto cycle, the work per unit mass will be maximum. Assume that the maximum and minimum temperatures are T_3 and T_1, respectively.

The net work done in the cycle per kg of air is given by

$$W = \eta Q_1 = \eta C_V (T_3 - T_2)$$

From Eq. (6.32a), we recall that

$$T_2 = T_1 \left(\frac{V_1}{V_2} \right)^{\gamma-1} = T_1 \rho_c^{\gamma-1}$$

Hence, on using this result and the value of η from Eq. (6.34), we can write

$$W = C_V \left(1 - \frac{1}{\rho_c^{\gamma-1}} \right) (T_3 - \rho_c^{\gamma-1} T_1) = C_V [T_3 - T_3 \rho_c^{-(\gamma-1)} - \rho_c^{\gamma-1} T_1 + T_1]$$

To calculate the maximum work for a given compression ratio, we differentiate this expression with respect to ρ_c and equate the resultant expression to zero:

$$\frac{\sigma W}{\sigma \rho_c} = (\gamma - 1) \rho_c^{-\gamma} T_3 - (\gamma - 1) \rho_c^{\gamma-2} T_1 = 0$$

$$T_3 \rho_c^{-\gamma} = T_1 \rho_c^{\gamma-2} \Rightarrow \rho_c^{2\gamma-2} = \frac{T_3}{T_1}.$$

Hence, compression ratio for maximum work can be expressed in terms of temperatures as

$$\rho_c = \left(\frac{T_3}{T_1} \right)^{\frac{1}{2(\gamma-1)}} = \left(\frac{T_3}{T_1} \right)^{1.25}$$

You can easily convince yourself that for this value of compression ratio, the second order derivative of W will be negative.

Example 6.13 In an air standard Otto cycle, the compression ratio is 8 and the pressure and temperature at the end of the compression stroke are 1.5 MPa and 550 K, respectively. The pressure at the end of the heat intake is 4.5 MPa. Calculate the net work done per kilogram of air. Take molecular weight of air as 28.97 and γ= 1.4 corresponding to a diatomic gas.

The net work done in the cycle per kg of air is given by

$$W = \eta Q_1$$

The heat taken reversibly per kg in the process $b \to c$ is given by $q_1 = C_V(T_3 - T_2)$. Since molar heat capacity at constant volume for a diatomic gas is $5R/2$, for 1 kg of air

$$C_V = \frac{5}{2} \times \frac{(8.314 \text{ kJ kmol}^{-1} \text{ K}^{-1})}{28.97 \text{ kg mol}^{-1}} = 0.7175 \text{ kJ K}^{-1}$$

Using ideal gas law for the isochoric process, we can calculate the value of T_3:

$$T_3 = \frac{p_3 T_2}{p_2} = \frac{(4.5\text{ MPa}) \times (550\text{ K})}{(1.5\text{ MPa})} = 1650\text{ K}$$

We recall that for the Otto cycle with $\rho_c = 8$, $\eta = 0.5647$. Hence

$$W = \eta C_V (T_3 - T_2) = 0.5647 \times (0.7175\text{ kJ K}^{-1}) \times (1100\text{ K}) = 445.7\text{ kJ}.$$

(b) The air standard diesel cycle The diesel engine is used extensively in heavy transport vehicles such as trucks, buses, tractors, cars and locomotives. The fuel—diesel—is less expensive and equally efficient. Now-a-days, diesel pump sets are also used to generate electricity at social functions or as substitute for power shut down in the business establishments. It also operates various devices used in agriculture such as water pumps, crop harvesting and thrashing machines, among others. In a diesel engine, only air is fed into the cylinder in the intake stroke and the fuel is injected into the air at the end of the compression stroke. This facilitates significant increase in compression ratio and combustion occurs spontaneously. The injection of oil is adjusted so that constant pressure is maintained. For this reason, the diesel engine is also called *compression-ignition engine.* The other operations are the same as those in a Otto engine. The ideal cycle for diesel engine is depicted in Fig. 6.13 on $p - V$ and $T - S$ diagrams.

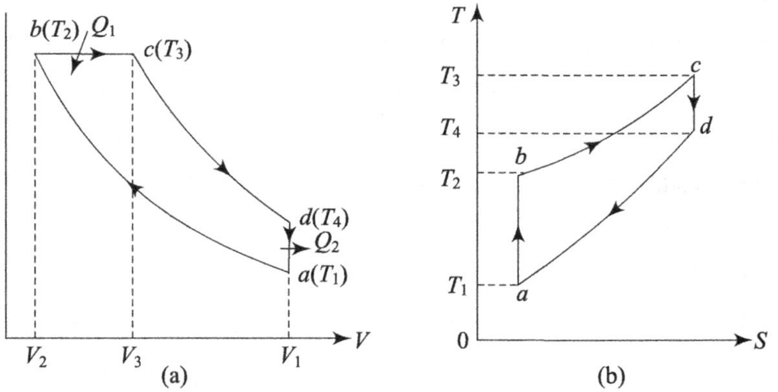

(a) $p - V$ diagram and (b) $T - S$ diagram of diesel cycle.

On comparing Figs. 6.12 and 6.13, you will note that the basic difference in the Otto cycle and diesel cycle is that in the latter case,

> The fuel burns at constant pressure. That is, the heat is absorbed reversibly and isobarically in the process $b \rightarrow c$ from a series of external reservoirs in the range T_2 to T_3. The compression and expansion ratios are now not equal.

For the analysis of diesel cycle, we define compression ratio as before:

$$\rho_c = \frac{V_1}{V_2}.$$

expansion ratio, also called *fuel cut-off ratio,* is defined as the ratio of the volume of air after heat intake at constant pressure to the volume at the end of the isentropic compression:

$$r_c = \frac{V_3}{V_2}$$

Note that the cut-off ratio is related to the actual operation of diesel engine. The efficiency of the diesel engine is given by

$$\eta = 1 - \frac{Q_2}{Q_1}$$

$$= 1 - \frac{nC_V(T_4 - T_1)}{nC_p(T_3 - T_2)}$$

$$= 1 - \frac{T_1[(T_4/T_1) - 1]}{\gamma T_2[(T_3/T_2\} - 1]} \tag{6.35}$$

Since compression and expansion occur adiabatically, the states a and b as well as c and will be connected by the equation of an adiabat. Hence, we can write

$$T_1 V_1^{\gamma-1} = T_2 V_2^{\gamma-1} \Rightarrow \left(\frac{T_1}{T_2}\right) = \left(\frac{V_2}{V_1}\right)^{\gamma-1} = \frac{1}{\rho_c^{\gamma-1}}$$

$$T_4 V_1^{\gamma-1} = T_3 V_3^{\gamma-1}$$

On combining these equations, we can write

$$\frac{T_4}{T_1} = \frac{T_3}{T_2}\left(\frac{V_3}{V_2}\right)^{\gamma-1} = r_c^{\gamma-1}\frac{T_3}{T_2}$$

Since heat is added at constant pressure, we can use Charle's law to write

$$\frac{T_2}{V_2} = \frac{T_3}{V_3} \Rightarrow \frac{T_3}{T_2} = \frac{V_3}{V_2} = r_c$$

$$\frac{T_4}{T_1} = r_c^{\gamma}$$

Using these results in Eq. (6.35), the expression for the efficiency of Diesel engine takes a simple form:

$$\eta = 1 - \frac{1}{\gamma \rho_c^{\gamma-1}}\left[\frac{r_c^{\gamma} - 1}{r_c - 1}\right] = 1 - \frac{A}{\rho_c^{\gamma-1}} \tag{6.36}$$

$\dfrac{r_c^{\gamma} - 1}{\gamma(r_c - 1)}$ depends only on the cut-off ratio; compression ratio does not affect it in any way whatsoever. For $A = 1$, Eqs. (6.34) and (6.36) become identical. It means gives information how rate of addition of heat influences the efficiency of a diesel engine. Further, since r_c as well as γ are greater than one, A will always be less than one. It means that for the same compression ratio, the efficiency of Otto cycle will be more. In practice, $7 < \rho_c < 9$, whereas $5 < r_c < 20$. As a result of this, an actual diesel engine has greater efficiency than the Otto (petrol) engine. Moreover, from Eq. (6.36), we note that the efficiency of a diesel engine will improve by

increasing the compression ratio;
decreasing the cut-off ratio; and
using a gas characterised by a high value of γ.

You should now go through the following example carefully to fix your ideas.

Example 6.14 A air-standard Diesel engine operates between 300 K and 1200 K. If the minimum pressure is 120 kPa and the heat intake is 600 kJ kg^{-1}, calculate the efficiency of the engine. Take $C_p = 29.2$ J mol^{-1} K^{-1} and molecular weight of air $M = 28.97$ kg.

We know that in a diesel cycle, the heat is absorbed reversibly and isobarically in the process $b \to c$.Therefore, we can write

$$Q_1 = C_p(T_3 - T_2) \Rightarrow 600\,\text{kJ kg}^{-1} = \frac{(29.2\,\text{Jmol}^{-1}\text{K}^{-1})}{28.97\,\text{kg}}(1200\,\text{K} - T_2)$$

$$= 1200\,\text{K} - \frac{(600\,\text{kJ kg}^{-1}) \times (28.97\,\text{kg})}{(29.2\,\text{Jmol}^{-1}\text{K}^{-1})} = 1200\,\text{K} - 295.3\,\text{K} = 904.7\,\text{K}$$

For the adiabatic process $a \to b$, we can write

$$\frac{p_2}{p_1} = \left(\frac{T_2}{T_1}\right)^{\frac{\gamma}{\gamma-1}} = \left(\frac{604.7\,\text{K}}{300\,\text{K}}\right)^{3.5} = 11.6 \Rightarrow p_2 = 1392\,\text{kPa}$$

The compression ratio $\rho_c = \dfrac{V_1}{V_2} = \left(\dfrac{p_2}{p_1}\right)^{\frac{1}{1.4}} = (11.6)^{\frac{1}{1.4}} = 5.75$

The cut-off ratio $r_c = \dfrac{V_3}{V_2} = \dfrac{T_3}{T_2} = \dfrac{1200}{604.7} = 1.98$

Since the process $b \to c$ is isobaric, the efficiency of the engine is given by Eq. (6.36):

$$= 1 - \frac{1}{{}_{c}^{\gamma-1}}\left[\frac{r_c^{\gamma} - 1}{r_c - 1}\right] = 1 - \frac{1}{1.4 \times (5.75)^{04}}\left[\frac{(1.98)^{1.4} - 1}{1.98 - 1}\right] = 1 - \frac{1}{1.4 \times (5.75)^{0.4}} \times 1.635$$

$$= 0.42$$

Note that this value is less than that obtained for the Otto engine. But in practise, it is not so. In general, the typical temperature attained at the end of the adiabatic process is much higher than 1200 K and minimum pressure is lower than 120 kPa.

Having discussed power cycles associated with different types of heat engines, we will now concentrate on another aspect of the second law of thermodynamics, the refrigeration

Refrigeration Cycles

You now know that a refrigerator is a device which also operates in a cycle and consumes

That is, heat is absorbed by the working substance from the object to be kept cold! The efficiency of a refrigerator is rated in terms of its coefficient of performance, which is the highest for a Carnot refrigerator. Most commercial refrigeration plants, which are used to store fruits and vegetables or maintain central air-conditioning in office complexes, and domestic refrigerators are of vapour-compression type. We will now discuss it in some

Vapour-Compression Refrigeration Cycle

Refer to Fig. 6.14, which schematically depicts vapour-compression refrigeration cycle. A fixed amount of the refrigerant follows the cyclic path involving phase changes.

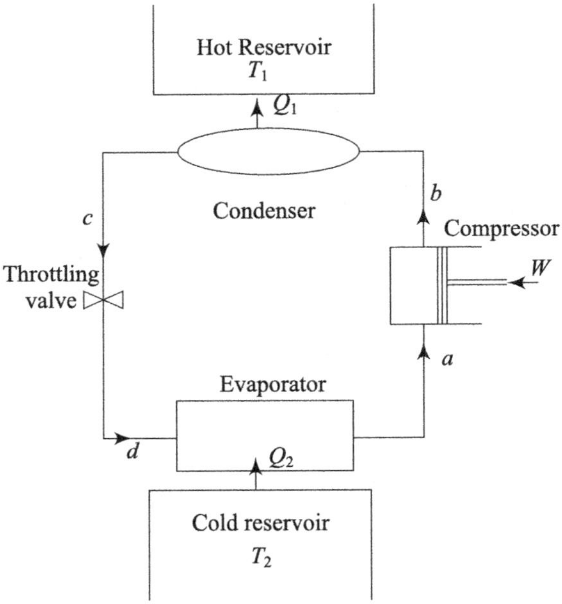

Schematic representation of vapour-compression refrigerator.

V and $T-S$ diagrams of the corresponding ideal refrigeration cycle are shown in Fig. 6.15. The ideal refrigeration cycle is defined by the following processes:

Reversible adiabatic compression of a saturated vapour defined by low pressure and low temperature (p_2, T_2) to an unsaturated vapour at high pressure-high temperature (p_1, T_3).

Reversible isobaric cooling of the vapour to temperature T_1. The superheated vapour is completely condensed to a saturated liquid defined by (p_1, T_1). Heat Q_1 is rejected by the refrigerant.

Irreversible adiabatic throttling (isenthalpic expansion) from pressure p_1 to p_2. This is accompanied by cooling to temperature T_2 and partial vaporisation. (A throttling process is indicated by the broken lines as it can not be plotted on a state diagram as a smooth curve.)

Reversible isothermal-isobaric vaporisation. Heat Q_2 is absorbed by the refrigerant at temperature T_2.

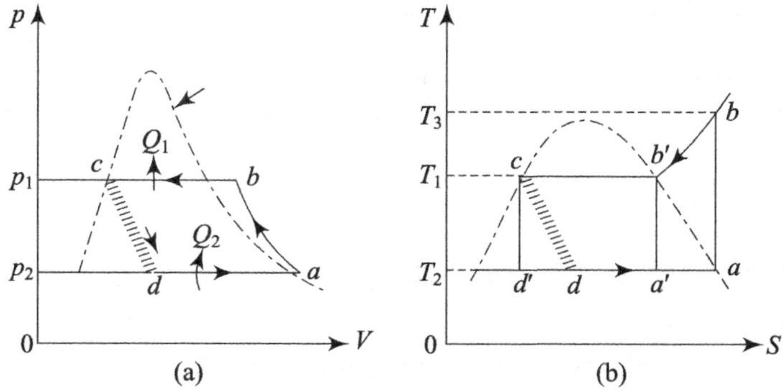

(a) $p-V$ and (b) $T-S$ diagrams of an ideal vapour-compression refrigeration cycle.

On comparison with Figs. 6.10 and 6.11, you will note that the refrigeration cycle is essentially a Rankine cycle in reverse, except that a pump in Rankine cycle has been substituted by the throttling process. The Carnot refrigeration cycle operating between temperatures T_1 and T_2 is shown as $a' \to b' \to c \to d' \to a'$ on the $T-S$ diagram. However, it is impractical since one has to handle a mixture of liquid and vapour in the process and an irreversible expansion process $c \to d'$, when the mixture of liquid and vapour is discharged.

You can easily convince yourself that the coefficient of performance is given by

$$\omega = \frac{Q_2}{W} = \frac{Q_1 - W}{W} = \frac{Q_1}{W} - 1$$

$$\frac{Q_1}{W} = 1 + \omega \tag{6.37}$$

This result shows that if $\omega \geq 2$, the heat liberated at the higher temperature is many times greater than the work done. This is the working principle of a heat pump, which is used to heat houses in cold climates. If the flow of refrigerant in the heat pump is reversed, the device acts as a air-conditioner to cool houses in summer.

For the air standard refrigeration cycle, the $p-V$ and $T-S$ diagrams are depicted in Fig. 6.16. You can easily calculate the expression for ω in this case:

$$\omega = \frac{T_1 - T_4}{T_2 - T_3 - T_1 + T_4} \tag{6.38}$$

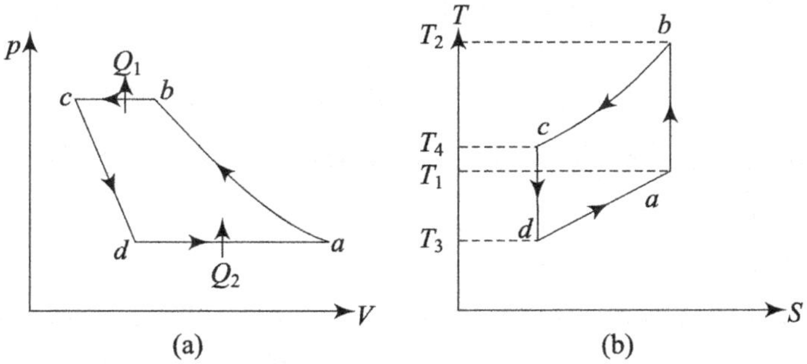

So far we have dealt only with convertibility of heat into work and vice versa. Although this is a very useful aspect and hastened industrialisation of Europe, we are many a times more interested in the evolution of thermodynamic processes. For instance, we would be interested in applying the second law to analyse combustion processes or chemical processes occurring in a system, an engine say. A detailed discussion of this aspect forms the subject matter of the next chapter.

DITIONAL EXAMPLES

Example 6.15 Consider four heat reservoirs *A*, *B*, *C* and *D*. The efficiency of a reversible engine working between *A* and *C* is half of the sum of the efficiencies of the same engine when it is operated between *A* and *B* and between *A* and *D*. Calculate the temperature of reservoir *C*.

Let us denote the temperature of heat reservoirs as T_A, T_B, T_C and T_D. For a reversible engine, we know that efficiency is given by Eq. (6.13):

$$\eta = 1 - \frac{T_2}{T_1}$$

is temperature of sink and T_1 is temperature of source. In the instant case, we are told that

$$\eta_{AC} = \frac{1}{2}[\eta_{AB} + \eta_{AD}]$$

In terms of temperatures, we can write

$$1 - \frac{T_C}{T_A} = \frac{1}{2}\left[1 - \frac{T_B}{T_A} + 1 - \frac{T_D}{T_A}\right]$$

$$T_C = \frac{1}{2}(T_B + T_D)$$

That is, temperature of the reservoir *C* is arithmetic mean of the temperatures of sinks of the reservoirs *B* and *D*.

Example 6.16 An inventor claims that his heat engine delivers work at the rate of 1 kW and absorbed energy at the rate 65 kJ/min from a source at 1127°C when the temperature of sink is 27°C. Is the claim justified?

The maximum possible efficiency of an engine is the Carnot efficiency:

$$\eta = 1 - \frac{T_2}{T_1}$$

In the instant case, $T_2 = 300$ K and $T_1 = 1400$ K. Hence,

$$\eta = 1 - \frac{300}{1400} = 0.7857$$

The efficiency claimed by inventor $= \dfrac{W}{Q_1} = \dfrac{10^3 \text{ Js}^{-1}}{\frac{65}{} \times 10^3 \text{ Js}^{-1}} = \dfrac{60}{65} = 0.9231$

Since the inventor claims efficiency higher than the Carnot efficiency, the claim is not

Example 6.17 A doctor practising in Delhi preserves some medicines at –23°C. (a) If a refrigerator used for this purpose has a 200 W compressor, how long will it run if the outside temperature is 27°C and the cooling rate is 5×10^4 kJ per day. (b) If the same refrigerator is used in Barmer in Thar desert where room temperature on a summer day is 47°C, estimate the fraction of time the compressor runs for the same cooling rate.

(a) We know that the coefficient of performance is given by

$$\omega = \frac{T_L}{T_H - T_L}$$

= (273 – 23) K = 250 K, T_H = (273 + 27) K = 300 K. Hence,

$$\omega = \frac{250}{300 - 250} = \frac{250}{50} = 5$$

The coefficient of performance is also defined as

$$\omega = \frac{Q_L}{W}$$

$$W = \frac{Q_L}{\omega} = \frac{5 \times 10^4 \text{ kJd}^{-1}}{5} = 10^4 \text{ kJd}^{-1}$$

We know that the work done by compressor in a day

$$(200 \text{ Js}^{-1}) \times (3600 \times 24)\text{s} = 200 \times 3600 \times 24 \text{ J}$$
$$= 17280 \text{ kJ}$$

Hence, the fraction of time the compressor runs in Delhi $= \dfrac{W}{17280 \text{ kJ}}$

$$= \frac{10^7 \text{ J}}{1728 \times 10^4 \text{ J}}$$

$$= 0.5787$$

(b) The coefficient of performance of the refrigerator in Barmer is

$$\omega = \frac{250}{320 - 250} = \frac{250}{70} = 3.571 = \frac{Q_L}{W}$$

$$W = \frac{5 \times 10^4 \text{ kJd}^{-1}}{3.571} = 1.40 \times 10^4 \text{ kJd}^{-1}.$$

Hence, the fraction of time for which compressor runs in Barmer

$$= \frac{1.40 \times 10^4 \text{ kJd}^{-1}}{17280 \text{ kJd}^{-1}} = 0.81$$

As expected, the refrigerator will run for greater time in Barmer than in Delhi.

Example 6.18 Calculate the efficiency of a Carnot engine working between 127°C and 27°C It absorbs 80 cal of heat from the source. How much heat is rejected?

We know that efficiency of a Carnot engine is given by

$$\eta = 1 - \frac{T_2}{T_1}$$

= (127 + 273) K = 400 K, T_2 = (27 + 273) K = 300 K. Hence,

$$\eta = 1 - \frac{300}{400}$$
$$= 0.25$$
$$= 25\%$$

Since efficiency is the ratio of work done and heat taken, we can write

$$\eta = \frac{W}{Q}$$

$$W = \eta Q$$
$$= 0.25 \times (80 \text{ cal})$$
$$= 20 \text{ cal}$$

Hence, heat rejected by the engine

$$Q_2 = Q_1 - W$$
$$= 80 \text{ cal} - 20 \text{ cal}$$
$$= 60 \text{ cal}$$

Note that a low efficiency engine cuts two ways; it causes wastage of fuel i.e., it is uneconomical, and degrades the environment.

Example 6.19 A Carnot engine has an efficiency of 30% when the temperature of the sink is 27°C Calculate the change in temperature of the source to raise its efficiency

We have to first calculate the temperature of the source for which efficiency is 30%. Therefore, we recall that

$$\eta = 1 - \frac{T_2}{T_1}$$

$$T_1 = \frac{T_2}{1 - \eta}$$

= (27 + 273) K = 300 K and η = 0.3. Hence,

$$T_1 = \frac{300 \text{ K}}{1 - 0.3} = 428.6 \text{ K}$$

Similarly, the temperature of the source which helps to raise efficiency to 50% is

$$T_1' = \frac{300 \text{ K}}{1 - 0.5} = 600 \text{ K}$$

Hence, change in temperature $= T_1' - T_1$

$$= 600\text{ K} - 428.6\text{ K}$$
$$= 171.4\text{ K}.$$

Example 6.20 A 100 kW engine is operating between 127°C and 27°C. Calculate the amount of heat absorbed, heat rejected and efficiency of the engine.

We know that work done by an engine is equal to the difference of heat absorbed and heat rejected:

$$W = Q_1 - Q_2 \tag{i}$$

Also ratio of heat absorbed to heat rejected is equal to the ratio of temperature of the source and the sink:

$$\frac{Q_1}{Q_2} = \frac{T_1}{T_2} \tag{ii}$$

= (127 + 273) K = 400 K and T_2 = (27 + 273) K = 300 K. Hence,

$$\frac{Q_1}{Q_2} = \frac{400}{300} = \frac{4}{3}$$

$$\frac{Q_1}{Q_2} - 1 = \frac{1}{3}$$

$$\frac{Q_1 - Q_2}{Q_2} = \frac{1}{3}$$

$$3W = Q_2$$

$$Q_2 = 3 \times 10^5\ \text{Js}^{-1}$$

Using this result in Eq. (i), we get

$$Q_1 = W + Q_2$$
$$= 4 \times 10^5\ \text{Js}^{-1}$$

Note that heat absorbed per second = 4×10^5 J and heat rejected per second = 3×10^5 J. The efficiency of the engine is given by

$$\eta = 1 - \frac{T_2}{T_1}$$
$$= 1 - \frac{300}{400} = 0.25$$
$$= 25\%$$

Example 6.21 A heat engine working between two temperatures converts one-eighth of the heat supplied into work. If the lower temperature is reduced by 95°C the efficiency is doubled. Calculate the initial temperatures of the source and the sink.

Suppose that the initial temperatures of the source and the sink are T_1 and T_2. Then efficiency of this engine

$$\eta = \frac{1}{8} = 1 - \frac{T_2}{T_1}$$

When the temperature of the sink is reduced by 95°C = 95 K, the efficiency is given by

$$\frac{1}{4} = 1 - \frac{T_2 - 95\text{ K}}{T_1}$$

$$= 1 - \frac{T_2}{T_1} - \frac{95\text{ K}}{T_1}$$

$$= \frac{1}{8} - \frac{95\text{ K}}{T_1}$$

$$\frac{95\text{ K}}{T_1} = \frac{1}{8}$$

$$T_1 = (95\text{ K}) \times 8 = 760\text{ K}$$

$$T_2 = \frac{7}{8} \times T_1 = (95\text{ K}) \times 7 = 665\text{ K}.$$

Example 6.22 A Carnot engine has an efficiency of 0.5 while working between a source at 400 K and a sink at T_2K. Calculate the efficiency of the engine, if temperatures of both the source and the sink are increased by 100 K. Calculate the efficiency, if the temperature of the source and the sink are reduced by 100 K.

We know that efficiency of a heat engine is given by

$$\eta = 1 - \frac{T_2}{T_1}$$

= 0.5 and T_1 = 400 K. Hence, we can write

$$0.5 = 1 - \frac{T_2}{400\text{ K}}$$

$$\frac{T_2}{400\text{ K}} = 0.5$$

$$T_2 = 200\text{ K}.$$

In the second stage, T_1 = 500 K and T_2 = 300. Hence,

$$\eta' = 1 - \frac{300\text{ K}}{500\text{ K}}$$

$$= 0.4$$

Note that by increasing temperatures of the sink and the source equally, we have lowered the efficiency of the engine. However, if the temperatures of the source and sink were reduced by 100 K, the efficiency will become

$$\eta'' = 1 - \frac{100\text{ K}}{300\text{ K}} = 0.67.$$

Example 6.23 Two Carnot engines A and B are operated in series. Engine A absorbs heat at 500 K and rejects heat to a sink at temperature T. Engine B absorbs half of the heat rejected by engine A and rejects heat to the sink at 200 K. If the work done in both the case is equal, calculate T.

Refer to Fig. 6.17, which is a schematic drawing of the process under consideration. The efficiencies of and B are given by

$$\eta_A = 1 - \frac{T}{T_1} = 1 - \frac{Q_2}{Q_1} \quad \text{(i)}$$

$$\eta_B = 1 - \frac{T_3}{T} = 1 - \frac{Q_3}{Q_2/2} \quad \text{(ii)}$$

$= W_B$, we can write

$$Q_1 - Q_2 = \frac{Q_2}{2} - Q_3$$

$$Q_1 - \frac{3}{2}Q_2 + Q_3 = 0 \quad \text{(iii)}$$

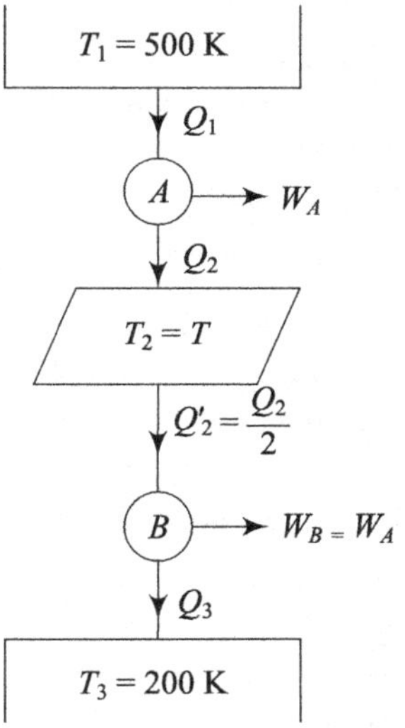

Fig. 6.17

On dividing throughout by Q_1, we get

$$1 - \frac{3}{2}\frac{Q_2}{Q_1} + \frac{Q_3}{Q_1} = 0$$

We rewrite it as

$$1 - \frac{3}{2}\frac{Q_2}{Q_1} + \frac{Q_3}{Q_2} \cdot \frac{Q_2}{Q_1} = 0 \quad \text{(iv)}$$

$$\frac{Q_3}{Q_2} = \frac{Q_3}{Q_2/2} \times \frac{1}{2}$$

$$= \frac{T_3}{2T}$$

Hence, in terms of temperatures, we can express it as

$$1 - \frac{3}{2}\frac{T}{T_1} + \frac{T_3}{2T} \cdot \frac{T}{T_1} = 0$$

$$1 - \frac{3}{2}\frac{T}{(500\text{ K})} + \frac{200\text{ K}}{2\times 500\text{ K}} = 0$$

This can be easily simplified to obtain

$$T = 400\text{ K}.$$

Example 6.24 An ideal monatomic gas occupies 2 litre at 30 K and 5×10^{-3} Pa. The internal energy of the gas is taken to be zero at this point. It undergoes the following changes. (a) The temperature is raised to 300 K at constant volume. (b) The gas is then expanded adiabatically till it attains the initial temperature. (c) Finally, it is compressed isothermally. Calculate the efficiency of the cycle.

Solution: Refer to Fig. 6.18 which is indicator diagram for the processes under consideration. Here

: isochoric compression
: adiabatic expansion
: isothermal compression

, internal energy is zero. To calculate the number of moles of the gas, we use the equation of state for an ideal gas:

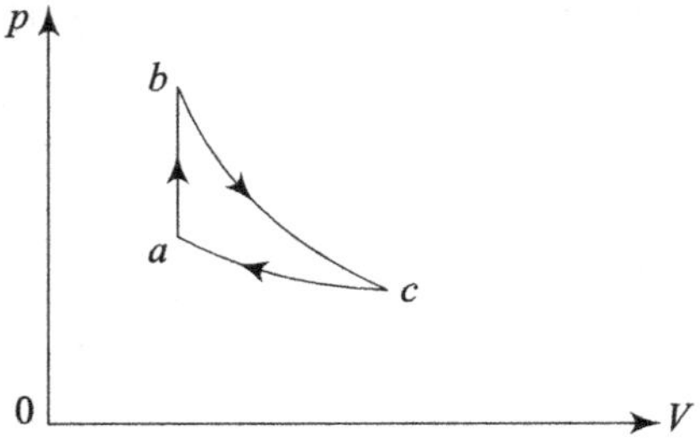

Fig. 6.18

$$pV = nRT$$

$= 2$ litre $= 2 \times 10^{-3}$ m^3, $p = 5 \times 10^2$ Nm^{-2} and $T = 30$ K. Using these values, we get

$$(5\times 10^2\text{ Nm}^{-2})\times(2\times 10^{-3}\text{ m}^3) = nR\times(30\text{ K})$$

$$30\,nR = 1\text{ Nm K}^{-1}$$

Hence, the number of moles of the gas can be expressed in terms of R:

$$n = \frac{1}{30\,R}\text{ JK}^{-1}$$

To calculate the efficiency, we need to know the work done as well as heat absorbed. So we consider the three processes one by one.

b, the process is isochoric and no work is done:

$$\delta W_{a\to b} = 0$$

$$Q_{a\to b} = nC_V(T_b - T_a)$$

For a monoatomic gas, $C_V = \frac{3}{2}R$

$$Q_{a\to b} = \frac{1\text{ JK}^{-1}}{30\,R}\times\frac{3}{2}R\times(300-30)\text{ K}$$

$$= 13.5\text{ J}$$

$$U_b - U_a = 13.5\text{ J} \quad \text{But} \quad U_a = 0 \Rightarrow U_b = 13.5\text{ J}$$

c, the process is adiabatic and no heat is exchanged:

$$\delta Q_{b\to c} = 0.$$

From the first law, we can write

$$W_{b\to c} = -\Delta U = -(U_c - U_b)$$

= 30 K and U is a function of only temperature for an ideal gas, we note that = 0 since $U_a = 0$.

$$W_{b\to c} = U_b = 13.5\text{ J}$$

$$W_{c\to a} = nRT\ln\left(\frac{V_a}{V_c}\right) = -nRT\ln(V_c/V_b)$$

V_b

To calculate the ratio $\frac{V_a}{V_c}$, we note that points c and b are located on the same adiabat. So we can write

$$TV^{\gamma-1} = \text{constant}$$

$$T_b V_b^{\gamma-1} = T_c V_c^{\gamma-1}$$

$$\left(\frac{V_c}{V_b}\right)^{\gamma-1} = \frac{T_b}{T_c}$$

= 1.67, we get

$$\left(\frac{V_c}{V_b}\right)^{0.67} = \frac{300\text{ K}}{30\text{ K}} = 10$$

$$\frac{V_c}{V_b} = (10)^{1/0.67} = (10)^{1.4925} = 31.08$$

$$W_{c\to a} = -\frac{1\text{ JK}^{-1}}{30R} \times R \times (30\text{ K})\ln(31.08)$$

$$= -3.4\text{ J}$$

$$Q_{c\to a} = -3.4\text{ J and } U_a = U_c = 0$$

Hence, the net work done = 13.5 J – 3.4 J = 10.1 J
and heat absorbed = 13.5 J

$$\eta = \frac{10.1\text{ J}}{13.5\text{ J}} = 0.748 = 74.8\%$$

Example 6.25 A Carnot engine is operated in reverse order. The work input to the unit is 12 kW and $COP = 2.5$. For cooling, calculate the ratio of temperatures within which the device operates.

We know that COP of a refrigerator is given by

$$COP = \frac{Q_2}{W} = \frac{\text{Heat extracted at low temperature}}{\text{Work input}}$$

$COP = 2.5$ and $W = 12$ kW, the heat extracted at low temperature

$$Q_2 = W \times COP$$
$$= (12 \text{ kW}) \times 2.5$$
$$= 30 \text{ kW}$$

$Q_2 = 10$ kW, we get

$$Q_1 = Q_2 + 10 \text{ kW}$$
$$= 40 \text{ kW}$$

Ratio of the temperatures $\dfrac{T_1}{T_2} = \dfrac{Q_1}{Q_2} = \dfrac{40}{30} = 1.33$

Example 6.26 A freezer is maintained at a temperature of –10°C and the room temperature is 30°C. To maintain the freezer temperature, heat is removed at the rate of . Calculate the COP and the amount of heat discarded to the surroundings.

The COP of a Carnot refrigerator is given by

$$COP = \frac{Q_2}{W} = \frac{Q_2}{Q_1 - Q_2}$$

In terms of temperatures, we can write

$$COP = \frac{T_2}{T_1 - T_2} = \frac{263 \text{ K}}{40 \text{ K}} = 6.575$$

To calculate the heat released to the surroundings, we have to calculate W:

$$W = \frac{Q_2}{COP} = \frac{1200 \text{ Js}^{-1}}{6.575} = 182.5 \text{ Js}^{-1}$$

$$Q_1 = W + Q_2$$
$$= (182.5 + 1200) \text{ Js}^{-1}$$
$$= 1382.5 \text{ Js}^{-1}$$

Example 6.27 A Carnot engine is operated in the reverse cycle between 200 K and 300 K. Calculate the power required to affect refrigeration by 150 W, if the refrigerator works at (a) 100% coefficient and (b) 60% coefficient of performance.

The maximum coefficient of performance

$$(COP)_{max} = \frac{T_2}{T_1 - T_2}$$

= 300 K and T_2 = 200 K

$$(COP)_{max} = \frac{200\ \text{K}}{100\ \text{K}} = 2$$

(a) For 100% coefficient, $COP = (COP)_{max} = 2$. But

$$COP = \frac{Q_2}{W}$$

$$W = \frac{150\ \text{W}}{COP}$$

$$= 75\ \text{W}$$

(b) For 60% coefficient, $COP = \frac{60}{100} \times 2 = 1.2$

$$W = \frac{150\ \text{W}}{1.2} = 125\ \text{W}$$

Example 6.28 A heat engine operating between 137°C and –43°C receives only 250 kcal of heat per minute. The inventor of the engine claims to develop 12 HP with this engine. Do you agree with his claim? Take 1 HP = 746 W.

We know that for a Carnot cycle

$$\frac{Q_1}{Q_2} = \frac{T_1}{T_2}$$

= (137 + 273) K = 410 K, T_2 = (–43 + 273) K = 230 K, Q_1 = 250 kcal min^{-1}.

$$Q_2 = Q_1 \times \frac{T_2}{T_1}$$

$$= (250\ \text{kcal min}^{-1}) \times \frac{230\ \text{K}}{410\ \text{K}}$$

$$= 140\ \text{kcal min}^{-1}$$

The work obtained from the engine per minute

$$W = Q_1 - Q_2$$

$$= (250 - 140)\ \text{kcal}$$

$$= 110\ \text{kcal}$$

Since 1 kcal = 4180 J and 1 minute = 60 s,

$$\text{Power developed} = \frac{P}{t}$$

$$= \frac{(110\ \text{kcal}) \times (4180\ \text{J kcal}^{-1})}{60\ \text{s}}$$

$$= 7663 \text{ W}$$

$$= \frac{7663 \text{ W}}{746 \text{ W HP}^{-1}}$$

$$= 10.3 \text{HP}$$

Thus, inventor's claim of 12 HP is incorrect.

Let us now sum up what you have learnt in this chapter.

MMARY

According to Kelvin–Planck statement of the second law of thermodynamics, no process is possible whose sole result is complete conversion of heat into work.

According to Clausius statement of the second law of thermodynamics, no process is possible whose sole result is to transfer heat from a body at a lower temperature to a body at a higher temperature.

The Kelvin–Planck and Clausius statements are equivalent; if one is violated, the other is also necessarily violated.

A heat engine converts heat into work. To be a useful device, a heat engine must operate continuously; absorb heat at a higher temperature and reject it at a lower temperature. Thus, a heat engine has to operate in a cycle between two heat reservoirs.

The efficiency of a reversible Carnot engine is given by

$$\eta = 1 - \frac{T_2}{T_1}$$

T_1 and T_2 are the temperatures of the source and the sink respectively.

If two or more engines operate between the same temperature limits, the efficiency of a Carnot engine is maximum.

The concept of thermodynamic temperature is based on Carnot cycle and is independent of the properties of the working substance. For this reason, it is also called absolute temperature scale.

The size of one degree on absolute scale is equal to that on the Celsius scale.

Irreversibility of natural processes is responsible for loss of useful energy.

The efficiency of the air standard Otto engine is given by

$$\eta = 1 - \left(\frac{V_2}{V_1}\right)^{\gamma-1} = 1 - \frac{1}{\rho_c^{\gamma-1}},$$

r_v signifies compression ratio.

The efficiency of the air standard diesel engine is given by

$$\eta = 1 - \frac{1}{\gamma \rho_c^{\gamma-1}}\left[\frac{r_c^{\gamma} - 1}{r_c - 1}\right] = 1 - \frac{A}{\rho_c^{\gamma-1}}$$

r_c signifies cut-off ratio. A diesel engine is more efficient than an Otto engine.

ERCISES

An ideal gas engine operates in a cycle, which when represented on a p-V diagram, is a rectangle. If we call p_1, p_2 the lower and higher pressures respectively and V_1, corresponding volumes,

Calculate the work done in one complete cycle.

Indicate in which parts of the cycle heat is absorbed and in which part librated.

Calculate the quantity of heat flowing into the gas in one cycle, and

Show that the efficiency of the engine is

$$\eta = \frac{\gamma - 1}{\dfrac{\gamma p_2}{p_2 - p_1} + \dfrac{V_1}{V_2 - V_1}}$$

Figure 6.19 shows the p-V diagram of an ideal engine. All processes are quasi-static and C_p is a constant. Prove that the efficiency of the engine is given by

$$\eta = 1 - \left(\frac{p_a}{p_b}\right)^{\frac{\gamma-1}{\gamma}}$$

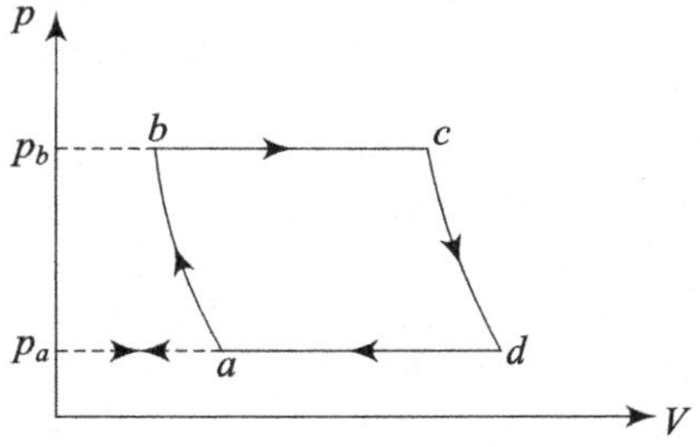

Fig. 6.19

Hydrogen is used in a Carnot cycle as a working substance. Calculate the efficiency of the cycle if in an adiabatic expansion

the volume of the gas is doubled, and

the pressure of the gas is halved (*Ans:* 0.25, 0.18)

An ideal gas goes through a cycle consisting of

isochoric, adiabatic and isothermal lines,

isobaric, adiabatic and isothermal lines

When the isothermal occurs at the minimum temperature, calculate the efficiency of each cycle if the absolute temperature varies n –1 fold within the cycle.

$$\left(Ans: \eta = 1 - \frac{\ln n}{n-1} \text{ for both}\right)$$

Calculate the minimum amount of work in joules required to freeze one gram of water at 0°C by means of an engine which operates in surroundings at 25°C. Given latent heat of ice is 80 cal g^{-1}. (*Ans:* 30.5J)

The cold reservoir of a Carnot engine is at 7°C. Its efficiency is 40% which is to be increased to 50%.

By how many degrees the temperature of the hot reservoir must be increased if the temperature of the cold reservoir is kept constant?

By how many degrees the temperature of the cold reservoir be decreased if the temperature of *hot* reservoir is kept constant? (*Ans:* 93°C, 47°C)

A thermal power plant generating 1000 MW of electrical power is located by the side of a river. The overall efficiency of the plant is 40%.

What is the total thermal power input to the plant?

At what rate is waste heat discharged from the plant?
If the waste heat is delivered to the river, and if a temperature rise of no more than 5°C is permissible, how much water must be available per second?
(*Ans:* (a) 2500MW (b) 1500MW (c) 7.2×10^4 kg s^{-1})

When the sun shines on a glass-roofed greenhouse, the temperature inside becomes higher than outside. Does this violate the second law of thermodynamics?

A reversible engine converts one-third of heat into work. When the temperature of the sink is reduced by 100°C, it converts one-half of heat input into work. Calculate the temperatures of the source and the sink. (*Ans:* T_1 = 600 K, T_2 = 400 K)

A refrigerator is fitted with a motor of 200 W. The temperature of its freezing compartment is –3°C and that of the room air is 27°C. Calculate the maximum amount of heat which the refrigerator can remove from the freezing compartment in 10 minutes. (*Ans:* 2.58×10^5 cal)

The temperature of a household refrigerator is 5°C and the room temperature is 30°C. Nearly 3×10^8 J heat flows from the room to the refrigerator in a day and has to be pumped out of the refrigerator to maintain its temperature. If the coefficient of performance of the refrigerator is 60% of the efficiency of a Carnot engine working between the same temperature limits, calculate the power required to run the refrigerator. (*Ans:* 520 W)

In an air standard Otto cycle, 1000 kJ kg^{-1} net work is done. The maximum temperature in the cycle is 3200 K and the temperature at the end of the compression stroke is 800 K. Calculate the compression ratio of the engine. (*Ans:* 8.8)

7 ENTROPY

arning Objectives

In this chapter, you will learn how to

e entropy and state the second law of thermodynamics in terms of entropy;

e expressions for change in entropy of a system undergoing reversible/ersible changes;

represent Carnot cycle on $T-S$ diagram and calculate the work done in stage; and

blish Clausius inequality.

TRODUCTION

In our discussion of the second law of thermodynamics in the preceding chapter, we confined ourselves to the problem of conversion of heat into work and observed that the thermodynamic state of the working substance changes continuously in a cycle. Moreover, the direction of operation of Carnot cycle determines whether a device acts as a heat engine or a refrigerator. Although this is a very interesting aspect with important economic implications through energy generation to power our homes/factories, from purely physics point of view, we are more concerned with the direction of evolution of a process. For instance, we would be interested in the analysis of processes such as expansion of a gas into vacuum (Joule expansion), cooling of a cup of tea or coffee (spontaneous heat flow), mixing of two gases or two solutions (diffusion), burning of coal or discharge of a battery during use (chemical processes), to mention a few.

We know that left to itself, a system has a tendency to change spontaneously towards equilibrium, if it is initially away from it. Now the question arises: What determines the direction of a process? Can we give a quantitative criterion by which we can predict the direction of evolution of a process? The answer to these questions was given by Clausius when he introduced the concept of *entropy*.* In Sec 7.2, we begin by answering the question "What is entropy" by considering a reversible process. You will discover that it is a mathematical device, which is a function of state. (However, it cannot be measured like temperature or pressure). This discussion is followed in Sec. 7.3 for calculation of entropy changes in a few typical reversible changes. But you may recall from Chapter 4 that most natural processes are irreversible. The search for a criterion for evolution of a natural change leads to *Clausius inequality*. This is the subject matter of discussion in

---

It has been derived from the Greek word *tropos*, which means *in change*. As such, it is an abstract concept

Sec. 7.4. Entropy changes in irreversible processes are discussed in Sec. 7.5. The principle of increase of entropy is discussed in Sec. 7.6. The entropy form of the first law of thermodynamics is discussed in Sec. 7.7 along with its application to calculate the entropy change of an ideal gas. Finally, we will discuss the connection between entropy and unavailable energy.

It may be mentioned here that so far we have not arrived at any mathematical formulation of the second law of thermodynamics. By introducing the concept of entropy, Clausius developed a general mathematical formulation of the second law and showed that only such processes can occur for which entropy does not decrease. Boltzmann showed that entropy is a measure of disorder in molecular arrangement of a system and, as such, it is an abstract concept. Since entropy characterises the approach to equilibrium, it can be related to the probability of occurrence of a state. Thus, it helps us to relate the results of thermodynamics to those of statistical mechanics. For this reason, it is said that thermodynamics became a powerful tool only after it divorced itself from the convertibility of heat into work and engaged to entropy.

HAT IS ENTROPY?

What is entropy? To answer this question, let us consider a system that undergoes a reversible process from state 1 to state 2 along path *A*. Let us assume that the cycle is completed along *B*, which is also reversible (Fig. 7.1a). We know that a Carnot cycle is reversible. And any reversible cycle can be represented by a series of miniature Carnot cycles, as shown in Fig. 7.1b. From Example 6.11, you may recall that for each reversible cycle, we can take

$$\frac{Q_1}{T_1} - \frac{Q_2}{T_2} = 0$$

We can rewrite it as follows:

$$\oint_{1A2B1} \frac{\delta Q_R}{T} = 0 = \int_1^2 \left(\frac{\delta Q_R}{T}\right)_{\text{via } A} + \int_2^1 \left(\frac{\delta Q_R}{T}\right)_{\text{via } B} \tag{7.1}$$

The subscript *R* denotes a reversible cycle.

Now refer to Fig. 7.1c which depicts another reversible cycle. The initial process is along the path *A* but is completed along the path *C*. For this cycle, we can write

$$\oint_{1A2C1} \frac{\delta Q_R}{T} = 0 = \int_1^2 \left(\frac{\delta Q_R}{T}\right)_{\text{via } A} + \int_2^1 \left(\frac{\delta Q_R}{T}\right)_{\text{via } C} \tag{7.2}$$

On combining Eqs. (7.1) and (7.2), we can write

$$\int_2^1 \left(\frac{\delta Q_R}{T}\right)_{\text{via } B} = \int_2^1 \left(\frac{\delta Q_R}{T}\right)_{\text{via } C} \tag{7.3}$$

This result strongly suggests that the integral $\int_2^1 \frac{\delta Q_R}{T}$ is same between states 2 and 1, no matter what reversible path we choose. That is, for any two coordinate systems, we can

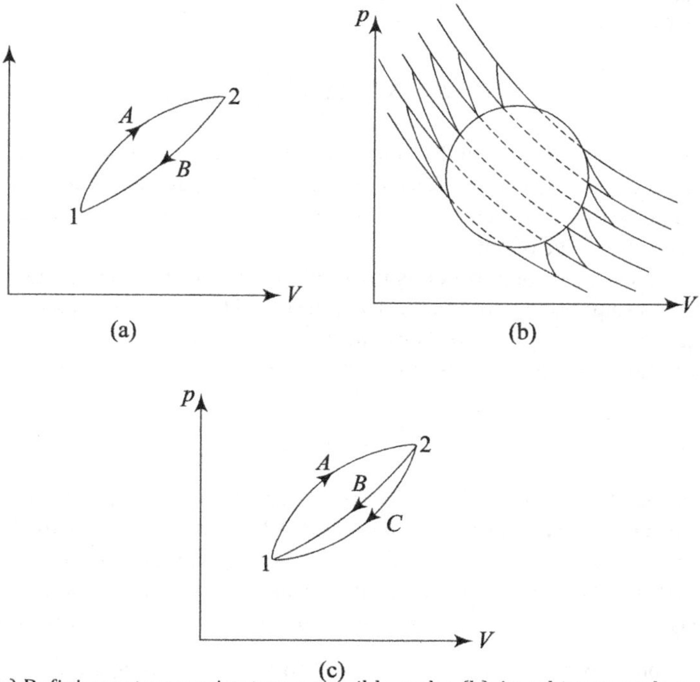

(a) Defining entropy using two reversible paths. (b) An arbitrary cycle may be approximated by a sequence of adiabates and isotherms. (c) Change in entropy between two states is independent of path for a reversible process.

cycles and whatever path we take, we will get the same result. Mathematically, we can say

is an exact differential* and write it as

$$\frac{\delta Q_R}{T} = \mathrm{d}S \tag{7.4}$$

This suggests that we can define a new quantity S which is a function of state and is independent of the path followed. This new function of state is called *entropy* of the

Put into words, Eq. (7.4) states if we supply an infinitesimal amount of heat δQ to a system reversibly at temperature T, the entropy of the system increases by $\dfrac{\delta Q}{T}$. Note that we have defined entropy in terms of a reversible process connecting states 1 and 2. Like volume and charge, entropy is also an extensive variable. The SI unit of entropy is joule per kelvin (JK^{-1}).

From Chapter 5, we recall that heat is a path function and δQ is an inexact differential. However, $(\delta Q_R/T)$ is an exact differential. From mathematical point of view, an inexact differential may be changed to an exact differential by introducing an integrating factor. Therefore, $(1/T)$ serves as the integrating factor in converting the inexact differential δQ to the exact differential $\delta Q/T$ for a reversible process.

In the so-called postulatory approach, the existence of entropy is assumed for all equilibrium states depending on *a posteriori* rather than *a priori* justification:

1. There exists a function called entropy which attains a maximum value in the equilibrium state.

The change in entropy of a system, as it undergoes a change of state, is obtained by integrating Eq. (7.4):

$$S_2 - S_1 = \Delta S = \int_1^2 \frac{\delta Q_R}{T} \tag{7.5a}$$

Equation (7.5a) enables us to determine changes in entropy along a reversible path in going from one state to another. (This will also be the magnitude of entropy change for an irreversible process between these two states.) However, this equation tells us nothing about the absolute value of entropy.

When no change of composition is involved, it is adequate to give values of entropy with respect to some arbitrarily chosen standard state. This, in turn, is chosen with respect to the state where entropy is zero. Thus, the entropy of a system in a given state, relative to some arbitrary reference state is indeterminate to the extent of an additive constant:

$$S_2 = S_n + \int_n^2 \frac{\delta Q_R}{T} \tag{7.5b}$$

denotes the entropy of the substance in some reference state n. This constant cannot be determined within the framework of the first and the second laws of thermodynamics. In fact, this aspect is contained in the third law of thermodynamics. (Entropy of a substance becomes zero at absolute zero of temperature. This is discussed in Chapter 10.)

We may now conclude that to measure entropy in any state, we have to take the substance along any reversible path to its standard state and compute the value of $\frac{\delta Q_R}{T}$ along this path. Obviously, for a cyclic reversible process, there will be no net entropy change. Thus, we can write

$$\oint_{\text{rev}} \frac{\delta Q}{T} = 0 \tag{7.6}$$

Having introduced the concept of entropy, let us now calculate entropy change in some typical reversible processes.

A common example of a reversible isothermal process is change of phase at constant pressure. The heat flow into the system per unit mass equals the latent heat. This is further brought out in the following example.

Example 7.1 10^{-2} kg of steam at 100°C and 1 atmosphere pressure condenses at the same temperature and pressure. Calculate the entropy change of steam. The latent heat of steam is 2.26×10^6 J kg^{-1}.

Condensation is an isothermal process. If it occurs slowly enough, it can be considered as a reversible process. Thus, Eq. (7.4) can be used to determine the change in entropy of steam when it condenses into water.

$$dS = \frac{\delta Q}{T} = -\frac{m \times L}{T}$$

The negative sign signifies the fact that heat is given out in the process under consideration.

$m = 10^{-2}$ kg, $T = 100°\text{C} = 373\text{K}$ and $L = 2.26 \times 10^6 \text{J kg}^{-1}$. Hence, change in entropy is given by

$$dS = -\frac{(10^{-2}\text{ kg})(2.26 \times 10^{6}\text{ J kg}^{-1})}{373\text{ K}}$$
$$= -60.6\text{ JK}^{-1}.$$

This result shows that entropy decreases when steam condenses into water and vice versa. You should expect a similar result when water condenses to form ice.

You may now like to solve a practise problem.

Problem 7.1 Calculate the increase in entropy of 1 kg of ice when it melts at 0°C. The latent heat of fusion of ice is 3.36×10^5 J kg^{-1}. Assume that melting is an isothermal reversible process.

1231 JK^{-1}

NTROPY CHANGE IN REVERSIBLE PROCESSES

Having established that entropy is a thermodynamic property of a system, we will now discuss its significance in various reversible processes and draw corresponding T–S diagrams. These diagrams are often useful in engineering and meteorology. We begin our discussion with the Carnot cycle.

he Carnot Cycle

Consider the Carnot cycle represented on the p-V diagram (Fig. 7.2(a)). The first process in this cycle is the isothermal transfer of heat to the working substance from the hot reservoir. The gain in entropy during this process is given by

$$\Delta S = \int_A^B \left(\frac{\delta Q}{T}\right)_{\text{rev}}$$

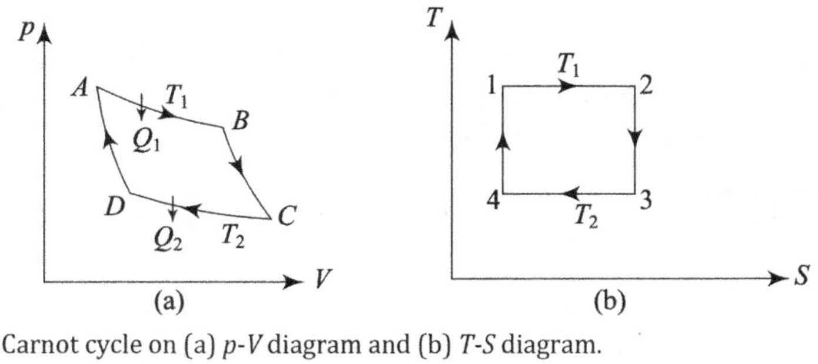

Carnot cycle on (a) p-V diagram and (b) T-S diagram.

Since the temperature of the working substance remains constant, equal to T_1, this equation can be integrated to give

$$\Delta S = \frac{1}{T_1}\int_A^B \delta Q = \frac{Q_1}{T} \qquad (7.7)$$

is the heat absorbed by the system. This is shown as horizontal line $1 \rightarrow 2$ on

The second process in Carnot cycle is a reversible adiabatic expansion. Since no heat is exchanged in this process ($\delta Q = 0$), it readily follows from Eq. (7.4) that entropy remains constant in a reversible adiabatic process. A constant entropy process is called an process. (This explains the use of the subscript 'S' in Chapter 5 to denote an adiabatic change.) However, this change is accompanied by a drop in temperature. This is represented by the vertical line $2 \rightarrow 3$ in Fig. 7.2 (b).

The third process is the reversible isothermal compression in which heat is transferred from the working substance to the cold reservoir. For this, we can write

$$\Delta S = \int_C^D \left(\frac{\delta T}{T} \right)_{\text{rev}} = -\frac{Q_2}{T_2} \tag{7.8}$$

The negative sign signifies that heat is rejected by the working substance during this process.

Equation (7.8) indicates that the entropy of the working substance decreases during isothermal compression. This is shown by the horizontal line $3 \rightarrow 4$. It is important to note that the final process $4 \rightarrow 1$ is a reversible adiabatic compression (and therefore isentropic). Therefore, the decrease in entropy in process $3 \rightarrow 4$ must exactly equal entropy increase in process $1 \rightarrow 2$, i.e., $S_2 - S_1 = S_3 - S_4$.

On comparing Figs. 7.2 (a) and (b), you will note that the curvilinear quadrilateral representing Carnot cycle on p-V diagram becomes a rectangle on the T–S diagram. From Sec. 6.4, we recall that the area $ABCD$ on the p–V diagram gives the useful work obtainable in one cycle. However, the corresponding area on the T–S diagram ($1 \rightarrow 2 \rightarrow 3 \rightarrow 4 \rightarrow 1$) gives the actual amount of heat energy converted into work in that cycle. Can you show it mathematically? That is why p–V diagram is also called the *work diagram* and the T–S diagram the *heat diagram*.

The efficiency of the Carnot engine can be readily calculated using the T–S diagram and Eq. (6.4):

$$\eta = \frac{Q_1 - Q_2}{Q_1}$$

$Q_2 = (T_1 - T_2)\,\Delta S$ and $Q_1 = T_1 \Delta S$, we obtain

$$\eta = 1 - \frac{T_2}{T_1}$$

This result is identical to Eq. (6.14). You will agree that the calculations based on the $T - S$ diagram are simpler as compared to the $p - V$ diagram.

We would like you to go through the following example carefully.

Example 7.2 Compare the efficiencies of the cycles $ABCA$ shown in Fig. 7.3.

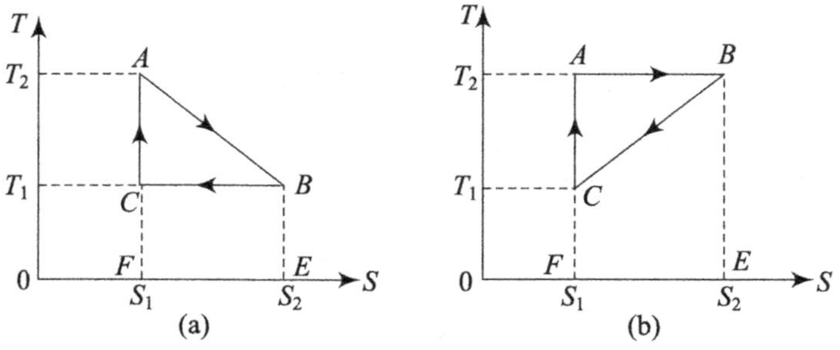

From Fig. 7.3(a), we note that the heat supplied is given by the area *ABEFA* and is equal to

$$Q_1 = T_1 (S_2 - S_1) + \frac{1}{2}(T_2 - T_1) \times (S_2 - S_1)$$

On the other hand, the work obtained is equal to the area of ΔABC and is equal tp

$) \times (S_2 - S_1)$.

$$\eta_a = \frac{\frac{1}{2}(T_2 - T_1)(S_2 - S_1)}{\frac{1}{2}(T_2 - T_1)(S_2 - S_1) + T_1 (S_2 - S_1)} = \frac{T_2 - T_1}{T_2 + T_1} \quad \text{(i)}$$

From Fig. 7.3(b), we note that heat supplied to the system is given by

$$Q_1 = (S_2 - S_1) \times T_2$$

and work obtained is (1/2) $(S_2 - S_1)(T_2 - T_1)$ so that

$$\eta_b = \frac{\frac{1}{2}(S_2 - S_1)(T_2 - T_1)}{(S_2 - S_1)T_2} = \frac{T_2 - T_1}{2T_2} \quad \text{(ii)}$$

From Eqs. (i) and (ii), we get

$$\frac{\eta_a}{\eta_b} = \frac{2T_2}{T_1 + T_2}$$

This result shows that efficiency of an engine depends on the order in which reversible cycles are carried out.

Let us next consider reversible heat transfer processes.

Reversible Heat Transfer

In most reversible heat transfer processes, the temperature is not constant. Then increase in entropy can readily be calculated from Eq. (7.4):

$$\Delta S = \int_{T_1}^{T_2} \left(\frac{\delta Q}{T}\right)_{\text{rev}}$$

If we assume that heat capacity of the system remains constant over the temperature interval of interest and changes in phase are excluded during this process, the entropy increase is given by

$$\Delta S = mc \int_{T_1}^{T_2} \frac{\mathrm{d}T}{T} = mc \ln(T_2/T_1) \quad (7.9)$$

This change is depicted by curve *AB* on the *T–S* diagram (Fig. 7.4). Note that to raise the temperature T_2 reversibly, we need a large number of heat baths of gradually increasing temperature with which the system is put in contact in succession.

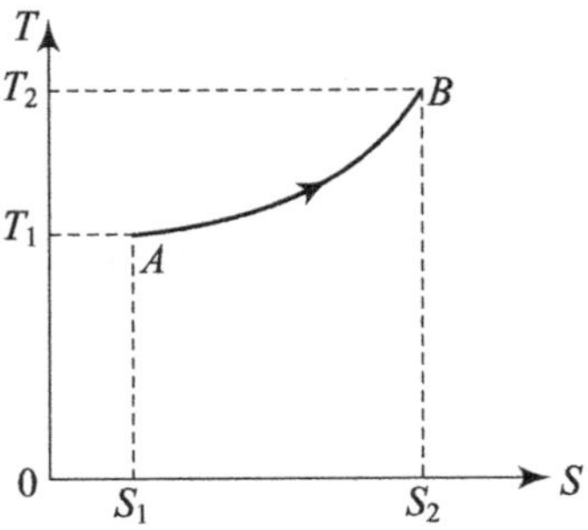

Fig. 7.4 *T–S* diagram for reversible heat transfer.

Example 7.3 1 kg water is heated from C to 100°C and converted into steam at the same temperature. Calculate the increase in entropy and draw a *T–S* diagram. Given that specific heat capacity is 4.18×10^3 J kg^{-1} K^{-1} and latent heat of vapourisation is 2.26×10^6 J kg^{-1}.

If heating is carried out reversibly, the increase in entropy when temperature of water is raised from 0°C to 100°C is given by

$$\ln (T_2/T_1)$$

$$= (1 \text{ kg}) \times (4.18 \times 10^3 \text{ J kg}^{-1} \text{ K}^{-1}) \times 2.303 \times \log_{10} \left(\frac{373 \text{ K}}{273 \text{ K}} \right)$$

$$= 1305 \text{ J K}^{-1}$$

This is shown by curve *AB* in Fig. 7.5.

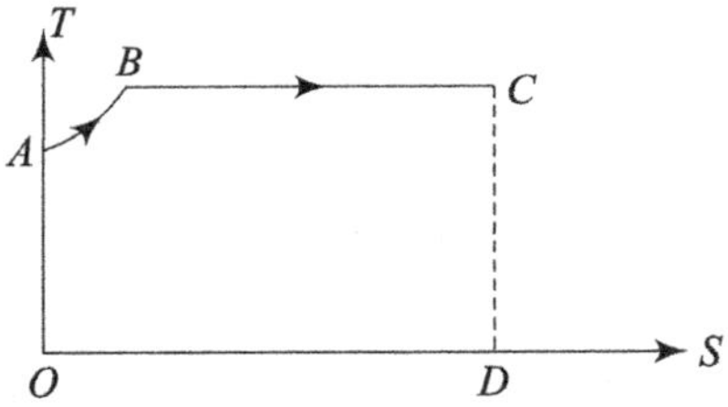

Fig. 7.5 *T–S* diagram for conversion of water at 0°C into steam at 100°C.

When a phase change takes place, it is accompanied by evolution or absorption of heat. In such a situation, we speak of entropy of fusion or evaporation. In the instant case, water at 100°C changes into steam. Therefore, increase in entropy when 1 kg water at 100°C changes into steam is given by

$$\Delta S_2 = \frac{\delta Q}{T} = \frac{mL}{T}$$

is latent heat of steam. (In case of melting of ice, we will use latent heat of fusion.) On substituting the numerical data, we get

$$\Delta S_2 = \frac{(1 \text{ kg}) \times (2.26 \times 10^6 \text{ Jkg}^{-1})}{373 \text{ K}} = 6.06 \times 10^3 \text{ JK}^{-1}$$

This is shown by the horizontal line *BC*.

Hence, total increase in entropy is given by

$$\Delta S = \Delta S_1 + \Delta S_2$$
$$= (1305 + 6060) \text{ JK}^{-1}$$
$$= 7365 \text{ JK}^{-1}$$

ABC is the *T–S* diagram and the area *ABCDOA* gives the heat absorbed by one kg of water when it changes from 0°C into steam at 100°C.

Example 7.4 10 g water at 60°C is mixed with 30g water at 20°C. Will the entropy of the system increase or decrease? Calculate the change. Take specific heat capacity as 1

Let t °C be the temperature of the mixture. Then, from the law of mixtures, we can write

$$(10\text{g}) \times (1 \text{ cal g}^{-1}\text{K}^{-1}) \times (60°\text{C} - t) = (30 \text{ g}) \times (1 \text{ cal g}^{-1}\text{K}^{-1}) \times (t - 20°\text{C})$$

This equation can be solved to give

$$t = 30°\text{C}$$

The change in entropy of 10g water when its temperature changes from 60°C (333 K) to C (303 K) is given by Eq. (7.9)

$$\Delta S_1 = mc \ln\left(\frac{T_2}{T_1}\right)$$
$$= (10 \text{ g}) \times (1 \text{ cal g}^{-1}\text{K}^{-1}) \ln\left(\frac{303}{333}\right)$$
$$= -0.944 \text{ cal K}^{-1}.$$

The change in entropy of 30 g water when its temperature changes from 20°C (293 K) to C (303 K) is

$$\Delta S_2 = (30 \text{ g}) \times (1 \text{ cal g}^{-1}\text{K}^{-1}) \times \ln\left(\frac{303}{293}\right) = 1.007 \text{ cal K}^{-1}.$$

Hence, net change in entropy is given by

$$\Delta S = \Delta S_1 + \Delta S_2$$
$$= (-0.944 + 1.007) \text{ cal K}^{-1}$$
$$= 0.063 \text{ cal K}^{-1}$$

This result shows that the entropy increases in this process by 0.063 cal K^{-1}.

Example 7.5 Calculate the change in entropy when one kg of water at 291 K is mixed with two kg of water at 300 K. Take specific heat capacity of water as 1cal g^{-1}K^{-1}.

Let the temperature of the mixture be t K. Then from the law of mixtures, we

$$(2000 \text{ g}) \times (1 \text{ cal g}^{-1}\text{K}^{-1}) \times (300 \text{ K} - t) = (1000 \text{ g}) \times (1 \text{ cal g}^{-1}\text{K}^{-1}) \times (t - 291 \text{ K})$$

On simplification, we get

$$3t = 891 \text{ K}$$
$$t = 297 \text{ K}$$

The change in entropy of 2000 g water when its temperature changes from 300 K to 297 K is given by Eq. (7.9):

$$\Delta S_1 = mc \ln\left(\frac{T_2}{T_1}\right)$$

$$= (2000 \text{ g})(1 \text{ cal g}^{-1}\text{K}^{-1}) \ln\left(\frac{297 \text{ K}}{300 \text{ K}}\right)$$

$$= -20.1 \text{ cal K}^{-1}$$

The change in entropy of 1000 g water when its temperature rises from 291 K to 297 K

$$\Delta S_2 = (1000 \text{ g})(1 \text{ cal g}^{-1}\text{K}^{-1}) \ln\left(\frac{297 \text{ K}}{291 \text{ K}}\right)$$

$$= 19.8 \text{ cal K}^{-1}$$

Therefore, the total change of entropy

$$\Delta S = \Delta S_1 + \Delta S_2 = -20.1 + 19.8$$
$$= -0.3 \text{ cal K}^{-1}$$

The negative sign shows that entropy of the mixture decreases.

Example 7.6 Calculate the increase in entropy when 50 g ice at −10°C is converted into steam at 100°C. Given that specific heat capacity of ice is 2090 J $\text{kg}^{-1}\text{K}^{-1}$, specific heat capacity of water is 4180 J $\text{kg}^{-1}\text{K}^{-1}$, latent heat of ice is 3.35×10^5 J kg^{-1}, and latent heat of steam is 2.26×10^6 J kg^{-1}.

(a) The change in entropy when 50 g ice at −10°C is heated to 0°C is given by

$$\Delta S_1 = mc \int_{263}^{273} \frac{dT}{T}$$

$$= (50 \times 10^{-3} \text{ kg}) \times (2090 \text{ J kg}^{-1}\text{K}^{-1}) \log_e \left(\frac{273 \text{ K}}{263 \text{ K}}\right)$$

$$= 50 \times 2.09 \times 2.3026 \times \log_{10}\left(\frac{273 \text{ K}}{263 \text{ K}}\right) \text{J K}^{-1}$$

$$= 50 \times 2.09 \times 2.3026 \times 0.0162 \text{ J K}^{-1}$$

$$= 3.90 \text{ J K}^{-1}$$

The change in entropy when 50 g ice at 0°C is converted into water 0°C is given by

$$\Delta S_2 = \frac{\delta Q}{T} = \frac{mL}{T} = \frac{(50 \times 10^{-3} \text{ kg}) \times (3.35 \times 10^5 \text{ J kg}^{-1})}{273 \text{ K}}$$

$$= \frac{50 \times 3.35}{2.73} \text{J K}^{-1}$$

$$= 61.35 \text{ J K}^{-1}$$

The change in entropy when 50 g water is heat from 0°C to 100°C is given by

$$\Delta S_3 = mc \int_{273}^{373} \frac{dT}{T}$$

$$= (50 \times 10^{-3}\,\text{kg}) \times (4.18 \times 10^{3}\ \text{J kg}^{-1}\,\text{K}^{-1}) \times 2.3026 \times \log_{10}\left(\frac{373\ \text{K}}{273\ \text{K}}\right)$$

$$= 50 \times 4.18 \times 2.3026 \times 0.1355\ \text{JK}^{-1}$$

$$= 65.21\ \text{J K}^{-1}$$

The change in entropy when 50 g water at 100°C is converted into steam at the same temperature is

$$\Delta S_4 = \frac{\delta Q}{T} = \frac{mL_{\text{vap}}}{T} = \frac{(50 \times 10^{-3}\ \text{kg}) \times (2.26 \times 10^{6}\ \text{J kg}^{-1})}{373\ \text{K}}$$

$$= \frac{50 \times 2.26 \times 10^{3}}{373}\ \text{JK}^{-1}$$

$$= 302.94\ \text{J K}^{-1}$$

Therefore, the total change (i.e., increase) in entropy is obtained by adding the values for each change:

$$\Delta S = \Delta S_1 + \Delta S_2 + \Delta S_3 + \Delta S_4$$

$$= (3.90 + 61.35 + 65.21 + 302.94)\ \text{J K}^{-1}$$

$$= 433.4\ \text{J K}^{-1}.$$

Problem 7.2 Calculate the increase in entropy when 5 kg of water is heated from °C. Given that specific heat capacity may be assumed to have the constant value ^{3} J kg^{-1} K^{-1}.

3.912×10^{3} J K^{-1}

Problem 7.3 10 g of ice is fully converted into steam at normal atmospheric pressure. Calculate the change in entropy in this process. Take latent heat of fusion as latent heat of steam as 540 cal g^{-1} and specific heat capacity of water as $^{-1}$.

20.5 cal K^{-1}

Example 7.7 m gram of water at temperature T_1 is isobarically and adiabatically mixed with an equal mass of water at temperature T_2. Show that the change in entropy is

$$2m\, c_p \ln\left(\frac{T_{av}}{\sqrt{T_1 T_2}}\right),$$

$_{av} = (T_1 + T_2)/2$.

Since the same mass of water is being mixed, the temperature of the mixture will be equal to $T_{av} = (T_1 + T_2)/2$. For the isobaric process, we denote the specific heat capacity of water by c_p. Then, the change in entropy, when the temperature of m gram of water changes from T_1 to T_{av}, is

$$\Delta S_1 = \int_{T_1}^{T_{av}} \frac{\delta Q}{T}$$
$$= m c_p \ln\left(\frac{T_{av}}{T_1}\right)$$

Similarly, the change in entropy when temperature of m gram of water falls from T_2

$$\Delta S_2 = mc_p \ln(T_{av}/T_2)$$

Hence, the net change in entropy is

$$\Delta S = \Delta S_1 + \Delta S_2$$

$$= mc_p \ln\left(\frac{T_{av}}{T_1}\right) + mc_p \ln\left(\frac{T_{av}}{T_2}\right)$$

$$= mc_p \ln\left(\frac{T_{av}}{T_1} \times \frac{T_{av}}{T_2}\right) \qquad (\because \ln a + \ln b = \ln(ab))$$

$$= mc_p \ln\left(\frac{T_{av}^2}{T_1 T_2}\right)$$

$$= 2 m c_p \ln(T_{av}/\sqrt{T_1 T_2}\,) \quad (\because \ln x^n = n \ln x)$$

Note that the arithmetic mean of two numbers is greater than their geometric mean. Therefore, entropy of mixing ΔS will be positive definite.

HE CLAUSIUS INEQUALITY

In the preceding section, we discussed the entropy changes for reversible processes. But a reversible process is a limiting case of a natural process. Therefore, to provide a criterion for the direction of natural change, we should generalise our earlier results for irreversible processes. The first step in this direction is to establish the Clausius inequality.

Let us consider an irreversible cycle engine working between temperatures T_1 and T_2. If a reversible engine were operating between the same temperatures, it follows from Carnot theorem (Sec. 6.4.1), that

$$\eta_{\text{irr}} < \eta_{\text{rev}}$$

$$\frac{Q_1^{\text{irr}} - Q_2^{\text{irr}}}{Q_1^{\text{irr}}} < \frac{Q_1^{\text{rev}} - Q_2^{\text{rev}}}{Q_1^{\text{rev}}}$$

This expression can be rearranged as

$$\frac{Q_2^{\text{irr}}}{Q_1^{\text{irr}}} > \frac{Q_2^{\text{rev}}}{Q_1^{\text{rev}}} = \frac{T_2}{T_1}$$

$$\frac{Q_2^{\text{irr}}}{T_2} > \frac{Q_1^{\text{irr}}}{T_1}$$

Consequently, for an *irreversible* cycle engine, we must have

$$\oint_{\text{irr}} \frac{\delta Q}{T} = \frac{Q_1^{\text{irr}}}{T_1} - \frac{Q_2^{\text{irr}}}{T_2} < 0 \tag{7.10}$$

In words, this inequality implies that the value of the cyclic integral $\oint_{\text{irr}} \frac{\delta Q}{T}$ will be less than zero for an irreversible process. If we make the engine more and more irreversible, the integral in Eq. (7.10) will progressively become smaller and smaller. Thus, for all irreversible heat engine cycles, we can write

$$\oint_{\text{irr}} \frac{\delta Q}{T} < 0 \tag{7.11}$$

On combining Eqs. (7.6) and (7.11), we find that for heat engines we can, in general, write

$$\oint \frac{\delta Q}{T} \leq 0 \tag{7.12}$$

This relation is known as the *Clausius inequality*. Note that the equality holds only for reversible heat engines.

We may mention here that Eq. (7.12) applies to reversible and irreversible refrigeration cycles as well. But we have left it as an exercise of you. (Problem 7.4)

Problem 7.4 Prove the inequality of Clausius for refrigeration cycles.

It is important to point out here that while dealing with reversible processes, the temperature at which heat is supplied is that of the system itself. This may not necessarily be true in irreversible processes.

Example 7.8 A heat engine receives 100 kcal of heat from a source at 1000 K. It rejects 50 kcal, 75 kcal and 25 kcal of heat to the surroundings at 500 K (Fig. 7.6). Investigate the nature of change in each case.

From Fig. 7.6a, we note that

$$\oint \frac{\delta Q}{T} = \left(\frac{100}{1000} - \frac{50}{500}\right) \text{kcal K}^{-1}$$
$$= (0.1 - 0.1) \text{ kcal K}^{-1}$$
$$= 0.0$$

That is, the changes in this case are reversible.

From Fig. 7.6(b), we note that

$$\oint \frac{\delta Q}{T} = \left(\frac{100}{1000} - \frac{75}{500}\right) \text{kcal K}^{-1}$$
$$= (0.1 - 0.15) \text{ kcal K}^{-1}$$

–1

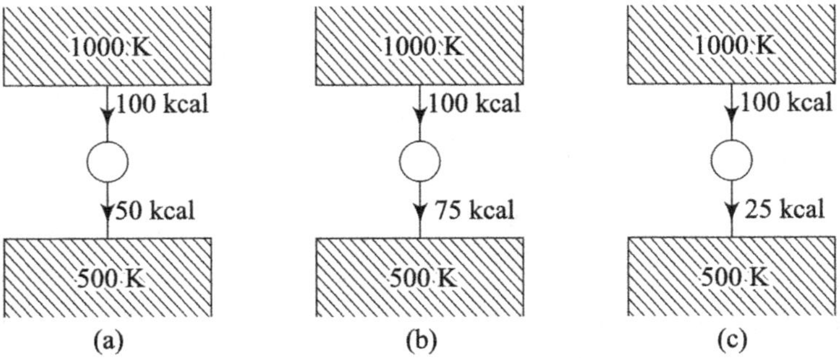

Demonstration of the inequality of Clausius.

$\frac{Q}{} < 0$ and the changes in this case are irreversible.

From Fig. 7.6(c), we note that

$$\oint \frac{\delta Q}{T} = \left(\frac{100}{1000} - \frac{25}{500}\right) \text{ kcal K}^{-1}$$
$$= (0.1 - 0.05) \text{ kcal K}^{-1}$$
$$= 0.05 \text{ kcal K}^{-1}$$

$\frac{\delta Q}{T} > 0$ and the change depicted in this case cannot occur.

NTROPY CHANGE IN AN IRREVERSIBLE PROCESS

Consider a system that undergoes the cycle shown in Fig. 7.7. For the irreversible cycle 1*A*2*B*1 made up of reversible processes *A* and *B*, we can write

$$\int \frac{}{T} = \int_1^2 \left(\frac{\delta Q_R}{T}\right)_{\text{via } A} + \int_2^1 \left(\frac{\delta Q_R}{T}\right)_{\text{via } B} = 0$$

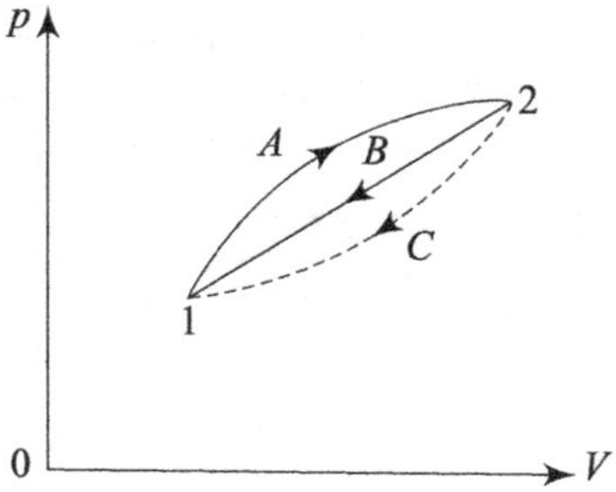

Fig. 7.7 Entropy change in an irreversible process.

Now consider the irreversible cycle 1*A*2*C*1 made up of reversible process *A* and irreversible process *C*. For this cycle, the Clausius inequality implies that

$$\int \frac{}{T} = \int_1^2 \left(\frac{\delta Q_R}{T}\right)_{\text{via } A} + \int_2^1 \left(\frac{\delta Q_R}{T}\right)_{\text{via } C} < 0$$

On combining these equations, we can write

$$\int_2^1 \left(\frac{\delta Q_I}{T}\right)_{\text{via } C} < \int_2^1 \left(\frac{\delta Q_R}{T}\right)_{\text{via } B}$$

is reversible and entropy is a function of state, we can write

$$S_1 - S_2 = \int_2^1 \left(\frac{\delta Q_R}{T}\right)_{\text{via } B} = \int_2^1 dS_B = \int_2^1 dS_C$$

Therefore, for an irreversible process

$$\int_2^1 dS_C > \int_2^1 \left(\frac{\delta Q_I}{T}\right)$$

Thus, for the general case we can write

$$\Delta S \geq \frac{\delta Q}{T} \tag{7.13a}$$

Or, for a change from state 1 to 2, the difference of entropies is given by

$$S_2 - S_1 \geq \int_1^2 \frac{\delta Q}{T} \tag{7.13b}$$

As before, the equality and greater than signs hold for reversible and irreversible processes, respectively.

This is an extremely important result and helps us in determining the direction in which a physical process evolves. In essence, it emphasises the influence of irreversibility on the entropy of a system. It may be pointed out here that the results contained in Eqs. (7.13a, b) hold, irrespective of the fact that δQ is equal to zero, less than zero or greater than zero. is negative, the entropy will tend to decrease as a result of heat transfer. However, the influence of irreversibility is still to increase the entropy of the system. That is why we say that entropy increases in all natural processes.

We would now like you to answer a practise problem.

Problem 7.5 Using the inequality of Clausius, show that the efficiency of a heat engine is limited to $\eta \leq \dfrac{T_1 - T_2}{T_1}$.

For an isolated system, there is no thermal interaction with the surroundings and in the Clausius inequality, we have $\delta Q = 0$. Therefore,

$$\delta S \geq 0 \tag{7.14}$$

That is, the entropy of an isolated system cannot decrease. For reversible changes in an isolated system, the entropy will remain constant, whereas for irreversible changes, the entropy will increase. A corollary of this result is that the entropy of an isolated system tends towards a maximum while it approaches equilibrium.

So far we have considered entropy changes in a system when it undergoes a change of state. We now consider the change in the entropy of a system and its surroundings; hence the whole universe*. This leads to the principle of increase of entropy.

Problem 7.6 Consider a system at temperature T_s in thermal contact with a heat ($T_s < T_h$). If the heat capacity of the system remains constant over the interval), show that temperature equalisation is accompanied by increase in entropy.

Strictly speaking, the surroundings of a system means everything in the universe other than the system. However, in our analysis, we will concentrate only on the local surroundings of the system.

HE PRINCIPLE OF INCREASE OF ENTROPY

Consider a process in which a quantity of heat δQ flows from the surroundings at temperature T_{surr} to the system at temperature T_{sys}. Let δW be the work done by the system during the process. Then from Clausius inequality, we can write

$$\Delta S_{\text{sys}} \geq \frac{\delta Q}{T_{\text{sys}}}$$

For the surroundings, δQ is negative and we can write

$$\Delta S_{\text{surr}} = -\frac{\delta Q}{T_{\text{surr}}}$$

The net change of entropy of the universe is, therefore,

$$\Delta S_u = \Delta S_{\text{sys}} + \Delta S_{\text{surr}} \geq \frac{\delta Q}{T_{\text{sys}}} - \frac{\delta Q}{T_{\text{surr}}}$$

$$= \delta Q\left(\frac{1}{T_{\text{sys}}} - \frac{1}{T_{\text{surr}}}\right) \tag{7.15}$$

$> T_{\text{sys}}$, the quantity within the brackets is positive and we conclude that

$$\Delta S_u \geq 0$$

$_{\text{surr}}$, the heat transfer will take place from the system to the surroundings, and and $\left(\frac{1}{T_{\text{sys}}} - \frac{1}{T_{\text{surr}}}\right)$ will be negative, yielding the same result.

We may therefore conclude that for all possible processes that a system in a given surroundings can undergo, the increase in entropy is given by

$$\Delta S_{\text{sys}} + \Delta S_{\text{surr}} = \Delta S_u \geq 0 \tag{7.16}$$

The equal to and greater than signs hold for reversible and irreversible processes, respectively. Since natural processes are irreversible, Eq. (7.16) implies that the entropy of the universe is increasing continuously. This result is known as the *principle of increase* . The significance of Eq. (7.16) is that it dictates unidirectionality for evolution of a process. This allows us to make a quantitative general statement of the second law of thermodynamics as: *The entropy of the universe can never decrease.* It applies to the cooling of a cup of coffee as also to all other unidirectional processes that occur in nature. You should now go through the following examples.

Example 7.9 Prove Eq. (7.16) for a system operating in a cycle.

Since our system is operating in a cycle, it will end up in exactly the same state as at the beginning. (In a real system, this cycle may involve irreversible processes.) Consequently, there will be no net change in its entropy, i.e., $\Delta S_{\text{sys}} = 0$. The Clausius inequality implies that it must be greater than or equal to $\oint \frac{\delta Q_{\text{sys}}}{T}$, i.e.,

$$0 = \Delta S_{\text{sys}} \geq \oint \frac{\delta Q_{\text{sys}}}{T}$$

While returning to the initial state, heat must flow to the system from the surroundings reversibly. Then, in each state of the cycle, $\delta Q_{\text{sys}} = -\,\delta Q_{\text{surr}}$ and hence,

$$0 = \Delta S_{\text{sys}} \geq \oint \frac{\delta Q_{\text{sys}}}{T} = \oint \frac{\delta Q_{\text{surr}}}{T}$$

We thus find that

$$0 \geq -\,\Delta\, S_{\text{surr}}$$

$$\Delta\, S_{\text{surr}} \geq 0$$

$$\Delta S_u = \Delta S_{\text{sys}} + S_{\text{surr}} \geq 0.$$

Example 7.10 A solid at temperature T_1 is brought in contact with a heat reservoir at temperature $T_2\ (> T_1)$. In due course of time, the system attains thermal equilibrium. Calculate the change in entropy of the universe.

The change in entropy of the solid is given by

$$\Delta S_s = mc_p \int_{T_1}^{T_2} \frac{dT}{T} = mc_p \ln\left(\frac{T_2}{T_1}\right)$$

is mass of the solid and c_p is specific heat capacity at constant pressure. The change in entropy of the reservoir is given by

$$\Delta S_r = -\frac{mc_p(T_2 - T_1)}{T_2}$$

The change in entropy of the universe is given by

$$\Delta S_U = \Delta S_s + \Delta S_r = mc_p \ln\left(\frac{T_2}{T_1}\right) - mc_p\left(1 - \frac{T_1}{T_2}\right)$$

To simplify this expression, we introduce a change of variable by defining $x = 1 - \left(\frac{T_1}{T_2}\right)$

$T \quad = \frac{1}{1-x}$. Hence, we can write

$$\frac{\Delta S_U}{mc_p} = -\ln(1-x) - x$$

T_1, $0 < x < 1$. If we now expand the logarithmic function, we can write

$$\frac{\Delta S_U}{mc_p} = \left(x + \frac{x^2}{2} + \frac{x^3}{3} + \cdot \frac{x^4}{4} + \ldots\right) - x = \frac{x^2}{2} + \frac{x^3}{3} + \frac{x^4}{4} + \ldots$$

> 0, $\Delta S_U > 0$. That is, the entropy of the universe will increase.

It is important to remark here that entropy change is not necessarily accompanied by heat flow. Since it is a measure of the irreversibility of a process, one would observe entropy change in Joule expansion of a perfect gas, intermixing of two or more gases as well as a throttling process. We now illustrate this for the Joule expansion.

Joule expansion Consider free expansion of a perfect gas into vacuum. Left to itself, a gas initially in the state depicted in Fig. 7.8a will end up in the state depicted in Fig. 7.8b. But experience tells that the reverse is not true. That is, free expansion is an irreversible process. Let us assume that the whole system is thermally insulated from the surroundings. We recall from Sec. 5.3 that in this case, the internal energy U remains constant. We wish to calculate the increase in entropy of the gas at temperature T when its volume changes from V_1 to V_2 but the pressure has decreased.

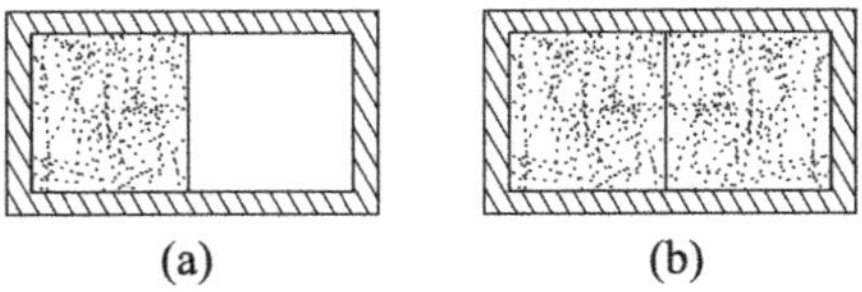

Fig. 7.8 Joule Expansion: an irreversible

Since the system has been thermally insulated, $\delta Q = 0$ and the change in entropy will be less than zero because work is done by the gas. But we must remember that the $= \delta Q/T$ does not hold true for an irreversible process. However, since entropy is a state function, we may replace this irreversible expansion by some reversible process connecting the initial and final states. (The basic distinction between these states in such an irreversible process is that in the final state, we have less knowledge of the position coordinates of molecules.) A convenient way is to choose an isothermal expansion from the initial volume to the final volume. From Eq. (7.13b), we recall that the increase in entropy will be given by

$$\Delta S = S_2 - S_1 = \int_{V_1}^{V_2} \frac{\delta Q}{T}$$

Substituting $\delta Q = \mathrm{d}U + p\mathrm{d}V$ for reversible process and writing $p = RT/V$ (for one mole of the gas), we get

$$\Delta S = \int_{V_1}^{V_2} \frac{\mathrm{d}U + p\,\mathrm{d}V}{T}$$

$= 0$ for an isothermal process, we can rewrite it as

$$\Delta S = \frac{1}{T}\int_{V_1}^{V_2} p\mathrm{d}V = R\ \ln\left(\frac{V_2}{V_1}\right) \tag{7.17}$$

Proceeding further, we note that $V_2 > V_1$. It means that ΔS is positive. That is, the entropy of the gas increases in isothermal expansion, although there is no heat flow. This would also be the entropy increase in Joule expansion.

Note that in this process, the entropy of the surroundings does not change as there is no thermal interaction. Hence, the increase in entropy of the universe in Joule expansion is also given by Eq. (7.17).

Example 7.11 One mole of an ideal gas expands isothermally to four times its initial volume. Calculate the entropy change in terms of R, the gas constant.

From Eq. (7.17), we have

$$\frac{\Delta S}{R} = \ln (V_2/V_1)$$

$$= 2.303 \log_{10} (V_2/V_1)$$

$V_1 = 4$, the entropy change in terms of the gas constant is

$$\frac{\Delta S}{R} = 2.303 \log_{10} 4$$

$$= 2.303 \times 0.6020$$

$$= 1.386.$$

HE ENTROPY FORM OF THE FIRST LAW

From the first law of thermodynamics, we were able to realise the existence of internal , a function of state. For any change of state in non-diffusive interactions, the U is given by Eq. (5.6):

$$dU = \delta Q - \delta W$$

For reversible changes, δW can in general, be written in the form $\delta W = \sum_i x_i \, dX_i$ where is the intensive variable parameter of state and X_i is conjugate extensive variable. Thus, we can rewrite the expression for internal energy as

$$dU = \delta Q - \sum_i x_i \, dX_i$$

For a reversible process between two equilibrium states, we have from the second law

$$\delta Q = T\,dS$$

These relations may be combined to give

$$dU = T\,dS - \sum_i x_i \, dX_i \tag{7.18}$$

Note that this equation combines the first and the second laws of thermodynamics for non-intensive interactions and all variables occurring here are functions of state. So Eq. (7.18) must be true for all changes, reversible and irreversible, taking place in a system. As we show, this result can be used to derive many important thermodynamical results. We will first use it to calculate the entropy change of an ideal gas.

ntropy of an Ideal Gas

For a hydrostatic system, $X = p$ and $x = V$ so that Eq. (7.18) takes the form

$$TdS = C_V\,dT + p\,dV \tag{7.19a}$$

where we have substituted $dU = C_V\,dT$ for an ideal gas.

This is one form of the entropy form of the first law of thermodynamics for a pVT system. Other equivalent forms are obtained by eliminating dV and dT using the equation of state for an ideal gas. This gives

$$T\mathrm{d}S = \frac{1}{R}(C_p p\mathrm{d}V + C_V V\mathrm{d}p) \tag{7.19c}$$

For other systems, such as a stretched wire, a surface film or a magnetic substance, the appropriate expression for δW may be used from Table 5.1, along with the equation of state.

On dividing by T and using the equation of state, Eqs. (7.19a, b, c) can respectively be rewritten as

$$\mathrm{d}S = C_V \frac{\mathrm{d}T}{T} + R\frac{\mathrm{d}V}{V}$$

$$\mathrm{d}S = C_p \frac{\mathrm{d}T}{T} - R\frac{\mathrm{d}p}{p}$$

$$\mathrm{d}S = C_p \frac{\mathrm{d}V}{V} + C_V \frac{\mathrm{d}p}{p}$$

Hence, the entropy change between any two states can be expressed as

$$S_2 - S_1 = \int_1^2 C_V \frac{\mathrm{d}T}{T} + R\int_1^2 \frac{\mathrm{d}V}{V} \tag{7.20a}$$

$$S_2 - S_1 = \int_1^2 C_p \frac{\mathrm{d}T}{T} - R\int_1^2 \frac{\mathrm{d}p}{p} \tag{7.20b}$$

$$S_2 - S_1 = \int_1^2 C_p \frac{\mathrm{d}V}{V} + \int_1^2 C_V \frac{\mathrm{d}p}{p} \tag{7.20c}$$

To integrate Eqs. (7.20a, b, c), one must know the temperature dependence of the heat capacities. If we assume that both C_p and C_V remain constant in the temperature interval of interest, we can perform the integration directly for reference states defined by) and (T_2, V_2) to obtain

$$S_2 - S_1 = C_V \ln\left(\frac{T_2}{T_1}\right) + R\ln\left(\frac{V_2}{V_1}\right) \tag{7.21a}$$

Similarly, for reference states signified by (T_1, p_1) and (T_2, p_2), we have

$$S_2 - S_1 = C_p \ln\left(\frac{T_2}{T_1}\right) - R\ln\left(\frac{p_2}{p_1}\right) \tag{7.21b}$$

and for states defined by (p_1, V_1) and (p_2, V_2), the entropy change is

$$S_2 - S_1 = C_p \ln\left(\frac{V_2}{V_1}\right) + C_V \ln\left(\frac{p_2}{p_1}\right) \tag{7.21c}$$

We can rewrite it as

$$S_2 - S_1 = C_V \ln\left(\frac{p_2}{p_1}\right) + \gamma C_V \ln\left(\frac{V_2}{V_1}\right)$$

$$= C_V \ln\left(\frac{p_2}{p}\frac{V_2^\gamma}{V^\gamma}\right) \tag{7.21d}$$

This result shows that for one mole of an ideal gas, the change in entropy is given by

$$\Delta S = C_V \ln(pV^{\gamma}) + \text{constant} \tag{7.21e}$$

From Eqs. (7.21a) and (7.21b), we note that for an isothermal process, the entropy change is given by

$$\Delta S_T = R\ln(V_2/V_1) = R\ln(p_1/p_2) \tag{7.22a}$$

This is in complete agreement with Eq. (7.17).
Similarly, for isobaric changes, we get

$$\Delta S_p = C_p \ln(T_2/T_1) = C_p \ln(V_2/V_1) \tag{7.22b}$$

and for isochoric changes, we have

$$\Delta S_V = C_V \ln(T_2/T_1) = C_V \ln(p_2/p_1) \tag{7.22c}$$

To appreciate the significance of these results, let us suppose that n moles of an ideal gas undergo isothermal expansion so that its volume becomes twice its value. Then Eq. (7.22a) implies that the change in entropy will be

$$\Delta S_T = nR \ln 2$$

From Eq. (7.22b) we note that if a gas is heated, its energy increases, whereas if it undergoes cooling, its entropy decreases. You may now ask: Does the total energy decrease in the cooling process? Recall that cooling is a natural process and entropy always increases in a natural process. To resolve this apparent contradiction, we note that when a body cools, the decrease in its entropy $\Delta S_1 = -\dfrac{Q}{T_2}$ is accompanied by an increase in entropy of the surroundings at a lower temperature and $\Delta S_2 = \dfrac{Q}{T_1}$.
Hence, the net change in entropy

$$\Delta S = \Delta S_1 + \Delta S_2 = Q\left(\frac{1}{T_1} - \frac{1}{T_2}\right)$$

T_2, $\Delta S > 0$. That is, total change in entropy of the system and its surroundings increases. However, for an adiabatic change, no change in entropy takes place since the system is isolated and no thermal energy is exchanged with the surroundings.
You should now go through the following examples carefully.

Example 7.12 Calculate the increase in entropy of 1 g of hydrogen when its temperature is raised from – 173°C to 27°C and its volume becomes four times its initial volume. Given $C_V = 4.86$ cal mol^{-1} K^{-1}, $R = 2$ cal mol^{-1} K^{-1} and $M_{H_2} = 2$ g mol^{-1}.

When n moles of an ideal gas are made to undergo changes in temperature and volume, the entropy change is obtained by multiplying Eq. (7.21a) by n. For 1 g of the gas,

$$n = \frac{1\,\text{g}}{2\,\text{g mol}^{-1}} = 0.5\ \text{mol}$$

$= (-173 + 273\ K) = 100$ K, $T_2 = (27 + 273)$ K $= 300$ K, $\dfrac{V_2}{V_1} = 4$,
$= 4.86$ cal mol^{-1} K^{-1} and $R = 2$ cal mol^{-1} K^{-1},

$$(0.5 \text{ mol}) \times 2.303\left(C_V \log_{10}\left(\frac{T_2}{T_1}\right) + R\log_{10}\left(\frac{V_2}{V_1}\right)\right)$$

$$(0.5 \text{ mol}) \times 2.303\left[(4.86 \text{ cal mol}^{-1}\text{ K}^{-1})\log_{10}\left(\frac{300 \text{ K}}{100 \text{ K}}\right) + (2 \text{ cal mol}^{-1}\text{ K}^{-1})\log_{10} 3\right]$$

$= 2.303\ [2.43 \log_{10}3 + \log_{10}4] \text{ cal K}^{-1}$

$= 2.303\ [(2.43 \times 0.4771) + 0.6021] \text{ cal K}^{-1}$

$= 4.06 \text{ cal K}^{-1}$.

You should now answer a practise problem.

Problem 7.7 Two moles of a gas occupy a volume of 10 litre at 4 atm. The gas is heated at constant volume till its pressure is 8 atm. Then, it is allowed to expand at constant pressure. If its final volume is 40 litre, calculate the change in entropy. Given $= 3 \text{ cal mol}^{-1}\text{ K}^{-1}$ and $R = 2 \text{ cal mol}^{-1}\text{ K}^{-1}$.

18.02 cal K^{-1}

Entropy of Mixing

Refer to Fig. 7.9, which shows two gases at constant temperature T and pressure p in two separate containers connected through a stopcock. The blank circles depict n_1 moles of one type of ideal gas, say, oxygen, and the solid circles depict n_2 moles of second type of ideal gas, say, nitrogen. (You could as well consider a more general case of a container having a number of ideal gases at the same temperature and pressure, separated from each other by suitable partitions.) Now suppose that the stopcock is opened and the two gases begin to mix. If no chemical reaction takes place, their pressures and temperature will not be affected. Let the partial pressures of individual gases in the mixture be p_1 and p_2. The entropy of mixing is equal to the sum total of the changes in entropy of each gas as it expands from the initial pressure to its partial pressure in the mixture.

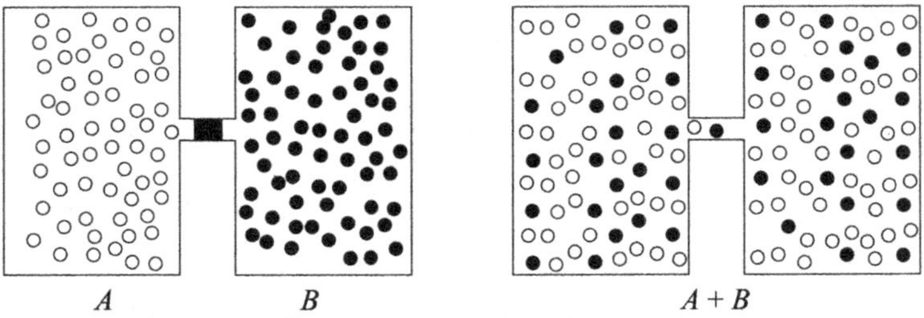

Entropy of mixing: Diffusion increases disorder.

Using Eq. (7.22a), we can express the changes in entropies of each gas for mixing at constant temperature T as

$$\Delta S_1 = n_1 R \ln\left(\frac{p}{p_1}\right)$$

$$\Delta S_2 = n_2 R \ln\left(\frac{p}{p_2}\right)$$

Hence, the entropy of mixing is given by

$$\Delta S_{\text{mix}} = \Delta S_1 + \Delta S_2 = n_1 R \ln\left(\frac{p}{p_1}\right) + n_2 R \ln\left(\frac{p}{p_2}\right) \tag{7.23}$$

$_2$ signify the mole fractions of these gases, we can express partial pressures in terms of total pressure as

$$p_1 = x_1 p$$

$$p_2 = x_2 p$$

Using these results in Eq. (7.23), we get

$$\Delta S_{\text{mix}} = n_1 R \ln\left(\frac{p}{x_1 p}\right) + n_2 R \ln\left(\frac{p}{x_2 p}\right)$$

$$= - n_1 R \ln x_1 - n_2 R \ln x_2 \tag{7.24}$$

The entropy of mixing per mole of the mixture is obtained by dividing both sides of Eq. (7.24) by $n_1 + n_2$. This gives

$$\frac{\Delta S_{\text{mix}}}{n_1 + n_2} = - R\left[\frac{n_1}{n_1 + n_2} \ln x_1 + \frac{n_1}{n_1 + n_2} \ln x_2\right]$$

Note that the ratios $\frac{n_1}{n_1 + n_2}$ and $\frac{n_2}{n_1 + n_2}$ define the mole fractions of two gases. Hence, the entropy of mixing for one mole of the gas is given by

$$\Delta s_{\text{mix}} = -R[x_1 \ln x_1 + x_2 \ln x_2] \tag{7.25}$$

x_1 and x_2 are less than one. It means that the entropy of mixing will be positive definite always.

Let us now suppose that two moles of the same gas are separated by a partition. If these are allowed to mix, Eq. (7.25) predicts that the entropy of mixing will increase. Physically this makes little sense because the state of the system cannot be changed by removing a partition. That is, there should be no change in the entropy of the gas. This is called *Gibb's* and arises because of the assumption that the two gases are indistinguishable. You will learn about it in detail in Chapter 13.

This concept is illustrated in the following example.

Example 7.13 11.2 litre of oxygen and 11.2 litre of hydrogen are allowed to mix at STP in an insulated chamber. If both gases show ideal gas behaviour, calculate the change

From Eq. (7.25), we recall that the entropy of mixing is given by

$$\Delta s_{mix} = -R[x_1 \ln x_1 + x_2 \ln x_2]$$

Since volume of both the gases is equal and these are at the same temperature and pressure, we can use Avogadro's hypothesis and take $n_1 = n_2$. Therefore, for one mole of the gas, n_2. Alternatively, you can convince yourself by noting that $n_1 = \dfrac{11.2}{22.4} = 0.5 = n_2$ $+ n_2 = 1$. Further, $x_1 = \dfrac{n_1}{n_1 + n_2} = 0.5 = x_2$. Hence,

$$\Delta s_{mix} = -R[0.5 \ln 0.5 + 0.5 \ln 0.5]$$

$$= -R[\ln 0.5] = -R \times (-0.693) = 0.693\,R.$$

NAVAILABLE ENERGY AND THERMAL DEATH OF NIVERSE

In Sec. 6.8, we discussed how irreversibility leads to loss of work output when a heat engine is operated under such conditions. The application of this idea to natural processes is of particular significance. We will now show that in every irreversible process, some energy becomes increasingly unavailable, i.e., quality of energy is degraded. Note that we have to express this result in terms of changes in entropy. To this end, from Eq. (6.30), we recall that the energy lost in an irreversible process is given by

$$LE = T_0\left(\frac{Q}{T_2} - \frac{Q}{T_1}\right)$$

In terms of entropy, we can rewrite this as

$$LE = T_0 \Delta S \tag{7.26}$$

is temperature of the coldest available reservoir in the given environment and $\dfrac{Q}{T_2} - \dfrac{Q}{T_1}$ is increase in entropy of the universe. From this, we may conclude that *energy that is not available for work during an irreversible process is T_0 times an increase in the entropy of the universe.* Since all natural processes are irreversible, energy available for work is continuously decreasing and entropy of the universe is piling up. You may now ask: How long will entropy continue to increase and what is the future of the natural world? We expect that ultimately, it will reach some upper limit and the principle of increase of entropy implies that the universe is marching towards a stagnant thermal state to suffer *thermal death.*

HYSICAL CONCEPT OF ENTROPY

It is likely that you may not have fully realised the significance of entropy yet. You may wonder as to what this quantity entropy is that is piling up in the universe. Now we wish to give an insight into the physical and philosophical aspects of the concept of entropy.

You may recall that entropy increases during an irreversible process. So, the word entropy serves as an effective vehicle for communicating our observations in respect of evolution of irreversible processes, like heat conduction along a bar or cooling of a cup

We also know that entropy of a system increases when heat is added to it. Since heat is chaos, from kinetic theory viewpoint, continuous withdrawal of heat makes the molecular arrangement increasingly ordered. That is, decrease in entropy implies a transition from chaotic (or more random state) to somewhat ordered state. Conversely, more disordered the state is, the larger is its entropy. This implies that solids have much less entropy than gases.

Another point to be made regarding entropy is that it can be related to the probability of occurrence of a state. For instance, increase in entropy in an irreversible process can be associated with a change of state from a less probable to a more probable state. If we re-examine Joule expansion of gas, we will have less information about the position coordinates of molecules after expansion. That is, the final state is a less ordered (or more random) state of the system. Recall that Joule expansion is associated with entropy increase. Similarly, the most probable state when a cup of coffee cools is that it will be at the same temperature as its surroundings. Therefore, as coffee cools, entropy increases and the system changes from less probable to more probable state.

In fact, when a system is left to itself, it passes from an ordered (least probable) state to a state of maximum disorder (most probable). That is, a system has a tendency to change in the direction of increasing entropy. This argument led Boltzmann to relate entropy with probability. Mathematically, he expressed it by the relation $S = f(W)$, where S and W respectively denote the entropy and thermodynamic probability of the state.

Lastly, you may enquire: Does the second law of thermodynamics apply to the universe as a whole? If this were true, how did it get in the state of low entropy? Are some processes going on in the universe elsewhere that compensate for increase in entropy? We cannot give conclusive answers to these questions on the basis of the second law of thermodynamics alone. In fact, a detailed discussion of these questions is beyond the scope of this book. However, you may like to think over these!

DDITIONAL EXAMPLES

Example 7.14 A lump of steel weighing 30 kg at a temperature of 627°C is dropped in 150 kg oil at 27°C. The specific heat capacities of steel and oil at constant pressure are K^{-1} and 2.5 kJ kg^{-1}K^{-1}, respectively. Calculate the total charge in entropy.

Let us assume that the final temperature of steel is T. By the law of mixtures, we can write

$$(mc_p \Delta T)_{\text{steel}} = (mc_p \Delta T)_{\text{oil}}$$

$$(30\text{ kg}) \times (0.5\text{ kJ kg}^{-1}\text{ K}^{-1}) \times (900\text{ K} - T) = (150\text{ kg}) \times (2.5\text{ kJ kg}^{-1}\text{ K}^{-1}) \times (T - 300\text{ K})$$

$$(15 \times 10^3\text{ J}) \times (900\text{ K} - T) = (375 \times 10^3\text{ J})(\text{T} - 300\text{ K})$$

or

$$900\text{ K} - T = 25(T - 300\text{ K})$$

$$26T = 8400\text{ K}$$

$$T = \frac{8400\text{ K}}{26} = 323\text{ K}$$

The change in entropy of steel lump when it is put in oil bath is given by

$$\begin{aligned}(\Delta S)_{\text{steel}} &= mc_p \ln(T/T_1)\\ &= (30\text{ kg}) \times (0.5\text{ kJ kg}^{-1}\text{K}^{-1}) \times \ln(323\text{ K}/900\text{ K})\\ &= -15.4\text{ kJ K}^{-1}\end{aligned}$$

Similarly, the change in entropy of oil when steel lump is put in it is

$$\begin{aligned}(\Delta S)_{\text{oil}} &= mc_p \ln(T/T_2)\\ &= (150\text{ kg}) \times (2.5\text{ kJ kg}^{-1}\text{K}^{-1}) \times \ln(323\text{ K}/300\text{ K})\\ &= 27.7\text{ kJ K}^{-1}\end{aligned}$$

Hence, total change in entropy of the system is given by

$$\begin{aligned}\Delta S &= (\Delta S)_{\text{steel}} + (\Delta S)_{\text{oil}}\\ &= -15.4\text{ kJ K}^{-1} + 27.7\text{ kJ K}^{-1}\\ &= 12.3\text{kJK}^{-1}.\end{aligned}$$

Example 7.15 m gram water at 0°C is mixed with an equal amount of water at 80°C. Calculate the total change in entropy.

Let the temperature of the mixture be t°C. Then from the law of mixtures, we

$$m_1 s(t - T_1) = m_2 s(T_2 - t)$$

$T_1 = 0°\text{C} = 273\text{ K}, T_2 = 80°\text{C} = (273 + 80)\text{ K} = 353\text{ K}$ and $m_1 = m_2 = m$. Hence, we

$$ms(t - 273\text{ K}) = ms(353\text{ K} - t)$$

On simplification, we get

$$2t = 626\text{ K}$$

$$t = 313\text{ K}$$

Increase in entropy when temperature of m gram of water rises from 273 K to 313 K

$$\begin{aligned}\Delta S_1 &= ms \ln\left(\frac{313\text{ K}}{273\text{ K}}\right)\\ &= m \times (1\text{ cal g}^{-1}\text{ K}^{-1}) \times 0.1367\\ &= 0.1367\, m\text{ cal g}^{-1}\text{ K}^{-1}\end{aligned}$$

Decrease in entropy when temperature of m gram of water drops from 353 K to 313 K

$$\begin{aligned}\Delta S_2 &= ms \ln\left(\frac{313\text{ K}}{353\text{ K}}\right)\\ &= m \times (1\text{ cal g}^{-1}\text{ K}^{-1}) \times (-0.1203)\\ &= -0.1203\, m\text{ cal g}^{-1}\text{ K}^{-1}\end{aligned}$$

Hence, net change in entropy

$$\Delta S = \Delta S_1 + \Delta S_2$$
$$= (0.1367m - 0.1203m) \text{ cal g}^{-1} \text{ K}^{-1}$$
$$= 0.0164m \text{ cal g}^{-1} \text{K}^{-1}$$

Example 7.16 10 kg water at 273 K is brought in contact with a heat reservoir at 373 K. Calculate the change in entropy of water if its temperature attains the temperature of reservoir. Also calculate the change in entropy of the reservoir and the universe. Take specific heat capacity of water = 4.18×10^3 J kg^{-1}K^{-1}.

The increase in entropy of water when its temperature increases from 273 K

$$dS_1 = ms \ln\left(\frac{373 \text{ K}}{273 \text{ K}}\right)$$
$$= (10 \text{ kg}) \times (4.18 \times 10^3 \text{ J kg}^{-1} \text{K}^{-1}) \times 0.312$$
$$= 13.05 \times 10^3 \text{ JK}^{-1}.$$

The reservoir can be assumed to have infinite heat capacity and its temperature remains unchanged. Therefore, the change in entropy of the reservoir

$$dS_2 = -\frac{\delta Q}{T} = -\frac{ms(T_2 - T_1)}{T_2}$$
$$= -\frac{(10 \text{ kg}) \times (4.18 \times 10^3 \text{ J kg}^{-1}\text{K}^{-1}) \times (100 \text{ K})}{373 \text{ K}}$$
$$= -11.21 \times 10^3 \text{ JK}^{-1}.$$

Hence, change in entropy of the universe

$$\Delta S = dS_1 + dS_2$$
$$= (13.05 \times 10^3 - 11.21 \times 10^3) \text{ JK}^{-1}$$
$$= 1.84 \times 10^3 \text{ JK}^{-1}.$$

Example 7.17 A metallic can of negligible heat capacity contains 2.5 kg water just above the freezing point. Another identical can contains the same quantity of water just below the boiling point. These cans are put in thermal contact. Calculate the change in the entropy of the system.

For contents of both cans, $\delta Q = ms\Delta T$. Here $m = 2.5$ kg, and $s = 4184$ JK^{-1}. The equilibrium temperature will be average of their initial temperatures, since their masses

$$T = \frac{273 \text{ K} + 373 \text{ K}}{2} = 323 \text{ K}$$

Hence, increase in entropy of the can containing cold water

$$\Delta S_1 = (2.5 \text{ kg}) \times (4184 \text{ JK}^{-1} \text{kg}^{-1}) \ln\left(\frac{323}{273}\right)$$
$$= 10460 \times \ln (1.18315)$$
$$= 1759.17 \text{ JK}^{-1}$$

Similarly, the drop in entropy of the can containing hot water

$$\Delta S_2 = (2.5 \text{ kg}) \times (4184 \text{ JK}^{-1} \text{kg}^{-1}) \ln\left(\frac{323}{373}\right)$$
$$= 10460 \times \ln (0.86595)$$
$$= -1505.49 \text{ JK}^{-1}$$

Hence, net change in entropy of the system

$$\Delta S = \Delta S_1 + \Delta S_2$$
$$= (1759.17 - 1505.49) \text{ JK}^{-1}$$
$$= 253.68.$$

Example 7.18 The temperature variation of molar specific heat capacity of a material at constant pressure, in the range 80 K to 100 K, is given in cal K^{-1} by

$$C_p = 0.07T - 0.0003T^2 - 0.15$$

The temperature is expressed in kelvin. If 2 mole of the substance is heated from 80 K to 100 K, calculate the change in entropy.

The change in entropy when 2 mole of the material is heated from 80 K to 100 K is given by

$$\Delta S = \int_{80}^{100} \frac{\delta Q}{T}$$
$$= \int_{80}^{100} \frac{2C_p \, dT}{T}$$
$$= 2\int_{80}^{100} \frac{(0.07T - 0.0003T^2 - 0.15)dT}{T} \text{ cal K}^{-1}$$
$$= 2\int_{80}^{100} \left[0.07 - 0.0003T - \frac{0.15}{T}\right] dT \text{ cal K}^{-1}$$
$$= 2\left[0.07T - 0.0003\frac{T^2}{2} - 0.15\ln T\right]_{80}^{100} \text{ cal K}^{-1}$$
$$= 2\left[0.07 \times 20 - 0.0003\frac{(10^4 - 6400)}{2} - 0.15 \ln\left(\frac{100}{80}\right)\right] \text{cal K}^{-1}$$
$$= 2[1.40 - 0.0003 \times 1800 - 0.15 \ln (1.25)] \text{ cal K}^{-1}$$
$$= 2[1.40 - 0.54 - 0.15 \times 0.2231] \text{ cal K}^{-1}$$
$$= 2[1.40 - 0.54 - 0.0335] \text{ cal K}^{-1}$$
$$= 1.65 \text{ cal K}^{-1}.$$

Example 7.19 Calculate the increase in entropy of 0.32 kg of oxygen when its temperature is raised from 27°C to 100°C. Assume that the volume of oxygen doubles in this process. Take $C_V = 20.8$ kJ mol^{-1}K^{-1} and $R = 8.314$ JK^{-1}mol^{-1}.

In this case, no change of phase takes place. The change in entropy, when 0.3 kg oxygen changes from 300 K to 373 K, is given by Eq. (7.21a):

$$\Delta S = nC_V \ln\left(\frac{T_2}{T_1}\right) + nR \ln\left(\frac{V_2}{V_1}\right)$$

$$= \frac{m}{M} C_V \ln(T_2/T_1) + \frac{m}{M} R \ln(V_2/V_1)$$

On substituting the given values, we get

$$\Delta S = \frac{0.032 \text{ kg}}{32 \text{ kg mol}^{-1}} \times (20.8 \times 10^3 \text{ J mol}^{-1}\text{K}^{-1}) \ln\left(\frac{373}{300}\right)$$

$$+ \frac{0.032 \text{ kg}}{32 \text{ kg mol}^{-1}} \times (8.314 \text{ JK}^{-1}\text{mol}^{-1}) \ln(2)$$

$$= (20.8 \text{ JK}^{-1}) \ln(1.2433) + 8.314 \times 10^{-3} \times 0.693 \text{ JK}^{-1}$$

$$= 4.530 \text{ JK}^{-1} + 5.763 \times 10^{-3} \text{ JK}^{-1}$$

$$= 4.536 \text{ JK}^{-1}.$$

Example 7.20 In an adiabatic process, the pressure of an ideal gas varies as αV, when p_0 and α are positive constants. Calculate the volume at which its entropy will be maximum.

We know that entropy of a gas remains constant in an adiabatic process. From Eq. (7.21e), we recall that

$$S = C_V \ln(pV^\gamma) + \text{constant}$$

$$= C_V \ln[(p_0 - \alpha V)V^\gamma) + \text{constant}$$

remains constant, we have to maximise the function $(p_0 - \alpha V)V^\gamma$, i.e., we put

$$\frac{d}{dV}[(p_0 - \alpha V)V^\gamma] = 0$$

On differentiation, we obtain

$$(p_0 - \alpha V)\gamma V^{\gamma-1} - \alpha V^\gamma = 0$$

$$\gamma p_0 V^{\gamma-1} - (\gamma + 1)\alpha V^\gamma = 0$$

Dividing throughout by $V^{\gamma-1}$, we can write

$$\gamma p_0 - (\gamma + 1)\alpha V = 0$$

$$V_{max} = \frac{\gamma}{\alpha(\gamma + 1)} p_0$$

Example 7.21 10 g steam at 100°C is blown into 90 g water at 0°C contained in a vessel of negligible heat capacity. If the whole steam condenses, calculate the increase in entropy of the system. Given latent heat of steam = 540 cal g^{-1}.

The change in entropy of steam during condensation

$$\Delta S_1 = -\frac{\delta Q}{T} = -\frac{(10\text{g})\times(540\text{ cal g}^{-1})}{373\text{ K}} = -14.777\text{ cal K}^{-1}$$

Note that heat is given out by steam during condensation as well as by hot water and it is absorbed by cold water. Obviously, cold water temperature will increase above its initial temperature. Let the final temperature be T. Using the method of mixtures, we can write

$$(10\text{ g})\times(540\text{ cal g}^{-1}) + (10\text{ g})\times(1\text{ cal g}^{-1}\text{K}^{-1})(373\text{ K} - T)$$
$$= (90\text{ g})\times(1\text{ cal g}^{-1}\text{K}^{-1})\times(T - 273\text{ K})$$

$$5400\text{ cal} + (3730\text{ K} - 10T)\text{ cal K}^{-1} = (90T - 273\times 90\text{ K})\text{ cal K}^{-1}$$

$$9130\text{ cal} - 10T\text{ cal K}^{-1} = 90T\text{ cal K}^{-1} - 24370\text{ cal}$$

$$100T(\text{K}^{-1}) = 33500$$

$$T = 335\text{ K}$$

Change in entropy when water at 373 K cools down to 335 K is given by

$$\Delta S_2 = ms\ln\left(\frac{T_2}{T_1}\right)$$
$$= (10\text{ g})(1\text{ cal g}^{-1}\text{K}^{-1})\times\ln\left(\frac{335}{373}\right)$$
$$= (10\text{ cal K}^{-1})\times\ln(0.8981)$$
$$= -1.0744\text{ cal K}^{-1}$$

Increase in entropy of cold water as its temperature rises from 273 K to 335 K

$$\Delta S_3 = (90\text{ g})\times(1\text{ cal g}^{-1}\text{K}^{-1})\times\ln\left(\frac{335}{273}\right)$$
$$= (90\text{ cal K}^{-1})\times\ln(1.227)$$
$$= 18.4188\text{ cal K}^{-1}$$

Hence, net change in entropy of the system is obtained by summing individual values:

$$\Delta S = \Delta S_1 + \Delta S_2 + \Delta S_3$$

$$= (-14.7770 - 1.0744 + 18.4188)\text{ cal K}^{-1}$$
$$= 2.5674\text{ cal K}^{-1} = 2.6\text{ cal K}^{-1}$$

Example 7.22 1 kg solid, whose melting point is 237°C is heated from 157°C to 330°C. Calculate the change in its entropy. Take latent heat of fusion = 14 cal g^{-1}, specific heat capacity of solid phase = 0.055 cal g^{-1}K^{-1} and specific heat capacity of liquid phase is 0.064 cal g^{-1}K^{-1}.

Note that the melting point of the given solid lies in-between the temperatures where heating is done. Therefore, calculation of entropy will consist of three steps: increase in temperature from 157°C (= 430 K) to 237°C (510 K), change of phase (melting) at 237°C, and increase of temperature of liquid phase from 230°C to 330°C.

Change in entropy when the solid is heated from 430 K to 510 K

$$\Delta S_1 = mc_V \ln\left(\frac{T_2}{T_1}\right)$$

$$= (10^3\,\text{g}) \times (0.055\,\text{cal g}^{-1}\text{K}^{-1}) \ln\left(\frac{510\text{ K}}{430\text{ K}}\right)$$

$$= (55\,\text{cal K}^{-1}) \ln(1.18605)$$

$$= 55 \times 0.17063\,\text{cal K}^{-1}$$

$$= 9.385\,\text{cal K}^{-1}.$$

The change in entropy when solid melts at 510 K is

$$\Delta S_2 = \frac{mL}{T} = \frac{(10^3\,\text{g}) \times (14\,\text{cal g}^{-1})}{510\text{ K}}$$

$$= 27.451\,\text{cal K}^{-1}$$

The change in entropy when molten solid is heated from 510 K to 603 K

$$\Delta S_3 = mc_V \ln\left(\frac{T_2}{T_1}\right)$$

$$= (10^3\,\text{g}) \times (0.064\,\text{cal g}^{-1}\text{K}^{-1}) \ln\left(\frac{603\text{ K}}{510\text{ K}}\right)$$

$$= (64\,\text{cal K}^{-1}) \ln(1.1824)$$

$$= 75.671$$

Hence, total change in entropy

$$\Delta S = \Delta S_1 + \Delta S_2 + \Delta S_3$$

$$= 112.51\,\text{cal K}^{-1}$$

Example 7.23 In Srinagar, temperature of ice on a cold winter night is measured as –20°C. Calculate the change in entropy when 1 kg of this ice is converted into steam at 100°C. Take specific heat capacity of ice = 500 cal kg^{-1}K^{-1}, latent heat of ice is J kg^{-1}, latent heat of steam = 2.26×10^6 J kg^{-1} and J = 4.2 J cal^{-1}.

Here the change in entropy occurs in four steps:

Increase in entropy when the temperature of 1 kg of ice increases from –20°C (253 K) to 0°C (273 K). It is given by

$$\Delta S_1 = ms \ln\left(\frac{T_2}{T_1}\right)$$

$$= (1\,\text{kg}) \times (500 \times 4.2\,\text{J kg}^{-1}\text{K}^{-1}) \ln\left(\frac{273\text{ K}}{253\text{ K}}\right)$$

$$= (2100 \text{ JK}^{-1}) \ln(1.07905)$$
$$= 159.77 \text{ JK}^{-1}.$$

Increase in entropy when 1 kg ice at 0°C is converted into water at 0°C can be calculated using the relation

$$\Delta S_2 = \frac{\delta Q}{T} = \frac{mL}{T}$$
$$= \frac{(1 \text{ kg}) \times (3.36 \times 10^5 \text{ J kg}^{-1})}{273 \text{ K}}$$
$$= 123.08 \text{ JK}^{-1}.$$

Increase in entropy when temperature of 1 kg of water at 0°C is raised to 100°C is

$$\Delta S_3 = m s_w \ln\left(\frac{373}{273}\right)$$
$$= (1 \text{ kg}) \times (10^3 \times 4.2 \text{ J kg}^{-1}\text{K}^{-1}) \ln\left(\frac{373}{273}\right)$$
$$= (4200 \text{ JK}^{-1}) \times \ln(1.3663)$$
$$= 1310.85 \text{ JK}^{-1}.$$

Increase in entropy when 1 kg of water at 100°C is converted into steam at the same temperature

$$\Delta S_4 = \frac{mL_{\text{steam}}}{T} = \frac{(1 \text{ kg}) \times (2.26 \times 10^6 \text{ J kg}^{-1})}{373 \text{ K}}$$
$$= 6058.98 \text{ JK}^{-1}$$

Hence, total change in entropy when ice at –20°C changes into steam at 100°C is

$$\Delta S = \Delta S_1 + \Delta S_2 + \Delta S_3 + \Delta S_4$$
$$= 159.77 \text{ JK}^{-1} + 123.08 \text{ JK}^{-1} + 1310.85 \text{ JK}^{-1} + 6058.98 \text{ JK}^{-1}$$
$$= 7652.68 \text{ JK}^{-1}.$$

Example 7.24 20 g water at 50°C is converted into ice at –10°C at constant atmospheric pressure. If specific heat of ice at constant pressure is 0.5 cal $g^{-1}K^{-1}$, calculate the total change in entropy of the system. Latent heat of ice = 80 cal g^{-1}.

The change in entropy will occur in three stages: drop in temperature of water from 50°C to 0°C, formation of ice at 0°C and decrease in temperature of ice from 0°C

Change in entropy when temperature of water falls from 50°C(= 323 K) to 0°C(273 K).

$$\Delta S_1 = ms \ln\left(\frac{T_2}{T_1}\right)$$
$$= (20 \text{ g}) \times (1.0 \text{ cal g}^{-1}\text{K}^{-1}) \times \ln\left(\frac{273 \text{ K}}{323 \text{ K}}\right)$$
$$= (20 \text{ cal K}^{-1}) \times \ln(0.8452)$$
$$= -0.3364 \text{ cal K}^{-1}.$$

Change in entropy when water at 273 K is converted into ice at the same temperature.

$$\Delta S_2 = -\frac{\delta Q}{T} = -\frac{m L_{\text{ice}}}{T}$$

$$= -\frac{(20\text{ g}) \times (80\text{ cal g}^{-1})}{273\text{ K}}$$

$$= -5.8608\text{ cal K}^{-1}.$$

Change in entropy when 10 g ice at 273 is cooled to 263 K.

$$\Delta S_3 = (20\text{ g}) \times (0.5\text{ cal g}^{-1}\text{K}^{-1}) \ln\left(\frac{263}{273}\right)$$

$$= (10\text{ cal K}^{-1}) \times \ln(0.9634)$$

$$= -0.3732\text{ cal K}^{-1}$$

Hence, total change in entropy

$$\Delta S = \Delta S_1 + \Delta S_2 + \Delta S_3$$

$$= -0.3364\text{ cal K}^{-1} - 5.8608\text{ cal K}^{-1} - 0.3732\text{ cal K}^{-1}$$

$$= -6.57\text{ cal K}^{-1}.$$

Example 7.25 The specific heat capacity of aluminium varies as

$$C_V = 0.77\text{J g}^{-1}\text{K}^{-1} + (0.46 + 10^{-3}\text{ J g}^{-2}\text{K}^{-2})T.$$

Calculate the increase in entropy, if an aluminium bar weighing 3 kg is heated from 300 K to 500 K.

The change in entropy when aluminium is heated from 300 K to 500 K is

$$\Delta S = m\int_{300}^{500}[0.77\text{ J g}^{-1}\text{K}^{-1} + (0.46 \times 10^{-3}\text{ J g}^{-2}\text{K}^{-2})T]\frac{dT}{T}$$

$$= (3 \times 10^3\text{ g})\left[(0.77\text{ J g}^{-1}\text{K}^{-1})\int_{300}^{500}\frac{dT}{T} + (0.46 \times 10^{-3}\text{ J g}^{-2}\text{K}^{-2})\right]\int_{300}^{500} dT$$

$$= (2.31 \times 10^3\text{ JK}^{-1}) \ln\left(\frac{500}{300}\right) + (1.38\text{ JK}^{-1}) \times 200$$

$$= 1180\text{ JK}^{-1} + 276\text{ JK}^{-1}$$

$$= 1456\text{ JK}^{-1}.$$

We now summarise what you have learnt in this chapter.

MMARY

The entropy of a system is defined through the relation

$$dS = \frac{\delta Q}{T}$$

Entropy is a function of state. However, we cannot measure it like temperature and pressure in a laboratory. As such, it is a mathematical device and gives the direction of evolution of a thermodynamic process.

Entropy is a measure of disorder in the system; more chaotic the system, greater will be its entropy.

The second law of thermodynamics is essentially the law of increase of entropy. It may be stated as: The entropy of the universe can never decrease. A more formal statement of the second law of thermodynamics in terms of entropy is: When an isolated system undergoes a change, its entropy cannot decrease; it either increases or remains constant. Mathematically, we write

$$\Delta S \geq 0$$

For a heat engine, the change in entropy is given by Clausius inequality

$$\oint \frac{\delta Q}{T} \leq 0$$

For all possible processes that a system in a given surroundings can undergo, the increase in entropy is given by

$$\Delta S_{\text{sys}} + \Delta S_{\text{surr}} = \Delta S_u \geq 0$$

The change in entropy of an ideal gas can be written in different forms as

$$\begin{aligned} S_2 - S_1 &= C_V \ln\left(\frac{T_2}{T_1}\right) + R \ln\left(\frac{V_2}{V_1}\right) \\ &= C_p \ln\left(\frac{T_2}{T_1}\right) - R \ln\left(\frac{p_2}{p_1}\right) \\ &= C_p \ln\left(\frac{V_2}{V_1}\right) + C_V \ln\left(\frac{p_2}{p_1}\right) \\ &= C_V \ln\left(\frac{p_2}{p_1}\frac{V_2^{\gamma}}{V_1^{\gamma}}\right) + \text{constant} \end{aligned}$$

The energy that is not available for work during an irreversible process is T_0 times the increase in the entropy of the universe.

ERCISES

Calculate the change in entropy when 100 g tin is heated from 20°C to 300°C. The melting point of tin is 232°C. Its latent heat of fusion is 14 cal g^{-1}. The specific heat capacities of solid and liquid tin are 0.055 cal g^{-1} K^{-1} and 0.064 cal g^{-1} K^{-1} respectively. (*Ans:* 6.60 cal K^{-1})

Calculate the change in entropy when 1 kg water at 27°C is converted into superheated steam at 200°C under constant atmospheric pressure. Specific heat capacity of liquid water is 4180 J kg^{-1} K^{-1}; temperature dependence of specific heat capacity of steam is given by the relation (1670 + 0.49 T) J kg^{-1} K^{-1} at T degree kelvin. Take latent heat of steam as 23×10^5 J kg^{-1}. (*Ans:* 7521 J kg^{-1})

55 g water at 0°C is mixed with an equal mass of water at 75°C. Calculate the resultant increase in entropy. (*Ans.* 0.825 cal K^{-1})

Calculate the change in entropy when 1g ice is converted into steam at atmospheric pressure. Given that specific heat capacity of water = 4200 $JK^{-1}kg^{-1}$, latent heat of ice = 3.36×10^5 J kg^{-1} and latent heat of steam = 2.26×10^6 J kg^{-1}.

(*Ans:* 8.6 JK^{-1})

The entropy of a liquid at 0.05 atm and 32°C is 0.52 cal K^{-1}. Assuming that the vapour of the liquid behaves ideally, calculate the entropy of its vapour per mole at C and 0.52 atm. It is given that the latent heat of vaporisation is 2422 cal mol^{-1}, $R = 2$ cal $mol^{-1}K^{-1}$. (*Ans:* 3.78 cal K^{-1})

10 A current is passed through a 25 Ω resistor for 1 s. Its temperature is maintained constant at 27°C. Calculate the entropy change of (a) the resistor and (b) the universe. The resistor is then insulated thermally. If its mass is 10 g and $C_p = 0.84$ J g^{-1} K^{-1}, compute the entropy change of (c) the resistor, and (d) the universe.

(*Ans:* (a) 0 (b) 8.33 JK^{-1} (c) 5.78 JK^{-1} (d) 5.78 JK^{-1})

Compute the entropy change of the universe in each of the following processes?

(a) A 1μF capacitor is connected to a 100 V reversible cell at 0°C.

(b) The same capacitor, after being charged to 100 V, is discharged through a 5 Ω resistor kept at 27°C. (*Ans:* (a) 0, (b) 1.7×10^{-5} J K^{-1})

At very low temperatures, the heat capacity of crystals is given by $C = aT^3$, where a is a constant. Calculate the entropy of a crystal as a function of temperature in this temperature interval. (*Ans:* $S = aT^3/3$)

One mole of a perfect gas undergoes a thermodynamic process so that its temperature changes from T_1 to T_2. Its entropy is related to temperature as $S = a/T$, where a is constant. Calculate the quantity of heat (ΔQ) transferred to and the work (ΔW) performed by the gas.

$$\left(Ans: \Delta Q = a \ln\left(\frac{T_1}{T_2}\right), \Delta W = a \ln\left(\frac{T_1}{T_2}\right) - C_V (T_1 - T_2) \right)$$

An ideal gas is expanded from a volume V_0 to a volume V. During this process, there occurs no heat exchange with the surroundings. Will the final pressure of the gas be the same in case of (a) a fast and (b) a very slow expansion process?

(*Ans:* Pressure will be higher in (a))

One mole of He and one mole of H_2 are mixed at a constant temperature. Calculate the entropy change. (*Ans:* 2.77 cal K^{-1})

Two globes of volume V_1 and V_2 contain n_1 and n_2 moles respectively of two ideal gases at the same temperature. A valve connecting the two is opened and the gases mix without any reaction. Show that

$$\Delta S = n_1 R \ln\left(\frac{V_1 + V_2}{V_1}\right) + n_2 R \ln\left(\frac{V_1 + V_2}{V_2}\right)$$

8 THERMODYNAMIC RELATIONS

arning Objectives

In this chapter, you will learn how to

e Maxwell relations from the first principle;

y Maxwell relations to discuss phase transitions of first order and explain malous expansion of water in the range 0°C –4°C;

obtain $T\,\mathrm{d}\,S$ equations, heat capacity equations and energy equations and y these to obtain pressure/volume dependence of heat capacities; and

derive Gibbs Helmholtz equation for a cell.

TRODUCTION

In the preceding chapters, you have learnt about several properties of a thermodynamic system. Some of these are pressure, temperature, compressibility, heat capacity, internal energy, enthalpy and entropy. Of these, internal energy and entropy cannot be measured in a laboratory. So you may like to know: How can we relate these with measurable properties? In order to answer this question, we require thermodynamic relations wherein these properties can be written as partial derivatives involving extensive or intensive variables. In general, the number of such derivatives is quite large but only four of these can be considered independent.

A thermodynamic relation is an exact rule which may be obtained by general thermodynamic reasoning and applies to almost every system. The usefulness of these relations lies in the fact that they frequently relate quantities which apparently seem unrelated. These enable us to link data obtained in different ways or replace a difficult measurement by an easier one. We can also use these to obtain values of one property from calculations of another, which may be more straightforward/convenient. In short, these relations are very general and extremely useful as they enormously simplify thermodynamic

We begin by deriving Maxwell thermodynamic relations in Sec. 8.2. Though there are many thermodynamic relations and even more possible routes to derive them, the most convenient way is to use the technique of partial differentiation. This is notwithstanding the fact that the calculations may be somewhat lengthy. However, to help you reproduce any relation, we have used simple mnemonic diagrams. We have also discussed applications of Maxwell's relations to first order phase transitions and anomalous expansion of water in the range 0°C –4°C. In Secs. 8.3–8.5, we have obtained heat capacity equations, T- $\mathrm{d}S$ equations and energy equations respectively using Maxwell relations. The results so derived have been used to study pressure/volume dependence of heat capacities. In Sec. 8.6, we

HE MAXWELL RELATIONS

We know that for a pVT system undergoing an infinitesimal reversible process, the combined form of the first and the second laws of thermodynamics is written as

$$dU = T\,dS - p\,dV \tag{8.1}$$

Note that this equation involves five functions of state: p, V, T, S and U. You may think that the state of the system can be determined by any two of these variables. For instance, we may choose pressure and temperature as independent variables. Then, U, V and S will be function of p and T. However, for generality, let us designate the independent variables . Then, we can write

$$U = U(x, y)$$
$$S = S(x, y)$$

$$V = V(x, y)$$

It is now possible to eliminate internal energy from Eq. (8.1) and write general relations linking the four thermodynamic variables. These relations are known as *Maxwell relations*. We may mention that these relations are not purely mathematical in content; they are alternative manifestations of the first and the second laws of thermodynamics.

To derive these relations, we note that since dU, dS and dV are exact differentials, we

$$dU = \left(\frac{\partial U}{\partial x}\right)_y dx + \left(\frac{\partial U}{\partial y}\right)_x dy \tag{8.2a}$$

$$dS = \left(\frac{\partial S}{\partial x}\right)_y dx + \left(\frac{\partial S}{\partial y}\right)_x dy \tag{8.2b}$$

$$dV = \left(\frac{\partial V}{\partial x}\right)_y dx + \left(\frac{\partial V}{\partial y}\right)_x dy \tag{8.2c}$$

On substituting these relations in Eq. (8.1) and comparing the coefficients of dx and dy,

$$\left(\frac{\partial U}{\partial x}\right)_y = T\left(\frac{\partial S}{\partial x}\right)_y - p\left(\frac{\partial V}{\partial x}\right)_y \tag{8.3a}$$

$$\left(\frac{\partial U}{\partial y}\right)_x = T\left(\frac{\partial S}{\partial y}\right)_x - p\left(\frac{\partial V}{\partial y}\right)_x \tag{8.3b}$$

) be a function of two variables x and y, which may be varied independently. Then, the basic theorem of partial differentiation states that simultaneous, arbitrary infinitesimal changes dx and dy in x produce a change dZ in Z which may be written as

$$dZ = \left(\frac{\partial Z}{\partial x}\right)_y dx + \left(\frac{\partial Z}{\partial y}\right)_x dy \tag{a}$$

$x)_y$ is the partial derivative of Z with respect to x at fixed y and $(\partial Z/\partial y)_x$ is the partial derivative with respect to y at fixed x

Now we differentiate Eq. (8.3a) with respect to y keeping x fixed. This gives

$$\left(\frac{\partial}{\partial y}\left(\frac{\partial U}{\partial x}\right)_y\right)_x = \left(\frac{\partial T}{\partial y}\right)_x \left(\frac{\partial S}{\partial x}\right)_y + T\left(\frac{\partial}{\partial y}\left(\frac{\partial S}{\partial x}\right)_y\right)_x$$

$$-\left(\frac{\partial p}{\partial y}\right)_x \left(\frac{\partial V}{\partial x}\right)_y - p\left(\frac{\partial}{\partial y}\left(\frac{\partial V}{\partial x}\right)_y\right)_x \qquad (8.4a)$$

Similarly, differentiation of Eq. (8.3b) with respect to x at fixed y gives

$$\left(\frac{\partial}{\partial x}\left(\frac{\partial U}{\partial y}\right)_x\right)_y = \left(\frac{\partial T}{\partial x}\right)_y \left(\frac{\partial S}{\partial y}\right)_x + T\left(\frac{\partial}{\partial x}\left(\frac{\partial S}{\partial y}\right)_x\right)_y$$

$$-\left(\frac{\partial p}{\partial x}\right)_y \left(\frac{\partial V}{\partial y}\right)_x - p\left(\frac{\partial}{\partial x}\left(\frac{\partial V}{\partial y}\right)_x\right)_y \qquad (8.4b)$$

To simplify these equations, we recall that for U, V and S to be perfect differentials of , we must have

$$\left(\frac{\partial}{\partial y}\left(\frac{\partial U}{\partial x}\right)_y\right)_x = \left(\frac{\partial}{\partial x}\left(\frac{\partial U}{\partial y}\right)_x\right)_y \qquad (8.5a)$$

$$\left(\frac{\partial}{\partial y}\left(\frac{\partial V}{\partial x}\right)_y\right)_x = \left(\frac{\partial}{\partial x}\left(\frac{\partial V}{\partial y}\right)_x\right)_y \qquad (8.5b)$$

$$\left(\frac{\partial}{\partial y}\left(\frac{\partial S}{\partial x}\right)_y\right)_x = \left(\frac{\partial}{\partial x}\left(\frac{\partial S}{\partial y}\right)_x\right)_y \qquad (8.5c)$$

Put in words, the order of differentiation does not alter the value of the double differential of a function of state.

Using Eq. (8.5a), we can equate the RHS of Eqs (8.4a) and (8.4b). To simplify the resultant expression, we use Eqs. (8.5b) and (8.5c). As a result, the double differentials of cancel out. This gives

$$\left(\frac{\partial T}{\partial y}\right)_x \left(\frac{\partial S}{\partial x}\right)_y - \left(\frac{\partial p}{\partial y}\right)_x \left(\frac{\partial V}{\partial x}\right)_y$$

$$= \left(\frac{\partial T}{\partial x}\right)_y \left(\frac{\partial S}{\partial y}\right)_x - \left(\frac{\partial p}{\partial x}\right)_y \left(\frac{\partial V}{\partial y}\right)_x$$

On rearranging terms, we can write

$$\left(\frac{\partial p}{\partial x}\right)_y \left(\frac{\partial V}{\partial y}\right)_x - \left(\frac{\partial p}{\partial y}\right)_x \left(\frac{\partial V}{\partial x}\right)_y = \left(\frac{\partial T}{\partial x}\right)_y \left(\frac{\partial S}{\partial y}\right)_x - \left(\frac{\partial T}{\partial y}\right)_x \left(\frac{\partial S}{\partial x}\right)_y$$

In the determinant form, this equation can be written as

$$\begin{vmatrix} \left(\frac{\partial p}{\partial x}\right)_y & \left(\frac{\partial p}{\partial y}\right)_x \\ \left(\frac{\partial V}{\partial}\right) & \left(\frac{\partial V}{\partial}\right) \end{vmatrix} = \begin{vmatrix} \left(\frac{\partial T}{\partial x}\right)_y & \left(\frac{\partial T}{\partial y}\right)_x \\ \left(\frac{\partial S}{\partial}\right) & \left(\frac{\partial S}{\partial}\right) \end{vmatrix} \qquad (8.6a)$$

In the Jacobian notation, this relation can be put in a rather elegant form:

$$\frac{\partial(p,V)}{\partial(x,y)}=\frac{\partial(T,S)}{\partial(x,y)} \tag{8.6b}$$

and y are different variables and may be chosen out of T, S, p and V. This can be done in six different ways but only four choices give useful thermodynamic relations. We will now derive these.

The First Relation Let us choose the temperature and the volume as independent variables and put $x = T$ and $y = V$ in Eq. (8.6b). Then we can write

$$\frac{\partial(T,S)}{\partial(T,V)}=\frac{\partial(p,V)}{\partial(T,V)}$$

We can show from the properties of the determinants that the above relation simplifies to*

$$\left(\frac{\partial S}{\partial V}\right)_T=\left(\frac{\partial p}{\partial T}\right)_V \tag{8.7}$$

This relation tells us that for any thermodynamic system, the increase in entropy per unit increase of volume at constant temperature is equal to the increase of pressure per unit increase of temperature at constant volume. We can apply Eq. (8.7) to study equilibrium between two states of the same substance, say phase change such as vaporisation of liquid or melting of a solid. (This is known as the first order phase transition.)

On multiplying both sides of Eq. (8.7) by T, we get

$$T\left(\frac{\partial S}{\partial V}\right)_T=T\left(\frac{\partial p}{\partial T}\right)_V$$

$$\left(\frac{\delta Q}{\partial V}\right)_T=T\left(\frac{\partial p}{\partial T}\right)_V \tag{8.8}$$

This result shows that in isothermal expansion, heat absorbed per unit volume is equal to the product of the absolute temperature and the rate of increase of pressure with temperature in an isochoric process.

Now, imagine a cylinder which contains a liquid in equilibrium with some of its vapour. (The pressure is called the *saturated vapour pressure*. It does not depend on the quantities of the liquid and the vapour present. So long as both are present and the system is in equilibrium, the saturated vapour pressure is a function of temperature only.) If we allow the system to expand at constant temperature, the vapour pressure will stay constant but

---

*The following rule will be helpful in this context:

Drop the variable occurring in the same position in the numerator as well as the denominator of the Jacobian. But if a variable occurs in a crossed position, it induces a negative sign. However, it signifies the constraint. Mathematically, we write

$$\frac{\partial(x,y)}{\partial(x,y)}=-\frac{\partial(y,x)}{\partial(x,y)}=1$$

$$\frac{\partial(q,y)}{\partial(x,y)}=\left(\frac{\partial q}{\partial x}\right)_y$$

liquid will evaporate to fill the extra space with vapour. Then we can write $\delta Q = Ldm$, is the specific latent heat of evaporation. The change in volume will be equal to)dm where v_{vap} and v_{liq} are the specific volumes for the vapour and the liquid, respectively. So we may write

$$\left(\frac{\delta Q}{\partial V}\right)_T = \frac{L}{v_{vap} - v_{liq}} = T\left(\frac{\partial p}{\partial T}\right)_V \tag{8.9}$$

Alternatively, if we hold the volume constant and increase the temperature somewhat, liquid will evaporate till the mixture reaches a new equilibrium state and hence a new saturated vapour pressure. Then we may identify

$$\left(\frac{\partial p}{\partial T}\right)_V = \left(\frac{\partial p}{\partial T}\right)_{sat}$$

On combining this result with Eq. (8.9), we can write

$$\left(\frac{\partial p}{\partial T}\right)_{sat} = \frac{L}{T\left[v_{vap} - v_{liq}\right]} \tag{8.10}$$

This is the well known *Clausius–Clapeyron equation.* It is one of the most important formulae in thermodynamics and gives the rate at which vapour pressure must change with temperature for two phases to coexist in equilibrium. (Inversely, we may investigate the effect of pressure on the boiling point.) In the next chapter, you will learn to derive it from the equality of Gibbs thermodynamic potentials, which govern the first order phase transitions. It may be remarked here that the Clausius-Clapeyron equation is obeyed by systems to a high degree of accuracy and over a wide range of experimental conditions. This provides strong evidence in favour of the second law of thermodynamics.

$_{vap} > v_{liq}$ always, $(\partial p/\partial T)_{sat}$ will be positive definite implying that increase in pressure raises boiling point and vice versa. This explains why food is cooked fast, and we save fuel, in a pressure cooker. This also explains why it is difficult to make boiled potatoes and cook food at high altitudes but more easily at the sea level. To give you a feel for the numbers, we may mention that the increase in temperature per atmosphere increase in pressure is about 27.5°C. On the other hand, at the top of the Mount Everest, an altitude of about 8 km above the sea level, water boils at about 80°C.

Although we have derived Eq. (8.10) for the evaporation process, the arguments can be extended to any phase change (solid-liquid, liquid-vapour and solid-vapour transition) involving latent heat. That is, Eq. (8.10) applies to all changes of phase in which there is discontinuity in entropy and volume at the transition temperature. These are known as *first order phase changes.*

The phase diagram* for a typical compressible system is shown in Fig. 8.1(a) and for water in Fig. 8.1(b). The curve on the p-T diagram represents the unique relationship which must hold for two phases to coexist. Note that the three curves intersect at one point. The – T diagram) where all three phases coexist is known as the *Triple Point.* For water, the triple point is defined by $T_{tr} = 0.0075$°C and $p_{tr} = 4.58$ mm of Hg. For CO_2, $= -56.6$°C and $p_{tr} = 5.11$ atm.

*A phase diagram is a plot of all pressures and corresponding temperatures at which a phase change occurs for a given substance.

From Fig. 8.1(a), we note that the slope of the solid-liquid curve is positive. This means that most substances expand on melting, i.e., ΔV, hence dp/dT is positive. So melting point of such materials will be raised by increase in pressure. However, for water, the solid-liquid curve has a negative slope implying that water expands on freezing and its melting point is lowered by increase in pressure (Example 8.1). Note that water is an exception in that it is possible to pass from vapour $\rightarrow$ solid $\rightarrow$ liquid phase at $T = T_{tr}$ by increasing pressure isothermally.

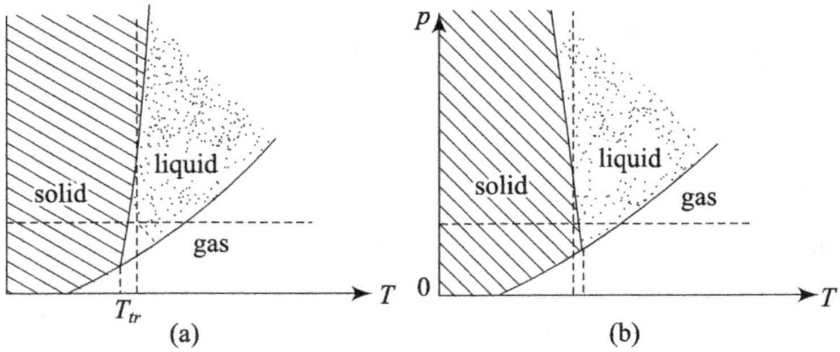

The phase diagrams for (a) a simple system, and (b) water. For water, if we follow the vertical dashed line, we observe that at constant temperature, increase in pressure makes the sequence of phase changes differ from that of most other materials.

We would now like you to go through the following examples.

Example 8.1 Calculate the change in the melting point of ice at 0°C when pressure is increased by 2 atm. How much pressure is required to lower the melting point by 1°C? Given, latent heat of fusion is 79.6 cal g^{-1} and the specific volumes of water and ice are g^{-1} and 1.0908 $cm^3\ g^{-1}$, respectively.

From Eq. (8.10), we have

$$\left(\frac{\partial p}{\partial T}\right)_{sat} = \frac{L}{T\left[v_{vap} - v_{liq}\right]}$$

On substituting $L = 79.6 \times 4.186 \times 10^7$ erg g^{-1}, $v_{water} = 1.0001\ cm^3\ g^{-1}$, $v_{ice} = 1.0908\ cm^3\ g^{-1}$ = 273.16 K, we get

$$\frac{dp}{dT} = \frac{79.6 \times (4.186 \times 10^7\ erg\,g^{-1})}{(273.16\ K) \times (1.0001 - 1.0908)\ cm^3 g^{-1}}$$

$$= -\ 13.45 \times 10^7\ dyne\ cm^{-2}\ K^{-1}$$

= 2 atm = 2.026×10^6 dyne cm^{-2}, the change in melting point is given by

$$dT = -\frac{2.026 \times 10^6\ dyne\ cm^{-2}}{13.45 \times 10^7\ dyne\ cm^{-2}\ K^{-1}} = -0.015\ K = -0.015°C$$

This result shows that the melting point of ice falls with pressure; the drop per atmosphere being 0.0075 °C. Thus, when enough pressure is applied, ice may melt—a significant fact to a skater as well as in the study of glaciers.

The pressure required to lower the melting point of ice by 1 °C is 1K/0.0075 K ≈

Example 8.2 A Carnot engine having 100 g water has steam as the working substance. At the beginning of the stroke, the volume of the working substance is 104 cm^3 at a pressure of 78.8 cm (B.P = 101°C). After a complete isothermal change, the volume of steam increases to 167,404 cm^3 and the pressure is lowered adiabatically to 73.37 cm (B.P = 99°C). If the engine is working between 99°C and 101°C, calculate the latent heat

From Clausius-Clapeyron equation, we know that change in pressure with temperature is given by

$$\frac{dp}{dT} = \frac{L}{T(v_2 - v_1)} \tag{i}$$

We can rearrange terms to write

$$L = T(v_2 - v_1)\frac{dp}{dT} \tag{ii}$$

$T = \left(\frac{99 + 101}{2} + 273\right)\text{K} = 373\ \text{K}$, volume of 100 g water is 104 cm^3, specific

$= \frac{104}{100} = 1.04\ \text{cm}^3\ \text{g}^{-1}$,

$= (78.80 - 73.37)$ cm of Hg $= 5.43$ cm of Hg $= 5.43 \times 13.6 \times 980$ dyne cm^{-2}

After the isothermal change, 100 g water turns into steam and occupies a volume 167,404 cm^3.

$$\text{Specific volume, } v_2 = \frac{167404}{100} = 1674.04\ \text{cm}^3\,\text{g}^{-1}$$

$$\text{Change in specific volume, } v_2 - v_1 = (1674.04 - 1.04)\,\text{cm}^3\,\text{g}^{-1} = 1673\ \text{cm}^3\,\text{g}^{-1}$$

$$dT = 101 - 99 = 2°\text{C} \equiv 2\ \text{K}$$

On inserting these values in Eq. (ii), we get the latent heat of steam

$$L = (373\ \text{K}) \times (1673\ \text{cm}^3\,\text{g}^{-1}) \times \frac{5.43 \times 13.6 \times 980\ \text{dyne cm}^{-2}}{2\ \text{K}}$$

$$= 2258.1 \times 10^7\ \text{erg g}^{-1} = 540.2\ \text{cal g}^{-1}$$

You may now like to answer a practise problem.

Problem 8.1 At 373.16 K, 1 cm^3 of water occupies 1674 cm^3 of space. Calculate the latent heat of vaporisation if $dp/dT = 27.12$ Torr* K^{-1}.

539 cal

---

*1 Torr = 1 mm of mercury

The Second Relation Let us now take temperature and pressure as independent variables and put $x = T$ and $y = p$. Then Eq. (8.6b) takes the form

$$\frac{\partial(T,S)}{\partial(T,p)} = \frac{\partial(p,V)}{\partial(T,p)}$$

Cancelling out common variables in the Jacobian, we get

$$\left(\frac{\partial S}{\partial p}\right)_T = -\left(\frac{\partial V}{\partial T}\right)_p \tag{8.11}$$

This relation implies that decrease in entropy per unit increase of pressure during an isothermal process is equal to the increase of volume per unit rise of temperature for an isobaric process.

To realise the significance of this relation, we relate it to thermal expansion of a substance. To this end, we multiply both sides of Eq. (8.11) by T and rewrite it as

$$\left(\frac{\delta Q}{\partial p}\right)_T = -T\left(\frac{\partial V}{\partial T}\right)_p = -TV\alpha \tag{8.12}$$

$\frac{1}{V}\left(\frac{\partial V}{\partial T}\right)_p$ is the volume expansion coefficient at constant pressure.

From Eq. (8.12), we note that if a substance expands on heating ($\alpha > 0$), $(\delta Q/\partial p)_T$ will be negative. This means that heat will have to be removed when such a substance is compressed isothermally. That is, heat is generated when a substance, which expands on heating, is compressed. What will happen in case of a substance which contracts on heating? An interesting example of this category is water below 4°C. If water at normal pressure and 0°C is heated isobarically, it shows anomalous expansion, i.e., its specific volume decreases until 4°C. Beyond this temperature, specific volume of water increases (Fig. 8.2). Thus $(\partial V/\partial T)_p$, the slope of the curve on V–T diagram, is negative at 0°C and hence $(\delta Q/\partial p)_T$ is positive. Conversely, if water below 4°C is compressed adiabatically, it cools, though internal energy has increased. Experiments have confirmed these observations, furnishing another evidence for the truth of the second law of thermodynamics.

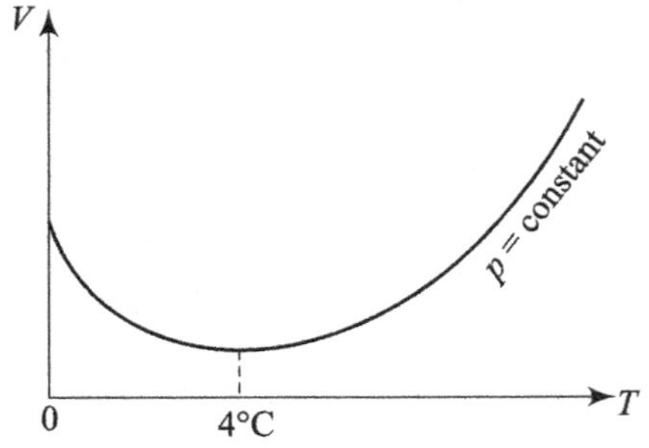

Fig. 8.2 Variation of specific volume of water with temperature. Volume decreases as temperature increases from 0°C to 4°C but increases thereafter.

We know that Indian rubber has a negative expansion coefficient. When stretched adiabatically, it shows heating effect, as expected. This can be felt by our lips.

The Third and the Fourth Relations Two other thermodynamic relations find applications in adiabatic-isochoric and adiabatic-isobaric processes. The first of these is obtained by taking the entropy and the volume as independent variables. So we put $x = S$ in Eq. (8.6b) and simplify. The result is

$$\left(\frac{\partial T}{\partial V}\right)_S = -\left(\frac{\partial p}{\partial S}\right)_V \tag{8.13}$$

In words, increase of temperature per unit increase of volume in an isentropic process is equal to the decrease of pressure per unit increase of entropy in an isochoric process.

Similarly, putting $x = S$ and $y = p$ in Eq. (8.6b) and simplifying the expression, we get the fourth Maxwell relation for a simple compressible system:

$$\left(\frac{\partial T}{\partial p}\right)_S = \left(\frac{\partial V}{\partial S}\right)_p \tag{8.14}$$

That is, increase of temperature per unit increase of pressure in an isentropic process is equal to the increase of volume per unit increase of entropy in an isobaric process.

The following points may be noted in respect of Maxwell relations:*

Cross multiplication of variables in any relation gives a form which has dimensions of energy.

The conjugate variables are constant.

appears with p in one of the partial derivatives, the sign is positive.

As already pointed out, we have written Maxwell relations for a simple compressible system. Evidently, similar Maxwell relations can be written for other systems. For instance, Eq. (8.1) can be generalised to

$$dU = TdS - pdV + FdL + \sigma dA + V_d dZ + BdM + \ldots \tag{8.15}$$

Thus, a magnetic system may not show significant variations in volume so that we

$$dU = TdS + BdM \tag{8.16}$$

and it readily follows that

$$\left(\frac{\partial T}{\partial M}\right)_{B,V} = \left(\frac{\partial B}{\partial S}\right)_{M,V} \tag{8.17}$$

*Maxwell relations may be recalled by translating the sentence: Special Programme on *TV* in a mnemonic diagram (Fig. 8.3). The first two relations are obtained by starting from *S* and going clockwise and anticlockwise, respectively. (A negative sign should be put in the latter case.) For instance, going clockwise , we have

$$\begin{array}{ccc} S & \rightarrow & p \\ \uparrow & & \downarrow \\ V & \leftarrow & T \end{array}$$

$$\left(\frac{\partial S}{\partial V}\right)_T = \left(\frac{\partial p}{\partial T}\right)_V$$

Similarly, for anticlockwise movement, we have

$$\begin{array}{ccc} S & \rightarrow & V \\ \uparrow & & \downarrow \\ p & \leftarrow & T \end{array}$$

$$\left(\frac{\partial S}{\partial p}\right)_T = -\left(\frac{\partial V}{\partial T}\right)_p$$

Other Maxwell relations can also be written in a similar manner.

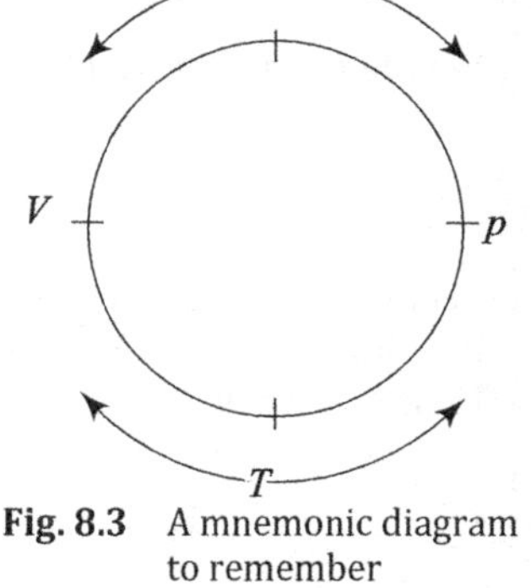

Fig. 8.3 A mnemonic diagram to remember Maxwell's relations.

Similarly for a stretched string and surface film, we can write

$$dU = TdS + FdL$$

$$dU = TdS + \sigma dA$$

By identifying p with $-F$ and V with L for a stretched string and p with $-\sigma$ and V with A for a surface film, you can easily write down the Maxwell relations applicable in these cases. We leave these as an exercise for you to develop problem solving skills. To make sure that you have understood the concepts discussed here, you must write out these. On doing so, you will realise that Maxwell relations are a vast treasure of information about thermodynamic behaviour of a system. In particular, these enable us to express experimentally indeterminate quantities in terms of measurable quantities. For instance, measurement of heat capacity at constant pressure for a solid/liquid is more convenient than the heat capacity at constant volume. But heat capacity equations resolve this problem quite naturally. Similarly, we cannot measure entropy and internal energy but express these in terms of measurable thermodynamic variables through $T\,d\,S$ equations and energy equations, respectively. In other words, Maxwell relations find very diverse applications and we will discuss some of these in this chapter.

ALTERNATIVE DERIVATION OF MAXWELL'S RELATION

Let us consider a thermodynamic function z, which is a function of state and depends on thermodynamic variables x and y. Then we can write

$$z = z(x, y)$$

$$dz = \left(\frac{\partial z}{\partial x}\right)_y dx + \left(\frac{\partial z}{\partial y}\right)_x dy$$

$$= Mdx + Ndy \qquad \text{(i)}$$

$= \left(\frac{\partial z}{\partial x}\right)_y$ and $N = \left(\frac{\partial y}{\partial y}\right)_x$.

is an exact differential, the order of differentiation does not matter. It means

$$\left[\frac{\partial}{\partial y}\left(\frac{\partial z}{\partial x}\right)_y\right]_x = \left[\frac{\partial}{\partial x}\left(\frac{\partial z}{\partial y}\right)_x\right]_y$$

We can rewrite this result in terms of M and N as

$$\left(\frac{\partial M}{\partial y}\right)_x = \left(\frac{\partial N}{\partial x}\right)_y \qquad \text{(ii)}$$

Let us now consider internal energy U and recall that $dU = TdS - pdV$. On comparing it with Eq. (i), we can identify M with T and N with $-p$. Similarly, x is S and y is V. Hence we can write

$$\left(\frac{\partial T}{\partial V}\right)_S = -\left(\frac{\partial p}{\partial S}\right)_V \qquad \text{(iii)}$$

This is identical to Eq. (8.13).

Similarly, if we consider enthalpy and write $dH = TdS + Vdp$, we obtain a relation identical to Eq. (8.14):

$$\left(\frac{\partial T}{\partial p}\right)_S = \left(\frac{\partial V}{\partial S}\right)_p$$

Next, by taking Helmholtz free energy and Gibbs free energy, you can easily write the other two Maxwell's relations.

You may now like to answer some practise problems.

Problem 8.2 A system involves magnetic and surface effects. Starting from Eq. (8.15) show that for an isentropic process

$$\left(\frac{\partial \sigma}{\partial M}\right)_{A,S,V} = \left(\frac{\partial B}{\partial A}\right)_{M,S,V}$$

Problem 8.3 The melting point of lead under normal pressure is 600 K. What will be the change in its value when pressure is increased to 100 atm. The density of lead in solid and liquid phases is 11.01 g cm^{-3} and 10.65 g cm^{-3}, respectively. The latent heat of fusion is 24.5×10^7 erg g^{-1}.

0.75°C

HERMODYNAMIC RELATIONS INVOLVING EAT CAPACITIES

An experimentalist prefers to work with heat capacity of a solid at constant pressure whereas a theoretician finds it more convenient to calculate heat capacity at constant volume. So if we wish to compare theory and experiment, the difference of heat capacities is of particular interest. This relation can easily be developed with the help of Maxwell relations as follows:

V be taken as independent variables. Since heat capacity can also be written in $(\partial S/\partial T)_x$, where x may be p or V, we write

$$S = S(T, V)$$

$$dS = \left(\frac{\partial S}{\partial T}\right)_V dT + \left(\frac{\partial S}{\partial V}\right)_T dV$$

Multiplying both sides of this relation by T while holding p constant, we can write

$$TdS_p = T\left(\frac{\partial S}{\partial T}\right)_V dT_p + T\left(\frac{\partial S}{\partial V}\right)_T dV_p$$

$$T\left(\frac{\partial S}{\partial T}\right)_p = T\left(\frac{\partial S}{\partial T}\right)_V + T\left(\frac{\partial S}{\partial V}\right)_T \left(\frac{\partial V}{\partial T}\right)_p$$

In terms of heat capacities, we can write

$$C_p = C_V + T\left(\frac{\partial S}{\partial V}\right)_T \left(\frac{\partial V}{\partial T}\right)_p$$

Using the first Maxwell relation (Eq. (8.7)) and rearranging terms, we can write

$$C_p - C_V = T\left(\frac{\partial p}{\partial T}\right)_V \left(\frac{\partial V}{\partial T}\right)_p = TV\alpha\left(\frac{\partial p}{\partial T}\right)_V \quad (8.18)$$

$\frac{1}{V}\left(\frac{\partial V}{\partial T}\right)_p$ is volume expansion coefficient at constant pressure.

To discover the physical significance of Eq. (8.18), we modify it by taking p as a function

$$p = p\,(T, V)$$

$$\mathrm{d}p = \left(\frac{\partial p}{\partial T}\right)_V \mathrm{d}T + \left(\frac{\partial p}{\partial V}\right)_T \mathrm{d}V$$

For an isobaric process (p = constant), we can rewrite it as

$$\left(\frac{\partial p}{\partial T}\right)_V \mathrm{d}T_p + \left(\frac{\partial p}{\partial V}\right)_T \mathrm{d}V_p = 0$$

$$\left(\frac{\partial p}{\partial T}\right)_V = -\left(\frac{\partial p}{\partial V}\right)_T \left(\frac{\partial V}{\partial T}\right)_p \quad (8.19)$$

Using this result in Eq. (8.18), we get

$$C_p - C_V = -T\left(\frac{\partial p}{\partial V}\right)_T \left(\frac{\partial V}{\partial T}\right)_p^2 \quad (8.20)$$

This is known as *heat capacity equation* and constitutes one of the most informative relations in thermodynamics. We can draw the following conclusions:

Since $(\partial p/\partial V)_T$ is negative for all known substances and $(\partial V/\partial T)_p^2$ is always positive, the difference $C_p - C_V$ will be always positive. That is, C_p can never be less than C_V. For liquids and solids, $(\partial V/\partial T)_p$ is relatively small and, therefore, the difference C_V is small. Further, when $(\partial V/\partial T)_p = 0$, as at 4°C for water (Fig. 8.2), C_p will be equal to C_V.

$\rightarrow$ 0K, $C_p \rightarrow C_V$, i.e., as we approach absolute zero, the constant pressure and constant volume heat capacities coalesce. This is confirmed by experiments. We have discussed it in detail in Chapter 10.

To perfect your analytical skills, you may now like to answer a simple problem.

Problem 8.4 Starting from Eq. (8.20) show that for a perfect gas

$$C_p - C_V = R$$

and for a van der Waals' gas

$$C_p - C_V = R\left(1 + \frac{2a}{RTV^2}\right).$$

Now go through the following solved examples.

Example 8.3 For a pure substance, show that $C_p - C_V$ can be expressed in terms of the volume expansivity and isothermal compressibility β_T as

$$C_p - C_V = \frac{T\alpha^2 V}{\beta_T}$$

From Eq. (8.20), we recall that

$$C_p - C_V = -T\left(\frac{\partial p}{\partial V}\right)_T \left(\frac{\partial V}{\partial T}\right)_p^2 \tag{i}$$

The volume expansivity of a substance is defined as

$$\alpha = \frac{1}{V}\left(\frac{\partial V}{\partial T}\right)_p$$

$$\left(\frac{\partial V}{\partial T}\right)_p^2 = \alpha^2 V^2 \tag{ii}$$

Similarly, by definition, isothermal compressibility is

$$\beta_T = -\frac{1}{V}\left(\frac{\partial V}{\partial p}\right)_T$$

$$\left(\frac{\partial p}{\partial V}\right)_T = -\frac{1}{V\beta_T} \tag{iii}$$

Using relations (ii) and (iii) in (i), we get the required result.
Note that both α and β_T can be measured experimentally. Therefore, the value of C_V can easily be obtained once we know C_p and vice versa.

Example 8.4 Starting from the relation $C_p - C_V = \frac{\alpha^2 VT}{\beta_T}$, show that the difference of adiabatic and isothermal compressibilities is given by

$$\beta_S - \beta_T = -\frac{\alpha^2 VT}{C_p}$$

We know that

$$C_p - C_V = \frac{\alpha^2 VT}{\beta_T} \tag{i}$$

Since the ratio of the adiabatic to isothermal elasticities is equal to the ratio of two heat capacities (Eq. (5.31)), we can write

$$\frac{C_p}{C_V} = \frac{E_S}{E_T} = \frac{\beta_T}{\beta_S} = \gamma \qquad \text{(ii)}$$

On dividing both sides of Eq. (i) by C_V and using (ii), we can write the expression involving difference of heat capacities as

$$\gamma - 1 = \frac{\alpha^2 VT}{\beta_T C_V}$$

$$\beta_T - \beta_S = \frac{\beta_S}{\beta_T} \frac{\alpha^2 VT}{C_V}$$

$$\frac{\beta_S}{\beta_T} = \frac{1}{\gamma} = \frac{C_V}{C_p}$$

we can write

$$\beta_S - \beta_T = -\frac{\alpha^2 VT}{C_p}$$

Example 8.5 Using Maxwell relations, show that the ratio of isentropic to isochoric pressure coefficients of expansion is $\gamma/(\gamma - 1)$.

The adiabatic pressure coefficient, also known as isentropic pressure coefficient, is defined as

$$\mu_S = \frac{1}{p}\left(\frac{\partial p}{\partial T}\right)_S$$

whereas the isochoric pressure coefficient is, by definition

$$\mu_V = \frac{1}{p}\left(\frac{\partial p}{\partial T}\right)_V$$

$$\frac{\mu_S}{\mu_V} = \frac{(\partial p/\partial T)_S}{(\partial p/\partial T)_V}$$

$$= \frac{1}{(\partial T/\partial p)_S\, (\partial p/\partial T)_V}$$

Using the fourth Maxwell relation (Eq. (8.14)), this expression can be rewritten as

$$\frac{\mu_S}{\mu_V} = \frac{1}{(\partial V/\partial S)_p\, (\partial p/\partial T)_V}$$

$$= \frac{1}{(\partial V/\partial T)_p\, (\partial T/\partial S)_p\, (\partial p/\partial T)_V}$$

By rearranging terms, we can write

$$\frac{\mu_S}{\mu_V} = \frac{(\partial S/\partial T)_p}{(\partial V/\partial T)_p\,(\partial p/\partial T)_V}$$

Multiplying the numerator as well as the denominator by T, noting that

$$C_p = T\left(\frac{\partial S}{\partial T}\right)_p$$

and using Eq. (8.18) $[C_p - C_V = T\,(\partial V/\partial T)_p\,(\partial p/\partial T)_V]$, we get

$$\frac{\mu_S}{\mu_V} = \frac{C_p}{C_p - C_V} = \frac{\gamma}{\gamma - 1}$$

Example 8.6 At temperature 6 K and pressure 19.7 atm, the specific volume of ^{4}He 10^{-2} m^3 kmol^{-1}. If isothermal compressibility is 9.42×10^{-8} m^2 N^{-1} and volume expansivity is 5.35×10^{-2} K^{-1}, calculate the difference in heat capacities of ^{4}He.

From Example 8.2, we recall that

$$C_p - C_V = \frac{TV\alpha^2}{\beta_T}$$

Using the data for ^{4}He given here, we get

$$C_p - C_V = \frac{(6\text{ K}) \times (2.64 \times 10^{-2}\text{ m}^3\text{ kmol}^{-1}) \times (5.35 \times 10^{-2})^2\text{ K}^{-2}}{9.42 \times 10^{-8}\text{ m}^2\text{ N}^{-1}}$$

$$= 4810\text{ J kmol}^{-1}\text{ K}^{-1}.$$

Example 8.7 Calculate the molar thermal capacity C_V, the ratio C_p/C_V and the adiabatic compressibility for mercury at 0°C. Take C_p = 27.96 J mol^{-1} K^{-1}, 10^{-4} K^{-1}, $\beta_T = 3.88 \times 10^{-11}$ m^{2}N^{-1}, and $V = 14.72 \times 10^{-6}$ m^3 mol^{-1}.

We know that the difference of heat capacities at constant pressure and at constant volume is given by

$$C_p - C_V = \frac{TV\alpha^2}{\beta_T}$$

On substituting the given values, we get

$$C_p - C_V = \frac{(273\text{ K}) \times (14.72 \times 10^{-6}\text{ m}^3\text{ mol}^{-1}) \times (3.28 \times 10^{-8}\text{ K}^{-2})}{3.88 \times 10^{-11}\text{ m}^2\text{ N}^{-1}}$$

$$= \frac{1.32 \times 10^{-10}}{3.88 \times 10^{-11}}(\text{Nm mol}^{-1}\text{ K}^{-1})$$

$$= 0.340 \times 10\text{ J mol}^{-1}\text{ K}^{-1}$$

$$= 3.40\text{ J mol}^{-1}\text{ K}^{-1}$$

$$C_V = (27.96 - 3.40)\text{ J mol}^{-1}\text{ K}^{-1}$$
$$= 24.56\text{ J mol}^{-1}\text{ K}^{-1}$$

The ratio of heat capacities

$$\frac{C_p}{C_V} = \frac{27.96\text{ Jmol}^{-1}\ K^{-1}}{24.56\text{ Jmol}^{-1}\ K^{-1}} = 1.14$$

Finally, we know that

$$\frac{C_p}{C_V} = \frac{E_S}{E_T} = \frac{\beta_T}{\beta_S}$$

is adiabatic compressibility. Therefore, we can write

$$\beta_S = \frac{\beta_T}{(C_p/C_V)} = \frac{3.88 \times 10^{-11}\text{ m}^2\text{ N}^{-1}}{1.14}$$
$$= 3.40 \times 10^{-1}\text{ m}^2\text{ N}^{-1}.$$

You may now like to answer a few practise problems.

Problem 8.5 Using Maxwell thermodynamic relations, show that the ratio of adiabatic to isobaric volume expansivity is $1/(1 - \gamma)$.

Problem 8.6 The pressure on a block of copper weighing 1 kg is increased reversibly to 100 atm at 15°C. Calculate $C_p - C_V$ for this process. Given $\alpha = 5 \times 10^{-5}\text{ K}^{-1}$, $10^{-12}\text{ m}^2\text{N}^{-1}$ and $V = 0.000114\text{ m}^3$.

9.55 J kg^{-1} K^{-1}

Usually, the values of C_p and C_V for pure substances are quoted as a function of p and T or respectively. So from a theoretician's viewpoint, one is quite frequently interested in studying pressure or volume variation of heat capacities at a given temperature. To accomplish this, we use the first Maxwell relation:

$$\left(\frac{\partial S}{\partial V}\right)_T = \left(\frac{\partial p}{\partial T}\right)_V$$

On differentiating both sides with respect to T at fixed V, we get

$$\left(\frac{\partial}{\partial T}\left(\frac{\partial S}{\partial V}\right)_T\right)_V = \left(\frac{\partial^2 p}{\partial T^2}\right)_V \tag{8.21}$$

We know that for a function of state, mixed second order differentials are independent of the order of differentiation. So we can rewrite Eq. (8.21) as

$$\left(\frac{\partial}{\partial T}\left(\frac{\partial S}{\partial V}\right)_T\right)_V = \frac{\partial}{\partial V}\left(\left(\frac{\partial S}{\partial T}\right)_V\right)_T = \left(\frac{\partial^2 p}{\partial T^2}\right)_V$$

$S/\partial T)_V = C_V$, we note that this relation takes a compact form:

$$\left(\frac{\partial C_V}{\partial V}\right)_T = T\left(\frac{\partial^2 p}{\partial T^2}\right)_V = T\frac{\partial}{\partial T}\left(\frac{\alpha}{\beta_T}\right) \tag{8.22a}$$

is isothermal compressibility and α is volume expansivity.

Alternatively, even if one knows isochoric pressure coefficient, μ_V, it is possible to calculate variation of C_V with volume since

$$\left(\frac{\partial C_V}{\partial V}\right)_T = T\,p\left(\frac{\partial \mu_V}{\partial T}\right)_V \tag{8.22b}$$

You can readily verify that for a perfect gas as well as a van der Waals' gas, = 0. This means that for an isothermal process, C_V is independent of volume, is a function of T only. In general, C_V increases initially with V to a maximum value and begins to decrease thereafter.

To study the pressure variation of constant pressure heat capacity for fixed T, we recall second Maxwell's relation:

$$\left(\frac{\partial S}{\partial p}\right)_T = -\left(\frac{\partial V}{\partial T}\right)_p$$

As before, we differentiate both sides of this relation with respect to T keeping p fixed and use the property of mixed differentials for entropy. Then, using the fact that $/\partial T)_p$, it readily follows that

$$\left(\frac{\partial C_p}{\partial p}\right)_T = -T\left(\frac{\partial^2 V}{\partial T^2}\right)_p = -TV\,\alpha^2 \tag{8.23}$$

This result shows that C_p always decreases as pressure increases since α^2, V and T are always positive for a pure substance. However, the change is usually small. For an ideal or a van der Waals' gas, we find from the equation of state that the right–hand side of Eq. (8.23) is zero, implying that for a given T, C_p is independent of p.

You should now go through the following example.

Example 8.8 Calculate the variation of C_p with pressure at constant temperature for a substance whose equation of state is given by the relation

$$V = \frac{RT}{p} - \frac{C}{T^3}$$

From Eq. (8.23), we have

$$\left(\frac{\partial C_p}{\partial p}\right)_T = -T\left(\frac{\partial^2 V}{\partial T^2}\right)_p$$

Since the equation of state of the given substance is given by the relation

$$V = \frac{RT}{p} - \frac{C}{T^3}$$

$$\left(\frac{\partial V}{\partial T}\right)_p = \frac{R}{p} + \frac{3C}{T^4}$$

$$\left(\frac{\partial^2 V}{\partial T^2}\right)_p = -\frac{12C}{T^5}$$

$$\left(\frac{\partial C_p}{\partial p}\right)_T = -T\left(-\frac{12C}{T^5}\right) = \frac{12C}{T^4}$$

This result shows that pressure variation of heat capacity is inversely proportional to the fourth power of temperature.

You should now answer the following practise problem.

Problem 8.7 The equation of state of a gas is $\frac{pV}{RT} = 1 + pB(T)$. Show that

$$C_p - C_{p0} = -RT\,p\left(\frac{\partial^2 (BT)}{\partial T^2}\right)_p$$

is the value of C_p at some reference temperature.

HE *TdS*- EQUATIONS

equations enable us to relate the entropy of a substance with directly measurable quantities, provided its equation of state and heat capacities are known. Depending on the choice of independent variables, we obtain three *TdS* equations. Since $\delta Q_{rev} = TdS$, one can also use these equations to calculate the heat flowing into a pure substance in a reversible process. To derive the first *TdS* equation, let us take *T* and *V* as independent variables and write

$$S = S(T, V)$$

$$dS = \frac{C_V}{T}dT + \left(\frac{\partial p}{\partial T}\right)_V dV$$

Multiplying throughout by *T* and noting that $C_V = T\left(\frac{\partial S}{\partial T}\right)_V$, we get

$$T\,dS = C_V\,dT + T\left(\frac{\partial S}{\partial V}\right)_T dV$$

Using first Maxwell relation (Eq. (8.7)), we can rewrite it as

$$T\,dS = C_V\,dT + T\left(\frac{\partial p}{\partial T}\right)_V dV \tag{8.24a}$$

This is known as the *first TdS-equation.*

On combining Eqs. (8.19) and (8.24a) we can rewrite the latter in a more useful form:

$$T\,dS = C_V\,dT - T\left(\frac{\partial p}{\partial V}\right)_T \left(\frac{\partial V}{\partial T}\right)_p dV$$

$$= C_V\,dT + T\,\alpha\,E_T\,dV \tag{8.24b}$$

is volume expansivity and E_T is isothermal elasticity.

If we take T and p as independent variables and follow the same steps as outlined above, use of Eq. (8.11) leads to the following equation:

$$T\,dS = C_p\,dT - T\left(\frac{\partial V}{\partial T}\right)_p dp \tag{8.25a}$$

second TdS-equation.

In terms of volume expansivity, we can rewrite it as

$$T\,dS = C_p\,dT - TV\alpha\,dp \tag{8.25b}$$

Similarly, by taking p and V as independent variables and writing $S = S(p, V)$, we get

$$T\,dS = T\left(\frac{\partial S}{\partial p}\right)_V dp + T\left(\frac{\partial S}{\partial V}\right)_p dV$$

To put it in a more meaningful form, we split the bracketed terms on the RHS of this equation and rewrite it as

$$T\,dS = T\left(\frac{\partial S}{\partial T}\right)_V \left(\frac{\partial T}{\partial p}\right)_V dp + T\left(\frac{\partial S}{\partial T}\right)_p \left(\frac{\partial T}{\partial V}\right)_p dV$$

$$T\,dS = C_V\left(\frac{\partial T}{\partial p}\right)_V dp + C_p\left(\frac{\partial T}{\partial V}\right)_p dV \tag{8.26}$$

This is known as the *third TdS-equation.*

As may be noted, TdS equations enable us to relate the entropy of a substance with directly measurable quantities, provided its equation of state and heat capacities are known. We will illustrate it for the case of a real gas. If we take T and V as independent variables, then from the first TdS equation, we can write

$$dS = \frac{C_V}{T}dT + \left(\frac{\partial p}{\partial T}\right)_V dV$$

If we assume that a real gas obeys van der Waals' equation of state, $\left(p + \frac{a}{V^2}\right)(V - b) = RT$, we can write

$$\left(\frac{\partial p}{\partial T}\right)_V = \frac{R}{V - b}$$

Using this result in the expression for dS and integrating the resultant expression, we get

$$S = \int_{T_0}^{T} C_V \frac{dT}{T} + R \ln \left| \frac{V-b}{V_0 - b} \right| + S_0 \tag{8.27}$$

is entropy in reference state defined by (V_0, T_0).
is independent of temperature in the region of interest, we can write

$$S = C_V \ln\left(\frac{T}{T_0}\right) + R \ln \left| \frac{V-b}{V_0 - b} \right| + S_0$$

Now you should go through the following examples.

Example 8.9 Consider a surface film defined by $TdS = dU - \sigma dA$. Write down equations.

By comparing the equation $TdS = dV - \sigma dA$ with that for a hydrostatic system, we note that p and V have been replaced by $-\sigma$ and A respectively. Using these in Eqs. (8.24a) and (8.24b), we can write

$$TdS = C_A dT - T\left(\frac{\partial \sigma}{\partial T}\right)_A dA$$

$$TdS = C_\sigma\, dT + T\left(\frac{\partial A}{\partial T}\right)_\sigma dp$$

and C_σ are heat capacities of the film at constant area and constant surface tension, respectively.

If the film is stretched isothermally and the area increases by dA, the heat transferred to the film is

$$\delta Q_T = TdS = -T\left(\frac{\partial \sigma}{\partial T}\right)_A dA$$

It is an experimental fact that surface tension is a function of temperature only. Therefore, we can write

$$\delta Q_T = -T\frac{d\sigma}{dT} dA$$

is negative for liquids, δQ_T will be positive if $dA > 0$.

Example 8.10 A wire under constant tension F is acted upon by an external stretching force. Express change in entropy in terms of changes in internal energy and change in length. Also calculate the change in tension for a small change in temperature and show that it is independent of the length of the wire.

When a wire under tension is acted upon by an external force F so that its length increases by dL, the work done in the infinitesimal process is $\delta W = -FdL$. Hence, the change in entropy in terms of changes in internal energy and change in length can be expressed using the combined form of the first and the second laws for this system:

$$TdS = dU - FdL \tag{i}$$

On comparing this expression with the corresponding relation for a hydrostatic system, we note that in this relation $p = -F$ and $V = L$. Hence, the second TdS equation in this case takes the form

$$TdS = C_F dT + T\left(\frac{\partial L}{\partial T}\right)_F dF \qquad \text{(ii)}$$

is the heat capacity of the wire under constant tension.

If the wire is stretched under adiabatic conditions, the change in temperature is obtained from Eq. (ii):

$$dT = -\frac{T}{C_F}\left(\frac{\partial L}{\partial T}\right)_F dF \qquad \text{(iii)}$$

To put it in a more physically acceptable form, we express it in terms of the properties of the wire: the coefficient of linear expansion α_L and the adiabatic elasticity E_S defined as

$$\alpha_L = \frac{1}{L}\left(\frac{\partial L}{\partial T}\right)_F$$

$$E_S = \frac{dF/A}{dL/L} = \frac{L\,dF}{A\,dL}$$

is area of cross section of the wire. Using these in Eq. (iii), we can write

$$dT = -\frac{T}{C_F}\alpha_L L \times \frac{A\,dL\,E_S}{L} = -\frac{AE_S T\alpha_L}{C_F}dL \qquad \text{(iv)}$$

This result shows that a wire cools on stretching since the coefficient of linear expansion is positive for most substances. However, India rubber is an exception in that α_L is negative and under moderate tension, it heats up.

Under isothermal conditions, heat will flow into the wire. The first term on the RHS of Eq. (ii) will drop out and it reduces to

$$TdS = \delta Q = T\left(\frac{\partial L}{\partial T}\right)_F dF = TAE_T\alpha_L\,dL \qquad \text{(v)}$$

is isothermal elasticity.

For a wire stretched between fixed supports, as in a sonometer or wired musical instruments, the length is constant. If it is heated, the change in tension is given by

$$\left(\frac{\partial F}{\partial T}\right)_L = -\frac{1}{\left(\frac{\partial L}{\partial F}\right)_T\left(\frac{\partial T}{\partial L}\right)_F} = -\left(\frac{\partial F}{\partial L}\right)_T\left(\frac{\partial L}{\partial T}\right)_F = -AE_T\alpha_L$$

This result shows that for a change in temperature ΔT_L, the change in tension is given by

$$\Delta F_L = -AE_T\alpha_L\Delta T_L \qquad \text{(vi)}$$

Note that the change in tension is independent of the length of the wire.

As another application of Maxwell relations, we now obtain the energy equations.

HE ENERGY EQUATIONS

In Sec. 5.3, we promised to show you that internal energy of a real gas varies with volume (as also pressure) even at constant temperature. We will first derive a general relation for change of internal energy for a pure substance. We start by writing

$$U = U(T, V)$$

So an infinitesimal change in internal energy induced by infinitesimal changes in temperature and volume can be expressed as

$$dU = \left(\frac{\partial U}{\partial T}\right)_V dT + \left(\frac{\partial U}{\partial V}\right)_T dV \quad (8.28)$$

system undergoing infinitesimal reversible change between two equilibrium states, the change in internal energy is given by

$$dU = T\,dS - p\,dV$$

From this relation, it readily follows that

$$\left(\frac{\partial U}{\partial V}\right)_T = T\left(\frac{\partial S}{\partial V}\right)_T - p$$

Using the first Maxwell relation, we can rewrite this expression for change in internal energy with volume as

$$\left(\frac{\partial U}{\partial V}\right)_T = T\left(\frac{\partial p}{\partial T}\right)_V - p \quad (8.29)$$

This result is known as the *first energy equation.*

On combining Eqs. (8.28) and (8.29), we obtain

$$dU = C_V\,dT + \left(T\left(\frac{\partial p}{\partial T}\right)_V - p\right)dV \quad (8.30)$$

This equation can be readily integrated to calculate the change in internal energy associated with a given change of state for a pure substance, provided C_V and the equation of state are known:

$$\Delta U = \int C_V\,dT + \int\left(T\left(\frac{\partial p}{\partial T}\right)_V - p\right)dV \quad (8.31)$$

To illustrate, we recall that for a perfect gas

$$p = \frac{RT}{V}$$

$$\left(\frac{\partial p}{\partial T}\right)_V = \frac{R}{V}$$

and the first energy equation implies that if temperature remains constant ($dT = 0$)

$$T\left(\frac{\partial p}{\partial T}\right)_V - p = \left(\frac{\partial U}{\partial V}\right)_T = \frac{RT}{V} - p = 0$$

That is, temperature remaining constant, internal energy of an ideal gas is independent of volume. Using this result in Eq. (8.31), we obtain

$$\Delta U = \int_{T_1}^{T_2} C_V \, \mathrm{d}T = C_V \, (T_2 - T_1) \tag{8.32}$$

where we have assumed that constant volume heat capacity does not change over the temperature range of interest.

For a real gas obeying van der Waals' equation of state

$$p = \frac{RT}{V-b} - \frac{a}{V^2}$$

$$\left(\frac{\partial p}{\partial T}\right)_V = \frac{R}{V-b}$$

Using this result in Eq. (8.29), we obtain

$$T\left(\frac{\partial p}{\partial T}\right)_V - p = \left(\frac{\partial U}{\partial V}\right)_T = \frac{RT}{V-b} - p$$

$$\left(\frac{\partial U}{\partial V}\right)_T = \frac{a}{V^2} \tag{8.33}$$

That is, $(\partial U/\partial V)_T$, the co-pressure, is finite and its genesis is in intermolecular interactions.

Using this result in Eq. (8.31), we obtain

$$U_2 - U_1 = \int_{T_1}^{T_2} C_V \, \mathrm{d}T + \int_{V_1}^{V_2} \frac{a}{V^2} \, \mathrm{d}V$$

$$= \int_{T_1}^{T_2} C_V \, \mathrm{d}T - a\left(\frac{1}{V_2} - \frac{1}{V_1}\right) \tag{8.34}$$

This result shows that internal energy of a real gas depends on volume. As volume increases, $(U_2 - U_1)$ also increases even at fixed temperature. This makes sense because as intermolecular separation increases, the potential energy of interaction decreases.

To investigate the dependence of internal energy on pressure, we write

$$U = U\,(T, p)$$

$$\mathrm{d}\,U = \left(\frac{\partial U}{\partial T}\right)_p \mathrm{d}\,T + \left(\frac{\partial U}{\partial p}\right)_T \mathrm{d}\,p \tag{8.35}$$

As before, from the relation

$$\mathrm{d}U = T\mathrm{d}S - p\mathrm{d}V$$

it follows that

$$\left(\frac{\partial U}{\partial p}\right)_T = T\left(\frac{\partial S}{\partial p}\right)_T - p\left(\frac{\partial V}{\partial p}\right)_T$$

Using the second Maxwell relation on the right-hand side, we have

$$\left(\frac{\partial U}{\partial p}\right)_T = -T\left(\frac{\partial V}{\partial T}\right)_p - p\left(\frac{\partial V}{\partial p}\right)_T \tag{8.36}$$

This result is known as the *second energy equation.*
Combining Eqs. (8.35) and (8.36) leads us to

$$dU = \left(\frac{\partial U}{\partial T}\right)_p dT - \left(T\left(\frac{\partial V}{\partial T}\right)_p + p\left(\frac{\partial V}{\partial p}\right)_T\right)dp \tag{8.37}$$

It may be remarked here that we have derived the energy equations using Maxwell relations. One can also accomplish this task using *TdS* equations and the reverse is also true (Problem 8.8) as well as from the first principle.

Problem 8.8 Derive *TdS* equations for a pure material using the energy equations.

You should now go through the following example:

Example 8.11 A gas obeys Dieterici's equation: $p = \frac{RT}{V-b}\exp\left[-\frac{a}{RTV}\right]$. Show that for an isothermal-isobaric expansion, the change in internal energy is equal to , if the final volume is 'e' times the initial volume.

For a *pVT* system undergoing infinitesimal reversible change between two equilibrium states, the first energy equation (Eq. (8.29) tells us that

$$\left(\frac{\partial U}{\partial V}\right)_T = T\left(\frac{\partial p}{\partial T}\right)_V - p \tag{i}$$

For a gas obeying Dieterici's equation, we have

$$p = \frac{RT}{V-b}\exp\left[-\frac{a}{RTV}\right]$$

$$\left(\frac{\partial p}{\partial T}\right)_V = \frac{R}{V-b}\exp\left(-\frac{a}{RTV}\right) + \frac{RT}{V-b}\frac{\partial}{\partial T}\left[\exp\left(-\frac{a}{RTV}\right)\right]_V$$

$$= \frac{R}{V-b}\exp\left(-\frac{a}{RTV}\right) + \frac{RT}{V-b}\left[\exp\left(-\frac{a}{RTV}\right)\right]\left(\frac{a}{RT^2V}\right)$$

$$T\left(\frac{\partial p}{\partial T}\right)_V = \frac{RT}{V-b}\exp\left(-\frac{a}{RTV}\right) + \frac{RT^2}{V-b}\left[\exp\left(-\frac{a}{RTV}\right)\right]\left(\frac{a}{RT^2V}\right)$$

The first term on the right-hand side is simply equal to *p*. Hence, we can write

$$T\left(\frac{\partial p}{\partial T}\right)_V - p = \frac{1}{V-b}\left[\exp\left(-\frac{a}{RTV}\right)\right]\left(\frac{a}{V}\right) = \frac{ap}{RTV}$$

$$\left(\frac{\partial U}{\partial V}\right)_T = \frac{ap}{RTV} \tag{ii}$$

For an isothermal-isobaric process, we can write

$$dU = k\frac{dV}{V} \tag{iii}$$

ap/RT is a constant. We can readily integrate Eq. (iii) to obtain

$$\Delta U = k \ln(V_2/V_1) = k \ln e = k = ap/RT$$

So far we have confined our discussion primarily to simple compressible substances. For pure paramagnetic materials, one can similarly write $T\mathrm{d}S$ equations, energy equations and heat capacity equations starting from the combined form of the first and the second laws of thermodynamics:

$$T\mathrm{d}S = \mathrm{d}U - B\mathrm{d}M$$

M and T as independent variables and using the relation $(\partial S/\partial M)_T = -(\partial B/\partial T)_M$, *first energy equation for a paramagnetic substance*:

$$\left(\frac{\partial U}{\partial M}\right)_T - B = -T\left(\frac{\partial B}{\partial T}\right)_M \tag{8.38}$$

Similarly, by writing $U = U(T, M)$ and using the above result we arrive at the first $T\mathrm{d}S$

$$T\mathrm{d}S = C_M\,\mathrm{d}T - T\left(\frac{\partial B}{\partial T}\right)_M \mathrm{d}M \tag{8.39}$$

By choosing B and T as independent variables and following the similar steps, we obtain

$$\left(\frac{\partial U}{\partial B}\right)_T - B\left(\frac{\partial M}{\partial B}\right)_T = T\left(\frac{\partial M}{\partial T}\right)_B \tag{8.40}$$

which is the *second energy equation*, and

$$T\mathrm{d}S = C_B\,\mathrm{d}T + T\left(\frac{\partial M}{\partial T}\right)_B \mathrm{d}B \tag{8.41}$$

second TdS equation. Equation (8.38) to Eq. (8.41) reveal the dependence of magnetic properties on temperature. This effect is known as *magneto-caloric effect*. From observed results, we know that for paramagnetic materials, $(\partial M/\partial T)_B$ is always negative. As a result, Eq. (8.41) implies that

a reversible isothermal increase in magnetic field causes rejection of heat, and vice versa; and

a reversible adiabatic decrease in magnetic field causes a drop in temperature. This process is widely used for production of temperatures below 1 K. We will discuss it in some detail in Chapter 10.

Maxwell thermodynamical relations can also be used to calculate *heat of reaction*, as in an electric cell. We now discuss this in detail.

EAT OF REACTION: GIBBS–HELMHOLTZ EQUATION

Chemists are frequently interested in knowing the heat of reaction. To measure it, one has to resort to heat flow methods, which are not always convenient. This problem is overcome in a somewhat roundabout way, wherein the chemical reaction of interest is made to occur reversibly in an electric cell.* Suppose we wish to estimate the heat of reaction when zinc is placed in a solution of copper salt. Normally, chemical reactions occur at constant pressure and are irreversible. So, the heat of reaction can be measured in terms of enthalpy

pV). In fact, the heat of reaction is numerically equal to the change in enthalpy.

Let us imagine that the reaction of interest is taking place nearly reversibly in a Daniel cell, which consists of zinc and copper electrodes dipped in acidified copper sulphate solution. If we ignore changes in volume, the parametric equation of state for the cell is

$$E = E\,(T, Z)$$

That is, emf is expressed as a function of temperature and charge. Then the first law of thermodynamics takes the form

$$\mathrm{d}U = T\mathrm{d}S + E\mathrm{d}Z$$

Expressing dS in terms of dT and dZ and combining with the first law, we get

$$\mathrm{d}U = T\left(\frac{\partial S}{\partial T}\right)_Z \mathrm{d}T + \left(E + T\left(\frac{\partial S}{\partial Z}\right)_T\right)\mathrm{d}Z \tag{8.42}$$

$T(\partial S/\partial Z)_T$ gives the rate at which heat is absorbed while charge increases reversibly and isothermally.

Using the thermodynamical relation

$$\left(\frac{\partial S}{\partial Z}\right)_{T,p} = -\left(\frac{\partial E}{\partial T}\right)_{Z,p} \tag{8.43}$$

in Eq. (8.42) and defining heat capacity at constant charge by the relation

$$C_Z = T\left(\frac{\partial S}{\partial T}\right)_Z$$

$$\mathrm{d}U = C_Z\,\mathrm{d}T + \left(E - T\left(\frac{\partial E}{\partial T}\right)_Z\right)\mathrm{d}Z$$

If the electrolyte is saturated, E will be a function of T only and we can rewrite the above equation is

$$\mathrm{d}U = C_Z\,\mathrm{d}T + \left(E - T\left(\frac{\partial E}{\partial T}\right)_{\mathrm{sat}}\right)\mathrm{d}Z \tag{8.44}$$

In an electric cell, the change in enthalpy arises partly from exchange of heat and partly from electric work. For an isobaric change at fixed volume, we find that enthalpy change is equal to the change of internal energy**:

$$\Delta H = \Delta U \tag{8.45}$$

---

*For an electric cell to be reversible, the Joule heat must be negligible and no gas should evolve in the reaction. How will you classify dry cells and lead accumulators?

$U + pV$, the change in H can be written as

$$\mathrm{d}H = \mathrm{d}U + p\mathrm{d}V + V\mathrm{d}p$$

fixed, it reproduces Eq. (8.45)

Using this result in Eq. (8.44), we obtain for isothermal charging:

$$\Delta H = \left(E - T\left(\frac{\partial E}{\partial T}\right)_{\text{sat}} \right)\Delta Z \tag{8.46}$$

This result constitutes what is known as the *Gibbs-Helmholtz equation* for a cell. It shows that to determine the heat of reaction, one has to just measure emf and its temperature dependence. Obviously, these measurements are much easier compared with calorimetric methods. Further, a positive value of ΔH means that an endothermic reaction is taking place, i.e., heat is absorbed, whereas a negative value of ΔH implies an exothermic reaction (heat is liberated).

The thermodynamic relations can also be used to explain temperature dependence of total as well as differential energy density of *black body radiation*. This is illustrated in Chapter 11 to bring out their intrinsic importance in physics.

DITIONAL EXAMPLES

Example 8.12 When ^{3}He melts, its volume increases. The melting curve of ^{3}He from 0.02 K to 1.2 K is shown below. Show a plot of change in entropy accompanying melting in this temperature range.

Refer to Fig. 8.4 (a)

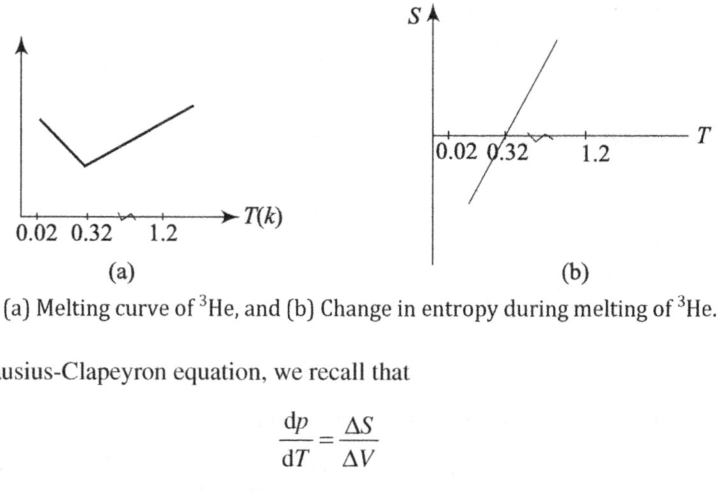

(a) Melting curve of ^{3}He, and (b) Change in entropy during melting of ^{3}He.

From Clausius-Clapeyron equation, we recall that

$$\frac{dp}{dT} = \frac{\Delta S}{\Delta V}$$

$$\Delta S = \frac{dp}{dT} \,.\, \Delta V$$

He melts, the volume increases. So $\Delta V > 0$ Therefore,

$$\text{For } 0.02\,K < T < 0.32\text{ K}, \frac{dp}{dT} < 0 \;\Rightarrow\; \Delta S < 0.$$

$$\text{For } 0.32\,K < T < 1.2\text{ K}, \frac{dp}{dT} > 0 \;\Rightarrow\; \Delta S < 0.$$

At $T = 0.32$ K, $\Delta S = 0$.

These results are depicted in Fig. 8.4 (b).

Example 8.13 Liquid ^{4}He has a normal boiling point of 4.2 K. However, at a pressure of 1 mm of Hg, it boils at 1.2 K. Calculate the average latent heat of vaporisation He in this temperature range. Take $R = 8.314$ Jmol^{-1} K^{-1}.

According to Clausius–Clapeyron equation, we have

$$\frac{dp}{dT} = \frac{1}{T(V_g - V_\ell)}$$

$$\approx \frac{L}{TV_g} \qquad \text{(i)}$$

and V_l signify the volumes of gaseous and liquid phases respectively. Also, if we approximate the behaviour of the gaseous phase by an ideal gas, we can write

$$pV_g = RT \qquad \text{(ii)}$$

Using this in Eq. (i), we can write

$$\frac{dp}{dT} = \frac{Lp}{RT^2}$$

On rearranging terms, we can write

$$\frac{dp}{p} = \frac{L}{R}\frac{dT}{T^2}$$

Hence, if temperature changes from T_0 to T and pressure changes from p_0 to p, we can write

$$\int_{p_0}^{p} \frac{dp}{p} = \frac{L}{R}\int_{T_0}^{T} \frac{dT}{T^2}$$

On integrating both sides, we get

$$\ln(p/p_0) = \frac{L}{R}\left(-\frac{1}{T} + \frac{1}{T_0}\right)$$

Hence, the expression for latent heat of vaporisation is given by

$$L = \frac{R \ln(p/p_0)}{\left(\frac{1}{T_0} - \frac{1}{T}\right)} = \frac{RT\,T_0}{(T - T_0)}\ln(p/p_0)$$

On substituting the values of different quantities, we get

$$L = \frac{(8.314 \text{ J mol}^{-1} \text{ K}^{-1}) \times (4.2 \text{ K}) \times (1.2 \text{ K})}{3 \text{ K}} \times$$

$$\ln\left[\frac{10^5 \text{ Nm}^{-2}}{(10^{-3} \text{ m})(13.6 \times 10^3 \text{ kg m}^{-3}) \times (9.8 \text{ ms}^{-2})}\right]$$

$$= (13.968 \text{ J mol}^{-1}) \ln(750.3)$$

$$= (13.968 \times 6.62) \text{ J mol}^{-1} = 92.5 \text{ J mol}^{-1}.$$

Example 8.14 Calculate the change in melting point of ice when it is subjected to a pressure of 100 atm. For ice, density = 0.92 g cm^{-3} and latent heat of fusion = 80 cal g^{-1}. Take = 981 cm s^{-2}.

From Clausius–Clapeyron equation we know that

$$\frac{dp}{dT} = \frac{L}{T(v_2 - v_1)}$$

On rearranging terms, w can write

$$dT = T\frac{(v_2 - v_1)dp}{L}$$

$$T = 0°\text{C} = 273 \text{ K}.$$

$$v_1 = \text{specific volume of ice at } 0°\text{C}$$

$$= \frac{1}{0.92} \text{cm}^3 \text{ g}^{-1}.$$

$$= 1.09 \text{ cm}^3 \text{ g}^{-1}$$

$$v_2 = \text{specific volume of water at } 0°\text{C}$$

$$= 1 \text{ cm}^3 \text{ g}^{-1}$$

$$(v_2 - v_1) = (1 - 1.09) \text{ cm}^3 \text{ g}^{-1} = -0.09 \text{ cm}^3 \text{ g}^{-1}$$

$$dp = (100 - 1) \text{ atm} = 99 \text{ atm}.$$

$$= 99 \times 1.013 \times 10^6 \text{ dyne cm}^{-2}$$

$$= 100.287 \times 10^6 \text{ dyne cm}^{-2}$$

$$L = 80 \text{ cal g}^{-1}$$

$$= 80 \times 4.2 \times 10^7 \text{ erg g}^{-1}.$$

$$dT = \frac{(273 \text{ K}) \times (-0.09 \text{ cm}^3 \text{ g}^{-1}) \times (100.287 \times 10^6 \text{ dyne cm}^{-2})}{(80 \text{ cal g}^{-1}) \times (4.2 \times 10^7 \text{ erg cal}^{-1})}$$

$$= -\frac{2464.05 \times 10^6}{336 \times 10^7} \text{K}$$

$$= -0.733 \text{ K} \equiv -0.733°\text{C}$$

Therefore, the change in melting point of ice at 100 atm pressure will be –0.733°C

Example 8.15 When 1 gram of water is converted into steam, the change in specific volume is 1676 cm^3. Calculate the pressure required to boil water at 250°C. Take latent heat of vaporization of steam = 540 cal g^{-1}, 1 atm = 10^6 dyne cm^{-2}, and 10^7 erg cal^{-1}.

From Clausius–Clapeyron equation, we have

$$\frac{dp}{dT} = \frac{L}{T(v_2 - v_1)}$$

Therefore, the change in pressure is given by

$$dp = \frac{dT \times L}{T(v_2 - v_1)}$$

$$dT = 250 - 100 = 150°C \equiv 150 \text{ K.}$$

$$T = 100°C = 373 \text{ K.}$$

$$(v_2 - v_1) = 1676 \text{ cm}^3$$

$$L = 540 \text{ cal g}^{-1} = 540 \times 4.2 \times 10^7 \text{ erg g}^{-1}$$

$$dp = \frac{(150 \text{ K}) \times (540 \times 4.2 \times 10^7 \text{ erg g}^{-1})}{(373 \text{ K}) \times (1676 \text{ cm}^3)}$$

$$= 0.544 \times 10^7 \text{ dyne cm}^{-2}$$

$$= 5.44 \text{ atm}$$

Pressure required = (1 + 5.44) atm = 6.44 atm.

Example 8.16 Water boils at 100°C under a pressure of 76 cm of Hg. In your laboratory, you observe that the atmospheric pressure is 74 cm. Calculate the change in boiling point using the following data: L(steam) = 540 cal g^{-1}, ρ(Hg) = 13.6 g cm^{-3}, specific volume of steam = 1677 cm^3, specific volume of water = 1 cm^3, and g = 980 cm s^{-2}.

From Clausius–Clapeyron equation, we have

$$\frac{dp}{dT} = \frac{L}{T(v_2 - v_1)}$$

We can rewrite it as

$$dT = \frac{T(v_2 - v_1)dp}{L}$$

$$T = 100°C = 373 \text{ K}, v_2 = 1677 \text{ cm}^3 \text{ g}^{-1}, v_1 = 1 \text{ cm}^3 \text{ g}^{-1}$$

$$dp = (76 - 74) \text{ cm of Hg} = 2 \text{ cm of Hg}$$

$$= (2 \text{ cm}) \times (13.6 \text{ g cm}^{-3}) \times 980 \text{ dyne cm}^{-2}$$

$$L = 540 \text{ cal g}^{-1} = 540 \times 4.2 \times 10^7 \text{ erg g}^{-1}$$

$$dT = \frac{(373 \text{ K}) \times (1676 \text{ cm}^3 \text{ g}^{-1}) \times (2 \times 13.6 \times 980 \text{ dyne cm}^{-2})}{(540 \text{ cal g}^{-1}) \times (4.2 \times 10^7 \text{ erg cal}^{-1})}$$

$$= 0.735 \text{ K}$$

$$= 0.735°C$$

Example 8.17 Water boils at a temperature of 101°C at a pressure of 787 mm of Hg. 1 g of water occupies 1601 cm^3 on evaporation. Calculate the latent heat of steam. $= 4.2 \times 10^7$ erg cal^{-1} and $g = 980$ cm s^{-2}.

From Clausius–Clapeyron equation, we recall that

$$\frac{dp}{dT} = \frac{L}{T(v_2 - v_1)}$$

On rearrangement, we can write

$$L = \frac{T(v_2 - v_1)dp}{dT}$$

$$T = 373\text{ K}, (v_2 - v_1) = (1601 - 1) = 1600\text{ cm}^3\text{ g}^{-1}$$

$$\begin{aligned} dp &= (787 - 760)\text{ mm of Hg} \\ &= 27\text{ mm of Hg} \\ &= 2.7\text{ cm of Hg} \\ &= (2.7\text{ cm}) \times (13.6\text{ g cm}^{-3}) \times (980\text{ cm s}^{-2}) = 35985.6\text{ dyne cm}^{-2} \end{aligned}$$

$$dT = (101 - 100)°\text{C} = 1°\text{C} \equiv 1\text{ K}.$$

$$L = \frac{(373\text{ K}) \times (1600\text{ cm}^3\text{ g}^{-1}) \times 35985.6\text{ dyne cm}^{-2}}{1\text{ K}}$$

$$= \frac{373 \times 1600 \times 35985.6}{4.2 \times 10^7\text{ erg cal}^{-1}}\text{ erg g}^{-1}$$

$$L = 511.3\text{ cal g}^{-1}.$$

Example 8.18 Minute droplets of water are slowly pushed out of an atomizer into air. The average radius of the droplets is 10^{-4} cm. If 1 kg of water is atomised isothermally at 25°C calculate the heat transferred. The specific volume of water at 25°C is 10^{-3} m^3 kg^{-1} and the rate of change of surface tension of water with temperature $\times 10^{-3}$ Nm^{-1} K^{-1}.

To calculate the heat transferred in atomising 1kg water, we have to determine the increase in surface area, which, in turn, requires the knowledge of number of droplets formed. The number of droplets formed is given by

$$N = \frac{\text{Volume of 1 kg of water}}{\text{Volume of one droplet}}$$

If the water drop can be regarded spherical, we can write

$$N = \frac{1.00187 \times 10^{-3}\text{ m}^3\text{ kg}^{-1}}{\frac{4\pi}{3}(10^{-6}\text{ m})^3}$$

$$= \frac{3 \times 1.00187 \times 10^{-3}}{4 \times 3.1417 \times 10^{-18}} \text{kg}^{-1}$$

$$= 2.392 \times 10^{14} \text{ kg}^{-1}$$

Hence, total surface area of droplets formed on atomisation of 1kg water is

$$A = 4\pi r^2 N$$

$$= 4 \times 3.1417 \times (10^{-6} \text{ m})^2 \times (2.392 \times 10^{14} \text{ kg}^{-1})$$

$$= 3.006 \times 10^3 \text{ m}^2 \text{ kg}^{-1}$$

From Example 8.9, we recall that the heat transferred to the droplets is given by

$$\delta Q_T = -T\frac{\mathrm{d}\sigma}{\mathrm{d}T}\mathrm{d}A$$

$$Q_T = -T\frac{\mathrm{d}\sigma}{\mathrm{d}T}A.$$

$= 25°\text{C} = 298 \text{ K}, \frac{\mathrm{d}\sigma}{\mathrm{d}T} = -0.152 \times 10^{-3} \text{ Nm}^{-1} \text{ K}^{-1}$ and

$\times 10^3 \text{ m}^2 \text{ kg}^{-1}$, we get

$$Q_T = (298 \text{ K}) \times (0.152 \times 10^{-3} \text{ Nm}^{-1} \text{ K}^{-1}) \times (3.006 \times 10^3 \text{ m}^2 \text{ kg}^{-1})$$

$$= 136.16 \text{ J kg}^{-1}.$$

Example 8.19 The pressure on one mole of mercury kept at STP is increased reversibly and isothermally to 101 atm. Calculate the heat transferred in this process and the change in internal energy of the system. Given $V = 14.7 \text{ cm}^3 \text{ mol}^{-1}$, coefficient of volume expansion $\alpha = 180 \times 10^{-6} \text{ K}^{-1}$ and isothermal elasticity $E_T = 2.6 \times 10^{10} \text{ Nm}^{-2}$.

In the first part, we have to calculate heat transfer in an isothermal process when pressure changes. So we consider the second $T\,\mathrm{d}S$ equation

$$T\,\mathrm{d}S = C_p\,\mathrm{d}T - T\left(\frac{\partial V}{\partial T}\right)_p \mathrm{d}p \qquad \text{(i)}$$

In an isothermal change, $\mathrm{d}T = 0$. Then Eq. (i) simplifies to

$$T\,\mathrm{d}S = \delta Q = -T\left(\frac{\partial V}{\partial T}\right)_p \mathrm{d}p \qquad \text{(ii)}$$

Hence, total heat transfer is obtained by integrating this expression:

$$Q = -T\int\left(\frac{\partial V}{\partial T}\right)_p \mathrm{d}p \qquad \text{(iii)}$$

The coefficient of volume expansion is defined as

$$\alpha = \frac{1}{V}\left(\frac{\partial V}{\partial T}\right)_p$$

Using this in Eq. (iii), we obtain

$$Q = -TV\alpha\int dp$$

$$= -TV\alpha(p_f - p_i) \qquad \text{(iv)}$$

= 273 K, $p_f - p_i$ = 100 atm = 100 × 1.013 × 10^5 Nm^{-2}, V = 14.7 × 10^{-6} m^3 mol^{-1} = 180 × 10^{-6} K. Using these values in Eq. (iv), we get

$$= -(273\text{ K}) \times (14.7 \times 10^{-6}\text{ m}^3\text{ mol}^{-1}) \times (180 \times 10^{-6}\text{ K}^{-1}) \times (1.013 \times 10^7\text{ Nm}^{-2})$$

7.317 J mol^{-1}

1.750 cal mol^{-1}

In the second part, we have to calculate change in internal energy. To do so, we recall that internal energy and heat are connected through the work done by/on the system. We, therefore, first obtain the expression for work done in an isothermal process:

$$W = \int p\text{d}V \qquad \text{(v)}$$

V as a function of T and p, we can write

$$V = V(T, p)$$

so that an infinitesimal change in V is given by

$$\text{d}V = \left(\frac{\partial V}{\partial T}\right)_p \text{d}T + \left(\frac{\partial V}{\partial p}\right)_T \text{d}p \qquad \text{(vi)}$$

For an isothermal change, dT = 0 and the first term on the RHS dropout. Hence, we can rewrite Eq. (vi) as

$$\text{d}V = \left(\frac{\partial V}{\partial p}\right)_T \text{d}p$$

We now recall that isothermal elasticity

$$E_T = -V\left(\frac{\partial p}{\partial V}\right)_T$$

$$\left(\frac{\partial V}{\partial p}\right)_T = -\frac{V}{E_T}$$

$$\text{d}V = -\frac{V}{E_T}\text{d}p \qquad \text{(vii)}$$

On combining Eqs. (v) and (vii), we can write

$$W = -\frac{V}{E_T}\int_{p_i}^{p_f} p\text{d}p = -\frac{V}{2E_T}\left(p_f^2 - p_i^2\right) = \frac{V}{2E_T}\left(p_i^2 - p_f^2\right) \qquad \text{(viii)}$$

On substituting the given values, we get

$$= \frac{(14.7 \times 10^{-6}\text{ m}^3\text{ mol}^{-1})}{10 \qquad 2} \times [(1.013 \times 10^5\text{ Nm}^{-2})^2] - [(101 \times 1.013 \times 10^5\text{ Nm}^{-2})^2]$$

$$= 2.827 \times 10^{-16}\ \text{N}^{-1}\ \text{m}^5\ \text{mol}^{-1} \times [(1.026 \times 10^{10}\ \text{N}^2\ \text{m}^{-4}) - 1.047 \times 10^{14}\ \text{N}^2\ \text{m}^{-4}]$$

$$= -2.96 \times 10^{-2}\ \text{Nm mol}^{-1}$$

$$= -2.96 \times 10^{-2}\ \text{J mol}^{-1}$$

$$= -0.708 \times 10^{-2}\ \text{cal mol}^{-1}$$

as well as W are negative, work is done on the system, which liberates energy. The change in internal energy is given by

$$U = Q - W$$

$$= (-1.75 + 0.0071)\ \text{cal mol}^{-1}$$

$$= -1.74\ \text{cal mol}^{-1}.$$

Example 8.20 Using Maxwell's relations, calculate the heat transferred when the pressure on 10 g of water at 0°C is increased reversibly by 1000 atm. Is heat absorbed or given out? The coefficient of volume expansion of water, $\alpha = -6.7 \times 10^{-5}$°C^{-1} and ^{5} dyne cm^{-2}.

From Maxwell's second relation, we have

$$\left(\frac{\partial S}{\partial p}\right)_T = -\left(\frac{\partial V}{\partial T}\right)_p$$

Since coefficient of volume expansion $\alpha = \frac{1}{V}\left(\frac{\partial V}{\partial T}\right)_p$, we can rewrite the above expression

$$\left(\frac{\partial S}{\partial p}\right)_T = -V\alpha$$

We know that heat transferred $\Delta Q = T\Delta S$. In terms of α, we can write

$$\Delta Q = -T V \alpha \Delta p$$

$$T = (0 + 273)\ \text{K} = 273\ \text{K}$$

$$V = \frac{10\ \text{g}}{1\text{g cm}^{-3}} = 10\ \text{cm}^3$$

$$\alpha = -6.7 \times 10^{-5}\ \text{C}^{-1} \equiv -6.7 \times 10^{-5}\ \text{K}^{-1} \text{ and } \Delta p = 1000\ \text{atm}$$

$$\Delta Q = -(273\ \text{K}) \times (10\ \text{cm}^3) \times (-6.7 \times 10^{-5}\ \text{K}^{-1}) \times (1000\ \text{atm})$$

$$= (182.91\ \text{cm}^3) \times 1\ \text{atm}$$

$$= (182.91\ \text{cm}^3) \times (10^5\ \text{dyne cm}^{-2})$$

$$= 1.83 \times 10^7\ \text{dyne cm}$$

$$= 1.83 \times 10^7\ \text{erg} = 1.83\ \text{J}$$

$$= \frac{1.83\ \text{J}}{4.2\ \text{J cal}^{-1}} = 0.44\ \text{cal}$$

> 0, we say that heat is absorbed.

Example 8.21 Prove that

$$\frac{\left(\frac{\partial p}{\partial T}\right)_V}{\left(\frac{\partial p}{\partial T}\right)_S} = 1 - \frac{1}{\gamma}$$

$$\text{RHS} = 1 - \frac{1}{\gamma} = \frac{\gamma - 1}{\gamma} = \frac{\frac{C_p}{C_V} - 1}{\frac{C_p}{C_V}} = \frac{C_p - C_V}{C_p}. \qquad \text{(i)}$$

From Eq. (8.18) we recall that

$$C_p - C_V = T\left(\frac{\partial p}{\partial T}\right)_V \left(\frac{\partial V}{\partial T}\right)_p$$

Also, by definition, $C_p = T\left(\frac{\partial S}{\partial T}\right)_p$

Using these results in Eq. (i), we can write

$$\text{RHS} = \frac{T\left(\frac{\partial p}{\partial T}\right)_V \left(\frac{\partial V}{\partial T}\right)_p}{T\left(\frac{\partial S}{\partial T}\right)_p} = \frac{\left(\frac{\partial p}{\partial T}\right)_V}{\left(\frac{\partial S}{\partial V}\right)_p} \qquad \text{(ii)}$$

From Maxwell's fourth relation, we know that

$$\left(\frac{\partial V}{\partial S}\right)_p = \left(\frac{\partial T}{\partial p}\right)_S$$

$$\left(\frac{\partial S}{\partial V}\right)_p = \left(\frac{\partial p}{\partial T}\right)_S$$

Using this result in Eq. (ii), we get

$$\text{RHS} = \frac{\left(\frac{\partial p}{\partial T}\right)_V}{\left(\frac{\partial p}{\partial T}\right)_S} = \text{LHS}.$$

We now summarise what you have learnt in this chapter.

MMARY

The four Maxwell relations are

$$\left(\frac{\partial S}{\partial V}\right)_T = \left(\frac{\partial p}{\partial T}\right)_V$$

$$\left(\frac{\partial S}{\partial p}\right)_T = -\left(\frac{\partial V}{\partial T}\right)_p$$

$$\left(\frac{\partial T}{\partial V}\right)_S = -\left(\frac{\partial p}{\partial S}\right)_V$$

$$\left(\frac{\partial T}{\partial p}\right)_S = \left(\frac{\partial V}{\partial S}\right)_p$$

The Clausius–Clapeyron equation describing a phase transition of first order is

$$\left(\frac{\partial p}{\partial T}\right)_{\text{sat}} = \frac{L}{T\,(v_{\text{vap}} - v_{\text{liq}})}$$

The heat capacity equation is readily obtained from Maxwell relations

$$C_p - C_V = -T\left(\frac{\partial p}{\partial V}\right)_T \left(\frac{\partial V}{\partial T}\right)_p^2 = TVE_T\,\alpha^2$$

d S equations relate changes in entropy to changes in temperature, volume and/or pressure. Therefore, these provide us a way to estimate changes in entropy in terms of experimentally measurable quantities:

$$T\,\mathrm{d}S = C_V\,\mathrm{d}T + T\left(\frac{\partial p}{\partial T}\right)_V \mathrm{d}V = C_V\,\mathrm{d}T + T\alpha E_T\,\mathrm{d}V$$

$$T\,\mathrm{d}S = C_p\,\mathrm{d}T - T\left(\frac{\partial V}{\partial T}\right)_p \mathrm{d}p = C_p\,\mathrm{d}T - TV\alpha\,\mathrm{d}p$$

$$T\,\mathrm{d}S = C_V\left(\frac{\partial T}{\partial p}\right)_V \mathrm{d}p + C_p\left(\frac{\partial T}{\partial V}\right)_p \mathrm{d}V$$

The energy equations enable us to study variation of internal energy with volume and pressure:

$$\left(\frac{\partial U}{\partial V}\right)_T = T\left(\frac{\partial p}{\partial T}\right)_V - p$$

$$\left(\frac{\partial U}{\partial p}\right)_T = -T\left(\frac{\partial V}{\partial T}\right)_p - p\left(\frac{\partial V}{\partial p}\right)_T$$

The Gibbs–Helmholtz equation gives us a measure' of heat of reaction:

$$\Delta H = \left(E - T\left(\frac{\partial E}{\partial T}\right)_{\text{sat}} \right) \Delta Z$$

ERCISES

Using Maxwell's relations, show that

$$C_p = T\left(\frac{\partial V}{\partial T}\right)_p \left(\frac{\partial p}{\partial T}\right)_S \text{ and } C_V = -T\left(\frac{\partial p}{\partial T}\right)_V \left(\frac{\partial V}{\partial T}\right)_S.$$

$$\frac{C_p}{C_V} = \gamma$$

A gas obeys the equation

$$p\,(V - b) = RT$$

where b is a constant. The gas has a constant value of C_V. Show that

U is a function of T only

During adiabatic process the gas obeys the relation

$$p(V - b)^{\gamma} = \text{constant}$$

where γ is a constant.

A system undergoes a phase transition at a constant temperature T and pressure p. If it suffers a change ΔV in its volume per unit mass, show that change in the internal energy is $mT^2 \dfrac{\partial}{\partial T}\left(\dfrac{p}{T}\right)_V \Delta V$, where m is the mass of the system.

A gas fills a cavity of volume V at a temperature T. The system expands adiabatically and reversibly to a volume equal to $8V$. By what factor does the temperature change?

(*Ans:* 0.5)

Calculate the boiling point of water at an altitude of 300 m. Assume that it is 100°C at the sea level. Given $g = 9.8$ ms^{-2}, $R = 2$ cal mol^{-1} K^{-1}, $T = 300$ K, molecular weight of air = 29 g mol^{-1} and the latent heat of evaporation of water at STP is 540 cal g^{-1}.

(*Ans:* ~ 99°C)

Calculate the change in the temperature of boiling water when the pressure is increased by 27.12 mm of Hg. The normal boiling point of water at atmospheric pressure is 100°C. Given latent heat of steam = 537 cal g^{-1}, specific volume of steam = 1674 cm^3 g^{-1}, $J = 4.2 \times 10^7$ erg cal^{-1}, density of mercury = 13.6 g cm^{-3} and = 980 cm s^{-2}.

(*Ans:* 1°C)

Calculate the pressure at which the freezing point of water would be –0.5°C, if change in specific volume when 1 g water freezes into ice is 0.091 cm^3 g^{-1}. Latent heat of fusion of ice is 80 cal g^{-1}.

(*Ans:* 68.6 atm)

Calculate the specific volume of solid sulphur from the following data: Melting point of sulphur = 115°C; latent heat of fusion of sulphur = 9.3 cal g^{-1}, volume of 1 g of liquid sulphur = 0.513 cm^3; rate of change of melting point with pressure is 0.025°C atm^{-1}. (1 atm. = 10^6 dyne cm^{-2})

(*Ans:* 0.488 cm^{3}g^{-1})

Calculate the change in the melting point of paraffin wax for a change in pressure by 5 atm. Given melting point of paraffin wax is 53.7°C. Latent heat of fusion is 36 cal g^{-1} and increase in volume of 1 g at melting point is 0.135 cm^3.

(*Ans:* 0.145 K)

Prove that

$$\left(\frac{\partial T}{\partial p}\right)_V \left(\frac{\partial S}{\partial V}\right)_p - \left(\frac{\partial T}{\partial V}\right)_p \left(\frac{\partial S}{\partial p}\right)_V = 1$$

$$\left(\frac{\partial p}{\partial T}\right)_S \left(\frac{\partial V}{\partial S}\right)_T - \left(\frac{\partial p}{\partial S}\right)_T \left(\frac{\partial V}{\partial T}\right)_S = 1$$

A 20 kg weight is placed on a large block of melting ice and it bears upon an area of 0.071 m^2. Ice floats on water with (11/12) of its volume submerged. To prevent melting and consequent penetration of weight, the temperature of ice is to be lowered. Calculate the change, assuming that the weight does not conduct heat. Take latent heat of fusion of ice as 333 kJ kg^{-1}. (*Ans:* -0.21×10^{-3} K)

Will a person weighing 60 kg be able to skate, if the temperature of ice is –2°C and the area of contact is 25 mm^2? Given L = 333 kJ kg^{-1} at 0°C and 1 atm pressure, change in specific volume of ice on melting is -9.05×10^5 m^3 kg^{-1} and = 9.81 ms^{-2}.

(*Ans:* No; The minimum weight required for skating is about 68.8 kg.)

In a cold storage, ice melts at the rate 3 kg per hour when the external temperature is 28°C. Calculate the minimum power output of the motor used to run the refrigerator just to prevent melting of ice. (*Ans:* 373 W)

= 800 – 0.705 T, calculate the specific heat capacity of steam using second latent heat equation. (*Ans:* –1.14 units)

Consider a gas obeying Berthelot's equation:

$$\left(p + \frac{a'}{TV^2}\right)(V - b) + RT$$

Show that the rate of change of internal energy with volume at constant temperature is inversely proportional to the square of volume.

9 FREE ENERGIES AND THERMODYNAMIC EQUILIBRIUM

arning Objectives

In this chapter, you will learn how to

- e thermodynamic potentials;
- e general condition for thermodynamic equilibrium and apply it to fic processes;
- y the condition determining phase equilibrium;
- e the first and second latent heat equations and explain different types uilibria;
- ribe second order phase transitions; and
- uss Gibbs phase rule for multi-component systems.

TRODUCTION

So far we have discussed four thermodynamic functions—temperature (T), internal energy), enthalpy (H) and entropy (S). You may recall that temperature helps us in fixing the equation of state of a thermodynamic system, and internal energy enables us to develop a mathematical formulation of the first law, which is essentially the law of conservation of energy. However it is inadequate as far as the issue of determining the direction of evolution of a process is concerned. This aspect is taken care of by the second law which emphasises that in all natural processes, entropy always increases. You may now ask the question: Can we say with certainty that processes where entropy increases will necessarily occur? The second law only implies that they are permissible. Further, these two laws of thermodynamics exclude the fact that all systems have an inherent tendency to approach equilibrium. It suggests that there is a definite need to supplement these laws when we wish to get complete information about a system.

The most general condition for a change to take place is discussed in Sec. 9.2. This condition is then used to arrive at the condition for thermodynamic equilibrium of physical and chemical systems. As we show, a change under specific constraints helps us to introduce a new function, called *free energy*. In Sec. 9.2, we will introduce two new functions of *Helmholtz free energy*, F and *Gibbs free energy*, G. The functions U, H, F and G are collectively called *thermodynamic potentials*. (This nomenclature seems to be borrowed from and analogous to the concept of potential in mechanics. A conservative mechanical system is in equilibrium when its potential energy has minimum value. You will learn that the minimum values of thermodynamic potentials determine the equilibrium states of a thermodynamic system under different constraints.) You may recall that enthalpy was introduced in Chapter 5 to express constant pressure heat capacity as its temperature derivative. However, its full significance will be realized in this and the next chapter. You will observe that each energy function has its own pair of natural variables. And these are fountainhead of all thermodynamic relations. This is illustrated in Sec. 9.3 for arriving at Maxwell relations. In fact, once any one of these energies is known explicitly, we can easily get complete information about the system.

We firmly believe that you must not memorise thermodynamic relations. Instead, you should develop capability and capacity to obtain these based on mnemonic diagrams. We illustrated this point in the previous chapter to obtain Maxwell relations and we have illustrated it again in Sec. 9.4 for free energies. It is a universal truth that 'Good Physicists Have Studied Under Very Fine Teachers'. We have chosen to highlight it to express infinitesimal changes in free energies in terms of thermodynamic variables (p, V,). In Sec. 9.5, we will discuss the general conditions for thermodynamic equilibrium and apply these in Sec. 9.6 to study the equilibrium between phases of one component as well as multi-component systems. We will revisit Clausius–Clapeyron equation for first order phase transitions derived earlier using Maxwell relations. However, the concepts developed here will also be used to derive Ehrenfest equations for second order phase transitions as well as Gibbs phase rule.

ENERAL CONDITION FOR A NATURAL CHANGE

Consider a system A of constant composition coupled thermally and mechanically to a heat reservoir (surroundings) at constant temperature T_{res} and pressure p_{res}. Let the composite system, shown in Fig. 9.1, be isolated from the universe. Suppose that the system and the reservoir are not in thermodynamic equilibrium. Let S, S_r and S_{total} be the entropies of the system A, heat reservoir and the composite system, respectively.

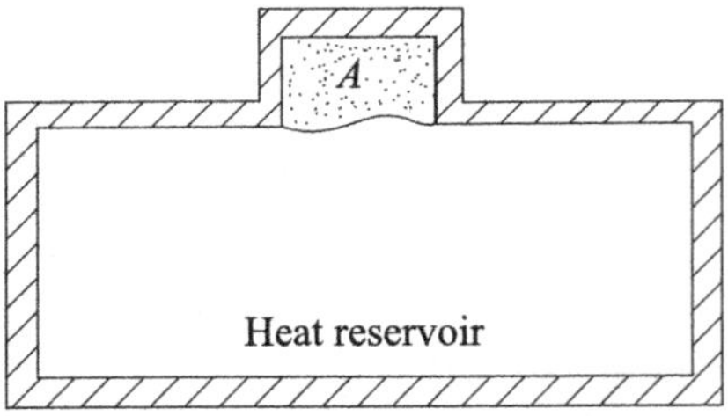

Fig. 9.1 A system in thermal and mechanical contact with heat reservoir.

For any change to occur in the composite system, the principle of increase of entropy implies that

$$\Delta S_{total} = \Delta S + \Delta S_{res} \geq 0 \tag{9.1}$$

Recall that the equality sign applies to a reversible process and greater than sign holds for an irreversible process.

Suppose that in this change, an amount of heat δQ is transferred reversibly to the system from the reservoir. Then, from Eq. (7.4), we can write

$$\Delta S_{res} = -\frac{\delta Q}{T_{res}}$$

Hence, Eq. (9.1) takes the form

$$\Delta S - \frac{\delta Q}{T_{res}} \geq 0 \tag{9.2a}$$

We can also rewrite it as

$$\Delta Q \leq T \quad \Delta S \tag{9.2b}$$

To proceed, we recall the entropy form of the first law of thermodynamics:

$$TdS = \delta Q = dU + \delta W \tag{9.3}$$

is increase in internal energy of system A in an infinitesimal process and δW is the work done by it. Combining Eqs. (9.2a) and (9.3), we obtain the general condition for a natural change to occur:

$$T_{res}\,\Delta S - (dU + \delta W) \geq 0 \tag{9.4}$$

In general, Eq. (9.4) depends on the system and its surroundings. In the special case where the temperature of the system and its surroundings are the same $T\,(= T_{res})$, it becomes a function of the system only:

$$T\Delta S - (dU + \delta W) \geq 0 \tag{9.5a}$$

or for an infinitesimal change

$$dU - TdS \leq -\,\delta W \tag{9.5b}$$

This is another formulation of the Clausius inequality. It may be mentioned here that the inequality condition in this equation specifies the requirement for a natural process to occur and takes the system towards equilibrium. The inequality holds for every infinitesimal part of the natural process but the equality sign will hold once the system has attained equilibrium. Thereafter, no natural process can take place as long as the physical conditions remain the same.

Note that as such, Eq. (9.5a, b) is so general that it is difficult to visualise its implications. We will, therefore, apply it to some important situations/systems subject to different constraints. You will note that it then takes simpler forms.

9.2.1 An Adiabatic Process

For an adiabatic process, $\delta Q = 0$ and Eq. (9.5a) reduces to

$$\Delta S \geq 0 \tag{9.6}$$

That is, the entropy of a thermally isolated system cannot decrease during a transformation. This requirement is consistent with the second law of thermodynamics. It may be remarked here that we will obtain the same condition if we consider a constant- energy ($dU = 0$) and constant-volume ($dV = 0$) configuration.

For an adiabatic-isobaric process, such as Joule–Thomson expansion, the system is not mechanically isolated. Then Eq. (9.5b) implies that

$$dU + \delta W \leq 0$$

system, $\delta W = p dV$. Further, for an isobaric process we can write $pdV = d(pV)$ so that the above inequality for an adiabatic-isobaric process reduces to

$$d(U + pV) \leq 0$$

We express this condition as

$$dH \leq 0 \tag{9.7}$$

$$H = U + pV \tag{9.8}$$

is enthalpy of the system (Eq. (5.26)).

Let us pause for a while and reflect on what we have achieved. We have discovered that if a natural change occurs in a thermally isolated system at constant pressure, its entropy increases but enthalpy decreases and becomes minimum in the equilibrium state. In fact, the entropy has maximum value at constant enthalpy and enthalpy is minimum at constant entropy.

By differentiating Eq. (9.8), we can write

$$dH = dU + p dV + V dp$$

$U + p dV = T dS$, we can relate the change in enthalpy in a hydrostatic system $V)$ to changes in entropy and pressure:

$$dH = T\, dS + V\, dp \tag{9.9}$$

Using this result, we can express temperature and volume in terms of enthalpy:

$$T = (\partial H / \partial S)_p \tag{9.9a}$$

$$V = (\partial H / \partial p)_S \tag{9.9b}$$

Note that Eq. (9.9a) suggests that enthalpy is heat content and Eq. (9.9b) implies that an increase in pressure raises the enthalpy in an isentropic (adiabatic) process. For a given increase in pressure, the larger the volume of the system, the greater is the increase in the enthalpy at constant entropy. And for a fixed mass, a gas will occupy greater volume than a solid. This suggests that at constant entropy, the rate of increase of enthalpy with pressure will be greater for a gas than that for a solid. It is important to note that entropy and pressure make up the natural pair for enthalpy and we can write $H = H(S, p)$.

Note that partial differentiation of enthalpy with respect to p and S gives thermodynamic variables (co-ordinates) V and T. This has mechanical analogy in that differentiation of mechanical potential with respect to the Cartesian co-ordinates gives components of force.

For an adiabatic-isochoric process, Eq. (9.5b) implies that

$$(dU)_{S,V} \leq 0 \tag{9.10}$$

That is, in an isolated system, a natural process at constant volume is accompanied by decrease in internal energy. In fact, internal energy becomes minimum in the equilibrium state. By comparing Eqs. (9.7) and (9.10), we can say that for isolated systems at constant volume, the role of internal energy is analogous to that of enthalpy for isolated systems at constant pressure.

9.2.2 An Isothermal Process

For an isothermal process, such as a phase change, entropy may change. Then we can write $TS)$ and Eq. (9.5b) takes the form

$$d\,(U - TS) \leq -\,\delta W$$

$$dF \leq -\,\delta W \tag{9.11a}$$

$$F = U - TS \tag{9.11b}$$

is called the *Helmholtz energy*. From this definition, we note that Helmholtz energy is expressed as a combination of functions of state. Therefore, F is also a property of state of a system and hence dF is an exact differential.

Equation (9.11a) signifies that when a system of constant composition undergoes an isothermal transformation, the work done by the system is less than or equal to the decrease in Helmholtz energy. The equality sign holds for reversible changes and the inequality sign applies to irreversible changes. Further, it follows from Eq. (9.11b) that internal energy is made up of two parts:

The free energy F, which is *available* for work in reversible isothermal processes, and

The latent energy TS which is *not available* for useful work. As entropy increases,

Example 9.1 Starting from the second law of thermodynamics, show that for a mechanically isolated system at constant temperature, the Helmholtz free energy never

Let us consider that a system undergoes an arbitrary isothermal transformation from state '1' to state '2'. Using the second law, we can write

$$\int_1^2 \frac{\delta Q}{T} \le S_2 - S_1 \tag{i}$$

The equality holds for a reversible transformation.
Since temperature remains constant, we can write

$$\frac{\Delta Q}{T} \le \Delta S \tag{ii}$$

From the first law of thermodynamics, we can write

$$\Delta Q = \Delta U + \Delta W$$

$$\Delta U + \Delta W \le T\Delta S \tag{iii}$$

Using Eq. (9.11b), we can write

$$\Delta F = \Delta U - T\Delta S - S\Delta T \tag{iv}$$

At constant temperature, the last term will drop and we get

$$\Delta F = \Delta U - T\Delta S \tag{v}$$

On combining Eqs (iii) and (iv), we can write

$$\Delta W \le -\Delta F$$

$$\Delta F \le -\Delta W$$

F can never increase.

Let us now consider some typical isothermal processes.

1. An Isothermal–Isochoric Process If a system is further constrained to a fixed volume, = 0, Eq. (9.11a) takes the form

$$\mathrm{d}F \le 0 \tag{9.12}$$

That is, at a fixed temperature, a system of constant volume is characterised solely by the Helmholtz energy and a spontaneous change will occur in the direction in which *F*

To show that thermal properties of a constant temperature, constant volume system are uniquely determined by *F*, we differentiate Eq. (9.11b) to obtain

$$\mathrm{d}F = \mathrm{d}U - T\,\mathrm{d}S - S\,\mathrm{d}T$$

If we use Eq. (9.3) as applied to a *pVT* system, we get

$$\mathrm{d}F = -S\,\mathrm{d}T - p\,\mathrm{d}V \tag{9.13}$$

This equation defines the dependence of *F* on independent variations of *T* and *V*. These therefore constitute the natural pair for Helmholtz energy and we can write $F = F(T, V)$. Further, it readily follows from Eq. (9.13) that entropy and pressure of a fixed *T* and fixed system are given by

$$S = -\left(\frac{\partial F}{\partial T}\right) \tag{9.14a}$$

$$p = -\left(\frac{\partial F}{\partial V}\right)_T \tag{9.14b}$$

This means that once F is known for a system under consideration, we can obtain complete information about its thermal properties. Further, Eq. (9.14a) shows that the Helmholtz energy decreases with rise in temperature, since entropy of any substance is positive definite. The rate of decrease of F will be greater, the higher the entropy of a substance. That is why, the rate of fall of F is maximum for gases and minimum for solids as temperature increases. Similarly, Eq. (9.14b) shows that an increase in volume decreases the Helmholtz energy; the rate of fall being greater at higher pressures.

It may be pointed out here that the Helmholtz energy is particularly important as it provides a link between statistical mechanics and thermodynamics.

Substituting for entropy in Eq. (9.11b) we obtain the *Gibbs–Helmholtz equation*, which relates the temperature dependence of F with internal energy:

$$U = F + TS = F - T\left(\frac{\partial F}{\partial T}\right)_V = -T^2\left(\frac{\partial}{\partial T}\left(\frac{F}{T}\right)\right)_V = \left(\frac{\partial(F/T)}{\partial(1/T)}\right)_V \tag{9.15}$$

$$d(1/T) = -\frac{1}{T^2}dT.$$

Since absolute determination of Helmholtz energy is not possible, we are usually interested in changes in F. So we rewrite Eq. (9.15) in a slightly different form by noting that the change in Helmholtz energy $F_2 - F_1 = -\Delta F$ is equal to the work done by the system, is heat of reaction, Q. Hence,

$$F_2 - F_1 = -\Delta F = \Delta U + T\left(\frac{\partial}{\partial T}(-\Delta F)\right)_V$$

$$w = Q + T\left(\frac{\partial w}{\partial T}\right)_V \tag{9.16}$$

We now illustrate the use of Gibbs–Helmholtz equation for a thin surface film.

Example 9.2 The surface energy of a surface film created isothermally with the help of a wire frame is given by

$$U = A\left(\sigma - T\frac{d\sigma}{dT}\right)$$

Calculate its entropy.

The Gibbs–Helmholtz equation for a surface film takes the form

$$U = F - T\left(\frac{\partial F}{\partial T}\right)_A$$

On comparing it with the given expression for the surface energy of the film, we note that

$$F = \sigma A$$

$$\left(\frac{\partial F}{\partial T}\right)_A = A\frac{d\sigma}{dT}$$

This result shows that the surface tension is equal to the free energy per unit area.

Using Eq. (9.14a) for a surface film, we can express the relation between entropy and free energy as

$$S = -\left(\frac{\partial F}{\partial T}\right)_A = -A\frac{d\sigma}{dT}$$

We would now like you to answer a practise exercise.

Problem 9.1 An enclosure with volume V is divided by a freely sliding piston into two parts having volume V_1 and V_2 and pressures p_1 and p_2. The piston is released at constant temperature and allowed to slide freely. Using the principle of minima of Helmholtz free energy, determine the equilibrium condition.

p_2

2. Isothermal–Isobaric Process A phase change, such as evaporation of a liquid, melting of ice or a chemical reaction, occurs under isothermal-isobaric conditions. From the preceding section, we recall that for systems held at constant temperature, the condition of change is given by Eq. (9.11a): $dF \le \delta W$. If such a system is further constrained to have constant pressure p, we may write $\delta W = p dV = d(pV)$. Then, the condition of change modifies to

$$d(F + pV) \le 0$$

$$dG \le 0 \tag{9.17}$$

$$G = F + pV = U - TS + pV = H - TS \tag{9.18}$$

Gibbs energy. Since all quantities occurring in Eq. (9.18) are functions of state, Gibbs energy G is also a function of state. That is why we have denoted dG as an exact differential. Note that Gibbs energy never increases in an isothermal-isobaric process.

Equation (9.17) implies that a system held at constant temperature and pressure is defined completely by G. Note that a spontaneous change under these conditions will occur in the direction of decreasing Gibbs energy. But Gibbs energy will be conserved in a reversible change.

On differentiating Eq. (9.18), we obtain

$$dG = dU - TdS + SdT + pdV - Vdp$$

Using Eq. (9.3) with $\delta W = pdV$, we get

$$dG = -SdT + Vdp \tag{9.19}$$

This equation specifies the dependence of G on (independent) variations of T and p. Hence, we can write $G = G(T, p)$.

It readily follows from Eq. (9.19) that the thermal properties of a hydrostatic system are given by

$$S = -\left(\frac{\partial G}{\partial T}\right)_p \tag{9.20a}$$

$$V = \left(\frac{\partial G}{\partial p}\right)_T \tag{9.20b}$$

The minus sign in Eq. (9.20a) shows that increase in temperature at constant pressure

On the other hand, Eq. (9.20b) implies that at constant temperature an increase in pressure increases Gibbs energy. Further, for a given increase in pressure, the larger the volume of the system the greater will be the increase in the Gibbs energy. This explains why the Gibbs energy of a gas increases more rapidly with pressure than that of a liquid or a solid.

For a pure substance, we can integrate Eq. (9.20b) and write

$$G = G_o(T) + \int_{p_0}^{p} V\,\mathrm{d}p \tag{9.21}$$

T) is the Gibbs energy at pressure p_0.

If the substance is in the condensed phase, the volume is nearly independent of pressure and we can write

$$G = G_0(T) + V(p - p_0) \tag{9.22a}$$

On the other hand, for n moles of an ideal gas

$$G = G_o(T) + nRT \int_{p_0}^{p} \frac{\mathrm{d}p}{p}$$

$$= G_o(T) + nRT \ln\left(\frac{p}{p_0}\right) \tag{9.22b}$$

mole of a gas, the Gibbs energy is called *chemical potential*. It is denoted by the

$$\mu = \frac{G}{n} \tag{9.23}$$

Hence, for an ideal gas we can write

$$\mu = \mu_o(T) + RT \ln\left(\frac{p}{p_0}\right) \tag{9.24}$$

We can now express Helmholtz free energy F in terms of Gibbs energy G by using Eq. (9.20b) in Eq. (9.18):

$$F = G - pV$$

$$= G - p\left(\frac{\partial G}{\partial p}\right)_T$$

$$= -p^2\left(\frac{\partial}{\partial p}\left(\frac{G}{p}\right)\right)_T = \left(\frac{\partial (G/p)}{\partial (1/p)}\right)_T \tag{9.25}$$

$$\mathrm{d}\left(\frac{1}{p}\right) = -\frac{1}{p^2}\,\mathrm{d}p.$$

Note that this relation is similar to Eq. (9.15).

Now you go through the following example which connects enthalpy and Gibbs energy.

Example 9.3 Show that $H = \left(\frac{\partial (G/T)}{\partial (1/T)}\right)_p$ and $U = G - T\left(\frac{\partial G}{\partial T}\right)_p - p\left(\frac{\partial G}{\partial p}\right)_T$

From Eq. (9.20a), we recall that

$$S = -\left(\frac{\partial G}{\partial T}\right)_p \tag{i}$$

Also, from Eq. (9.18) we can write

$$S = -\frac{G-H}{T} \tag{ii}$$

On combining Eqs. (i) and (ii), we can write

$$\left(\frac{\partial G}{\partial T}\right)_p = \frac{G-H}{T}$$

On rearranging terms, we can write the required expression depicting temperature dependence of enthalpy:

$$H = G - T\left(\frac{\partial G}{\partial T}\right)_p = -T^2\left(\frac{\partial}{\partial T}\left(\frac{G}{T}\right)\right)_p = \left(\frac{\partial (G/T)}{\partial (1/T)}\right)_p \tag{iii}$$

We know that internal energy can be expressed in terms of Gibbs energy as

$$U = H - pV = G + TS - pV = G - T\left(\frac{\partial G}{\partial T}\right)_p - p\left(\frac{\partial G}{\partial p}\right)_T \tag{iv}$$

where we have used Eqs. (9.20a, b).

It may be mentioned here that Eqs. (iii) and (iv) are also known as Gibbs–Helmholtz

Recall the statement that thermodynamic potentials are the fountain-head of all important thermodynamic relations. We illustrate this by first deriving Maxwell relations.

REE ENERGIES AND MAXWELL RELATIONS

We now know that in addition to two mechanical properties p and V, a hydrostatic system has three fundamental properties T, U and S. You may recall that each one of these has genesis in a law of thermodynamics. Moreover, such a system has three composite H, F and G. Of these, any one quantity may be expressed in terms of the other two. For instance, a hydrostatic system can be characterised by the following differential

$$dU = TdS - pdV \tag{9.26a}$$

$$dH = TdS + Vdp \tag{9.26b}$$

$$dF = -SdT - pdV \tag{9.26c}$$

$$dG = -SdT + Vdp \tag{9.26d}$$

It is important to recognise that these four relations are essentially four different ways of looking at one fundamental equation (Eq. (9.26a)). Further, since free energies are state functions, the expressions on the right-hand side of these equations are exact differential expressions. Hence, the cross derivatives will be equal. This immediately leads us to the four Maxwell relations which connect difficult to measure quantities to measurable quantities. We discuss it now.

Suppose z is a function of state and x and y are two state variables, i.e., $z = z(x, y)$. Then an infinitesimal change dz in z can be expressed as

$$dz = \left(\frac{\partial z}{\partial x}\right)_y dx + \left(\frac{\partial z}{\partial y}\right)_x dx$$

$$= Mdx + Ndy \tag{9.27a}$$

$= \left(\frac{\partial z}{\partial x}\right)_y$ and $N = \left(\frac{\partial z}{\partial y}\right)_x$. If we differentiate M with respect to y keeping x fixed with respect to x keeping y fixed, we get

$$\left(\frac{\partial M}{\partial y}\right)_x = \left(\frac{\partial}{\partial y}\left(\frac{\partial z}{\partial x}\right)_y\right)_x$$

$$\left(\frac{\partial N}{\partial x}\right)_y = \left(\frac{\partial}{\partial x}\left(\frac{\partial z}{\partial y}\right)_x\right)_y$$

Since the order of differentiation does not matter for a perfect differential, we can write

$$\left(\frac{\partial M}{\partial y}\right)_x = \left(\frac{\partial N}{\partial x}\right)_y \tag{9.27b}$$

We are now in a position to use Eqs. (9.27a) and (9.27b) to obtain Maxwell relations from thermodynamic potentials. Compare Eqs. (9.26a) and (9.27a). These equations will become identical if we identify U with z, T with M, $-p$ with N, S with x and V with y. (This means and p are now functions of S and V.) Therefore, using Eq. (9.27b), we can write

$$\left(\frac{\partial T}{\partial V}\right)_S = -\left(\frac{\partial p}{\partial S}\right)_V \tag{9.28a}$$

Note that this is the second Maxwell relation.
Similarly, if we identify H with z, T with M, S with x, V with N and p with y in Eq. (9.26b), then using Eq. (9.27b), we get the fourth Maxwell relation:

$$\left(\frac{\partial T}{\partial p}\right)_S = \left(\frac{\partial V}{\partial S}\right)_p \tag{9.28b}$$

You can similarly write down other two relations. We strongly advise you to attempt this.

In the preceding chapter, we mentioned that there is no need to memorise Maxwell relations. Instead, we could easily write down these based on a simple sentence. The same holds true for differential changes in free energies. These relations can readily be obtained from a simple mnemonic diagram. For this purpose, many schemes have been suggested. We illustrate our point in two different ways.

HERMODYNAMIC MNEMONIC DIAGRAMS

Refer to Fig. 9.2, which depicts a square with diagonal arrows pointing upwards. It is based on a magic sentence due to Born: *Valid Facts and Theoretical Understanding Generate Solutions to Hard Problems*. The sides of the square are labelled with four energies in alphabetical order proceeding in clockwise direction. The Helmholtz energy is at the top. The two corners at the left are labelled with extensive variables V and S. The two corners at the right are labelled with intensive variables T and p.

Note that each of the four energies appearing on the square is flanked by its natural variables. The diagonal arrows determine the algebraic sign in the differential expression for free energies. An arrow pointing away from a thermodynamic variable implies a positive coefficient whereas an arrow pointing towards a natural variable implies a negative coefficient. Therefore, we

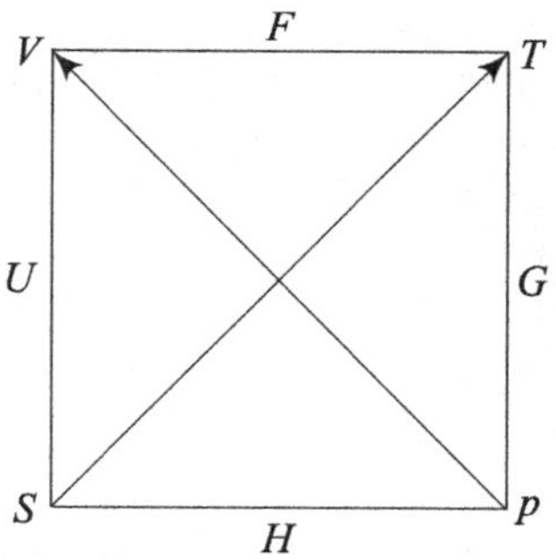

Fig. 9.2 A simple mnemonic diagram to recall Maxwell relations based on *Valid Facts and Theoretical Understanding Generate Solutions to Hard Problems.*

$$\mathrm{d}U = (\ )\,\mathrm{d}S - (\ )\,\mathrm{d}V \qquad (9.29)$$

The parentheses should be filled so that the product has dimensions of energy.

The Maxwell relations can also be read from the diagram. To this end, we deal only with the corners of the square. For instance, consider the first of the squares shown below. It is suggestive of the relation

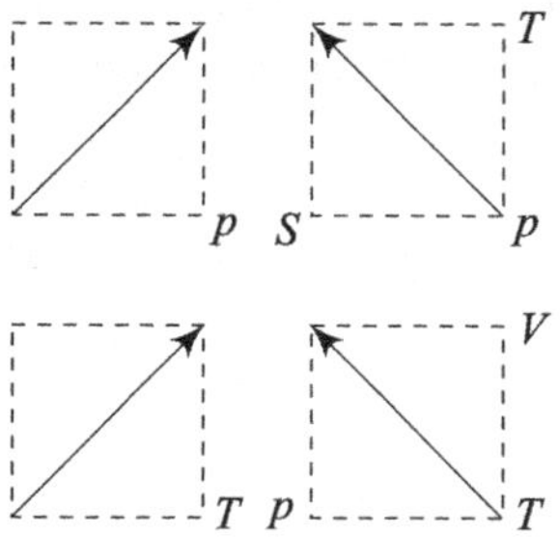

$$\left(\frac{\partial V}{\partial S}\right)_p = \left(\frac{\partial T}{\partial p}\right)_S \qquad (9.30a)$$

To obtain the second Maxwell relation, the square in Fig. 9.2 may be mentally rotated on its side. Then, by exactly the same construction, we obtain the second of these squares which readily gives

$$\left(\frac{\partial S}{\partial p}\right)_T = -\left(\frac{\partial V}{\partial T}\right)_p \qquad (9.30b)$$

The negative sign on the right-hand side arises due to the unsymmetrical placement of arrows in

The remaining two relations can be obtained by other two rotations. Can you construct mnemonic diagrams for these relations? You should try these on your own. This will encourage learning by doing and discourage passive reading; active learning will help you to develop mastery in problem solving and enjoy the subject.

The second magic sentence we illustrate is due to Gilvarry: *Good Physicists Have Studied Under Very Fine Teachers*. The mnemonic diagram, given in Fig. 9.3, consists of a square whose upper right

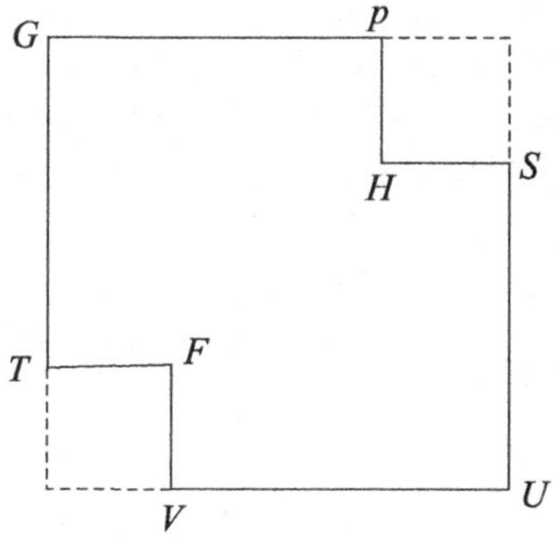

Fig. 9.3 A mnemonic diagram based on *Good Physicists Have Studied Under Very Fine Teachers.*

and lower left corners have been clipped. Starting from the upper left corner of the square and proceeding clockwise, the first letter of each word of Gilvarry's sentence is placed successively at each corner. As before, each energy is flanked by its respective set of natural variables. The expressions for dG, dH, dU and dF may be written by taking the energy corner as origin and noting the position of natural variables. When a variable is below the respective origin or to the left of it, a negative sign precedes it. Thus, we can write

$$dF = -(\ )\,dT - (\ )\,dV \tag{9.31a}$$

$$dG = (\ )\,dp - (\ )\,dT \tag{9.31b}$$

As before, the variable in brackets will ensure dimensions of energy for the product.

ENERAL CONDITIONS FOR HERMODYNAMIC EQUILIBRIUM

In the preceding section, we established the most general condition for a natural change to occur. If this condition is not satisfied, no change can take place. That is, the system will be in equilibrium, if Eqs. (9.5a, b) do not hold. Thus, the most general condition for equilibrium may be stated as follows:

If for any infinitesimal change in a system, the condition for natural change given by the inequality in Eq. (9.5) is not satisfied, the system is in stable thermodynamic equilibrium.

We now consider some special cases.

9.5.1 An Adiabatic Process

If a system is thermally isolated but it is not mechanically isolated, work-like interactions with the surroundings can take place. The condition for a natural adiabatic process to

$$dS > 0$$

This implies that the condition to be satisfied by a system in equilibrium may be

$$dS \le 0 \tag{9.32}$$

We know that any infinitesimal change can be reversed if the direction of external changes bringing about this change is reversed. This means that if the change dS < 0 implies equilibrium, then dS > 0 also suggests equilibrium. But the latter possibility is ruled out by Eq. (9.6). Hence, the condition for equilibrium reduces to

$$dS = 0 \tag{9.33}$$

For the equilibrium to be stable, the entropy must be maximum. It cannot be a minimum because a system, left to it has a tendency to change in the direction of increasing entropy. Hence, the further constraint for stable equilibrium is

$$d^2S < 0 \tag{9.34}$$

We will now consider two different types of systems.

1. **Thermally Isolated-Isochoric System** The appropriate set of conditions on a thermally isolated-isochoric system in stable equilibrium is

$$dS = 0, \quad dV = 0, \quad dU = 0 \quad \text{and} \quad d^2S < 0 \tag{9.35}$$

2. **Thermally Isolated-Isobaric System** The general condition for a natural change to occur in a thermally isolated isobaric system is given by Eq. (9.7):

$$dH \leq 0$$

Following the argument outlined in the preceding paragraphs, we can say that the set of conditions for stable equilibrium of a thermally isolated-isobaric system is

$$dH = 0, \quad dS = 0, \quad dp = 0 \tag{9.36}$$

9.5.2 An Isothermal Process

During an isothermal process, entropy may change but temperature remains constant. So one of the conditions of equilibrium obviously is $dT = 0$. The system may further be held at constant volume or constant pressure.

1. **Isothermal-Isochoric System** For an isothermal-isochoric system, the conditions of stable equilibrium are

$$dT = 0, \quad dV = 0, \quad dF = 0 \tag{9.37}$$

That is, of the various states that a system may have by isothermal processes without doing any useful work, only those are stable which correspond to a minimum of Helmholtz energy.

2. **Isothermal-Isobaric System** In an isothermal-isobaric system, the condition of stable equilibrium requires a minimum value of Gibbs energy. So the appropriate set of conditions in this case is

$$dT = 0, \quad dp = 0, \quad dG = 0 \tag{9.38}$$

We note that in every set of conditions for stable equilibrium, each free energy appears with its natural variables. It is important to emphasise here that these four set of conditions (Eqs. (9.35) to (9.38)), rest on the law of increase of entropy and are equivalent in the sense that they lead to identical physical results; the choice of a particular set is a matter of convenience under given constraints.

QUILIBRIUM BETWEEN PHASES

From Sec. 8.2.1, we recall that phase transition is a process in which a thermodynamic system changes from one state to another with different physical properties. The phase transition from solid to liquid state is marked by a discontinuous change in the shear strength of the substance, whereas the change to the gaseous phase involves a discontinuous change in density. It is possible to bring about a phase transition in most cases by suitable changes in temperature and/or pressure. The melting of ice to form water or condensation of water into ice (fusion) and evaporation of water to form water vapour are the best known examples of phase transitions. Another familiar example is sublimation in which a solid directly changes from the solid state to the gaseous state. The existence of these changes can be satisfactorily explained on the basis of the equation of state (Chapter 3). Some more interesting and technologically important phase transitions involve abrupt changes in electrical conductivity (gas to plasma, normal conductor to superconductor), magnetism (paramagnetism to ferromagnetism below Curie point) and viscosity (normal helium to superfluid helium).

All phase transitions are caused by the interplay of attractive forces between particles making up the system and their thermal motion. Whereas attractive forces tend to hold them together and trap them in bound states, thermal energy contributes to random and free motion. If enough thermal energy is not available, particles are not able to break free and ultimately remain trapped. This happens in plasma-gas, gas-liquid and liquid-solid phase transitions. Such transitions are called *first order phase transitions* and are marked by discontinuous changes in entropy and volume.

Once trapped, particles get distributed amongst various discrete bound states. If some more thermal energy is taken away from the system, the trapped particles may begin to condense into the ground state. This *thermal imprisonment* of the system is accompanied by a distinct change in a physical property revealing a phase transition. These so-called *second order phase transitions* are gradual and occur over an extremely small range of temperature and/or pressure or any other intensive variable.

It is important to point out here that when a system consists of more than one phase,* each phase should be considered as a separate system within the whole. For simplicity, we first restrict ourselves to pVT systems with uniform chemical composition, i.e., consider one component system which can exist in more than one phases.

One Component System

Consider a system of two phases of the same substance (one component system) comprising a liquid in contact with its vapour. We can imagine that the arrangement is kept in a cylinder fitted with a piston, as shown in Fig. 9.4. The cylinder is in thermal contact with a heat reservoir at constant temperature T and is maintained at constant pressure p.

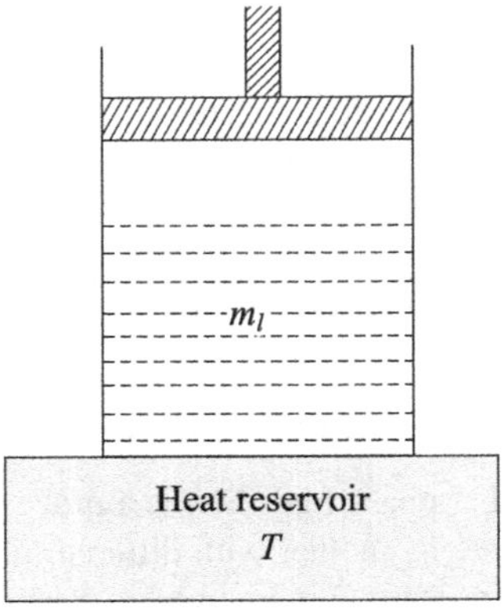

A liquid and its vapour for a one component system in equilibrium.

Suppose that the masses of the liquid and vapour phases are m_l and m_v, respectively for the equilibrium state. The $p - V$ diagram for this system is shown in Fig. 9.5 for two different temperatures. Note that

at temperature T, the portions ab and cd respectively represent the liquid and gaseous states; the corresponding states at temperature $T + \Delta T$ are represented by $a'b'$ and $c'd'$, respectively;

the portions bc and $b'c'$ respectively signify liquid-vapour phase equilibrium at temperatures T and $T + \Delta T$;

the volume of the liquid-vapour phase decreases as temperature increases.

*A phase is defined as a system which is homogeneous in physical structure, i.e., has definite boundaries, and chemical composition. It may be a chemically pure substance or may contain more than one component

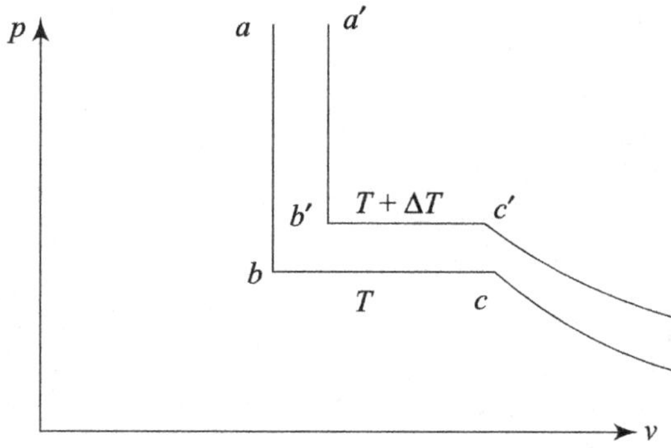

Isotherms for a liquid-vapour system. The horizontal portions represent liquid-vapour phase equilibrium.

g_v be specific Gibbs energies of these phases. Then the Gibbs energy for the whole system in state 1 is given by

$$G_1 = m_\ell g_\ell + m_v g_v \tag{9.39}$$

is a function of p, T, m_ℓ and m_v.

Let us think of a reversible change in the system consistent with external conditions whereby some liquid is converted into the vapour at constant temperature and pressure. This essentially means that some mass of the liquid transfers to the vapour state. Suppose m_ℓ of the liquid vaporises reversibly and the systems approaches a neighbouring equilibrium state 2. If the mass of the vapour produced is $\mathrm{d}m_v$, the Gibbs energy of the system in new equilibrium state 2 is

$$\begin{aligned} G_2 &= (m_\ell - \mathrm{d}m_\ell)g_\ell + (m_v + \mathrm{d}m_v)g_v \\ &= m_\ell g_\ell + m_v g_v + \mathrm{d}m_v g_v - \mathrm{d}m_\ell g_\ell \\ &= G_1 + \mathrm{d}G \end{aligned}$$

Recall the condition of equilibrium for a system at constant temperature and pressure; the Gibbs energy has a minimum value and for a reversible change under these conditions, there will be no change in Gibbs free energy and we can write

$$\mathrm{d}G = g_\ell\,\mathrm{d}m_\ell - g_v\,\mathrm{d}m_v = 0 \tag{9.40}$$

We know that for a closed system, mass is conserved. Therefore, we must have

$$\mathrm{d}m_\ell = \mathrm{d}m_v \tag{9.41}$$

Using this result in Eq. (9.40), we obtain

$$g_\ell = g_v \tag{9.42}$$

That is, when two (or more) phases of a one component system are in equilibrium, the specific Gibbs energies have the same value in both phases. It may be pointed out here that although Eq. (9.42) has been derived for a system at constant T and p, it is quite general. That is, it holds for all conditions of external constraints for the whole system. We now illustrate the use of Eq. (9.42) for first order phase transitions.

1. First Order Phase Transitions: Clausius–Clapeyron Equation We now know that first order phase transitions occur at a particular temperature, pressure or for some other intensive parameter. You may ask: Is it possible to use this information to predict other temperatures and pressures at which phase transitions may occur? For instance, water boils at about 100°C under atmospheric pressure. Can we predict the boiling point if pressure were reduced to half? Or can we predict the pressure under which boiling point may be raised to 200°C? To answer these questions, we must know how p and T are connected when two phases are in equilibrium. The required information is contained in the Clausius–Clapeyron equation (Eq. (8.10)). We will now deduce this important result starting from Eq. (9.42).

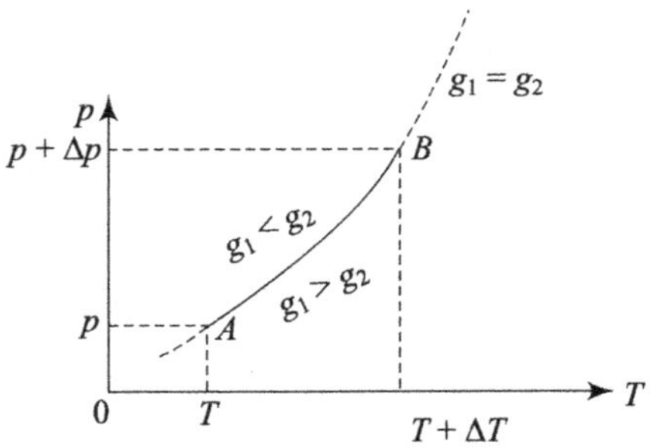

Fig. 9.6 A portion of the curve separating two phases of a typical substance

Now refer to Fig. 9.6. The curve AB shows a portion of the line separating two phases for a typical substance. Along this curve, the two phases will be in equilibrium and, therefore, the specific Gibbs energies of two phases will be the same. In particular, let correspond to temperature T and pressure p. Then, we must have

$$g_1^A = g_2^A \tag{9.43a}$$

If we now increase the temperature to $T + \Delta T$, pressure will change to $p + \Delta p$. Then the saturated vapour pressure and the Gibbs energy will also change. For the two phases to remain in equilibrium, we must have

$$g_1^B = g_2^B \tag{9.43b}$$

so that in passing from A to B, the change in specific Gibbs energies is

$$g_1^B - g_1^A = g_2^B - g_2^A$$

$$\mathrm{d}g_1 = \mathrm{d}g_2 \tag{9.44}$$

$$\mathrm{d}g_1 = \left(\frac{\partial g_1}{\partial p}\right)_T \mathrm{d}p + \left(\frac{\partial g_1}{\partial T}\right)_p \mathrm{d}T$$

$$= v_1\,\mathrm{d}p - s_1\,\mathrm{d}T \tag{9.45a}$$

and s_1, respectively denote specific volume and entropy in phase 1. Similarly, we can write

$$\mathrm{d}g_2 = \left(\frac{\partial g_2}{\partial p}\right)_T \mathrm{d}p + \left(\frac{\partial g_2}{\partial T}\right)_p \mathrm{d}T$$

$$= v_2\,\mathrm{d}p - s_2\,\mathrm{d}T \tag{9.45b}$$

and s_2 are specific volume and entropy in phase 2.

On combining Eqs. (9.44) and (9.45a, b), we can write

$$v_1\,\mathrm{d}p - s_1\,\mathrm{d}T = v_2\,\mathrm{d}p - s_2\,\mathrm{d}T$$

$$\frac{\mathrm{d}p}{\mathrm{d}T} = \frac{s_2 - s_1}{v_2 - v_1} = \frac{\Delta s}{\Delta v} \tag{9.46}$$

$= s_2 - s_1$ and $\Delta v = v_2 - v_1$, respectively denote changes in specific entropy and specific volume in going from one phase to another. This means that first order phase transition is accompanied by changes in entropy and volume. We may therefore conclude that in such transitions, the specific Gibbs energy is continuous but its first order derivatives show discontinuity at the transition temperature (Fig. 9.7). is the latent heat of vaporisation per unit mass required to accomplish the transition, then we

$$L_{\text{vap}} = T\Delta s$$

Hence, we can rewrite Eq. (9.46) as

$$\left(\frac{p}{dT}\right)_{\text{vapour}} = \frac{L_{\text{vap}}}{T(v_2 - v_1)} = \frac{L_{\text{vap}}}{T\Delta v} \qquad (9.47a)$$

Do you recognise this equation? It is *Clausius–Clapeyron equation* for liquid-vapour equilibrium. (Eq. (8.10)). It gives the rate at which pressure must change with temperature for two phases to co-exist in equilibrium. From Sec. 8.2, we recall that an increase in pressure decreases the melting point of water. This lowering of melting point under pressure exerted by the weight of a skater is partly responsible for making ice skating possible.

If water is studied under very high pressures, several crystalline forms of ice are observed. In fact, under very high pressure, melting ice is very hot! (Under a pressure of 25,000 atm, ice melts at about

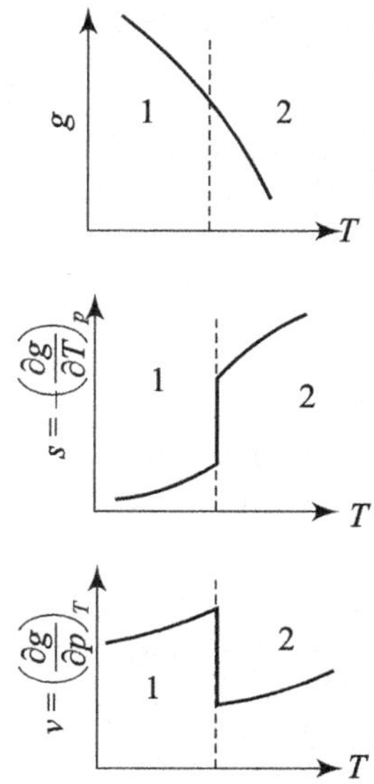

Fig. 9.7 First order phase transition. Temperature variation of (a) specific Gibbs function, (b) specific entropy and (c) specific volume.

In Fig. 8.1, we had depicted the phase diagram of a simple substance as well as water. You may recall that water has only one triple point. But there can be more than one triple point even for one component system. To illustrate that, in Fig. 9.8, we have shown the phase diagram for sulphur. At STP, the stable form of sulphur is rhombic sulphur. If heated slowly, it changes to solid monoclinic sulphur at 95.4°C. It is stable up to 119°C. On further heating, monoclinic sulphur melts and passes into liquid phase. As will be noted, there are three triple points in Fig. 9.8. The equilibrium conditions are

$$= g_{\text{mono}} = g_{\text{gas}} \quad \text{at } 95.4°\text{C}$$
$$= g_{\text{liq}} = g_{\text{gas}} \quad \text{at } 119°\text{C}$$
$$= g_{\text{mono}} = g_{\text{liq}} \quad \text{at } 151°\text{C}$$

During a phase transition, p depends only and is independent of V. This important result may be established if one derives Eq. (9.47a) from the first TdS

Following the same argument, we can express the Clausius–Clapeyron equations for solid-liquid and solid-vapour phases as

$$\left(\frac{dp}{dT}\right)_{\text{fusion}} = \frac{L_{\text{fusion}}}{T\Delta v} \qquad (9.47b)$$

$$\left(\frac{dp}{\quad}\right. \qquad L_{\text{sub}} \qquad (9.47c)$$

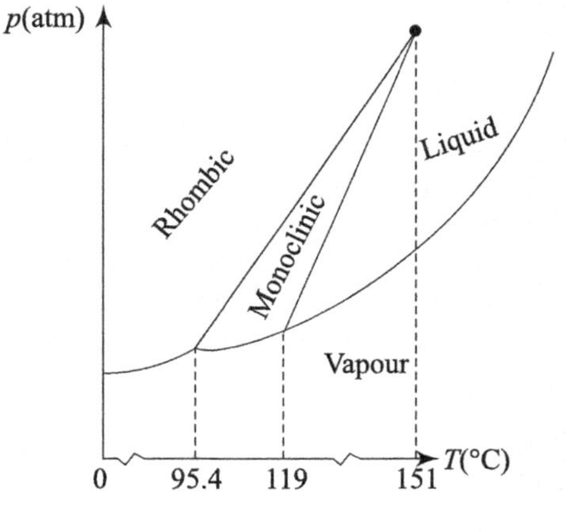

GEOMETRICAL PROOF FOR THE EXISTENCE OF TRIPLE POINT

We now know that phase transitions such as fusion (liquid to solid), melting (solid to liquid), vaporization (liquid to vapour) and sublimation (solid to vapour) are isobaric-isothermal processes. These are characterised by the fact that specific Gibbs energies for both phases are equal (9.42):

$$g_{\text{solid}} - g_{\text{lid}} = 0 \quad \text{(i)}$$

$$g_{\text{liq}} - g_{\text{vap}} = 0 \quad \text{(ii)}$$

$$g_{\text{solid}} - g_{\text{vap}} = 0 \quad \text{(iii)}$$

In coordinate geometry, you may have learnt that the general equation of any curve passing through the intersection of the curves $\sigma_1 = 0$ and $\sigma_2 = 0$ is given by

$$\sigma_1 + C\sigma_2 = 0 \quad \text{(iv)}$$

is an arbitraty constant.

Suppose that Eqs (i) and (ii) are represented by curves 1 and 2, respectively, as shown the Fig. B.1. It then readily follows from Eq. (iv) that the equation of any curve passing through the intersection P of the curves 1 and 2 will be given by

$$(g_{\text{solid}} - g_{\text{liq}}) + C(g_{\text{liq}} - g_{\text{vap}}) = 0$$

We can rewrite it as

$$g_{\text{solid}} + (C - 1)g_{\text{liq}} - Cg_{\text{vap}} = 0$$

Note that for $C = 1$, it reduces to Eq. (iii). It means that the curve representing Eq. (iii) passes through the point of intersection of Eqs. (i) and (ii). In other words, the point is a triple point.

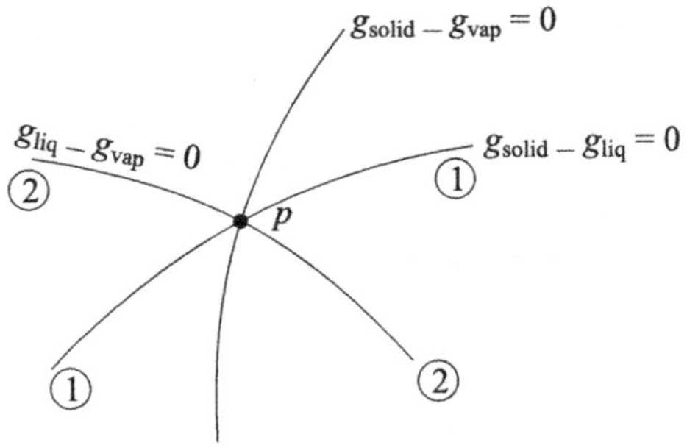

Although the latent heat varies with temperature, it is always positive. Moreover, the specific volume of the vapour phase is always greater than the specific volumes of the liquid or the solid phases. This implies that the slope of the vapour pressure curve and sublimation pressure curve will always be positive. In general, the specific volume of the liquid phase is greater than the specific volume of the solid phase so that the slope of the melting curve will also be positive. (As discussed earlier, water below 4°C is an exception.)

You should now go through the following examples.

Example 9.4 For a two phase system in equilibrium, p is a function of T only, so

$$\left(\frac{\partial p}{\partial T}\right)_V = \left(\frac{\partial p}{\partial T}\right)_S = \frac{\mathrm{d}p}{\mathrm{d}T}$$

$$\frac{C_V}{\beta_S} = TV\left(\frac{\mathrm{d}p}{\mathrm{d}T}\right)^2$$

irrespective of the type of transition.

Let us take T and V as independent variables and write

$$S = S(T, V)$$

$$\mathrm{d}S = \left(\frac{\partial S}{\partial T}\right)_V \mathrm{d}T + \left(\frac{\partial S}{\partial V}\right)_T \mathrm{d}V$$

For an adiabatic process ($\mathrm{d}S = 0$). Therefore, we can write

$$\left(\frac{\partial S}{\partial T}\right)_V = -\left(\frac{\partial S}{\partial V}\right)_T \left(\frac{\partial V}{\partial T}\right)_S$$

Using first Maxwell relation, we can rewrite it as

$$\left(\frac{\partial S}{\partial T}\right)_V = -\left(\frac{\partial p}{\partial T}\right)_V \left(\frac{\partial V}{\partial T}\right)_S$$

Since heat capacity at constant volume $C_V = T(\partial S / \partial T)_V$, we can write

$$C_V = -T\left(\frac{\partial p}{\partial T}\right)_V \left(\frac{\partial V}{\partial T}\right)_S$$

$$= -T\left(\frac{\partial p}{\partial T}\right)_V \left(\frac{\partial V}{\partial p}\right)_S \left(\frac{\partial p}{\partial T}\right)_S$$

$$= TV\,\beta_S \left(\frac{\partial p}{\partial T}\right)_V \left(\frac{\partial p}{\partial T}\right)_S$$

$$= TV\,\beta_S \left(\frac{\mathrm{d}p}{\mathrm{d}T}\right)^2$$

is adiabatic compressibility.

Example 9.5 Discuss the temperature dependence of vapour pressure for first order phase transition. You may assume that variation of the latent heat with temperature is negligible and the vapour behaves like an ideal gas.

Suppose that v_1 and v_2 respectively denote the molar volume of the liquid and the vapour. Obviously, $v_1 << v_2$ and can be ignored in Eq. (9.47a). Then using the ideal gas equation, we can put $v_2 = \frac{RT}{p}$. Hence, we can write

$$\left(\frac{dp}{dT}\right)_{\text{vap}} = \frac{L_{\text{vap}}}{RT^2}p \qquad \text{(i)}$$

This can be rewritten as

$$\frac{dp}{p} = d(\ln p) = \frac{L_{\text{vap}}}{RT^2}dT$$

This equation can be integrated to obtain

$$\ln p = -\frac{L_{\text{vap}}}{RT} + K \qquad \text{(ii)}$$

is the constant of integration.

On taking antilogarithms on both sides of Eq. (ii), we get

$$p = K_0 \exp\left[-\frac{L_{\text{vap}}}{RT}\right] \qquad \text{(iii)}$$

= exp (K). This result shows that the vapour pressure decreases exponentially with temperature and the exact increment is determined by the magnitude of the latent heat of vaporisation.

Example 9.6 An experimentalist discovers that the vapour pressure curve of ammonia on the $p - T$ diagram near the triple point can be fitted with the equation $\ln \; = 24.38 - \frac{3063\text{ K}}{T}$, where ln p is in pascal. Similarly, the solid-vapour equilibrium curve can be fitted with the equation $\ln p = 27.92 - \frac{3754\text{ K}}{T}$. Calculate the triple point pressure and temperature.

We know that the vaporisation and sublimation curves on a $p - T$ diagram meet at the triple point. Therefore, we can write

$$24.38 - \frac{3063\text{ K}}{T_{tr}} = 27.92 - \frac{3754\text{ K}}{T_{tr}}$$

On rearranging terms, we can write

$$\frac{(3754 - 3063)\text{ K}}{T_{tr}} = 27.92 - 24.38$$

$$T_{tr} = \frac{(3754 - 3063)\text{ K}}{27.92 - 24.38} = \frac{691\text{ K}}{3.54} = 195.2\text{ K}$$

Triple point temperature = 195.2 K

On using this result in the given equation, we can determine the triple point pressure:

$$\ln p_{tr} = 27.92 - \frac{3754}{T} = 27.92 - \frac{3754}{195.2} = 8.69$$

$$p_{tr} = \exp(8.69) = 5943.2$$

Triple point pressure = 5493.2 Pa

By comparing the vaporisation and sublimation curves with Eq. (ii) in Example 9.5, you can calculate the latent heats of vaporisation and sublimation in terms of $R : L_{\text{vap}} = 3063R$ $= 3754R$.

Example 9.7 A first order phase transition occurs at constant temperature and pressure. Derive second latent heat equation $\frac{dL}{dT} - \frac{L}{T} = (C_2)_{\text{sat}} - (C_1)_{\text{sat}}$ where $(C_1)_{\text{sat}}$ and are heat capacities of two phases at temperature T and L is the latent heat. (This is also known as *Clausius equation.*) Discuss its implications for specific heat capacity of a

We know that for a first order phase transition

$$s_2 - s_1 = \frac{L}{T} \tag{i}$$

$$\frac{d}{dT}\left(\frac{L}{T}\right) = \frac{ds_2}{dT} - \frac{ds_1}{dT}$$

We can rewrite it as

$$\frac{1}{T}\frac{dL}{dT} - \frac{L}{T^2} = \frac{ds_2}{dT} - \frac{ds_1}{dT}$$

$$\frac{dL}{dT} - \frac{L}{T} = T\frac{ds_2}{dT} - T\frac{ds_1}{dT} = (C_2)_{\text{sat}} - (C_1)_{\text{sat}}\,. \tag{ii}$$

Note that molar heat capacities of two phases are defined for *constant saturation* rather than constant temperature or constant pressure. This equation gives us a method of calculating *specific heat capacity of saturated vapour.*

The specific heat capacity of saturated vapour is defined as the amount of heat required to raise the temperature of a unit mass of the vapour through 1°C, keeping the vapour always saturated. In Example 9.5, we have seen that the vapour pressure of a liquid rises exponentially with temperature. When the temperature of a saturated vapour is raised, the vapour becomes unsaturated. Therefore, to keep it saturated, we have to decrease its volume by applying pressure from outside. It means that both pressure and volume undergo a change and may have to be adjusted so as to keep vapour pressure constant. That is, specific heat capacity of saturated vapour is defined neither with reference to constant volume nor to constant pressure. Moreover, due to application of pressure on the vapour, an external work is done on it and the energy appears as heat inside it and tends to raise its temperature.

Suppose that Q_2 amount of heat is required to raise the temperature of the vapour by 1°C. The increase in temperature will cause the vapour to expand and become unsaturated. To restore saturation, the vapour is compressed and Q_1 amount of heat generated in this process has to be removed from the system, if only, no further rise in temperature is allowed. Now three possibilities can arise:

Q_1 That is, the heat of compression of the vapour is equal to the total heat required to raise the temperature of the vapour through 1°C and no net heat is supplied to the system. The *specific heat capacity of the vapour is then zero.*

Q_1 That is, the heat of compression of the vapour is less than the total heat required to raise the temperature of the vapour through 1°C and a net amount of heat is supplied to the system. *The specific heat capacity of the vapour is then said to be positive.*

Q_1 In this case, the heat of compression of the vapour is greater than the total heat required to raise the temperature of the vapour through 1°C and some heat has to be removed from the system to conserve vapour pressure. The vapour is then said to have *negative specific heat capacity.*

The last prediction appears to be surprising but it was supported by experimental data due to Hirn in the year 1862. To give you an idea about the numerical values, let us consider the case of boiling water. In this case, $T = 100°\text{C} = 373$ K, $L = 540$ cal g^{-1},

0.64 cal g^{-1} K^{-1}, and $(C_1)_{\text{sat}} = 1.01$ cal g^{-1} K^{-1}. Hence,

$$(C_2)_{\text{sat}} = \left(1.01 - 0.64 - \frac{540}{373}\right) \text{cal g}^{-1}\ \text{K}^{-1} = -1.077 \text{ cal g}^{-1}\ \text{K}^{-1}$$

You should now answer a few practise problems.

Problem 9.2 Starting from Clausius–Clapeyron equation, show that internal latent heat can be expressed as $T^2 \dfrac{\partial}{\partial T}\left(\dfrac{p}{T}\right)_V \, dV$.

Problem 9.3 The temperature variation of heat capacity at constant pressure and isobaric expansivity shows that their values become infinite at the transition temperature where both phases are present. Starting from the basic definitions, discuss how far it is correct. What are its possible implications?

Example 9.8 The weight on the steam exhaust opening of a domestic pressure cooker allows the pressure to rise only up to 202 kPa. Calculate the temperature at which water will boil in the cooker. Given $(L_{\text{vap}})_{\text{water}} = 2257$ kJ kg^{-1} at 100°C, = 0.4619 kJ kg^{-1} K^{-1} and 1 atm = 1.01×10^5 Pa.

Using Eq. (ii) in Example 9.5, we can write

$$\ln\left(\frac{p_2}{p_1}\right) = \frac{L_{\text{vap}}}{R}\left[\frac{1}{T_1} - \frac{1}{T_2}\right] \tag{i}$$

$= 202 \times 10^3$ Pa, $p_1 = 1.01 \times 10^5$ Pa $= 101 \times 10^3$ Pa, $R = 0.4619 \times 10^3$ J kg^{-1}K^{-1}, $= 2257 \times 10^3$ J kg^{-1} and $T_1 = 373$ K. On substituting these values in (i), we get

$$\ln\left(\frac{202 \times 10^3 \text{ Pa}}{101 \times 10^3 \text{ Pa}}\right) = \frac{2257 \times 10^3 \text{ J kg}^{-1}}{0.4619 \times 10^3 \text{ J kg}^{-1}\text{K}^{-1}}\left[\frac{1}{373\text{ K}} - \frac{1}{T_2}\right]$$

$$\ln 2 = 0.6920 = \left[\frac{4886.3}{373} - \frac{4886.3\text{ K}}{T_2}\right]$$

$$T_2 = \frac{4886.3\text{ K}}{12.407} = 393.8\text{ K} = 120.8°\text{C}$$

2. Second Order Phase Transitions: Ehrenfest Equations We now know that in a pure substance, first order phase transitions (melting, vapourisation, sublimation) occur at constant temperature and constant pressure. These are marked by continuity of Gibbs energy but finite changes are observed in entropy and volume at the transition temperature. However it has been observed, particularly with liquid helium, that during a phase transition, specific Gibbs energy, entropy and volume remain the same but discontinuities appear in constant pressure heat capacity and isobaric volume expansivity. In the language of mathematics, we can say that the specific Gibbs energy and its first derivatives are continuous but its second derivatives show discontinuous changes as a substance passes from one phase to another (Fig. 9.9). (Since s remains constant, there is no latent heat.) Such transitions are termed second order transitions.*

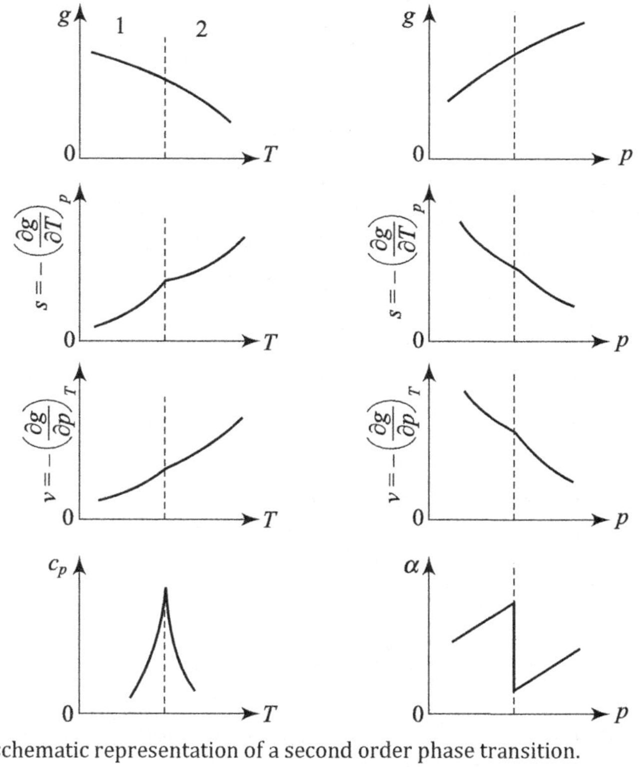

A schematic representation of a second order phase transition.

If we apply Clausius–Clapeyron equation to transitions of order higher than one, we would obtain an indeterminate result. It is because both numerator and denominator become zero. This led Ehrenfest to use the equalities of the entropies or volumes instead of the

*Following Ehrenfest, we define the order of a transition as order of the lowest differential of the specific

specific Gibbs energies at the transition. Following him, we assume that if the transition occurs at temperature T and pressure p, we first take*

$$s_1 = s_2 \tag{9.48}$$

As temperature is raised to $T + \mathrm{d}\,T$ and pressure to $p + d\,p$, we can similarly write

$$s_1 + \mathrm{d}\,s_1 = s_2 + \mathrm{d}\,s_2$$

$$\mathrm{d}\,s_1 = \mathrm{d}\,s_2 \tag{9.49}$$

As before, we can choose s to be a function of T and p so that

$$s = s\,(T, p)$$

An infinitesimal change in entropy can be expressed as

$$\mathrm{d}s = \left(\frac{\partial s}{\partial T}\right)_p \mathrm{d}T + \left(\frac{\partial s}{\partial p}\right)_T \mathrm{d}p$$

Using second Maxwell relation (Eq. (8.11)), we can write

$$\mathrm{d}s = \left(\frac{\partial s}{\partial T}\right)_p \mathrm{d}T - \left(\frac{\partial v}{\partial T}\right)_p \mathrm{d}p$$

$$= \frac{c_p}{T}\,\mathrm{d}T - v\alpha\,\mathrm{d}p \tag{9.50}$$

$T\,(\partial s / \partial T)_p$ is specific heat capacity at constant pressure and $\alpha = \frac{1}{v}\left(\frac{\partial v}{\partial T}\right)_p$ is isobaric volume expansivity.

Since the specific volume v is the same in both phases, we can write

$$\mathrm{d}s_1 = \frac{c_{p_1}}{T}\,\mathrm{d}T - v\alpha_1\,\mathrm{d}p \tag{9.51a}$$

$$\mathrm{d}s_2 = \frac{c_{p_2}}{T}\,\mathrm{d}T - v\alpha_2\,\mathrm{d}p \tag{9.51b}$$

On combining Eqs. (9.49) and (9.51 a, b), we obtain

$$\frac{c_{p_1}}{T}\,\mathrm{d}T - v\alpha_1\,\mathrm{d}p = \frac{c_{p_2}}{T}\,\mathrm{d}T - v\alpha_2\,\mathrm{d}p$$

$$\frac{\mathrm{d}p}{\mathrm{d}T} = \frac{c_{p_2} - c_{p_1}}{Tv(\alpha_2 - \alpha_1)} \tag{9.52}$$

-transition for He I and He II, is characterised by the following experimental values of various quantities occurring in Eq. (9.52):

$$c_{p_1} = 1.2 \text{ cal g}^{-1},\ c_{p_2} = 2.9 \text{ cal g}^{-1},\ \alpha_1 = 0.2 \text{ K}^{-1},$$
$$\alpha_2 = -0.04 \text{ K}^{-1},\ v = 6.84 \text{ cm}^3\,\text{g}^{-1} \quad \text{and} \quad T = 2.19 \text{ K}.$$

* Note that we are working with specific entropy because in a second order phase transition, specific Gibbs energy and specific entropy remain constant.

On substituting these in Eq. (9.52), we get dp/dT = –78.4 atm K^{-1}. This agrees fairly well with the observed value of – 81.9 atm K^{-1}.

Similarly, if we start by assuming $v_1 = v_2$, we get for an infinitesimal change

$$\mathrm{d}v_1 = \mathrm{d}v_2 \tag{9.53}$$

Expanding dv instead of ds, in terms of T and p, we can write

$$\mathrm{d}v = \left(\frac{\partial v}{\partial T}\right)_p \mathrm{d}T + \left(\frac{\partial v}{\partial p}\right)_T \mathrm{d}p$$

$$= v\alpha\, \mathrm{d}T - v\beta_T\, \mathrm{d}p \tag{9.54}$$

$$= -\frac{1}{v}\left(\frac{\partial v}{\partial p}\right)_T$$ is isothermal compressibility.

Proceeding as before, we can write

$$v\alpha_1\, \mathrm{d}T - v\beta_{T_1}\, \mathrm{d}p = v\alpha_2\, \mathrm{d}T - v\beta_{T_2}\, \mathrm{d}p$$

$$\frac{\mathrm{d}p}{\mathrm{d}T} = \frac{\alpha_2 - \alpha_1}{\beta_{T_2} - \beta_{T_1}} \tag{9.55}$$

Equations (9.52) and (9.55) are known as *Ehrenfest's equations* for second order phase

It may be remarked here that the superconducting transition of a metal in the absence of the magnetic field provides the only example of an ideal second order phase transition. The lambda transition in liquid ^{4}He, transition from ferromagnetism to paramagnetism at the Curie temperature, structural transitions from one solid phase to another are examples of nearly second-order phase transitions in the sense that g, s and v are continuous across the transition point but c_p, β_T, and α do not exhibit the behaviour expected theoretically. In all these transitions, the new phase is characterised by greater order. For this reason, second order phase transitions are also called *order-disorder transitions.*

The phase diagram for ^{4}He, the abundant isotope of helium, is shown in Fig. 9.10. Note that helium has two triple points but neither of these refers to the co-existence of solid, liquid and vapour phases. It is because helium remains liquid down to absolute zero at atmospheric pressure. This is due to extremely weak forces between helium atoms. (In a sense, helium is an exceptional substance. Scientists are of the view that helium is present in abundance in lunar rocks and at some future time, it may be a source of relatively inexpensive energy.)

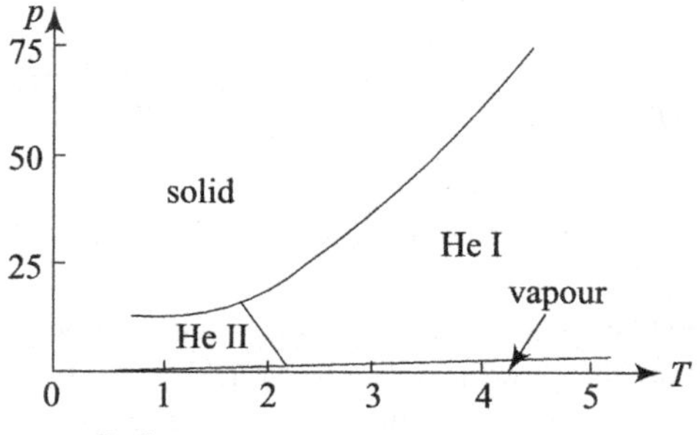

Phase diagram of helium.

At atmospheric pressure, helium condenses into liquid at 4.2 K. As temperature is lowered further, it undergoes another phase change at 2.17 K to a liquid of very surprising properties. This liquid is known as helium-II, abbreviated as He II to differentiate it from normal liquid, which we call helium I (He I). The transition temperature of 2.17 K is referred to as the lambda (λ) point because of the shape of the specific heat capacity curve (Fig. 9.9). As mentioned earlier, in He I $\rightarrow$ He II transition, no latent heat is evolved. That is, it is not accompanied by bubbling or boiling.

He II has zero viscosity. That is, it behaves as a *super-fluid*. This means that He II will flow through the smallest crack without any impediment. If you put it in circular motion around the container, it will keep on going forever. Similarly, the infinite thermal conductivity means that regardless of where you heat the liquid, it will evaporate from the top surface.

The transition from normal to superfluid state is well understood now and you will learn more about it in Chapter 15.

So far we have derived conditions for equilibrium of a one component system which may exist in two or more phases. You will recall that mass was conserved in this case. But when two or more phases are in equilibrium, there is inter-phase flow of matter. We will now generalise these ideas to multi-component systems.

Multi-component Systems: Gibbs Phase Rule

Consider a general system of P phases and C components. For the sake of generality, we assume that each of the C components can exist in all the P phases. We shall indicate phases by superscripts and components by subscripts. We assume that

No surface effect or chemical reaction takes place in the system, and
The system is at constant temperature T and pressure p.

We know that at a given temperature and pressure, the condition for thermodynamic equilibrium of any system is that its Gibbs energy is a minimum. Further, since the surface effects have been ignored, G for the entire system can be expressed as a sum of Gibbs energies of all the phases, i.e.,

$$G = G^{(1)} + G^{(2)} + \dots + G^{(P)} \tag{9.56}$$

For any phase, the Gibbs energy depends upon T, p and the masses of the different components in this phase. Hence, for the ith phase we can write

$$G^{(i)} = G^{(i)}(T, p, m_1^{(i)}, m_2^{(i)}, \dots m_C^{(i)}) \tag{9.57}$$

We can express the differential change in $G^{(i)}$ as

$$\mathrm{d}G^{(i)} = \left(\frac{\partial G^{(i)}}{\partial T}\right)_{p,\, m_j^{(i)}} \mathrm{d}T + \left(\frac{\partial G^{(i)}}{\partial p}\right)_{T,\, m_j^{(i)}} \mathrm{d}p + \sum_{j=1}^{C}\left(\frac{\partial G}{\partial m_j^{(i)}}\right)_{T,\, p} \mathrm{d}m_j^{(i)}$$

$$= -S^{(i)}\,\mathrm{d}T + V^{(i)}\,\mathrm{d}p + \sum_{j=1}^{C} g_j^{(i)}\,\mathrm{d}m_j^{(i)} \tag{9.58}$$

The quantities $g_j^{(i)}$ are known as *chemical potentials* of the jth component in the ith phase. It is evident from Eq. (9.58) that these are intensive variables. Further, they must be functions of state since they can only depend on temperature, pressure and composition of the phase.

Suppose we increase the total mass of a phase while keeping T, p and composition constant. Then, Eq. (9.58) reduces to

$$\mathrm{d}G^{(i)} = \sum_{j=1}^{C} g_j^{(i)}\, \mathrm{d}m_j^{(i)} = \sum_{j=1}^{C} g_j^{(i)}\, m_j^{(i)}\, \mathrm{d}x \tag{9.59}$$

$\mathrm{d}x = \mathrm{d}m_j^{(i)} / m_j^{(i)}$ is the fraction by which mass is increased. Eq. (9.59) may now be integrated to give

$$G^{(i)} = \sum_{j=1}^{C} g_j^{(i)}\, m_j^{(i)} \tag{9.60}$$

This equation shows that the chemical potential is analogous to the specific Gibbs energy of a pure substance. In fact, in a phase containing a single component, they are identical but in a mixture chemical potential depends on concentration and proportions of other components.

In the above discussion, we have used m_j as the extensive parameter to bring out the connection between chemical potential and the specific Gibbs energy of a substance. Often, it is more convenient to work in molar quantities and use the number of moles n_j of the components rather than masses. Then, Eq. (9.58) takes the form

$$\mathrm{d}G = -S\,\mathrm{d}T + V\,\mathrm{d}p + \sum_{j=1}^{C} \mu_j\, \mathrm{d}n_j \tag{9.61}$$

is known as the *molar potential*. It is defined as

$$\mu_j = M_j g_j \tag{9.62}$$

is molar mass of the jth component. In terms of the molar potential, one $G^{(i)}$ as

$$G^{(i)} = \sum_{j=1}^{C} \mu_j^{(i)}\, n_j^{(i)} \tag{9.63}$$

so that Gibbs energy of the multi-phase, multi-component system is given by

$$G = \sum_{i=1}^{P} G^{(i)} = \sum_{j=1}^{C} \sum_{i=1}^{P} \mu_j^{(i)}\, n_j^{(i)} \tag{9.64}$$

For equilibrium at constant temperature and pressure, G will be minimum, i.e., we must

$$\mathrm{d}G = 0$$

Substituting for G from Eq. (9.64), we get

$$\sum_{j=1}^{C} \sum_{i=1}^{P} \mu_j^{(i)}\, \mathrm{d}n_j^{(i)} = 0 \tag{9.65}$$

As there are no chemical reactions, the total number of moles of each component must remain constant. Therefore, the system is further subject to the constraint

$$\sum_{i=1}^{P} n_j^{(i)} = \text{constant}$$

$$\sum^{P} \mathrm{d}n_j = 0 \quad \text{for each } j \tag{9.66}$$

To relax this condition, we use the method of undetermined multipliers. So we multiply Eq. (9.66) by λ_j for each component and add the resultant to Eq. (9.65). Thus, the general condition of equilibrium for a multiphase, multi-component system can be expressed as

$$\sum_{j=1}^{C}\sum_{i=1}^{P}(\mu_j^{(i)}+\lambda_j)\,\mathrm{d}n_j^{(i)}=0 \tag{9.67a}$$

We can rewrite it as

$$\sum_{=1}^{C}\Big((\mu_j^{(1)}+\lambda_j)\,\mathrm{d}n_j^{(1)}+(\mu_j^{(2)}+\lambda_j)\,\mathrm{d}n_j^{(2)}+\ldots+(\mu_j^{P}+\lambda_j)\,\mathrm{d}n_j^{(P)}\Big)=0 \tag{9.67b}$$

In the expanded form, we have

$$\begin{aligned}
&(\mu_1^{(1)}+\lambda_1)\,\mathrm{d}n_1^{(1)}+(\mu_1^{(2)}+\lambda_1)\,\mathrm{d}n_1^{(2)}+\ldots+(\mu_1^{(P)}+\lambda_1)\,\mathrm{d}n_1^{(P)}\\
+\;&(\mu_2^{(1)}+\lambda_2)\,\mathrm{d}n_2^{(1)}+(\mu_2^{(2)}+\lambda_2)\,\mathrm{d}n_2^{(2)}+\ldots+(\mu_2^{(P)}+\lambda_2)\,\mathrm{d}n_2^{(P)}\\
&\vdots\\
&\vdots\\
&\vdots\\
+\;&(\mu_C^{(1)}+\lambda_C)\,\mathrm{d}n_C^{(1)}+(\mu_C^{(2)}+\lambda_C)\,\mathrm{d}n_C^{(2)}+\ldots+(\mu_C^{(P)}+\lambda_C)\,\mathrm{d}n_C^{(P)}=0
\end{aligned} \tag{9.67c}$$

For this equation to hold, the coefficients of the various dn terms must vanish separately:

$$\begin{aligned}
&\mu_1^{(1)}+\lambda_1=0,\quad \mu_1^{(2)}+\lambda_1=0,\ldots\mu_1^{(P)}+\lambda_1=0\\
&\mu_2^{(1)}+\lambda_2=0,\quad \mu_2^{(2)}+\lambda_2=0,\ldots\mu_2^{(P)}+\lambda_2=0\\
&\vdots\\
&\vdots\\
&\mu_C^{(1)}+\lambda_C=0,\quad \mu_C^{(2)}+\lambda_C=0,\ldots\mu_C^{(P)}+\lambda_C=0
\end{aligned} \tag{9.68}$$

For these relations to hold for any component, we must have

$$\begin{aligned}
&\mu_1^{(1)}=\mu_1^{(2)}=\ldots=\mu_1^{(P)}=-\lambda_1\\
&\mu_2^{(1)}=\mu_2^{(2)}=\ldots=\mu_2^{(P)}=-\lambda_2\\
&\vdots\qquad\qquad\vdots\\
&\mu_C^{(1)}=\mu_C^{(2)}=\ldots=\mu_C^{(P)}=-\lambda_C
\end{aligned} \tag{9.69}$$

This result shows that molar partial potential of any component is conserved in all phases at equilibrium. However, different components have different potentials. It is important to note that as expected, these results are generalisation of the condition for equilibrium between phases of a single component.

We will now determine the total number of variables of equilibrium.

In every phase, we have C components, all of which may have different concentrations. Since the total number of moles is fixed for a closed system, only $(C-1)$ of these will be independent for each phase. Hence, for all the P phases, the number of composition variables would be $P(C-1)$. In addition, we have a choice of temperature and pressure

to fix the state of the system. Therefore, the total number of variables at our disposal is*
– 1) + 2. Of these, some are already fixed. From the set of Eqs. (9.69) we note that component exists in two phases, we have *one* restriction ($\mu_j^{(1)} = \mu_j^{(2)} = \mu_j^{(3)}$) and P phases of any component, we have (P – 1) restrictions. Therefore, the total number of restrictions for the system to be in equilibrium are $C(P – 1)$. Hence, the number of degrees of freedom f of the system, which signify the number of independent variables that can be arbitrarily chosen without affecting equilibrium, is given by

$$\begin{aligned} f &= \text{number of variables} - \text{number of restrictions} \\ &= P(C-1) + 2 - C(P-1) \\ &= C - P + 2 \end{aligned}$$

$$P + f = C + 2 \tag{9.70}$$

This is known as *Gibbs phase rule*. It holds when no chemical reaction takes place in the

Example 9.9 For a one component pVT system, show that the maximum number of phases that can co-exist at equilibrium is three.

From the Gibbs phase rule, we have

$$P = C - f + 2$$

= 1. Further, for maximum number of phases to co-exist at equilibrium, the degrees of freedom should be minimum, which we may take to be zero. (Such a system is said to be non-variant.) Hence,

$$P = 1 + 2 = 3$$

That is, a pure substance cannot have more than three phases co-existing at equilibrium. This fact has been experimentally confirmed by Tammann and Bridgman. In case of water, they even observed five varieties of ice but in no case more than three phases co-existed. Similarly, we know that sulphur can exist in four phases—vapour, liquid, rhombic and monoclinic forms—but only three can co-exist at a time.

Moreover, since $C = 1$, the phase rule tells us that for one phase $f = 2$, i.e., one can vary p without disturbing equilibrium. Similarly, for two phases $f = 1$, i.e., one can vary either pressure or temperature and yet keep equilibrium. However, for three phases to co-exist we must have $f = 0$. The system is then said to be *bivariant, monovariant* or , respectively. As an example of an invariant system, we take the case of water at the triple point defined by $T_{tp} = 0.0075$°C and $p = 4.58$ mm (Fig. 8.1b). If we raise the temperature, ice disappears and if we lower the temperature, water disappears.

Next we consider another example in three parts.

A bell jar is inverted over a water beaker. The air in the jar will get saturated in due course of time. It means that here water and air are the two components, i.e., $C = 2$. You can easily visualise that these can exist in two phases as liquid water and air-vapour mixture, i.e., $P = 2$. Hence, Gibbs phase rule dictates that the number of degrees of

*If extra factors, such as gravitational energy, magnetic field and electric field influence the nature of phase, the numerical factor of two will increase accordingly.

freedom $f = 2 - 2 + 2 = 2$. It means that one can vary two parameters, which could be p. The pressure can be varied by connecting the bell jar to a pump.

If air in point 1 above is kept at atmospheric pressure and alcohol is added to water, the system will have three components (water, air and alcohol) and two phases (water-alcohol mixture and air-vapour mixture), i.e., $C = 3$ and $P = 2$. Hence, $f = 3 - 2 + 2 = 3$. Therefore, the variables could be temperature, pressure and concentration of alcohol in water.

Suppose some ice is in equilibrium with water and alcohol mixture. Then $C = 3$ (water, alcohol and air), $P = 3$ (ice, water-alcohol mixture and air-vapour mixture). So $= 3 - 3 + 2 = 2$. In this case, the variables are pressure and alcohol concentration. It means that temperature will not be an independent variable in this case.

Problem 9.4 (a) Calculate the number of degree of freedom for a pVT system consisting of a vapour in equilibrium with a liquid of (i) a single component, and (ii) two non-interacting chemical components.

(b) Common salt is added gradually in a beaker containing ice, and water in equilibrium under a bell jar, which has air at atmospheric pressure. Determine the degrees of freedom and identify the variables when some salt remains undissolved.

(a) (i) 1, (ii) 2; (b) $f = 1$, pressure

DITIONAL EXAMPLES

Example 9.10 The latent heat of vaporisation of ammonia, expressed in Jg^{-1} at t°C is represented by the equation

$$L = 138.5(135 - t)^{1/2} - 2.48(135 - t)$$

Calculate the specific heat capacity of ammonia vapour at constant saturation at 35°C. The specific heat capacity of the saturated liquid is 4.82 $\text{Jg}^{-1}\ \text{K}^{-1}$.

From the second latent heat equation, we recall that

$$(C_{\text{vapour}})_{\text{sat}} - (C_{\text{liquid}})_{\text{sat}} = \frac{dL}{dT} - \frac{L}{T}.$$

$= [138.5\,(135 - t)^{1/2} - 2.48(135 - t)]\ \text{Jg}^{-1}$, the first derivative w.r.t. T is

$$\frac{dL}{dT} \equiv \frac{dL}{dt} = \left[\frac{1}{2}\frac{138.5}{(135 - t)^{1/2}}(-1) + 2.48\right] \text{Jg}^{-1}\text{K}^{-1}$$

$$\left.\frac{dL}{dt}\right|_{t=35^\circ\text{C}} = \left(-\frac{1}{2} \times \frac{138.5}{10} + 2.48\right) \text{Jg}^{-1}\text{K}^{-1}$$

$$= (-6.93 + 2.48)\ \text{Jg}^{-1}\text{K}^{-1} = -4.45\ \text{Jg}^{-1}\text{K}^{-1}$$

Similarly, we can calculate latent heat at 35°C using the given relation:

$$L = (138.5 \times 10 - 2.48 \times 100)\ \text{Jg}^{-1}$$

$$= (1385 - 248)\ \text{Jg}^{-1} = 1137\ \text{Jg}^{-1}$$

$$\left.\frac{L}{T}\right|_{308\text{ K}} = \frac{1137}{308}\text{J g}^{-1}\text{K}^{-1}$$
$$= 3.69 \text{ J g}^{-1}\text{K}^{-1}$$

$$(C_{\text{vapour}})_{\text{sat}} = 4.82\,\text{Jg}^{-1}\text{K}^{-1} - 4.45\,\text{Jg}^{-1}\text{K}^{-1} - 3.69\,\text{Jg}^{-1}\text{K}^{-1}.$$
$$= -3.32\,\text{Jg}^{-1}\text{K}^{-1}$$

You may now ask: What is the significance of negative sign with specific heat capacity. It implies that heat must be withdrawn to maintain saturation conditions at a higher temperature.

Example 9.11 Water is converted into steam at atmospheric pressure and 100°C. If entropies of water and steam are 0.31 cal g^{-1} K^{-1} and 1.76 cal g^{-1} K^{-1} respectively, calculate (a) the heat of vaporisation. If the enthalpy of steam is 640 cal g^{-1}, (b) calculate the enthalpy of water under the given conditions, and (c) the Gibbs free energy of water

(a) The heat of vaporisation is given by

$$L = T\Delta S = (373\text{ K})\times(1.76-0.31)\text{ cal g}^{-1}\text{K}^{-1}$$
$$= 373\times 1.45\text{ cal g}^{-1}$$
$$= 541\text{ cal g}^{-1}.$$

(b) To calculate the enthalpy of water, we use the relation $dH = TdS + Vdp$. Since water is converted into steam under isobaric-isothermal conditions, we can write

$$H_{\text{steam}} - H_{\text{water}} = TdS$$

$$H_{\text{water}} = H_{\text{steam}} - TdS$$

On substituting the given values, we get

$$H_{\text{water}} = (640-541)\text{ cal g}^{-1}$$
$$= 99\text{ cal g}^{-1}$$

(c) Recall that the Gibbs free energy is defined as

$$G = H - TS$$

$$G_{\text{water}} = H_{\text{water}} - (373\text{ K})\times S_{\text{water}}$$
$$= 99\text{ cal g}^{-1} - (373\text{ K})\times(0.31\text{ cal g}^{-1}\text{K}^{-1})$$
$$= (99-115.6)\text{ cal g}^{-1}$$
$$= -16.6\text{ cal g}^{-1}$$

Similarly, Gibbs free energy of steam is given by

$$G_{\text{steam}} = H_{\text{steam}} - (373\text{ K})\times S_{\text{steam}}$$
$$= (640\text{ cal g}^{-1}) - (373\text{ K})\times(1.76\text{ cal g}^{-1}\text{K}^{-1})$$

$$= (640 - 656.5) \text{ cal g}^{-1}$$
$$= -16.5 \text{ cal g}^{-1}.$$

Example 9.12 The vapour pressure, in mm of Hg, of a substance in solid state is given by the relation $\ln p = 23.03 - \frac{3754\text{ K}}{T}$, where T in degree Kelvin. The vapour pressure, in mm of Hg, of the substance in liquid state is given by the relation $\ln p = 19.49 - \frac{3063\text{ K}}{T}$. Calculate (a) the coordinates of the triple point, (b) latent heat of vaporisation at the triple point and (c) latent heat of melting at the triple point, if the latent heat of sublimation at the triple point is 7508 cal mol^{-1}. Also, comment on the nature of the gas.

(a) We are told that on the $p - T$ diagram, the triple point satisfies the equation

$$23.03 - \frac{3754\text{ K}}{T_{tr}} = 19.49 - \frac{3063\text{ K}}{T_{tr}} \tag{i}$$

On rearranging terms, we can write

$$\frac{(3754 - 3063)\text{ K}}{T_{tr}} = 23.03 - 19.49$$

$$T_{tr} = \frac{691\text{ K}}{3.54} = 195.2\text{ K} \tag{ii}$$

This corresponds to ammonia.
On substituting this value of T_{tr} in the equation defining solid state of the substance, we get

$$\ln p = 23.03 - \frac{3754\text{ K}}{195.2\text{ K}}$$
$$= 23.03 - 19.23$$
$$= 3.8$$

$$p = \exp(3.8) = 44.7 \text{ mm of Hg.}$$

Thus the coordinates of the triple point are T_{tr} = 195.2 K and p = 44.7 mm of Hg.
(b) From the equation relating the vapour pressure and temperature of liquid ammonia

$$\ln p = 19.49 - \frac{3063\text{ K}}{T} \tag{iii}$$

we get on differentiation

$$\frac{dp}{p} = \frac{3063\text{ K}}{T^2} dT$$

$$\frac{dp}{dT} = \frac{3063\, p}{T^2}\text{ K} \tag{iv}$$

From Classius–Clapeyron equation, we know that

$$\frac{dp}{dT} = \frac{L}{TV} \tag{v}$$

On rearrangement, we can write

$$L = TV\frac{dp}{dT}$$

$$= TV \times \frac{(3063\ K)\,p}{T^2}$$

$$= \frac{pV}{T} \times (3063\ K)$$

$$= (3063\ K)\ R$$

$$= (3063\ K) \times 8.314\ J\ mol^{-1}\ K^{-1}$$

$$= 2.55 \times 10^4\ J\ mol^{-1}.$$

(c) To calculate latent heat of melting, we use the relation

$$L_{melting} = L_{sublimation} - L_{vapourisation}$$

$$= (7508\ cal\ mol^{-1} \times 4.2\ J\ cal^{-1}) - 2.55 \times 10^4\ J\ mol^{-1}$$

$$= (3.15 \times 10^4 - 2.55 \times 10^4)\ J\ mol^{-1}$$

$$= 6 \times 10^3\ J\ mol^{-1}$$

$$= 1.43 \times 10^3\ cal\ mol^{-1}.$$

Example 9.13 The melting point of ice increases by 0.0075°C for one atmosphere decrease in pressure. If pressure of its saturated vapours at 0°C and 1°C are 4.60 mm and 4.94 mm of Hg, calculate the coordinates of the triple point.

Suppose that the coordinates of the triple point are t°C and p mm of Hg. From the given data, we note that increase in saturated vapour pressure for 1°C rise in temperature is (4.94 – 4.60) mm of Hg = 0.34 mm of Hg. Hence, increase in vapour pressure for t°C rise in temperature = 0.34 t. Thus, the equation defining saturated vapour pressure (in mm of Hg) at triple point is given by

$$p = 4.60 + 0.34\,t \tag{i}$$

We know that the melting point of ice at 760 mm of Hg pressure is 0°C and the pressure at the triple point is given by Eq. (i). Hence, decrease in pressure in mm of Hg is

$$\Delta p = 760 - (4.60 + 0.34\,t)$$

$$= 755.4 - 0.34\,t$$

We are told that decrease in pressure by 760 mm (= 1atm) increases the melting point of ice by 0.0075°C. Hence, increase in melting point for decrease of pressure by (755.4 – 0.34t) mm of Hg will be

$$\Delta T = \frac{(0.0075°\,C) \times (755.4 - 0.34\,t)\ mm\ of\ Hg}{760\ mm\ of\ Hg}$$

$$t = 0°\,C + \frac{(0.0075°\,C) \times (755.4 - 0.34\,t)\ mm\ of\ Hg}{760\ mm\ of\ Hg}$$

$$= \frac{0.0075 \times (755.4 - 0.34\,t)}{760}\,{}^\circ\text{C}$$

$$760\,t = 5.67 - 2.55 \times 10^{-3}\,t$$

$$t = 0.00746^\circ\text{C}$$

On substituting this value of t in (i) we get

$$p = (4.60 + 0.34 \times 0.00746)\,\text{mm of Hg}$$
$$= 4.603\,\text{mm of Hg}$$

Example 9.14 Calculate the specific heat capacity of saturated steam at 100°C using the following data:

$L(90^\circ\text{C}) = 545.25$ cal, $L(100^\circ\text{C}) = 539.30$ cal and $L(110^\circ\text{C}) = 533.17$ cal.

Specific heat capacity of water at 100°C is $1.013\ \text{cal g}^{-1}\,\text{K}^{-1}$.

From second latent heat equation, we recall that

$$(C_{\text{steam}})_{\text{sat}} - (C_{\text{water}})_{\text{sat}} = \frac{dL}{dT} - \frac{L}{T}$$

Since we are given values of latent heat at three different temperatures, we can calculate the rate of change of latent heat with temperature as

$$\frac{dL}{dT} = \frac{L(110^\circ\text{C}) - L(90^\circ\text{C})}{20\ \text{K}}\ \text{cal g}^{-1}$$

$$= \frac{(533.17 - 545.25)\ \text{cal g}^{-1}}{20\ \text{K}}$$

$$= -\frac{12.08\ \text{cal g}^{-1}}{20\ \text{K}}$$

$$= -0.604\ \text{cal g}^{-1}\,\text{K}^{-1}$$

$$(C_{\text{steam}})_{\text{sat}} = (1.013\ \text{cal g}^{-1}\text{K}^{-1}) - 0.604\ \text{cal g}^{-1}\text{K}^{-1} - \frac{539.30\ \text{cal g}^{-1}}{373\ \text{K}}$$

$$= -1.037\ \text{cal g}^{-1}\,\text{K}^{-1}.$$

The negative specific heat capacity of saturated steam implies that heat is given out when liquid and saturated vapour are in contact.

Example 9.15 For a first-order phase transition, show that

$$\Delta U = \Delta H\left(1 - \frac{d\ln T}{d\ln P}\right)$$

From Eq. (9.18), we recall that enthalpy can be expressed in terms of Gibbs energy, temperature and entropy as

So an infinitesimal change in enthalpy is given by

$$\Delta H = \Delta G + T\Delta S + S\Delta T. \tag{i}$$

For a first-order phase transition, $\Delta G = 0$ and $\Delta T = 0$. Hence, Eq. (i) reduces to

$$(\Delta H)_{T,\,p} = T(\Delta S)_{T,\,p} \tag{ii}$$

From Clausius–Clapeyron equation, we can write

$$\frac{\Delta p}{\Delta T} = \frac{\Delta S}{\Delta V} = \frac{(\Delta H)_{T,\,p}}{T\Delta V} \tag{iii}$$

In terms of U, p and V, enthalpy is defined as

$$H = U + pV$$

so that an infinitesimal change in it can be expressed as

$$\Delta H = \Delta U + p\Delta V + V\Delta p$$

For first order transition,

$$(\Delta H)_p = \Delta U + p\Delta V$$

$$\Delta V = \frac{(\Delta H)_p - \Delta U}{p} \tag{iv}$$

On combining Eqs. (iii) and (iv), we get

$$\frac{\Delta p}{\Delta T} = \frac{(\Delta H)_{T,\,p}}{T} \cdot \frac{p}{(\Delta H)_p - \Delta U}$$

$$\frac{\Delta T}{T} \cdot \frac{p}{\Delta p} = \frac{\Delta H - \Delta U}{\Delta H}$$

Considering ΔT and Δp as differentials of T and p, respectively, we can write

$$\frac{\mathrm{d}(\ln T)}{\mathrm{d}(\ln p)} = 1 - \frac{\Delta U}{\Delta H}$$

$$\Delta U = \Delta H\left(1 - \frac{\mathrm{d}\ln T}{\mathrm{d}\ln p}\right)$$

Example 9.16 The specific Gibbs energy of a gas is given as

$$g = RT\ln\left(\frac{p}{p_0}\right) - Bp$$

is a function of T only. Obtain the equation of state of the gas.

We know that

$$g = RT\ln\left(\frac{p}{p_0}\right) - Bp \tag{i}$$

For small change in T and p, we can write

$$dg = R\ln\left(\frac{p}{p_0}\right)dT + RT\frac{dp}{p} - p\frac{dB}{dT}\cdot dT - Bdp \qquad \text{(ii)}$$

From Eq. (9.20b), we recall that

$$v = \left(\frac{\partial g}{\partial p}\right)_T$$

Hence, using (ii) we get

$$v = \frac{RT}{p} - B$$

$$p(v+B) = RT.$$

is the required equation of state of the gas.

We will now summarise what you have learnt in this chapter.

MMARY

When a system can be subject to work by application of pressure only, we can define four thermodynamic potentials functions:

Internal energy	U
Enthalpy	$H = U + pV$
Helmholtz free energy	$F = U - TS$
Gibbs free energy	$G = U - TS + pV = F + pV = H - TS$

An infinitesimal change in a thermodynamic potential can be expressed in terms of the corresponding natural thermodynamic variables as

$$dU = TdS - pdV$$

$$dH = TdS + Vdp$$

$$dF = -SdT - pdV$$

$$dG = -SdT + Vdp$$

The conditions for thermodynamic equilibrium of a system under different conditions

Adiabatic-isochoric process	$dS = 0$, $dV = 0$, $dU = 0$
Adiabatic-isobaric process	$dS = 0$, $dp = 0$, $dH = 0$
Isothermal-isochoric process	$dT = 0$, $dV = 0$, $dF = 0$
Isothermal-isobaric process	$dT = 0$, $dp = 0$, $dG = 0$

The condition for equilibrium of two phases to co-exist at constant temperature T and pressure p is

$$g_1 = g_2$$

For first order phase transitions, the entropy and volume show discontinuities, whereas for second order phase transitions, heat capacity at constant pressure and isobaric volume expansivity/ isothermal compressibility show discontinuity.

The Clausius–Clapeyron equation governing the temperature variation of pressure for first order phase transition is

$$\left(\frac{dp}{dT}\right)_{\text{vapour}} = \frac{L_{\text{vap}}}{T(v_2 - v_1)} = \frac{L_{\text{vap}}}{T\Delta v}$$

L_{vap} is latent heat of vaporisation and v_1 and v_2 are specific volumes of vapour and liquid phases.

The co-existence of phases in a second order phase transition is governed by Ehrenfest equations:

$$\frac{dp}{dT} = \frac{c_{p_2} - c_{p_1}}{Tv(\alpha_2 - \alpha_1)}$$

$$\frac{dp}{dT} = \frac{\alpha_2 - \alpha_1}{\beta_{T_2} - \beta_{T_1}}.$$

When equilibrium between various phases is influenced by temperature, pressure and concentration, the number of degrees of freedom f of the system is related to the components C and phases P by the Gibbs phase rule:

$$f = C - P + 2$$

ERCISES

Show that for n moles of an ideal gas

$$F = \int C_V \, dT - T\int \frac{C_V}{T} \, dT - nRT \ln V - a_1 T + a_2$$

$$G = \int C_p \, dT - T\int \frac{C_p}{T} \, dT + nRT \ln p - b_1 T + b_2$$

where a_1, a_2, b_1 and b_2 are constants.

Prove that

$$dU = (C_p - pV\alpha)\, dT + V(\beta_T p - \alpha T)\, dp$$

$$dH = C_p \, dT + V(1 - \alpha T)\, dp$$

$$dF = -(pV\alpha + S)\, dT + pV\beta_T \, dp$$

A thermodynamic system is described in terms of the parameters p, V, T, m and n, through the relation $TdS = dU + pdV - mdn$. If we define $G = U + pV - TS$, prove

$dG = -S\,dT + V\,dp + m\,dn$ **(b)** $-\left(\frac{\partial S}{\partial p}\right)_{n,T} = \left(\frac{\partial V}{\partial T}\right)_{n,p}$

$-\left(\frac{\partial S}{\partial n}\right) = \left(\frac{\partial m}{\partial T}\right)$ **(d)** $\left(\frac{\partial V}{\partial}\right) = \left(\frac{\partial m}{\partial}\right)$

Obtain the Clausius–Clapeyron equation starting from the p–V diagram.

Starting from second latent heat equation, show that

$$\frac{dL}{dT} = C_{p_2} - C_{p_1} + L\left[\frac{1}{T} - \frac{v_2\beta_2 - v_1\beta_1}{v_2 - v_1}\right]$$

where β_1 and β_2 are isothermal compressibilities of phase 1 and 2 respectively.

At 1 atm, mercury melts at –38.87°C and its density is 13.59 g cm^{-3}. The density of the solid is 14.19 g cm^{-3}. The heat of fusion is 2.33 cal g^{-1}. Calculate the melting point of mercury at 1000 atm. (*Ans:* –31.3°C)

Calculate the reduction in melting point of ice due to the pressure exerted on it by a skater who is standing still. Latent heat of ice = 3.3×10^5 J kg^{-1}.

(*Ans:* 0.066 K if the skater is equivalent to a uniform pressure of 10^6 N m^{-2})

The weight on the steam exhaust opening of a pressure cooker can maintain the highest pressure of 220 kPa inside it. Calculate the temperature at which water will boil in the cooker. Given, latent heat of vaporisation of water is 2257 kJ kg^{-1} at 100°C and 1atm = 1.01325×10^5 Pa. (*Ans:* 396 K)

The latent heat of vaporisation of ammonia (expressed in J g^{-1}) at temperature t°C is characterised by the equation $L = 138.5(133 - t)^{1/2} - 2.476(133 - t)$. If specific heat capacity of saturated liquid at 306 K is 4.822 Jg^{-1} K^{-1}, calculate the specific heat capacity of ammonia vapour. (*Ans:* –3.342 Jg^{-1}K^{-1})

The Helmholtz free energy of a system is given by

$$F = A + BT(1 - \ln T) - CT\ln V$$

where A, B, C are constant. Obtain expressions for pressure, entropy, internal energy, enthalpy and Gibbs energy. Also discuss the significance of the constants B and C.

The Gibbs free energy of a system is given by $S = -\left(\frac{\partial G}{\partial T}\right)_p$. If $G = RT\ln\left[\frac{ap}{(RT)^{5/2}}\right]$ show that $C_p = \frac{5}{2}R$.

10 PRODUCTION OF LOW TEMPERATURES AND THEIR APPLICATIONS

arning Objectives

In this chapter, you will learn how to

explain ordinary methods to produce low temperatures;

discuss Linde's method of regenerative cooling for liquefaction of gases;

n expression for Integral Joule–Thomson coefficient from thermodynamic considerations;

describe the method of adiabatic demagnetisation for production of ultra-low temperatures;

state the third law of thermodynamics and discuss its consequences.

NTRODUCTION

In the previous chapter, we observed that second order phase transitions occur at temperatures of a few kelvin and a lot of good physics is involved in their study. Therefore, it is important to learn how we can go below room temperature of about 300 K down to such low temperatures. Do you know that our quest for low temperatures is essentially the story of liquefaction of gases such as oxygen, nitrogen, hydrogen and helium and which unfolded a lot of new physics? (From Chapter 3, you may recall that a gas cannot be liquefied by application of pressure alone howsoever high, if the temperature of the gas is not below its critical temperature.) Today, liquefied gases are used as refrigerants, as fire extinguisher, as fuel (LPG and CNG) to drive automobiles and generate power. These also find many industrial, scientific and medical applications. As research tools, liquefied gases catalysed the discovery of many unusual phenomena such as superconductivity, superfluidity and Bose–Einstein condensate. (In the latter-half of the 20th century, high temperature superconductivity was thought of as a possible solution to varied energy needs of the world population but it is yet to materialise commercially.) Although liquefaction of gases requires a great deal of sophistication, physicists found this field increasingly interesting because of such exciting discoveries. In fact, this has led to the development of a fascinating branch of physics known as *Low Temperature Physics*, also referred to as *Cryogenics*. (*Cryogenics* is a Greek word and means "the production of freezing cold".) However, it is not well defined at what point on the temperature scale refrigeration ends and cryogenics begins. But according to The National Institute of Standards and Technology at Boulder, Colorado, the field of cryogenics involves temperatures below –180°C (93.15 K), since the normal boiling points of the so-called permanent gases (helium, hydrogen, nitrogen, oxygen and normal air) lie below –180°C while the Freon refrigerants, have boiling points above –180°C.

The knowledge of heating (fire) had been acquired long before the dawn of recorded history. (And this was probably the first brilliance and one of the best innovations of human intellect, which paved the way for development.) But the secret of refrigeration is barely older than the oldest living man. It is common knowledge that we can cool things by the method of mixtures as well as by evaporation. Did you know that water can be cooled in a clay pitcher by evaporation through its pores? But we cannot go below about 20°C by this method. A refrigerator helps us to achieve 0°C or so and that is why we can enjoy cold water/chilled soft drinks. (We discussed the working of a refrigerator in Chapter 6.) The common methods of cooling are discussed in Sec. 10.2.

As a child, you may have seen your grandmother preparing ice-cream by churning a vessel filled with sweetened milk and other necessary ingredients in a box. You may have wondered but the trick was a freezing mixture of ice and common salt put in the box. From Chapter 3, you may recall that if a gas is cooled below its critical temperature, it liquefies by application of pressure alone. However, but for CO_2, the critical temperatures of gases such as oxygen, hydrogen and helium are much below room temperature. Ordinary methods of cooling carry us nowhere near the critical temperatures of these gases. Therefore, we have to look for other methods to produce low temperatures.

The earliest and the simplest method to produce low temperatures and liquefy gases *Cascade process*, also known as *Pictet method*. In this process, cooling is produced in stages by a series of isothermal compressions and evaporation under reduced pressure. Under isothermal compression, a gas kept below its critical temperature liquefies when pressure reaches its saturation vapour pressure at that temperature. The liquid, kept in an insulated container, is made to evaporate by reducing pressure on it. Since latent heat of vaporisation is supplied at the expense of internal energy, a drop in temperature is observed. It may be mentioned here that theoretically, the minimum temperature that can be attained in a vapour compression machine is its triple point, i.e., the lowest temperature at which the refrigerant in use boils under reduced pressure. (The liquid phase of the gas exists in-between its critical temperature and the triple point.) In this cascade, care has to be taken in choosing the refrigerants in that the triple point temperature of any gas has to be below the critical temperature of the gas in the immediately next stage, and so on for successive stages.

So if we start with a gas such as ammonia, methyl chloride, carbon dioxide or ethylene, whose critical temperature is above room temperature, it is possible to obtain cooled liquid after the compression-evaporation process and use it as a low-temperature bath to cool another gas below its critical temperature. By repeating this process through a cascade of stages with a series of gases arranged successively in order of decreasing critical temperatures, it is possible to attain very low temperatures. For instance, if methyl chloride, ammonia, ethylene, methane, oxygen and nitrogen are used in a sequence, nitrogen can be liquefied. But it is not possible to extend the cascade process further as there is no gas which, when evaporated under reduced pressure, may produce a temperature low enough to attain the critical temperature of hydrogen (33.3 K); it is much below the triple point temperature of nitrogen (63.15 K). To bridge this gap and liquefy hydrogen, other methods had to be devised. In this search, adiabatic expansion presented itself as an attractive alternative. It is discussed in Sec. 10.3. We have discussed the method of regenerative cooling with particular emphasis on Linde's method for liquefaction of air, Dewar's method for hydrogen and Onnes' method for helium.

Joule–Kelvin irreversible adiabatic expansion, initially designed to confirm the existence of intermolecular forces of attraction (Chapter 3), is a slight variation of adiabatic reversible expansion. Coupled with the concept of regenerative cooling, it enabled Linde to attain liquid oxygen temperature. In a slightly modified format, Dewar attained liquid hydrogen temperature and Kamerlingh Onnes reached liquid helium temperature. In Sec.10.4, we revisit this process from thermodynamical viewpoint to discuss how the journey of low temperatures unfolded.

For temperatures below 1 K, adiabatic demagnetization of a paramagnetic salt is frequently used. A detailed discussion of this method forms the subject matter of Sec. 10.5. He cryostats, $^3He/^4He$ dilution refrigerator, Pomeranchuk cooling and nuclear demagnetisation are some other techniques used for attaining temperatures below 0.1 K. Some of these more convenient and efficient methods of producing such ultra-low temperatures are discussed in Sec. 10.6. We may mention that our emphasis will be on the physics of these methods and discussion of their detailed features or technical aspects will be limited due to space requirements.

Do you know that the lowest natural temperature recorded in astronomical observations in 2003 was at the Boomerang Nebula and the average temperature of the universe due to cosmic microwave background radiation is 2.73 K? Scientists reported a record low of 100 pico-kelvin in 1999 by cooling a piece of rhodium metal.

As you proceed, you will recognise that the journey towards absolute zero becomes progressively difficult. Even with an infinite series of operations, it is not possible to attain absolute zero. This observation supports the second law of thermodynamics and was postulated by Nernst in the form of the *third law of thermodynamics*. We have discussed it in Sec.10.7, along with its consequences. In particular, this law enables us to fix the absolute value of entropy of a system. The practical uses and applications of low temperatures are discussed in brief in Sec. 10.8.

Ordinary Methods of Cooling

We live in an environment whose temperature is around 310 K. In everyday experience, we may need to cool things a few degrees below this value. For this, we deprive the body of its heat energy. To achieve this, we put ice cubes in lemonade. In the household refrigerator also, heat is extracted from the eatables kept inside. That is, the heat of the body to be cooled is transferred to other objects. But this technique cannot be used to cool things much below 0°C. Use of freezing mixtures (salt + ice) in ice cream freezers helps to attain a temperature of about –21°C. In fact, it has been the oldest method for producing subzero temperatures By adding salts such as NH_4Cl, $NaNO_3$, $CaCl_2$, or KOH to ice, it is possible to attain temperatures as low as –65°C. This method was used by Faraday to , H_2S, and NO_2.

Cooling can also be produced by evaporation. That is how in summer earthen pots cool drinking water. We get a feel of diffusive cooling when we stop running or step out of a hot shower and the moisture on our skin begins to evaporate. The amount of cooling however depends on the nature of the liquid and the rate of evaporation, etc. The rate of evaporation can be increased by using high speed suction pumps or by getting the vapour absorbed in another liquid. For instance, at low temperatures, we can put liquid nitrogen, hydrogen or helium in partial vacuum and vaporisation of some of the liquid cools the rest.

ADIABATIC COOLING

If the object we wish to cool is colder than every other thing around, we have to first insulate it from its environment. Otherwise, heat will flow in the opposite (wrong) direction. From Chapter 5, we recall that when a system is made to expand adiabatically, some of its (internal) energy is lost and results in drop in temperature. This is how we condense gases into the liquid form (Fig. 10.1). For an ideal gas undergoing a reversible adiabatic expansion, the temperature and pressure are connected through the relation (Eq. (5.19c)):

$$\frac{T^{\gamma}}{p^{\gamma-1}} = \text{Constant}$$

Therefore, for two states characterised by temperatures (T_1, T_2) and pressures (p_1, p_2), we

$$\frac{T_2}{T_1} = \left(\frac{p_2}{p_1}\right)^{\frac{\gamma-1}{\gamma}}$$

If we subtract both sides from 1, we obtain expression for drop in temperature:

$$\Delta T = T_1 - T_2 = T_1\left[1 - \left(\frac{p_2}{p_1}\right)^{\frac{\gamma-1}{\gamma}}\right] \qquad (10.1)$$

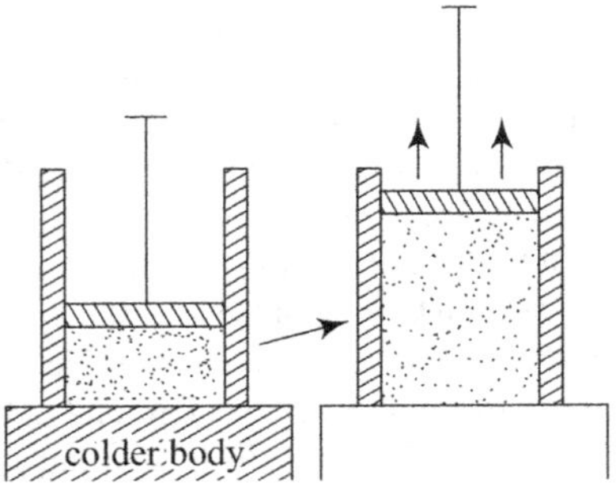

Fig. 10.1 Schematic representation of cooling by adiabatic expansion.

This result shows that adiabatic expansion can be used gainfully to produce low temperatures through either diffusive or mechanical interactions. You may recall how the refrigerant in a refrigerator cools; it is first compressed and then made to undergo adiabatic expansion. In principle, we may allow compressed ice cooled air to expand adiabatically and make nitrogen to condense. Then we can allow compressed hydrogen gas, kept at liquid nitrogen temperature, to undergo adiabatic expansion and get cooled so much that some hydrogen condenses. The process may be repeated with helium, held initially at liquid hydrogen temperature. Liquid helium gives us access to about 4 K. To reduce the temperature to about 1 K, liquid helium is made to evaporate under low pressure (Fig. 10.2).

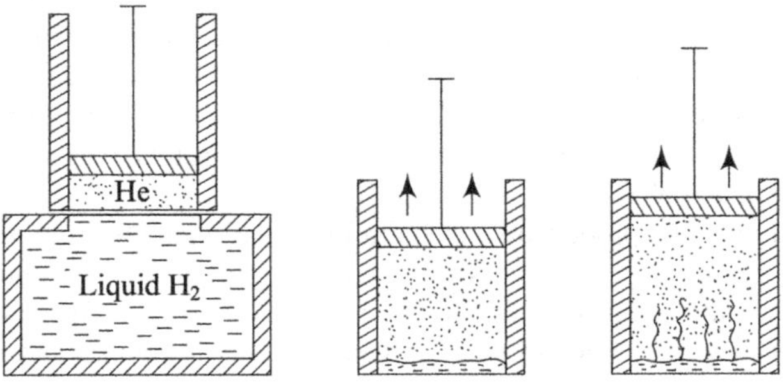

(a) Compressed helium gas in thermal equilibrium with liquid hydrogen,
(b) Adiabatic expansion cools helium and it begins to condense and
(c) As pressure is continuously lowered, vaporisation of some liquid produces further cooling.

This process has some obvious limitations which have genesis in the motion (friction) of pistons and use of lubricants. To overcome the difficulties arising due to lubricants, Claude used petroleum ether and succeeded in liquefying air. But it did not work for hydrogen.

The question now arises: How can we go down to liquid hydrogen temperatures? To answer this question, it is important to note that in this temperature range, some energy loss invariably occurs due to friction, lack of perfect insulation and absorption. A part of this may even return to the gas through the walls of the container. Therefore, one should look for methods which do not require bulky equipment or moving parts. A slight variation of the adiabatic expansion is the so-called Joule–Kelvin expansion. We discussed it in some detail in Chapter 3 and revisit its theoretical understanding here from thermodynamical viewpoint (Sec. 10.4). Debye and Giauque independently suggested the use of magnetic substances to accomplish it. The process is called *adiabatic demagnetisation* (Sec 10.5).

OULE–THOMSON EFFECT

In Chapter 3, you have learnt that if a gas at a temperature below its inversion temperature is allowed to expand adiabatically through a porous plug, it cools. However, the drop in temperature is quite small; a few tenths of a degree per atmosphere. Obviously, this is not sufficient to liquefy a gas all by itself. But it should in principle be possible to accumulate the total effect by using the cooled gas to pre-cool the incoming gas in a heat inter-exchanger before it is made to expand through the porous plug at a lower initial temperature and experience gradually steeper fall in temperature. Repetition of this cycle of operations *accumulated* or *regenerative cooling* and ultimately the gas temperature may fall so much that on being subject to further Joule–Thomson expansion, it readily liquefies. Linde harnessed *Joule–Thomson effect* successfully to produce temperatures sufficient for liquefaction of air. Later on, Dewar used it for liquefaction of hydrogen and Onnes for helium. As such, this method affords considerable economy over Cascade process because the gas has to be initially cooled below its inversion temperature rather than the critical temperature; the former being about 6.75 times higher than the latter (Eq. (3.36)). We will therefore first obtain expression for Joule–Kelvin coefficient and then discuss regenerative cooling giving technical details of machines used by Linde and others.

Joule–Kelvin Effect: An Isenthalpic Process

In the porous plug experiment, the pressures on the two sides of the porous plug are kept constant, say at values p_1 and $p_2 (< p_1)$. Let us assume that the initial internal energy and volume of the gas are respectively U_1 and V_1. After passing through the porous plug, let the final internal energy and volume of the gas be U_2 and V_2, respectively. Since the system is completely isolated thermally, the first law of thermodynamics implies that

$$\delta Q = 0 = U_2 - U_1 + p_2 V_2 - p_1 V_1$$

$$U_1 + p_1 V_1 = U_2 + p_2 V_2$$

$$H_1 = H_2 \tag{10.2}$$

That is, the Joule–Kelvin expansion can be regarded as a quasistatic isenthalpic process. It means that if a series of experiments are performed on a gas keeping the initial temperature and pressure constant for different final pressures, the values of temperature will be determined by the properties of the gas. On a $p - T$ diagram, we obtain a series of discrete points, which indicate the final equilibrium states with constant enthalpy. The smooth curve obtained on joining these points is shown in Fig. 10.3(a) and is known as the *isenthalpic* . (Note that it is not a plot of the throttling process, which is highly irreversible.) As may be seen, between states (1) and (4) the gas heats up but between (1) and (7), the gas cools down. In fact, the cooling will be more if the initial state corresponds to the maximum of the isenthalpic curve.

Next, if another series of experiments is carried out by changing the initial temperature of the gas but keeping the same initial pressure, we will obtain another isenthalpic curve with a different value of H. In this way, we can obtain a family of isenthalps, i.e., curves with $H(T, p)$ = constant for the gas under study. The isenthalpic curves shown in Fig. 10.3(b) are typical of most gases.

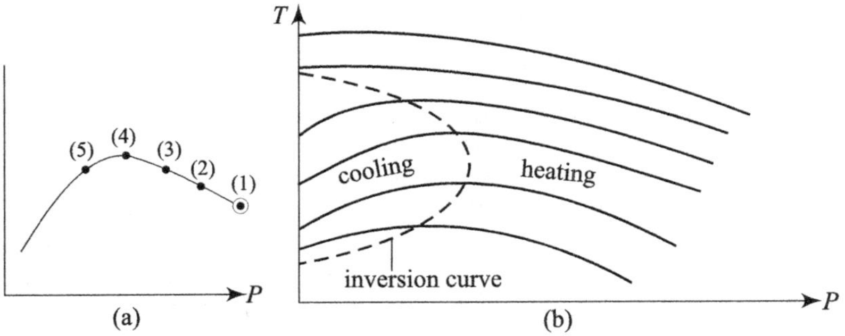

(a) An isenthalpic curve and (b) a family of isenthalpic curves for a typical gas.

The change in temperature associated with change in pressure is expressed as $(\partial T/\partial p)_H$. To calculate the change in temperature with change in pressure, we write H as a function of temperature and pressure:

$$H = H(T, p)$$

An infinitesimal change in enthalpy corresponding to infinitesimal changes in temperature and pressure can be expressed as

$$\mathrm{d}H = \left(\frac{\partial H}{\partial T}\right)_p \mathrm{d}T = \left(\frac{\partial H}{\partial p}\right)_T \mathrm{d}p \tag{10.3}$$

By definition, heat capacity at constant pressure is

$$C_p = \left(\frac{\partial H}{\partial T}\right)_p$$

Hence, we can rewrite Eq. (10.3) as

$$\mathrm{d}H = C_p \mathrm{d}T + \left(\frac{\partial H}{\partial p}\right)_T \mathrm{d}p \tag{10.4}$$

To express this relation in a more compact form, we recall Eq. (9.9):

$$\mathrm{d}H = T\mathrm{d}S + V\mathrm{d}p$$

This equation shows that in Joule–Kelvin expansion, the entropy change is finite, even though the process is adiabatic. Can we say that it violates the second law of thermodynamics? It is not so because in this process mechanical interaction (expansion) takes place, i.e., the system is not isolated.

Proceeding further, we express the change in enthalpy with pressure as

$$\left(\frac{\partial H}{\partial p}\right)_T = T\left(\frac{\partial S}{\partial p}\right)_T + V = -T\left(\frac{\partial V}{\partial T}\right)_p + V \tag{10.5}$$

where we have made use of the second Maxwell relation given by Eq. (8.11).

On combining Eqs. (10.4) and (10.5), we get*

$$\mathrm{d}H = C_p \,\mathrm{d}T + \left(V - T\left(\frac{\partial V}{\partial T}\right)_p\right)\mathrm{d}p \tag{10.6}$$

Since Joule–Kelvin expansion is isenthalpic ($\mathrm{d}H = 0$), the change in temperature with pressure is given by

$$\left(\frac{\partial T}{\partial p}\right)_H = \mu = \frac{1}{C_p}\left(T\left(\frac{\partial V}{\partial T}\right)_p - V\right) = \frac{T^2}{C_p}\left(\frac{\partial}{\partial T}\left(\frac{V}{T}\right)\right)_p \tag{10.7}$$

Joule–Kelvin coefficient.

$_2 < p_1$, $\mathrm{d}p$ will be negative. So μ will have a positive value only if ΔT is negative, i.e., cooling is produced. Similarly, a negative value of μ implies that Joule–Kelvin expansion leads to heating and $\mu = 0$ means that there is neither heating nor cooling. The physical reason for this may be understood by noting from Eq. (10.3) that

$$\mu = \left(\frac{\partial T}{\partial p}\right)_H = -\left(\frac{\partial T}{\partial H}\right)_p \left(\frac{\partial H}{\partial p}\right)_T = -\frac{1}{C_p}\left(\frac{\partial H}{\partial p}\right)_T$$

Further, since enthalpy

$$H = U + pV$$

we can write

$$\left(\frac{\partial H}{\partial p}\right)_T = \left(\frac{\partial U}{\partial p}\right)_T + \left(\frac{\partial}{\partial p}(pV)\right)_T$$

This result shows that the Joule–Kelvin coefficient can be expressed as a sum of two terms:

$$\mu = -\frac{1}{C_p}\left(\left(\frac{\partial U}{\partial p}\right)_T + \left(\frac{\partial}{\partial p}(pV)\right)_T\right) \tag{10.8}$$

The first term on the RHS of Eq. (10.8) is a measure of deviation from Joule's law:), whereas the second term signifies deviation from Boyle's law: pV = constant. For a perfect gas, both these terms are zero. From this we may conclude that if Joule–Kelvin coefficient vanishes for a gas, it may be regarded as perfect. For a real gas, the first term always contributes to cooling but the second term may contribute to heating or cooling, depending on the temperature and pressure of the gas. So, the net result of Joule–Kelvin expansion is determined by the interplay of these two terms and can have either positive or negative sign. It will vanish when cooling due to deviations from Joule's law exactly cancels heating due to deviations from Boyle's law, i.e. when

$$\left(\frac{\partial U}{\partial p}\right)_T + \left(\frac{\partial}{\partial p}(pV)\right)_T = 0$$

From Eq. (10.5) we can rewrite this condition as

$$-V + T\left(\frac{\partial V}{\partial T}\right)_p = 0 \tag{10.9}$$

*We can arrive at this result directly from Eq. (9.9) by using the second $T\mathrm{d}S$ equation (Eq. (8.25a)).

This equation defines the *inversion curve* in the (T, p) plane (Fig. 10.4). That is, it represents the locus of points for which $\mu = 0$. We note that for a given initial temperature , the gas will cool only if the pressure is chosen so that the initial state lies within the inversion curve, i.e., $(\partial T/\partial p)_H$ is positive. This means that if a gas is allowed to expand when it corresponds to points A or B (i.e., temperatures and pressures T_1, p_1 and T_2, p_2) it will cool. Outside the inversion curve, $(\partial T/\partial p)_H$ is negative and the gas warms up on expansion.

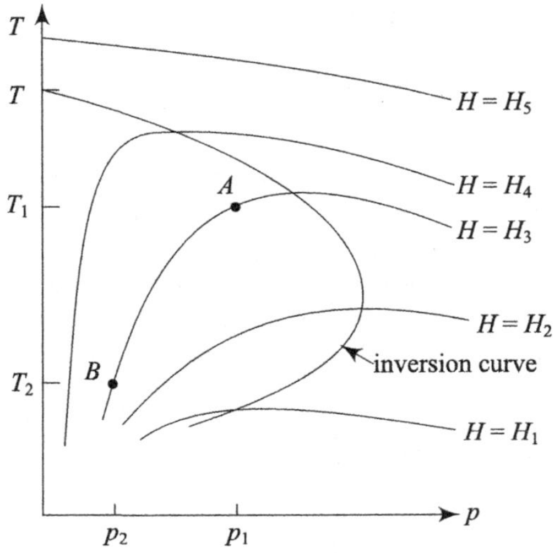

The inversion curve. Inside it, the gas cools on undergoing Joule–Thomson expansion.

Recall the important findings of Joule and Kelvin for a real gas:

No real gas behaves perfectly since $\mu \neq 0$.

Except hydrogen and helium, all gases show cooling effect at room temperature. In fact these gases show slight heating effect. (This paradox was resolved when it was recognised that to obtain any cooling, the initial temperature of the gas must be below its inversion temperature.)

To understand this, we recall Eq. (3.6) for a van der Waals' gas:

$$\left(p + \frac{a}{V^2}\right)(V - b) = RT \tag{10.10}$$

and note that

$$\left(\frac{\partial V}{\partial T}\right)_p = \frac{R}{\left(p + \frac{a}{V^2}\right) - \frac{2a}{V^3}(V - b)} \tag{10.11}$$

Multiplying the numerator as well as the denominator on the right-hand side of this expression by $(V - b)$ and using Eq. (10.10), we can write

$$\left(\frac{\partial V}{\partial T}\right)_p = \frac{R(V - b)}{RT - \frac{2a(V - b)^2}{V^3}} = \frac{(V - b)}{T\left(1 - \frac{2a(V - b)^2}{RTV^3}\right)} \tag{10.12}$$

V, we can write, using binomial expansion

$$\left(\frac{\partial V}{\partial T}\right)_p = \frac{V}{T}\left(1-\frac{b}{V}\right)\left(1-\frac{2a}{RTV}\right)^{-1}$$

$$= \frac{V}{T}\left(1-\frac{b}{V}\right)\left(1+\frac{2a}{RTV}+\cdots\right)$$

$$= \frac{V}{T}\left(1-\frac{b}{V}+\frac{2a}{RTV}-\frac{2ab}{RTV^2}\right) \tag{10.13}$$

where we have neglected the quadratic or higher powers of van der Waals' constants.

If we now ignore the term containing the product of van der Waals' constants in Eq. (10.13), we can write it in a compact form as

$$T\left(\frac{\partial V}{\partial T}\right)_p - V = \frac{2a}{RT} - b \tag{10.14}$$

On substituting this result in Eq. (10.7), we get

$$\mu = \frac{1}{C_p}\left(\frac{2a}{RT} - b\right) \tag{10.15}$$

Note that this equation is analogous to Eq. (3.34).

In terms of inversion temperature $T_\text{I} = (2a/Rb)$, we can rewrite it as

$$\mu = \frac{b}{C_p}\left(\frac{T_i}{T} - 1\right) \tag{10.16}$$

Integral Joule-Thomson Effect In the above discussion, we assumed that the pressure gradient across the porous plus is gradual. However, if pressure drop is finite, we should integrate Eq. (10.7) to obtain the change in temperature of the gas. This gives

$$\int_{T_1}^{T_2} C_p \, \mathrm{d}T = \int_{p_1}^{p_2} \left(T\left(\frac{\partial V}{\partial T}\right)_p - V\right) \mathrm{d}p \tag{10.17}$$

Using the cyclic relation $\left(\frac{\partial V}{\partial T}\right)_p = -\left(\frac{\partial V}{\partial p}\right)_T \left(\frac{\partial p}{\partial T}\right)_V$, we can write for constant T

$$\left(\frac{\partial V}{\partial T}\right)_p \mathrm{d}p_T = -\left(\frac{\partial p}{\partial T}\right)_V \mathrm{d}V_T \tag{10.18}$$

Further, since

$$\int V \mathrm{d}p = \int \mathrm{d}(pV) - \int p \mathrm{d}V \tag{10.19}$$

we can combine Eqs. (10.17) – (10.19) to obtain

$$T_2 - T_1 = \frac{1}{C_p}\left(-\int_{V_1}^{V_2} T\left(\frac{\partial p}{\partial T}\right)_V \mathrm{d}V - \int_1^2 \mathrm{d}(pV) + \int_{V_1}^{V_2} p\mathrm{d}V\right)$$

$$= \frac{1}{C_p}\left(p_1V_1 - p_2V_2 + \int_{V_1}^{V_2} p\mathrm{d}V - \int_{V_1}^{V_2} T\left(\frac{\partial p}{\partial T}\right)_V \mathrm{d}V\right) \tag{10.20}$$

is the heat capacity over the temperature range T_1 to T_2.

For a van der Waals gas $p = \frac{RT}{V-b} - \frac{a}{V^2}$ and $\left(\frac{\partial p}{\partial T}\right)_V = \frac{R}{V-b}$ so that

$$T_2 - T_1 = \frac{1}{C_p}\left(p_1 V_1 - p_2 V_2 + \int_{V_1}^{V_2} p\,\mathrm{d}V - \int_{V_1}^{V_2} \frac{RT}{V-b}\mathrm{d}V\right)$$

$$= \frac{1}{C_p}\left(p_1 V_1 - p_2 V_2 - \int_{V_1}^{V_2} \frac{a}{V^2}\mathrm{d}V\right)$$

$$= \frac{1}{C_p}\left(p_1 V_1 - p_2 V_2 + a\left[\frac{1}{V_2} - \frac{1}{V_1}\right]\right) \tag{10.21}$$

Do you recall this equation? If we identify the work done by the gas as $W = C_p(T_2 - T_1)$, Eq. (10.21) becomes identical to Eq. (3.29).

Now go through the following example to appreciate the connection between Joule-coefficient and Joule–Thomson coefficient.

Example 10.1 The Joule coefficient is defined as $\eta = (\partial T/\partial p)_U$. Show that it is connected to the Joule–Kelvin coefficient μ by the relation.

$$\eta\left(C_p - \left(\frac{\partial(pV)}{\partial T}\right)_p\right) = \mu C_p + \left(\frac{\partial(pV)}{\partial p}\right)_T$$

Let us take internal energy as a function of temperature and pressure, i.e.,

$$U = U(T, p)$$

Then, we can write

$$\mathrm{d}U = \left(\frac{\partial U}{\partial T}\right)_p \mathrm{d}T + \left(\frac{\partial U}{\partial p}\right)_T \mathrm{d}p$$

$$\left(\frac{\partial T}{\partial p}\right)_U = \eta = -\frac{\left(\frac{\partial U}{\partial p}\right)_T}{\left(\frac{\partial U}{\partial T}\right)_p} \tag{i}$$

By definition, enthalpy $H = U + pV$ so that

$$\left(\frac{\partial U}{\partial T}\right)_p = \left(\frac{\partial H}{\partial T}\right)_p - \left(\frac{\partial}{\partial T}(pV)\right)_p$$

$$= C_p - \left(\frac{\partial}{\partial T}(pV)\right)_p$$

Substituting this result in Eq. (i), we obtain

$$\eta = -\frac{\left(\frac{\partial U}{\partial p}\right)_T}{C_p - \left(\frac{\partial}{\partial T}(pV)\right)_p} \tag{ii}$$

For a perfect gas, the numerator in Eq. (ii) is zero implying that $\eta = 0$.

If we now take H as a function of T and p, we can write

$$H = H(T, p)$$

$$\mathrm{d}H = \left(\frac{\partial H}{\partial T}\right)_p \mathrm{d}T + \left(\frac{\partial H}{\partial p}\right)_T \mathrm{d}p$$

For an isenthalpic process,

$$\left(\frac{\partial T}{\partial p}\right)_H = \mu = -\left(\frac{\partial H}{\partial p}\right)_T \Big/ \left(\frac{\partial H}{\partial T}\right)_p$$

$$= -\frac{1}{C_p}\left(\frac{\partial H}{\partial p}\right)_T \tag{iii}$$

By definition, $H = U + pV$ and we can write

$$\left(\frac{\partial H}{\partial p}\right)_T = \left(\frac{\partial U}{\partial p}\right)_T + \left(\frac{\partial}{\partial p}(pV)\right)_T$$

On combining this result with Eq. (iii), we get

$$\left(\frac{\partial U}{\partial p}\right)_T + \left(\frac{\partial}{\partial p}(pV)\right)_T = -\mu C_p$$

$$\left(\frac{\partial U}{\partial p}\right)_T = -\mu C_p - \left(\frac{\partial}{\partial p}(pV)\right)_T \tag{iv}$$

Using this result in Eq. (ii) we get the required expression connecting the Joule coefficient with Joule–Kelvin coefficient:

$$\eta = \frac{\mu C_p + \left(\frac{\partial}{\partial p}(pV)\right)_T}{C_p - \left(\frac{\partial}{\partial T}(pV)\right)_p}$$

$$\eta\left(C_p - \left(\frac{\partial}{\partial T}(pV)\right)_p\right) = \mu C_p + \left(\frac{\partial}{\partial p}(pV)\right)_T$$

Example 10.2 A gas obeying van der Waals' equation is made to undergo Joule–Kelvin expansion from an initial constant pressure 41.2 atm to final constant pressure 1.2 atm. It is initially at 273 K. Calculate the Joule–Kelvin coefficient and change in temperature. Given a = 1.34 × 10^{-6} atm m^6 mol^{-2}, b = 36.5 × 10^{-6} m^3 mol^{-1} and × 10^{-5} atm m^3 K^{-1} mol^{-1}. Take $R = 8.2 \times 10^{-5}$ atm m^3 K^{-1} mol^{-1}.

From Eq. (10.15), we know that

$$\mu = \frac{1}{C_p}\left(\frac{2a}{RT} - b\right)$$

On substituting the given data, we get

$$= \frac{1}{(28.7 \times 10^{-5}\ \text{atm m}^3\ \text{K}^{-1}\ \text{mol}^{-1})} \times \left(\frac{(2 \times 1.34 \times 10^{-6}\ \text{atm m}^6\ \text{mol}^{-2})}{(8.2 \times 10^{-5}\ \text{atm m}^3\ \text{K}^{-1}\ \text{mol}^{-1}) \times (273\ \text{K})} - (36.5 \times 10^{-6}\ \text{m}^3\ \text{mol}^{-1})\right)$$

$$= \frac{1}{(28.7 \times 10^{-5}\ \text{atm m}^3\ \text{K}^{-1}\ \text{mol}^{-1})} [(119.7 \times 10^{-6}\ \text{m}^3\ \text{mol}^{-1} - (36.5 \times 10^{-6}\ \text{m}^3\ \text{mol}^{-1})]$$

$$= \frac{83.2 \times 10^{-6}}{28.7 \times 10^{-5}}\ \text{K atm}^{-1} = 0.29\ \text{K atm}^{-1}$$

Hence, drop in temperature ΔT is given by

$$\begin{aligned} \Delta T &= \mu \Delta p \\ &= (0.29\ \text{K atm}^{-1}) \times 40\ \text{atm} \\ &= 11.6\ \text{K} \end{aligned}$$

That is, the gas will be cooled by 11.6 K.

Having discussed Joule–Thomson effect for production of low temperatures, we will now discuss its practical applications for liquefaction of gases. As mentioned earlier, Linde used it coupled with the principle of regenerative cooling for liquefaction of air. We now discuss it in brief.

Producing Liquid Air Temperatures: Linde's Liquefier

Refer to Fig. 10.5, which shows schematic diagram of the Linde's liquefier for attaining liquid air temperatures. The entire apparatus is thermally insulated to avoid loss of energy due to conduction and radiation. The main feature of the liquefier is that a closed circulating system is maintained at a high pressure with the help of two compressors. Initially air is compressed to a pressure of about 25 atm by compressor C_1 and then made to pass through a copper spiral tube immersed in a cold water jacket. To free air from water vapour and carbon dioxide, the compressed air is passed through a KOH solution. This is a necessary requirement as otherwise the circulatory system may choke due to solidification of water vapour and CO_2.

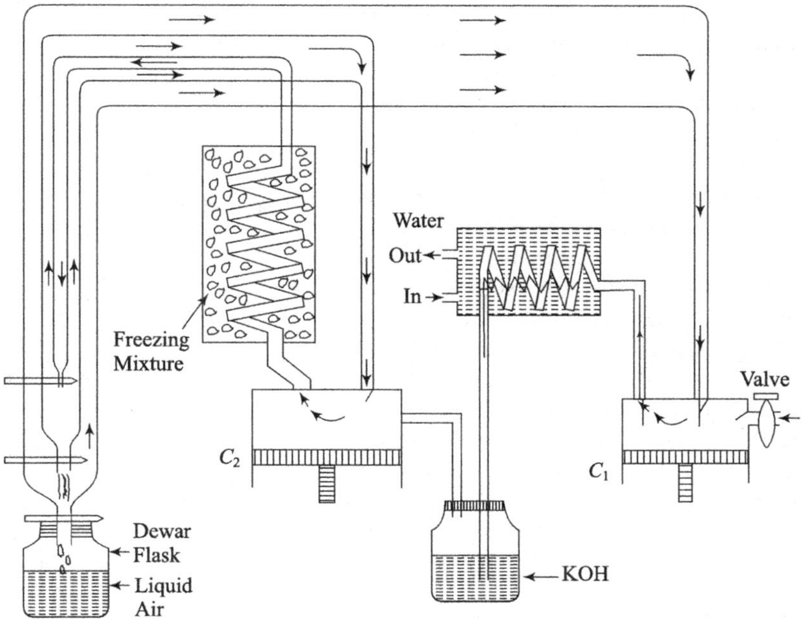

Schematic diagram of Linde's liquefier.

Once freed from water vapour and CO_2, air is fed to the second compressor C_2, which raises its pressure to about 200 atm. The highly compressed air is passed through a spiral tube surrounded by a jacket containing a freezing mixture. As a result, it is cooled to about –20°C. This high pressure cooled air is made to undergo Joule–Thomson expansion against a pressure of about 20 atm as it passes through porous plug V_1. As a result of sudden drop of pressure, the temperature of the gas goes down to about –70°C. This low pressure cooled air returns to compressor C_2 through the outer tube and in so doing exchanges heat with the air flowing through the inner pipe and cools it.

This process is repeated a few times and results in significant lowering of temperature of the incoming air. At this stage, the cool air is again made to expand through the second against a pressure of about one atmosphere. This results in partial liquefaction at a temperature of –188°C and saturated liquid air collects in a Dewar flask through valve while the saturated vapour returns to compressor C_1. The entire process is repeated all

It may be mentioned here that Linde's liquefier is not very efficient. Therefore, this apparatus was modified by Linde and Hampson and they succeeded in producing liquid hydrogen temperatures. We now briefly discuss it.

Producing Liquid Hydrogen Temperatures: Linde-Hampson Method

In this liquefier also, use is made of Joule–Thomson expansion with counter-current heat exchangers. However, liquid nitrogen is used as pre-coolant. (For air, the pre-coolant could be liquid ammonia.) A schematic representation of Linde-Hampson liquefier is shown in Fig. 10.6. Initially, hydrogen gas is compressed isothermally in stages. High pressure gas is made to pass through a heat exchanger where it exchanges heat with low pressure cool gas moving in the opposite direction. For efficient heat transfer, the heat exchanger is well insulated and the coils carrying opposite streams of the gas are put in good thermal contact. The chamber *B* contains liquid nitrogen and cools hydrogen further. After passing through the second heat exchanger, the gas is made to undergo Joule–Thomson expansion, which cools the gas further. The low pressure gas returns to the compressor via the heat exchanger. After several such cycles, the gas attains sufficiently low temperature (about –250°C) and on being subject to Joule–Thomson expansion, it liquefies partially. The saturated liquid hydrogen collects in the Dewar flask while the saturated vapour is made to return to compressor. The amount of hydrogen that has liquefied is replaced by the hydrogen gas, which is allowed to enter through the compressor.

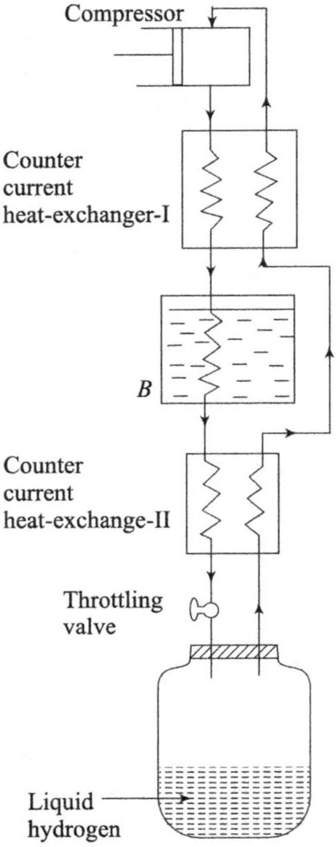

Fig. 10.6 Schematic representation of Linde-Hampson Liquefier.

Attaining Liquid Helium Temperatures

After attaining temperature which made liquefaction of hydrogen possible, researchers refined their methods, improved the efficiency of their machines and tried different ways to go further down towards absolute zero. In the series of these classical experiments, Kamerlingh Onnes, Dutch physicist, succeeded in liquefying helium in 1908 and attain temperature of about 1 K. His method has since been modified by Kapitza, Collins, Simon and others. However, for its historical importance in our continuing march towards absolute zero, we discuss how Onnes attained liquid helium temperature. We have also discussed Collins method, which is based on Claude process, wherein adiabatic expansion is used in the higher temperature range and Joule–Thomson expansion is used in the low energy range.

Refer to Fig .10.7. It depicts a schematic diagram of the apparatus used by Onnes to liquefy helium. Helium gas compressed to a pressure of about 35 atm by the compressor is forced into the spiral tube *AB* at *A*. It is divided into two streams and is made to flow through spiral coils C_1 and C_2. These streams reunite at *B* and then again divide into two, flowing through coils C_3 and C_4. The coils C_1 and C_3 are enclosed in the double-walled , whereas coils C_2 and C_4 are enclosed in another identical vessel *F*. The gas passing through coils C_1 and C_3 is cooled by hydrogen vapour and the coils C_2 and C_4 are cooled by helium gas kept in vessel *F* (The liquid hydrogen is made to boil continuously under reduced pressure.) The helium gas is therefore cooled in two stages in vessels *E* and *F* so that by the time these reunite again, the gas cools down considerably.

The cooled gas is then made to undergo Joule–Thomson expansion at the nozzle *N* and cools down further. The cold helium vapour rising up in vessel *F* cools the down coming

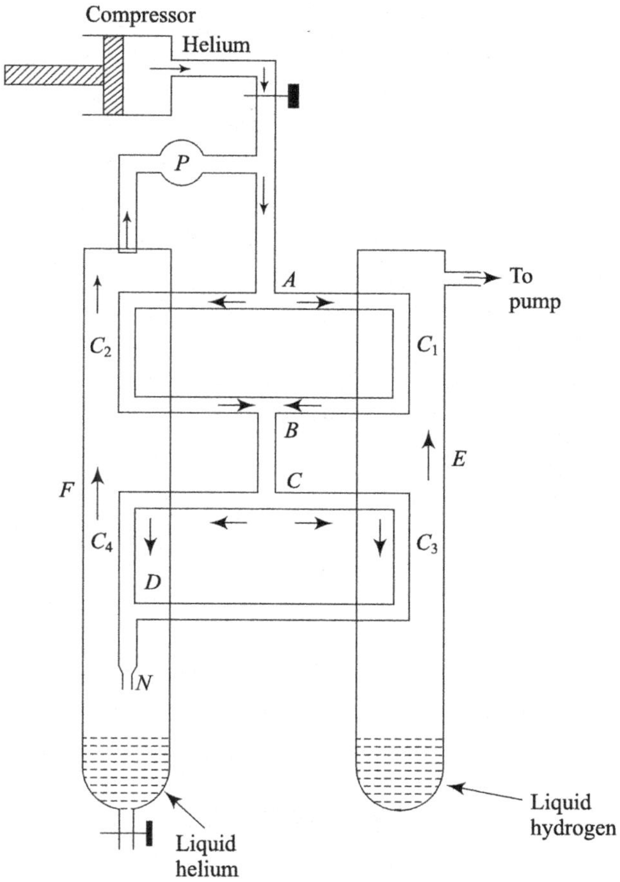

Schematic diagram of Onnes liqufier.

streams of the gas in coils C_2 and C_4. This process is repeated and after a few cycles, the temperature of helium becomes so low that after another Joule–Thomson expansion, it liquefies and is collected at the bottom of the vessel F. The unliquefied helium vapours are compressed again by P and subject to the entire process all over again.

It may be mentioned here that the entire arrangement described above is surrounded by Dewar flasks to ensure perfect insulation.

Collin's Liquefier The Collin's helium liquefier is an extension of Onnes' method and makes it possible to liquefy helium directly from high pressure gas at room temperature. The schematic diagram of Collins liquefier is shown in Fig. 10.8. The gas is initially compressed and made to exchange heat in counter-current heat exchanger I. As it comes out, it is divided into two streams, about 80% is diverted to an expansion engine E_1 where it is made to expand to 1 atm. As a result, it gets cooled and is fed to compressor C through heat exchangers II and I, where it cools the incoming gas. The other stream, after cooling in heat exchanger III, is again divided into two streams. One part of this gas is fed to the second expansion engine E_2 and is similarly cooled before reuniting with the low-pressure stream. The other part is made to undergo Joule–Thomson expansion, where it ultimately liquefies at 4.2 K at atmospheric pressure as a normal liquid and is collected in the vessel.

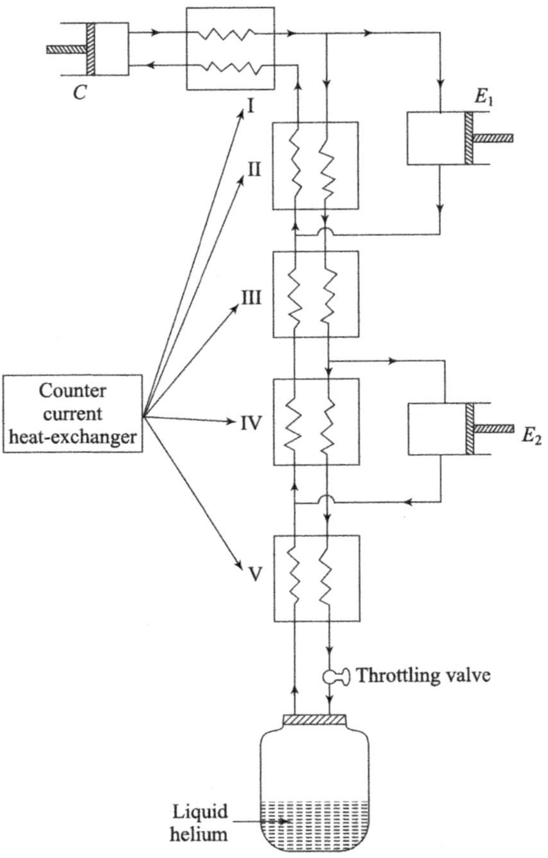

Schematic diagram of Collin's helium liquefier.

Note that in this method, the requirement of liquid hydrogen is done away with for cooling the incoming gas.

Liquid Helium: A Unique Substance It may be mentioned here that liquid helium exhibits very remarkable properties as temperature is lowered below 4.2 K. For instance, it undergoes a phase transition at 2.17 K. Instead of solidifying at this temperature, liquid helium changes into another liquid of vastly different and extremely surprising properties. To differentiate these, the phase above 2.17 K is termed as Helium I and that below 2.17 K as Helium II. The transition temperature is accompanied by an anomalous behaviour of the heat capacity. As mentioned in Chapter 9, the heat capacity curve exhibits a shape similar to the Greek letter lambda (λ), as shown in Fig. 10.9a. (For this reason, the transition temperature is referred to as the *lambda point*.) The He I to He II phase transition is visually characterised by the disappearance of bubbles and boiling (Fig. 10.9 b, c).

The viscosity of He II is apparently zero and it can flow rapidly through the smallest crack without any impediment whatsoever. For this reason, He II is called *superfluid* and characterises a new state of matter. This property of He II is known as *superfluidity*.

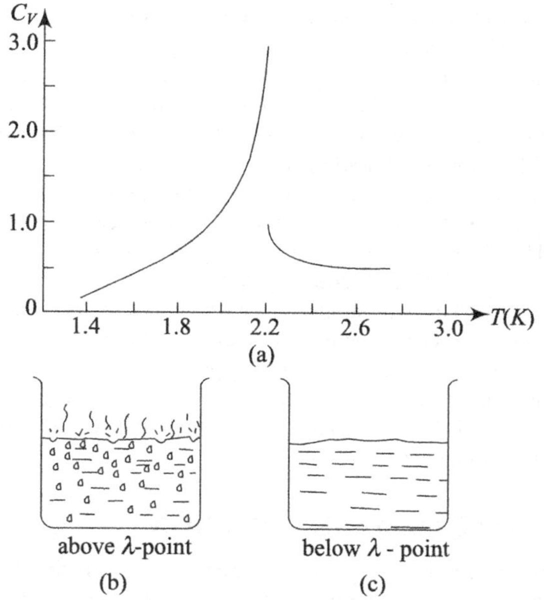

(a) Temperature variation of the heat capacity of helium about the λ-point, (b) He I shows normal properties of a liquid and (c) The He I-He II transition is accompanied by disappearance of bubbles and boiling.

He II has extraordinarily high thermal conductivity; nearly 13.5×10^6 times more than He I and about 800 times more than copper. This means that irrespective of where we heat the liquid, it will evaporate from the top surface. (This is unlike ordinary fluids such as water, which vaporise from wherever the heat input is.)

Some of the most spectacular effects arise due to superfluidity of He II. These include *fountain effect*, the *second sound* and formation of rolling film.

To achieve the fountain effect, we plug a U-tube whose one end is fitted with a capillary tube, with emery powder and cotton at its bottom. It is then immersed in liquid He-II as shown in Fig. 10.10. If now light from a torch is flashed on the powder (or helium inside the tube is given some heat), a fountain of liquid helium shoots out of the capillary end. It is found to rise as high as 30 cm.

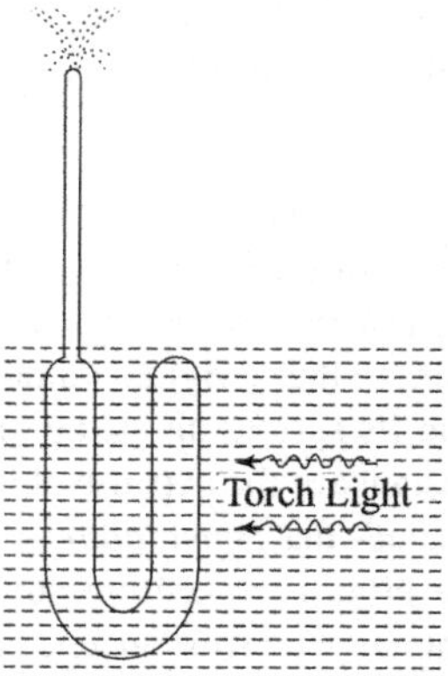

Fig. 10.10 The fountain effect exhibited by helium II.

The second sound is a temperature or entropy wave, unlike the familiar (first) sound wave, which is a pressure wave. It was predicted by Tisza in 1940 and observed by Peshkov in 1944. We will not go into details here.

He II forms rolling film on all solid surfaces and flows from one vessel to another through this. (This film is several hundred atoms thick.) If a tube containing He II is dipped in a He II bath, it is observed that

the liquid He II from the bath enters the tube, if the level of He II in the tube is lower than that of the He II bath.

the liquid He II from the tube enters the bath, if the level of He II in the tube is higher than that of the He II bath.

If the tube containing He II is completely out of the He II bath, it is observed that He II begins to creep out of the tube along the sides of the tube and the process continues till such time that the tube is completely empty. It collects at its bottom of the tube in the form of drops and falls into the liquid bath.

Another amazing property of helium is that it has *no* triple point. It has genesis in the fact that liquid helium does not solidify as temperature is lowered below 4.2 K. Solid helium was first obtained by Keesom, a student of Onnes, at a pressure of about 250 atm at 4.2 K and at nearly 23 atm at 1.1 K. Subsequently, solid helium was obtained at a pressure of about 5800 atm at 42 K. This was strange as helium could not be liquefied so long as the temperature was above the critical temperature of 5.2 K but it could exist in solid form at much higher temperatures. It means that at high pressures, the melting point exceeds the critical temperature and when solid helium melts, it turns into gas.

A lot of good physics is involved in the various phenomena exhibited by He II. We will discuss some of these in Chapter 15.

ADIABATIC DEMAGNETISATION

Adiabatic demagnetisation is a very powerful method of producing low temperatures. It was suggested independently by Debye in 1926 and by Giauque in 1927. From theoretical considerations they predicted that if a paramagnetic salt is first cooled to about 1 K by means of liquid helium in the presence of strong magnetic field and then subject to adiabatic demagnetisation, the energy spent in doing magnetic work leads to a fall in temperature below 1 K. These predictions were taken up for experimental verification by Simon and Kurti, and Giauque and McDougall in 1931. We now discuss the principle of adiabatic demagnetisation in some detail.

We know that atoms/molecules in a paramagnetic substance act as small magnetic dipoles, which are randomly oriented, pointing in all possible directions. As a result, the net intensity of magnetisation is ordinarily zero, i.e., $M = 0$. But when a paramagnetic substance is placed in a magnetic field (**B**), these small magnets tend to align in the direction of the magnetic field. From Chapter 5, we recall that the work done the substance is given by

$$dW = -B dM \tag{10.22}$$

is the change in the intensity of magnetisation. This work increases the internal energy of the substance and temperature of the paramagnetic substance rises. However, when such a substance is demagnetised by switching off the magnetic field, these dipoles return to their original random distribution. However, if the system is in thermal isolation from its surroundings, i.e., if the demagnetisation is adiabatic, the work is done by the substance at the expense of its internal energy and leads to drop in temperature. So we can conclude that *temperature of a paramagnetic substance drops in adiabatic demagnetisation.* The process of heating due to magnetisation and cooling due to demagnetisation of a paramagnetic substance is known as *magneto-caloric effect* or *thermo-magnetic effect.*

According to *Curie's law*, the susceptibility of a paramagnetic substance varies inversely with absolute temperature, i. e., $\chi \propto \frac{1}{T}$. So if this law is assumed to be valid up to the liquid helium temperature, a paramagnetic substance will get highly magnetised at this temperature and will therefore show a pronounced cooling effect on adiabatic demagnetisation. This helps us to obtain a temperature much below the liquid helium temperature.

We can also understand how adiabatic demagnetisation produces cooling in terms of the behaviour of entropy, which is the most significant thermal property of a paramagnetic substance at low temperatures. The entropy of a solid is made up of several components. For instance, in a paramagnetic crystal, the lattice sites are occupied by ions and their thermal motion contributes to entropy. Similarly, contributions also arise due to thermal motion of free electrons. However, some of the ions may have permanent dipole moments oriented randomly and this disorder contributes to entropy. (The dipole moments arise from uncompensated electron spins in incomplete shells.) At very low temperatures, the thermal effects are almost negligible and if we ignore these, it is sufficient to consider contribution to entropy arising due to magnetic moment only. This is illustrated in Fig. 10.11, where we have plotted entropy of a paramagnetic substance as a function of temperature for different values of magnetic field. We note that

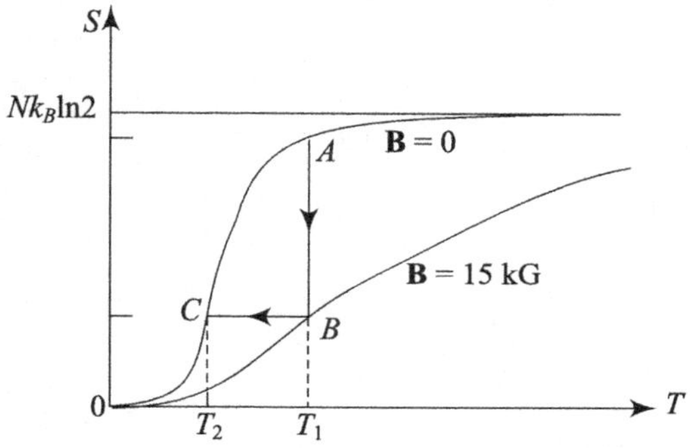

Fig. 10.11 T–S curves for a paramagnetic salt for different values of magnetic field.

In the process $A \rightarrow B$, entropy decreases, whereas in the process $B \rightarrow C$, the entropy remains constant. (These processes are analogous to those in which a gas is first compressed isothermally and reversibly and then allowed to expand to original volume adiabatically and reversibly, respectively.)

The entropy of the specimen is more when there is no magnetic field. Physically, we can understand it as follows: Due to thermal energy, the spins of a paramagnetic substance are oriented randomly. But when a magnetic field is applied, these spins are partially aligned along the field and entropy of the specimen decreases.

In Fig. 10.12, we have shown a system of four spin ½ particles, each having the same magnetic moment. There are six different arrangements having two up and two down magnetic moments giving no net magnetic moment. But there is only one arrangement having them all aligned in the same direction. That is why the entropy of the ordered system is smaller.

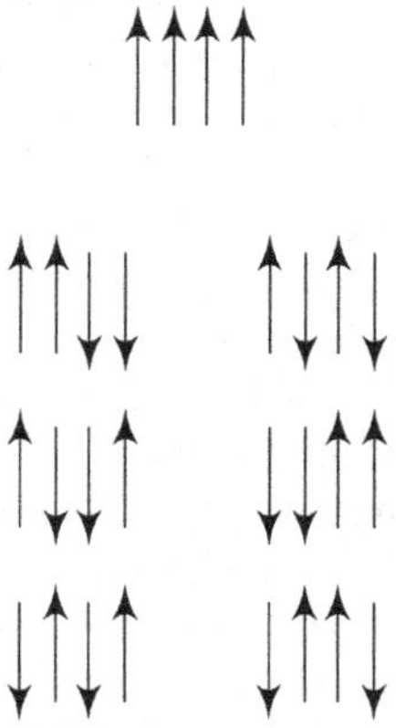

Fig. 10.12 A system of four spin 1/2 particles.

Experimental Procedure Refer to Fig 10.13, which shows the experimental arrangement for measurement of temperature in adiabatic demagnetisation. The paramagnetic specimen P is suspended by a silk thread inside a cylindrical glass tube (A) which is surrounded by a liquid helium bath (B) under a reduced pressure of about 1 mm of mercury. The space in A is connected to a pump or to a supply of helium gas. The liquid helium

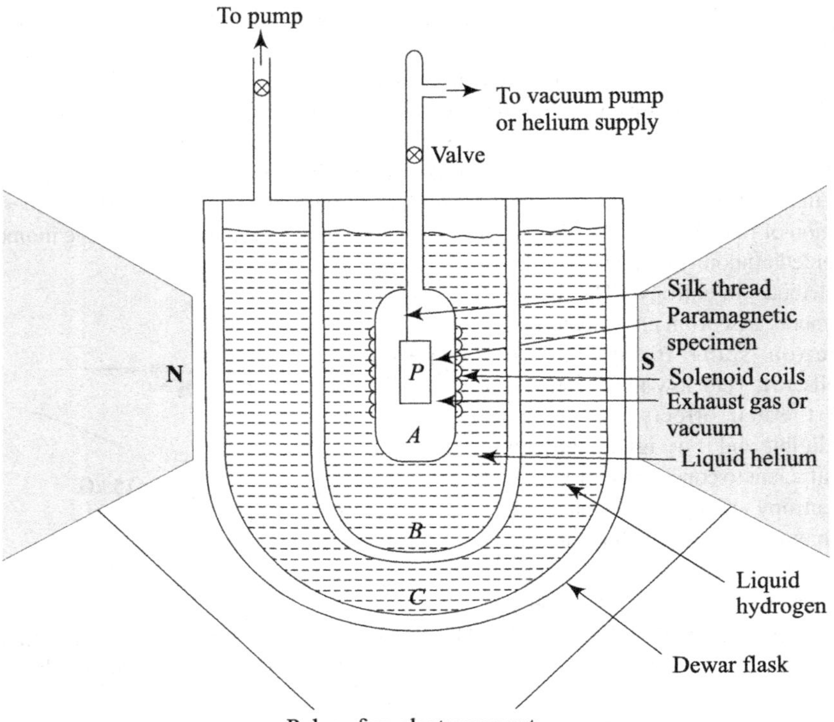

Schematic diagram of the arrangement used for adiabatic demagnetisation of a

is kept in a Dewar vessel which is further surrounded by another bath (*C*) of liquid hydrogen (or liquid nitrogen). The whole assembly is placed between the poles of a strong electromagnet.

To begin with, helium gas is admitted into bulb *A* to put the salt specimen in thermal contact with liquid helium in bulb *B*. As a result, the paramagnetic specimen acquires the temperature of liquid helium (1.2 to 1.5 K). Next, the electromagnet is switched on and a magnetic field of about 3 tesla is applied to the specimen. As a result, it is highly magnetised and the heat produced in this process is carried out by helium gas in bulb *A* to the liquid helium in bulb *B*. The specimen in *A* is thus both cold and highly magnetised. The exchange gas in *A* is now pumped out so that the specimen is thermally isolated and *C*. The magnetic field is then switched off suddenly so that the specimen is demagnetised adiabatically. The temperature of the specimen falls and it is estimated by fitting a coaxial solenoid coil around *A* and susceptibility is measured at the beginning and at the end of the experiment.

It may be mentioned here that direct measurement of the low temperature attained after demagnetisation is not possible. It is for this reason that we measure susceptibility χ of the paramagnetic substance. Then by assuming the validity of Curie's law, the low temperature is calculated. This is termed as the 'Curie Temperature' or the magnetic temperature.

Note that like adiabatic expansion, adiabatic demagnetisation is also a two stage process. In the first stage the paramagnetic salt is isothermally magnetised by switching on a magnetic field. In the second stage, the specimen is thermally isolated and the applied magnetic field is switched off adiabatically and reversibly. To ensure reversibility, this switching off is done very slowly. In practise, this stage may be carried out in a few seconds.

A complete theory of adiabatic demagnetisation is rather complicated. So we give a very elementary treatment to calculate the final temperature in terms of the properties of the paramagnetic substance. From Chapter 5, we recall that the thermodynamic state of a paramagnetic substance, when placed in a magnetic field of strength B, depends on only. To calculate the temperature drop, we recall the second TdS equation for a paramagnetic substance given by Eq. (8.41):

$$TdS = C_B\,dT + T\left(\frac{\partial M}{\partial T}\right)_B dB \qquad (10.23)^*$$

Since the process is adiabatic, $TdS = 0$ and we can rewrite Eq. (10.23) as

$$dT = -\frac{T}{C_B}\left(\frac{\partial M}{\partial T}\right)_B dB \qquad (10.24)$$

We know from Curie's law that for a paramagnetic substance, $(\partial M/\partial T)_B$ is always negative. Since dB is also negative, the adiabatic demagnetisation is accompanied by fall in temperature. To determine this drop, we calculate $(\partial M/\partial T)_B$ using Curie's law:

$$M = \frac{kVB}{T}$$

is molar volume, i.e., volume of one mole and k is Curie's constant. Then we

$$\left(\frac{\partial M}{\partial T}\right)_B = -\frac{kVB}{T^2}$$

Using this result in Eq. (10.24), we obtain

$$TdT = \frac{kV}{C_B} B\,dB$$

If magnetic field decreases from B_0 to 0, and T_0 and T_f respectively denote initial and final temperatures, we can rewrite above equation as

$$\int_{T_0}^{T_f} TdT = \frac{kV}{C_B}\int_{B_0}^{0} B\,dB \qquad (10.25)$$

In writing Eq. (10.25), we have assumed that heat capacity at constant magnetic field and Curie's constant are independent of temperature.

---

*The first term will become zero during the first stage.

The integrals in Eq. (10.25) can now be readily evaluated to obtain*.

$$T_f^2 - T_0^2 = -\frac{kV}{C_B} B_0^2$$

$$T_f - T_0 = \Delta T = -\frac{kV}{2\, C_B T_{\text{av}}} B_0^2 \qquad (10.26)$$

= $(T_f + T_0)/2$ is the average temperature of the specimen. Note that kV is Curie constant per mole. If one gram of paramagnetic salt is taken, kV will stand for Curie constant per gram.

Equation (10.26) shows that

Adiabatic demagnetisation produces cooling.
The fall in temperature would be more if the initial field is high and/or the initial temperature is low; at very low temperatures, the drop is very steep since $C_B \to 0$ $\to 0$.

Note that this process is isentropic. So the degree of order in the orientation of magnetic dipoles can not change suggesting that the magnetic moment will be constant if the final field is non-zero. For this reason, some scientists think that it would be more appropriate to name the process as '*adiabatic reduction of the magnetic field*'.

It may be pointed out here that this method has been known for over 80 years now and is used extensively. But it has some disadvantages. Some of these are:

It is a one shot experiment with the temperature changing continuously so that steady-state experiments are difficult.
The cooling capacity is small and one cannot cool any large sized material.
One cannot perform experiments which require the absence of magnetic fields.

In view of these serious limitations, one uses adiabatic demagnetisation cautiously.
You should now go through the following example:

Example 10.3 Calculate the fall in temperature produced by adiabatic demagnetisation of a paramagnetic salt at an initial temperature of 2 K when the field is switched off from 10,000 oersted to zero. Given $C_B = 0.2$ J g^{-1} K^{-1} and Curie constant per gram mole per cm^3 = 0.042 erg K g^{-1}oe^{-2}.

We know that fall in temperature in adiabatic demagnetisation is given by

$$\Delta T = -\frac{kV}{2 C_B T_{\text{av}}} B_0^2$$

= 10,000 Oe, $C_B = 0.42$ J g^{-1}K^{-1} = 0.42×10^7 erg g^{-1}K^{-1} and Curie constant = 0.042 erg K g^{-1}oe^{-2}. Hence,

$$\Delta T = -\frac{0.042 \,\text{erg K g}^{-1}\text{oe}^{-2}}{2 \times (0.42 \times 10^7 \text{ erg g}^{-1}\text{ K}^{-1}) \times 2\,\text{K}} \times (10{,}000\,\text{oe})^2$$

$$= -0.25\,\text{K}$$

$_B$ varies with temperature as

$$C_B = aT^3$$

Using this relation in Eq. (10.25) and carrying out the integration, you will get

$$T_f^5 - T_0^5 = -\frac{5k}{2a} B_0^2$$

He Cryostats

He cryostats have been widely used for studies below 1 K since the early sixties. We know ^{3}He has a critical temperature of 3.3 K. At about 1.2 K, its vapour pressure is 10 Torr, which decreases to about 10^{-2} torr at 0.3 K. Since the latent heat of vaporisation is quite appreciable, a very convenient closed cycle operation of a ^{3}He cryostat is possible.

He gas is condensed into a liquid at 1.3 K, using a ^{4}He bath. This liquid is put into a thermally insulated vessel, where it is made to boil at a reduced pressure and cooled to a low temperature of 300 mK. The evaporated ^{3}He is re-condensed at 1.3 K and the cycle is completed. So a continuous cooling of substances down to 300 mK is possible.

We may remark here that ^{3}He cryostat has the obvious advantage in that it is a steady-state cooling. So one can keep the temperature constant at any value for any length of time. Moreover, the cooling capacity is large and even complicated experiments can be done. Also, since no magnetic field is present, one can perform experiments which are spoiled by magnetic fields. However, liquid ^{3}He cryostats have a lower limit of about 300 mK. To produce cooling below this value, we have to use the Pomeranchuk cooling or nuclear demagnetisation.

Pomeranchuk Cooling

The Pomeranchuk cooling of a mixture of solid and liquid ^{3}He depends upon a peculiar property of the phase diagram of ^{3}He, namely, a minimum in the melting curve at 0.32 K and 29 atm (Fig. 10.14). The melting curve reaches 34 atm as $T \to 0$, with a very faint maximum at about 0.5 mK. The Clausius–Clapeyron equation implies that the entropy of a solid is greater than the entropy of the liquid when dp/dT is negative. (This very unusual situation arises due to quantum effects exhibited by liquid ^{3}He.) Pomeranchuk suggested

He is pre-cooled to a temperature somewhat below 0.3 K and compressed beyond 29 atm, it is converted to solid ^{3}He. If compression of liquid-solid mixture of ^{3}He is carried out adiabatically along the melting curve below T_{min}, temperature drops producing cooling. With reference to the phase diagram, the liquid-solid mixture co-exists along the negative slope until all the helium solidifies at about 2 mK.

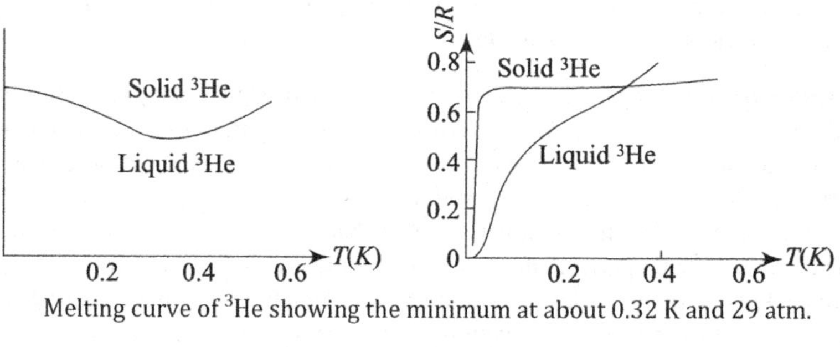

Melting curve of ^{3}He showing the minimum at about 0.32 K and 29 atm.

In ordinary adiabatic demagnetisation, alignment of electronic spins determines the entropy of the system. Encouraged by the success of this method, it was suggested to employ nuclear demagnetisation using nuclear spins. But it was only in 1956 that the associated technical difficulties could be overcome successfully. The nuclear demagnetisation experiments are intrinsically difficult because nuclear moments are 5×10^{-4} times that due to electronic spins. This necessitates us to start with a temperature as low as 30 mK and a magnetic field of 3 tesla. In this way, it became possible to cool the system to about 100 pK in 1990. Because of the interest in the phenomena occurring about such temperatures, nuclear demagnetisation is being used extensively these days.

By now you must have realised that a good deal of experimental sophistication is required to attain and maintain low temperatures. An equally impressive degree of sophistication is required to measure them. We cannot think of buying a thermometer that works in this region. Magnetic thermometry has been widely used for many years now. Here one measures the magnetic susceptibilities or conductivities and compares them with recorded data at different temperatures. But the question arises: How was the first such thermometer calibrated? If investigations are made at low temperatures that have never been probed before or where no previous data is available and properties of materials are unknown, how can we know the temperature?

A discussion of these and other similar questions will carry us too far and is of no interest to us here. However, we now reflect on some of the practical applications of low temperatures.

Practical Uses and Applications of Low Temperatures

Low temperatures find several industrial uses and research applications. Particular mention may be made of the discovery of a wide variety of new phenomena, such as superconductivity and superfluidity as we approach absolute zero. All of us know what an important role liquid oxygen plays as a life-support system for artificial respiration in case of critically ill patients. Some of the emerging applications of low temperatures are in preparation of fuels for rockets used for exploration of outer space and study of cosmological phenomena. We now discuss some of these.

Industrial Applications

Liquefied gases such as liquid nitrogen and liquid helium are commonly used in cryogenics and allow for the lowest attainable temperatures to be reached.

Cryogenic transfer pumps are used on LNG piers to transfer Liquefied Natural Gas from LNG carriers to LNG storage tanks, as cryogenic valves.

Cryogenic processing The field of cryogenics made significant progress during World War II when scientists found that metals frozen to low temperatures showed more resistance to wear and tear and the life of metallic tools could be increased by more than 200% of the original life expectancy. This evolved in the late 1990s into the treatment of amplifier valves (for improved sound quality), baseball bats and golf clubs (greater sweet spot), racing engines (greater performance under stress), firearms (less warping after continuous shooting), knives, razor blades, and brake rotors. This field was put on a higher pedestal in the later part of the twentieth century when cryogenic processor was coupled with microprocessor-based industrial controls.

Cryogens, like liquid nitrogen, are also used for specialty chilling and freezing applications such as freezing of foods, life saving vaccines, semen samples to improve fertility of animals and many such other applications.

Another use of cryogenics is as cryogenic fuels; mainly liquid hydrogen has fuel as its primary means of getting into orbit, as did all of the rockets built for the Soviet Space Programme. However, all efforts made by ISRO to use cryogenic fuels have so far been disastrous for lack of perfection in indigenous technology and non-availability of imported know-how.

Some other industrial applications of cryogenics include:

Magnetic Resonance Imaging (MRI) In MRI, objects are imaged using a strong magnetic field to detect relaxation of protons that have been perturbed by a radio-frequency pulse. This magnetic field is generated by electromagnets, and high field strengths can be achieved by using superconducting magnets. Traditionally, liquid helium is used as coolant and metallic superconductors can be used for the coil wiring.

Electric Power Transmission It is neither economical nor safe to transmit power by overhead cables in big cities. For these reasons, underground cables are being increasingly used. But underground cables get heated and increase in resistance of the wire leads to wastage of power. To minimise such wastage, liquefied gases are sprayed on the cables to keep them cool and at low resistance.

Frozen Food Cryogenic gases are used in transportation of large masses of frozen food. When very large quantities of food must be transported to regions like war fields, earthquake hit regions, etc., they must be stored for a long time. So cryogenic food freezing is used. Cryogenic food freezing is also helpful for large scale food processing industries, which are coming up in a big way now in India.

Blood Banking The blood groups are stored at low temperatures, nearly –165°C. In fact, rapid progress in biotechnology and genetic engineering has now made it possible to store animal semen to improve the quality of the cattle.

Scientific Applications

Gases at low temperatures are gainfully used to produce high vacuum. If the system to be evacuated contains gases such as CO_2, N_2 and CH_3, it is connected to a tube containing some freshly prepared charcoal and surrounded by liquid air. It is observed that these gases are absorbed by the charcoal. In case the container has hydrogen, one uses palladium black, instead of charcoal. However, water vapour or sulphuric acid vapours can be removed from a vessel by putting it in a bath of liquid air. Similarly, air can be removed by surrounding it by liquid hydrogen and so on.

The low temperature physics has opened up the possibilities of improving our understanding of upper atmosphere since those conditions can be recreated in the laboratory.

By fractional distillation, oxygen, nitrogen, argon and neon are obtained commercially in pure form from liquid air.

Low temperature researches led to the discovery of the amazing phenomenon of superconductivity in metals and superfluidity in helium.

The Third Law of Thermodynamics

From the preceding discussion, it is clear that march towards absolute zero becomes progressively difficult. Essentially, this is a simple manifestation of the fact that lower the temperature, less useful the heat is. This means that to attain absolute zero, an infinite number of adiabatic/nuclear demagnetisation operations will be needed. This observation is contained in the third law of thermodynamics:

It is not possible to reduce any assembly to the absolute zero of temperature by any

You will recognise that this statement is along the lines somewhat parallel to the second law of thermodynamics. In fact, just like the second law, the third law has a number of equivalent statements. For instance, we can enunciate it in terms of entropy as well. To arrive at this formulation, we re-examine Eq. (7.5b) and recall that even for a reversible process, entropy is indeterminate to the extent of an arbitrary additive constant. This means that within the framework of the first and the second laws of thermodynamics, we can calculate entropy difference only, i.e., we cannot fix the absolute value of entropy. This aspect is contained in the third law, which deals with the entropy of a system as its temperature tends to absolute zero. It was discovered experimentally by Nernst in 1906. He observed that the net change in entropy of a system is very small when we go from one low temperature equilibrium state to another state. For instance, we may change liquid helium into solid helium by applying pressure or convert separate samples of sodium and chlorine at, say, 4 K into NaCl at the same temperature. This means that near absolute zero, all systems are highly ordered and the entropy of all states (of every substance) is the same; in fact extremely small or essentially zero. Thus, we may also state the third law as follows:

The entropy changes in all reversible isothermal processes tend to zero as temperature approaches absolute zero.

This was called the *Nernst heat theorem* and later on as the *Nernst–Simon statement of the third law*. Mathematically, we write

$$\lim_{T \to 0} \Delta S \to 0 \tag{10.27}$$

The importance of this result arises from the fact that the zero temperature entropy does not depend on any other property of the system, such as volume or pressure. Various futile attempts made to obtain this law from the second law have only helped in establishing its independent status. In fact, unlike the zeroth, first and second laws, which constitute an essential part of the basic structure of thermodynamic theory, the third law of thermodynamics has implications only in the low temperature range near absolute zero.

As such, the Nernst–Simon statement (Eq. (10.27)) is a macroscopic manifestation of quantum effects, which become prominent at low temperatures. According to Planck, the entropy of every solid or liquid substance in internal equilibrium at absolute zero is

$$\lim_{T \to 0} S \to 0 \tag{10.28}$$

In other words, a reversible isothermal process at $T = 0$ is isentropic.

Today, this law is supported by experimental evidence and finds wide applications in low temperature physics as well as in the study of chemical reactions.

Taking the validity of this law for granted, we now discuss some of its consequences and give experimental evidence, wherever possible. For simplicity, we will confine ourselves to ideas applicable to perfectly crystalline substances. However, the real significance of this law can be appreciated only in the context of quantum statistics.

Consequences of the Third Law

The third law of thermodynamics has several important consequences. We now discuss

1. Heat Capacities For changes at constant volume, the change in entropy can be written as

$$S(T_2, V) - S(T_1, V) = \int_{T_1}^{T_2} C_V \frac{dT}{T} \tag{10.29a}$$

Similarly, for changes at constant pressure, the change in entropy at constant pressure can be written as

$$S(T_2, p) - S(T_1, p) = \int_{T_1}^{T_2} C_p \frac{\mathrm{d}T}{T} \tag{10.29b}$$

If we take the limit $T_1 \to 0$ in Eq. (10.29a, b), the validity of third law demands that the integral in these equations should be finite. This requires that C_V and C_p should tend $T \to 0$. Otherwise, the integrals will diverge. This property of heat capacities, that they tend to zero as $T \to 0$, was experimentally established by Nernst. Some typical results for gold are shown in Fig. 10.15. Note how heat capacity varies with temperature. This result is not compatible with the law of equipartition of energy. It means that the third law cannot be explained within the framework of classical physics; it brings out the connection of third law with quantum theory.

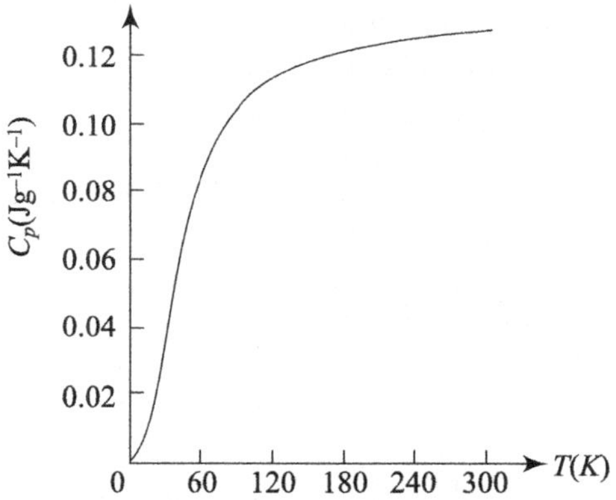

Temperature variation of heat capacity of gold at constant pressure.

2. Expansion Coefficients Let us consider thermal expansion of a solid at low temperatures. We know that volume expansion coefficient is defined as

$$\alpha = \frac{1}{V}(\partial V/\partial T)_p$$

The second Maxwell relation (Eq. (8.11)) tells us that

$$\left(\frac{\partial V}{\partial T}\right)_p - \left(\frac{\partial S}{\partial p}\right)_T$$

so that we can write

$$\alpha = -\frac{1}{V}(\partial S/\partial p)_T \tag{10.30}$$

But according to the third law, the changes in entropy tend to vanish in an isothermal process for all values of p at low temperatures. So, the RHS in Eq. (10.30) will tend to zero 0. This means that α must tend to zero as we approach absolute zero.

For a solid, the volume expansion coefficient is three times its linear expansion α_ℓ. Thus at low temperatures, both α and α_ℓ should tend to vanish. Experimental results confirm this prediction of the third law. Similarly, we can show that the result

$$\lim_{T \to 0} \left(\frac{\partial p}{\partial T} \right)_V = 0 \tag{10.31}$$

follows from Eq. (8.7). That is, the pressure expansion coefficient at constant volume

We may thus conclude that all expansion coefficients must tend to zero as temperature approaches absolute zero.

Corresponding results follow for other properties. For instance, for a surface film

$$\left(\frac{\partial S}{\partial A} \right)_T = -\frac{d\sigma}{dT}$$

so that by third law, *surface tension becomes constant as* $T \to 0$:

$$\lim_{T \to 0} \frac{d\sigma}{dT} \to 0 \tag{10.32}$$

This is found to be so for both liquid ^{4}He and liquid ^{3}He.
You may now like to answer a practise problem.

Problem 10.1 For a paramagnetic salt show that a necessary consequence of the third law is that susceptibility becomes constant as $T \to 0$.

3. Continuance of C_p and C_V Using Eq. (8.18) we can relate the difference of heat capacities to volume expansion coefficient:

$$C_p - C_V = TV\alpha \left(\frac{\partial p}{\partial T} \right)_V \tag{10.33}$$

From the above discussion, we know that $\to 0$. This means that $C_p - C_V \to 0$ as 0. Thus the validity of the third law implies confluence of C_p and C_V (Fig. 10.16). This is in conformity with experimental findings.

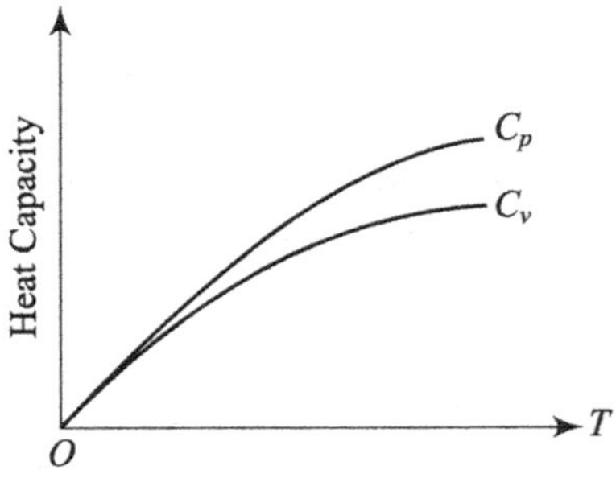

Fig. 10.16 As temperature is lowered, C_p and C_V coalesce.

4. Degeneracy of an Ideal Gas From Chapter 7, we recall that entropy of an ideal gas is given

$$S = C_V \ln T + R \ln V + S_0 \tag{10.34}$$

0, $S \to -\infty$, which is physically not meaningful. This means that at low temperatures, an ideal gas will not behave according to Eq. (10.34). This departure from ideal behaviour implies degeneracy of an ideal gas, which increases as T decreases. You will learn more about degeneracy in Chapters 13-15.

5. Unattainability of Absolute Zero As a final exercise, we show that unattainability of absolute zero is equivalent to entropy tending to zero as $T \to 0$. Let us assume the contrary so that we can operate a Carnot engine between two reservoirs, one maintained at absolute zero and the other at some finite temperature T as shown in Fig. 10.17.

For a cyclic process, we recall that

$$\Delta S = \oint \frac{\delta Q}{T} = 0$$

But we can write

$$= \Delta S_{12} + \Delta S_{23} + \Delta S_{34} + \Delta S_{41}$$

$$\Delta S_{12} = \frac{Q}{T}$$

is heat absorbed at temperature T. For an adiabatic process, $\Delta S_{23} = \Delta S_{41} = 0$ and by the $\Delta S_{34} = 0$. Hence,

$$dS = \oint \frac{\delta Q}{T} = \Delta S_{12} \neq 0 \qquad (10.35)$$

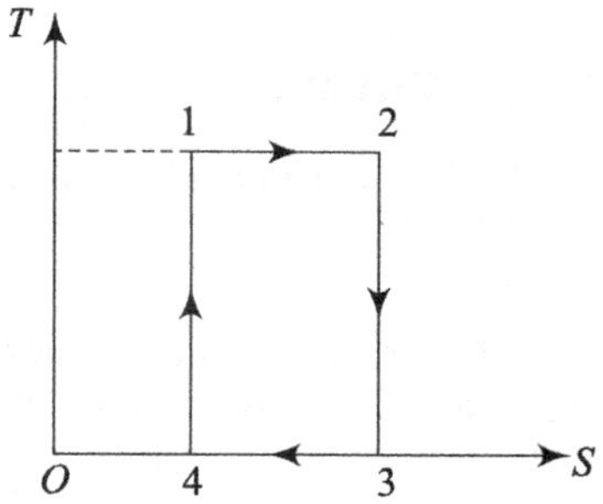

Fig. 10.17 T-S diagram of Carnot reversible engine operating between absolute zero and a finite temperature: Unattainability of absolute zero.

But this contradicts the second law of thermodynamics. This inconsistency demands that we cannot operate a Carnot engine using a single reservoir, i.e., *it is not possible to attain absolute zero by means of any process howsoever idealised in a finite number of operations.*

It is important to point out here that although the third law of thermodynamics is well supported by experiments, there are exceptions to it also. Some of these are apparent in the sense that discrepancy can be traced to non-existence of true equilibrium states. This is particularly true for gases, H_2O, etc., where all the molecules are not oriented in the same direction near absolute zero. But differences in the values of entropy for processes involving substances like H_2 which can exist in forms differing in their nuclear spins are difficult to explain. We may thus conclude that the third law enjoys the status of an independent law and provides a useful check both for experiments and for new theories at low temperatures.

MMARY

Cascade process is used to liquefy gases. In this process, cooling is produced in stages by a series of isothermal compression and evaporation under reduced pressure.

Coupled with the principle of regenerative cooling, Joule–Thomson effect is used for liquefaction of air in Linde's liquefier.

Liquid helium exhibits amazing properties as temperature is lowered below 4.2 K: it undergoes a phase transition at 2.17 K and instead of solidifying, liquid helium changes into another liquid of vastly different and extremely surprising properties. The phase above 2.17 K is termed as Helium I and that below 2.17 K as Helium II.

He II exhibits superfluidity and superconductivity.

If a paramagnetic salt is first cooled to about 1 K in the presence of strong magnetic field and then subject to adiabatic demagnetisation, there is fall in temperature.

The third law of thermodynamics states that it is not possible to reduce any assembly to the absolute zero of temperature by any process, however idealised, in a finite number of operations.

Nernst heat theorem states that the entropy changes in all reversible isothermal processes tend to zero as temperature approaches absolute zero: $\lim_{T \to 0} \Delta S \to 0$

ERCISES

In the region of moderate pressures, the equation of state of one mole of a gas may be written as

$$pV = RT + c_2 p + c_3 p^2$$

where c_2 and c_3 are the second and third virial coefficients. Taking these to be functions of T only, show that as the pressure approaches zero

$$\mu c_p \to T\frac{dc_2}{dT} - c_2$$

Also show that the inversion curve is described by equation

$$p\left(c_3 - T\frac{dc_3}{dT}\right) + \left(c_2 - \frac{dc_2}{dT}\right) = 0$$

We know that μ provides us with a measure of temperature change during a throttling process. A similar measure of the temperature change produced by an isentropic change of pressure can be obtained from the coefficient μ_S defined as

$$\mu_S = \left(\frac{\partial T}{\partial p}\right)_s$$

Prove that

$$\mu_S - \mu = \frac{V}{C_p}$$

Calculate the pressure corresponding to an inversion temperature of 20 K for helium assuming that it behaves as a van der Waals' gas. Use the data given in Exercise 3.11.

(*Ans:* 19 atm)

Using the fact that $\left(\frac{\partial S}{\partial V}\right)_{T=0} = \left(\frac{\partial^2 S}{\partial V^2}\right)_{T=0} = 0$ at $T = 0$,

prove that $\lim_{T\to 0}\left(\frac{\partial(1/\beta_T)}{\partial T}\right)_V = 0$ where β_T is isothermal compressibility.

Calculate the fall in temperature produced by adiabatic demagnetisation of a paramagnetic salt, if magnetic field is reduced from 8400 Oe to zero at 3 K. Take Curie constant per cm^3 = 0.6 erg K g^{-1} Oe^{-2} and C_B = 0.2 cal g^{-1} K^{-1}.

11 RADIATION: CLASSICAL AND QUANTUM THEORIES

arning Objectives

In this chapter, you will learn how to

- e a blackbody and discuss Kirchoff's laws;
- n expression for radiation pressure and apply it to derive Stefan–mann law;
- uss how blackbody radiation behaves when subject to isothermal and batic expansion;
- e expression for spectral density based on Wien's and Rayleigh–Jeans
- uss Planck's theory to explain observed spectral distribution of kbody radiation;
- w that Planck's radiation law is the most general law of blackbody tion and all other laws are contained in it; and
- n expression for entropy, pressure and other thermodynamic meters for blackbody radiation.

NTRODUCTION

From your school physics, you may recall that transmission of thermal energy (heat) takes place from one point to another by conduction, convection and radiation. We experience these in many ways. On a hot summer day, a metallic handle feels hotter than a wooden rod, brisk stirring helps to heat water faster and the less privileged give themselves warmth by standing near a fire and so on. As you know, these are examples of conduction, convection and radiation, respectively. In *conduction*, heat flows in a differentially heated body from regions of higher temperature to regions of lower temperature. For instance, when we place one end of a metal rod in fire, thermal energy is communicated to the other end through collisions of successive vibrating atoms/molecules. Although this mode of transmission is, in principle, permissible in all the three states of matter (solids, liquids and gases), it is the dominant mode in solids only but for a few exceptions such as mercury. We usually characterise the conduction in a solid through thermal conductivity. The analytical discussion of conduction for good as well as not so good conductors and different methods for the determination of thermal conductivity are given in Appendix IV.

When we heat water in a vessel, water becomes less dense and tends to rise at the point of heat input under the buoyancy force. The space created by the rising water in this process is taken by colder (and heavier) fluid from other regions and gets heated in turn, rises to top, and so on the process continues. The process in which each particle of the substance participates and carries thermal energy is known as *convection*. This is the common mode of transmission of heat in fluids. Note that in convection, energy is transferred by physical movement of matter, whereas in conduction, the thermal energy is transmitted by molecules/atoms via collisions in the local regions. You are familiar with the notion that wind is convection current in the atmosphere caused by unequal heating. Other natural processes in which convection plays an important role include setting up monsoons in India and adiabatic lapse rate in the stratosphere (Chapter 5). When hot air from the Indian landmass rises up and moves towards the Indian Ocean, moisture laden less hot air moves into the subcontinent leading to formation of south-west monsoon. Convection is also responsible for ventilation in rooms, the central heating in buildings and the cooling system in automobiles, among others. However, we will not into these details here. Note *conduction and convection involve actual participation of the material medium.*

While standing near a fire place, we feel the heat. You may be aware that in this case, energy is being transported without active participation of the intervening medium. The process in which energy can be transferred directly from one place/body to another is *radiation*. In fact, radiation does not require a medium and that is how solar energy reaches us on this planet. Do you know that radiation is the main mechanism for energy transfer in large systems such as the sun, the solar system, interstellar space and the galaxy?

It is now well accepted that thermal radiations are electromagnetic in nature and akin to light but for the wavelength, which lies in the infra-red region (> 800 nm) and produces sensation of warmth in us. Accordingly, heat radiations exhibit all the phenomena associated with light such as reflection, refraction, absorption, interference, diffraction, and polarisation. An enclosure maintained at a constant temperature is filled with electromagnetic radiation in thermal equilibrium with the walls. The electromagnetic radiation trapped in a cavity is called *blackbody radiation* and constitutes a simple thermodynamic system having a well-defined temperature, a definite volume and finite pressure. We will first use the laws of thermodynamics to study the behaviour of blackbody radiation.

We know that all bodies emit thermal radiation by virtue of temperature at the expense of their internal energy. However, the intensity, wavelength and the rate of emission change, in fact increase, with temperature. For instance, at room temperature, most of the energy is radiated in the far infra-red region, whereas at 6000 K, the temperature of the outer surface of the Sun, it lies in the visible region. You may have also seen a blacksmith heating a piece of iron in his furnace to create different implements. The colour of iron changes from dull red to reddish yellow and 'white' as he continues to heat it to give the desired shape. It is important to mention here that the quality of radiation also depends on the nature of the surface of the body. A highly polished surface reflects most of the radiation incident on it and emits very little.

The development of complete theory of blackbody radiation is one of the most exciting stories in intellectual history of physics. A correct understanding required the scientific creativity of the highest order; it completely changed physicists' view of nature and opened totally new perspectives. In fact, many new branches of study were born as our understanding of blackbody radiation improved. Starting with Kirchhoff's laws, in this chapter, we will give you a glimpse of the applications of the principles of thermodynamics to electromagnetic radiations trapped in a cavity, before moving into newer territory of discrete energy exchanges for gaining correct insight into the nature of spectral distribution in the entire range.

To facilitate this journey, in Sec. 11.2, we have introduced a few useful terms so as to be able to express theoretical results in physically acceptable forms. This is followed by a discussion of Kirchhoff's laws in Sec. 11.3. Blackbody radiation as a thermodynamic system is discussed in Sec. 11.4. In particular, we have discussed Stefan–Boltzmann law and how blackbody radiation behaves when subject to isothermal and adiabatic expansions. Spectroscopic analysis of blackbody radiation shows that it consists of a distribution spread over a range of wavelengths rather than discrete spectral lines. Using thermodynamics, Wien and Rayleigh–Jeans obtained expressions for the energy density of equilibrium blackbody radiation. These form the subject matter of Sec. 11.5. However, you will discover that their efforts were not enough to explain the observed results on spectral distribution; the greatest physicists struggled to arrive at a correct explanation. This inspired Planck, who made a drastic deviation from the then accepted concepts of energy exchange between bodies. He proposed the concept of quanta as carrier of energy in emission or absorption of blackbody radiation and obtained excellent agreement with observed results. In fact, he showed that all other laws of radiation then known could be obtained from Planck's law. This is also discussed in Sec 11.5.

You will agree that physics is an experimental science and our vast knowledge pool has genesis in the confirmation of theoretical predictions by experimental findings. Theory is invariably modified in the light of experimental evidences; new experiments are designed to test newer theories. Therefore, for completeness, we have presented experiments used to verify some of the radiation laws in Sec. 11.6. This is followed by a brief discussion of solar constant in Sec. 11.7. It is a measure of the energy received per minute per unit area by a perfectly black surface held at right angles to the solar radiations at the mean distance of the Earth from the Sun, provided there were no absorption in the atmosphere.

OME USEFUL DEFINITIONS AND CONCEPTS

1. Total energy density (u) at any point denotes the total radiant energy for all wavelengths ∞ per unit volume around that point. Its unit is Jm^{-3}.

2. Spectral energy density (u_λ) for the wavelength λ is a measure of the energy per unit volume per unit wavelength. Therefore, $u_\lambda d\lambda$ denotes the energy per unit volume in the wavelength range between λ and λ + dλ. It is related to total energy density through the

$$u = \int_0^\infty u_\lambda \, d\lambda \tag{11.1}$$

3. Emissivity (e), also known as *total emissive power* of a body, is defined as the total radiant energy of all wavelengths from 0 to ∞ emitted per second per unit surface area of the body. Its unit is $\text{J m}^{-2}\text{s}^{-1}$ or W m^{-2}.

4. Spectral emissive power (e_λ) of a body for the wavelength λ signifies the radiant energy per second per unit surface area per unit range of wavelength. Therefore, $e_\lambda d\lambda$ denotes the energy per unit area per second in the wavelength range between λ and λ + dλ. It is related to emissivity through the relation

$$e = \int_0^\infty e_\lambda \, d\lambda \tag{11.2}$$

5. Spectral absorptivity (a_λ) is defined as the fraction of incident energy absorbed per unit surface area per second at wavelength λ. Suppose that δQ_λ radiation of wavelength and $\lambda + d\lambda$ is incident on a unit area of the surface of the body per second from all possible directions. If $a_\lambda \delta Q_\lambda$ is the amount of radiation absorbed, then a_λ signifies the absorptivity of the body for wavelength λ. Note that a_λ has no dimensions; it is a fraction whose maximum value is unity. (Such a body is said to be perfectly blackbody.)

It may be mentioned here that e and e_λ characterise the properties of a body as emitter describes the properties as an absorber of radiation. However, all the three depend on the temperature of the body and on the nature of its surface.

A blackbody absorbs all radiations incident on it. When radiation of a particular wavelength λ is incident on a body, it may be partially reflected, partially absorbed and partially transmitted. Then we can write

$$r_\lambda + a_\lambda + t_\lambda = 1$$

and t_λ respectively denote energy reflection and transmission coefficients of the body for wavelength λ. If $r_\lambda = t_\lambda = 0$, and $a_\lambda = 1$ for all wavelengths, i. e., when a body absorbs all radiations incident on it, regardless of frequency, the body is said to be perfectly black. (This is analogous to the colour that we see due to selective absorption of light; white cloth appears so because it reflects all colours in the visible range and a red one absorbs all except red, which it reflects. Similarly, the letters in the text of this book appear black because these absorb all light falling on them.) In practise, no surface or body satisfies this ideal definition strictly. Even for lamp black and platinum black, a_λ is less than unity; these respectively absorb nearly 96% and 98% of visible light. But a small hole in a large enclosure or cavity is a better practical approximation to blackbody (Fig. 11.1). Suppose the hole is sufficiently small and the radiation enters the cavity. It suffers multiple reflections at the inner surface of the wall and a fraction of it will be absorbed at each reflection with the result that only a negligible fraction, if at all, can escape through the hole. So we can say that a hole in a cavity is a blackbody in that it absorbs the entire radiation incident on it. (Do not confuse it with a black hole, which is a heavenly body at the end

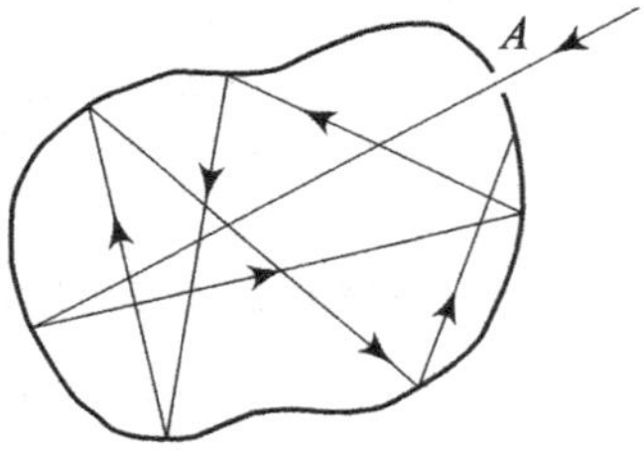

Fig. 11.1 A small hole in a large cavity acts as a blackbody.

11.2.1 Prevost's Theory of Exchanges

In pre-1792 period, physicists had little knowledge about the nature of radiant energy and its behaviour; they used terms like hot radiations and cold radiations. But Prevost argued against this nomenclature. He stated that all bodies spontaneously emit radiation, without any regard for their surroundings, as long as they are above absolute zero. Moreover, the amount of energy radiated is more at a higher temperature. And when two bodies at different temperatures are put in a large enclosure, they attain thermal equilibrium with the enclosure by exchanging heat radiations and the rate is determined by the temperature of the body rather than the surroundings. That is, rise or fall in temperature of a body occurs due to continuous exchange of radiation with the surroundings. According to Prevost, bodies exchange radiation even when they are in equilibrium. However, the amount of radiation emitted equals the amount of radiation absorbed. In other words, spontaneous emission and absorption of radiation are mutually independent, though each is a function of temperature. This is known as *Prevost's Theory of Exchanges*.

To illustrate, consider a child sitting near a fire in winter. He gets sensation of warmth because his body receives more energy from the fire than it loses by its own radiation. Similarly, if you stand near a air conditioner, you will get a sensation of cold because your body loses more energy by radiation than it receives from the air conditioner, which is much below the body temperature.

11.2.2 Radiation Pressure

As mentioned earlier, thermal radiations are electromagnetic in nature and akin to light but for the wavelength. It is therefore natural to assume that like light, radiation also exerts a small but finite pressure. You may have studied in science magazines that as comets approach the Sun, their tail continuously veers round so as to be always away from the Sun. Kepler explained this phenomenon on the assumption that solar light exerts pressure on all material bodies. Initially, experiments designed to measure light pressure failed probably because it was too small to be detected. However, on the basis of electromagnetic theory of light, Maxwell predicted that light pressure should be equal to the energy density of light for a parallel beam. This provided firm theoretical confirmation for the existence of light pressure.

For a parallel beam, the intensity of radiation I is the radiant energy incident per second on a unit area held normal to the incident beam. In one second, radiation will travel a , equal to velocity of light, and will fill a cylinder of unit cross-sectional area c so that the volume of the cylinder is c. Hence, energy density of radiation u is amount of radiation contained in a unit volume and we can write $u = I/c$ Using the result of electromagnetic theory, we can write the pressure exerted by the radiation as

$$p = u = \frac{I}{c} \tag{11.3}$$

To get an idea of the magnitude of radiation pressure exerted by solar radiation, it may be mentioned here that the amount of thermal energy absorbed by a unit area of a blackbody per minute placed at a mean distance of the Earth and the Sun is 1.35 kWm^{-2}. Hence,

$$I = (1.35 \times 10^3/60)\ \mathrm{Js^{-1}m^{-2}}$$

$$p = \frac{135}{6 \times 3 \times 10^8}\ \mathrm{Nm^{-2}} = 7.5 \times 10^{-8}\mathrm{Nm^{-2}}$$

This result shows that the pressure exerted by solar radiation on 1 m^2 of a blackbody is equal to the weight of 7.6×10^{-9} kg matter.

11.2.3 Radiation in an Enclosure

Imagine an enclosure of arbitrary shape and material. Suppose its inner walls are maintained at a constant temperature T say. We now know that all bodies emit radiation by virtue of their temperature. According to Prevost's theory of exchange of radiant energy, all bodies continuously emit and absorb radiation irrespective of their surroundings. Thus, if two bodies at different temperatures are kept in a large enclosure, they attain thermal equilibrium with the enclosure by exchanging radiations. Notwithstanding thermal equilibrium, the exchange process continues with the difference that now each body emits as much radiation as it absorbs. So we can say that *spontaneous emission of radiation and its absorption are independent of each other.*

The enclosure will be filled up with radiation of all wavelengths due to continuous *isothermal enclosure, the radiation will be in thermal equilibrium with the walls*; its quality and quantity remain constant with time. Further, it will not be influenced if we introduce any material body, provided it is small in size compared to the enclosure. And *the cavity radiation is homogeneous and isotropic*, i.e., same at all points and in all directions. In fact, all enclosures at the same temperature will have the same quality of radiation, irrespective of their shape and the nature of the inner walls.

You should now go through the following example carefully.

Example 11.1 Cavity radiation is isotropic and homogeneous. Using this fact, calculate the expression for pressure exerted by blackbody radiation on the walls of the

Refer to Fig. 11.2. Let QR be the blackbody. The radiation incident on it will consist of a large number of parallel beams, each of equal intensity with directions of propagation distributed evenly. One such beam incident at an angle θ with the normal to QR is shown. If the energy density of the radiation is u, the energy incident on the per unit area per second is uc. If surface area of the plane PR is s, the energy incident on the surface PQ per second will be ucs. But surface area of the plane PR is equal to the surface area of the plane QR times $\cos\theta$. Let us denote it as $s'\cos\theta$ where s' is surface area of the plane QR. Hence, energy incident on the surface PR per second will be $uc\, s'\cos\theta$.

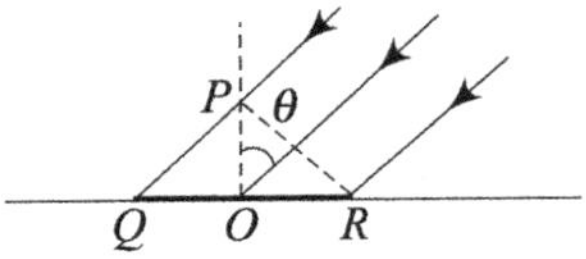

Fig. 11.2 Calculation of pressure of cavity radiation.

From Fig. 11.2, it is obvious that the total energy crossing the plane PR per second is equal to the energy incident on the plane QR and will be $uc\, s'\cos\theta$. Hence, the energy incident on the plane QR per unit area per second $\theta/s') = uc \cos\theta$.

The momentum incident on the plane QR per unit area per second is $= (uc \cos\theta/c) = u \cos\theta$.

The component of this momentum in the direction normal to plane $QR = (u \cos\theta)\cos\theta = u\cos^2\theta$.

Note that this momentum, if completely absorbed, gives the total rate of change of momentum per second per unit area normal to the surface QR, which, by definition, is pressure exerted by the incident radiation on surface QR. Hence, we can write

$$p = u\cos^2\theta \tag{i}$$

Since the radiation in a cavity will be incident from all directions with equal probability, the correct value of radiation pressure will be obtained by averaging the value of $\cos^2\theta$ over the hemisphere. Hence, we can write

$$p = u\overline{\cos^2\theta} \tag{ii}$$

In a hollow cone between angles θ and $\theta + d\theta$, the solid angle is $2\pi \sin\theta\, d\theta$. Hence,

$$\overline{\cos^2\theta} = \int_0^{\pi} \frac{2\pi\sin\theta\cos^2\theta\, d\theta}{4\pi}$$

To evaluate this integral, we introduce a change of variable by defining $\cos\theta = x$ so that $= dx$ and the limits of integration change from 0 to π to 1 to -1. Hence, we can write

$$\overline{\cos^2\theta} = -\frac{1}{2}\int_1^{-1} x^2\, dx = \frac{1}{2}\int_1^{1} x^2\, dx = \frac{1}{2}\times\frac{x^3}{3}\bigg|_1^1 = \frac{1}{3}$$

Using this result in Eq. (ii), we get the desired expression for pressure of cavity radiation:

$$p = \frac{u}{3} \tag{iii}$$

This result shows that the pressure exerted by radiation on any surface is one-third of its energy density. This is a fundamental result for blackbody radiation and we will obtain it again in Chapter 15. However, it is important to mention here that this result implies an analogy between blackbody radiation and a perfect gas [$p = (2/3)\rho$]. If we extend this analogy further, we can consider blackbody radiation as a thermodynamic system and calculate its energy as well as entropy.

You should now answer a practise problem.

Problem 11.1 Our earth receives 10^3 Jm^{-2}s^{-1} solar energy. Calculate the total force experienced on the earth due to solar radiations. Take the radius of the earth
$\times 10^6$ m.
$\times 10^9$ N

IRCHHOFF'S LAW: RELATION BETWEEN e_λ AND a_λ

The Kirchhoff's law states that *the ratio of the spectral emissive power e_λ to the spectral absorptivity a_λ for a particular wavelength λ is the same for all bodies at the same temperature and is equal to the emissive power of a perfectly black body at that temperature*. Mathematically, we write

$$\frac{e_\lambda}{a_\lambda} = E_\lambda \tag{11.4}$$

is emissive power of a perfectly blackbody. Note that the ratio e_λ/a_λ is a universal function of λ and T.

You can easily arrive at Eq. (11.4) by noting that if δQ denotes the thermal energy between wavelengths λ and λ + dλ incident on the unit area of a surface in one second, the quantity of thermal energy absorbed by the surface will be $a_\lambda \delta Q$, where a_λ is spectral absorptivity of the body. The remaining energy will be either reflected or/and transmitted.

If the spectral emissive power is e_λ, the amount of energy emitted per second per unit surface area between wavelengths λ and λ + dλ by virtue of temperature dλ. Hence, the total energy given out per second per unit surface area is .+ e_λdλ.

In the equilibrium state, we can write

$$\delta Q = (1 - a_\lambda)\delta Q + e_\lambda \mathrm{d}\lambda$$

$$a_\lambda \delta Q = e_\lambda \mathrm{d}\lambda \tag{11.5}$$

For a perfectly blackbody, $a_\lambda = 1$ and $e_\lambda = E_\lambda$ so that we can rewrite Eq. (11.5) as

$$\delta Q = E_\lambda \mathrm{d}\lambda$$

Using this result in Eq. (11.5) and simplifying the resultant expression we reproduce

$$a_\lambda E_\lambda \mathrm{d}\lambda = e_\lambda \mathrm{d}\lambda$$

$$E_\lambda = \frac{e_\lambda}{a_\lambda}$$

A rigorous proof of Kirchhoff's law is given in Appendix III. However, the original proof given by Kirchhoff is given in the box.

ORIGINAL PROOF OF KIRCHHOFF'S LAW

Consider two bodies A and B of infinite extent in all directions. Suppose that body A emits some energy and body B absorbs a part of it and reflects the rest. Of the energy reflected by B, some fraction is absorbed by A and the rest is reflected. This process continues till such time that thermal equilibrium is attained and the amount of energy emitted by one is equal to the energy absorbed by it. If emissive power is denoted by e and the absorptive power by , then the series of processes described above can be mathematically modelled as follows:

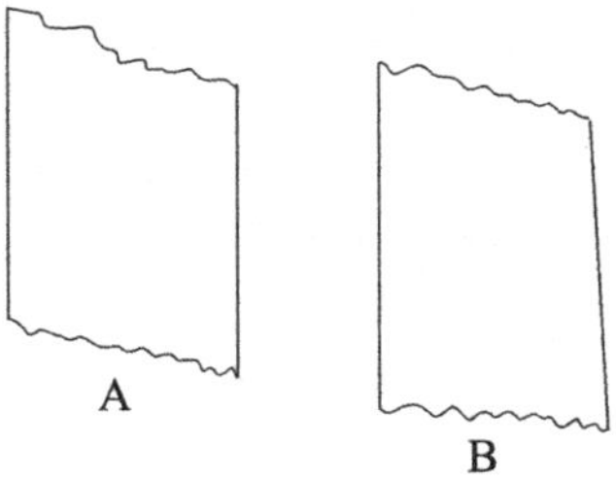

Fig. B.11.1

Amount of energy emitted by A e_1	Amount of energy absorbed by B a_2e_1	Amount of energy reflected by B $(1 - a_2)e_1$	Amount of energy absorbed by A $(1 - a_2)e_1a_1$
Amount of energy reflected by A $)(1 - a_2)e_1$	Amount of energy absorbed by B $(1 - a_1)(1 - a_2)\, a_2e_1$	Amount of energy reflected by B $(1 - a_1)(1 - a_2)^2e_1$	Amount of energy absorbed by A $(1 - a_1)(1 - a_2)^2e_1a_1$

Hence, the amount of energy absorbed by A till thermal equilibrium is attained can be expressed as

$$\begin{aligned} E &= (1-a_2)e_1a_1 + (1-a_1)(1-a_2)^2 e_1a_1 + (1-a_1)^2(1-a_2)^3 e_1a_1 + \ldots \\ &= (1-a_2)e_1a_1\,[1 + x + x^2 + x^3 + \ldots] \text{ where we have put } x = (1-a_1)(1-a_2). \end{aligned}$$

The term within the square brackets is an infinite series with common ratio x (<1)and its sum is $(1 - x)^{-1}$. Hence, the expression for the amount of energy absorbed by A till thermal equilibrium is attained takes a compact form:

$$E_A = (1-a_2)e_1a_1\left(\frac{1}{1-x}\right) = \frac{(1-a_2)e_1a_1}{1-(1-a_1)(1-a_2)} = \frac{(1-a_2)e_1a_1}{a_1 + a_2 - a_1a_2} \quad \text{(i)}$$

Similarly, the amount of energy absorbed by B till thermal equilibrium is attained can be written as

$$E_B = a_2e_1 + (1-a_1)(1-a_2)a_2e_1 + (1-a_1)^2(1-a_2)^2 a_2e_1 + \ldots$$

As before, we can sum up this series to obtain

$$E_B = \frac{e_1a_2}{\;+\;-\;} \quad \text{(ii)}$$

Next, starting from the emission e_2 by B and proceeding as above, you can easily convince yourself that the energy absorbed by A is given by

$$E'_A = \frac{a_1 e_2}{a_1 + a_2 - a_1 a_2} \tag{iii}$$

On combining Eqs. (i) and (ii), we find that the total energy absorbed by A is given by

$$U_A = \frac{(1 - a_2) e_1 a_1}{a_1 + a_2 - a_1 a_2} + \frac{a_1 e_2}{a_1 + a_2 - a_1 a_2} = \frac{e_1 a_1 + a_1 e_2 - e_1 a_1 a_2}{a_1 + a_2 - a_1 a_2}$$

Under equilibrium condition, we can write

$$e_1 = \frac{e_1 a_1 + a_1 e_2 - e_1 a_1 a_2}{a_1 + a_2 - a_1 a_2}$$

$$e_1 a_1 + e_1 a_2 - e_1 a_1 a_2 = e_1 a_1 + e_2 a_1 - e_1 a_1 a_2$$

On simplification, we get the expression for Kirchhoff's law:

$$e_1 a_2 = e_2 a_1$$

$$\frac{e_1}{a_1} = \frac{e_2}{a_2}$$

Applications of Kirchhoff's Law

Kirchhoff's law implies that if a body can emit radiation of a particular wavelength, it will absorb the same radiation when incident on it. Moreover, if e_λ is more, a_λ will also be correspondingly high so that their ratio is constant. It means that a good absorber of radiation is a good emitter also and vice versa. From your school physics, you may be aware of many examples, which illustrate the validity of Kirchhoff's law. For instance, in a dark room, the decorations on a piece of china heated to about 1000°C appear much brighter than the white part. Similarly, a black spot coated on a polished metal ball using platinum black shines more brilliantly than the polished surface, if heated to about 1000°C.

Have you seen greenhouses being used for protecting plants in severe cold conditions/climates? These are based on Kirchhoff's law. We know that glass is a good transmitter and poor absorber of visible light but it is good absorber of longer wavelengths. The greenhouses protect the plants by freely transmitting sunlight which is absorbed by the plants. But a greenhouse acts as a radiation shield to outgoing longer wavelength thermal radiation emitted by the plants.

The line-reversal experiment of Kirchhoff and Bunsen is a beautiful experiment which can also be demonstrated in physics laboratory. (In some universities, students are required to use it to determine the wavelength of sodium doublet.) When a white light source is seen through a sodium flame by means of a spectrometer, two fine dark lines in the yellow region in the positions corresponding to D_1 and D_2 lines of sodium are observed. This is because of selective absorption by the sodium flame from the white light precisely the same wavelengths which it can emit. Oxides of erbium and didymium also exhibit selective absorption. On heating, these oxides emit bright bands along with the continuous spectrum. When this light is made to pass through a solution of these oxides, the very same bands appear in the absorption spectrum.

An excellent illustration of Kirchhoff's law is in the explanation of the origin of Fraunhofer dark lines in solar spectrum. (The light emitted by the glowing mass in the core of the sun is in the form of continuous spectrum. When it passes through the surrounding atmosphere, which is comparatively cooler, the atoms of elements such as sodium, copper, caesium, etc., which exist in the gaseous state, absorb those wavelengths which they can emit. As a result those wavelengths are seen as dark lines. However, in a total solar eclipse, when the sun's core is completely covered, the Fraunhofer lines appear bright.) It is important to mention here that Kirchhoff's law has far reaching importance than providing natural explanation of Fraunhofer lines; its true worth was realised as it gave birth to two new fields of study: Astrophysics and Spectroscopy. We shall however not into further details of these applications here.

Proceeding further, we shall focus on blackbody radiation and explain how far classical theories help us to understand its behaviour.

LACKBODY RADIATION AS A THERMODYNAMIC YSTEM

We now know that the quality of blackbody radiation is determined only by the temperature of the walls of the enclosure. Like a gas, it can be completely specified in terms of (T, p,) and can therefore be treated as a thermodynamic system. It means that all the relations derived for a gas will apply to blackbody radiation. Since energy density is a function of temperature, it follows that pressure of cavity radiation is also a function of temperature (Example 11.1). But as of now, we have not determined the nature of dependence of energy density on temperature explicitly. This information is contained in Stefan's law and we discuss it now.

The Stefan–Boltzmann Law

To know how the energy density of radiation varies with temperature, we perform a thought experiment whereby we trap blackbody radiation in a cylinder (container) with perfectly reflecting walls (to prevent exchange of thermal energy between the walls and the radiation) and compress it. If compression is sufficiently slow, the radiation may be considered as being in equilibrium at temperature T. We consider it as a simple thermodynamic fluid and use the first energy equation:

$$u = \left(\frac{\partial U}{\partial V}\right)_T = T\left(\frac{\partial p}{\partial T}\right)_V - p \tag{11.6}$$

$\frac{u}{3}$, we can write

$$\left(\frac{\partial p}{\partial T}\right)_V = \frac{1}{3}\left(\frac{\partial u}{\partial T}\right)_V$$

Using these results in Eq. (11.6), we get

$$u = \frac{T}{3}\left(\frac{\partial u}{\partial T}\right)_V - \frac{u}{3}$$

Note that we are now left with a relation between u and T only. So we may replace the partial differential with an exact one and write

$$\frac{du}{dT} = 4\frac{u}{T}$$

$$\frac{du}{u} = 4\frac{dT}{T}$$

We can readily integrate it to obtain

$$\ln u = 4\ln T + \ln a$$
$$= \ln(aT^4)$$

Hence, on taking antilog of both sides, we obtain

$$u = aT^4 \tag{11.7}$$

is a constant which does not depend on the properties of the body.

Equation (11.7) shows that energy density is proportional to the fourth power of temperature. It is called *Stefan's law.*

The total rate of emission of radiant energy per unit area is proportional to energy density. That is,

$$E \; \alpha \; u$$

Therefore, we can write the total rate of emission of energy by a body per unit area as proportional to the fourth power of its temperature:

$$E = \sigma T^4 \tag{11.8a}$$

is known as *Stefan's constant.* It is connected to a through the relation $\sigma = ca/4$, is speed of light. The value of σ is 5.672×10^{-8} J m^{-2} K^{-4} s^{-1}. It means that at room temperature (T = 300 K), a blackbody radiates 459.4 Wm^{-2} per unit area per unit time. From Eqs. (11.7) and (11.8), we note that *the energy density as well as total emissive power of a blackbody are proportional to the fourth power of absolute temperature.* This is the statement of the *Stefan–Boltzmann law.*

Note that if a blackbody is placed in an enclosure at a temperature T_0, Eq. (11.8a) modifies as

$$E = \sigma(T^4 - T_0^4) \tag{11.8b}$$

You should now go through the following examples.

Example 11.2 Obtain expressions for pressure, entropy, Helmholtz free energy and Gibbs function of the blackbody radiation.

Since radiation pressure is one-third of the energy density, we can write

$$p = \frac{u}{3} = \frac{1}{3}aT^4 \tag{i}$$

To obtain expression for entropy, we must know C_V, which, in turn, requires knowledge of the total internal energy of the cavity radiation enclosed in volume V.

Using Eq. (11.7), we can write

$$U = uV = aVT^4 \tag{ii}$$

The heat capacity at constant volume

$$C_V = \left(\frac{\partial U}{\partial T}\right)_V = 4aVT^3 \tag{iii}$$

To calculate entropy, we assume that the temperature of the walls of the enclosure is increased in an isochoric process from 0 to T. Then, we can write

$$S=\int_0^T \frac{1}{T} C_V \,\mathrm{d}T=\int_0^T 4\,aVT^2\,\mathrm{d}T=\frac{4}{3}aVT^3 \qquad \text{(iv)}$$

The Helmholtz free energy is

$$F=U-TS=aVT^4-\frac{4}{3}aVT^4=-\frac{1}{3}aVT^4 \qquad \text{(v)}$$

The Gibbs function is

$$G=F+pV=-\frac{1}{3}aVT^4+\frac{1}{3}aVT^4=0 \qquad \text{(vi)}$$

We shall revisit some of these relations in Chapter 15.

Example 11.3 A blackbody is placed in an enclosure whose walls are kept at 300 K. Compare the rate at which the heat is gained or lost by the body when its temperature 500 K and **(b)** 200 K.

(a) In this case, the blackbody will lose energy and the rate at which heat is radiated is given by Eq.(11.8b):

$$E_1=\sigma(T^4-T_0^{\,4})=\sigma(500^4-300^4)\mathrm{K}^4=[\sigma(625-81)\times 10^8]\,\mathrm{K}^4=(544\times 10^8\,\sigma)\,\mathrm{K}^4$$

In the second case, the blackbody will gain energy and the rate at which energy is gained is given by

$$E_2=\sigma(300^4-200^4)\mathrm{K}^4=[\sigma(81-16)\times 10^8]\,\mathrm{K}^4=(65\,\sigma\times 10^8)\,\mathrm{K}^4$$

$$\frac{E_1}{E_2}=\frac{(544\times 10^8\,\sigma)K^4}{(81\times 10^8\,\sigma)K^4}=6.72$$

Example 11.4 A blackened cubical metal container of side 5 cm and negligible heat capacity is filled with water. It is placed in an evacuated enclosure whose walls are kept at 300 K. Calculate the time in which water temperature will drop from 310 K to the temperature of the enclosure. Take $\sigma = 5.67 \times 10^{-8}$ Wm^{-2} K^{-4} and $J = 4.2$J cal^{-1}.

The mass of water in the container, $m = 125$ g since density of water is 1 g cm^{-3}.

Heat lost by water in cooling from 310 K to 300 K = $ms(T_2 - T_1)$ = (125 g) × (10 K) × K^{-1}) = 1250 cal

Suppose that time t second elapses in cooling from 310 K to 300 K. Hence,

$$\text{Rate of cooling, } R=\frac{1250}{t}\text{cal}=\frac{1250\times 4.2}{t}\text{J}$$

The average temperature of water in the container = 305 K

Using Stefan's law we can write

$$R=A\sigma(T^4-T_0^{\,4})$$

On substituting the values of various physical quantities, we get

$$\frac{1250 \quad 4.2}{} \text{J s}^{-1} = 6\times(25\times10^{-4}\,\text{m}^2)\times(5.67\times10^{-8}\ \text{Wm}^{-2}\,\text{K}^{-4})\times\left[(305)^4-(300)^4\right]K^4$$

$$= (1.5 \times 10^{-2}) \times (5.67 \times 5.54)\ \text{s}^{-1} = 0.47\ \text{s}^{-1}$$

$$t = 3 \text{ hr } 6 \text{ min.}$$

The fourth power law refers to the total power emitted per unit area of a blackbody without any reference whatsoever to the surroundings with which it exchanges energy. To account for energy exchange, let us assume that a body *X* at temperature is enclosed in a chamber *Y* at temperature , as shown in Fig. 11.3. If we assume that all surfaces are black, the power emitted by outwards is $A_X \sigma T_1^4$, where A_X is surface . Similarly, the power emitted by inwards is $A_Y \sigma T_1^4$, where A_Y is surface area of *Y*. Note that all the power emitted by *X* strikes the walls of *Y* and is absorbed there; only a small part, $A_X \sigma T_2^4$, of it falls on and is absorbed by *X*.

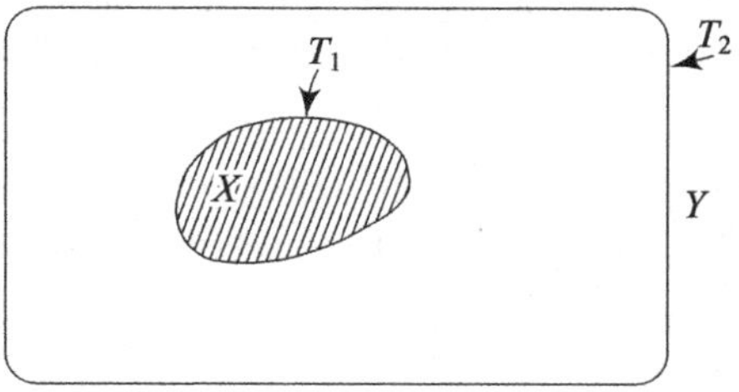

Fig. 11.3 Exchange of radiation between a blackbody and its enclosure.

The net energy exchanged by *X*, according to Stefan-Boltzmann law, is given by

$$E_{ex} = A_X \sigma (T_1^4 - T_2^4) \tag{11.9}$$

body *X* loses power and vice versa.

For small temperature difference, we can rewrite Eq. (11.9) as

$$E_{ex} = A_X \sigma (T_1 - T_2)(T_1 + T_2)(T_1^2 + T_2^2)$$

$$= A_X (T_1 - T_2) 4T^3 \tag{11.10}$$

is mean temperature.

Equation (11.10) shows that the net rate of energy loss is proportional to the temperature difference between the body and its surroundings. This is *Newton's law of cooling*.

You may now like to answer a couple of practise problems.

Problem 11.2 A blackbody weighing 60 g and at an initial temperature of 327°C is allowed to cool inside an evacuated enclosure surrounded by melting ice at the rate of 0.35°C per second. If the specific heat capacity and surface area of the body are $^{-1}$K^{-1} and 12 × 10^{-2}m^2, respectively, calculate Stephan's constant.

5919 × 10–8 Jm^{-2}s^{-1} K^{-4}

Problem 11.3 The power emitted by a filament is given by $P = e\sigma T^4 A$, where is emissivity. The mean temperature of the tungsten filament of a 40 W electric bulb is 2500°C. Its radius is 50 μm and length is 0.1 m. If all heat is lost by radiation, calculate emissivity of the filament. Take σ = 5.672 × 10^{-8} Jm^{-2}K^{-4}s^{-1}. Calculate emissivity of the

Isothermal and Adiabatic Expansion of Blackbody Radiation

As before, we consider an enclosure (a cylinder fitted with a piston) of perfectly reflecting walls and capable of expanding filled with radiation of energy density u at the temperature of the enclosure. Suppose that this radiation is in equilibrium with a tiny black piece of matter of negligible thermal capacity. The equation of state for equilibrium blackbody radiation is given by the relation between pressure and temperature of the system. Recall that radiation pressure is one-third of the energy density, which in turn is proportional to fourth power of absolute temperature. Mathematically, we can write

$$p = \frac{u}{3} = \frac{aT^4}{3} \tag{11.11}$$

To discuss the effects of changes in volume on blackbody radiation, we recall the first equation:

$$T\mathrm{d}S = C_V\,\mathrm{d}T + T\left(\frac{\partial p}{\partial T}\right)_V \mathrm{d}V \tag{11.12}$$

Since internal energy $U = uV = aT^4V$, the heat capacity at constant volume $C\ = \left(\frac{U}{\ }\right)_V = 4aT^3V$. Similarly from Eq. (11.11), we can write $\left(\frac{\partial p}{\partial T}\right)_V = \frac{4}{3}aT^3$. Using these results in Eq. (11.12), we get

$$T\mathrm{d}S = 4aT^3V\mathrm{d}T + \frac{4}{3}aT^4\,\mathrm{d}V \tag{11.13}$$

For an isothermal expansion, the first term on the RHS of Eq. (11.13) vanishes so that the heat exchanged by a blackbody with an external reservoir is given by

$$T\mathrm{d}S = \delta Q = \frac{4}{3}aT^4\,\mathrm{d}V$$

If volume changes from V_i to V_f, the total heat exchange is given by

$$Q = \frac{4}{3}aT^4\,(V_f - V_i) \tag{11.14}$$

Suppose now that the inner walls of the cylinder as well as the face of the piston fitted in it are perfectly reflecting. The radiation is made to undergo *reversible adiabatic expansion.* Since there is no exchange (neither absorption nor emission) of energy between the walls and the radiation, the work on the surroundings is accomplished at the cost of internal energy of the radiation. As a result, the energy density of the radiation will drop. Suppose it decreases from u_1 to u_2, which is characteristic of some lower temperature. Note that the radiation is always in equilibrium with the tiny black speck. So its temperature will also fall. To calculate the final temperature, we put $\mathrm{d}S = 0$ in Eq. (11.13) and obtain

$$4aT^3V\mathrm{d}T + \frac{4}{3}aT^4\,\mathrm{d}V = 0$$

$$4aT^3V\mathrm{d}T = -\frac{4}{3}aT^4\,\mathrm{d}V$$

$$\left(\frac{dV}{V}\right) = -3\left(\frac{dT}{T}\right)$$

This can be readily integrated to obtain

$$\ln V = -3\ln T + \ln K = \ln(T^{-3}) + \ln K$$

is a constant.

Expressing this in numbers, we can write

$$(VT^3)_S = K \tag{11.15}$$

is new temperature of the radiation. This result shows that if the volume of the blackbody radiation is increased eight times, it will be in equilibrium with matter at temperature one-half of the initial temperature. It also suggests that the quality of radiation does not change due to adiabatic expansion (or compression); only temperature changes and the corresponding energy density at the end of the process will be lower (higher): $u_2 = aT^4$.

Besides its importance in Wien's theory of blackbody radiation (Sec. 11.5), Eq. (11.15) finds useful applications in cosmology and radio-astronomy. According to the theory of the universe, it all began with a very dense and hot cloud of matter and radiation, which subsequently expanded and cooled. According to theoretical predictions of Gamow, the correctness of big bang theory demands that the universe must be flooded with blackbody radiation whose present temperature should be about 3 K. Recent measurement on cosmic background radiations have shown that the temperature of the universe is indeed 2.7 K, confirming Gamow's theory. This is of great significance to astrophysicists because the density of radiation at 2.7 K contributes to the energy of the universe.

You should now go through the following example.

Example 11.5 Blackbody radiation in a cavity at 2000 K is subject to isothermal-reversible expansion through 10^3 cm^3. Calculate **(a)** the heat transferred, and **(b)** the work done. If the expansion had been adiabatic, calculate the change in temperature of the radiation. Given that $\sigma = 5.672 \times 10^{-8}$ J m^{-2}K^{-4}s^{-1}.

From Eq. (11.14), we recall that

$$Q = \frac{4}{3}aT^4\,(V_f - V_i) = \frac{16}{3c}\sigma T^4\,(V_f - V_i)$$

σ/c. Hence,

$$= \frac{16}{3\times(3\times10^8\ \text{ms}^{-1})}(5.672\times10^{-8}\ \text{J m}^{-2}\,\text{K}^{-4}\,\text{s}^{-1})\times(2\times10^3\ \text{K})^4\times(10^{-3}\ \text{m}^3)$$

$$= 1.61\times10^{-5}\ \text{J}$$

$$W = \frac{aT^4}{3}(V_f - V_i) = 4.03\times10^{-5}\ \text{J}$$

For reversible adiabatic change, VT^3 = constant. Hence, we can write

$$\frac{T_f}{T_i} = \left(\frac{V_i}{V_f}\right)^{1/3} = \left(\frac{100}{1100}\right)^{1/3} = 0.45$$

$$T_f = (2000\text{ K}) \times 0.45 = 900\text{ K}.$$

We now know that Stefan–Boltzmann law relates the total energy density of blackbody radiation with temperature. However, it does not give any information about the actual distribution of energy in different parts of the spectrum. You will agree that our knowledge about blackbody radiation is incomplete if we do not know the spectral distribution of energy at an arbitrary temperature and how it changes with temperature. We discuss it now.

PECTRAL DISTRIBUTION OF RADIANT ENERGY

Refer to Fig. 11.4, which shows the results of spectroscopic analysis of black body radiation. We note that

For a given wavelength λ, u_λ increases with temperature.
For each temperature, the spectral energy density versus wavelength plot shows a maximum, which shifts to shorter wavelengths as temperature increases.
The energy density goes smoothly to zero as $\lambda \rightarrow 0$ or as $\lambda \rightarrow \infty$.

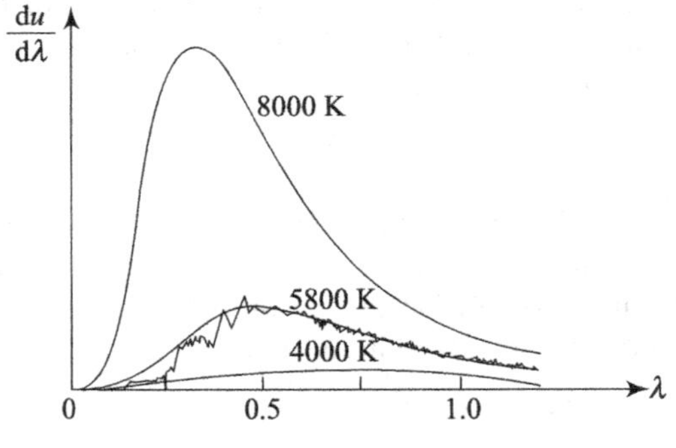

Plot of spectral energy density with wavelength at different temperatures.

Several leading physicists calculated energy distribution over different wavelengths to explain observed results on the basis of thermodynamic reasoning. The first notable success in this direction was achieved by Wien. We will now discuss Wien's law.

Wien's Law

To arrive at Wien's law, we revisit the reversible adiabatic expansion of blackbody radiation contained in a cylinder whose walls as well as the face of the piston are perfectly reflecting and elastic. We assume that as the piston moves outward slowly with a uniform speed v_s(<<
is speed of light), due to Doppler effect, the frequency (wavelength) of each spectral component changes on reflection at the moving surface of the piston (or walls of

the enclosure). Wien regarded radiation filling the enclosure as electromagnetic waves and expected that the wavelength of any particular spectral component may change. Therefore, to establish Wien's law we must work out the connection of the size of the enclosure with (i) the wavelength of electromagnetic waves and (ii) temperature of enclosure.

For simplicity, we consider a spherical enclosure, whose walls can move outward and are perfectly reflecting. Suppose that an electromagnetic. wave of frequency ν is incident normally and suffers multiple reflections from a wall moving slowly outwards. Note that the waves inside the enclosure will be incident on the walls at all angles. If the walls of the enclosure move with uniform speed $v_s (<< c)$, the frequency of the wave to an observer on the walls will appear to be $\nu(1 - v_s/c)$, in accordance with Doppler's effect. This wave will be reflected by the outward moving wall, which then acts as a moving source so that the frequency of the reflected wave will decrease further by a factor of $\left(1+\frac{v_s}{c}\right)^{-1}$. The resultant frequency after one reflection will therefore be given by

$$\nu + \mathrm{d}\,\nu = f\left(1-\frac{v_s}{c}\right)\left(1+\frac{v_s}{c}\right)^{-1} \tag{11.16}$$

$<< c$, this relation can be simplified using binomial expansion as

$$\nu + \mathrm{d}\,\nu \approx \nu\left(1-\frac{v_s}{c}\right)^2 \approx \nu\left(1-\frac{2\,v_s}{c}\right)$$

Hence, on simplification, we can write

$$\frac{\mathrm{d}\nu}{\nu} = -\frac{2\,v_s}{c} \tag{11.17}$$

$/\lambda$ we find that $\mathrm{d}\nu = -\frac{c}{\lambda^2}\mathrm{d}\lambda$ so that $\frac{\mathrm{d}\nu}{\nu} = -\frac{\mathrm{d}\lambda}{\lambda}$. Hence, in terms of wavelength, Eq. (11.17) can be expressed as

$$\frac{\mathrm{d}\lambda}{\lambda} = \frac{2\,v_s}{c} \tag{11.18}$$

For oblique incidence, the effective part of the velocity v_s is the component $v_s \cos\theta$ along the direction of incidence so that the change in wavelength in one reflection is given by

$$\frac{\mathrm{d}\lambda}{\lambda} = \frac{2\,v_s \cos\theta}{c} \tag{11.19}$$

is the angle between the perpendiculars to the wavefront and the surface of the container, as shown in Fig. 11.5.

From the geometry shown in Fig. 11.6 we note that a wave propagating at velocity c may suffer a number of reflections in one second. So the total change in wavelength will be much greater than that given by Eq. (11.19). Since $v_s << c$, we can assume that θ remains the same for all reflections while the enclosure expands by dr.

To calculate the total change in wavelength, we note that between two successive reflections, the wave travels a distance $2r\cos\theta$ and therefore the time between reflections $\frac{\theta}{\quad}$. It means that the number of reflections suffered by a wave in one second is

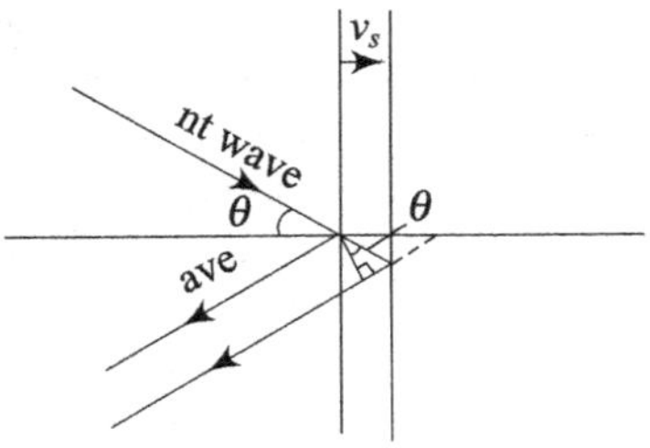

Change in wavelength of an electromagnetic wave reflected by a moving surface.

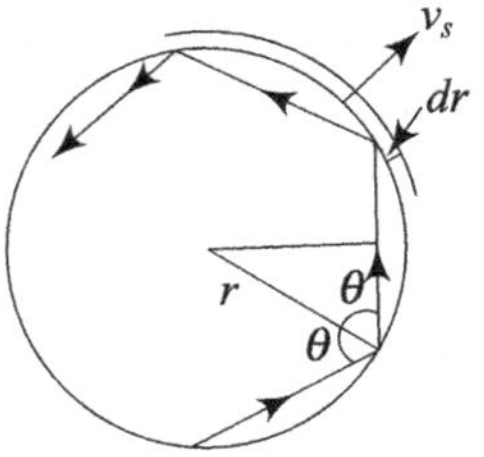

Fig. 11.6 Reflection of an electromagnetic wave inside an expanding spherical enclosure.

. The fractional change in wavelength $\frac{d\lambda}{\lambda}$ per second is obtained by multiplying the right-hand side of Eq. (11.19) with the total number of reflections. This gives

$$\frac{d\lambda}{\lambda} = \left(\frac{2 v_s \cos\theta}{c}\right)\left(\frac{c}{2r\cos\theta}\right) = \frac{v_s}{r}$$

$$\frac{d\lambda}{\lambda} = \frac{v_s}{r} = \frac{dr}{r} \tag{11.20}$$

= dr signifies the increase in radius of the enclosure in one second.
On integrating Eq. (11.20), we get

$$\frac{\lambda}{r} = \text{constant}$$

$$\lambda \propto r \tag{11.21}$$

That is, the wavelength of a wave is linearly proportional to the (size) radius of the spherical

We will now derive the relation connecting the temperature of enclosure to its radius. This follows directly from the first law of thermodynamics:

$$\delta Q = dU + p\,dV$$

In an adiabatic process $\delta Q = 0$. Also we know that for cavity radiation $U = uV$ and /3, where u is the energy density. Hence,

$$d(uV) + \frac{u}{3} dV = 0$$

We rewrite it as

$$V\,du + u\,dV + \frac{u}{3} dV = 0$$

$$\frac{du}{u} = -\frac{4}{3}\frac{dV}{V}$$

Integrating, we get

$$uV^{4/3} = K, \text{ a constant} \tag{11.22}$$

From Stefan's law, we know that $u = aT^4$. Hence, we can rewrite Eq. (11.22) as

$$TV^{1/3} = K_1 \tag{11.23}$$

is some other constant.

For a spherical enclosure, $V = \frac{4\pi}{3}r^3$. Using this expression in Eq. (11.23), we get

$$rT = \text{constant} \tag{11.24}$$

That is, temperature is inversely proportional to the radius of the spherical container. The infinitesimal changes in r and T are connected through the relation

$$rdT + Tdr = 0$$

$$\frac{dT}{T} = -\frac{dr}{r} \tag{11.25}$$

On combining Eqs. (11.20) and (11.25), we get

$$\frac{d\lambda}{\lambda} = -\frac{dT}{T}$$

On integrating both sides, we obtain

$$\lambda T = \text{constant} \tag{11.26}$$

This means that if blackbody radiation at a certain temperature is adiabatically expanded, wavelength of each spectral component changes such that it is inversely proportional to temperature. That is, *the wavelength of radiation is displaced as temperature changes so that the product λT remains constant.* In particular, if we follow the maximum of the radiation spectrum, it follows T-inverse law. This is *Wien's displacement law.*

We now consider the effect of adiabatic expansion on the spectral components between + $d\lambda$. Let the energy density of this band be $u_\lambda\, d\lambda$. We know that in adiabatic expansion, the spectral density, wavelength and the width of this band will change. Applying the first law of thermodynamics to this band, we can write

$$d(U_\lambda d\lambda) + pdV = 0 \tag{11.27}$$

$_\lambda = u_\lambda V$ and $p = u_\lambda d\lambda/3$, we obtain

$$d(u_\lambda V d\lambda) = -\frac{u_\lambda d\lambda}{3} dV$$

$$d(u_\lambda d\lambda)V + u_\lambda d\lambda\, dV = -\frac{u_\lambda d\lambda}{3} dV$$

On rearranging terms and dividing throughout by $Vu_\lambda d\lambda$, we obtain

$$\frac{d(u_\lambda d\lambda)}{u_\lambda d\lambda} = -\frac{4}{3}\frac{dV}{V} \tag{11.28}$$

For a spherical enclosure, $V = (4\pi/3)r^3$ so that $dV = 4\pi r^2 dr$ and $\frac{dV}{V} = \frac{3dr}{r}$. Using this result in Eq. (11.28), we get

$$\frac{d(u_\lambda d\lambda)}{u_\lambda d\lambda} = -4\frac{dr}{r}$$

On integration, we obtain

$$\ln(u_\lambda \, d\lambda) = -4 \ln r + \ln A$$

$$\ln(u_\lambda \, d\lambda r^4) = \ln A$$

is a constant.

On taking antilog of both sides, we get

$$r^4 u_\lambda \, d\lambda = \text{constant}$$

From Eq. (11.21) we recall that $\lambda \propto r$. So $d\lambda$ will also be proportional to r. This implies that

$$u_\lambda \lambda^5 = \text{constant} \tag{11.29}$$

Using Eq. (11.26), we can correlate u_λ with temperature:

$$u_\lambda T^{-5} = \text{constant} \tag{11.30}$$

From Eq. (11.30), it is clear that u_λ is a function of temperature T. Hence, we expect that the constant in these equations will involve T. But the constant is such that it remains the same throughout the adiabatic expansion. In view of Eq. (11.26) we expect this constant to be a function of the product λT. It means that the radiation law must have the form

$$u_\lambda \lambda^5 = A' F(\lambda T)$$

is any constant and $F(\lambda T)$ is a differentiable function which depends only on the T rather than on individual values of these variables. Thus we can write

$$u_\lambda \, d\lambda = A' \lambda^{-5} F(\lambda T) \, d\lambda \tag{11.31a}$$

Since emissive power E_λ is proportional to the energy density, we can write

$$E_\lambda \, d\lambda \propto u_\lambda \, d\lambda$$

$$= \frac{A}{\lambda^5} F(\lambda T) \, d\lambda \tag{11.31b}$$

$E_\lambda d\lambda$ denotes the amount of energy contained in the spectral range from λ to emitted by a black body at temperature T.

Equation (11.31a) is *Wien's distribution law* in terms of wavelength. Note that the form of the function $F(\lambda T)$ is essentially undefined so far and we can go no further based on thermodynamic reasoning. In fact, determination of this functional form posed a very serious challenge to physicists for a long time. That it is exponential became clear when Planck gave a completely satisfactory theory of blackbody radiation.

We can express Wien's law in different forms. To this end, we write energy density in the spectral range from λ to $\lambda + d\lambda$ emitted by a blackbody at temperature T as

$$u_\lambda \, d\lambda = A' \lambda^{-5} \frac{T^5}{T^5} F(\lambda T) \, d\lambda = A' T^5 G(\lambda T) \, d\lambda \tag{11.31c}$$

$T) = (\lambda T)^{-5} F(\lambda T)$

As before, we can express the amount of energy contained in the spectral range from λ emitted by a blackbody at temperature T as

$$E_\lambda \, d\lambda \propto u_\lambda \, d\lambda = A \lambda^{-5} F(\lambda T) \, d\lambda \tag{11.31d}$$

Sometimes, it is more convenient to write it in terms of frequencies rather than wavelength. If frequencies in the range ν and $\nu + d\nu$ correspond to wavelengths in the range λ and

$$|u_\lambda \, d\lambda| = |u_\nu \, d\nu| \qquad (11.32)$$

$\lambda = \frac{c}{\nu}$, $d\lambda = -\frac{c}{\nu^2} d\nu$. On using this relation in Eq. (11.31d), we get

$$u_\nu \, d\nu = A' \left(\frac{c}{\nu}\right)^{-5} F(\nu/T) \left(-\frac{c}{\nu^2} d\nu\right)$$

$$u_\nu \, d\nu = B\nu^3 F(\nu/T) d\nu \qquad (11.33a)$$

$A'c^{-4}$ and $F(\nu/T)$ is any differentiable function which depends only on the ratio rather than on the individual values of these variables.

The energy contained in the spectral region from frequency ν to $\nu + d\nu$ emitted by a blackbody at temperature T is given by

$$E \, d\nu = A'T^5 G(\lambda T)(-\frac{c}{\nu^2} d\nu) = CT^3 \left(\frac{\nu}{T}\right)^{-2} G\left(\frac{\nu}{T}\right) d\nu = CT^3 \phi\left(\frac{\nu}{T}\right) d\nu \qquad (11.33b)$$

It is true that thermodynamic reasoning can not predict the form of the function F/F', but, if we plot $u_\lambda T^{-5}$ versus λT, the curve so obtained resembles the blackbody radiation curves shown in Fig. 11.4. From Eq. (11.31b), we can show that for a given temperature, u_λ will be maximum when the function $F(\lambda T)$ becomes maximum, corresponding to a constant value of the product λT. To show this, we differentiate u_λ with respect to λ and obtain

$$\frac{du_\lambda}{d\lambda} = A'[\lambda^{-5} TF'(\lambda T) - 5\lambda^{-6} F(\lambda T)]$$

$$= A'\lambda^{-6} [(\lambda T) F'(\lambda T) - 5 F(\lambda T)]$$

is derivative of F with respect to the argument λT. The quantity in brackets can now be seen to be simply another function of λT. Let us denote it by $Q(\lambda T)$ so that the above expression can be written as

$$\frac{du_\lambda}{d\lambda} = DQ(\lambda T)$$

$A'\lambda^{-6}$. The maximum value of u_λ occurs at wavelength λ_{max} at temperature such that $Q(\lambda T) = 0$. That is, whatever its form, the function Q must be zero when its argument equals $\lambda_{max} T$. Hence, we can write

$$\lambda_{max} T = \lambda'_{max} T' = \lambda''_{max} T'' = b \qquad (11.34)$$

is a constant and has value 2.892×10^{-3} m-K. This result is known as *Wien's displacement law*. It predicts that as temperature is raised, the maximum of spectral energy density will shift towards lower wavelengths. It has been used to correctly predict the temperature of celestial objects including solar surface. (We have to insert the value of wavelength corresponding to the maximum of the spectral energy density curve obtained for the Sun.) It is for the same reason that the colour of iron piece put in blacksmith furnace changes continuously from red dull to cherry red to orange to yellow to white.

In an effort to reproduce experimental results, Wien made some assumptions about the mechanism of emission and absorption of radiation and empirically concluded that the function $F(\lambda T)$ should be exponential: $a_1 \exp(-b/\lambda T)$, where a_1 and b are adjustable

Then we can write Wien's law as

$$u_\lambda \, d\lambda = a\lambda^{-5} \exp(-b/\lambda T) d\lambda \tag{11.35}$$

The Wien's law was put to a series of experimental tests by a number of researchers, including Rubens and Kurlbaum. It explained the qualitative features of observed spectral distribution of blackbody radiation available then at high frequencies (low wavelengths) and seemed acceptable. But when more accurate experimental results were reported in the low frequency (high wavelength) region, there was disagreement in details. In fact, Wien's law was completely inadequate at low frequencies and high temperatures (Fig. 11.7). In particular, Rubens and Kurlbaum predicted that high temperature spectral distribution should be proportional to temperature but Wien's law failed to explain it. Moreover, the value of b had to be adjusted for different wavelengths and the variation was quite

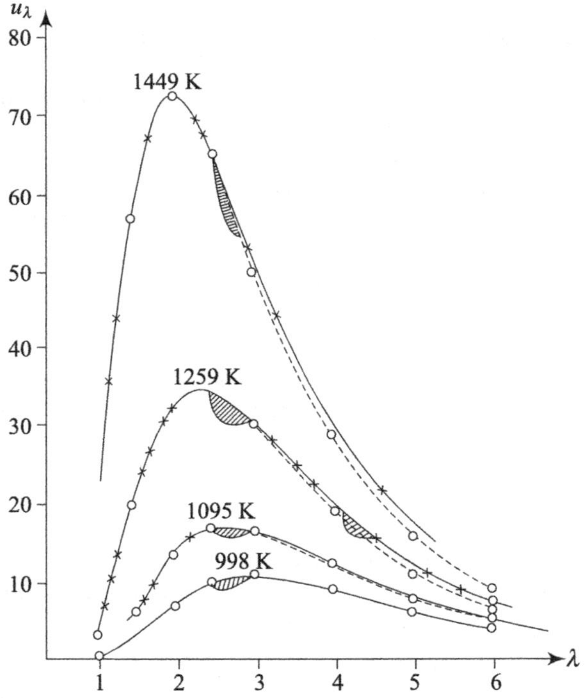

Comparison of observed results with Wien's law.

To account for the inconsistencies in Wien's law, some other alternative was needed. But before we discuss other alternatives, you may like to go through the following examples.

Example 11.6 The spectral energy curve of the moon shows maxima at 470 nm m. What conclusions can you draw from this? Calculate the energy density and radiation pressure.

Using Eq. (11.34), we can calculate the temperature corresponding to the maximum in the curve.

For the first maxima, the temperature is given by

$$T_1 = \frac{2.892\times10^{-3}\ \text{mK}}{470\times10^{-9}\ \text{m}} = 6.153\times10^3\ \text{K}$$

T_1 corresponds to the temperature of the surface of the Sun. It means that the first maxima coincides with the maximum in solar spectrum and suggests that it arises due to reflection of solar light by the lunar surface.

The energy density of solar radiation is

$$u = aT_1^4 = \frac{4\sigma}{c}T_1^4 = \frac{4\times(0.567\times10^{-7}\ \text{Jm}^{-2}\text{s}^{-1}\text{K}^{-4})}{(3\times10^8\ \text{ms}^{-1})}\times(6153\ \text{K})^4 = 1.084\ \text{J m}^{-3}$$

The pressure exerted by solar radiation is given by

$$p = \frac{u}{3} = 0.361\ \text{J m}^{-3} = 0.361\ \text{N m}^{-2}$$

The temperature corresponding to the second maximum is given by

$$T_2 = \frac{2.892\times10^{-3}\ \text{mK}}{14\times10^{-6}\ \text{m}} = 2.066\times10^2\ \text{K}$$

T_2 corresponds to the temperature of the lunar surface temperature.

The energy density of lunar radiation is

$$u = aT_2^4 = \frac{4\sigma}{c}T_2^4 = \frac{4\times(0.567\times10^{-7}\ \text{Jm}^{-2}\text{s}^{-1}\text{K}^{-4})}{3\times10^8\ \text{ms}^{-1}}\times(206.6\ \text{K})^4 = 1.377\times10^{-6}\ \text{J m}^{-3}$$

The pressure exerted by lunar radiation is given by

$$p = \frac{u}{3} = 0.459\times10^{-6}\ \text{N m}^{-2}.$$

Example 11.7 A spherical blackbody of radius 4 cm is kept at a temperature of 227 K. Calculate **(a)** the power radiated and **(b)** the wavelength at which maximum energy is radiated. Take $\sigma = 5.67\times10^{-8}\text{W m}^{-2}\ \text{K}^{-4}$.

(a) The rate of emission of energy is obtained using Stefan-Boltzmann law:

$$R = \sigma T^4 A = (5.67\times10^8\ \text{W m}^{-2}\ \text{K}^{-4})\times(500\ \text{K})^4\times4\pi(16\times10^{-4}\ \text{m}^2)$$

$$= 71.2\ \text{W}$$

To calculate the wavelength at which maximum energy is radiated, we use Wien's displacement law:

$$\lambda_{\text{max}} = \frac{2898\times10^{-6}\ \text{mK}}{500\ \text{K}} = 5.79\times10^{-6}\ \text{m}.$$

As mentioned earlier, Wien's law does not conform to observed results at longer wavelengths and high temperatures even with exponential form of the function $F(\lambda T)$ conjectured by Wien. This inadequacy of thermodynamics stimulated Lord Rayleigh to work on the theory of blackbody radiation. Rayleigh analysed the problem by applying the principle of equipartition of energy to electromagnetic field in terms of standing waves.

However, he made a numerical error, which was corrected by Jeans in a paper to *Nature* in 1906 and that is why his name was included. We now discuss formulation of Rayleigh–

Rayleigh–Jeans Law

Rayleigh–Jeans formula for the distribution of energy in the blackbody radiation is based on the principle of equipartition of energy for all possible modes of standing electromagnetic waves. Each mode of vibration, on the basis of equipartition of energy, is assigned energy (which is equally shared by kinetic and potential energy components). To determine the number of modes, we consider the radiation to be enclosed in a hollow cubic enclosure of and having perfectly reflecting walls. Since blackbody radiation is electromagnetic in character, Rayleigh considered that blackbody radiation in the enclosure consists of a number of electromagnetic waves, which travel in all directions, 'bang' against the walls, undergo multiple reflections and superpose. This results in the formation of standing waves with nodes at the walls. The wave equation for such a system is

$$\nabla^2 \psi = \frac{\partial^2 \psi}{\partial x^2} + \frac{\partial^2 \psi}{\partial y^2} + \frac{\partial^2 \psi}{\partial z^2} = \frac{1}{c_s} \frac{\partial^2 \psi}{\partial t^2} \tag{11.36}$$

is velocity of the waves.

Since walls of the cubical enclosure are fixed, these will act as nodal points, i. e., the amplitude of the waves ψ will be zero at $x, y, z = 0$ and $x, y, z = L$. We now assume a solution of Eq. (11.36) in the form

$$\psi(x, y, z, t) = C \exp(-i\omega t) \sin\left(\frac{n_x \pi x}{L}\right) \sin\left(\frac{n_y \pi y}{L}\right) \sin\left(\frac{n_z \pi z}{L}\right) \tag{11.37}$$

n_y, n_z) are integers and ω is angular frequency of the wave. Note that each combination of (n_x, n_y, n_z) constitutes a mode of oscillation of the waves in the box. In the absence of damping or coupling between the modes, the waves will oscillate independently and indefinitely with constant amplitude.

We now substitute the expression for $\psi(x, y, z, t)$ in Eq. (11.36). On simplification, we obtain the relation between the values of (n_x, n_y, n_z) and the angular frequency of the wave:

$$\frac{\pi^2}{L^2}(n_x^2 + n_y^2 + n_z^2) = \frac{\omega^2}{c_s^2}.$$

$$(n_x^2 + n_y^2 + n_z^2) = \frac{\omega^2 L^2}{\pi^2 c_s^2} = \left(\frac{2 \nu L}{c_s}\right)^2 = \left(\frac{2L}{\lambda}\right)^2 \tag{11.38}$$

c_s/ν defines the wavelength of the standing wave.

Note that Eq. (11.38) gives the allowed frequencies or modes of vibration inside the cube for different positive integral values of (n_x, n_y, n_z). The total number of modes of vibration are specified by the total number of possible sets (n_x, n_y, n_z).

If we now put $\frac{2L}{\lambda} = n$, Eq. (11.38) simplifies to

$$(n_x^2 + n_y^2 + n_z^2) = n^2 \tag{11.39}$$

This result suggests that n is the radius of a sphere in (n_x, n_y, n_z) space and the number of allowed modes can be obtained by plotting n_x, n_y, n_z and counting the number of points corresponding to positive integral values. These will obviously lie in the positive octant of a sphere of radius n. (In other octants, at least one value either of n_x, n_y or n_z will be negative.)

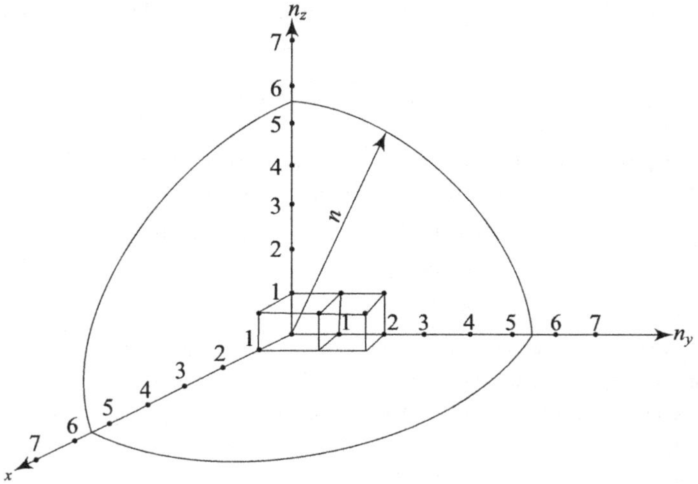

Calculating the number of allowed modes of standing waves in a box.

Refer to Fig. 11.8. It shows the values corresponding to all positive integral values of n_x, . An allowed set of values of n_x, n_y, and n_z is specified by their intersection and gives a mesh of small cubes. For sufficiently large values of n, each point will correspond to one unit cube in this octant. Therefore, the total number of allowed modes of vibration will be equal to the volume of the octant and we can write

$$N = \frac{1}{8}\left(\frac{4\pi}{3}n^3\right) = \frac{1\times 4}{8\times 3}\pi\left(\frac{2L}{\lambda}\right)^3 = \frac{4\pi L^3}{3\lambda^3}$$

Hence, the number of modes between λ and λ + dλ is obtained by differentiating this expression for total number of modes. Thus,

$$|N_\lambda\, d\lambda| = \left|\frac{4\pi V}{3}(-3\lambda^{-4})\, d\lambda\right| = \frac{4\pi V}{\lambda^4} d\lambda \tag{11.40}$$

L^3 is volume of the enclosure.

You may recall that we are dealing with e. m. waves, which are transverse in nature and there are two independent polarisation states for a given value of wave vector. We therefore have to multiply Eq. (11.40) by two so that the correct number of allowed modes will be twice as many. Hence

$$N_\lambda\, d\,\lambda = \frac{8\pi V}{\lambda^4} d\lambda \tag{11.41a}$$

Using the relation $\lambda = c/\nu$, you can easily convince yourself that for blackbody radiation, the number of modes in the frequency range ν to $\nu + d\nu$ is given by

$$N_\nu\, d\,\nu = \frac{8\pi V}{c^3}\nu^2\, d\,\nu \tag{11.41b}$$

To give you an idea about the number of modes, go through the following example.

Example 11.8 Calculate the number of modes in a chamber of volume 100 cm^3 in the frequency range 4×10^{14} Hz to 4.01×10^{14} Hz.

From Eq. (11.41b), we recall that the number of modes in the frequency range ν is given by

$$N_\nu \mathrm{d}\,\nu = \frac{8\pi V}{c^3}\nu^2 \mathrm{d}\,\nu$$

= 100 cm^3 = 10^{-4}m^3, $\nu = 4 \times 10^{14}$ Hz, dν= 0.01×10^{14} Hz, $c = 3 \times 10^8$ ms^{-1}. On substituting these values in the above expression, we get

$$\nu = \frac{8\times3.14\times(4\times10^{14}\ \text{Hz})^2\times(10^{-4}\ \text{m}^3)\times(0.01\times10^{14}\ \text{Hz})}{(3\times10^8\ \text{ms}^{-1})^3} = 1.49\times10^{13}.$$

Now we have all the wherewithal that we need to arrive at the Rayleigh–Jeans formula for blackbody radiation based on the principle of equipartition of energy, which assigns an amount of energy k_BT to each mode of vibration. Therefore, on multiplying Eq. (11.41b) , we obtain total energy of blackbody radiation in the enclosure between frequencies + dν:

$$E_\nu \mathrm{d}\,\nu = \frac{8\pi V k_B T}{c^3}\nu^2 \mathrm{d}\,\nu$$

Hence, the energy density within frequency range ν to $\nu + \mathrm{d}\nu$ is given by

$$u_\nu \mathrm{d}\,\nu = \frac{8\pi k_B T}{c^3}\nu^2 \mathrm{d}\,\nu \tag{11.42}$$

This is Rayleigh–Jeans law.

To write Rayleigh–Jeans law in terms of wavelength, we use the relation $\lambda = c/\nu$ in the equality

$$|u_\lambda \mathrm{d}\lambda| = |u_\nu \mathrm{d}\,\nu|$$

$$= \frac{8\pi k_B T}{\lambda^4}\mathrm{d}\lambda \tag{11.43}$$

You will agree that Rayleigh's analysis is brilliant. It reproduced experimental curve at long wavelengths but turned out to be a poor representation of experimental data, particularly at shorter wavelengths (Fig. 11.9). Moreover, the total energy density $u = \int_0^\infty u_\lambda \mathrm{d}\lambda \to \infty$ 0 or $\nu \to \infty$, since there is no maximum. An infinite energy density in a field is clearly impossible and physically unacceptable. The implications of this failure were so serious that the behaviour of the theory at low wavelengths (high frequencies) was termed *ultraviolet catastrophe*. This behaviour was obviously undesirable; the disaster really upset Rayleigh and Jeans. To resolve the syndrome, they added a decaying exponential term: T) where c_2 is a constant. We may therefore conclude that blackbody radiation is Rayleigh-like in the low frequency region and Wien-like in the high frequency region. However, a general law for the complete energy range was yet to come.

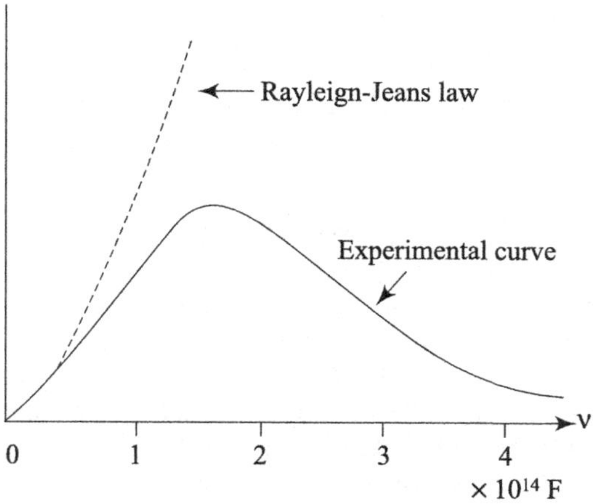

Comparison of observed results with Rayleigh–Jeans law.

As such, the failure of Rayleigh–Jeans formula was a serious chink in the applicability of classical physics to blackbody radiation because, as such, the derivation was flawless and it did not involve any undetermined constant. Therefore, doubts were expressed about the applicability of the principle of equipartition of energy, which favours continuous energy exchange, to radiation. Planck was aware of this work through Rubens. It is therefore quite possible that he conjectured, albeit correctly, that the emission and absorption of radiation is a discontinuous process. We now discuss Planck's law for blackbody radiation.

As we now know, Planck's law is the most general law of blackbody radiation and explains all experimental results in the complete energy range. In fact, all other laws of blackbody radiation follow from Planck's law. We can go about discussion of Planck's law in many different ways: Discuss the developments in chronological order or start by taking the theory as we now know and present the subject accordingly. While most authors choose to follow the latter approach as it is more convenient, we will follow the former as it is more informative and learning-friendly; it will help you develop better understanding of the physics involved in blackbody radiation. It will also give you a feel of how scientists handle difficult unknown situations, particularly when their results do not conform to experimental results, and modify their ideas/theories. This law can also be obtained by treating radiation as an assembly of photons, which obey Bose–Einstein statistics. We have discussed it in Chapter 15.

Planck's Law

Planck presented the following formula for energy density empirically to fit the experimentally observed blackbody spectrum:

$$u_\nu \, d\nu = \frac{8\pi\nu^2}{c^3}\left(\frac{h\nu}{\exp(h\nu/k_B T)-1}\right) d\nu \tag{11.44}$$

We can rewrite it as

$$u_\nu \, d\nu = n_\nu \, \varepsilon_\nu \, d\nu = \frac{8\pi\nu^2 \varepsilon_\nu}{c^3} d\nu \tag{11.45a}$$

$$\varepsilon_\nu = \left(\frac{h\nu}{\exp(h\nu / k_B T) - 1} \right) \tag{11.45b}$$

defines the energy of a Planck resonator and

$$n_\nu \, d\nu = \frac{8\pi\nu^2}{c^3} d\nu \tag{11.45c}$$

defines the number of modes per unit volume in the frequency range ν to $\nu + d\nu$. (This result was rederived by Indian physicist S N Bose based on statistical considerations and communicated to Einstein on June 4, 1924 with a request to translate it in German and get it published in *Zeitschrift fur Physik*. Einstein read his paper carefully, recognised its importance and got it published. Later he generalised it for material particles leading to formulation of Bose-Einstein statistics.)

If we compare Eqs. (11.41b) and (11.45c), we note that these are identical, though these have been arrived at in two different ways. This coincidence is remarkable. Further, $_\nu = k_B T$ in Eq. (11.45a), we obtain Rayleigh–Jeans law, which is based on the law of equipartition where energy is considered an infinitely divisible continuous variable. (It suggests that ε_ν is the average energy of a mode of oscillation in Planck's theory. Also, it is truly amazing that two different approaches led to the same number of modes.) However, Planck was convinced about the inappropriateness of the classical theories and to his pleasant surprise; he was informed that his formula was in excellent agreement with observed results. This convinced Planck that there must be some fundamental significance to his formula and he set out to discover what it was. Unfortunately, he found it extremely difficult and frustrating to give a logical justification; it is believed that out of desperation he concluded—*energy is not a continuous variable.* In fact, on December 18, 1900, he declared in a paper presented to the German Physical Society that the only way to derive the correct blackbody radiation formula was to postulate as follows:

Exchange of energy between matter (walls) and radiation (cavity) could take place only in bundles of a certain size.

The quantum of exchange is directly proportional to its frequency. That is, the energy of an oscillator having frequency ν could only be an integral multiple of $h\nu$, where is a small constant whose value may be obtained by fitting the observed results.

These postulates marked a fundamental departure from the then existing ideas. The is now known as *Planck's constant* in his honour. Its value is 6.62618×10^{-34} Js. Planck was awarded Nobel Prize in physics in 1918 for his work on blackbody radiation.

To realise the significance of Planck's postulates, let us consider the following: Suppose two litre of milk is to be distributed between two persons. Since milk is an infinitely indivisible quantity, you can divide it between two persons in an infinite number of ways. Next you are asked to distribute milk in units of a litre. Now both the persons can receive 0,1 and 2 litre meaning thereby that the number of ways reduces to three. The number of ways will be five if the unit (quantum) of distribution is half-a-litre. This example shows how discretisation introduces a drastic change. You may expect the same in regard to blackbody radiation due to the concept of discrete energy quanta for energy exchange.

In order to unfold the significance of his hypothesis, Planck imagined that blackbody radiation chamber was filled up not only with radiation but also with the molecules of a perfect gas, which exchanged energy via resonators of molecular dimensions.

(Matter-radiation interaction was necessary to introduce the notion of temperature.) The resonators were assumed to absorb energy from the radiation and transfer the same wholly or partially to gas molecules when they collide with them. This helped to establish thermodynamic equilibrium. (You may think that the process is somewhat roundabout but this was the only one possible consistent with accepted ideas at that time.)

Let us now suppose that the total number of Planck resonators is N and their total E. We now calculate the average energy of a Planck resonator. This can be done by classical methods, using Maxwell's formula, according to which the probability that a resonator will possess energy $h\nu$ is $\exp(-h\nu/k_BT)$, where k_B is Boltzmann constant and is absolute temperature. Let $N_0, N_1, N_2, \ldots N_p, \ldots$ be the number of oscillators having $h\nu, 2h\nu, \ldots, ph\nu, \ldots$ respectively. Then we can write

$$N_1 = N_0 \exp(-h\nu/k_BT)$$

$$N_2 = N_0 \exp(-2h\nu/k_BT)$$

$$N_3 = N_0 \exp(-3h\nu/k_BT)$$

$$\ldots\ldots$$

$$N_p = N_0 \exp(-ph\nu/k_BT)$$

Hence, the total number of resonators

$$\begin{aligned} N &= N_0 + N_1 + N_2 + \ldots. \\ &= N_0 + N_0 \exp(-h\nu/k_BT) + N_0 \exp(-2h\nu/k_BT) + \ldots \\ &= N_0 \, [1 + \exp(-h\nu/k_BT) + \exp(-2h\nu/k_BT) + \ldots] \end{aligned}$$

Note that the terms within the square brackets form an infinite geometric series with common ratio $r = \exp(-h\nu/k_BT)$. Since the sum of an infinite series is $(1 - r)^{-1}$, the expression for the total number of resonators takes the form

$$N = \frac{N_0}{1 - \exp(-h\nu/k_BT)} \tag{11.46}$$

Similarly, the total energy ε of resonators is

$$\varepsilon = (N_0 \times 0) + (N_1 \times h\nu) + (N_2 \times 2h\nu) + (N_3 \times 3h\nu) + \ldots$$

On substituting the values of $N_1, N_2, N_3, \ldots$, we can rewrite the expression for total energy

$$\varepsilon = (N_0 \times 0) + [N_0 h\nu \exp(-h\nu/k_BT)] + [2N_0 h\nu \exp(-2h\nu/k_BT)] + [3N_0 h\nu \exp(-3h\nu/k_BT)] + \ldots$$

To simplify this, we rewrite it as

$$\begin{aligned} \varepsilon &= [N_0 h\nu \exp(-h\nu/k_BT)]\left[1 + 2\exp(-h\nu/k_BT) + 3\exp(-2h\nu/k_BT) + \ldots\right] \\ &= [N_0 h\nu \exp(-h\nu/k_BT)] \times S \end{aligned} \tag{11.47}$$

$$S = 1 + 2\exp(-h\nu/k_BT) + 3\exp(-2h\nu/k_BT) + \ldots \tag{11.48}$$

To simplify notation, we put $x = \exp(-h\nu/k_BT)$ Then Eq. (11.48) takes a very elegant form:

$$S = 1 + 2x + 3x^2 + 4x^3 + \ldots \tag{11.49}$$

To evaluate the sum, we multiply both sides of Eq. (11.49) by x. This gives

$$xS = x + 2x^2 + 3x^3 + 4x^4 + ... \quad (11.50)$$

On subtracting Eq. (11.50) from Eq. (11.49), we get

$$S(1-x) = 1 + x + x^2 + x^3 + ... \quad (11.51)$$

Note that the right-hand side of Eq. (11.51) is an infinite geometric series with common and has the value $\frac{1}{1-x}$. Hence, Eq. (11.51) takes the form

$$S(1-x) = \frac{1}{1-x}$$

$$S = \frac{1}{(1-x)^2} = \frac{1}{[1-\exp(-h\nu/k_B T)]^2} \quad (11.52)$$

Using this result in Eq. (11.47), we get the desired expression for the total energy of Planck resonators:

$$\varepsilon = \frac{N_0 h\nu \exp(-h\nu/k_B T)}{[1-\exp(-h\nu/k_B T)]^2} \quad (11.53)$$

Hence, the mean energy of a resonator is obtained by dividing the expression for total energy of Planck resonators [Eq. (11.53)] by the total number of resonators [Eq. (11.46)]:

$$\bar{\varepsilon} = \frac{\varepsilon}{N} = \frac{N_0 h\nu \exp(-h\nu/k_B T)}{[1-\exp(-h\nu/k_B T)]^2} \times \frac{[1-\exp(-h\nu/k_B T)]}{N_0}$$

On simplification, we can write

$$\bar{\varepsilon} = \frac{\varepsilon}{N} = \frac{h\nu \exp(-h\nu/k_B T)}{[1-\exp(-h\nu/k_B T)]} = \frac{h\nu}{[\exp(h\nu/k_B T)-1]} \quad (11.54)$$

Note that the average energy of a Planck resonator is $\frac{h\nu}{[\exp(h\nu/k_B T)-1]}$ rather than $k_B T$. However, in the limit $\nu \to 0$, the Planck factor reduces to $k_B T$:

$$\frac{h\nu}{[\exp(h\nu/k_B T)-1]} \to \frac{h\nu}{\left(1 + \frac{h\nu}{k_B T} + ... - 1\right)} = k_B T$$

This shows that when $h\nu$ is small compared to $k_B T$, the discreet nature of energy in not important. However, as frequency increases, the average energy of an oscillator decreases.

In terms of wavelength, we can express Planck's formula using the relation $\nu = c/\lambda$ in Eq. (11.44). Note that $|d\nu| = \left|-\frac{c}{\lambda^2} d\lambda\right|$ and use the fact that $u_\lambda d\lambda$ corresponds to $u_\nu d\nu$. Hence Planck's law in terms of wavelength can be expressed as

$$u_\lambda \, d\lambda = u_\nu \, d\nu$$

$$= \frac{8\pi h}{c^3}\left(\frac{c}{\lambda}\right)^3 \left(\frac{1}{\exp(hc/\lambda k_B T)-1}\right)\left|-\frac{c}{\lambda^2} d\lambda\right|$$

$$= \frac{8\pi hc}{\lambda^5}\left(\frac{1}{\exp(hc/\lambda k_B T)-1}\right)\mathrm{d}\lambda \tag{11.55}$$

Now refer to Fig. 11. 10, which shows a plot of the Planck's law based on Eq. (11.44). We have also depicted the Rayleigh–Jeans law and Wien's law to facilitate comparison. As mentioned earlier, Planck's law explained all the observed results in the entire spectral range for blackbody radiation, notwithstanding several inner contradictions in the Planck's method. To that extent, it is the complete law and thereby the idea of discreteness of energy, which was truly revolutionary, came to stay. In fact, it led to most interesting developments in photoelectricity, spectral emission and ultimately gave birth to quantum mechanics.

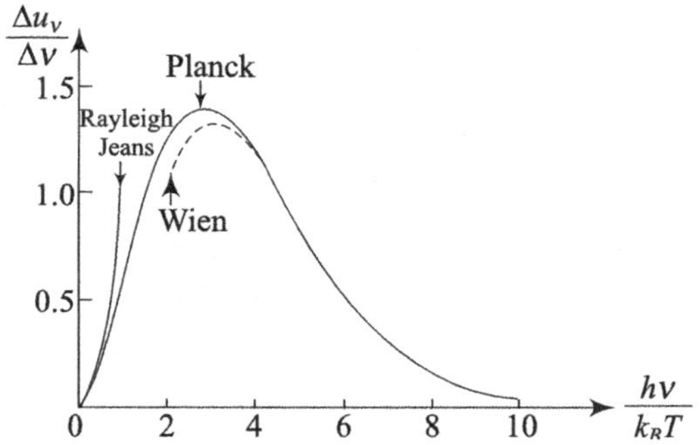

Graphs of Planck's law based on Eq. (11.44), the Wien's law and the Rayleigh–Jeans law.

Deductions from Planck's law

1. Rayleigh–Jeans law We now know that Planck's law provides us with the most general description of blackbody radiation. We can use it to obtain all other laws of blackbody radiation. In particular, Wien's law and Rayleigh–Jeans law are its special cases in the region of shorter and longer wavelengths (high and low frequencies), respectively. To show this, we note that when $\lambda >> hc/k_BT$, the exponential term in Eq. (11.55) can be approximated as

$$\exp(hc/\lambda k_B T) \approx 1 + \frac{hc}{\lambda k_B T} + \ldots$$

$$\exp(hc/\lambda k_B T) - 1 = \frac{hc}{\lambda k_B T}$$

Hence, Eq. (11.55) reproduces Rayleigh–Jeans law for $\lambda >> hc/k_BT$:

$$u_\lambda \,\mathrm{d}\lambda = \frac{8\pi hc}{\lambda^5} \times \left(\frac{\lambda k_B T}{hc}\right)\mathrm{d}\lambda = \frac{8\pi k_B T}{\lambda^4}\mathrm{d}\lambda \tag{11.56}$$

2. Wien's law For $\lambda << hc/k_BT$, the exponential term will be much greater than unity. Therefore, by ignoring one in comparison to the exponential, we find that Eq. (11.55)

$$u_\lambda \, \mathrm{d}\lambda = \left(\frac{8\pi hc}{\lambda^5} \right) \exp(-hc/\lambda k_B T)\mathrm{d}\lambda \qquad (11.57)$$

Wien's law.

3. Wien's displacement law The wavelength at which maximum in the curve occurs can be obtained from Eq. (11.55) using the condition

$$\left[\frac{\partial u_\lambda}{\partial \lambda} \right]_{\lambda=\lambda_{\max}} = 0 \qquad (11.58)$$

This leads to

$$\frac{}{\partial\lambda} = 8\pi hc \frac{\partial}{\partial\lambda}\left[\frac{\lambda^{-5}}{\exp(hc/\lambda k_B T)-1} \right]$$

$$= 8\pi hc \left[\frac{-5\lambda^{-6}\,[\exp(hc/\lambda k_B T)-1] - \lambda^{-5}\left(-\frac{hc}{\lambda^2 k_B T}\exp(hc/\lambda k_B T) \right)}{\left(\exp(hc/\lambda k_B T)-1\right)^2} \right]$$

$$= \frac{8\pi hc}{\lambda^5}\frac{1}{\left(\exp(hc/\lambda k_B T)-1\right)}\left[-\frac{5}{\lambda} + \frac{hc}{\lambda^2 k_B T}\frac{\exp(hc/\lambda k_B T)}{\left(\exp(hc/\lambda k_B T)-1\right)} \right]$$

$$= u_\lambda \left[-\frac{5}{\lambda} + \frac{hc}{\lambda^2 k_B T}\frac{\exp(hc/\lambda k_B T)}{\left(\exp(hc/\lambda k_B T)-1\right)} \right]$$

Suppose the value of u_λ is maximum for $\lambda = \lambda_{\max}$. Therefore, we equate the right-hand side of the above expression equal to zero and put $\lambda = \lambda_{\max}$. This gives

$$\left[-\frac{5}{\lambda_{\max}} + \frac{hc}{\lambda_{\max}{}^2 k_B T}\frac{\exp(hc/\lambda_{\max} k_B T)}{\left(\exp(hc/\lambda_{\max} k_B T)-1\right)} \right] = 0$$

$$\frac{hc}{\lambda_{\max} k_B T}\frac{\exp(hc/\lambda_{\max} k_B T)}{\left[\exp(hc/\lambda_{\max} k_B T)-1\right]} = 5 \qquad (11.59)$$

We will now introduce a new variable by defining $x = hc/\lambda_{\max} k_B T$. Then we can rewrite Eq. (11.59) in an elegant form:

$$x\frac{\exp(x)}{\exp(x)-1} = 5$$

$$x = 5(1-\mathrm{e}^{-x}) \qquad (11.60)$$

This is a transcendental equation and can be solved graphically or numerically. However, we expect a root in the neighbourhood of 5. By applying the method of approximation, the exact value of x is found to be 4.965. Hence, the result is

$$x = \frac{hc}{\lambda_{\max} k_B T} = 4.965$$

$$\lambda_{\max} T = b = \frac{hc}{k_B \times 4.965} \tag{11.61}$$

Wien's displacement law.

On substituting for h, c and k_B, we get

$$\lambda_{\max} T = 2.897 \times 10^{-6} \text{ mK} \tag{11.62}$$

4. Stefan's law The total energy density for photons of all wavelengths (frequencies) is obtained by integrating Eq. (11.57). This gives

$$u(T) = \int_0^\infty u_\lambda \, d\lambda = 8\pi hc \int_0^\infty \frac{d\lambda}{\lambda^5 [\exp(hc/\lambda k_B T) - 1]} \tag{11.63}$$

To evaluate this integral, we introduce a change of variable by defining $x = \dfrac{hc}{\lambda k_B T}$ so that and $d\lambda = -\dfrac{hc}{x^2 k_B T} dx$. The limits of integration change to $-\infty$ to 0. Using these results in Eq. (11.63), we get

$$u(T) = 8\pi hc \int_{-\infty}^{0} \frac{\left(-\dfrac{hc}{x^2 k_B T}\right) dx}{\left(\dfrac{hc}{x k_B T}\right)^5 [\exp(x) - 1]}$$

To absorb the negative sign, we change the limits of integration. Hence, we get

$$u(T) = \frac{8\pi k_B{}^4 T^4}{c^3 h^3} \int_0^\infty \frac{x^3 \, dx}{\exp(x) - 1}$$

This integral is a standard one and has the value $\Gamma(4)\,\zeta(4) = \pi^4/15$. So, the total energy density at temperature T is given by

$$u(T) = \frac{8\pi^5 k_B^4}{15 h^3 c^3} T^4$$

$$= aT^4 \tag{11.64}$$

$$= \frac{8\pi^5 k_B^4}{15 h^3 c^3} = 7.56 \times 10^{-16} \text{ Jm}^3 \text{ K}^{-4}.$$

If we consider the interior of the Sun as consisting of a photon gas at constant temperature 3×10^6 K, we find that its energy density is equal to

$$u = (7.56 \times 10^{-14} \text{ Jm}^{-3} \text{ K}^{-4}) \times (3 \times 10^6 \text{ K})^4$$

$$= 6.1 \times 10^{10} \text{ Jm}^{-3}$$

The volume of the sun is nearly equal to 1.4×10^{27} m^3. So the total energy of photons inside the sun is

$$E = uV = 8.6 \times 10^{37} \text{ J}$$

If there were a small opening in a cavity, photons will **effuse** through it. The net rate of flow of radiation per unit area of the opening is given by

$$R = \frac{1}{4} u\, c = \frac{2\pi^5 k_B^4}{15 h^3 c^2} T^4$$

$$R = \sigma T^4 \tag{11.65}$$

$$= \frac{2\pi^5 k_B^4}{15 h^3 c^2} = 5.67 \times 10^{-8} \text{ J m}^{-2} \text{ K}^{-4} \text{ s}^{-1}$$ is Stefan–Boltzmann constant.

Since radiation pressure exerted on the walls of the enclosure is one-third of the energy density, we can write

$$p = \frac{u}{3} = \frac{8\pi^5 k_B^4}{45 c^3 h^3} T^4 \tag{11.66}$$

This result shows that radiation pressure is independent of volume of the enclosure and varies as the fourth power of temperature.

You may now like to solve a practise problem.

Problem 11.4 In a nuclear explosion, a temperature of 10^6K is produced in a sphere of radius 0.12 m. Calculate the energy density of radiation in the sphere and wavelength corresponding to the maxima in the spectral energy distribution curve.

$\times 10^8$ Jm^{-3}; 2.9 nm

You will agree that physics is an experimental science and theoretical predictions have to stand the test of experimental findings. Experience shows that theory is invariably modified in the light of experimental evidences. Therefore, for completeness, we now present simple experiments used to verify some of the radiation laws.

XPERIMENTAL VERIFICATION OF LAWS OF RADIATION

Kirchhoff's Law

The experimental verification of Kirchhoff's law was done using Rotche's apparatus (Fig. 11.11). It essentially consists of two identical cylindrical metallic vessels V_P and V_B. (The and B signify 'Polished' and 'Blackened' surfaces.) These are connected by a U-tube filled with some coloured liquid. Another cylindrical metallic vessel V, which can be rotated, is placed between V_P and V_B. The flat surface of V facing the polished surface is blackened. The other flat surface which faces the blackened face of V_B is polished.

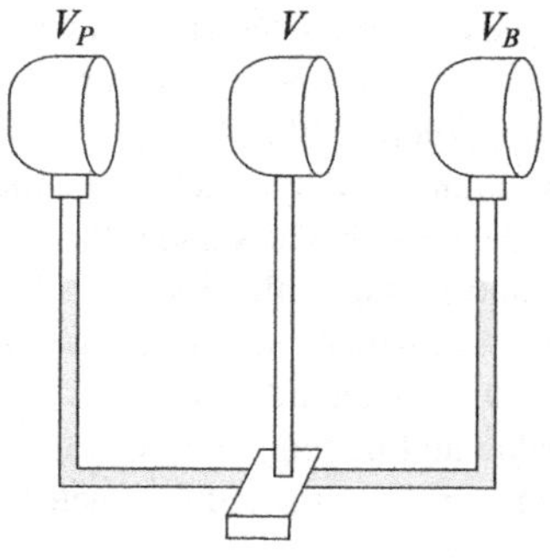

Fig. 11.11 Rotche's apparatus used for verification of Kirchhoff's law.

To begin with, V is filled with boiling water so that the levels of the liquid in the two limbs of the U-tube are at the same height. It indicates that V_P and V_B are at the same temperature.

Now, the cylinder V is rotated so that its blackened surface faces the blackened surface of V_B and the polished surfaces of V and V_P face each other. It is observed that there is a difference in the levels of the liquid in the limbs. The observation essentially establishes that the liquid columns in the limbs are indicative of the temperatures of the vessels V_B and , which are subject to change depending on the quantum of the radiated energy.

Now, let us analyse the situation when the liquid columns in the limbs were at the same height. Suppose that the area of the flat surfaces of the cylindrical vessels is α. Let us denote the emissive and absorptive powers of the polished and the blackened by e_p and e_b, respectively. Then energy emitted by the blackened

$_b\alpha$.

If a fraction f of this energy falls on the polished surface, the energy absorbed by it will , where a_p is the absorptive power of the polished surface.

Similarly, the energy incident on the blackened surface is $fe_p\alpha$ and the same is absorbed by it. Therefore, we can write

$$fa_p e_b \alpha = fe_p \alpha \tag{11.67}$$

On simplification, we obtain the desired result:

$$\frac{e_p}{a_p} = e_b$$

This result shows that by measuring e_p, a_p and e_b, we can easily verify Kirchhoff's law.

Stefan's Law

Stefan's law was experimentally verified by Lummer and Pringsheim in 1897. They carried out a series of experiments over a wide range of temperatures—from 100°C to 1300°C. The apparatus used by them is shown in Fig. 11.12. An interesting feature of the experiment was that the blackbody was taken in two different forms in two different ranges—a copper sphere blackened from inside by platinum black was placed in a bath of fused nitre in the temperature range 100°C – 600°C and an iron cylinder with platinum black coating was placed in a gas furnace in the temperature range 600°C – 1300°C.

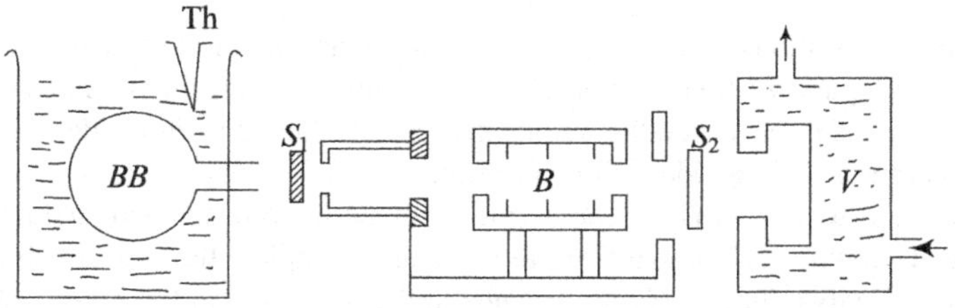

Lummer and Pringsheim apparatus used for verification of Stefan's law.

is a water cooled shutter, which allows the radiation from *BB* to fall on the grid of the *B*. The bolometer is used to measure the radiant energy. It is standardised by using another blackbody *V*, which is a vessel coated from inside with platinum black. It is maintained at a fixed temperature by means of a boiling water heat bath. Here S_2 acts as the shutter. The standardisation of *B* was done by locating *V* at different distances from and noting deflections, *d*, of the galvanometer in the bolometer. It was observed that , which established that *d* was proportional to the incident radiation.

Once standardisation of *B* was achieved, the blackbody *BB* was heated to any desired value and its temperature was measured by a thermocouple 'Th' and the corresponding deflections of the galvanometer in *B* was noted. It was observed that

$$d = k(T^4 - T_0^4) \tag{11.68}$$

and T_0 are the absolute temperatures of *BB* and S_1 respectively, and *k* is a constant.

Note that the constant *k* is not Stefan's constant but proportional to it depending on the calibration of the bolometer. However, you can appreciate that the Stefan's law stands

We shall now discuss a simple experiment that you can perform in your college physics laboratory to determine Stefan's constant.

Determination of Stefan's constant in a physics laboratory

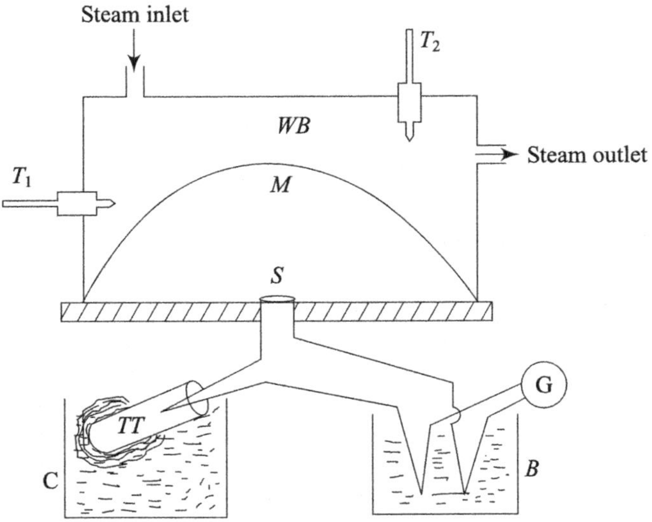

Schematics of the Stefan constant apparatus used in a physics laboratory.

Refer to Fig. 11.13. *WB* is a wooden box lined inside with tin. It is heated by passing steam. Thermometers T_1 and T_2 are used to record temperatures. *WB* is fitted on to a table which has a hole at the centre. A metallic hemisphere *M*, blackened from inside is enclosed by such that the hole coincides with its centre.

The hole is provided with a vulcanite ring. A silver disc *S* blackened on the top is fitted to the ring. Two wires are drawn from the disc and complete the electric circuit, which also includes a galvanometer *G* and thermocouples as shown in the figure. The hot end of the thermocouple is placed in a test tube (*TT*) containing an oil. *TT* is surrounded by a

You can easily make out that the hemisphere acts as a blackbody. It emits radiation, say at temperature, T_1, which is received by the disc *S*. At any instant the disc will also emit radiation. If the temperature of the disc at any instant is *T*, the rate at which energy is gained by it is given by $\sigma A(T_1^4 - T^4)$, where *A* is area of the disc and, σ is Stefan's constant. Let *m* be the mass and *s* be the specific heat capacity of the material of disc. Then the rate at which energy is gained by the disc can also be expressed as $ms\frac{dT}{dt}$. On equating these expressions, we get

$$ms\frac{dT}{dt} = \sigma A(T_1^4 - T^4)$$

$$\sigma = \frac{ms}{A(T_1^4 - T^4)}\frac{dT}{dt} \qquad (11.69)$$

This expression shows that σ can be determined from the knowledge of the quantities on the RHS of Eq. (11.69). Of these $\frac{dT}{dt}$ is determined in two stages. To this end, we write

$\frac{d}{dt} \quad \frac{d}{d} \frac{d\theta}{dt}$ where θ denotes galvanometer deflection. In the first stage we determine

while in the second stage, we determine $\frac{d\theta}{dt}$. These are respectively called the

and the *static* parts.

In the dynamical part, the galvanometer deflection is calibrated against the temperature of the disc, when the process of emission and absorption of radiation takes place. In the static part, the disc is first removed and the hemisphere is heated to a steady temperature . Thereafter, the disc is replaced and keeping the cold end at 0°C, galvanometer deflection is recorded at intervals of 10s. These measurements are used to draw a graph θ. On this graph, we locate the time *t* corresponding to a value of θ which, in turn, corresponds to temperature *T* on the calibration curve. Then we calculate the slope of the tangent at that value of θ. The nature of *T* vs θ and *t* vs θ graphs are shown in Fig. 11.14. It may be mentioned here that the typical tangent plays a crucial role in obtaining a correct measured value of Stefan's constant.

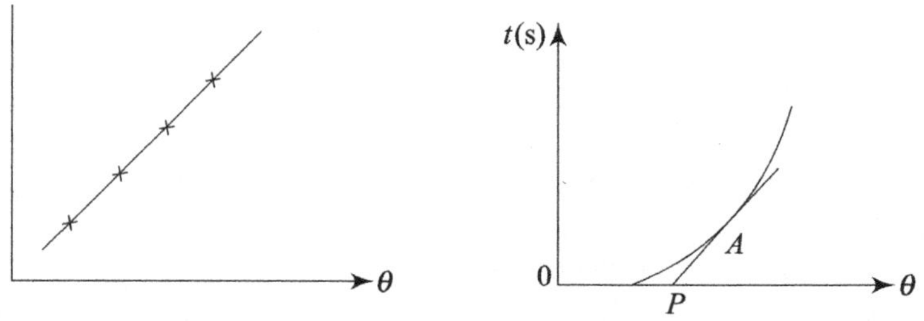

(a) (b)

Expected plots of (a) *T* vs θ and (b) *t* vs θ.

•lar Constant

We now know that energy radiated by the sun supports life on the Earth. From your school of hydrogen atoms at millions of degrees temperature. Moreover, the amount of energy released is huge and is radiated continuously in space. The upper atmosphere of earth receives about 4.1×10^{18} W-hr (thermal) of solar radiations every day. (This is more than times the energy used by the human population of the earth.) And out of this incoming radiation, a considerable part is lost by reflection, scattering and absorption in earth's atmosphere and is sent back into interstellar space. Moreover, absorption is variable; it depends on the time of day and the season of the year. (The solar radiations are reflected by water, snow and clouds, whereas scattering is due to dust particles and air molecules. Nearly, 20%–40% radiation is absorbed.)

To get a measure of the energy received by the earth, it is advisable to look for a more stable quantity. This leads us to the so-called *solar constant*, which is defined *as the amount of energy received per minute per unit area by a perfectly black surface held at right angles to the solar radiations at the mean distance of the earth from the sun, provided there were no absorption in the atmosphere*. It is denoted by the symbol S. Initially, the solar constant was expressed in calories per minute and after making allowance for the losses in earth's atmosphere, measurements yielded a value $S = 1.933$ cal min^{-1} cm^{-2}. The presently accepted value of solar constant is $S = 1.35$ kW m^{-2}.

The principle of measurement of solar constant is very simple. The solar energy absorbed by a known area in a fixed time is measured with the help of pyrheliometers. To minimise the effect of absorption by the atmosphere, the observations are repeated at different altitudes of the sun on the same day under similar sky conditions. The observed solar constant is related to actual solar constant through the relation

$$S' = Sa^{\sec Z} \tag{11.70}$$

is altitude (angular elevation) of the sun.

A plot of $\ln S'$ versus $\sec Z$ gives a straight line. If this straight line is produced backward, the intercept on the y-axis gives $\ln S$. The value so obtained easily yields the solar constant.

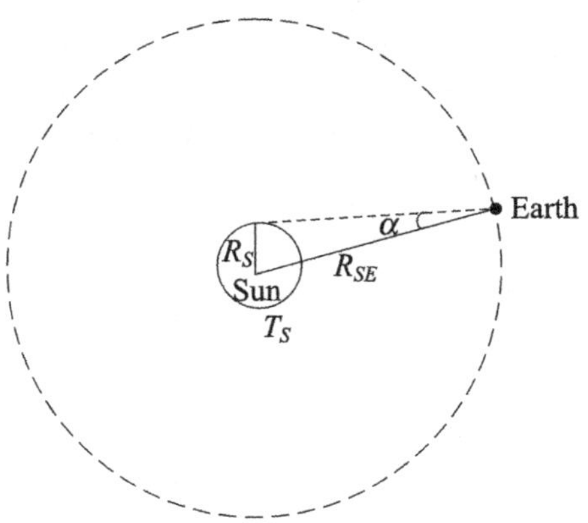

Calculation of blackbody radiation of the sun.

The concept of solar constant can also be used to obtain blackbody temperature of the sun. It is defined as the temperature of a blackbody having the dimensions and position of the core of the sun, free from any atmosphere, which emits radiation at the same rate as is actually received on the surface of the earth. Refer to Fig. 11.15, which depicts the sun-earth configuration. If the radius of the sun is R_S and its blackbody temperature is T_S, the total energy emitted per second is given by $U = 4\pi R^2 \times \sigma T^4$. If the mean distance between the Sun and the Earth is R_{SE}, as depicted by the broken circle, the energy incident per second on unit area of the Earth equals the solar constant. Therefore, we can write

$$S = \frac{4\pi R_S^2 \sigma T_S^4}{4\pi R_{SE}^2} = \left(\frac{R_S}{R_{SE}}\right)^2 \sigma T_S^4$$

Hence, the blackbody temperature of the sun can be obtained by inverting this relation:

$$s = \left(\frac{R_{SE}}{R_S}\right)^{1/2} \left(\frac{S}{\sigma}\right)^{1/4} \tag{11.68}$$

= 1.35 kW m^{-2}, $R_S = 7 \times 10^8$ m, $R_{SE} = 15 \times 10^{10}$ m and $\sigma = 5.7 \times 10^{-8}$ Wm^{-2}K^{-4},

$$T_S = \left(\frac{15\times 10^{10}\ \text{m}}{7\times 10^{8}\ \text{m}}\right)^{1/2} \times \left(\frac{1.35\times 10^{3}\ \text{Wm}^{-2}}{5.7\times 10^{-8}\ \text{Wm}^{-2}\,\text{K}^{-4}}\right)^{1/4}$$

$$= 14.64 \times 392.3\ \text{K}$$

$$= 5743\ \text{K}$$

Problem 11.5 The surface temperature of the Sun is 6000 K, its radius $R_S = 7 \times$ m, and the mean distance from the Earth $R_{SE} = 15 \times 10^{10}$ m. If the radius of the Earth 10^6 m, calculate the temperature of the Earth. Assume that both Sun and the Earth behave as black bodies.

289.8 K

DDITIONAL EXAMPLES

Example 11.9 A body of mass 1 kg is put in an enclosure at 300 K. If the body is at 500 K, calculate the rate at which it cools. The specific heat capacity and area of the emitting surface of the body are 0.1 kcal kg^{-1}K^{-1} and 0.1 m^2 respectively. Take $\times 10^{-8}$ Jm^{-2}K^{-4}s^{-1}.

The energy emitted by area A of a body at temperature T in time t is given by

$$Q = A\sigma(T^4 - T_0^4)t$$

is temperature of the surroundings. Hence, the rate of loss of heat is given by

$$\frac{dQ}{dt} = A\sigma(T^4 - T_0^4)$$

But we know that $Q = msT$ for a body of mass m and specific heat capacity s. Therefore, we can write

$$\frac{d(msT)}{dt} = A\sigma(T^4 - T_0^4)$$

$$\frac{dT}{dt} = \frac{A\sigma}{ms}(T^4 - T_0^4)$$

On substituting the given values, we get

$$\frac{dT}{} = \frac{(10^{-1}\ \text{m}^2)\times(5.672\times10^{-8}\ \text{J m}^{-2}\text{K}^{-4}\text{s}^{-1})}{(1\ \text{kg})\times(100\times4.2\ \text{J kg}^{-1}\text{K}^{-1})}\times[(500\ \text{K})^4-(300\ \text{K})^4]$$

$$= \frac{5.672\times10^{-9}\ \text{JK}^{-4}\text{s}^{-1}}{420\ \text{J K}^{-1}}\times(625-81)\times10^8\ \text{K}^4$$

$$= \frac{3.086\times10^2\ \text{K s}^{-1}}{420}$$

$$= 7.35\ \text{K s}^{-1}.$$

Example 11.10 The heat radiated by a planet of radius r obeys Stefan's law. You may also assume that the surface temperature of the planet attains a steady value when the loss of heat from its surface gets compensated exactly by the heat received from the sun. If every other thing remains invariant, show that $T \propto d^{-1/2}$, where d is distance of the planet from the centre of the sun. Also, calculate the temperature of the planet if 10^7 km, the radius of the sun is 7.5×10^5 km and the surface temperature of sun

We assume that the sun behaves as a perfect blackbody of radius R and surface temperature T_1. Using Stefan's law, we can write the expression for heat radiated per second by the sun as

$$Q = 4\pi R^2\sigma T_1^4 \tag{i}$$

Therefore, the amount of solar energy incident on unit area of the planet per second is

$$\frac{Q}{4\pi d^2} = \frac{4\pi R^2\sigma T_1^4}{4\pi d^2} = \frac{R^2\sigma T_1^4}{d^2} \tag{ii}$$

The total energy incident upon the entire surface of the planet can be expressed as

$$E = \frac{R^2\sigma T_1^4}{d^2}\times\pi r^2 \tag{iii}$$

Under steady state, we can write

$$\frac{R^2\sigma T_1^4}{d^2}\cdot\pi r^2 = 4\pi r^2\sigma T^4$$

$$T^4 = \frac{R^2\, T_1^4}{4d^2}$$

$$T = T_1\sqrt{\frac{R}{2d}} \tag{iv}$$

T_1 are supposed to be invariant, we get the required result:

$$T \propto \frac{1}{}$$

On substituting the given values in Eq. (iv), we get

$$T = (6000\ \text{K}) \times \sqrt{\frac{7.5\times10^5\ \text{km}}{2\times15\times10^7\ \text{km}}} = \frac{6000\ \text{K}}{20} = 300\ \text{K}.$$

Example 11.11 A solid iron sphere of radius 0.1 m kept at 200 K is suspended inside a closed chamber whose walls are at almost 0 K. The sphere acts as a blackbody. Calculate the time required for its temperature to drop to 100 K. It is given that density of iron is 7.8 × 10^3 kg m^{-3}, its specific heat capacity is 0.11 J kg^{-1}K^{-1} and × 10^{-8} J s^{-1}m^{-2}K^{-4}.

As the chamber is closed, the iron ball will act as a blackbody. Here = 200 K and $T_0 = 0$ K. Let the radius of the iron sphere be r. So, its surface area $A = 4\pi r^2$ and its mass = $\frac{4}{3}\pi r^3 \rho$, where ρ denotes density. The net rate of loss of heat $= -\frac{dQ}{dt}$. (The negative sign appears as Q decreases with t). Hence, we can write

$$-\frac{dQ}{dt} = \sigma A(T^4 - T_0^4) = \sigma A T^4 \qquad (\because T_0 = 0\ \text{K}) \tag{i}$$

$$\frac{dQ}{dt} = ms\frac{dT}{dt} \tag{ii}$$

is mass of the sphere and s is its specific heat capacity. On combining Eqs. (i) and (ii), we can write

$$-ms\frac{dT}{dt} = \sigma A T^4$$

$$dt = -\frac{ms}{\sigma A}\frac{dT}{T^4}$$

By integrating this expression over t between given time limits, we get

$$t = \int_0^t dt = -\frac{ms}{\sigma A}\int_{200}^{100} T^{-4}\,dT$$

$$= -\frac{ms}{\sigma A}\frac{T^{-3}}{(-3)}\Bigg|_{200}^{100}$$

$$= \frac{ms}{3\sigma A}\left(\frac{1}{100^3\ \text{K}^3} - \frac{1}{200^3\ \text{K}^3}\right)$$

$$= \frac{7ms}{24\sigma A}\times10^{-6}\ \text{K}^{-3} = \frac{7\times\frac{4}{3}\pi r^3\rho\times s}{24\sigma\times(4\pi r^2)}\times10^{-6}\ \text{K}^{-3} = \frac{7r\rho s}{72\times10^6\sigma}$$

By putting the values of r, ρ, s and σ, we get

$$t = \frac{7 \times (0.1\ \text{m}) \times (7.8 \times 10^3\ \text{kg m}^{-3}) \times (0.11\ \text{J kg}^{-1}\text{K}^{-1})\ \text{K}^{-3}}{72 \times 10^6 \times (5.672 \times 10^{-8}\ \text{J s}^{-1}\text{m}^{-2}\text{K}^{-4})}$$

$$= \frac{0.6006 \times 10^3}{4.084}\ \text{s}$$

$$= \frac{600.6}{4.084}\ \text{s}$$

$$= 147\ \text{s}.$$

Example 11.12 An oscillator vibrates with frequency 1.5×10^{14} Hz at $T = 1800$ K Calculate its average energy treating it as **(a)** a classical oscillator and **(b)** Planck's

(a) The average energy of a classical oscillator

$$\bar{\varepsilon} = k_B T = (1.38 \times 10^{-23}\ \text{J K}^{-1}) \times (1800\ \text{K})$$

$$= 2.48 \times 10^{-20}\ \text{J}$$

The average energy of Planck's oscillator

$$\bar{\varepsilon} = \frac{h\nu}{e^{h\nu/k_B T} - 1} = \frac{k_B T (h\nu/k_B T)}{e^{h\nu/k_B T} - 1}$$

$$\frac{h\nu}{k_B T} = \frac{(6.62 \times 10^{-34}\ \text{Js}) \times (1.5 \times 10^{14}\ \text{s}^{-1})}{(1.38 \times 10^{-23}\ \text{JK}^{-1}) \times (1800\ \text{K})}$$

$$= \frac{9.93 \times 10^{-20}\ \text{J}}{2.48 \times 10^{-20}\ \text{J}}$$

$$= 4$$

$$\bar{\varepsilon} = \frac{(2.48 \times 10^{-20}\ \text{J}) \times (4)}{e^4 - 1}$$

$$= \frac{9.92 \times 10^{-20}\ \text{J}}{53.6}$$

$$= 1.85 \times 10^{-20}\ \text{J}.$$

Example 11.13 Calculate the number of modes of vibration in a 100 cm^3 chamber in the wavelength region **(a)** 500.0 nm – 500.2 nm and **(b)** frequency range 1.5×10^{14} Hz 10^{14} Hz.

(a) The number of modes per unit volume in the wavelength region λ to is given by

$$n(\lambda) = \frac{8\pi}{\lambda^4} d\lambda$$

Therefore, the number of modes within wavelength range λ to λ + dλ in a chamber of is

$$N(\lambda) = \frac{8 \times 3.14 \times (100\ \text{cm}^3)}{(5 \times 10^{-5}\ \text{cm})^4} \times (2 \times 10^{-8}\ \text{cm})$$

$$= \frac{50.24 \times 10^{-6}}{625 \times 10^{-20}}$$

$$= 8.014 \times 10^{12}$$

The number of modes in the frequency range v to v + dv in chamber of volume V is

$$N(v) = V \frac{8\pi v^2}{c^3} dv$$

= 1.5 × 10^{14} Hz, dv = 0.01 × 10^{14} Hz, c = 3 × 10^{+10} cm s^{-1}, and V = 100 cm^3

$$N(v) = \frac{8 \times 3.14 \times (100\ \text{cm}^3) \times (1.5 \times 10^{14}\ \text{s}^{-1})^2}{(3 \times 10^{10}\ \text{cm s}^{-1})^3} \times (0.01 \times 10^{14}\ \text{s}^{-1})$$

$$= \frac{56.52 \times 10^{42}\ \text{cm}^3\,\text{s}^{-3}}{27 \times 10^{30}\ \text{cm}^3\,\text{s}^{-3}}$$

$$= 2.09 \times 10^{12}.$$

Example 11.14 For the $u_v - v$ curve based on Planck's law, obtain the value of v corresponding to the peak of u_v at 1000 K.

From Eq. (11.44), we recall that the energy density of blackbody radiation according to Planck's law is given by

$$u_v\, dv = \frac{8\pi v^2}{c^3} \left(\frac{hv}{e^{hv/k_B T} - 1} \right) dv$$

For maximum value of u_v, we differentiate it with respect to v and equate the resultant expression to zero:

$$\frac{\partial u_v}{\partial v} = \frac{8\pi h}{c^3} \left[3 \frac{v^2}{e^{hv/k_B T} - 1} + \frac{v^3 \left(-\frac{h}{k_B T} \right) e^{hv/k_B T}}{(e^{hv/k_B T} - 1)^2} \right]_{v = v_m} = 0$$

$$3-\frac{h\nu_m}{k_B T}\,\frac{e^{h\nu_m/k_B T}}{e^{h\nu_m/k_B T}-1}=0.$$

$\nu_m/k_B T = x$, this expression simplifies to

$$3=\frac{xe^x}{e^x-1}$$

This is a transcendental equation and its solution is

$$x=\frac{h\nu_m}{k_B T}\approx 3$$

$$\nu_m=\frac{3k_B T}{h}.$$

= 1000 K, the frequency corresponding to the peak of spectral density is given by

$$\nu_m=\frac{3\times(1.38\times10^{-23}\ \text{JK}^{-1})\times(10^3\ \text{K})}{6.62\times10^{-34}\ \text{Js}}=6.25\times10^{13}\ \text{s}^{-1}$$

Note that this frequency lies in the infra-red region.

Example 11.15 An aluminium sphere of radius 5 cm and kept at 77°C is suspended by a fine thread inside an evacuated jar, which is maintained at a constant temperature of 27°C. It means that heat is lost by radiation only. Calculate the rate of loss of heat by the sphere and how long will it take to cool through 10°C. The density of aluminium is $^{-3}$ and surface emissivity of the metal is 0.4. Take $\sigma = 5.672\times10^{-8}$ J m^{-2}K^{-4}s^{-1} = 920 J kg^{-1}K^{-1}.

The rate of loss of heat by the sphere is given by

$$Q=e\sigma A(T_{\text{al}}^4-T_{\text{jar}}^4) \qquad \text{(i)}$$

$e=0.4,\ \sigma=5.672\times10^{-8}\ \text{Jm}^{-2}\text{K}^{-4}\text{s}^{-1},\ T_{\text{al}}=(273+77)\text{K}=350\ \text{K},\ \ T_{\text{jar}}=300\ \text{K},$ $=4\pi\ ^2=4\pi\times(0.05\ \text{m})^2=0.0314\ \text{m}^2$. On substituting these values in Eq. (i), we get

$$=(0.4)\times(5.672\times10^{-8}\ \text{Jm}^{-2}\text{K}^{-4}\text{s}^{-1})\times(0.0314\ \text{m}^2)\times(350^4-300^4)\ \text{K}^4$$

$$=(0.071\times10^{-8})\times(1.5\times10^{10}-0.81\times10^{10})\ \text{Js}^{-1}$$

4.9 Js^{-1}

We know that

$$Q\Delta t = m s_{\text{al}}\,\Delta T$$

$920\,\text{J}\,\text{kg}^{-1}\text{K}^{-1}$,

$$=\rho_{\text{al}}\left(\frac{4\pi r^3}{3}\right)=(2700\ \text{kg m}^{-3})\times\frac{4\times3.1417}{3}\times(0.05\ \text{m}^3)$$

1.414 kg,

$\Delta\ = 10°\text{C} = 10\ \text{K}$

$$\Delta t = \frac{m s_{al} \Delta T}{Q}$$

$$= \frac{(1.414 \text{ kg}) \times (920 \text{ J kg}^{-1}\text{K}^{-1}) \times (10 \text{ K})}{4.9 \text{ J s}^{-1}}$$

$$= 2654 \text{ s}$$

Example 11.16 A light bulb filament is made from 2 cm tungsten wire of diameter m and is enclosed in an evacuated glass bulb. Calculate the temperature of the filament when it is operated at a power of 1 W. Take emissivity of the tungsten surface as 0.38.

The power radiated by a body of emissivity e and surface area A is given by

$$P = e\sigma T^4 A$$

When the surroundings of the body are at a temperature T_0, we can write

$$P = e\sigma(T^4 - T_0^4)A$$

$$e = 0.38,\ \sigma = 5.672 \times 10^{-8} \text{ Jm}^{-2}\text{K}^{-4}\text{s}^{-1},\ P = 1 \text{ Js}^{-1} \text{ and}$$

$$= 2\pi \quad = 2\pi(25 \times 10^{-6} \text{ m}) \times (0.02 \text{ m}) = 3.1417 \times 10^{-6} \text{ m}^2$$

On substituting the given values, we get

$$1 \text{ Js}^{-1} = (T^4 - T_0^4) \times 0.38 \times (5.672 \times 10^{-8} \text{ Js}^{-1}\text{m}^{-2}\text{K}^{-4}) \times (3.1417 \times 10^{-6} \text{ m}^2)$$

$$= 6.77 \times 10^{-14}(T^4 - T_0^4)\text{K}^{-4}\text{Js}^{-1}$$

$$- \quad = \frac{1 \text{ Js}^{-1}}{6.77 \times 10^{-14} \text{ Js}^{-1}\text{K}^{-4}}$$

$$= 1.48 \times 10^{13} \text{ K}^4$$

$T = 300$ K, $T_0^4 = 8.1 \times 10^9$ and this is almost insignificant in comparison to T^4. Hence, we can write

$$T^4 \approx 14.8 \times 10^{12} \text{ K}^4$$

$$T = 1961 \text{ K}.$$

Example 11.17 The volume of an enclosure is increased from 10^{-5} m^3 to ${}^{-3}$ m^3, while its walls are kept at a temperature of 2500 K. Calculate **(a)** the radiation pressure on the enclosure, **(b)** the work done during the expansion, and **(c)** final temperature when the volume of the enclosure is reduced reversibly and adiabatically to one-hundredth of its initial value. Take $\sigma = 5.672 \times 10^{-8}$ Jm^{-2}K^{-4}s^{-1}.

(a) We know that enclosure can be treated as a blackbody and radiation pressure is one-third of energy density:

$$p = \frac{u}{3} \tag{i}$$

To calculate u, we recall that it is connected to emissive power through the relation

$$u = \frac{4R}{c} \tag{ii}$$

Hence, we can rewrite Eq. (i) as

$$p = \frac{4}{3}\frac{R}{c} \tag{iii}$$

Using Stefan's law for a blackbody, we can write

$$R = \sigma T^4$$

On combining this result with Eq. (iii), we get

$$p = \frac{4}{3c}\sigma T^4 \tag{iv}$$

$= 3 \times 10^8$ ms^{-1}, $\sigma = 5.672 \times 10^{-8}$ Jm^{-2} K^{-4} s^{-1} and T = (2500 + 273) K = 2773 K. On inserting these values in Eq. (iv), we get

$$p = \frac{4}{3 \times (3 \times 10^8 \text{ ms}^{-1})} \times (5.672 \times 10^{-8} \text{ Jm}^{-2}\text{ K}^{-4}\text{ s}^{-1})(2773 \text{ K})^4$$

$$= 1.49 \times 10^{14} \times 10^{-16} \text{ Jm}^{-3}$$

$$= 1.49 \times 10^{-2} \text{ Nm}^{-2}.$$

The work done at constant pressure is given by

$$W = p\Delta V$$

$$= (1.49 \times 10^{-2} \text{ Nm}^{-2}) \times (1.01 \times 10^{-3} - 10^{-5}) \text{ m}^3$$

$$= 1.49 \times 10^{-2} \times 10^{-3} \text{ Nm}$$

$$= 1.49 \times 10^{-5} \text{ J}.$$

During an adiabatic change, the temperature and volume are connected by Eq. (11.15):

$$VT^3 = \text{constant}$$

$$\left(\frac{T_f}{T_i}\right) = \left(\frac{V_i}{V_f}\right)^{1/3}$$

$$T_f = T_i\left(\frac{V_i}{V_f}\right)^{1/3}$$

$= 2500$ K and $\left(\frac{V_i}{V_f}\right) = 100$. Hence,

$$T_f = (2500 \text{ K}) \times (100)^{1/3}$$

$$= 11604 \text{ K}.$$

Example 11.18 A metal ball of mass 5 kg and radius 0.1 m, at a temperature of 127°C is suspended in a closed container whose walls are maintained at a temperature of 27°C. The maximum rate of fall of temperature of the ball is found to be 3.0×10^{-2}°Cs^{-1}.

Calculate the specific heat capacity of the metal. Take $\sigma = 5.7 \times 10^{-5}$ erg $s^{-1}cm^{-2}K^{-4}$.

As the ball is suspended in a closed container, it will behave like a blackbody and its temperature will fall at the maximum rate. So using Stefan's law, we can write

$$\frac{dQ}{dt} = \sigma A(T^4 - T_0^4) \qquad \text{(i)}$$

$\frac{Q}{\ }$ may be related to $\frac{d\theta}{dt}$ as

$$\frac{dQ}{dt} = ms\frac{d\theta}{dt} \qquad \text{(ii)}$$

On combining Eqs (i) and (ii), we can write

$$\sigma A(T^4 - T_0^4) = ms\frac{d\theta}{dt}$$

$$s = \frac{\sigma A(T^4 - T_0^4)}{m\frac{d\theta}{dt}}$$

Here $T = (127 + 273)$ K $= 400$ K and $T_0 = (27 + 273)$ K $= 300$ K. On substituting the given σ, m, $d\theta/dt$ and $A = 2\pi r^2$, we get

$$= \frac{(5.7 \times 10^{-5}\ \text{erg s}^{-1}\text{cm}^{-2}\text{K}^{-4})4\pi \times 10^2\ \text{cm}^2\{(400)^4 - (300)^4\}\ \text{K}^4}{(5 \times 10^3\ \text{g}) \times (3.0 \times 10^{-2}\ {}^\circ\text{C s}^{-1})}$$

$$= \frac{5.7 \times 175 \times 4\pi \times 10^5\ \text{erg}}{150 \times 4.2 \times 10^7\ \text{erg cal}^{-1}\ \text{g}\ {}^\circ\text{C}} = 0.2\ \text{cal g}^{-1}\,{}^\circ\text{C}^{-1}$$

Example 11.19 Two bodies A and B are characterised by thermal emissivities 0.01 and 0.81, respectively. The outer surface areas of the two bodies are same and their rate of emission is also same. The wavelength λ_A corresponding to the maximum spectral radiance is less than λ_B by 10,000 Å. Calculate λ_B and the temperature of A, if the temperature is 1661°C.

From Stefan's law, we know that the rate of emission of energy is proportional , where e is the emissivity, α is surface area, and T is the temperature in kelvin. is same, we can write

$$0.01\,T_A^4 = 0.81\,T_B^4$$

$$T_B = (1661 + 273)\ \text{K} = 1934\ \text{K}$$

$$\left(\frac{T_A}{T_B}\right)^4 = \frac{0.81}{0.01} = 81 = 3^4$$

$$\frac{T_A}{T_B} = 3$$

$$T_A = 3T_B = 3 \times 1934 \text{ K} = 5802 \text{ K} = (5802 - 273)°\text{C} = 5529°\text{C}$$

From Wien's Displacement law, we know that

$$\lambda T = \text{constant}$$

$$\lambda_A T_A = \lambda_B T_B$$

$$\lambda_A = (\lambda_B - 10^4)\,\text{Å}$$

$$(\lambda_B - 10^4) T_A = \lambda_B T_B$$

$$\frac{T_A}{T_B} = \frac{\lambda_B}{\lambda_B - 10^4}$$

On using this result in (i), we rewrite it as

$$\frac{\lambda_B}{\lambda_B - 10^4} = 3$$

$$\lambda_B = 3\lambda_B - 3 \times 10^4$$

$$2\lambda_B = 3 \times 10^4$$

$$\lambda_B = \frac{3}{2} \times 10^4 \text{ Å} = 1.5 \times 10^4 \text{ Å}.$$

Let us now summarise what you have learnt in this chapter.

MMARY

The pressure exerted by radiation on any surface is one-third of its energy density.

Kirchhoff's law states that the ratio of the spectral emissive power e_λ to the spectral absorptivity a_λ for a particular wavelength λ is the same for all bodies at the same temperature and is equal to the emissive power of a perfect blackbody at that temperature:

$$\frac{e_\lambda}{a_\lambda} = E_\lambda$$

E_λ is emissive power of a perfect blackbody. It explains the existence of Fraunhoffer lines in solar spectrum and gave birth to two new fields of study: Astrophysics and Spectroscopy.

Stefan–Boltzmann law states that the total rate of emission of radiant energy by a body per unit area is related to energy density as fourth power of its temperature:

$$E = \sigma T^4$$

σ is known as Stefan's constant. Its value is 5.672×10^{-8} Jm^{-2}K^{-4}s^{-1}.

When blackbody radiation is made to undergo adiabatic expansion (or compression), the quality of radiation does not change; only temperature changes and the corresponding energy density at the end of the process will be lower (higher).

Spectroscopic analysis of blackbody radiation shows that

- for a given wavelength λ, u_λ increases with temperature;
- for each temperature, the spectral energy density versus wavelength plot shows a maximum, which shifts to shorter wavelengths as temperature increases; and
- the energy density goes smoothly to zero as $\lambda \to 0$ or as $\lambda \to \infty$.

Wien's displacement law states that if blackbody radiation at a certain temperature is adiabatically expanded, the wavelength of each spectral component follows T-inverse law. That is, the product $\lambda_m T$ remains constant.

Wien's distribution law states that when blackbody radiation is subject to adiabatic change, the energy density corresponding to wavelength band λ and λ + dλ is given by

$$u_\lambda \, d\lambda = A'\lambda^{-5} F(\lambda T)\, d\lambda$$

Rayleigh–Jeans formula for the distribution of energy treats blackbody radiation as standing electromagnetic waves which arise due to multiple reflections at the walls of the enclosure and each mode (of standing electromagnetic waves) is assigned energy according to the principle of equipartition of energy. The energy density within wavelength range λ and λ + dλ is given by

$$u_\lambda \, d\lambda = \frac{8\pi k_B T}{\lambda^4} d\lambda$$

According to Planck, blackbody radiation chamber is filled up not only with radiation but also with the molecules of a perfect gas, which exchange energy via resonators of molecular dimensions. The energy density of blackbody spectrum is given by

$$u_\nu \, d\nu = \frac{8\pi\nu^2}{c^3}\left(\frac{h\nu}{\exp(h\nu/k_B T)-1}\right) d\nu$$

Planck's radiation formula explains all experimental results satisfactorily in the complete energy range.

Solar constant is defined as the amount of energy received per minute per unit area by a perfectly black surface held at right angles to the solar radiations at the mean distance of the Earth from the Sun, provided there were no absorption in the atmosphere. The presently accepted value of solar constant is 1.35 kW m^{-2}.

ERCISES

Calculate the wavelength at which human body radiates maximum energy. Take body temperature as 37°C and Wien's constant $b = 2.898 \times 10^{-3}$ m K.

(*Ans:* 9.35×10^{-6} m)

A blackbody at 500 K is put in an evacuated enclosure whose walls are at 300 K. Calculate the rate of loss of energy of the blackbody per unit area.

(*Ans:* 3.086×10^{3} Wm^{-2})

The operating temperature of a 60 W filament lamp is 2000 K and its total emissivity is 0.30. Calculate the surface area of the filament. (*Ans*: 2.19×10^{-4} m^2)

A vacuum tube has a cathode in the form of a long cylinder of radius 0.05 cm. The cathode heater dissipates 1 W per cm of its length. If the temperature of the anode is 500 K, calculate the temperature of the cathode, which radiates like a blackbody.

(*Ans:* 1433 K)

A blackbody with an initial temperature of 300°C cools inside an evacuated chamber surrounded by melting ice at a rate 0.35°C per second. If the mass, specific heat capacity and surface area of the body are 32 g, 0.418 $Jg^{-1}K^{-1}$, and 8 cm^2, respectively, calculate Stefan's constant. (*Ans:* $5.75 \times$ 10–5 erg s^{-1} cm^{-2} K^{-4})

Calculate the average energy of a Planck oscillator of frequency 0.06×10^{14} Hz at 2000 K. How does it compare with the energy of a classical oscillator?

Ans: 1.24×10^{-20} J; Nearly half of classical oscillator, which is 2.76×10^{-20} J)

Calculate the number of modes in a chamber of volume 1m³ in the frequency range $\times 10^{14}$ Hz to 0.61×10^{14} Hz. (*Ans:* 33.5×10^{14})

A spherical satellite is moving around the sun in a circular orbit. The sun subtends an angle of 6.5×10^{-5}. If the emissivity of the satellite is independent of wavelength and calculate the temperature of satellite. Take temperature of the sun as 6000 K.

(*Ans:* 286 K)

Calculate the amount of energy emitted by the human body in a day. If one cold drink provides 200 kcal, how many cold drinks would you need to take per day to compensate for this energy? Assume that the body does not absorb any radiation. Take $\sigma = 5.672 \times 10^{-8}$ $Jm^{-2}K^{-4}s^{-1}$, T 310 K and $A = 0.66$ m^2. (*Ans:* 149.4)

(a) If our earth were to act like a blackbody at 275 K, calculate the rate of emission of energy by it. Take $R = 6400$ km and $\sigma = 5.672 \times 10^{-8}$ $Jm^{-2}K^{-4}s^{-1}$.

(b) The earth receives 10^3 $Jm^{-2}s^{-1}$ solar energy. Calculate the total force experienced on the earth due to solar radiations. Take $R = 6400$ km.

(*Ans:* 1.67×10^{17} Js; 1.72×10^{9} N)

12 BASIC CONCEPTS OF STATISTICAL MECHANICS

arning Objectives

In this chapter, you will learn how to

- explain the terms microstates and macrostates, phase space and quantum states and energy states and energy levels;
- establish Boltzmann entropy relation $S = k_B \ln \Omega$ and discuss statistical interpretation of entropy;
- obtain expressions for thermodynamic probabilities and distribution functions for particles obeying Maxwell–Boltzmann, Fermi–Dirac and Bose–Einstein statistics; and
- highlight differences between different types of statistics.

NTRODUCTION

So far we have developed the subjects of kinetic theory and thermodynamics. In kinetic theory, one applies the laws of mechanics to individual molecules of the system, while in thermodynamics, one deals with directly observable macroscopic properties without any reference whatsoever to microscopic structure. The laws of thermodynamics provide us with extremely powerful methods to develop meaningful relations between macroscopic variables, such as pressure, volume, internal energy, entropy and temperature of a system. Yet, it is phenomenological and one has to rely on observation.

Macroscopic variables can be perceived through our senses. But we cannot do the same with variables such as the positions and velocities of individual atoms/molecules constituting a system. In statistical mechanics, we deal with systems having a large number of particles ($N \approx 10^{25}$) and attempt to relate the macroscopic properties in terms of *statistical* with no reference to the details of molecular motion/collisions. As an analogy, consider the statement: According to Census-2011, the life expectancy in India is about 64 years, up from around 30 years in pre-independent India. Do you know that this figure, though useful, was arrived at without any consideration for the state of health and nutrition available to the individuals in the country?

A physical system is governed by the laws of classical or quantum mechanics, depending on the level of description and the extent of analysis. As such, statistical mechanics cannot help us determine the life history of any particular particle but it helps us to know its probable behaviour. Accordingly, we construct models and make assumptions about microscopic dynamics. For instance, many a times, we choose a discrete model of a gas comprising N particles within an enclosure of volume V which can be divided into cells that may be either empty or contain a definite number of particles. When statistical methods are mechanics. The origin of this subject can be traced back to the work of Maxwell. It then grew in the hands of Boltzmann, Gibbs and Fowler. The advent of quantum mechanics and the insights of Fermi, Einstein, Dirac and Bose, among others, gave it a new dimension. Due to generality of its arguments, the subject of statistical mechanics is mathematically elegant and very versatile. In fact, it has emerged as the most powerful tool of theoretical physics. This chapter is devoted to a discussion of the basic concepts of statistical mechanics and to the development of some of its necessary tools.

In Sec. 12.2, we begin by elaborating the vital link between the probable behaviour of the elements of a physical system and the properties of the system as a whole. In Sec. 12.3, we have introduced the concept of phase space and quantum states. You will learn that to describe a general system we need six-dimensional phase space. However, it cannot be represented in a plane and is used as a purely mathematical device. In Sec. 12.4, we will discuss the states of a particle in a three-dimensional box and generalise these N particles. You will learn that the number of quantum states of such a system can be specified by $3N$ quantum numbers. This is followed by a discussion of macrostates and microstates of a statistical system in Sec. 12.5.

The probability calculations and elementary ideas about statistical ensembles are introduced in Sec. 12.6 and 12.7. Boltzmann entropy relation forms the subject matter of discussion of Sec. 12.8. This relation—entropy of a system is proportional to the logarithmic of its thermodynamic probability—is fundamental and provides a connection between thermodynamics and statistical mechanics. These results are used in Sec. 13.9 to obtain expressions for thermodynamic probabilities for Maxwell–Boltzmann (classical) as well as Fermi–Dirac and Bose–Einstein (quantum) particles making up a system in equilibrium. Since thermodynamic probabilities are very large numbers, the classical and quantum distribution functions are obtained by maximising the logarithm of thermodynamic probabilities. Physically acceptable forms of these functions demand that we introduce a new mathematical device—the partition function. You will also learn that spin of a particle plays a very prominent role in determining its behaviour, particularly at low temperatures.

RIDGING MICROSCOPIC AND MACROSCOPIC EHAVIOURS

Suppose you toss a coin and want to predict the outcome. Whichever way you may call, the odds of being right as well as wrong are rather high (50%). But if a large number of coins are tossed, you can safely predict that roughly half of them will show *heads* and half . The larger the number of coins, the more dependable is your prediction.

The situation is similar to the rolling of a single die. Irrespective of the prediction, (i.e., 1, 2, 3, 4, 5 or 6), 5/6 of them will be wrong. But if a larger number of identical dice are rolled, you can quite correctly predict that roughly 1/6 of them will show 1, 1/6 will show 2 and so on. This again suggests that larger the number of dice, more reliable will be your prediction.

The predictions can be directed the other way too. In the foregoing examples, we used the known probabilities for a single coin or die to predict the outcome of tossing or rolling a large number of them. You could have observed the outcomes first and used that to ascertain the probabilities in the case of a single coin or a die. For example, if you had observed that significantly more than half of a large number of tosses of identical coins landed with 'head' up, you could have readily inferred that the coin is biased, giving a larger probability of the outcome 'head'.

The constituents of actual systems where we shall apply our knowledge of probability are not as simple as those of a large number of tossed coins or rolled dice. Yet, they resemble these in many ways. For instance, from the foregoing examples, you can say that when a system consists of a large number of identical particles, one can use the observed behaviour of an individual element to predict the behaviour of the entire system. Conversely, the probable behaviour of individual elements can be obtained from the observed properties of the entire system. In the following sections, we elaborate upon this vital link between the behaviour of the elements of a physical system and the properties of the system as a whole. To develop an appreciation for the concepts discussed here, go through the following example carefully.

Example 12.1 **(a)** Suppose you flip ten thousand identical coins and find that two of them end up standing on edge. If you flip one more such coin, what is the per cent likelihood that it will end up standing on its edge?

If you select 7,000 dates at random how many of them are likely to be a Sunday?

(a) The observed probability of a coin standing on its edge for the whole system

$= 2 \times 10^{-4}$ So the per cent likelihood of one more such coin standing on its edge

when flipped is $2 \times 10^{-4} \times 100 = 0.02$.

Since there are seven days in a week, the probability of a particular date being a Sunday is 1/7. So if 7000 dates are selected at random roughly 1000 of them will be

A similar conclusion, however, cannot be drawn if the total number of selected dates is a small number.

We know that actual systems are far more complicated than those considered above. The number of equally probable configurations will be far more than that of a coin or a die. An element may have a molecule with a very large number of possible positions, orientations and kinds of motion. Again, these probabilities may vary with temperature or pressure. Any change in the behaviour of individual elements will be reflected in changes in the properties of the system as a whole. For instance, if we consider a tumbler full of ice cubes and another filled with water, the differences in the probable behaviour of individual water molecules in the two tumblers will be reflected in their thermodynamic properties. From thermodynamic reasoning, we expect that molecular chaos will be more in the tumbler filled with water than in the one filled with ice. We will now analyse this and similar other questions using the tools of statistical mechanics.

You should now answer a practise exercise.

Problem 12.1 **(a)** Suppose you are dealing one card from each of one million well-shuffled decks of 52 playing cards. How many of the dealt cards will be (i) queens, (ii) spades and (iii) queen of spades.

If you flip 10 coins, would it be wise to bet on exactly 5 outcomes being head?

(i) 7.7×10^4 (ii) 2.5×10^5 (iii) 1.9×10^4; **(b)** No.

HASE SPACE AND QUANTUM STATES

Let us consider the motion of a particle along a straight line (Fig. 12.1a). The mechanical state of the particle at any instant is given by its position x from a fixed point on the straight line and its velocity $v_x = \frac{dx}{dt}$ at that instant. However, (in view of uncertainty principle) it is more desirable to work with momentum $p_x(= mv_x)$ instead of velocity. As the particle moves along the straight line, the values of x and p_x change. So the state of the particle at any instant is completely specified classically at a particular instant if its position and momentum are known. It may be represented by any point P on a two-dimensional hypothetical space, whose coordinate axes are x and p_x (Fig. 12.1b). With the passage of time, the point P traces out a certain trajectory in the $x - p_x$ plane.

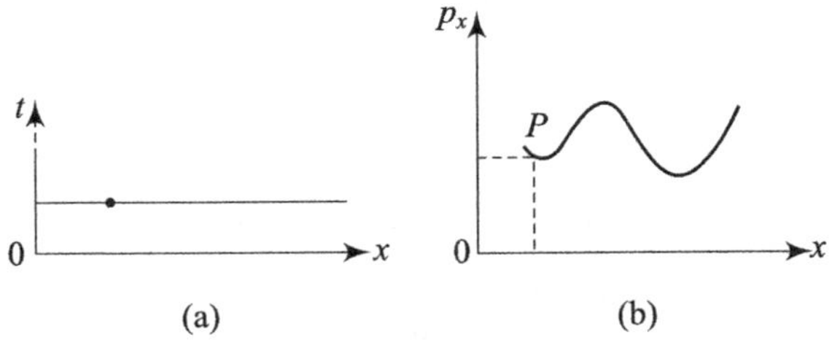

(a) Motion of a particle along a straight line and (b) Phase space for one-dimensional motion.

The state of the particle is referred to as the *phase*, the point P as the *phase point*, the trajectory as the *phase path* and the hypothetical two-dimensional plane as the *phase space*. These concepts are illustrated in the following example.

Example 12.2 Determine the phase path for a linear harmonic oscillator.

Suppose the mass and the force constant of the oscillator are m and k, respectively. From your school physics, you may recall that the total energy E of a linear harmonic oscillator is given by

$$E = \frac{1}{2}mv_x^2 + \frac{1}{2}kx^2$$

mv_x, we can write $\frac{1}{2}mv_x^2 = \frac{p_x^2}{2m}$

so that we can write

$$E = \frac{p_x^2}{2m} + \frac{1}{2}kx^2$$

$$\frac{p_x^2}{2mE} + \frac{x^2}{2(E/k)} = 1$$

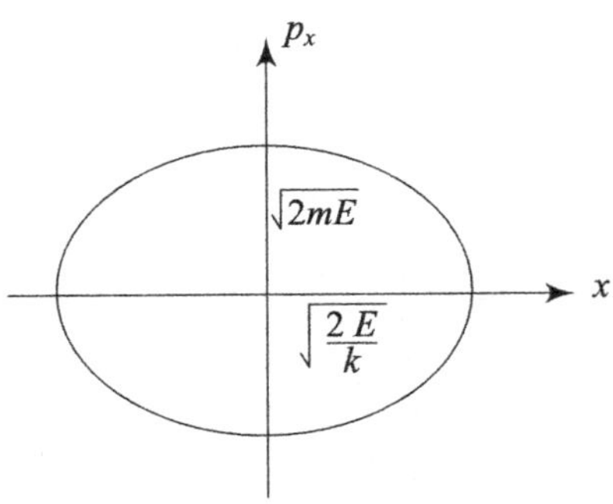

Fig.12.2 Phase path for a linear harmonic oscillator.

is constant, $2mE$ and $2E/k$ are also constant. So this relation indicates that the phase point of the oscillator traces out an ellipse (Fig.12.2) in the x–p_x plane with semi-major and semi-minor axes equal to $\sqrt{2E/k}$ and $\sqrt{2mE}$ respectively.

Any point on this ellipse represents the phase at that time. As the oscillator moves to and fro along a straightline periodically, the same positions and momenta are repeated. So the phase point describes the ellipse of Fig. 12.2 over and over again.

In this example, the trajectory is a closed path. It may not always be so (Problem 12.2). But the curve will not intersect itself in any case.

Problem 12.2 A stone at rest falls freely. Determine its phase trajectory.

A parabola

You will note that the actual motion of a linear harmonic oscillator takes place along a straightline. An actual mechanical system is, in general, more complex than a particle moving along a straightline. A general treatment demands three position coordinates) and three components of momentum (p_x, p_y, p_z). In other words, we require six numerical quantities to specify the state of a system at a particular instant. That is, we need six-dimensional phase space. (Figure 12.1 only gives a symbolic representation of such a space.) It is referred to as μ–space. The state of translational motion of a molecule at any instant is completely specified by a representative point in this hypothetical space and the state of a system of particles corresponds to a certain distribution of points in phase space. For a system of N molecules, the instantaneous state is represented by a set of N points. As the position and momentum change with time, all these points may undergo extremely complicated motions in this space.

Note that the notion of phase space provides geometrical framework of some sort to statistical mechanics and helps to minimise abstraction. You will agree that it is not possible to draw such a space on a plane and for this reason, phase space should be considered a purely mathematical concept.

The uncertainty principle helps us to elaborate what we mean by a point in phase space. Suppose we divide the phase space into small six-dimensional cells of sides Δx, Δy, Δz, p_z. If we reduce the size of the cells, we approach more and more closely to the limit of a point in phase space. However, the volume of each of these cells is

$$H = \Delta x \Delta y \Delta z \Delta p_x \, \Delta p_y \, \Delta p_z$$

According to the uncertainty principle, the uncertainty in position and momentum coordinates in two-dimensional $x - p_x$ phase space is connected as

$$\Delta x \Delta p_x \geq \hbar \tag{12.1a}$$

$\frac{h}{2\pi}$; h is Planck's constant.

Identical relations hold for y and z-components:

$$\Delta y \Delta p_y \geq \hbar \tag{12.1b}$$

$$\Delta z \Delta p_z \geq \hbar \tag{12.1c}$$

On combining Eqs. (12.1 a–c), we can say that

$$H \geq \hbar^3 \tag{12.2}$$

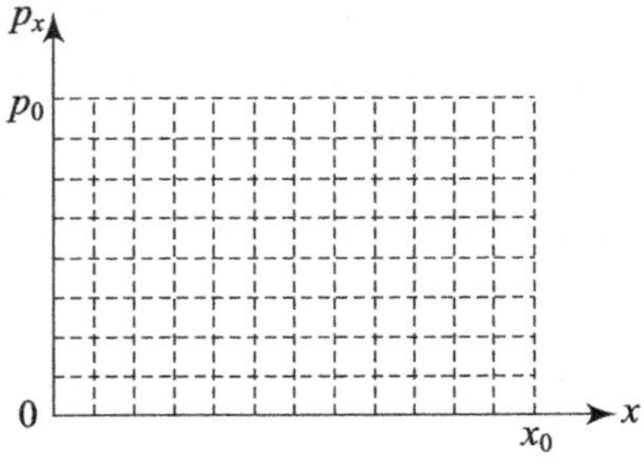

Fig. 12.3 The total number of states available to a particle confined between $0 \le x \le x_0$ and $0 \le p_x \le p_0$.

We can now conclude that a point in phase space is a six-dimensional cell (or 'quantum box') whose volume is of the order of $\hbar^3$. A more detailed analysis shows that each cell in phase space has volume h^3, which is in conformity with the uncertainty principle argument since $h^3 >$. It means that a particle in phase space will be located somewhere in such a cell centred at some location (x, y, z, p_x, p_y, p_z) instead of being precisely at the point itself.

For simplicity, let us consider a particle moving along the x-direction. Let it be confined between $x = 0$ and $x = x_0$ and have its – component of momentum between $p_x = 0$ $_0$ (Fig. 12.3). The number of states available to the particle in two-dimensional space is given by

$$n = \frac{\text{Total area}}{\Delta x\ \Delta p_x} = \frac{x_0\, p_0}{h} \tag{12.3}$$

Similarly, we can say that the number of states available in six-dimensional phase space

$$n = \frac{\text{Total six-dimensional volume}}{\Delta x \Delta y \Delta z\ \Delta p_x\, \Delta p_y\, \Delta p_z} = \frac{V_r\, V_p}{h^3} \tag{12.4}$$

and V_p denote the volumes in coordinate and momentum space, respectively. The number of states available in the six-dimensional volume element $d^3r\, d^3p$ is given by

. Mathematically, we can write

$$\text{Number of states} = \frac{dx\, dy\, dz\, dp_x\, dp_y\, dp_z}{h^3} \tag{12.5}$$

From this we can say that the number of quantum states included in any interval of any of the coordinates is directly proportional to the length of the interval.

To give you an appreciation of the numbers, we give below a solved example. Go through it carefully.

Example 12.3 Calculate the number of quantum states available to the following:

A particle is moving in one dimension. It is confined to 10^{-5}m of space and its momentum lies between -10^{-25} kg ms^{-1} and 10^{-25} kg ms^{-1}.

A proton inside a nucleus (radius = 10^{-14}m) whose momentum cannot exceed kg ms^{-1}.

(a) From Eq. (12.3), we recall that the number of quantum states available to a particle moving in one dimension is given by

$$n = \frac{x_0\, p_0}{h} \tag{i}$$

$= 10^{-5}$ m, and $p_0 = 2 \times 10^{-25}$ kg ms^{-1}. Hence,

$$n = \frac{(10^{-5}\ \text{m}) \times (2 \times 10^{-25}\ \text{kg m s}^{-1})}{(6.626 \times 10^{-34}\ \text{Js})} = 3{,}000$$

Since a proton inside a nucleus is free to move in 6 – D space, from Eq. (12.4), we can write

$$n = \frac{V_r V_p}{h^3} \tag{ii}$$

We know that $V_r = \frac{4\pi}{3} r_0^3$ with $r_0 = 10^{-14}$ m and $V_p = \frac{4\pi}{3} p_0^3$ with $p_0 = 10^{-19}$ kg m s^{-1}. On substituting this data in Eq. (ii), we get

$$n = \frac{[(4\pi/3) \times 10^{-42}\ \text{m}^3] \times [(4\pi/3) \times 10^{-57}\ \text{kg}^3\,\text{m}^3\,\text{s}^{-3}]}{(6.626 \times 10^{-34}\ \text{Js})^3} = 600$$

Now you should answer a practise problem.

Problem 12.3 Calculate the number of quantum states available to the following: An electron moving in one dimension is confined within 10 Å. Its speed does not m s^{-1}.

A particle within a rectangular box (1 cm × 1 cm × 2 cm) whose momentum does not exceed 4g cm s^{-1}.

(a) 27 (b) 4.45×10^{80}

In our study of statistical mechanics, we are very often required to obtain a summation over all the states available to a particle. In general, quantum states are discrete, and the spectra of accessible states for actual systems are vastly different. Yet it is possible to group quantum states in broad categories depending on the similarities exhibited by them.

For many physical systems which include photons, vibrations of solids (phonons), electrons in the conduction band, helium atoms in liquid state, etc., the separation between neighbouring states is much less than $k_B T$ (= 2.6×10^{-2} eV at 300 K). For air molecules, the separation between neighbouring states is about 10^{-23} eV. In such cases, it is possible to replace summation over discrete states by integration:

$$\sum_{\text{states}} \rightarrow \int \frac{d^3 r\, d^3 p}{h^3} \tag{12.6}$$

In actual applications, it often becomes difficult to ascertain the region of phase space over which the above integration has to be carried out. Several constraints on the position and momentum coordinates of the particle come into play due to the presence of interactions. For example, an atom in a crystalline solid is held to the particular site where its potential energy (due to interactions) is minimum. Similarly, interactions may constrain the momenta of some particles.

When the spectra of accessible states occur in bands, the separation of neighbouring states in each band is extremely small but the bands may be separated by rather large gaps. In such cases, we use the above result for the width of individual band only.

If the interactions with the neighbours can be ignored, as in a perfect gas, the particles will have access to the entire domain of the phase space. If the system is confined to V and integration is carried out over all momenta directions (i.e., with p sinθdθdϕ where θ varies from 0 to π and ϕ varies from 0 to 2π, we get

$$\iiiiiint \frac{d^3 p\, d^3 r}{h^3} = \left(\frac{4\pi V}{h^3}\right)\int p^2\, dp \tag{12.7}$$

is volume in Cartesian space. For non-relativistic particles

$$\varepsilon = \frac{p^2}{2m}$$

$= m$dε and

$$p^2\, dp = \sqrt{2m\varepsilon} \times m d\varepsilon$$
$$= \sqrt{2}\, m^{3/2}\, \varepsilon^{1/2}\, d\varepsilon$$

Then we get

$$\sum_{\text{state}} = \frac{2\pi V}{h^3}(2m)^{3/2}\int \varepsilon^{1/2}\, d\varepsilon \tag{12.8}$$

At this stage, we introduce another important factor called the *density of states*, which is defined as the number of accessible states per unit energy. It is denoted by $D(\varepsilon)$. So we have

$$\sum_{\text{states}} = \int D(\varepsilon)\, d\varepsilon \tag{12.9}$$

where the range of integration is defined over all values of available energies. On comparing Eqs. (12.8) and (12.9), we can say that the density of states for a system of non-interacting, non-relativistic particles is given by

$$D(\varepsilon)\, d\varepsilon = \frac{2\pi V}{h^3}(2m)^{3/2}\, \varepsilon^{1/2}\, d\varepsilon \tag{12.10}$$

Note that the form of the function $D(\varepsilon)d\varepsilon$ may differ considerably from that given by Eq. (12.10) if the mutual interactions between the particles are not negligible. We will consider these in detail a little later.

PECIFICATION OF THE STATE OF A SYSTEM

The specification of the state of a system demands that we should be able to describe the possible outcomes of an experiment involving it. For example, the state of four coins after any toss can be described by the head-tail configuration of each. The spin state of four electrons may be described by identifying the electrons in spin-up and spin-down states. The state of an electron in hydrogen atom can be specified by indicating the in the equation $E_n = -13.6/n^2$. (n is called the *principal quantum number* and has integral values, which define the energy level/state in which an electron can exist. $= 1$ corresponds to the lowest energy state called ground state. All other values correspond to excited states.) Let us now consider some other specific examples.

12.4.1 A Particle in One-dimensional Box*

Let us consider a particle of mass m, which is free to move in one dimension in the range $0 \leq x \leq L$, as shown in Fig. 12.4. It means that within this range, the potential energy of this particle is zero and elsewhere the potential energy is infinite. This motion is described by 1-D Schrödinger equation:

$$\frac{d^2\psi(x)}{dx^2} + \frac{2mE}{\hbar^2}\psi(x) = 0$$

whose solution is given by

$$\psi(x) = A \sin kx + B \cos kx \qquad (12.11)$$

$$= \frac{2\,mE}{\hbar^2}.$$

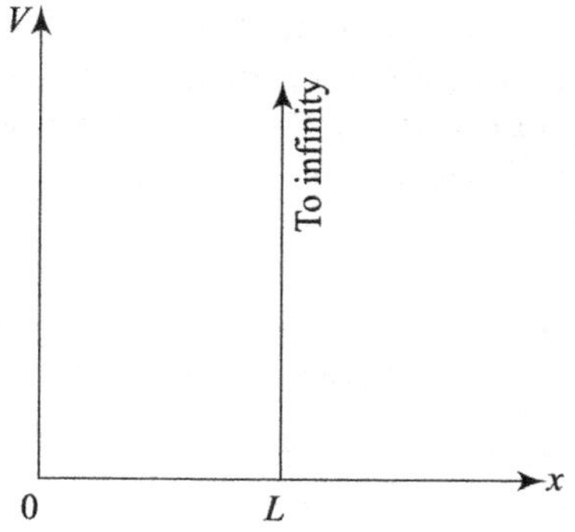

Fig. 12.4 A one-dimensional box.

We now use boundary conditions to determine the constants A and B. Since the movement of the particle is forbidden outside the box, $\psi(x)$ is zero outside it**. Hence, we must have

$$\psi(x) = 0 \text{ at } x = 0 \text{ and } x = L$$

On applying these conditions to Eq. (12.11), we get

$$B = 0$$

$$k = \frac{n\pi}{L}, n = 1, 2, 3, 4 \ldots \qquad (12.12)$$

$$\psi(x) = A \sin\left(\frac{n\pi}{L}x\right) \qquad (12.13a)$$

$$E = \frac{n^2\pi^2\hbar^2}{2\,mL^2} \qquad (12.13b)$$

This result shows that the state of a particle in a 1-D box is specified by Eqs. (12.13a) and (12.13b) with n as a positive integer. Now refer to Fig. 12.5. An electron moving through a crystalline solid can be considered almost free between consecutive lattice points. However, it experiences a strong field at the lattice points. So the state of an electron between the lattice points can be described using the above model.

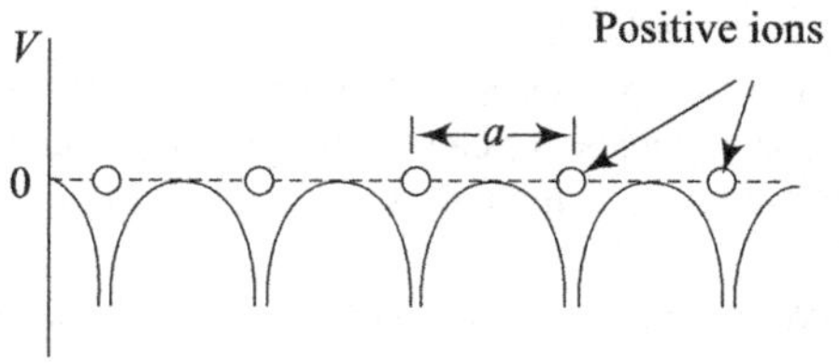

Fig. 12.5 Potential energy of an electron in

---

plot takes the shape of a box, hence, the name one-dimensional box.

ψ^2dx represents the probability that the particle is found in the range x to x + dx. Moreover, it is essential that ψ be continuous everywhere.

12.4.2 A Particle in a Three-dimensional Box

The one-dimensional box model can be extended to the case of a particle in a three-dimensional box (Fig. 11.6). The particle is free to move in the region defined by

$$0 \le x \le L_x, 0 \le y \le L_y, \text{and } 0 \le z \le L_z$$

By solving the 3-D Schrödinger equation, you must satisfy yourself that its state is specified by

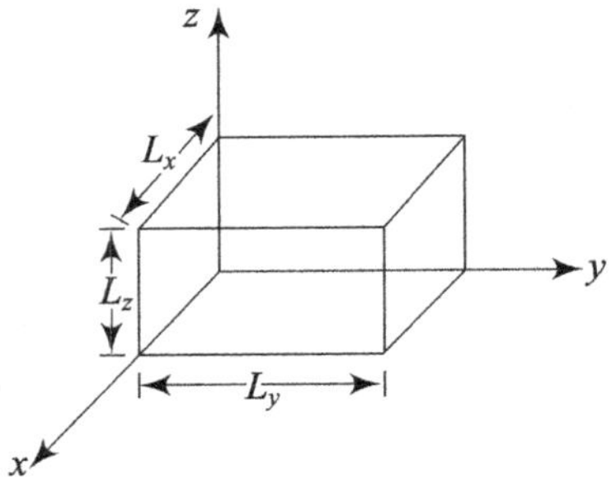

Fig. 12.6 A particle in a three-dimensional box.

$$A \sin\frac{\pi n_x}{L_x} \sin\frac{\pi n_y}{L_y} \sin\frac{\pi n_z}{L_z} \qquad (12.14a)$$

$$= \frac{\pi^2 \hbar^2}{2m}\left(\frac{n_x^2}{L_x^2} + \frac{n_y^2}{L_y^2} + \frac{n_z^2}{L_z^2}\right) \qquad (12.14b)$$

n_y, n_z take values 1, 2, 3, 4 …

V, we can write $L^2 = V^{2/3}$. Hence, the expression for energy level of a free particle in a cubical box can be rewritten as

$$E = \frac{\pi^2 \hbar^2}{2mV^{2/3}}(n_x^2 + n_y^2 + n_z^2) = Cn_j^2 \qquad (12.14c)$$

where constant $C = \frac{\pi^2 \hbar^2}{2mV^{2/3}} = \frac{h^2}{8mV^{2/3}}$ and $n_j^2 = (n_x^2 + n_y^2 + n_z^2)$. Note that energy level of a free particle depends on the quantum number n_j as well as the volume. If volume decreases, the value of E increases. The lowest energy level (j = 1) will correspond to

$= 1$. Then $n_1^2 = 3$ and $E_1 = \frac{3h^2}{8mV^{2/3}}$. Note that there is only one set of quantum

n_x, n_y, n_z) which correspond to this energy level. Such a state is said to be *non-*

. However, j = 2 could correspond to any one state defined by the following three combinations of n_x, n_y, n_z: (2, 1, 1), (1, 2, 1) (1, 1, 2). In each state, $n_2^2 = 6$ and

$\frac{2}{mV^{2/3}}$. It means that three different states correspond to the same energy level. Such a level is said to be *three-fold degenerate*. From this discussion of energy levels and degeneracies, we can say that a free particle in a box is an example of energy quantisation. Think of a schematic way of depicting energy levels, energy states and degeneracy of a

We would now like you to go through the following example carefully.

Example 12.4 Assuming $n_x = n_y = n_z$, estimate the value of n_x for an atom of helium in a cube of volume $V = 0.0224 \text{m}^3$* at STP. The mass of a helium atom is 6.65×10^{-27} kg.

From Eq. (12.14c), we recall that the energy of a particle of mass m in a cubical box of side L and volume V is given by

* At STP, a volume equal to 0.0224m^3, i.e., 22.4 litres contains one mole of helium atoms.

$$E = \frac{\pi^2 \hbar^2}{2\,mL^2}\, n_j^2 \tag{i}$$

V, we can write $L^2 = V^{2/3}$. Hence the energy of jth level can be rewritten as

$$E_j = C\, n_j^2 \tag{ii}$$

where constant $C = \dfrac{\pi^2 \hbar^2}{2\,mV^{2/3}} = \dfrac{h^2}{8\,mV^{2/3}}$.

The mass of a helium atom is 6.65×10^{-27} kg and $V = 0.0224\ \text{m}^3$. On substituting the , m and V, we get $C = 1.04 \times 10^{-40}$ J.

From Sec. 1.3, we recall that the mean energy of a particle at an absolute temperature $_BT$. Since $k_B = 1.38 \times 10^{-23}\ \text{JK}^{-1}$, the mean energy of an atom of helium gas at

$$E = \frac{3}{2} \times (1.38 \times 10^{-23}\ \text{J K}^{-1}) \times (273\ \text{K}) = 5.65 \times 10^{-21}\ \text{J}$$

So, on substituting the value of C and E in Eq. (ii), we can write

$$n_x^2 + n_y^2 + n_z^2 = \frac{5.65 \times 10^{-21}\ \text{J}}{1.04 \times 10^{-40}\ \text{J}} = 5.43 \times 10^{19}$$

$n_y = n_z$, we get

$$n_x^2 = 1.81 \times 10^{19}$$

$$n_x = 4.3 \times 10^{9}$$

This result shows that a helium atom in a cubical box of volume $V = 0.0224\ \text{m}^3$ at STP is likely to be in a single particle quantum state having quantum numbers in the range

.

You may now like to answer a practise problem.

Problem 12.4 For a single particle in a cubical box, calculate energies of the ground state and the first two excited states in terms of C.

, $6C$, $9C$

N-Particles in a Three-dimensional Box

Now let us consider a system of N particles, each of mass m, confined in the 3-D box. If the interaction between the particles is negligible, the total energy E of the system is equal to the sum of the energies of individual particles, i.e. equal to the internal energy and we can write

$$E = E_1 + E_2 + E_3 + \ldots E_N \tag{12.15}$$

is energy of the ith particle. The *state* of each particle is specified by its three quantum numbers n_{ix}, n_{iy} and n_{iz} and the energies corresponding to different possible values.

Its energy will then be given by an expression analogous to Eq. (11.14b). This implies that the number of possible quantum states of the entire system will be specified by the N quantum numbers

$$(n_{1x}, n_{1y}, n_{1z}, n_{2x}, n_{2y}, n_{2z}, \dots n_{Nx}, n_{Ny}, n_{Nz})$$

The corresponding energy is given by Eq. (12.15) where for each i, E_i has the form given by Eq. (12.14b).

We may now conclude that the state of an ideal gas consisting of N non-interacting particles can be described by this model with a fair degree of confidence.

The total number of particles in an energy state corresponding to a particular level is *occupation number* and the sum of occupation numbers N_j over all levels gives the total number of particles N:

$$N = \sum_j N_j \tag{12.16}$$

Since all particles in different states corresponding to any level have the same energy ε_j, the total energy of the particles in the jth level will be $\varepsilon_j N_j$ so that the total energy of the

$$E = \sum_j \varepsilon_j N_j \tag{12.17}$$

In statistical mechanics, we are concerned with the determination of the possible distributions of particles, in*distinguishable* as in a gas or *distinguishable* as in a crystal lattice, among energy levels and energy states. This brings us to the discussion of macrostates and microstates.

MACROSTATES AND MICROSTATES

macrostate of a system is that state which can be described by specifying only those quantities which can be determined by macroscopic measurement, without any reference whatsoever to the microscopic details. For instance, simple count of the number of particles/phase points in each energy level/cell defines a macrostate of the system. It essentially means that a macrostate does not require any information about the identity of constituent atoms/molecules of the system and we concentrate on their number in different levels: N_1 in level 1, N_2 in level 2, and so on.

The state of a system described in the minutest possible details of each atom/molecule/particle is said to be *microstate*. That is, we have to specify to which sub-level each molecule belongs temporarily. For indistinguishable particles, *a specification of the total number of particles in each energy state defines a* . To visualise this, refer to Fig. 12.7, which shows a set of three energy $= 1, 2, 3\dots$; their degeneracies and their occupation numbers N_j. The

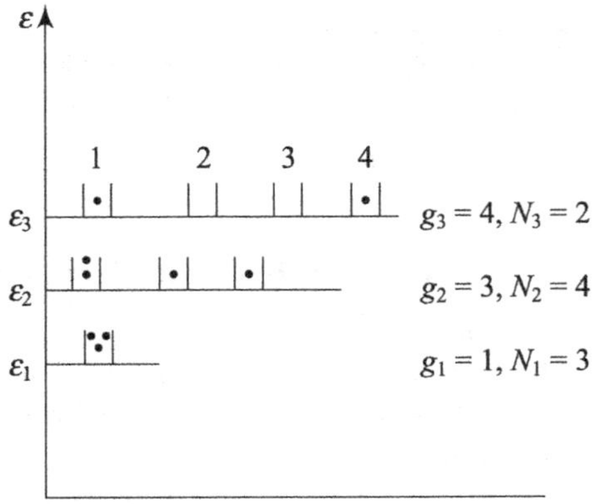

Fig. 12.7 Schematic representation of energy levels, their degeneracies and occupation numbers.

energy levels are shown as shelves at different elevations and energy states are shown by open boxes on each self. These are depicted as (1), (2), (3),.... The degeneracy of a level is equal to the number of boxes on the corresponding shelf.

Suppose the particles are indistinguishable and we first concentrate on level 3. The microstate of the assembly is specified by saying that there is one particle in state (1) as well as in state (4) and there is no particle in states (2) and (3). In level 2, there are two particles in state (1) and one particle in each of the states (2) and (3) and so on. If one or both the particles in level 3 were in states other than (1) and (4), the microstate would have been different but the macrostate would be unchanged since $N_3 = 2$. In fact, *many microstates may correspond to the same macrostate.*

For a system of distinguishable particles, a specification of the energy state and energy level of each particle defines a microstate. That is, we have to specify the number of particles in each state as also which particles they are. Thus, if distinguishable particles are denoted as *a*, *b*, *c*,... , we must specify whether *a* or *b* or *c* is in states (1) or (4) of level 3. Similarly, which particle(s) is (are) in states (1), (2) and (3) and so on. Recall that in a system of indistinguishable particles, microstate remains the same irrespective of the particle occupying a state in a particular level, whereas for a system of distinguishable particles, microstate is considered different if two particles switch their states/levels. However, if an energy state is occupied by more than one particle, the change in their ordering does not induce change in the microstate. Obviously, the number of microstates that are considered different, for a given set of occupation numbers, is significantly larger for an assembly of distinguishable particles than indistinguishable particles.

These concepts are further illustrated by considering a system of four distinguishable *b*, *c* and *d*. We distribute these into two exactly identical compartments in an open box. Since the compartments are alike, the *a priori* probability of particles going into either compartment will be the same, equal to one-half. (The word *a priori* implies a notion which exists in our mind prior to and independent of the outcome of the observations we are about to make. The assumption of equal probability for equally likely events constitutes the principle of equal a priori probability.) The five possible ways in which these particles can be distributed in two compartments are: (0, 4), (1, 3), (2, 2), (3, 1) and (4, 0). Each of these distributions constitutes a macrostate. We can generalise this result to *n* distinguishable gas molecules and the various macrostates will be (0, *n*), (1, *n* – 1), (2, *n* – 2), ..., (*n* – 1, , 0). That is, the total number of macrostates for *n* particles is (*n* + 1).

Proceeding further, let us examine the arrangement (combination) of particles in each compartment corresponding to different macrostates. Refer to Table 12.1. You will note that the distribution (0, 4) has only one combination, whereas distribution (1, 3) has four distinct combinations. Similarly, the distribution (2, 2) has six distinct combinations and so on. (You can easily compute these numbers using the relation ${}^nC_r$, where $n = 4$ and r can take values 0, 1, 2, 3, 4.) Each of these combinations for a particular macrostate constitutes *microstate*. Note that the compartment occupied by a particular set of particles leads to a different microstate. We can, therefore, conclude that *a given macrostate may consist of a number of microstates.* The total number of microstates when four particles are to be distributed in two identical compartments is $16 = 2^4$ For *n* distinguishable gas molecules, the number of microstates will be 2^n.

Table 12.1 *Combination of particles in each compartment corresponding to different macrostates.*

Macrostate	Possible arrangements		No. of microstates
	Compartment 1	Compartment 2	
0, 4	0	*abcd*	1
1, 3	*a*	*bcd*	4
	b	*cda*	
	c	*dab*	
	d	*abc*	
2, 2	*ab*	*cd*	6
	ac	*bd*	
	ad	*bc*	
	bc	*ad*	
	bd	*ac*	
	cd	*ab*	
3, 1	*bcd*	*a*	4
	cda	*b*	
	dab	*c*	
	abc	*d*	
4, 0	*abcd*	0	1

You may now like to answer a practise problem.

Problem 12.5 **(a)** 3 particles are to be distributed in two compartments with equal a priori probability. Calculate the number of macrostates and microstates, if the particles are (i) distinguishable, and (ii) indistinguishable.

A public school has a total enrolment of 1500 students in its secondary and senior secondary classes. Define its macrostates and microstates.

(i) macrostates 4; microstates 8; (ii) macrostates 4, microstates 4; **(b)** Macrostates are determined by the total number of students in different grades, degeneracy by the classes/sections in each grade and microstates are defined by the combination of number of students in each section.

In Chapter 9, we learnt that a system has an inherent tendency to approach equilibrium. Let us now redefine equilibrium from the point of view of probability. We begin by considering the familiar example of opening a bottle of perfume, say at one end of a room. In view of the preceding discussion, we can say that on being released, the quantum states located near the mouth of the bottle have relatively higher probabilities of containing the molecules of the perfume than those at the other end of the room. However, with passage of time, the perfume molecules move on and fill the entire room. In other words, the probabilities keep changing till such time that the molecules are equally likely to be found anywhere in the room. Then the system is said to be in equilibrium. So, *an isolated system is said to be in equilibrium when the probabilities of it being in the various accessible states do not vary with time.*

Probability Calculations

We now know that the concept of equilibrium of a state and the probability of its occurrence are related. It is therefore important to know how the probabilities are calculated. This calculation forms the backbone of the analysis of all statistical problems. We will make probability calculations on the basis of the following fundamental postulate:

An isolated system in equilibrium is equally likely to be in any of its accessible states.

Note that accessible states are those states which are consistent with the given constraints of a system. If we put the constraint that no compartment should remain empty in the example discussed in Table 12.1, the macrostates (0,4) and (4,0) will not be accessible. The system will then have only three macrostates with their associated microstates.

As such, this postulate seems to be quite reasonable. The predictions made on the basis of this postulate have been put to test for different systems and every time the postulate has met with phenomenal success. This has confirmed our faith in this postulate.

If we denote the number of states accessible to a system by Ω_0, the probability of finding it in any one of them is (Fig. 12.8)

$$P = \frac{1}{\Omega_0}$$

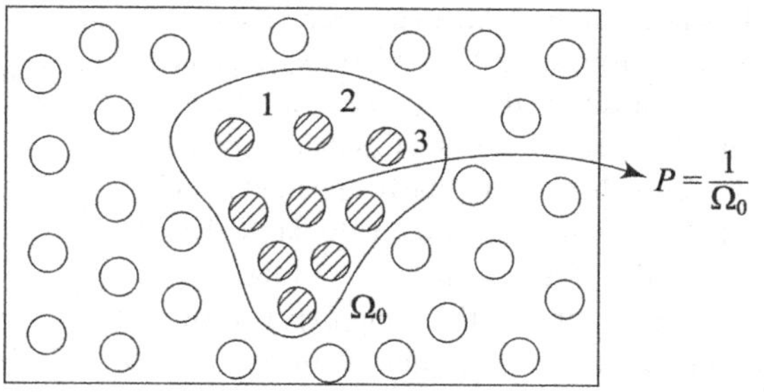

Schematic representation of probability.

From this we can say that the probability of finding the system in a certain set of accessible states is given by

$$P_i = \frac{\Omega_i}{\Omega_0} \tag{12.18}$$

is the number of states in the set.

$\sum\limits_{=1}^{n} P_i$ of the probabilities over all states is equal to one:

$$\sum_{i=1}^{n} P_i = 1 \tag{12.19}$$

In statistical theory, we wish to know the mean value of a physical quantity. The mean value of a parameter y is given by

$$\bar{y} = \sum_{i=1}^{n} P_i\, y_i = \frac{1}{\Omega_0} \sum_{i=1}^{n} \Omega_i\, y_i \tag{12.20}$$

It is important to note that $\bar{y}$ represents the (central) value about which the several values) of the parameter y are distributed.

Problem 12.6 The electrons in plasma are confined within a fixed volume. The upper limit of their momenta is 10^{-22} kg ms^{-1}. What is the probability that the momentum of a particular electron is less than 10^{-24} kg ms^{-1}? (*Hint*: $\Omega_0 = V_r V_p / h^3$)

The average in Eq. (12.20) is referred to as the ensemble average. In fact, the concept of ensemble is very important in statistical mechanics and in the paragraphs that follow, we will discuss it in detail.

YPES OF ENSEMBLES

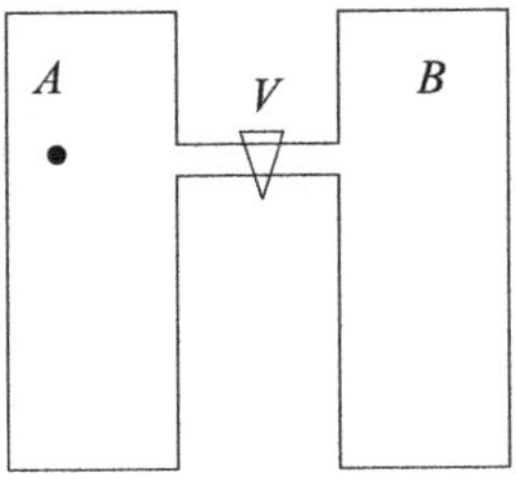

Fig. 12.9 Two identical containers joined by a narrow tube fitted with a valve.

The concept of ensemble was put forward by Gibbs. Following him, let us consider a simple situation wherein two identical A and B are connected by a narrow tube fitted with (Fig. 12.9). Let us assume that A contains a single molecule and B is empty. Suppose that one second after the is opened, the probability of finding the molecule in B is 0.4. What does this claim imply? Obviously, this probability varies with time. It is not even possible to ascertain the time-dependent behaviour of probability by performing repeated trials with a single apparatus to find out whether the molecule or not. So this claim essentially implies that if you had set up a large number of replicas of the above apparatus, each with the molecule in A, then on opening the valve at $t = 0$, the molecule would be found in B at $t = 1$ s in 40% of the cases. (Alternatively, it can be said that if the same experiment is performed several times, we expect to find the molecule in B at $t = 1$ s, on opening the $= 0$, in 40% of the trials.) Such a set of mental constructs of the replicas (or repeats) of the experiment was termed an *ensemble* by Gibbs. An ensemble average thus refers to the average calculated over the population of a suitably chosen ensemble at the start of the experiment.

Note that an ensemble is a mathematical device; there is nothing physical about it. However, it provides us with an opportunity to follow a probabilistic approach to describe the state of the system at a given time. Resorting to this approach implies that we do not require detailed knowledge of the state of the system. If we knew exactly the position and velocity of a molecule at $t = 0$ and the nature of forces acting on it, we could work out the trajectory of the molecule and easily locate its position (i.e., whether in A or B) at any point of time. In practise, we rarely have access to such a detailed knowledge about a system. So we represent its state by a suitably chosen ensemble of systems which corresponds to our knowledge of microstates.

Suppose you know that at $t = 0$ the speed of the molecule is v. It would then be reasonable to assume that in the starting ensemble ($t = 0$), the molecule is uniformly distributed over the volume of A. Its direction of motion is also distributed uniformly. Then, in principle, the trajectory can be calculated for each state in the starting ensemble. Hence, we can also calculate the number of cases in which the molecule would be in B after one second. In other words, we make an assumption about the initial probabilities. Then we use arguments, based on the laws of mechanics, to find how the ensemble, and hence the probabilities, vary with time.

We adopt that procedure in statistical thermodynamics where a thermodynamic state is described in terms of probabilities, and the theoretical value of a thermodynamic variable is obtained in terms of its mean value averaged over a suitably chosen ensemble.

Ensembles may be classified according to the behaviour of their constituents with respect to the surroundings. From Chapter 5, you will recall that a system can interact with its surroundings in three ways: (a) thermal, (b) mechanical and (c) diffusive. During the interaction, the member of an ensemble may be involved in none or all or any combination of these. Of the various possibilities, three are most likely: (a) *micro-canonical*, (b) *canonical* *grand-canonical* ensemble*. The members of a micro-canonical ensemble do not interact at all with their surroundings. An example of such an ensemble is a large number of identical tea-pots stored in perfectly insulating thermos flasks (Fig. 12.10a). The members of a canonical ensemble can exchange thermal energy and work with a reservoir. That is, they can interact mechanically as well as thermally with the reservoir. The water molecules in a teapot (Fig. 12.10b) are good examples of a canonical ensemble. The members of a grand canonical ensemble can exchange heat, work as well as matter (particle) with a reservoir. In other words, they can interact thermally, mechanically and diffusively with the reservoir. A large number of identical grains of sugar in a tea-pot undergo all these interactions (Fig. 12.10c).

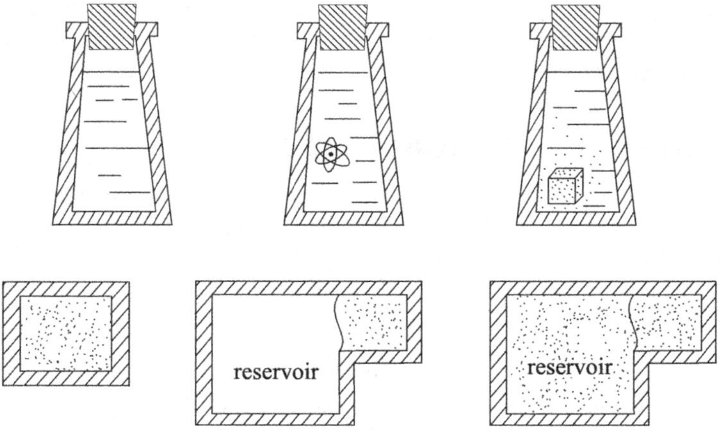

(a) (b) (c)

Types of ensembles: (a) Micro-canonical, (b) Canonical and (c) Grand-canonical.

A micro canonical ensemble is an idealised concept and does not exist in practice. The systems that we come across generally interact mechanically and thermally with their surroundings. Hence, they constitute canonical ensembles. An ice cube in water or common salt in water solution constitutes a grand-canonical ensemble. However, we will not go into further details of canonical ensembles here.

Problem 12.7 Comment on the following state: "Liquid nitrogen in a closed laboratory Dewar flask approximates to being a member of a micro-canonical ensemble."

*The micro-canonical, canonical and grand-canonical ensembles respectively correspond to isolated, closed and open systems in thermodynamics. As such, statistical ensembles are mathematical idealisations.

At the beginning of Sec. 12.5, we stated the fundamental postulate of statistical mechanics. To that, we add the following three:

A given macrostate is realised through an enormously large number of microstates.
The principle of conservation of energy holds. In this connection the energy-time uncertainty relation should be kept in mind. According to it, if the lifetime of a state , the energy of that state is uncertain by a factor $\Delta E(= \hbar/\Delta t)$.
*The equilibrium values of the macroscopic variables are given by the ensemble average of the corresponding quantities**.

Now that we are familiar with the calculation of probability we shall attempt to relate it to some thermodynamic quantities. To this end, we recall that the states of a system are very conveniently represented by its energy. So we first seek a connection between the value of energy and the corresponding number of accessible states of a system.

Thus we have seen that in *equilibrium two interacting macroscopic systems will have such a distribution of energy between them that the number of available states is a* . This result is completely independent of the nature of interactions. We can now state the conditions of attainment of equilibrium as follows:

When two interacting macroscopic systems are in equilibrium, the system parameter will take values such that the number of states available to the combined system is a maximum. This condition suggests a connection between entropy and probability. In Chapter 7, we defined entropy on the basis of Carnot's reversible cycle. We shall now define entropy from the standpoint of statistical mechanics.

NTROPY AND PROBABILITY

From Chapter 7, we recall that entropy characterises disorder in a system. Since equilibrium is the most disordered state, the entropy of a system is maximum when it is in equilibrium. From the preceding discussion, we can also say that equilibrium is the most probable state. This suggests a connection between entropy and thermodynamic probability. Let us express it mathematically as

$$S = f(\Omega_0) \tag{12.21}$$

The nature of the function f is unknown at this stage but we shall soon discover it.

Let us now consider two completely independent systems having entropies S_1 and S_2. We know that entropy is an extensive (additive) quantity. This means that the entropy of the combined system will be sum of the entropies of individual systems:

$$S_0 = S_1 + S_2 \tag{12.22a}$$

On the other hand, if Ω_1 and Ω_2 are probabilities of the individual systems, the probability of the combined system is obtained as a product of probabilities of the individual systems and we can write

$$\Omega_0 = \Omega_1 \Omega_2 \tag{12.22b}$$

Physically, we can say that for every microstate of one system, the other may be in any one of its possible microstates.

We determine the form of the function f using the information provided by Eqs. (12.22 a, b). We expect the function f to be such that S increases with Ω and translates a product

*This is also known as the *ergodic hypothesis*

into a sum. Mathematically, we can write

$$S_1 + S_2 = f(\Omega_1) + f(\Omega_2) = f(\Omega_1 \Omega_2) \tag{12.23}$$

Proceeding further, we take partial derivatives of both sides of Eq. (12.23) with respect , keeping Ω_2 fixed. Since $f(\Omega_1)$ is a function of Ω_1 only, its partial derivative with Ω_1 will be equal to its total derivative

$$\left(\frac{\partial f(\Omega_1)}{\partial \Omega_1}\right)_{\Omega_2} = \frac{df(\Omega_1)}{d\Omega_1} \tag{12.24}$$

The partial derivative of $f(\Omega_2)$ with respect to Ω_1 is zero, since Ω_2 is constant. The partial derivative of $f(\Omega_1\Omega_2)$ with respect to Ω_1 is equal to the total derivative of $f(\Omega_1\Omega_2)$ with respect to its argument, multiplied by the partial derivative of its argument with respect , which is simply the constant Ω_2. Thus, if we denote the derivative of $f(\Omega_1\Omega_2)$ with respect to its argument $\Omega_1\Omega_2$ by $f'(\Omega_1\Omega_2)$, we can write

$$\frac{df(\Omega_1)}{d\Omega_1} = \Omega_2 f'(\Omega_1 \Omega_2) \tag{12.25}$$

Similarly, we differentiate Eq. (12.23) with respect to Ω_2 keeping Ω_1 fixed and following the same line of reasoning, we can write

$$\frac{df(\Omega_2)}{d\Omega_2} = \Omega_1 f'(\Omega_1 \Omega_2) \tag{12.26}$$

We now multiply Eq. (12.25) by Ω_1 and Eq. (12.26) by Ω_2. This gives

$$\Omega_1 \frac{df(\Omega_1)}{d\Omega_1} = \Omega_2 \frac{df(\Omega_2)}{d\Omega_2} = \Omega_1 \Omega_2 f'(\Omega_1 \Omega_2)$$

Note that in this expression, the first term is a function of Ω_1 only and the second term is a function of Ω_2 only. Hence, we can write

$$\Omega_1 \frac{df(\Omega_1)}{d\Omega_1} = \Omega_2 \frac{df(\Omega_2)}{d\Omega_2} = \text{Constant} \tag{12.27}$$

We can readily integrate it to obtain

$$f(\Omega_1) = k \ln \Omega_1 + C_1$$

$$f(\Omega_2) = k \ln \Omega_2 + C_2$$

In general, we can write

$$f(\Omega) = S = k \ln \Omega + C \tag{12.28}$$

This result shows that the function f is logarithmic and constant k is same for all systems. This relation was derived by Boltzmann but he did not determine the constants k and C. Subsequently, Planck used the fact that at absolute zero, the entropy is zero and $\Omega = 1$ so that the constant C can be taken to be zero. He identified Ω as thermodynamic probability. The constant k was identified as Boltzmann constant, k_B by considering an equivalent situation in thermodynamics. Let us now understand how it was done.

Identification of Constant *k*

Refer to Fig. 7.8 again and consider *irreversible free expansion* (Joule expansion) of one mole of an ideal gas at pressure p_1 and volume V_1 into an evacuated chamber of volume V_2. Let the final pressure be p_2 and final volume be $V_1 + V_2$. From Chapter 7, you may recall that no change in temperature occurs in Joule expansion, i.e., the expansion is isothermal but we wish to calculate change in entropy of the gas. After the expansion has occurred, the probability of finding any molecule in the container of volume V_1 will be $\dfrac{V_1}{V_1 + V_2}$. You may recall that the number of molecules in one mole of any gas is equal to Avogadro , and for independent events, the probability of a composite event is equal to the product of individual events. Therefore, the probability of finding a molecule in the container of volume V_1 will now be

$$\Omega_1 = \left(\frac{V_1}{V_1 + V_2}\right)^{N_A}$$

$_A$ molecule will necessarily be inside the chamber, we can say that the probability of finding any one molecule somewhere in the entire container of volume $V_1 + V_2$ will be

$$\Omega_2 = \left(\frac{V_1 + V_2}{V_1 + V_2}\right)^{N_A} = (1)^{N_A}$$

Using Eq. (12.28), we can write change in entropy in this case as

$$\Delta S = S_2 - S_1 = k \ln \Omega_2 - k \ln \Omega_1$$

$$= k \ln\left(\frac{\Omega_2}{\Omega_1}\right) \tag{12.29}$$

On substituting the values of Ω_1 and Ω_2 in Eq. (12.29), we get

$$\Delta S = k \ln\left(\frac{1}{V_1/(V_1 + V_2)}\right)^{N_A} = N_A k \ln\left(\frac{V_1 + V_2}{V_1}\right) \tag{12.30}$$

From Eq. (7.21a), we recall that change in entropy of a gas when it changes from the state defined by (V_1, T_1) to another state defined by $(V_1 + V_2, T_2)$ can be written as

$$\Delta S = C_V \ln\left(\frac{T_2}{T_1}\right) + R \ln\left(\frac{V_1 + V_2}{V_1}\right)$$

For an isothermal change, the first term on the right-hand side of this equation will drop out and we get

$$\Delta S = R \ln\left(\frac{V_1 + V_2}{V_1}\right) \tag{12.31}$$

On comparing Eqs. (12.30) and (12.31), we get

$$N_A k = R$$

$$k = \frac{R}{N_A} \tag{12.32}$$

From Chapter 1, you may recall that the ratio $\frac{R}{N_A}$ defines Boltzmann constant, which has $\times 10^{-23}$ JK^{-1}. Hence, we can now identify k with k_B and write Eq. (12.28) as

$$S = k_B \ln \Omega \tag{12.33}$$

Equation (12.33) is called *Boltzmann's entropy relation* and is one of the most fundamental relations of statistical mechanics. (This relation is engraved on the final resting place of Boltzmann in the central cemetery in Vienna.) It states that the entropy of a system is proportional to the logarithm of its thermodynamic probability. Therefore, statistically speaking, the universe always tends to change towards statistically more probable state. It may be mentioned here that Eq. (12.33) provides a bridge between statistical mechanics and thermodynamics. In fact, it opens the way for a more direct and elegant application of the techniques of probability theory to study widely diverse systems and obtain their thermodynamic properties.

It may be remarked here that for a macroscopic system, Ω is usually a very large $10^{10^{24}}$ but $\ln \Omega$ can be conveniently managed. To appreciate this, answer the following problem.

Problem 12.8 A system is made of three interacting systems A_1, A_2 and A_3 for $= 10^{10^{20}}$, $\Omega_2 = 3 \times 10^{10^{20}}$ and $\Omega_3 = 4 \times 10^{3\times10^{20}}$. Calculate Ω_0, S_1, S_2, S_3 and in terms of k_B.

$= 12 \times 10^{5\times10^{20}}$, $S_1 = S_2 = (2.3 \times 10^{20})\, k_B$, $S_3 = (6.9 \times 10^{20})\, k_B$,
$\times 10^{20})\, k_B$

Example 12.5 The volume of a system changes but the number of particles and internal energy remain constant. Show that the number of accessible states changes exponentially with volume.

We know that

$$p = T\left(\frac{\partial S}{\partial V}\right)_{E,\,N}$$

$k_B \ln \Omega$, we can write

$$p = k_B T\left(\frac{\partial \ln \Omega}{\partial V}\right)_{E,\,N}$$

If the number of accessible states changes from Ω_1 to Ω_2 corresponding to a change in V with fixed E and N, we get

$$p = k_B T \frac{\Delta(\ln \Omega)}{\Delta V}$$

$$\frac{p\Delta V}{k_B T} = \Delta(\ln \Omega) = \ln \Omega_1 - \ln \Omega_2 = \ln\left(\frac{\Omega_2}{\Omega_1}\right)$$

$$\left(\frac{\Omega_1}{\Omega_2}\right) = \exp[\, p\Delta V / k_B T \,]$$

Statistical Interpretation of Entropy

Statistically speaking, the entropy of a system consisting of a very large number of particles is proportional to the natural logarithm of the total number of microstates available to the system. Therefore, if only one microstate is available to an assembly, we have $\Omega = 1$, $= 0 = S$. The state of each particle can be uniquely specified and the system is said to be perfectly ordered. However, if more than one energy states become available to the system, $\Omega > 1$, and $S > 0$. Consequently, we can not specify the state of each particle uniquely. It means that the system has become disordered. Therefore, entropy of a system may be construed as a measure of disorder in the system.

The statistical interpretation of entropy helps us to gain greater insight into the meaning of absolute zero. According to Planck's statement of the third law of thermodynamics, the equilibrium entropies of all systems tend to zero as temperature approaches absolute $S \to 0$ as $T \to 0$. Therefore, we can say that systems in equilibrium are more perfectly ordered at absolute zero.

Classical and Quantum Statistics

A microscopic system can be statistically described in two different ways:

A single particle or a group of particles can occupy any of the several possible states (classical statistics).

A quantum state can be occupied by any of the several particles (quantum statistics).

The first of these alternatives caters to distinguishable particles whereas the second applies to indistinguishable particles. To understand this, imagine that a football match is being played. If footballer *A* (Messi/Maradona/Pele) is a striker and foot baller *B* (Platini/Falcao Beckenbauer) is a midfielder, the combination has a particular sense (striking power). But interchange positions, we get a different sense (reduced striking power). This is the classical way of looking at things. If we consider a gaseous system, then in this approach the focus is on the particle which can occupy several states* (Fig. 12.11a). But this cannot be said about electrons in an atom or photons in a cavity. For example, there is nothing to distinguish the configuration "electron–1 in $n = 1$ orbit and electron–2 in $n = 2$ orbit"

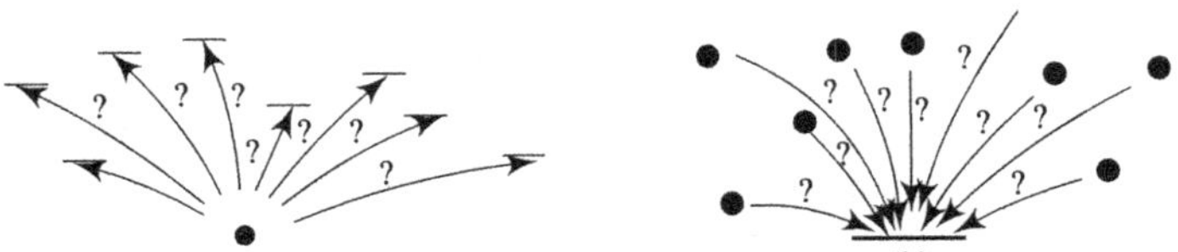

(a) (b)

Bars indicate states and solid circles indicate particles. (a) Classical picture and (b) Quantum picture.

*In this case, we assume that we are capable of identifying all particles individually.

from the configuration "electron–2 in $n = 1$ orbit and electron–1 in $n = 2$ orbit". That is, these particles are indistinguishable and this depicts the quantum picture, where the focus of attention is the state which can be occupied by any of the several particles (Fig. 12.11b). Before we consider classical and quantum distributions, go through the following example.

Example 12.6 A rubber band can be modelled as a polymer involving N molecules linked together end-to-end. The angle between successive links is equally likely to be 0° or 180°. Calculate the number of arrangements that give the overall length of $L = 2md$.

Let us assume that there are N^+ links of 0° and N^- links of 180°. Then,

$$N^+ + N^- = N$$

$$N^+ - N^- = 2m$$

On combining these relations, we can write

$$N^+ = \frac{N}{2} + m$$

$$N^- = \frac{N}{2} - m$$

Hence, the number of arrangements that give an overall length of $2md$ is given by

$$G(N, m) = \frac{N!}{N^+!N^-!} = \frac{2N}{\left(\frac{N}{2} + m\right)!\left(\frac{N}{2} - m\right)!}$$

12.9.1 Distribution Functions

Now that we have developed the necessary tools of statistical mechanics, we can apply them to obtain expressions for distribution functions for particles obeying classical (Maxwell–Boltzmann) and quantum (Bose–Einstein and Fermi–Dirac) statistics. For simplicity, we begin with classical distribution.

1. Calculation of Ω_{MB}: Maxwellian gas To obtain expression for Ω_{MB}, we consider N *distinguishable** particles enclosed in a volume V, each of which can have any energy. However, for mathematical convenience, we assume that these particles can be put in G (discrete) groups having energies $\varepsilon_1, \varepsilon_2, \varepsilon_3, \ldots, \varepsilon_G$ and let the number of particles in these $n_1, n_2, n_3, \ldots, n_G$, respectively. (Note that the assumption about discrete energy groups is made to simplify calculations, and should not be confused with any quantum mechanical significance, because ε can be made as small as we wish.) We divide the phase space into cells of volume H and the numbers of cells in which particles of various energy groups are distributed are $g_1, g_2, g_3, \ldots, g_G$. (Alternatively, we can say that there is more than one state of differing momenta or spin for each energy level; i. e. the energy levels *degenerate*. Though the terminology sounds quantum mechanical, it can be applied to classical theory as well. Mechanically, we can consider the phase space as a compartment, which has been divided into several cubicles.) This arrangement is tabulated as follows:

*The atoms of different elements and atoms localised in a crystal are obvious examples. However, even identical particles in an ideal gas are, in principle, supposed to be told apart.

1	2	3	...	i	...	G
ε_1	ε_2	ε_3	...	ε_i	...	ε_G
n_1	n_2	n_3	...	n_i	...	n_G
g_1	g_2	g_3		g_I	...	g_G

In Maxwell–Boltzmann statistics, we deal with

distinguishable non-interacting/weakly interacting particles, say atoms of a perfect monatomic gas,

(b) calculate the number of ways Ω in which we can distribute all particles among the groups and cells, and

(c) maximise ln Ω to obtain the most probable distribution subject to the following constraints:

(i) The total number of particles is conserved: $N = \sum_{i=1}^{G} n_i = \text{constant}$

(ii) The total energy of the system is conserved: $E = \sum_{i=1}^{G} \varepsilon_i n_i = \text{constant}$

(iii) There is no limit on the number of particles that can occupy a particular cell.

In classical statistics, we focus our attention on the particles and determine the number of ways in which the total number of particles, N can be distributed into various groups. The number of ways in which we can do so is equal to the number of permutations of things out of which n_1 are alike, n_2 others are alike, n_3 others are alike, and so on. To understand this, we note that the number of ways in which n_1 particles out of N will occupy the first group is

$$^{N}C_{n1} = \frac{N!}{n_1\,!(N-n_1)!} \tag{12.34}$$

Now we are left with $(N - n_1)$ particles. Out of these, let n_2 occupy the second group. This can be done in $N - {}^{n1}C_{n2}$ ways:

$$N - {}^{n1}C_{n2} = \frac{(N-n_1)!}{n_2\,!(N-n_1-n_2)!} \tag{12.35}$$

Hence, the total number of ways in which n_1 particles occupy the first group, n_2 particles occupy the second group and n_i particles occupy the ith group is obtained by multiplying terms like those contained in Eqs. (12.34), (12.35) and so on, i.e.,

$$\frac{N!}{n_1\,!(N-n_1)!}\,\frac{(N-n_1)!}{n_2!(N-n_1-n_2)!}\,\frac{(N-n_1-n_2)!}{n_3!(N-n_1-n_2-n_3)!}\cdots\frac{(N-n_1-n_2-\ldots-n_i)!}{n_i\,!(N-n_1-n_2\ldots n_i)!}\cdots$$

$$\frac{N!}{n_1\,!n_2\,!n_3\,!\ldots n_i\,!\ldots} = \frac{N!}{\prod_i n_i\,!} \tag{12.36}$$

$N \quad \sum_{i=1}^{G} n_i$, $0! = 1$ and $\prod_i$ denotes the product of n_i for all values of i. (It corresponds to the summation sign for the sum of terms of a series.)

Now, we have to determine the number of ways in which the particles can be distributed in the cells associated with different groups. Let us denote these probabilities by P_{MB} (for classical particles). Then $\Omega_{Cl} = W \times P_{MB}$.

To calculate the number of ways in which n_i particles can be distributed among g_i cells when there is no restriction on the number of particles occupying a cell, we recall that classical particles are taken as distinguishable. Since each particle of ith group has equal a priori probability of occupying any of the g_i cells, we can say that it can be distributed in ways. (Seen from this angle, g_i's can be considered the a priori probability.) So the total number of ways in which n_i particles can occupy g_i cells is $g_i^{n_i}$. Hence, the total number of ways in which g_1 cells are occupied by n_1 particles, g_2 by n_2, g_3 by n_3 and so on is given by

$$P_{MB} = g_1^{n_1} g_2^{n_2} g_3^{n_3} \ldots = \prod_i g_i^{n_i}$$

Hence, the total number of ways in which N distinguishable particles can be distributed groups and the particles in various groups are distributed into different cells is

$$\Omega_{MB} = \frac{N!}{\prod_i n_i!} \left(\prod_i g_i^{n_i}\right) \tag{12.37}$$

This defines thermodynamic probability of N distinguishable particles when there is no limit on the number of particles occupying a cell. The most probable distribution is obtained by maximising $\ln \Omega_{MB}$ subject to the conditions on the total number of particles and total energy of the system. However, we will illustrate it a little later.

If the particles are treated as indistinguishable, there is no limit on the number of particles occupying a cell and the energy is regarded a continuous variable, the expression for thermodynamic probability is obtained by dividing the expression contained in Eq. $N!$ This leads us to what is called classical statistics:

$$\Omega_{cl} = \prod_i \frac{g_i^{n_i}}{n_i!} \tag{12.37a}$$

In quantum statistics, we treat

particles as indistinguishable;
energy as quantised and defines various levels; and
occupancy of a state as fundamental.

Before we calculate thermodynamic probability for a quantum system, we would like you to read the following anecdote, which demonstrates the dedication of great physicists in their convictions, acts and words.

> Prof. and Mrs. Dirac visited India in 1950 and Prof. S. N. Bose invited them to Kolkata (then Calcutta). Prof. Dirac delivered a lecture and as a matter of courtesy, Prof. Bose accompanied Prof. and Mrs. Dirac to their hotel after the lecture. Having made sure that the couple was comfortably seated in the back seat of their limousine, Bose got into the front seat. Two of Bose's pupils also wanted to accompany him and they too squeezed into the front seat.
>
> *"Couldn't Professor Bose come to the back seat?"* asked Mrs. Dirac. *"He will be much more comfortable here."* As a good host, Bose politely declined.
>
> *"It is a matter of statistics, my dear!"* Dirac offered an explanation in his brief but cogent style.

From this anecdote, you may have got the hint that the number of particles in an energy level is limited in Fermi–Dirac statistics, whereas there is no such restriction in Bose–Einstein statistics. Fermions are very reserved but bosons like staying together!

2. Calculation of Ω_{FD} Fermions To obtain the expression for Ω_{FD}, we consider N *indistinguishable* half-integral spin particles such as electrons, protons, neutrons, ^{3}He, neutrino, muons, etc. The wave functions associated with Fermions are anti-symmetric and they obey Pauli Exclusion Principle (no two particles having all four same quantum numbers can occupy the same sublevel). For mathematical convenience, we assume that these particles can be divided into G quantum groups or levels having energies ε_1, ε_2, , ..., and the number of particles in these levels are $n_1, n_2, n_3, \dots, n_G$, respectively. Let us assume that each level is degenerate, i.e., each level is associated with several states/ $_1, g_2, g_3, \dots g_i, \dots, g_G$. (Each quantum state corresponds to a cell in phase space.) This arrangement is tabulated below:

Group/Energy level	1	2	3	...	i	...	G
	ε_1	ε_2	ε_3	...	ε_I	...	ε_G
Occupation number	n_1	n_2	n_3	...	n_i	...	n_G
Degeneracy	g_1	g_2	g_3		g_i	...	g_G

We wish to determine the number of ways in which n_i fermions can occupy g_i sub-levels subject to following conditions:

The particles are indistinguishable from each other. Therefore, there is no distinction between the different ways in which n_i particles can be chosen.
Since fermions are governed by the Pauli Exclusion Principle, the number of particles in each sublevel will be one or none. This implies that g_i must be greater than or equal , i.e., $g_i \geq n_i$.
The sum of the energies of all the particles in different quantum groups taken together gives the total energy of the system.

The number of ways in which n_i fermions can occupy g_i sublevels is equal to the number of ways in which n_i things can be taken at a time from g_i different things. We arrive at this result by noting that we can place the first particle in the ith sublevel in g_i different ways. According to Pauli Exclusion Principle, no particle can be assigned to a filled state. Thus, we are left with $(g_i - 1)$ sublevels, which can be filled with the second particle in $(g_i - 1)$ different ways and so on. Thus, the total number of ways in which n_i fermions can occupy sublevels is equal to the product

$$g_i(g_i - 1)(g_i - 2)\dots(g_i - n_i + 1) = \frac{g_i\,!}{(g_i - n_i)!} \tag{12.38}$$

We know that permutations among identical particles do not give distinct distributions and such permutations must be excluded from Eq. (12.38). This can be done by dividing Eq. (12.38) by $n_i\,!$. Hence, the number of ways in which n_i fermions can be distributed sub-levels is

$$(\Omega_{FD})_i = \frac{g_i\,!}{(g_i - n_i)!\,n_i\,!} = {}^{g_i}C_{n_i}\,. \tag{12.39}$$

Similar expressions can be obtained for other quantum groups/states. Considering them all, we can write the total number of distinguishable arrangements for a Fermi–Dirac system as

$$\Omega_{FD} = \prod_i \frac{g_i\,!}{n_i\,!(g_i - n_i)!} \tag{12.40}$$

This defines the thermodynamic probability for an assembly of fermions. The most probable distribution is obtained by maximising ln Ω subject to given constraints.

3. Calculation of Ω_{BE}: Bosons To obtain the expression for Ω_{BE}, we consider N *indistin-* integral spin particles such as photons, ^{4}He atoms, phonons, magnons, π-meson, etc. The wave functions associated with Bosons are symmetric. It is worthwhile to mention here that Bose communicated his work on statistics of photon gas to Einstein, who immediately recognised its fundamental importance. He generalised Bose's work for all integral spin particles, translated it in German and communicated for publication. At that time, Bose was working in Dhaka University, Dhaka (now in Bangladesh).

As before, for mathematical convenience, we assume that these indistinguishable particles can be divided into G distinguishable quantum groups or levels. Suppose that there $n_3, \ldots, n_G$ particles in these groups whose energies are $\varepsilon_1, \varepsilon_2, \varepsilon_3, \ldots$, respectively. Again, let us assume that each level is degenerate, i.e., each level is associated with several sublevels/states $g_1, g_2, g_3, \ldots, g_G$. This situation is depicted below:

Group/Energy level	1	2	3	...	i	...	G
	ε_1	ε_2	ε_3	...	ε_i	...	ε_G
Occupation number	n_1	n_2	n_3	...	n_i	...	n_G
Degeneracy	g_1	g_2	g_3		g_i	...	g_G

We wish to know the number of ways in which n_i bosons can occupy g_i sublevels subject to following conditions:

The particles are indistinguishable from each other. Therefore, there is no distinction between the different ways in which n_i particles can be chosen.

Each sublevel of the ith quantum group may contain any number of particles from 0, 1, 2, ... n_i. Bosons love company; they are gregarious!

The sum of the energies of all the particles in different quantum groups taken together gives the total energy of the system.

To compute Ω_{BE}, we should know the number of ways in which n_i bosons can occupy g_i sublevels when there is no restriction on the number of particles occupying a sublevel. But you must remember that the particles are indistinguishable. So if we represent each of the sublevels by one compartment, there will be nothing to distinguish a particle occupying compartment number 1 from the one occupying compartment number 2. What matters is the number of particles going into a compartment. The number of particles in a particular compartment may vary between 0 and n_i. We have illustrated this situation by dividing a compartments (Fig. 12.12), where dots represent the particles.

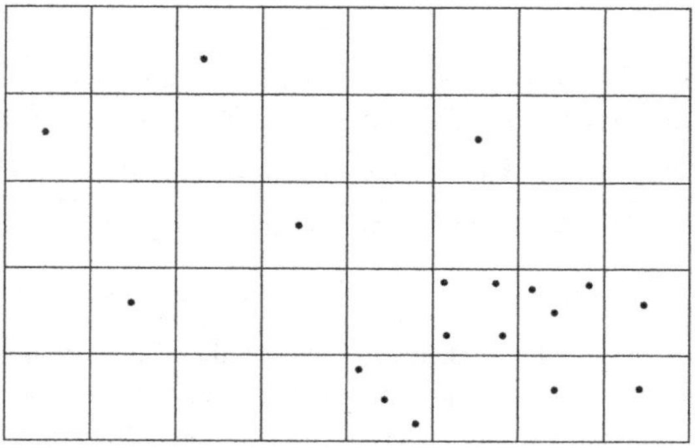

Division of a box into compartments.

Since the particles are indistinguishable, all of them can go in any one of the g_i sublevels and obviously the choice as to which of these sublevels will lead the sequence can be made ways and in each of these sequences, $[n_i + (g_i - 1)]$ sublevels and particles have to be arranged in any order. Obviously, this can be done in $[(n_i + g_i - 1)!]$ ways. Hence, the total number of ways in which n_i particles can be distributed in g_i sublevels is $g_i[(n_i + g_i - 1)!]$

As in case of F–D statistics, we recall that the permutations among identical particles do not lead to distinct arrangements and such permutations must be excluded. This can be done by dividing the above expression by n_i !.

Further, the distributions which arise by mere permutation of sublevels among themselves also do not produce different states. Obviously, there are g_i! such permutations. Hence, the number of ways in which n_i indistinguishable particles can be distributed among sublevels of the ith quantum level without any restriction on the number of particles occupying a sublevel is

$$\frac{g_i\,[(n_i + g_i - 1)!]}{n_i\,!g_i\,!} = \frac{[(n_i + g_i - 1)!]}{n_i\,!(g_i - 1)!} \tag{12.41}$$

Similar expressions can be obtained for other quantum groups/states. Considering them all, we can write the total number of distinguishable arrangements for a B–E system as

$$\Omega_{BE} = \frac{(N_1 + g_1 - 1)!}{n_1\,!(g_1 - 1)!} \times \frac{(N_2 + g_2 - 1)!}{n_2\,!(g_2 - 1)!} \times \ldots \frac{[(n_i + g_i - 1)!]}{n_i\,!(g_i - 1)!}$$

$$= \prod_i \frac{[(n_i + g_i - 1)!]}{n_i\,!(g_i - 1)!} \tag{12.42}$$

Before proceeding further, you may like to go through the following example.

Example 12.7 Consider a system consisting of two particles, which can exist in three different energy states. Calculate the number of ways in which these particles can be distributed in the given states.

Here $n_i = 2$ and $g_i = 3$. Hence, it readily follows from Eqs. (12.37), (12.40) and (12.42) that

$$\Omega_{MB} = 3^2 = 9$$

$$\Omega_{FD} = \frac{3!}{2! \times 1!} = 3$$

$$\Omega_{BE} = \frac{4!}{2! \times 2!} = 6$$

We know that in a M–B and B–E system, all particles can occupy a given energy state whereas only one fermion can occupy a given state. We now define a parameter ζ as

$$\zeta = \frac{\text{Probability that two particles can occupy the same state}}{\text{Probability that particles occupy different states}}$$

$= \dfrac{3}{9} = \dfrac{1}{3}$, $\zeta_{FD} = 0$ and $\zeta_{BE} = \dfrac{3}{6} = \dfrac{1}{2}$. This result shows that $\zeta_{BE} > \zeta_{MB} > \zeta_{FD}$. It means that bosons have a greater tendency to stick together. This explains why the condition for realising lasing action in actual practise is statistically favourable. In fact, we expect these values of ζ to make a striking difference in these statistics.

The Most Probable Distribution To know the most probable distribution, it is more convenient to maximise ln Ω instead of Ω. Therefore, we set

$$\delta(\ln \Omega) = 0 \tag{12.43}$$

subject to the constraints that the total number of particles in and total energy of the system are conserved:

$$N = n_1 + n_2 + n_3 + \ldots = \sum_i n_i = \text{constant}$$

$$E = n_1 \varepsilon_1 + n_2 \varepsilon_2 + n_3 \varepsilon_3 + \ldots = \sum_i n_i \varepsilon_i = \text{constant}$$

In the differential form, we can rewrite these constraints as

$$\sum_i \delta_i n_i = 0 \tag{12.44a}$$

$$\sum_i \varepsilon_i \delta n_i = 0 \tag{12.44b}$$

1. Maxwell–Boltzmann Statistics We first illustrate the procedure of obtaining the most probable distribution for Maxwell–Boltzmann statistics. From Eq. (12.37), we recall that thermodynamic probability is given by

$$\Omega_{MB} = \frac{N!}{\prod_i n_i!}\left(\prod_i g_i^{n_i}\right)$$

Taking natural logarithm on both sides, we get

$$\ln \Omega_{MB} = \ln N! + \sum_i [n_i \ln g_i - \ln n_i!]$$

To simplify this expression, we use Stirling's approximation* (ln $x! = x \ln x - x$) since g_i n_i are large. This gives

$$\ln \Omega_{MB} = N \ln N - N + \sum_i [n_i \ln g_i - n_i \ln n_i + n_i]$$

If we now invoke the relation $N = \sum_i n_i$, this expression simplifies to

$$\ln \Omega_{MB} = N \ln N + \sum_i [n_i \ln g_i - n_i \ln n_i] \tag{12.45}$$

We now use the fact that N is constant and g_i is not subject to variation. Therefore, on differentiating Eq. (12.45) with respect to n_i we can write

$$\delta(\ln \Omega_{MB}) = \sum_i \delta[n_i \ln g_i - n_i \ln n_i]$$

$$= \sum_i \left[\delta n_i \ln g_i - n_i\left(\frac{1}{n_i}\right)\delta n_i - \delta n_i \ln n_i\right]$$

$$= -\sum_i \left[\ln\left(\frac{n_i}{g_i}\right) + 1\right]\delta n_i$$

*For a statistical system, the relative error for $n = 100$ is just 1%. For $n = 6 \times 10^{26}$ the relative error will be almost insignificant.

For most probable state, we equate $\delta(\ln \Omega_{MB}) = 0$. This gives

$$\sum_i \left[\ln\left(\frac{n_i}{g_i}\right) + 1 \right] \delta n_i = 0 \tag{12.46}$$

Note that this equation is subject to the conditions expressed in Eqs. (12.44a,b) on the number of particles in and energy of the system. To relax these constraints, we use Lagrange's method of undetermined multipliers, wherein we multiply the two constraints with constants α and β, respectively. The resultant expressions are added to Eq. (12.46). The result is

$$\sum_i \delta n_i \left[\ln\left(\frac{n_i}{g_i}\right) + \alpha + \beta \varepsilon_i \right] = 0 \tag{12.47}$$

It may be mentioned here that we have absorbed the integer 1 in α.

The coefficients δn_i occurring in Eq. (12.47) are arbitrary and non-zero. Therefore, this result will hold for each i, only if we have

$$\ln\left(\frac{n_i}{g_i}\right) + \alpha + \beta\varepsilon_i = 0$$

$$\ln\left(\frac{n_i}{g_i}\right) = -(\alpha + \beta\varepsilon_i)$$

This can be rewritten as

$$\frac{n_i}{g_i} = \exp[-(\alpha + \beta\varepsilon_i)]$$

$$n_i = g_i \exp[-(\alpha + \beta\varepsilon_i)] = \frac{g_i}{\exp(\alpha + \beta\varepsilon_i)} \tag{12.48}$$

This result is known as *Maxwell–Boltzmann distribution law*. Note that we intended to know the set (n_i) which characterises the equilibrium state. But in this process, we have obtained two unknown constants. We will evaluate these in terms of known quantities a

Before proceeding further, we would like you to work out a practise problem.

Problem 12.9 Starting from Eq. (12.37a), derive Eq. (12.48).

2. Fermi–Dirac Statistics For F–D statistics, we write Eq. (12.40) as

$$\ln \Omega_{FD} = \sum_i [\ln g_i! - \ln n_i! - \ln(g_i - n_i)!]$$

Using Stirling's formula, we can rewrite it as

$$\ln \Omega_{FD} = \sum_i [g_i \ln g_i - g_i - (g_i - n_i)\ln(g_i - n_i) + (g_i - n_i) - n_i \ln n_i + n_i]$$

On simplification, we get

$$\ln \Omega_{\text{FD}} = \sum [g_i \ln g_i - (g_i - n_i)\ln(g_i - n_i) - n_i \ln n_i] \tag{12.49}$$

As before, we use the fact that N is constant and g_i is not subject to variation. Therefore, on differentiating Eq. (12.49) with respect to n_i, we get

$$\delta(\ln \Omega_{FD}) = \sum_i \delta[g_i \ln g_i - n_i \ln n_i + (n_i - g_i)\ln(g_i - n_i)]$$

$$= \sum_i [\delta n_i \ln(g_i - n_i) + \frac{(n_i - g_i)}{(g_i - n_i)}(-\delta n_i) - n_i\left(\frac{1}{n_i}\right)\delta n_i - \delta n_i \ln n_i]$$

The second and third terms on the RHS cancel out. Then the expression for $\delta(\ln \Omega_{FD})$ simplifies to

$$\delta(\ln \Omega_{FD}) = \sum_i -\delta n_i [\ln n_i - \ln(g_i - n_i)]$$

$$= -\sum_i \left[\ln\left(\frac{n_i}{g_i - n_i}\right)\right]\delta n_i$$

For most probable state, we equate $\delta(\ln \Omega_{FD}) = 0$. This gives

$$\sum_i \left[\ln\left(\frac{n_i}{g_i - n_i}\right)\right]\delta n_i = 0 \tag{12.50}$$

This equation is subject to the conditions expressed in Eqs. (12.44a, b) on the number of particles in the system and energy of the system. To relax these constraints, as before, we use Lagrange's method of undetermined multipliers, wherein we multiply the two constraints with constants α and β, respectively. The resultant expressions are added to Eq. (12.50). This leads us to the following expression:

$$\sum_i \delta n_i \left[\ln\left(\frac{n_i}{g_i - n_i}\right) + \alpha + \beta\varepsilon_i\right] = 0 \tag{12.51}$$

Note that the coefficients δn_i occurring in Eq. (12.51) are arbitrary and non-zero. Therefore, this result will hold for each i, only if the coefficients of δn_i vanish identically. Thus, we obtain

$$\ln\left(\frac{n_i}{g_i - n_i}\right) + \alpha + \beta\varepsilon_i = 0$$

This can be rewritten as

$$\frac{n_i}{g_i - n_i} = \exp[-(\alpha + \beta\varepsilon_i)]$$

$$\frac{g_i - n_i}{n_i} = \exp(\alpha + \beta\varepsilon_i)$$

$$\frac{g_i}{n_i} - 1 = \exp(\alpha + \beta\varepsilon_i)$$

$$\frac{g_i}{n_i} = \exp(\alpha + \beta\varepsilon_i) + 1$$

Hence, the expression for the most probable distribution of particles among various energy levels of a system obeying Fermi–Dirac statistics is given by

$$n_i = \frac{g_i}{e^{\alpha+\beta\varepsilon_i}+1} \tag{12.52}$$

This result is known as *Fermi–Dirac distribution law*.

3. Bose–Einstein Statistics For–B-E statistics, we take logarithm of Eq. (12.42) and write

$$\ln \Omega_{BE} = \sum_i [\ln (n_i + g_i - 1)! - \ln n_i! - \ln(g_i - 1)!]$$

Using Stirling's formula, we can rewrite it as

$$\ln \Omega_{BE} = \sum_i [(n_i + g_i - 1)\ln(n_i + g_i - 1) - (n_i + g_i - 1) - n_i \ln n_i + n_i$$
$$- (g_i - 1)\ln(g_i - 1) + (g_i - 1)]$$

and g_i are very large, we can take $n_i + g_i - 1 \approx n_i + g_i$ and $g_i - 1 \approx g_i$. Hence, the expression for $\ln \Omega_{BE}$ takes the form

$$\ln \Omega_{BE} = \sum_i [(n_i + g_i)\ln(n_i + g_i) - (n_i + g_i) - n_i \ln n_i + n_i - g_i \ln g_i + g_i]$$

You will note that the second, fourth and sixth terms on the RHS of this expression cancel out. Then, the expression for $\ln \Omega_{BE}$ simplifies to

$$\ln \Omega_{BE} = \sum_i [(n_i + g_i)\ln(n_i + g_i) - n_i \ln n_i - g_i \ln g_i] \tag{12.53}$$

We recall that n_i varies continuously but g_i is not subject to variation. Therefore, on differentiating Eq. (12.53) with respect to n_i we get

$$\delta(\ln \Omega_{BE}) = \sum_i \delta[(g_i + n_i)\ln(g_i + n_i) - n_i \ln n_i - g_i \ln g_i]$$
$$= \sum_i [\delta n_i \ln(g_i + n_i) + \frac{(n_i + g_i)}{(g_i + n_i)}\delta n_i - n_i\left(\frac{1}{n_i}\right)\delta n_i - \delta n_i \ln n_i]$$

The second and third terms on the RHS cancel out. Then the expression for $\delta(\ln \Omega_{BE})$ simplifies to

$$\delta(\ln \Omega_{BE}) = \sum_i \delta n_i [\ln(n_i + g_i) - \ln n_i]$$
$$= -\sum_i \left[\ln\left(\frac{n_i}{g_i + n_i}\right)\right]\delta n_i$$

For most probable state, we equate $\delta(\ln \Omega_{BE}) = 0$. This gives

$$\sum_i \left[\ln\left(\frac{n_i}{g_i + n_i}\right)\right]\delta n_i = 0 \tag{12.54}$$

As before, this equation is subject to the conditions expressed in Eqs. (12.44a, b) on the number of particles in and energy of the system. To relax these constraints, as before, we use Lagrange's method of undetermined multipliers, wherein we multiply the two constraints with constants α and β, respectively. The resultant expressions are added to Eq. (12.54). The result is

$$\sum_i \delta n_i \left[\ln\left(\frac{n_i}{g_i + n_i}\right) + \alpha + \beta\varepsilon_i \right] = 0 \tag{12.55}$$

Note that the coefficients δn_i occurring in Eq. (12.55) are arbitrary and non-zero. Therefore, this result will hold for each i, only if the coefficients of δn_i vanish identically. Thus, we obtain

$$\ln\left(\frac{n_i}{g_i + n_i}\right) + \alpha + \beta\varepsilon_i = 0$$

This can be rewritten as

$$\frac{n_i}{g_i + n_i} = \exp[-(\alpha + \beta\varepsilon_i)]$$

$$\frac{g_i + n_i}{n_i} = \exp(\alpha + \beta\varepsilon_i)$$

$$\frac{g_i}{n_i} = \exp(\alpha + \beta\varepsilon_i) - 1$$

Hence, the expression representing the most probable distribution of particles among various energy levels of a system obeying Bose–Einstein statistics is given by

$$n_i = \frac{g_i}{e^{\alpha + \beta\varepsilon_i} - 1} \tag{12.56}$$

This result is known as *Bose–Einstein distribution law*.
Before proceeding further, you may like to solve a practise problem.

Problem 12.10 Equation (12.49) and (12.53) can be combined into one expression by introducing a parameter κ which takes the value +1 for F–D statistics and –1 for B–E

$$\ln\Omega = \sum_i [\kappa g_i \ln g_i - \kappa(g_i - \kappa n_i)\ln(g_i - \kappa n_i) - n_i \ln n_i]$$

Following the procedure outlined above, show that

$$n_i = \frac{g_i}{e^{\alpha + \beta\varepsilon_i} + \kappa}$$

Determination of Constants You must have noted that the constants α and β occurring in different statistical distributions are unknown so far. We now evaluate these by referring to M–B distribution (Eq. 12.48):

$$n_i = g_i \exp[-(\alpha + \beta\varepsilon_i)]$$

To determine α, we make use of the fact that $\sum_i n_i = N$. Therefore, we can write

$$N = \exp(-\alpha)\sum_i g_i \exp(-\beta\varepsilon_i) = \exp(-\alpha) Z, \tag{12.57}$$

$$\exp(-\alpha) = \frac{N}{Z} \tag{12.58}$$

and we can write the expression for the number of particles in the ith group as

$$\frac{n_i}{g_i} = \frac{N}{Z}\exp(-\beta\varepsilon_i) \tag{12.59}$$

$$Z = \sum_i g_i \exp(-\beta\varepsilon_i) \tag{12.60}$$

denotes sum over all states and is known as *partition function*. (This nomenclature is due to Darwin and Fowler. However, Planck used symbol Z and derived it from the German *Zustandssummae*.) It may be mentioned here that partition function is a mathematical device. As such, it has no latent physical significance. However, it occupies a pivotal position and provides us an elegant tool to derive expressions for various thermodynamic variables/quantities.

Proceeding further, we evaluate the constant β and recall Eq. (12.45) for $\ln(\Omega_{MB})$:

$$\ln \Omega_{MB} = N \ln N + \sum_i [n_i \ln g_i - n_i \ln n_i]$$

$$= N \ln N + \sum_i n_i \ln\left(\frac{g_i}{n_i}\right)$$

On combining this result with Eq. (12.59), we obtain

$$\ln \Omega_{MB} = N \ln N + \sum_i n_i \ln\left(\frac{Z}{N}\exp(\beta\varepsilon_i)\right)$$

$$= N \ln N + \sum_i n_i \ln Z - \sum_i n_i \ln N + \beta \sum_i n_i \varepsilon_i$$

Note that the first and the third terms on the RHS of this expression cancel out. Moreover, n E, the total energy of the system. Then, we can write

$$\ln \Omega_{MB} = \sum_i n_i \ln Z + \beta E$$

Using Boltzmann's entropy relation, we get

$$S = k_B \ln \Omega_{MB} = N k_B \ln Z + \beta k_B E$$

For a system of non-interacting particles, E denotes internal energy, U. So this equation can be rewritten as

$$S = N k_B \ln Z + \beta k_B U \tag{12.61}$$

We use this relation to introduce the concept of temperature by relating entropy and internal energy of a system at constant volume (Eq. (7.18):

$$\frac{1}{T} = \left(\frac{\partial S}{\partial U}\right)_V = \frac{N k_B}{Z}\left(\frac{\partial Z}{\partial \beta}\right)_V \left(\frac{\partial \beta}{\partial U}\right)_V + k_B \beta + k_B U \left(\frac{\partial \beta}{\partial U}\right)_V \tag{12.62}$$

$Z = \sum_i g_i \exp(-\beta\varepsilon_i)$, we note that

$$\left(\frac{\partial Z}{\partial \beta}\right)_V = -\sum_i g_i \varepsilon_i \exp(-\beta\varepsilon_i) \tag{12.63a}$$

Similarly, since $U = \sum_i n_i \varepsilon_i = \frac{N}{Z}\sum_i g_i\, \varepsilon_i \exp(-\beta\varepsilon_i)$, we can write

$$U = -\frac{N}{Z}\left(\frac{\partial Z}{\partial \beta}\right)_V$$

$$\left(\frac{\partial Z}{\partial \beta}\right)_V = -\frac{ZU}{N} \tag{12.63b}$$

On using this result in Eq. (12.62), we get

$$\frac{1}{T} = -\frac{Nk_B}{Z}\frac{ZU}{N}\left(\frac{\partial \beta}{\partial U}\right)_V + k_B\beta + k_B U\left(\frac{\partial \beta}{\partial U}\right)_V$$

Note that the first and the third terms on the right-hand side of this expression cancel out and we get the desired result:

$$\beta = \frac{1}{k_B T} \tag{12.64}$$

Comparison of Different Distribution Functions To give you an idea about how particles obeying different statistics behave, we note that the ratio (n_i/g_i) defines the occupation number of the ith group. If we define chemical potential through the relation T, we can express occupation numbers for different distributions by a single expression:

$$\frac{n_i}{g_i} \equiv f = \frac{1}{\exp[(\varepsilon_i - \mu)/k_B T] + \kappa} \tag{12.65}$$

The constant κ takes values 0, 1 and –1 for M–B (classical), F–D and B–E statistics, respectively.

Note that in spite of the vast differences in the assumptions used to derive expressions for distribution functions, they appear reasonably similar. Yet, as we discover later, the presence of constant κ in the denominator has enormous consequences, particularly at low temperatures and high densities.

n_i/g_i vs. $(\varepsilon_i - \mu)/k_B T$ at two different temperatures T_1 and T_2 $(> T_1)$ for fermions and bosons are shown in Fig. 12.13(a, b). As may be noted, for $\varepsilon = \mu$, $f_{FD} = 1/2$ whereas . For $\varepsilon_i - \mu >> k_B T$, κ can be ignored compared to the exponential term so that $f \to \exp[-(\varepsilon_i - \mu)/k_B T]$.

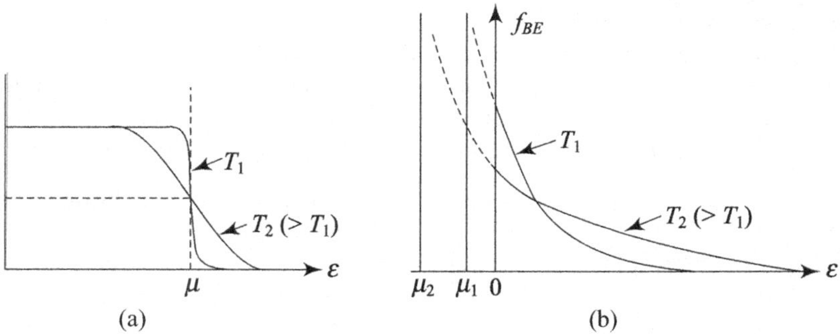

(a) Plot of f for fermions at two temperatures T and T (> T) and

From this, we can say that

distribution of bosons is skewed towards lower energy states, whereas fermions are skewed towards higher energy states, That is, there is higher probability of finding bosons in low level energy states and fermions in higher energy states; and
at sufficiently high energies classical and quantum results are identical.

This is illustrated in Fig. 12.14.

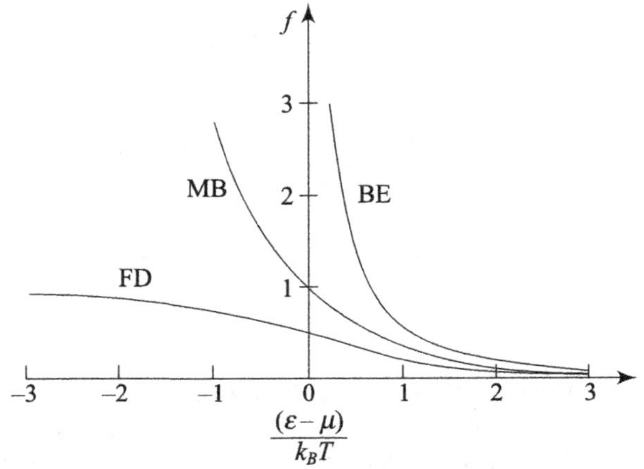

Plots of distribution function for M–B, F–D and B–E statistics versus $(\varepsilon_i - \mu)/k_BT$ for fixed T and μ.

However, at lower energies, we cannot ignore κ in the denominator in comparison with $\mu)]$. This implies that the classical and quantum results will be different at low energies. In this range of energies, we have

$$f_{BE} > f_{MB} > f_{FD} \tag{12.66}$$

It may be mentioned here that all elementary particles in nature are either bosons or fermions. However, both fermions and bosons behave as classical particles at high temperatures.

DDITIONAL EXAMPLES

Example 12.8 Six distinguishable particles are distributed over three non-degenerate levels of energies 0, ε and 2ε.

Calculate the total number of microstates of the system.
Determine the thermodynamic probabilities (TP) for all possible distributions and obtain the total energy of the distribution for which the probability is maximum.
Using the results of the TP calculations and knowledge of permutation, verify (a)

(a) We are told that six distinguishable particles are to be distributed in three non-degenerate levels. So the total no. of microstates = $3^6 = 729$.

The distribution of the 6 particles can be in following combinations:

1. (6, 0, 0); 2. (5, 1, 0); 3. (4, 2, 0); 4. (4, 1, 1); 5. (3, 2, 1);
6. (3, 3, 0); 7. (2, 2, 2)

The thermodynamic probabilities for these cases are as follows:

$$\frac{6!}{6!0!0!} = \frac{6!}{6!(1)(1)} = 1$$

$$\frac{6!}{5!1!0!} = \frac{6!}{5!(1)(1)} = 6$$

$$\frac{6!}{4!2!0!} = \frac{6!}{4!(2)(1)} = \frac{5 \times 6}{2} = 15$$

$$\frac{6!}{4!1!1!} = \frac{6!}{4!(1)(1)} = 5 \times 6 = 30$$

$$\frac{6!}{3!2!1!} = \frac{6!}{3!(2)(1)} = \frac{6 \times 5 \times 4}{2} = 60$$

$$\frac{6!}{3!3!0!} = \frac{6!}{6 \times 6 \times 1} = \frac{6 \times 5 \times 4 \times 3 \times 2 \times 1}{6 \times 6} = 20$$

$$\frac{6!}{2!2!2!} = \frac{720}{2 \times 2 \times 2} = 90$$

Obviously, the most probable distribution is (7).
And the total energy = $2 \times 0 + 2 \times \varepsilon + 2 \times 2\varepsilon = 6\varepsilon$.

The number of microstates in each case is

$$1 \times — = 3$$

$$3! = 36$$

$$3! = 90$$

$$30 \times \frac{3!}{2!} = 90$$

$$3! = 360$$

$$20 \times \frac{3!}{2!} = 60$$

$$90 \times \frac{3!}{3!} = 90$$

Total = 729

Example 12.9 Consider two identical particles. Each particle can be in one of the three possible quantum states of energies 0, ε and 3ε. Calculate the number of microstates of the system for M–B, B–E and F–D statistics. Also determine the ratio of the probability that the two particles are found in the same state to the probability that two particles are found in different states in each of the three cases.

The number of microstates can be obtained as under:

M–B Statistics

For non-identical particles, the number = $3^2 = 9$. Out of these $\frac{3!}{2!} = 3$ cases will be common. So the required number of microstates is = $9 - 3 = 6$.

B–E Statistics

The situation is depicted schematically below:

States → (columns); Occupancy (rows)

0	e	3e
..		
	..	
		..
.	.	
	.	.
.		.

Thus, the required number = 6 (It is the same as obtained for M–B statistics).

You can also work it out directly as $\frac{(2+3-1)!}{2!(3-1)!} = \frac{4!}{2!2!} = \frac{24}{2\times 2} = 6.$

F–D Statistics

The required number $= {}^3C_2 = \frac{3!}{2!((3-2)!} = \frac{6}{2\times 1} = 3$

Schematic Representation

.	.	
	.	.
.		.

Second Part

The required ratios are:

M–B: 3:3 = 1

B–E: 3:3 = 1

F–D: 0:3 = 0

Example 12.10 If three identical particles are distributed over three single particle states, how many possibilities are allowed if the particles are **(a)** spin-zero pions and electrons?

(a) Note that spin-zero pions are bosons. So

$$\text{the number distribution} = \frac{(3+3-1)!}{3!(3-1)!}$$

$$= \frac{5!}{3!2!} = \frac{4\times 5}{1\times 2} = 10$$

Recall that electrons are fermions. So

$$\text{the required number of distribution} = \frac{3!}{3!(3-3)!} = \frac{3!}{3!0!} = 1.$$

Example 12.11 The energy levels of a rigid rotator are given by $E = \frac{m^2\hbar^2}{2I}$, where $= \pm 1, \pm 2, \pm 3, \ldots$ Determine the trajectories in the phase space. Show that the phase space is divided into cells of area h.

Let us first understand the origin of the result, $E = \frac{m^2 \hbar^2}{2I}$. Recall that a rigid rotator is rotational analogue of a free particle whose energy is given by $E = \frac{p^2}{2m}$, where is the linear momentum and m is the mass of the free particle. For the rigid rotator (Fig. 12.15(a)), p is replaced by L_z and m by the moment of inertia I. Thus, we can write

$$E = \frac{L_Z^2}{2I}$$

In quantum mechanics course, you will learn that $L_z = m\hbar$. Therefore, we can write

$$E = \frac{m^2 \hbar^2}{2I}, \qquad m = \pm 1, \pm 2, \pm 3, \ldots$$

takes positive as well as negative integral values, which are indicative of spin up and spin down.

The reference coordinates of the phase space are usually taken as p_ϕ and ϕ. Note that ϕ is azimuthal angle, which varies in the range 0 to 2π and p_ϕ is the generalised momentum, which we have denoted by L_z, i.e., $p_\phi = m\hbar$. So the phase space will be as shown in Fig. 12.15 (b).

Thus the phase space is divided into cells, each of which is a rectangle having sides $\hbar$. So area of each such cell is $2\pi \times \frac{h}{2\pi} = h$.

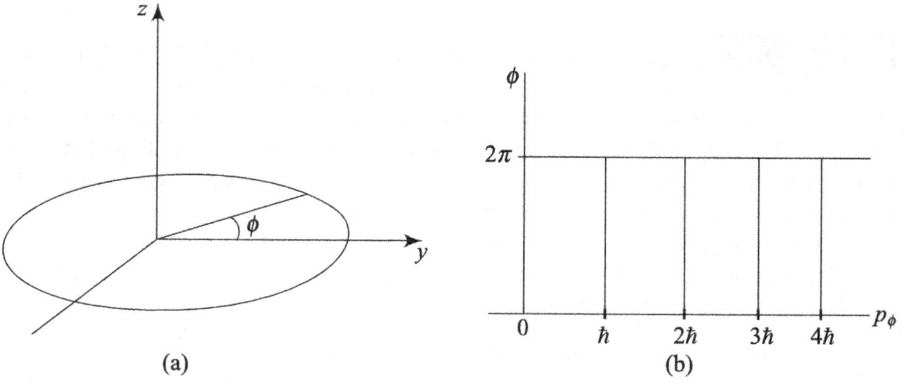

Fig. 12.15

Example 12.12 Starting from Eq. (12.37a) show that entropy of a clasical system is

$$S = Nk_B[\ln Z - \ln N + 1] + \frac{U}{T}$$

From Eq. (12.37a) we know that thermodynamic probability of a classical system is given

$$\Omega_{cl} = \prod_i \frac{g_i^{n_i}}{n_1!}$$

On taking natural logarithm of both sides and using Stirling's approximation, we obtain

$$\ln \Omega_{cl} = \sum_i [n_i \ln g_i - n_i \ln n_i + n_i]$$

We now use Boltzmann entropy relation to obtain

$$S = k_B \ln \Omega_{cl}$$

$$= k_B \left[N + \sum_i n_i \ln\left(\frac{g_i}{n_i}\right)\right]$$

g $= \frac{N}{Z}\exp(-\beta\varepsilon_i)$, we can rewrite the expression for entropy as

$$S = k_B \left[N + \sum n_i \ln\left(\frac{Z}{N}\exp(\beta\varepsilon_i)\right)\right]$$

$$= k_B \left[N + \sum_i n_i \ln Z - \sum_i n_i \ln N + \sum_i n_i \beta\varepsilon_i \right]$$

$$= k_B [N + N \ln Z - N \ln N + \beta E]$$

$= \frac{1}{k_B T}$ and $E \equiv U$, the expression for entropy takes the form

$$S = Nk_B[\ln Z - \ln N + 1] + \frac{U}{T}$$

$$U - TS = F = -Nk_B T[\ln Z - \ln N + 1].$$

Example 12.13 Consider a lattice containing N_a atoms of type A and N_b atoms of . These are placed one by one on available lattice sites. The structure so formed is a configuration. The entropy of formation of the structure is known is configurational entropy. (The configurational entropy is determined solely by the number of different ways in which the atoms may be arranged over the available number of lattice sites.) Obtain the expression for the configurational entropy S_{cf} for such a structure.

What assumption do you need to make?

Use the result to obtain the value of S_{cf} for a lattice having N atoms and n vacancies. Hence show that for a perfect lattice $S_{cf} = 0$

The required thermodynamic probability, $\Omega = \frac{(N_a + N_b)!}{N_a!N_b!}$

$$S_{cf} = k_B \ln \Omega = k_B \ln\left\{\frac{(N_a + N_b)!}{N_a!N_b!}\right\}$$

$$= k_B \{\ln(N_a + N_b)! - \ln N_a! - \ln N_b!\}$$

Here we have assumed that the lattice sites are all equivalent. It means that it is equally likely for a site to be occupied by A or B.

In the above result, we put $N_a = N$, $N_b = n$ so as to get

$$S_{cf} = k_B \ln\left[\frac{(N+n)!}{N!n!}\right]$$

For a perfect lattice $n = 0$. $\quad \therefore\ S_{cf} = k \ln\left(\frac{N!}{N!}\right)$

$$= k \ln 1! = 0$$

Example 12.14 For a gaseous molecule, the wave number is 540 cm^{-1}. Calculate the value of $h\nu/k_BT$ at $T = 100$ K. Also determine the relative probabilities of the first three vibrational modes.

We know that

$$\frac{h\nu}{k_B T} = \frac{hc}{k_B T \lambda}$$

and wave number $k = 2\pi/\lambda_0$. Hence,

$$\frac{h\nu}{k_B T} = \frac{hc}{k_B T}\frac{k}{2\pi} = \frac{(6.62\times 10^{-34}\text{ Js})\times(3\times 10^{8}\text{ ms}^{-1})}{(1.38\times 10^{-23}\text{ JK}^{-1})(1000\text{ K})\times 6.28}\times 54000\text{ m}^{-1}$$

$$= \frac{107.24\times 10^{-22}}{8.67\times 10^{-20}}$$

$$= 0.124$$

Hence, probabilities of three successive vibrational states are

$$e^0, e^{-h\nu/k_B T} \text{ and } e^{-2h\nu/k_B T}$$

$$P_1 = 1, P_2 = e^{-0.124} = 0.88 \text{ and } P_3 = e^{-0.248} = 0.78$$

Example 12.15 A system of two energy levels E_0 and E_1 is populated by N particles at temperature T. The particles populate the energy levels according to the classical distribution law. **(a)** Obtain an expression for the average energy per particle and discuss its behaviour at low temperature. **(b)** Also derive an expression for the specific heat capacity.

(a) The average energy of a particle that can exist in two energy states is given

$$\bar{\varepsilon} = \frac{E_0 e^{-E_0/k_B T} + E_1 e^{-E_1/k_B T}}{e^{-E_0/k_B T} + e^{-E_1/k_B T}} \tag{i}$$

Suppose that $E_1 > E_0 > 0$ and $\Delta E = E_1 - E_0$. Then on dividing throughout by $e^{-E_0/k_B T}$,

$$\bar{\varepsilon} = \frac{E_0 + E_1 e^{-\Delta E/k_B T}}{1 + e^{-\Delta E/k_B T}} \tag{ii}$$

0, $(k_BT)^{-1} \to \infty$, so that using binomial theorem we can write

$$\bar{\varepsilon} = (E_0 + E_1 e^{-\Delta E/k_B T})(1 - e^{-\Delta E/k_B T} + \ldots)$$

$$= E_0 + (E_1 - E_0)e^{-\Delta E/k_B T} - E_1 e^{-2\Delta E/k_B T}$$

If we retain terms only up to first order in $e^{-\Delta E/k_B T}$, we get in the low temperature limit

$$(\bar{\varepsilon})_{T\to 0} = E_0 + \Delta E\, e^{-\Delta E/k_B T}. \tag{iii}$$

The molar specific heat capacity is given by

$$C = N_A \frac{\partial \bar{\varepsilon}}{\partial T}$$

Using Eq. (ii), we get

$$= N_A k_B \left(\frac{\Delta E}{k_B T}\right)^2 \frac{e^{-\Delta E/k_B T}}{(1 + e^{-\Delta E/k_B T})^2}$$

$\rightarrow$, $e^{-\Delta E/k_BT} << 1$ so that expression for molar heat capacity simplifies to

$$C = R\left(\frac{\Delta E}{k_B T}\right)^2 e^{-\Delta E/k_B T}$$

Example 12.16 A system of particles occupying single-particle levels and obeying M–B statistics is in thermal contact with a heat reservoir at temperature T. If the population distribution in the non-degenerate energy levels is as tabulated below, calculate the temperature of the system.

10^3	30.1	21.5	12.9	4.3
(%)	3.1	8.5	23	63

According to M–B statistics, the population distribution is given by

$$\frac{n_2}{n_1} = \exp\left[(\varepsilon_1 - \varepsilon_2)/k_B T\right]$$

$$T = \frac{\varepsilon_1 - \varepsilon_2}{k_B} \ln\left(\frac{n_1}{n_2}\right)$$

Using the given values, we get

$$T_1 = \frac{[(30.1 - 21.5)\times 10^{-3}\,\text{eV}]\times(1.6\times 10^{-19}\,\text{J eV}^{-1})}{1.38\times 10^{-23}\ \text{JK}^{-1}} \cdot \frac{1}{\ln\left(\dfrac{8.5}{3.1}\right)}$$

$$= 9.97\times 10\times 0.9914\ \text{K}$$

$$= 98.8\ \text{K}$$

$$T_2 = \frac{[(21.5 - 12.9)\times 10^{-3}\,\text{eV}]\times(1.6\times 10^{-19}\,\text{J eV}^{-1})}{1.38\times 10^{-23}\ \text{JK}^{-1}} \cdot \frac{1}{\ln\left(\dfrac{23}{8.5}\right)}$$

$$= 9.97\times 10^1 \times 0.9955$$

$$= 99.3\ \text{K}$$

Following the same steps, you can calculate the temperatures corresponding to other distributions.

Example 12.17 The average kinetic energy of hydrogen atoms in a certain stellar atmosphere, assumed to be in thermal equilibrium, is 1.2 eV. Calculate the ratio of the number of atoms in the second excited state ($n = 3$) to the number in the ground state.

The energy levels for hydrogen atom are defined by

$$E_n = -\left(\frac{13.6}{n^2}\right)\text{eV}$$

According to M–B statistics, the ratio of the number of hydrogen atoms in the second excited state to the ground state can be expressed as

$$\frac{N_3}{N_1} = \exp\left[\frac{E_1 - E_3}{k_B T}\right]$$

$= -13.6$ eV, $E_3 = -13.6$ eV/9 $= -1.51$ eV and $k_B T = \frac{2}{3} \times 1.2$ eV $= 0.8$ eV. Hence,

$$\frac{N_3}{N_1} = \exp\left[\frac{-13.6 + 1.51}{0.8}\right]$$
$$= \exp(-15.1)$$
$$= 2.74 \times 10^{-7}$$

Let us now summarise what you have learnt in this chapter for quick recapitulation.

MMARY

Phase space is a purely mathematical device, which provides some sort of geometrical framework to statistical mechanics and helps to minimise abstraction.

The macrostate of a system can be described by specifying only those quantities which can be determined by macroscopic measurement, without any reference whatsoever to the microscopic details.

The state of a system described in the minutest possible details of each atom/molecule/particle is said to be microstate.

For indistinguishable particles, a specification of the total number of particles in each energy state defines a microstate. For a system of distinguishable particles, a specification of the energy state and energy level of each particle defines a microstate.

A given macrostate may consist of a number of microstates.

The entropy and thermodynamic probability are connected through the relation ln W, where k_B is Boltzmann constant.

The thermodynamic probability of distributing N distinguishable particles in various energy groups and the particles in different groups into cells is given by

$$\Omega_{MB} = \frac{N!}{\prod_i n_i!}\left(\prod_i g_i^{n_i}\right)$$

For a Fermi–Dirac system, the thermodynamic probability of distributing N indistinguishable particles into various energy states subject to Pauli principle is given by

$$\Omega_{FD} = \prod_i \frac{g_i!}{n_i!(g_i - n_i)!}$$

For a Bose–Einstein system, the thermodynamic probability of distributing N indistinguishable particles into various energy states is given by

$$\Omega_{BE} = \prod_i \frac{[(n_i + g_i - 1)!]}{n_i!(g_i - 1)!}$$

To determine the most probable state, we maximise ln Ω rather than Ω, since the latter is very large and inconvenient to handle.

The most probable distribution function for MB, FD and BE statistics can be expressed as

$$\frac{n_i}{g_i} = f = \frac{1}{\exp[(\varepsilon_i - \mu)/k_B T] + \kappa}$$

The constant κ takes values 0, 1 and –1 for M–B (classical), F–D and B–E statistics, respectively.

ERCISES

As the internal energy of a system is doubled, the density of states is also doubled. What is the number of degrees of freedom of the system? Is the system microscopic or macroscopic? (*Ans:* 2, Microscopic)

By defining temperature through the relation $\frac{1}{T} = \left(\frac{\partial S}{\partial E}\right)_{V,N}$, obtain the zeroth law of thermodynamics.

A system has a net charge of 10^{-12}C at a temperature of 20°C. Its potential is lowered by 1V, keeping the temperature and the number of particles unchanged. Will the number of states accessible to the system increase? If so, by what factor?

(*Ans:* Yes, $10^{1.04 \times 10^8}$)

A container of volume V_0 contains N_0 non-interacting molecules of a gas. Assuming that each molecule is equally likely to be located anywhere within the container, calculate

the probability that exactly N molecules are located within a sub-volume V of the container

the mean number, N, of molecules within the sub-volume V.

$$\left[Ans: \text{(a)} \frac{N!}{N!(N_0 - N)!} \left(\frac{V}{V_0}\right)^N \left(1 - \frac{V}{V_0}\right)^{N_0 - N}, \text{(b)}\ N_0 \frac{V}{V_0} \right]$$

Consider a sample of N non-interacting magnetic atoms. The atoms may point either up or down. Suppose N_1 atoms are pointing up and N_2 are pointing down. Calculate the entropy of the system assuming N_1 and N_2 to be very large.

$$\left[Ans: k_B \left[N_1 \ln\left(\frac{N}{N_1}\right) + N_2 \ln\left(\frac{N_1}{N_2}\right) \right] \right]$$

Three particles have to be accommodated in four available states. Calculate the number of ways in which this can be done if the particles obey

Classical statistics **(b)** F–D statistics **(c)** B–E statistics

(*Ans:* (a) 64 (b) 4 (c) 20)

particles obeying classical statistics are distributed among three states having energies $\varepsilon_1 = 0$, $\varepsilon_2 = k_B T$ and $\varepsilon_3 = 2k_B T$. If the total equilibrium energy of the system is 1000 $k_B T$, what is the value of N? (*Ans:* 2354)

13

MAXWELL–BOLTZMANN STATISTICS

arning Objectives

In this chapter, you will learn how to

- express thermodynamic functions for a classical system in terms of partition function;
- evaluate partition function for a monatomic gas and obtain expressions for thermodynamic functions;
- explain Gibbs paradox and discuss its genesis;
- derive Sackür–Tetrode equation and show that it predicts entropy of mixing correctly;
- evaluate partition function for a diatomic gas and obtain expressions for heat capacity of hydrogen;
- discuss Einstein's and Debye's theories of heat capacity of solids;
- uss thermodynamic behaviour of finite energy level systems; and
- explain the concept of negative temperatures, and action of lasers.

NTRODUCTION

In Chapter 12, we obtained expressions for distribution functions of a classical system (of distinguishable) as well as a quantum system (of indistinguishable) particles. While evaluating Lagrange multiplier α for a Maxwell–Boltzmann distribution, we discovered that we have to perform sum over all states and introduce a new mathematical device—the partition function, Z. Its importance stems from the fact that thermodynamic functions such as pressure, internal energy, entropy and thermodynamic potentials in M–B statistics can be expressed in terms of Z or its partial derivatives. This is illustrated in Sec. 13.2. You will therefore agree that to apply the methods of statistics to a system of interest, we must first evaluate Z for it. We will illustrate this in Sec. 13.3 for a monatomic gas of non-interacting particles and use this information to obtain expression for entropy, apart from other thermodynamic functions. In so doing, we come across what is termed Gibbs paradox in that even self-diffusion of a gas causes increase in entropy of the system. The genesis of the paradox lies in treating all the identical molecules of a gas as distinguishable and over-estimating the number of accessible states. The paradox was resolved by recognising that molecules of a monatomic gas are identical and hence indistinguishable. This led to the Sackür–Tetrode equation for absolute entropy.

In Sec. 13.4, we have applied M–B statistics to a few typical gaseous systems. In particular, we have obtained distribution law for molecular speeds and expressions for heat capacities of monatomic as well as diatomic gaseous molecules such as HCl and H_2. We show that the observed temperature variation of heat capacity of hydrogen can be explained only by including rotational, vibrational and spin contributions. In Sec. 13.5, we have considered heat capacity of solids within the framework of Einstein and Debye theories.

In Sec. 13.6, we have discussed thermodynamic behaviour of systems of finite number of energy levels as exist in paramagnetic substances. You will discover that the nature of variation of C_B is very different from that of the specific heat capacity of a solid. This analysis has been extended to introduce the concept of negative temperatures. You will learn that this nomenclature has origin in population inversion and, in fact, negative temperatures are higher than positive temperatures. This is followed by a discussion of Einstein's A and B coefficients and operation of a laser—a highly directional, intense source of monochromatic light. As you are aware, lasers have revolutionised the way we communicate in the present day world across oceans and continents.

At sufficiently high temperatures as occurring in the interior of stars, all substances exist in ionised state. The theory of thermal ionisation was developed by eminent Indian physicist M.N. Saha and forms the subject matter of discussion of Sec. 13.7. It successfully explained ionisation in solar chromosphere. The most spectacular success of ionisation formula was in categorisation of the vast number of stars into a few groups, which form a continuous and linear series.

ARTITION FUNCTION AND THERMODYNAMIC ROPERTIES OF A SYSTEM

Consider an ideal monatomic gas made up of N identical yet distinguishable particles enclosed in volume V and having total energy E. The entropy of such a system is given by Eq. (12.61):

$$S = Nk_B \ln Z + \beta k_B U \tag{13.1}$$

We now know that $\beta = (k_B T)^{-1}$. Therefore, the expression for entropy takes the form

$$S = Nk_B \ln Z + \frac{U}{T} \tag{13.2}$$

We can now express all thermodynamic functions in terms of Z. To begin with, we recall that Helmholtz free energy is given by

$$F = U - TS$$

Using Eq. (13.2), we can write

$$F = -Nk_B T \ln Z \tag{13.3}$$

To express internal energy in terms of Z, we recall that

$$S = -\left(\frac{\partial F}{\partial T}\right)_{N,V} = Nk_B \ln Z + Nk_B T\left(\frac{\partial \ln Z}{\partial T}\right)_{N,V} \tag{13.4}$$

On comparing this expression for entropy with that given in Eq. (13.2), we get

$$U = Nk_B T^2 \left(\frac{\partial (\ln Z)}{\partial T}\right)_{N,V} \tag{13.5}$$

If the property of interest is pressure exerted by the system, we recall that

$$p = -\left(\frac{\partial F}{\partial V}\right)_{T,N}$$

On substituting for F from Eq. (13.3), we get

$$p = Nk_B T\left(\frac{\partial \ln Z}{\partial V}\right)_{T,N} = \frac{Nk_B T}{Z}\left(\frac{\partial Z}{\partial V}\right)_{T,N} \tag{13.6}$$

Similarly, chemical potential, μ, is defined as

$$\mu = \left(\frac{\partial F}{\partial N}\right)_{T,V}$$

On substituting for F from Eq. (13.3), we can write

$$\mu = \left[\frac{\partial}{\partial N}(-Nk_B T \ln Z)\right]_{T,V} = -k_B T \ln Z \tag{13.7}$$

We have now seen that all thermodynamic functions of a monatomic gas have been expressed in terms of the partition function Z. It implies that once we evaluate Z, we can easily know the thermodynamic state of the system.

Before we illustrate how to evaluate partition function for an ideal monatomic gas, go through the following solved examples.

Example 13.1 Consider a one level system having energy

$$\varepsilon = -k_B T \ln\left(\frac{V}{V_0}\right)$$

is a constant.

Write down the partition function for this system,

Calculate the average pressure for this system as a function of volume and temperature.

(a) For simplicity, we assume that the only accessible state is non-degenerate. Then, we can write the partition function for this system as

$$Z = \exp\left[-\beta\left\{-k_B T \ln\left(\frac{V}{V_0}\right)\right\}\right] = \exp\left[\ln\left(\frac{V}{V_0}\right)\right]$$

From Eq. (13.6), we recall that

$$p = Nk_B T\left(\frac{\partial \ln Z}{\partial V}\right)_{T,N} = \frac{Nk_B T}{Z}\left(\frac{\partial Z}{\partial V}\right)_{T,N}$$

$= \ln(V/V_0) = \ln V - \ln V_0$, we find that $\left(\frac{\partial}{\partial V}\ln Z\right)_{T,N} = \frac{1}{V}$. Hence, the expression for pressure exerted by the system is given by

$$p = \frac{Nk_B T}{V}$$

Do you recognise this equation? It is analogous to the equation of state for an ideal gas and follows readily by the application of methods of statistical mechanics.

Example 13.2 Consider a system of N particles and a phase space consisting of only two states with energies 0 and ε (> 0). Calculate the partition function and the internal

Since the N particle system can have only two energy states, 0 and ε, we can write the partition function as

$$Z = \sum_{i=1}^{2} \exp(-\varepsilon_i / k_B T) = 1 + \exp(-\varepsilon / k_B T) \qquad \text{(i)}$$

= 1 and $\varepsilon_2 = \varepsilon$. The occupation numbers are given by

$$n_1 = \frac{N \exp(-\varepsilon_1 / k_B T)}{Z} = \frac{N}{1 + \exp(-\varepsilon / k_B T)} = \frac{N}{1 + \exp(-\theta / T)} \qquad \text{(ii)}$$

where we have put $\theta = \dfrac{\varepsilon}{k_B}$.

Similarly, the occupation number in the second level is given by

$$n = \frac{N \exp(-\varepsilon_2 / k_B T)}{Z} = \frac{N \exp(-\varepsilon / k_B T)}{1 + \exp(-\varepsilon / k_B T)} = \frac{N \exp(-\theta / T)}{1 + \exp(-\theta / T)} = n_1 \exp(-\theta / T) \qquad \text{(iii)}$$

Hence, internal energy of the two-level system is

$$U = n_1 \varepsilon_1 + n_2 \varepsilon_2 = n_2 \varepsilon_2 = \frac{N \varepsilon \exp(-\theta / T)}{1 + \exp(-\theta / T)}$$

Note that if $T << \theta$, the exponential term in the denominator of (ii) will drop out giving . That is, at low temperatures, all the particles will be confined to the lower energy level. However, if $T >> \theta$, $n_1 \cong n_2 \cong \dfrac{N}{2}$.

In general, we note that

$$\frac{n_2}{n_1} = \exp(-\theta / T)$$

$n_2 < n_1$, as expected. However, $n_1 > n_2$ implies population inversion, which forms the operating principle of a laser. We discuss it in detail a little later.

We would like you to answer a practise exercise.

Problem 13.1 Show that

$$G \quad N k_B T \left[\ln Z - \left(\frac{\partial \ln Z}{\partial \ln V} \right)_{T,N} \right]$$

$$H = N k_B T \left[\left(\frac{\partial \ln Z}{\partial \ln T} \right)_{V,N} + \left(\frac{\partial \ln Z}{\partial \ln V} \right)_{T,N} \right]$$

The energy of a simple harmonic oscillator is given by $\left(n+\frac{1}{2}\right)h\nu$. Assume that these are non-degenerate and show that the partition function in this case is

$$Z = \frac{1}{\exp\left(\frac{h\nu}{2k_BT}\right) - \exp\left(-\frac{h\nu}{2k_BT}\right)}$$

Show that the partition function of a non-degenerate N-level system is given by

$$Z = \frac{1-\exp(-Nh\nu/k_BT)}{1-\exp(-h\nu/k_BT)}$$

HE PARTITION FUNCTION FOR AN IDEAL MONATOMIC GAS

Consider an ideal monatomic gas contained in a box. We know that all gaseous particles are identical and their motion is completely random. However, we assume that these can be distinguished from each other. Let the number of particles be N. We now wish to obtain the partition function of this system of N non-localised and distinguishable particles. Since molecules of an ideal monatomic gas do not interact among themselves, except at the instant of collision, it is reasonable to assume that

each particle is essentially independent and has the same set of energy levels as does a single particle in a box;

there is no potential energy due to interatomic interactions; and

at moderate temperatures, the higher energy electronic and nuclear states of individual particles are not excited and we consider only the translational motion of any one molecule*.

In view of the above, it is sufficient to consider single particle partition function.

Single Particle Partition Function

Let us consider one particle in a cubical box of side L, i.e., having volume $V = L^3$. The box is placed in thermal contact with a heat reservoir at a constant temperature T. Note that the particle has translational degrees of freedom only and classically speaking, energy can be treated as a continuous variable. From Eq. (12.60) we recall that the partition function

$$Z = \sum_i g_i \exp(-\beta\varepsilon_i) \tag{13.8a}$$

For a continuous energy distribution, the discrete sum is replaced by integral, with a lower limit of zero and an upper limit of infinity. This gives

$$Z = \int_0^\infty D(\varepsilon)\exp(-\beta\varepsilon)\, d\varepsilon \tag{13.8b}$$

*For most gases, the electronic and nuclear excited states of atoms and molecules lie more than 1 eV above the ground state and are hardly excited at room temperature.

where density of states is given by Eq. (12.10): $D(\varepsilon)\,\mathrm{d}\varepsilon = \dfrac{2\pi V}{h^3}(2m)^{3/2}\,\varepsilon^{1/2}\,\mathrm{d}\varepsilon$.

However, for a non-degenerate, non-relativistic system, Eq. (13.8 a) takes the form

$$Z = \int \exp(-\beta\varepsilon)\,\mathrm{d}\varepsilon = \int \exp\left(-\frac{\beta p^2}{2m}\right)\mathrm{d}p \tag{13.9}$$

The probability that a molecule has momentum between p and $p + \mathrm{d}p$ is equal to the number of cells in 6-D phase space within which such a molecule may exist. If each cell has volume h^3, the single particle partition function can be written as

$$Z = \frac{1}{h^3}\iiint \mathrm{d}x\,\mathrm{d}y\,\mathrm{d}z \iiint_{-\infty}^{\infty} \exp\left[-\frac{\beta}{2m}(p_x^2 + p_y^2 + p_z^2)\right]\mathrm{d}p_x\,\mathrm{d}p_y\,\mathrm{d}p_z \tag{13.10}$$

Note that the limits of integration over momentum coordinates vary from $-\infty$ to ∞.

Proceeding further, we note that integration over space variables gives V so that the expression for single particle partition function simplifies to

$$Z = \frac{V}{h^3}\int_{-\infty}^{\infty}\mathrm{d}p_x \int_{-\infty}^{\infty}\mathrm{d}p_y \int_{-\infty}^{\infty}\mathrm{d}p_z \exp\left(-\frac{\beta}{2m}(p_x^2 + p_y^2 + p_z^2)\right) \tag{13.11}$$

Note that all the three integrals are identical and it will be sufficient to evaluate any one of them. Let us therefore consider

$$I_x = \int_{-\infty}^{\infty} \exp\left(-\frac{\beta p_x^2}{2m}\right)\mathrm{d}p_x$$

$$= \int_{-\infty}^{0} \exp\left(-\frac{\beta p_x^2}{2m}\right)\mathrm{d}p_x + \int_{0}^{\infty} \exp\left(-\frac{\beta p_x^2}{2m}\right)\mathrm{d}p_x \tag{13.12}$$

If we now put $p_x = q$ in the first integral, $dp_x = dq$ and the limits of integration change to). So you can easily convince yourself that both the integrals in Eq. (13.12) will be identical and we can write

$$I_x = 2\int_{0}^{\infty} \exp\left(-\frac{\beta p_x^2}{2m}\right)\mathrm{d}p_x \tag{13.13}$$

To evaluate this integral, we introduce a change of variable by defining $\xi = \dfrac{\beta p_x^2}{2m}$ so that d $\dfrac{m}{\beta}\mathrm{d}\xi$ and $\mathrm{d}p_x = \sqrt{\dfrac{m}{2\beta}}\,\xi^{-1/2}\,\mathrm{d}\xi$. Hence the integral in Eq. (13.13) takes the form

$$I_x = \sqrt{\frac{2m}{\beta}}\int_{0}^{\infty} \exp(-\xi)\,\xi^{-1/2}\,\mathrm{d}\xi \tag{13.14}$$

This is a standard gamma function integral of order (1/2) and has the value $\sqrt{\pi}$ so that

$$I_x = \int_{-\infty}^{\infty} \exp\left(-\frac{\beta p_x^2}{2m}\right)\mathrm{d}p_x = \sqrt{\frac{2m\pi}{\beta}} = \sqrt{2m\pi k_B T} \tag{13.15}$$

We will obtain the same values for integrals over dp_y and dp_z. Using these results in Eq. (13.11), we get a compact expression for single particle partition function:

$$Z = \frac{V}{h^3}\left(\frac{2m\pi}{\beta}\right)^{3/2} = \frac{V}{h^3}(2m\pi k_B T)^{3/2} \tag{13.16}$$

From Eq. (12.58), we recall that

$$\exp(-\alpha) = \frac{N}{Z}$$

Let us now define

$$A = \exp(\alpha) = \frac{Z}{N} = \frac{V(2\pi m k_B T)^{3/2}}{Nh^3} \tag{13.17}$$

From this we note that for high temperatures and/or low densities, A will be large. We can relate it to de Broglie wavelength λ_{dB} and inter-particle distance by noting that

$$\lambda_{dB} = \frac{h}{p}$$

$= k_B T$, the expression for de Broglie wavelength takes the form

$$\lambda_{dB} = \frac{h}{\sqrt{2mk_B T}}$$

Also, inter-particle distance is given by

$$r_0 = \left(\frac{V}{N}\right)^{1/3}$$

Using these expressions for λ_{dB} and r_0 in Eq. (13.17), we get

$$A = \left(\frac{r_0}{\lambda_{dB}}\right)^3 \pi^{3/2} \tag{13.18}$$

This result shows that when de Broglie wavelength is much less than inter-molecular $A >> 1$. We then say that we are in the classical regime. In terms of number $= N/V$, we can express this condition as $n\lambda_{dB}^3 << 1$.

N-Particle Partition Function and Thermodynamic Variables

For a system of indistinguishable N-particles characterised by non-degenerate energy states (cells), the partition function is given by

$$Z_N = \sum_i e^{-\beta E_i} \tag{13.19}$$

$\varepsilon_1 + \varepsilon_2 + \ldots + \varepsilon_N$. In expanded form, we can write

$$Z_N = \sum_i e^{-\beta(\varepsilon_1 + \varepsilon_2 + \ldots + \varepsilon_N)} = \sum (e^{-\beta\varepsilon_1})(e^{-\beta\varepsilon_2})\ldots(e^{-\beta\varepsilon_N})$$

Within the framework of classical statistics, we assume that these particles are distinguishable and independent. So we can write the expression for Z_N in a compact form

$$Z_N = \left(\sum_i \exp(-\beta \varepsilon_i) \right)^N = Z^N \qquad (13.20a)$$

is given by Eq. (13.16). Hence, the N – particle partition function is given by

$$Z_N = V^N \left(\frac{mk_B T}{2\pi \hbar^2} \right)^{3N/2} \qquad (13.20b)$$

In the preceding section, we expressed various thermodynamic functions in terms of the partition function. To proceed with calculations for the special case of an ideal monatomic gas consisting of N molecules, we take natural logarithm on both sides of Eq. (13.20b).

$$\ln Z_N = N\left[\ln V + \frac{3}{2}\ln(k_B T) + \frac{3}{2}\ln\left(\frac{m}{2\pi\hbar^2} \right) \right] \qquad (13.21)$$

We now recall that internal energy of a gas consisting of N-distinguishable particles is given by Eq. (13.5). But if we wish to use N-particle partition function, we should rewrite the expression for internal energy of a gas as

$$U = k_B T^2 \left(\frac{\partial \ln Z_N}{\partial T} \right)_{N,V} \qquad (13.22)$$

From Eq. (13.21) we note that

$$\left(\frac{\partial \ln Z_N}{\partial T} \right)_{N,V} = \frac{3\,Nk_B}{2\,k_B T} = \frac{3}{2}\frac{N}{T}$$

On using this result in Eq. (13.22), the expression for internal energy of a gas made up of -identical particles, which are however considered distinguishable, is given by

$$U = k_B T^2 \left(\frac{3N}{2T} \right) = \frac{3}{2} N k_B T \qquad (13.23)$$

This result agrees with that obtained on the basis of kinetic theory for a monatomic gas having three degrees of translational freedom.
The energy per molecule is given by

$$u = \frac{U}{N} = \frac{3}{2} k_B T \qquad (13.24)$$

At 300 K, the average energy of the molecule of an ideal gas is

$$u = \frac{3}{2} \times (1.38 \times 10^{-23}\ \text{JK}^{-1}) \times 300\ \text{K} = 6.21 \times 10^{-21}\ \text{J}$$

The heat capacity at constant volume is, by definition

$$C_V = \left(\frac{\partial U}{\partial T} \right)_V$$

Using Eq. (13.23), we find that heat capacity is independent of temperature:

$$C_V = \frac{3}{2} Nk_B = \frac{3}{2} nR \tag{13.25}$$

denotes the number of moles of the gas and R is universal gas constant.
The pressure exerted by a gas composed of N-identical molecules is given by

$$p = Nk_B T \left(\frac{\partial \ln Z}{\partial V} \right)_{T,N}$$

In terms of N-particle partition function, we express it as

$$p = k_B T \left(\frac{\partial \ln Z_N}{\partial V} \right)_{T,N} \tag{13.26a}$$

Using Eq. (13.21), we can write

$$\left(\frac{\partial \ln Z_N}{\partial V} \right)_{T,N} = \frac{N}{V}$$

so that Eq. (13.26a) takes the form

$$p = \frac{Nk_B T}{V} \tag{13.26b}$$

Do you recognise this equation? It is equation of state of an ideal gas. It is important to mention here that neither thermodynamics form nor kinetic theory of gases enabled us to establish its exact form. It means that natural explanation of molecular chaos lies in statistical arguments, which are more profound.
On combining this result with Eq. (13.23), we can write

$$p = \frac{2U}{3V} \tag{13.27}$$

That is, the pressure exerted by a gas of N-identical indistinguishable molecules is two-third of energy density.
Similarly, in terms of N-particle partition function, the Helmholtz free energy is given by

$$F = -k_B T \ln Z_N = -Nk_B T \ln \left[\frac{V}{h^3} (2\pi m k_B T)^{3/2} \right] \tag{13.28}$$

From Eq. (13.2), we know that in terms of N-particle partition function, the entropy of a system is expressed through by the relation

$$S = k_B \ln Z_N + \frac{U}{T}$$

On inserting the expressions for $\ln Z_N$ and U from Eqs. (13.21) and (13.23) respectively,

$$S(T,V,N) = Nk_B \left[\ln(VT^{3/2}) + \frac{3}{2} \ln k_B + \frac{3}{2} \ln \left(\frac{m}{2\pi\hbar^2} \right) \right] + \frac{3}{2} Nk_B$$

$$= Nk_B \left[\ln(VT^{3/2}) + \ln \left(\frac{2\pi m k_B}{h^2} \right)^{3/2} + \frac{3}{2} \right]$$

$$= Nk_B \left[\ln(VT^{3/2}) + \ln \left(\frac{2\pi m k_B e}{h^2} \right)^{3/2} \right]$$

$\ln\left(\frac{2\pi m k_B e}{h^2}\right)^{3/2}$ is a constant for a system.

In terms of de Broglie wavelength, we can rewrite the expression for entropy as

$$S(T, V, N) = Nk_B \left[\ln\left(\frac{V}{\lambda_{dB}{}^3}\right)(\pi e)^{3/2} \right] \tag{13.30}$$

Equation (13.29) gives us the entropy of an ideal monatomic gas in terms of T, V and . Note that it is in conformity with the thermodynamic expression for S in its dependence . However, according to Eq. (13.29), as $T \to 0$, $S \to -\infty$, which is not physically meaningful. In fact, it contradicts the third law of thermodynamics. This implies that M-B statistics should not be used to describe the behaviour of a system of distinguishable particles in the vicinity of absolute zero. In fact, at low temperatures quantum effects tend to become pronounced and the right choice will be to use quantum statistics.

1. Mixing of Two Different Ideas Gases From Chapter 7, you may recall that mixing of two different gases by diffusion is an irreversible process and is accompanied by an increase of entropy.

Let us now consider two perfect gases at the same temperature contained in two chambers having volumes V_1 and V_2 and separated by a rigid partition, as shown in Fig. 13.1. Suppose that both gases are in thermal equilibrium at temperature T. Then the entropy of the gas in each chamber can be written as

$$S_1 = N_1 k_B [\ln(V_1 T^{3/2}) + C_1]$$

$$S_2 = N_2 k_B [\ln(V_2 T^{3/2}) + C_2]$$

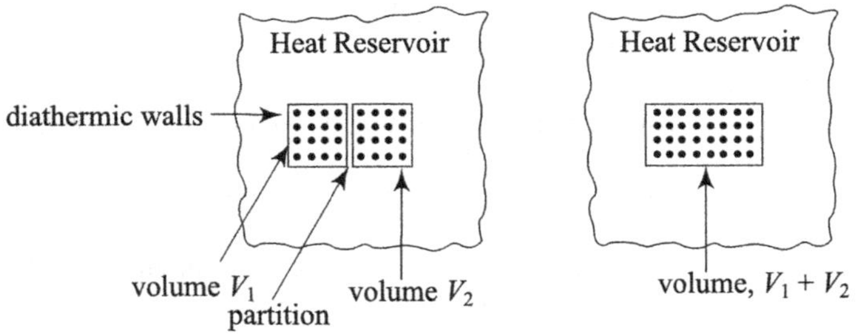

Mixing of two different gases leads to increase in entropy.

So the total initial entropy of the two gases is given by

$$S_i = S_1 + S_2 = N_1 k_B [\ln(V_1 T^{3/2}) + C_1] + N_2 k_B [\ln(V_2 T^{3/2}) + C_2] \tag{13.31}$$

Now let the gases mix by removing the partition. This can be regarded as the expansion of each gas to a volume $V = V_1 + V_2$. The final entropy of the system can now be expressed as

$$S_f = N_1 k_B [\ln(V T^{3/2}) + C_1] + N_2 k_B [\ln(V T^{3/2}) + C_2] \tag{13.32}$$

Hence, the entropy of mixing is obtained by taking the difference in entropy between the final and the initial states. It is given by

$$\Delta S = S_f - S_i = N_1 k_B [\ln(V/V_1)] + N_2 k_B [\ln(V/V_2)] \tag{13.33}$$

$\frac{V}{V_1}$ as well as $\frac{V}{V_2}$ are greater than one, $\Delta S > 0$. That is, entropy of mixing for two different gases is positive since it is an irreversible process and is in conformity with the results of thermodynamics. However, suppose the initial densities are also the same in the two sub-systems, i.e.,

$$\frac{N_1}{V_1} = \frac{N_2}{V_2}$$

In this case, we obtain

$$\Delta S = N_1 k_B\,[\ln(V/V_1)] + N_2 k_B\,[\ln(V/V_2)] = k_B\left[N_1 \ln\left(\frac{N_1 + N_2}{N_1}\right) + N_2 \ln\left(\frac{N_1 + N_2}{N_2}\right)\right] \quad (13.34)$$

which is also positive.

For the particular case of $N_1 = N_2 = \frac{N}{2}$ and $V_1 = V_2 = \frac{V}{2}$, Eq. (13.34) gives

$$\Delta S = 2Nk_B \ln 2 \quad (13.35)$$

2. Gibbs Paradox Let us now consider that the same monatomic gas is put in two chambers separated by a rigid partition. In this case, we expect the mixing to be reversible since there is no difference in the two subsystems. Moreover, physical removal or insertion of the partition should not influence the macroscopic behaviour of the gas or distribution of particles in various accessible microstates. Therefore, we expect the entropy of mixing for self-diffusion to be zero. But this contradicts Eq. (13.35). That is, even for self-mixing, Eq. (13.35) predicts the same increase in entropy as for inter-mixing of two different gases. (It is due to the fact that derivation of Eq. (13.35) does not depend on the identity of a gas or its molecules and Eq. (13.29) is not a homogeneous function of U, V and N.) Obviously, this is not physically tenable; it implies that entropy depends on the history of the system and is not a function of the thermodynamic state alone. That is, we can manage to change entropy of a system by factors extraneous to it. This result poses a paradox *Gibbs Paradox.*

This failure is rooted in the inability of earlier workers to appreciate the significance of the identity of particles in a system. Molecules are either identical or different. When we consider the intermixing of two different gases, the interchange of the position of the molecules of one gas with those of the other creates a new state. This is responsible for the increase of entropy. But in case of identical gases, no new state is created. It means that the number of accessible states is over-estimated. This is because we have considered the molecules to be distinguishable rather than indistinguishable.

3. Sackür-Tetrode Equation In 1911, Sackür suggested that if we divide the expression for thermodynamic probability (Eq. (12.37)) by !, Gibbs paradox could be resolved. But this was just an *ad hoc* correction, since he did not appreciate the real reason behind it—the indistinguishability of particles. To illustrate this point, let us consider a three particle system shown in Fig. 13.2. Among the possible states of the combined system, the six states illustrated

A	B	C		B	C	A
A	C	B		C	A	B
B	A	C		C	B	A

Fig. 13.2 Arrangement of three distinguishable particles

here are all different if the three particles are distinguishable. But they will be identical for indistinguishable particles. Therefore, in calculating Z, the different arrangements in Fig. 13.2 will correspond to six separate terms if the particles A, B and C are distinguishable, but only one term if they are indistinguishable.

You may convince yourself about this number 'six' by noting that three different things can be arranged between themselves in $3!(= 3 \times 2 \times 1 = 6)$ ways. So we can say that N distinguishable particles can be arranged in $N!$ ways. But if they are indistinguishable, all such permutations will be identical and do not lead to a new configuration. While arriving at Eq. (13.31) we have not accounted for these $N!$ identical permutations. This means that for an ideal gas made up of N indistinguishable particles, the expression for thermodynamic probability should be modified to

$$(\Omega_{\text{MB}})_{\text{modified}} = \prod_i \frac{g_i^{n_i}}{n_i!} \tag{13.36}$$

This is known as *Boltzmann counting* and leads to *classical statistics*.
Taking natural logarithm of both sides and using Stirling approximation, we get

$$(\ln \Omega)_{\text{Classical}} = \sum_i [n_i \ln g_i - n_i \ln n_i + n_i]$$

$$= N + \sum_i n_i \ln\left(\frac{g_i}{n_i}\right)$$

Hence, the correct expression for entropy is

$$S^c = k_B\,(\ln \Omega)_{\text{Classical}} = k_B\left[N + \sum_i n_i \ln\left(\frac{g_i}{n_i}\right)\right]$$

On using the result $\dfrac{n_i}{g_i} = \dfrac{N}{Z}\exp(-\beta\varepsilon_i)$ in this expression, we get

$$S^c = k_B\left[N + \sum_i n_i \ln\left(\frac{Z}{N}\exp(\beta\varepsilon_i)\right)\right]$$

$$= k_B\left[N + \sum_i n_i \ln Z - \sum_i n_i \ln N + \beta\sum_i n_i\varepsilon_i\right]$$

$$= k_B[N + N\ln Z - N\ln N + \beta U]$$

$$= Nk_B[\ln Z - \ln N + 1] + \frac{U}{T} \tag{13.37}$$

If we compare this expression for entropy obtained using Boltzmann counting with that obtained without it, we can conclude that Boltzmann counting influences the partition function in the same way as it does thermodynamic probability. Mathematically, we can express correct single particle and N particle partition functions as

$$Z^c = \frac{Z}{N!} \tag{13.38a}$$

$$Z_N^c = \frac{Z_N}{N!} \tag{13.38b}$$

Since internal energy U and pressure p depend on the derivative of the partition function, the presence of the factor $N!$ does not lead to any change in their expressions.

For a gas of N molecules, the correct expression for partition function is given by Eq. (13.20b) and its logarithm by Eq. (13.21). On combining Eqs. (13.21) and (13.37), we obtain correct expression for entropy based on Boltzmann counting:

$$S = k_B \ln Z_N = Nk_B \left[\ln V + \frac{3}{2}\ln(k_B T) + \frac{3}{2}\ln\left(\frac{m}{2\pi\hbar^2}\right) - \ln N + 1 \right] + \frac{U}{T}$$

$$= Nk_B \left[\ln\left(\frac{V}{N}\right) + \frac{3}{2}\ln(k_B T) + \frac{3}{2}\ln\left(\frac{m}{2\pi\hbar^2}\right) + 1 \right] + \frac{3}{2}Nk_B$$

$$= Nk_B \left[\ln\left(\frac{V}{N}\right) + \frac{3}{2}\ln(k_B T) + \frac{3}{2}\ln\left(\frac{m}{2\pi\hbar^2}\right) + \frac{5}{2} \right]$$

We can rewrite it in a more compact form as

$$S_N^c = Nk_B \left[\ln\left(\frac{VT^{3/2}}{N}\right) + C_0 \right] \tag{13.39}$$

$$= \ln\left[\left(\frac{2\pi m k_B}{h^2}\right)^{3/2} e^{5/2} \right].$$

In terms of de Broglie wavelength, Eq. (13.39) takes the form

$$S_N^c = Nk_B \ln\left[\left(\frac{V}{N\lambda_{dB}{}^3}\right) \pi^{3/2} e^{5/2} \right] \tag{13.40}$$

Equation (13.39) is known as the *Sackür–Tetrode equation*. Since there is no undetermined constant in this equation, it gives absolute value of entropy of a monatomic gas. It has been established experimentally that Sackür–Tetrode equation gives the correct expression for the entropy of an ideal monatomic gas.

Let us now show that Sackür–Tetrode equation is free from Gibbs paradox and calculate the entropy of self-mixing for a perfect gas. According to Eq. (13.39), the entropies of the gas in the two chambers are

$$S_1 = N_1 k_B [\ln(V_1/N_1) + \sigma]$$

$$S_2 = N_2 k_B [\ln(V_2/N_2) + \sigma]$$

The total initial entropy of the gas (contained in two separate chambers) is given by

$$S_i = S_1 + S_2 = N_1 k_B [\ln(V_1/N_1) + \sigma] + N_2 k_B [\ln(V_2/N_2) + \sigma] \tag{13.41a}$$

On removing the partition, there will be N particles in volume V. Then, from Eq. (13.39) the final entropy of the gas is

$$S_f = Nk_B [\ln(V/N) + \sigma] \tag{13.41b}$$

Hence, the change in entropy is

$$\Delta S = S_f - S_i = Nk_B \ln(V/N) - N_1 k_B \ln(V_1/N_1) - N_2 k_B \ln(V_2/N_2) \quad (13.42)$$

Since the molecules in the two chambers are at the same pressure and temperature, the number densities of the two samples must be the same:

$$\frac{N_1}{V_1} = \frac{N_2}{V_2} = \frac{N}{V} = \rho$$

Using this fact in Eq. (13.42), we find that the entropy of mixing is zero :

$$\Delta S = k_B\left[(N_1 + N_2)\ln\left(\frac{1}{\rho}\right) - N_1 \ln\left(\frac{1}{\rho}\right) - N_2 \ln\left(\frac{1}{\rho}\right)\right] = 0 \quad (13.43)$$

This means that removal of the partition does not change the accessible microstates and accounting for the indistinguishable particles of an ideal monatomic gas is the key to the resolution of Gibbs' paradox.
You should now solve a practise problem.

Problem 13.2 Calculate the entropy of one mole of neon at STP. Take atomic weight of neon as 20.

4.74 J kg^{-1} K^{-1}

OME DEDUCTIONS FROM M–B STATISTICS

In Chapter 1, we derived expressions for distribution of molecular speeds and energies for an ideal gas based on kinetic theory. We also obtained expression for mean translational energy per degree of freedom for the molecules of a Maxwellian gas. We can obtain the same expressions in a more elegant way using the M–B statistics. We now illustrate this.

Distribution law for Molecular Speeds

From Eq. (12.59), we recall that according to M–B statistics, the number of particles in state is given by

$$n_i = g_i \frac{N}{Z} \exp(-\beta\varepsilon_i) \quad (13.44)$$

For a non-relativistic ideal gas made up of N molecules, it is convenient to consider a continuous distribution of molecular energies rather than a discreet set of energies ε_1, ε_2, , ... ε_G. Then we can rewrite Eq. (13.44) as

$$n(\varepsilon)\mathrm{d}\varepsilon = \frac{N}{Z} g(\varepsilon)\exp(-\varepsilon/k_B T)\,\mathrm{d}\varepsilon \quad (13.45)$$

$d\varepsilon$ gives the number of particles having energies in the range ε and $\varepsilon + d\varepsilon$. Using the relation $\varepsilon = p^2/2\mathrm{m}$, we can rewrite Eq. (13.45) in terms of momentum as

$$n(p)\mathrm{d}p = \frac{N}{Z} g(p)\exp(-p^2/2mk_B T)\,\mathrm{d}p \quad (13.46)$$

The number of cells in phase space within which molecules having momentum between + dp may exist can be expressed as

$$g(p)\,dp = \frac{\iiint dx\,dy\,dz \iiint dp_x\,dp_y\,dp_z}{h^3}$$

is the volume of one cell.

As before, we note that integration over space variables gives volume V occupied by the gas in Cartesian space: $\iiint dx\,dy\,dz = V$. To integrate over p_x, p_y and p_z, we consider only the resultant momentum p. For this, we draw a sphere of radius p about the origin in the momentum space defined by taking p_x, p_y and p_z along three mutually perpendicular axes. Then the surface of the sphere will contain all the points defined by $p_x^2 + p_y^2 + p_z^2 = p^2$. If we draw another concentric sphere of radius p + dp, the volume of a spherical shell of and thickness dp in momentum space will be given by

$$\iiint dp_x\,dp_y\,dp_z = 4\pi p^2\,dp$$

$$g(p)\,dp = \frac{4\pi V p^2\,dp}{h^3} \tag{13.47}$$

$$n(p)\,dp = \frac{4\pi VN}{Zh^3} p^2 \exp(-p^2/2mk_B T)\,dp \tag{13.48}$$

If we now use the relation $p = mv$ so that $p^2dp = m^3v^2dv$. Then on recalling from Eq. (13.16) that $Z = \frac{V}{h^3}(2m\pi k_B T)^{3/2}$, we obtain the expression for the Maxwell–Boltzmann distribution function:

$$n(v)\,dv \equiv dn_v = 4\pi N\left(\frac{m}{2\pi k_B T}\right)^{3/2} v^2 \exp(-mv^2/2k_B T)\,dv \tag{13.49}$$

This equation gives the number of molecules having speeds in the range v to v + dv in an ideal gas at temperature T.

It may be mentioned here that continuous distribution of energies occurs only in the translational motions of molecules. In quantum mechanics, you will learn that molecular rotations and vibrations are quantised where only a few specific energies are permitted. The relative probability that a quantum state of energy E_i will be occupied at temperature is given by Boltzmann factor: $\exp(-E_i/k_BT)$ and the number of modes is given by

$$n_i = g_i \exp(-E_i/k_B T) \tag{13.50}$$

g_i defines the number of quantum states that have the same energy E_i. All quantum states are assumed to have the same a priori probability, i.e., the cells that represent them in phase space have the same volume.

You should now answer a practise problem.

Problem 13.3 The energy of a one-dimensional harmonic oscillator is given by

$$\varepsilon_n = \left(n + \frac{1}{2}\right)\hbar\omega$$

Show that its entropy $S \to 0$ as $T \to 0$.

As another exercise in the applications of statistical methods, we consider specific heat capacity of gases. You are familiar with the difficulties encountered in understanding their observed behaviour from Chapter 1. We will first recapitulate the essence of that discussion and then explain how the problem is overcome using the concepts of quantum mechanics.

Specific Heat Capacity of Gases

From Sec. 1.4, you may recall the statement of the principle of equipartition of energy: The energy of the molecule of an ideal gas is equally divided among its various (translational, rotational, vibrational) degrees of freedom (d.f) and has the value $k_BT/2$. For an ideal gas, we obtained the principle of equipartition of energy from Maxwell's law (Example 1.9). Using this law, we obtained the following results for heat capacities of gases:

The molar heat capacities at constant pressure and at constant volume as also their ratio are the same and independent of temperature for all gases.
The ratio of molar heat capacities decreases with increasing atomicity.
The molar heat capacities vary with temperature.

For monatomic gases such as argon, helium, and neon, experimental results are supported by Eq. (13.25) as well. However, for diatomic gases, it is possible to reproduce only qualitative features of temperature variation of specific heat capacity (Fig. 13.3) and that too by including the contributions of rotational and vibrational modes based on the principle of equipartition of energy. However, the agreement is very poor in

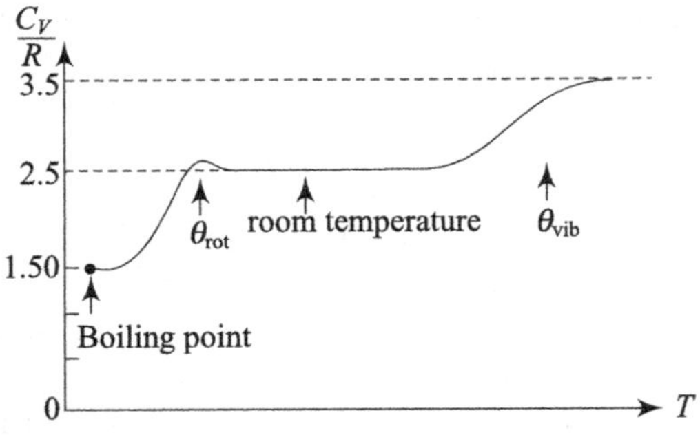

Fig. 13.3 Temperature variation of molar heat capacity of an ideal diatomic gas.

To understand the details, let us consider the specific case of hydrogen. For H_2, C_V is found to be less than $\frac{5R}{2}$ even at room temperature. Eucken observed that from 300 K to 500 K, the value of $C_V \approx \frac{5R}{2}$ but gradually falls to $\frac{3R}{2}$ at about 60 K. (It means that below 60 K, H_2 behaves as a monatomic gas and rotational as well as vibrational modes are 'frozen'.) The observed variation of the specific heat capacity of hydrogen with temperature (Fig.13.4) indicates serious departure from the classical theory.

The experimental data of Eucken suggested that may fall below $\frac{3R}{2}$; a result known as *degeneration of gases*.

Ehrenfest attempted a general explanation of these curves by suggesting that the rotational motion as well as vibrational motion is quantised. Due to quantisation, the contribution of rotational and vibrational modes to specific heat capacity tends to disappear at lower temperatures. The basic argument is very simple and can be understood as

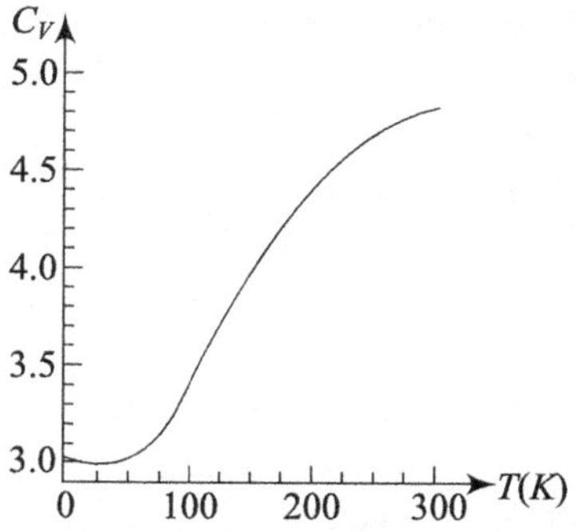

Fig.13.4 Observed temperature variation of specific heat

In quantum description of a system, only discrete energy levels are allowed. If the separation between the levels ΔE around the mean energy is much less than k_BT, the discrete nature of the spectrum is not important and the equipartition theorem constitutes a reasonable approximation. This certainly holds at high temperatures. $\approx k_BT$, the discrete nature of the spectrum becomes important.

From Sec. 13.2, you may recall that all thermodynamic properties, including specific heat capacity can be determined once we know the partition function. Therefore, we now calculate the partition function of a diatomic molecule.

Partition Function of a Diatomic Molecule

Refer to Fig. 13.5. It depicts a diatomic molecule wherein two identical atoms are separated by a finite distance r but held together by interatomic forces along the line joining the particles. (If the atoms are dissimilar, we get a dumb-bell shaped molecule.) If we choose -axis along the line joining the atoms, the molecule can rotate about the y- and z-axes, which are perpendicular to the x-axis. This imparts rotational energy to the molecule. Similarly, if the bonding between the atoms is not very rigid, these atoms can vibrate about their respective equilibrium positions. It means that a diatomic molecule will have vibrational energy as well.

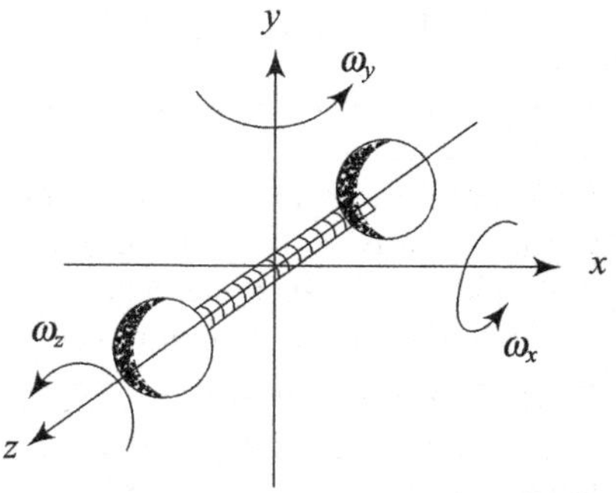

Fig. 13.5 A diatomic molecule having

If we consider a diatomic molecule inside a box at temperature T, the energy of the ith microstate of this molecule can, in general, be expressed as

$$\varepsilon_i = \varepsilon_{\text{tr}} + \varepsilon_{\text{rot}} + \varepsilon_{\text{vib}} + \varepsilon_e + \varepsilon_n \tag{13.51}$$

is the translational energy of the centre of mass of the molecule, ε_{rot} is the energy associated with the rotation of the constituent atoms in the molecule about the centre of is the energy associated with the vibrations of the two atoms along the line joining is the energy of the atomic electrons, and ε_n is the energy of the atomic nucleus. So the partition function of a single diatomic molecule can be written as

$$Z_D = \sum e^{-\beta\varepsilon_i}$$

Substituting for ε_i from Eq. (13.51), we get

$$Z_D = \sum_{(\text{States})} (e^{-\beta\varepsilon_{tr}})(e^{-\beta\varepsilon_{rot}})(e^{-\beta\varepsilon_{vib}})(e^{-\beta\varepsilon_e})(e^{-\beta\varepsilon_n})$$

If we now assume that the various contributions to the total energy given in Eq. (13.51) can vary independently, Z_D may be factorised as

$$Z_D = (\Sigma e^{-\beta\varepsilon_{tr}})(\Sigma e^{-\beta\varepsilon_{rot}})(\Sigma e^{-\beta\varepsilon_{vib}})(\Sigma e^{-\beta\varepsilon_e})(\Sigma e^{-\beta\varepsilon_n})$$

$$Z_D = Z_{tr}\, Z_{rot}\, Z_{vib}\, Z_e\, Z_n \tag{13.52}$$

Z_{rot}, Z_{vib}, Z_e and Z_n respectively denote the translational, rotational, vibrational, electronic and nuclear partition functions.

Taking natural logarithms on both sides, we can write Eq. (13.52) as

$$\ln Z_D = \ln Z_{tr} + \ln Z_{rot} + \ln Z_{vib} + \ln Z_e + \ln Z_n \tag{13.53}$$

Equation (13.53) indicates that the contributions of various partition functions to the thermodynamic variables are additive.

The partition function associated with translational motion of the centre of mass of a diatomic molecule is given by Eq. (13.16):

$$Z_{tr} = \left(\frac{m_D}{2\,\beta\pi\hbar^2}\right)^{3/2} V \tag{13.54}$$

is mass of the diatomic molecule.

Specific Heat Capacity of a Diatomic Molecule

We now know that a diatomic molecule can have translational, rotational and vibrational motions. In addition, electronic and nuclear (spin) effects may also influence it. While the principle of equipartition of energy holds well for translational motion at low temperatures, we have to use the laws of quantum mechanics to obtain expressions for energy of the molecules due to other internal modes (rotational, vibrational, nuclear and electrical). We first consider quantisation of rotational motion.

1. Quantisation of Rotational Motion and its Contribution to Specific Heat

We now know that a diatomic molecule can rotate about the two axes passing through the centre of mass of the two atoms. The rotational energy of a diatomic molecule can be expressed as

$$\varepsilon_{rot} = \frac{1}{2} I\omega^2$$

I $\frac{1}{2}mr^2$ is moment of inertia about an axis at right angles to the line joining the

$\omega^2 = \omega_y^2 + \omega_z^2$. Since angular momentum $J = I\omega$, we can write

$$\varepsilon_{rot} = \frac{J^2}{2I}$$

is the total angular momentum of the system about an axis passing through its centre of mass. This axis is perpendicular to the line joining the atoms.

According to quantum theory, J^2 can take only discrete values given by

$$J^2 = j(j+1)\hbar^2,$$

where rotational quantum number *j* can take integral values, 0, 1, 2, ... So we can expresss the rotational energy in terms of rotational quantum number as

$$\varepsilon_{\text{rot}} = j(j+1)\frac{\hbar^2}{2I} = j(j+1)\frac{h^2}{8\pi^2 I} \tag{13.55}$$

This result shows that unlike translational energy, which varies continuously, rotational energy can take only discrete values.

We would now like you to go through the following example.

Example 13.3 A cubical vessel contains diatomic H_2 gas at a temperature of 300 K. Each H_2 molecule consists of two hydrogen atoms of mass 1.66×10^{-27}kg and separated by about 10^{-10}m. Calculate the root mean square speed of rotation of the molecules about an axis which is perpendicular bisector of the line joining the two atoms.

The number of rotational degrees of freedom is 2. Hence, we can write

$$\frac{1}{2} I \overline{\omega^2} = 2(k_B T/2)$$

$mr^2/2$ is the moment of inertia of the molecule, *m* is mass of hydrogen atom and is the distance between the two hydrogen atoms. Hence, we get

$$\overline{\omega^2} = \frac{2k_B T}{I} = \frac{4k_B T}{mr^2}$$

$$= \frac{4 \times (1.38 \times 10^{-23} \text{ JK}^{-1}) \times 300 \text{ K}}{(1.66 \times 10^{-27} \text{ kg}) \times 10^{-20} \text{ m}^2}$$

$$= 9.976 \times 10^{26} \text{ s}^{-2}$$

$$\sqrt{\overline{\omega^2}} = 3.16 \times 10^{13} \text{ s}^{-1}$$

Due to space quantisation of angular momentum, the rotational levels are degenerate. For a given value of *j*, the angular momentum may be oriented in (2*j* + 1) directions, i. e., degeneracy is given by (2*j* + 1). Hence, the partition function for the rotational motion of a diatomic molecule may be written as

$$Z_{\text{rot}} = \sum_{j=0}^{\infty} (2j+1)\exp[-j(j+1)\hbar^2/2Ik_B T]$$

$$= \sum_{j=0}^{\infty} (2j+1)\exp[-j(j+1)\theta_{\text{rot}}/T] \tag{13.56}$$

$= \dfrac{\hbar^2}{2Ik_B} = \dfrac{h^2}{8\pi^2 Ik_B}$ is the *rotational temperature* and is characteristic of a

θ_{rot} is low for heavy molecules and vice versa. For example, $\theta_{\text{rot}} = 1.52$ K for HCl, 2.9 K for N_2, 2.1K for O_2 and 0.3K for Cl_2. On the other hand, $\theta_{\text{rot}} = 85.5$ K for H_2, $_2$ and 64 K for HD.

Case I High Temperature Limit

>> θ_{rot}, the spacing between the consecutive levels is very small compared to k_BT. Hence the energy may be treated as continuous. Then we can replace the summation in Eq. (13.56) by integration and write

$$Z_{rot} = \int_0^\infty (2j+1)\exp[-j(j+1)\theta_{rot}/T]\,dj \tag{13.57}$$

To evaluate this integral we introduce a change of variable by defining $u = j(j+1)\dfrac{\theta_{rot}}{T}$, so $\dfrac{\theta_{rot}}{T}(2j+1)\,dj$. On substituting these results in the expression for Z_{rot}, we obtain

$$Z_{rot} = \frac{T}{\theta_{rot}}\int_0^\infty \exp(-u)du = -\frac{T}{\theta_{rot}}\exp(-u)\Big|_0^\infty \tag{13.57a}$$

The exponential term vanishes at the upper limit and gives –1 at the lower limit so that

$$Z_{rot} = \frac{T}{\theta_{rot}}$$

On substituting for θ_{rot}, we have, in the limit $T >> \theta_{rot}$

$$Z_{rot} = \frac{2Ik_BT}{\hbar^2}$$

$$\ln Z_{rot} = \ln(2Ik_B/\hbar^2) + \ln T$$

$T >> \theta_{rot}$, the energy for rotational motion of one kmol is given by

$$U_{rot} = N_Ak_BT^2\left(\frac{\partial(\ln Z_{rot})}{\partial T}\right)_{V,N} = N_Ak_BT$$

This result shows that when $T >> \theta_{rot}$, the rotational energy of a diatomic molecule is k_BT. The rotational specific heat capacity of one kmol of a diatomic gas is given by

$$(C_V)_{rot} = \left(\frac{\partial U}{\partial T}\right)_V = N_Ak_B = R \tag{13.58}$$

is universal gas constant. Note that this result is consistent with the principle of equipartition of energy.

To obtain a more accurate expression for specific heat capacity, we use the Euler–Maclaurin formula for evaluating the integral contained in Eq. (13.57a):

$$\int_0^\infty f(x)dx = \frac{f(0)}{2} - \frac{f'(0)}{12} + \frac{f'''}{720}$$

From Eq. (13.57), we note that

$$f(x) = (2x+1)\exp\left(-\frac{x(x+1)\theta_{rot}}{T}\right)$$

(0) = 1

$$f'(x) = 2\exp\left(-\frac{x(x+1)\theta_{\text{rot}}}{T}\right) + (2x+1)\exp\left(-\frac{x(x+1)\theta_{\text{rot}}}{T}\right) \times \left(-\frac{\theta_{\text{rot}}}{T}\right)(2x+1)$$

$$f'(0) = 2 - \frac{\theta_{\text{rot}}}{T}$$

Hence, if we retain terms only up to first derivative of the function, we get

$$Z_{\text{rot}} = \frac{T}{\theta_{\text{rot}}} + \frac{1}{2} - \frac{1}{6} + \frac{1}{12}\frac{\theta_{\text{rot}}}{T}$$

$$= \frac{T}{\theta_{\text{rot}}} + \frac{1}{3} + \frac{1}{12}\frac{\theta_{\text{rot}}}{T}$$

$$= \frac{T}{\theta_{\text{rot}}}\left[1 + \frac{1}{3}\left(\frac{\theta_{\text{rot}}}{T}\right) + \frac{1}{12}\left(\frac{\theta_{\text{rot}}}{T}\right)^2\right]$$

If we include the contribution of third derivative also, this expression will be slightly modified to

$$Z_{\text{rot}} = \frac{T}{\theta_{\text{rot}}}\left[1 + \frac{1}{3}\frac{\theta_{\text{rot}}}{T} + \frac{1}{15}\left(\frac{\theta_{\text{rot}}}{T}\right)^2 + \cdots\right] \tag{13.59}$$

so that the expression for specific heat capacity of one kmol diatomic gas takes the form

$$(C_V)_{\text{rot}} = N_A k_B \frac{\mathrm{d}}{\mathrm{d}T}\left(T^2 \frac{\mathrm{d}}{\mathrm{d}T}\ln Z_{\text{rot}}\right) = N_A k_B\left[1 + \frac{1}{45}\left(\frac{\theta_{\text{rot}}}{T}\right)^2 + \ldots\right] \tag{13.60}$$

Before proceeding further, you must convince yourself about this result (Problem 13.4).

Problem 13.4 Starting from Eq. (13.59), verify Eq. (13.60).

At high temperatures, this expression reduces to Eq. (13.58). Since $(C_V)_{\text{rot}}$ must approach $\to 0$, we expect that the $(C_V)_{\text{rot}}$ versus T curve should show a maximum.

Case II Low Temperature Limit

$_{\text{rot}}$, the exponential term in Eq. (13.56) will drop rapidly. Therefore, it is sufficient to retain only a few terms in Eq. (13.56). So we can write the rotational partition function as

$$Z_{\text{rot}} = 1 + 3e^{-2\theta_{\text{rot}}/T} + 5e^{-6\theta_{\text{rot}}/T} + \ldots \tag{13.61}$$

At low temperatures, the thermal energy of the system, which is of the order of $k_B T$, is not sufficient to take the molecule to higher rotational levels. So it is very likely that the diatomic molecule stays in its ground state of rotational motion. (When $T \approx \theta_{\text{rot}}$ the rotational states of higher energy begin to get excited.) It means that at low temperatures, the series in Eq. (13.61) can be directly used to calculate $(C_V)_{\text{rot}}$. To this end, we note that

$$\frac{d\,Z_{rot}}{dT} = \frac{\theta_{rot}}{T^2}(6e^{-2\theta_{rot}/T} + 30e^{-6\theta_{rot}/T} + \ldots)$$

$$\frac{1}{Z_{rot}}\frac{d\,Z_{rot}}{dT} = \frac{\theta_{rot}}{T^2}\frac{(6e^{-2\theta_{rot}/T} + 30e^{-6\theta_{rot}/T} + \ldots)}{[1 + 3e^{-2\theta_{rot}/T} + 5e^{-6\theta_{rot}/T} + \ldots]}$$

Hence, the rotational energy of one kmol diatomic gas at low temperatures is

$$U_{rot} = N_A k_B T^2 \frac{1}{Z_{rot}}\left(\frac{d\,Z_{rot}}{dT}\right)$$

$$= \frac{6N_A k_B \theta_{rot}[e^{-2\theta_{rot}/T} + 5e^{-6\theta_{rot}/T} + \ldots]}{1 + 3e^{-2\theta_{rot}/T} + 5e^{-6\theta_{rot}/T} + \ldots}$$

$$\approx 6N_A k_B \theta_{rot} e^{-2\theta_{rot}/T}$$

This result shows that energy due to rotational motion decreases exponentially as $T \to 0$. And in the limit $T << \theta_{rot}$, the rotational heat capacity of one kmol diatomic gas is given by

$$(C_V)_{rot} = \frac{dU_{rot}}{dT} = 12N_A k_B\left(\frac{\theta_{rot}}{T}\right)^2 e^{-2\theta_{rot}/T} \tag{13.62}$$

From Eq. (13.62) we note that contribution of rotational specific heat capacity decreases exponentially as $T \to 0$. Now refer to Fig. 13.6, which depicts rotational heat capacity for a typical hetro-nuclear molecule. We note that

For one kilo-mole, $(C_V)_{rot} \to N_A k_B (= R)$.
$\to 0$, $(C_V)_{rot} \to 0$.
$\approx 0.81\theta_{rot}$, $(C_V)_{rot} \to 1.1R$.

These features are satisfactorily explained by Eqs. (13.61) and (13.62). We may now conclude that *rotational motion disappears at low temperatures, whereas classical and quantum theories lead to the same result at higher temperatures.*

It may be mentioned here that the results discussed above are valid only for hetro-nuclear diatomic molecules such as HCl, HI, etc. However, for diatomic molecules with

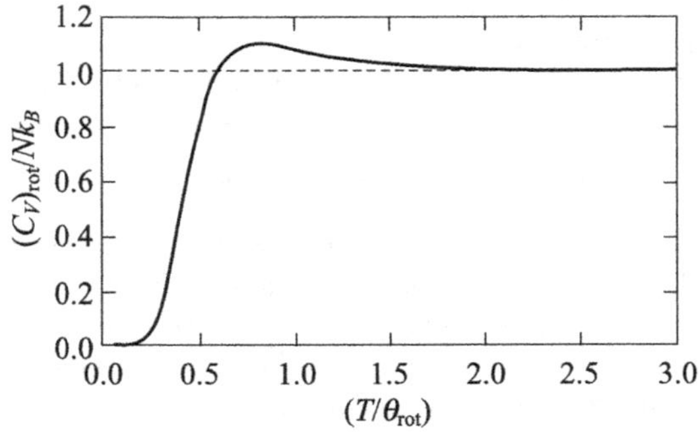

Variation of rotational specific heat capacity of a diatomic gas with (T/θ_{rot}) for a

identical atoms (homo-nuclear), these results have to be modified. At high temperatures, the required modification is quite simple. When a homo-nuclear diatomic molecule is made to rotate through 180°, we can not differentiate its orientation after the rotation from that before the rotation. It means that a homo-nuclear diatomic molecule has only half as many distinguishable orientations as a hetro-nuclear diatomic molecule. Consequently, the rotational partition function for homo-nuclear diatomic molecule is one-half of the rotational partition function for hetro-nuclear diatomic molecule. This difference is expressed in terms of the *symmetry number*, denoted as σ. It specifies the number of indistinguishable orientations that a molecule may have. For a homo-nuclear diatomic $\sigma = 2$ whereas for a hetro-nuclear diatomic molecule, $\sigma = 1$. Accordingly, we divide the expression of rotational partition function by σ.

It may be mentioned here that the effect of symmetry is of particular interest for a homo-nuclear diatomic molecule at low temperatures.

2. Quantisation of Vibrational Motion and its Contribution to Specific Heat The vibrational motion of atoms bound in a molecule can be taken to be nearly simple harmonic. The energy levels of a linear harmonic oscillator are non-degenerate and the vibrational energy of a diatomic molecule is given by

$$\varepsilon_n = (n+1)h\nu; \quad n = 0, 1, 2, 3, \ldots \tag{13.63}$$

is Planck's constant, ν is frequency of oscillation and n is vibrational quantum number. Equation (13.63) shows that the vibrational energy levels of an oscillator are separated by equal steps of $h\nu$. It is important to note that the minimum vibrational energy of an oscillator is not zero; it is $h\nu/2$.

We can now write the vibrational partition function as

$$Z_{\text{vib}} = \sum_{n=0}^{\infty} \exp(-\varepsilon_n / k_B T) = \sum_{n=0}^{\infty} \exp\left[-\left(\frac{h\nu}{k_B T}\right)\left(n + \frac{1}{2}\right)\right]$$

$$= \exp\left(-\frac{h\nu}{2k_B T}\right) \sum_{n=0}^{\infty} \exp\left(-\frac{nh\nu}{k_B T}\right) \tag{13.64}$$

$\exp(-nx) = 1 + e^{-x} + e^{-2x} + \ldots = \dfrac{1}{1 - e^{-x}}$, , the expression for the vibrational partition function simplifies to

$$Z_{\text{vib}} = \frac{\exp\left(-\dfrac{h\nu}{2k_B T}\right)}{1 - \exp\left(-\dfrac{h\nu}{k_B T}\right)} \tag{13.65}$$

If we now define the *vibrational temperature* as $\theta_{\text{vib}} = \dfrac{h\nu}{k_B}$, the vibrational partition function can be rewritten as

$$Z_{\text{vib}} = \exp\left(-\frac{\theta_{\text{vib}}}{2T}\right)\left[1 - \exp\left(-\frac{\theta_{\text{vib}}}{T}\right)\right]^{-1} \tag{13.66}$$

For an HCl molecule, $\theta_{\text{vib}} = 4130$ K, whereas for H_2 molecules, $\theta_{\text{vib}} = 6140$ K. This explains why vibrational states of diatomic molecules are not excited at room temperature and do

To calculate vibrational contribution to heat capacity, we first obtain expression for vibrational energy of one kmol diatomic gas using the relation

$$U_{\text{vib}} = N_A k_B T^2 \left(\frac{\partial (\ln Z_{vib})}{\partial T} \right)_{N,V}$$

On combining this with Eq. (13.66), we get

$$U_{\text{vib}} = N_A k_B T^2 \frac{\partial}{\partial T} \left[-\frac{\theta_{\text{vib}}}{2T} - \ln(1 - \exp(-\theta_{vib}/T)) \right]$$

$$= N_A k_B T^2 \left[\frac{\theta_{\text{vib}}}{2T^2} + \frac{\exp(-\theta_{\text{vib}}/T)}{1 - \exp(-\theta_{\text{vib}}/T)} \left(\frac{\theta_{\text{vib}}}{T^2} \right) \right]$$

$$= N_A k_B \theta_{\text{vib}} \left[\frac{1}{2} + \frac{1}{\exp(\theta_{\text{vib}}/T) - 1} \right] \tag{13.67}$$

$_{\text{vib}}$ we can write $\exp(\theta_{\text{vib}}/T) \approx 1 + \frac{\theta_{\text{vib}}}{T}$. Then Eq. (13.67) reduces to

$$U_{\text{vib}} = N_A k_B \theta_{\text{vib}} \left[\frac{1}{2} + \frac{T}{\theta_{\text{vib}}} \right] \approx N_A k_B T \tag{13.67a}$$

The vibrational specific heat capacity is obtained by differentiating Eq. (13.67) with respect
. For one kmol diatomic gas, we obtain

$$(C_V)_{\text{vib}} = N_A k_B \left(\frac{\theta_{\text{vib}}}{T} \right)^2 \frac{\exp(\theta_{\text{vib}}/T)}{[\exp(\theta_{\text{vib}}/T) - 1]^2} = N_A k_B f\left(\frac{\theta_{\text{vib}}}{T} \right) \tag{13.68a}$$

$$f\left(\frac{\theta_{\text{vib}}}{T} \right) = \left(\frac{\theta_{\text{vib}}}{T} \right)^2 \frac{\exp(\theta_{\text{vib}}/T)}{[\exp(\theta_{\text{vib}}/T) - 1]^2} \tag{13.68b}$$

From Eq. (13.68b), we note that for $T >> \theta_{\text{vib}}$, $f\left(\frac{\theta_{\text{vib}}}{T} \right) \to 1$ and we recover the classical

$$(C_V)_{\text{vib}} \to N_A k_B$$

On the other hand, for $T << \theta_{\text{vib}}$

$$(C_V)_{\text{vib}} \to 0$$

This result shows that the vibrational motion gets frozen at low temperatures. These features are depicted in Fig. 13.7, where we have
$C_V)_{\text{vib}}/R$ as a function of
). As may be noted, at $T = \theta_{\text{vib}}$,
≈ 0.93.

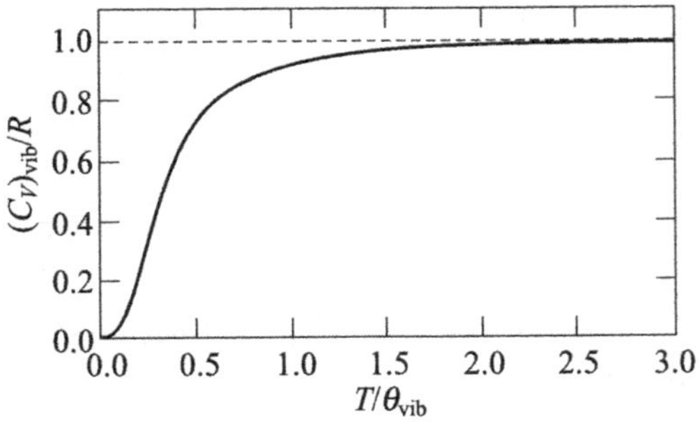

Fig. 13.7 Plot of vibrational specific heat capacity of a typical diatomic molecule

Specific Heat Capacity of Hydrogen

So far we have confined our discussion to hetro-nuclear diatomic molecules like HCl, NO, OH, etc. We now extend our discussion to H_2, which is a homo-nuclear molecule. As mentioned earlier, Eucken investigated the temperature dependence of C_V of H_2, particularly at low temperatures. He observed that the value of C_V gradually falls to $3R/2$ at about 60 K. Ehrenfest attempted a general explanation by including the contribution of rotational motion but he failed to explain the observed behaviour. Then Hori carried out a complete rotational analysis of the Lyman band of hydrogen atom and obtained a value of moment of inertia $I = 0.467 \times 10^{-40}$ g cm^{-2}. He used this value to compute rotational temperature, θ_{rot}, and rotational partition function, Z_{rot}. However, he too failed to explain the observed variation of specific heat capacity.

The anomaly was successfully resolved by Denison, who pointed out that contribution of nuclear spins to the partition function should also be included. We know that hydrogen nuclei have spin ½. Since the hydrogen molecule has two nuclei, their spins may be parallel) or anti-parallel (↑↓). In the former case, the resulting spin will be one whereas in the latter case, it will be zero. It suggests that H_2 gas consists of two varieties of hydrogen corresponding to $j = 1$ and $j = 0$, i.e., we should consider an *ortho-form* of hydrogen corresponding to $j = 1$ and *para-form* of hydrogen corresponding to $j = 0$. The statistical $J + 1$) of ortho- and para-forms of hydrogen are therefore 3 and 1, respectively. gas must be treated as a mixture of two distinct components—ortho-hydrogen and para-hydrogen in meta-stable equilibrium such that ortho-molecules are three times as numerous as those of para-form. Moreover, their relative proportions do not alter during experiments on the measurement of heat capacity.

It can be shown from Pauli's exclusion principle that ortho-molecules can exist only in odd values ($j = 1, 3, 5, \ldots$) quantum states while para-molecules can exist only in even- $= 0, 2, 4, \ldots$) quantum states. So we need to evaluate rotational partition functions for ortho and para-states separately. These are given by

$$(Z_n)_{ortho} = \sum_{j=1,3,5} (2j+1)e^{-j(j+1)\theta_{rot}/T}$$

$$= 3e^{-2\theta_{rot}/T} + 7e^{-12\theta_{rot}/T} + 11e^{-30\theta_{rot}/T} + \ldots \quad (13.69)$$

$$(Z_n)_{para} = \sum_{j=0,2,4} (2j+1)e^{-j(j+1)\theta_{rot}/T}$$

$$= 1 + 5e^{-6\theta_{rot}/T} + 9e^{-20\theta_{rot}/T} + \ldots \quad (13.70)$$

Hence, the equilibrium partition function for 'normal' hydrogen should be written as

$$(Z_{rot})_{eqm} = 3(Z_{rot})_{ortho} + (Z_{rot})_{para}$$

be the total number of molecules. Then,

$$N_{para} = N_0\,(Z_n)_{para} \quad \text{and} \quad N_{ortho} = N_0\,(Z_n)_{ortho} \quad (13.71)$$

Thus, the ratio of hydrogen molecules in ortho form to those in para form can be expressed

$$\frac{N_{\text{ortho}}}{N_{\text{para}}} = \frac{(Z_n)_{\text{ortho}}}{(Z_n)_{\text{para}}} = \frac{3 \sum\limits_{j=1,3,5,\ldots} (2j+1)\exp(-(\varepsilon_j)_{\text{rot}}/k_B T)}{\sum\limits_{j=0,2,4,\ldots} (2j+1)\exp(-(\varepsilon_j)_{\text{rot}}/k_B T)}$$

$$(\varepsilon_j)_{\text{rot}} = j(j+1)k_B\theta_{\text{rot}} = \frac{j(j+1)h^2}{8\pi^2 I}.$$ Hence,

$$\frac{\;}{N_{\text{para}}} = \frac{3\left[3\exp(-\varepsilon_1/k_B T) + 7\exp(-\varepsilon_3/k_B T) + 11\exp(-\varepsilon_5/k_B T) + \ldots\right]}{1 + 5\exp(-\varepsilon_2/k_B T) + 9\exp(-\varepsilon_4/k_B T) + \ldots} \qquad (13.72)$$

Note that energies ε_1, ε_2, ε_3 correspond to $j = 1, 2, 3, \ldots$ respectively.

The calculated percentages of ortho- and para-hydrogen at different temperatures are shown in Table 13.1 and the ratio $N_{\text{para}}/(N_{\text{para}} + N_{\text{ortho}})$ calculated for the equilibrium mixture at various values of T is shown in Fig. 13.8. At high temperatures, the proportion of ortho-molecules to para-molecules is 3:1. But as temperature decreases, the equilibrium proportion of para-molecules increases gradually. It becomes very nearly 100% at 20 K. However, it has been found in usual laboratory experiments that when ordinary hydrogen gas is cooled to low temperatures for heat capacity measurements, this equilibrium percentage is not reached. The time required for the attainment of equilibrium is about a

Table 13.1 *Percentage of ortho and para hydrogen At different temperatures.*

T(K)	O-H_2 (%)	p-H_2 (%)
20	0.18	99.82
40	11.39	88.61
50	23.11	76.89
170	72.01	27.99
273	74.87	25.13
Very high temperatures	75	25.00

Thus, we can write

$$(C_V)_{\text{rot}} = \frac{1}{4}(C_V)_{\text{para}} + \frac{3}{4}(C_V)_{\text{ortho}} \qquad (13.73)$$

We know that for one kmol diatomic gas, internal energy is given by

$$U = N_A k_B T^2 \left(\frac{\partial \ln Z}{\partial T}\right)_{N,V}$$

$$C_V = \left(\frac{\partial U}{\partial T}\right) = \frac{\partial}{\partial T}\left(N_A k_B T^2 \frac{\partial \ln Z}{\partial T}\right)$$

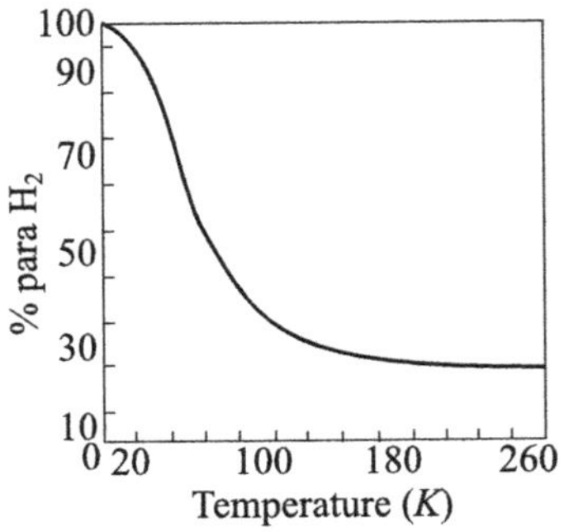

Fig. 13.8 Plot of $N_{\text{para}}/(N_{\text{para}} + N_{\text{ortho}})$ as a function of temperature.

$$(C_V)_{\text{para}} = \frac{\partial}{\partial T}\left\{ N_A k_B T^2 \frac{\partial (\ln Z_{\text{rot}})_{\text{para}}}{\partial T} \right\} = 180 R \left(\frac{\theta_{\text{rot}}}{T} \right)^2 \exp(-6\theta_{\text{rot}}/T) + \ldots \quad (13.74)$$

$$(C_V)_{\text{ortho}} = \frac{\partial}{\partial T}\left\{ N_A k_B T^2 \frac{\partial (\ln Z_{\text{rot}})_{\text{ortho}}}{\partial T} \right\} = \frac{700}{3} R \left(\frac{\theta_{\text{rot}}}{T} \right)^2 \exp(-10\theta_{\text{rot}}/T) + \ldots \quad (13.75)$$

The calculated values of $(C_V)_{\text{para}}$, $(C_V)_{\text{ortho}}$ and $(C_V)_{\text{rot}}$ are shown in Fig. 13.9 by curves (a), (b) and (c), respectively in terms of R. These curves are found to agree favourably with the experimental observations of several investigators.

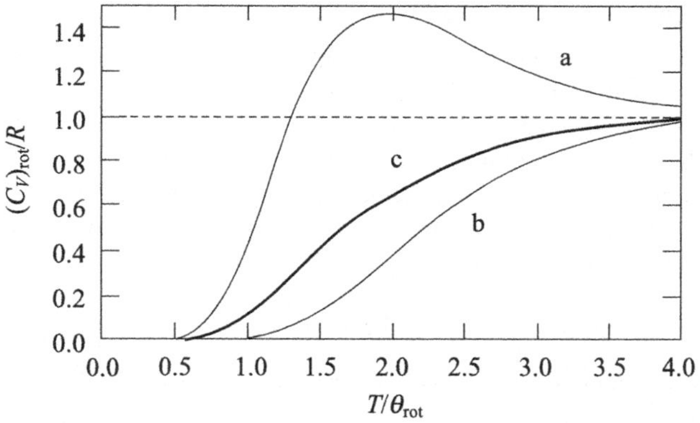

Plot of $(C_V)_{\text{para}}$, $(C_V)_{\text{ortho}}$ and (C_{rot}) with T/θrot is shown by curves a, b, and c

So far we have applied statistical methods to obtain thermodynamic functions of a large number of non-interacting indistinguishable particles. We have also explained temperature dependence of diatomic hetro-nuclear as well as homo-nuclear gas molecules. We now extend this study to understand the temperature variation of specific heat capacities of

PECIFIC HEAT CAPACITY OF SOLIDS

In your earlier physics classes, you have learnt that solids can be regarded as a collection of atoms arranged in a periodic pattern. We know that the atoms in a solid are held together at their equilibrium sites by atomic forces and these execute simple harmonic motion about their respective equilibrium positions. (We refer to this motion as lattice vibration.) That is, atoms can be regarded as independent harmonic oscillators. This suggests that internal energy of a solid resides in the vibrations of its constituent atoms.

To analyse the motion of atoms in a solid, we resolve their vibrational motion in three mutually perpendicular components. That is each atom can be taken to be equivalent to three linearly independent harmonic oscillators, each having an energy equal to k_BT. This result follows directly from Eq. (13.67a). If we consider I kmol of a substance, the energy associated with it will be

$$U = 3N_A k_B T = 3RT \tag{13.75a}$$

is Avogadro's number and R(=8.31 Jkmol^{-1}K^{-1}) is universal gas constant. The molar specific heat capacity at constant volume is given by

$$C_V = \left(\frac{\partial U}{\partial T}\right)_V = 3R = 24.9\ \text{Jkmol}^{-1}\text{K}^{-1} \tag{13.76}$$

This is known as *Dulong–Petit law.* This result was arrived at empirically. They showed that for most crystalline solids at room temperature, $C_V \approx 3R$. However, the Dulong–Petit law fails for light elements such as boron, beryllium and carbon. The specific heat capacities for these elements are 14.03 Jkmol^{-1}K^{-1}, 16.17 Jkmol^{-1}K^{-1} and 6.13 Jkmol^{-1}K^{-1}, respectively at room temperature. Moreover, this law is not consistent with the third law of thermodynamics.

Now refer to Fig. 13.10, which shows the temperature variation of the specific heat capacity of a few typical substances. We note that

the specific heat capacity drops very rapidly as temperature is lowered below room temperature and approaches zero as $T \to 0$;
the value of specific heat capacity is less than the classical value for most substances even at the room temperature; and
all curves show a similar general character. It suggests that there is something seriously wrong with the principle of equipartition of energy used to arrive at the Dulong–Petit law (Eq. (13.76)).

At very low temperatures, $\frac{\theta_{\text{vib}}}{T} >> 1$ and we can ignore the factor 1 in comparison to the exponential in the denominator of the second term in Eq. (13.67). Then, the expression for vibrational energy reduces to

$$\varepsilon_{\text{vib}} = \frac{U_{\text{vib}}}{N_A} = h\nu\left[\frac{1}{2} + \exp\left(-\frac{h\nu}{k_B T}\right)\right] \tag{13.77}$$

which is quite different from the equipartition result.

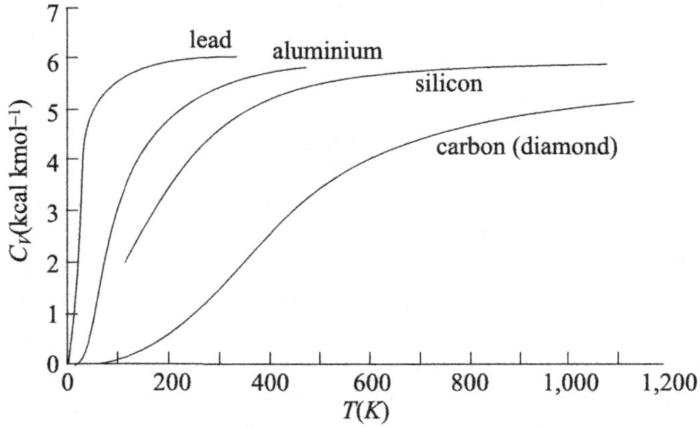

0, $\exp(-\theta_{vib}/T) \rightarrow 0$, so that $\varepsilon_{vib} \rightarrow \frac{1}{2} k_B \theta_{vib} = \frac{1}{2} h\nu$. This is referred to as the *zero-point energy*. This result is completely unknown in the realm of the classical theory.

The first successful attempt to explain the behaviour of the heat capacity of solids at low temperatures was made by Einstein. He recognised that the basic error in the derivation of Eq. (13.76) is the same as that responsible for the failure of Rayleigh–Jeans law to explain variation of spectral density for a blackbody radiation. That is, the spectrum of energies is not continuous but quantised in multiples of $h\nu$, the ideas developed by Planck to explain blackbody radiation. We will now discuss Einstein's theory of the heat capacity of solids.

Einstein's Theory

Einstein assumed that

> Each atom vibrates with the same angular frequency ω_E and may be treated as an independent oscillator. It means that all $3N_A$ modes of oscillation of a solid have the same frequency.
>
> From the standpoint of a single oscillator, the rest of the atoms in the solid act as a heat reservoir at constant temperature, T.

For one kmol of a solid, which is equivalent to an aggregate of $3N_A$ independent linear harmonic oscillators, we can rewrite Eq. (13.67) as

$$U_{vib} = 3N_A h\nu \left[\frac{1}{2} + \frac{1}{\exp(h\nu/k_B T) - 1} \right]$$

Hence, molar specific heat capacity is given by

$$C_V = \left(\frac{\partial U_{vib}}{\partial T} \right)_V = 3N_A h\nu \times \frac{\exp(h\nu / k_B T)}{\left[\exp\left(\frac{h\nu}{k_B T} \right) - 1 \right]^2} \times \frac{h\nu}{k_B T^2}$$

$$= 3N_A k_B \left(\frac{h\nu}{k_B T} \right)^2 \frac{\exp(h\nu/k_B T)}{\left[\exp\left(\frac{h\nu}{k_B T} \right) - 1 \right]^2}$$

$$= 3N_A k_B \left(\frac{\theta_E}{T} \right)^2 \frac{\exp(\theta_E /T)}{\left[\exp\left(\frac{\theta_E}{T} \right) - 1 \right]^2} \qquad (13.78)$$

$= \frac{h\nu_E}{k_B}$ is called the *Einstein temperature*. This is an expression for the heat capacity of solids. For copper, $\theta_E = 230$ K and you can easily convince yourself that Einstein frequency $\nu_E = 4.8 \times 10^{12} s^{-1}$.

At high temperatures ($T >> \theta_E$), Eq. (13.78) reproduces the Dulong–Petit law.

Problem 13.5 Show that at high temperatures, i.e., for $T >> \theta_E$, Eq. (13.78) reproduces Dulong–Petit value of molar specific heat capacity.

At low temperatures, $\frac{\theta_E}{T} >> 1$ and $\exp(\theta_E / T) - 1 \cong \exp(\theta_E / T)$ so that Eq. (13.78)

$$C_V = 3R\left(\frac{\theta_E}{T}\right)^2 \exp(-\theta_E / T) \tag{13.79}$$

Equation (13.79) shows that C_V falls off exponentially at low temperatures and approaches zero as $T \to 0$. These features are shown in Fig. 13.11 for aluminium with 10^{12}s^{-1}. There is a fairly good agreement between theoretical and experimental results, except at the low temperatures. It means that Einstein's theory is also not strictly valid in the complete temperature range and we must look beyond it for complete explanation of specific heat capacity of solids. A more refined theory was put forward by Debye and we discuss it now.

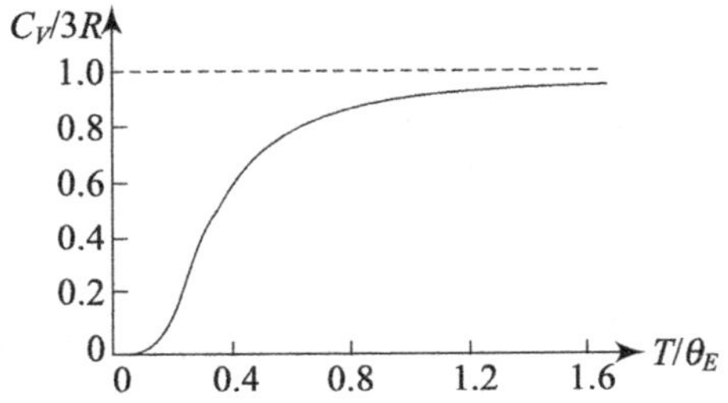

Debye Theory

Debye argued that the discrepancy at low temperatures is largely due to the oversimplified assumption made by Einstein in that the motion of each atom in a solid is independent of the motion of its neighbours. He suggested that the single frequency model of Einstein be replaced by a model where a spectrum of vibrational frequencies is considered. At low temperatures, if an atom is displaced from its mean position, it affects atoms in its neighbourhood. This means that many different modes can exist in a solid, each mode corresponding to a group of atoms vibrating in phase at the same frequency. So to give a satisfactory theory of the specific heat capacity of solids, Debye suggested that it is necessary to know the frequency distribution of these modes. We can do so by regarding the solid as a continuous elastic body and drawing an analogy between the theory of radiation in an enclosure and the wave nature of the motion of an atom in a solid. We will now discuss it in detail.

Suppose that all the $3N$ modes have different frequencies. Then, the vibrational partition function will be modified to

$$Z_{\text{vib}} = \sum_{n_1, n_2 \ldots n_{3N}} \exp\left[-\beta\left\{\left(n_1 + \frac{1}{2}\right)\hbar\omega_1 + \left(n_2 + \frac{1}{2}\right)\hbar\omega_2 + \ldots + \left(n_{3N} + \frac{1}{2}\right)\hbar\omega_{3N}\right\}\right] \tag{13.80}$$

$$e^{-\beta K_0}\left(\sum_{n_1} e^{-\beta n_1 \hbar\omega_1}\right)\left(\sum_{n_2} e^{-\beta n_2 \hbar\omega_2}\right)\cdots\left(\sum_{n_{3N}} e^{-\beta n_{3N} \hbar\omega_{3N}}\right)$$

$= (k_B T)^{-1}$ and $K_0 = \frac{\hbar}{2}\sum_{r=1}^{3N}\omega_r$.

The terms within each parentheses form a geometric series and we can evaluate their sum easily to obtain

$$Z_{\text{vib}} = e^{-\beta K_0}\left(\frac{1}{1-e^{-\beta\hbar\omega_1}}\right)\left(\frac{1}{1-e^{-\beta\hbar\omega_2}}\right)\cdots\left(\frac{1}{1-e^{-\beta\hbar\omega_{3N}}}\right)$$

$$= e^{-\beta K_0}\prod_{r=1}^{3N}(1-e^{-\beta\hbar\omega_r})^{-1}$$

$$\ln Z_{\text{vib}} = -\beta K_0 - \sum_{r=1}^{3N}\ln(1-e^{-\beta\hbar\omega_r}) \tag{13.81}$$

If we define the density of modes $D(\nu)d\nu$ as the number of modes lying in the frequency $\nu + d\nu$, we can rewrite Eq. (13.81) as

$$\ln Z_{\text{vib}} = -\frac{1}{k_B T}K_0 - \int_0^\infty \ln\left[1-\exp\left(-\frac{h\nu}{k_B T}\right)\right] D(\nu)d\nu \tag{13.82}$$

Hence, vibrational energy of the solid is given by

$$U_{\text{vib}} = k_B T^2\frac{\partial \ln Z_{\text{vib}}}{\partial T} = K_0 + \int_0^\infty \frac{h\nu\exp(-h\nu/k_B T)}{1-\exp(-h\nu/k_B T)}D(\nu)d\nu$$

On rearranging terms, we can rewrite it as

$$U_{\text{vib}} = K_0 + \int_0^\infty \frac{h\nu}{\exp(h\nu/k_B T)-1}D(\nu)d\nu \tag{13.83}$$

Thus, the molar specific heat capacity of the solid is given by

$$C_V = \left(\frac{\partial U_{\text{vib}}}{\partial T}\right)_V = k_B\int_0^\infty\left(\frac{h\nu}{k_B T}\right)^2\frac{\exp(h\nu/k_B T)}{[\exp(h\nu/k_B T)-1]^2}D(\nu)d\nu \tag{13.84}$$

To evaluate this integral, we must know the density of modes in accordance with Debye's model. In this model, the solid is treated as a continuous elastic medium and each normal mode* of vibration of this elastic medium has a characteristic wavelength λ. This approximation is good if $\lambda >> a$, the interatomic separation.

To obtain $D(\nu)$, we look for the relation between ν and the wave vector **k**, so as to relate and the momentum vector **p** associated with the wave. You may recall from your school physics classes that $p = h/\lambda = 2\pi\hbar/\lambda = \hbar k$. Once we are able to relate ν to p, we can use the fact that the number of translational states for the momentum range between **p** and

---

*If motion of a coupled vibrating system is not simple harmonic, it can be analysed in terms of normal modes, which are fundamental modes of vibration. That is, any particular state of vibration can be expressed as a linear combination of *normal modes*. For example, in the case of a coupled pendulum having bobs of equal mass and strings of equal length, there are two normal modes. The first mode corresponds to the individual amplitudes being equal and in the same direction. The second mode corresponds to the amplitudes being equal but in opposite directions.

$Vd^3\mathbf{p}/h^3$ where V is the volume of the solid. Thus, the number of modes having k and lying between k and $k+dk$ is

$$\frac{V}{h^3}4\pi p^2\,\mathrm{d}p = \frac{V}{h^3}4\pi(\hbar k)^2\,\mathrm{d}(\hbar k) = \frac{V}{(2\pi)^3}4\pi k^2\,\mathrm{d}k$$

$=\dfrac{2\pi}{\lambda}=\dfrac{2\pi}{c/\nu}$, where $\nu\left(=\dfrac{c}{\lambda}\right)$ is the natural frequency, we can write

$$k = \frac{2\pi\nu}{c}$$

$$\mathrm{d}k = 2\pi\frac{\mathrm{d}\nu}{c}$$

So, the number of modes between ν and $\nu + \mathrm{d}\nu$ is given by

$$D(\nu)\mathrm{d}\nu = \frac{V}{(2\pi)^3}4\pi\left(\frac{2\pi\nu}{c}\right)^2 2\pi\frac{\mathrm{d}\nu}{c} = \frac{4\pi V}{c^3}\nu^2\,\mathrm{d}\nu \tag{13.85}$$

At this point, we also need to count the states of polarisation of elastic waves in a solid. Incidentally, there are three such states, two for transverse waves and one for longitudinal . Moreover, the velocities of longitudinal and transverse waves (c_ℓ and c_t) are

Hence, we obtain $D(\nu)$ by modifying Eq. (13.85) as

$$D(\nu)\mathrm{d}\nu = \frac{4\pi V}{c_\ell^3}\nu^2\,\mathrm{d}\nu + 2\frac{4\pi V}{c_t^{\,3}}\nu^2\,\mathrm{d}\nu$$

$$= 4\pi V\left(\frac{1}{c_\ell^3}+\frac{2}{c_t^3}\right)\nu^2\,\mathrm{d}\nu \tag{13.86}$$

The plot of $D(\omega)$ vs. ω is shown in Fig. 13.12.

Now, we have to remember the restriction imposed by the fact that the total number of vibrational modes, known *phonons*, is $3N$, where N is the total number of atoms. In other words, there must be an upper limit on the frequency spectrum so that we can write

$$\int_0^{\nu_D} D(\nu)\mathrm{d}\nu = 3N \tag{13.87}$$

is called the *Debye cut-off frequency*. This is the same for transverse (t) and longitudinal (ℓ) modes (Fig. 13.13).

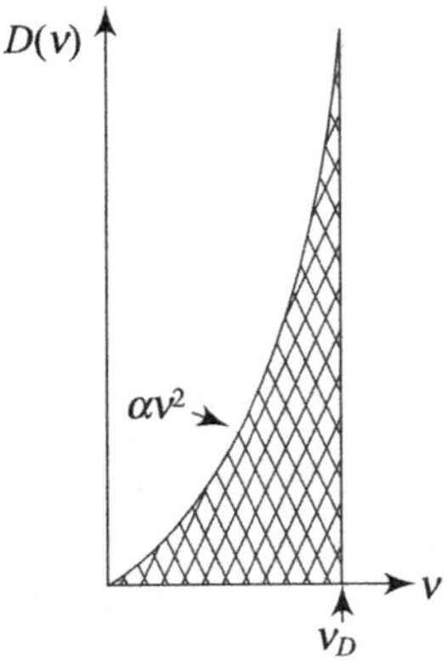

Fig. 13.12 Plot of $D(\nu)$ vs ν for a Debye

*In the longitudinal mode, the vibration of an atom is along the direction of propagation, whereas in the transverse modes the vibrations are perpendicular to the direction of propagation. Thus, the transverse mode has two independent components.

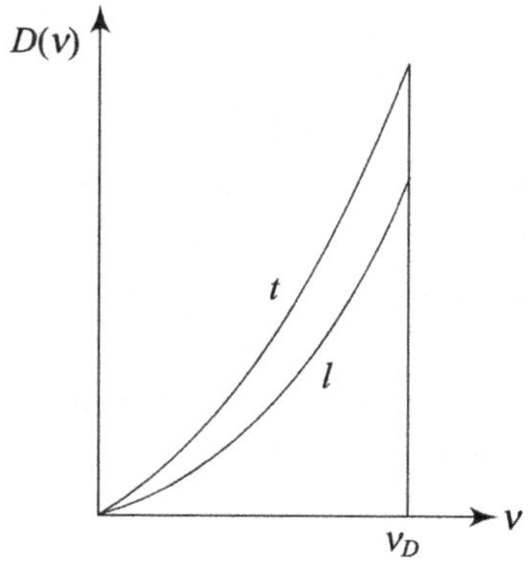

Debye cut-off frequency for transverse and longitudinal modes.

From Eqs. (13.86) and (13.87), we get

$$\nu_D^3 = \frac{9N}{4\pi V}\left(\frac{1}{c_\ell^3}+\frac{2}{c_t^3}\right)^{-1} \tag{13.88}$$

On inserting the value of $D(\nu)$ from Eq. (13.86), we can rewrite Eq. (13.84) as

$$C_V = 4\pi V k_B \int_0^{\nu_D}\left(\frac{h\nu}{k_B T}\right)^2 \frac{\exp(h\nu/k_B T)}{[\exp(h\nu/k_B T)-1]^2}\left(\frac{1}{c_\ell^3}+\frac{2}{c_t^3}\right)\nu^2\,\mathrm{d}\nu \tag{13.89}$$

To evaluate this integral we introduce a change of variable by defining $x = \frac{h\nu}{k_B T}$. $\mathrm{d} = \frac{h}{k_B T}\mathrm{d}\nu$ and $\nu^4\,\mathrm{d}\nu = \left(\frac{k_B T}{h}\right)^5 x^4\,\mathrm{d}x$ and the limits of integration vary from 0 $\frac{\nu_D}{k_B T}$. Substituting these results in Eq. (13.89), we get

$$C_V = 4\pi V k_B\left(\frac{k_B T}{h}\right)^3\left(\frac{1}{c_\ell^3}+\frac{2}{c_t^3}\right)\int_0^{x_D}\frac{x^4\exp(x)}{[\exp(x)-1]^2}\mathrm{d}x$$

x_D is related to the Debye temperature, θ_D of a solid through the relation

$$x_\mathrm{D} = \frac{h\nu_\mathrm{D}}{k_B T} = \frac{\theta_\mathrm{D}}{T}$$

Using Eq. (13.88), we can rewrite this expression for specific heat capacity in a compact

$$C_V = 9Nk_B\left(\frac{T}{\theta_\mathrm{D}}\right)^3\int_0^{\theta_\mathrm{D}/T}\frac{\mathrm{e}^x x^4}{(\mathrm{e}^x-1)^2}\mathrm{d}x \tag{13.90}$$

Integrating by parts, we obtain

$$C_V = 9Nk_B\left(\frac{T}{\theta_\mathrm{D}}\right)^3\left[-\frac{x^4}{\exp(x)-1}\Bigg|_0^{\theta_\mathrm{D}/T} + 4\int_0^{\theta_\mathrm{D}/T}\frac{x^3}{\mathrm{e}^x-1}\mathrm{d}x\right]$$

$$= 3Nk_B\left[12\left(\frac{T}{\theta_D}\right)^3 \int_0^{\theta_D/T} \frac{x^3}{e^x - 1}\,dx - \frac{3(\theta_D/T)}{e^{\theta_D/T} - 1}\right]$$

, the molar specific heat capacity of the solid is given by

$$C_V = 3R\left[12\left(\frac{T}{\theta_D}\right)^3 \int_0^{\theta_D/T} \frac{x^3}{e^x - 1}\,dx - \frac{3(\theta_D/T)}{e^{\theta_D/T} - 1}\right] \tag{13.91}$$

The term within the square brackets is termed as the *Debye function*. Note that the molar specific heat capacity of a solid is a function of (θ_D/T). The values of Debye temperature for a few substances are given in Table 13.2.

Table 13.2 *Debye temperatures for a few substances.*

Substance	θ_D(K)	Substance	θ_D(K)
Hg	75	Pb	95
H	100	Na	160
Fe	360	Al	375
Be	1200	C	1850
AgBr	150	NaCl	280

The integral occurring in the Debye function cannot be evaluated analytically. For this reason, the Debye function is computed numerically and its tabulated values are available. However, we can integrate it in the high and low temperature limits.

High Temperature Limit ($T >> \theta_D$) At high temperatures, $x(= h\nu/k_BT) << 1$ so that + ... or $e^x - 1 \approx x$. Hence, the integral in Eq. (13.91) reduces to

$$\int_0^{\theta_D/T} \frac{x^3\,dx}{e^x - 1} = \int_0^{\theta_D/T} x^2\,dx = \frac{1}{3}\left(\frac{\theta_D}{T}\right)^3$$

and the second term becomes equal to 3. This means that at high temperatures, the Debye function becomes unity and the molar specific heat capacity of a solid equals $3R$. This is in conformity with the Dulong–Petit law.

Low Temperature Limit ($T << \theta_D$) At very low temperatures, x is very large and we can neglect 1 in comparison with the exponential term. This makes the second term in the Debye function almost negligible.

Moreover, the upper limit of the integral in the Debye function may be taken as infinity. Then the expression for C_V simplifies to

$$C_V = 36R\left(\frac{T}{\theta_D}\right)^3 \int_0^{\infty} \frac{x^3\,dx}{e^x - 1}$$

The integral is a standard zeta function and is given by

$$\int_0^{\infty} \frac{x^3\,dx}{e^x - 1} = 3!\zeta(4) = \frac{4\pi}{15} \quad \left(\because \zeta(4) = \frac{\pi^4}{90}\right)$$

(4) is the Riemann Zeta function of order 4.

Hence, the molar specific heat capacity of a solid at low temperatures varies according

$$C_V = 36R\left(\frac{T}{\theta_D}\right)^3 \frac{\pi^4}{15} = \alpha T^3 \tag{13.92}$$

$\left(= \dfrac{12\pi^4 R}{5\theta_D^3}\right)$ is a constant for a given solid.

Equation (13.92) is the famous Debye's T^3-law. It shows that as $T \to 0$, $C_V \to 0$, in conformity with observed results. A plot of C_V vs T^3 for solid argon ($\theta_D = 93$ K) is shown in Fig. 13.14. It shows quite good agreement between the experimental and theoretical

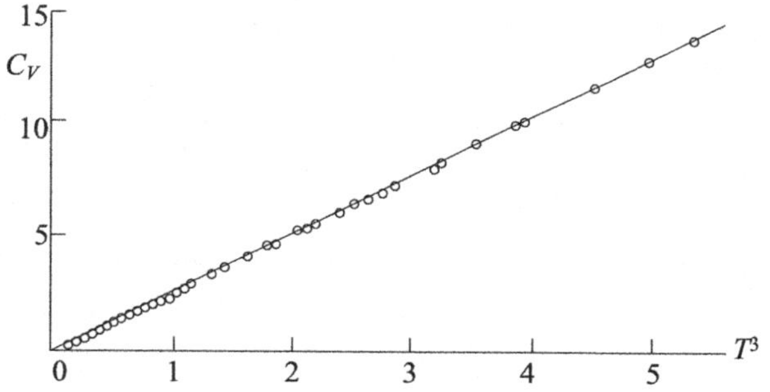

Plot of C_V vs. T^3 for solid argon below 5 K. The circles are the experimental points. The straight line corresponds to Eq. (13.92) with $\theta_D = 93$ K.

It is obvious now that Debye's theory is superior to Einstein's theory at low temperatures. But the assumption of an isotropic solid made by Debye may not hold always. For example, graphite has anisotropic layered structure and Debye's theory fails completely. Even for isotropic solids, the assumption of the existence of a threshold frequency made by Debye is open to discussion. To have a feel for this, let us estimate the order of magnitude of ν_D. By taking $(N/V) = 10^{28}$ per m^3 and the velocity of sound to be of the order of 10^3 ms^{-1}, $\sim 10^{13}$ per second. This corresponds to a minimum wavelength of the order of 1Å, indicating that the continuous model may be erroneous, particularly in the high frequency region.

A very rigorous analysis of the specific heat capacity of solids was done by Born and Kerman. An excellent account of this analysis can be found in Lattice Dynamics by Ghatak and Kothari (1971).

So far we have considered systems having closely spaced energy levels. An important class of phenomena is exhibited by systems having a few discrete energy levels. In particular, a two-level system is of great importance in physics. Such a system is characterised by only two microstates. A typical example of such a system is an electron which can be either in a spin-up or a spin-down state. In some cases, we may have several particles with only two energy states available to them. We shall now discuss the physics of such systems in some detail.

HERMODYNAMIC FUNCTIONS OF SYSTEMS WITH INITE NUMBER OF ENERGY LEVELS

For simplicity, we consider a system having two non-degenerate microstates, with energies ε_1 and ε_2, as shown in Fig. 13.15. The energy difference between the levels is

$$\varepsilon = \varepsilon_2 - \varepsilon_1$$

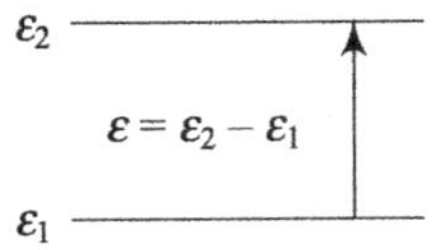

Fig. 13.15 A non-degenerate

Now, let us assume that this system is in thermal equilibrium with a heat reservoir at temperature T. The partition function of this system is

$$Z = \exp(-\varepsilon_1/k_B T) + \exp(-\varepsilon_2/k_B T)$$

The probability of occupancy of these states is given by

$$P_1 = \frac{\exp(-\varepsilon_1/k_B T)}{Z} = \frac{\exp(-\varepsilon_1/k_B T)}{\exp(-\varepsilon_1/k_B T) + \exp(-\varepsilon_2/k_B T)} \tag{13.93a}$$

$$P_2 = \frac{\exp(-\varepsilon_2/k_B T)}{Z} = \frac{\exp(-\varepsilon_2/k_B T)}{\exp(-\varepsilon_1/k_B T) + \exp(-\varepsilon_2/k_B T)} \tag{13.93b}$$

For convenience, we introduce a characteristic temperature θ, by defining

$$\theta = \frac{\varepsilon_2 - \varepsilon_1}{k_B} \tag{13.94}$$

In terms of θ, we can rewrite Eqs. (13.93a, b) as

$$P_1 = \frac{1}{1 + \exp(-\varepsilon/k_B T)} = \frac{1}{1 + \exp(-\theta/T)} \tag{13.95a}$$

$$P_2 = \frac{\exp(-\varepsilon/k_B T)}{1 + \exp(-\varepsilon/k_B T)} = \frac{\exp(-\theta/T)}{1 + \exp(-\theta/T)} \tag{13.95b}$$

From these equations we note that P_1 as $_2$ depends on T. If $T = 0$, $P_1 = 1$ and $= 0$, which means that the system is in its ground state. A plot of P_2 vs. T/θ is shown in Fig. 13.16. When $T << \theta$, i.e., when P_2 is negligible; it begins to increase significantly when $T > 0.2\theta$. The rate of increase is maximum in the range $0.3\theta <$. As T increases further, P_2 approaches a limiting value of 0.5. For example, when $T = 20$ K, $= 0.4875$.

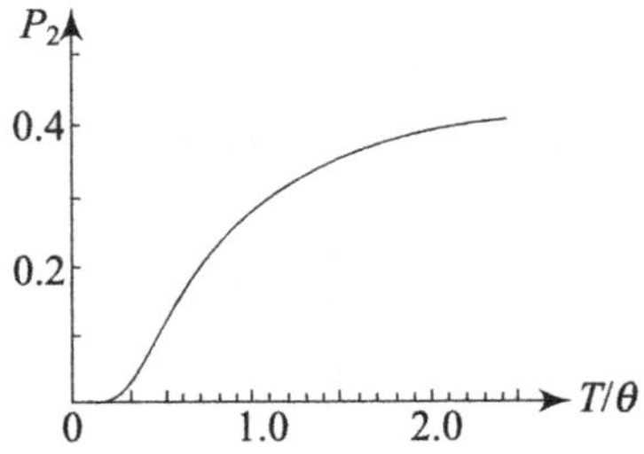

Problem 13.6 (**a**) Calculate P_2 when $T = 0.01\theta$ and 0.1θ.

Prove that the average energy of the two-level system defined above is given by

$$\bar{\varepsilon} = \frac{\varepsilon_1 + \varepsilon_2 \exp(-\theta/T)}{1 + \exp(-\theta/T)}$$

Hence show that for $\theta = T$, $\bar{\varepsilon} = 0.731\,\varepsilon_1 + 0.269\,\varepsilon_2$.

3.72×10^{-44}, 4.54×10^{-5}

We now consider a particular case where a system has two magnetic potential energy levels. You will discover that such a system exhibits paramagnetism.

Consider a paramagnetic substance having N magnetic atoms per unit volume placed in an external magnetic field **B**. Let us assume that each atom has spin ½ and an intrinsic magnetic moment μ_B. Classically, the magnetic moment can point along any direction in space. So the energy of an atom in this field will be $-\mu_B B\cos\theta$, where θ is the angle between the directions of magnetic moment and the field. However, owing to the restrictions imposed by the principles of quantum mechanics, θ can have only two values: 0 and 180°. Accordingly, only two energy levels, with energies $-\mu_B B$ and $+\mu_B B$ will be available to the system. We assume that the energy levels are non-degenerate. That is, there is only one state in each level. However, there is no restriction on the number of atoms per state. The energy of the lower state in which μ_B is parallel to the magnetic field is $\varepsilon_1 = -\mu_B B$ and the energy of the higher state in which μ_B is parallel to the magnetic field is $\varepsilon_2 = \mu_B B$. Thus, if the system is in thermal equilibrium at temperature T, the partition function of the system can be written as

$$Z = \exp\left(\frac{\mu_B B}{k_B T}\right) + \exp\left(-\frac{\mu_B B}{k_B T}\right) \tag{13.96}$$

If we define $x = \dfrac{\mu_B B}{k_B T}$, the two level partition function can be expressed as

$$Z = 2\cosh x = 2\cosh\left(\frac{\mu_B B}{k_B T}\right) \tag{13.97}$$ *

Proceeding further, let us assume that the number of atoms whose moments are parallel and anti-parallel to **B** are N_1 and N_2, respectively. Then we can write

$$N_1 = \frac{N}{Z}\exp\left(-\frac{\varepsilon_1}{k_B T}\right)$$

Strictly speaking, the total energy of the system should be expressed as

$$E = E_{vib} + E_{int} + E_p$$

is the vibrational energy of the system, E_{int} is the energy associated with internal magnetic and electric interactions and E_p is the (magnetic) potential energy.

Accordingly, the partition function of the system should be

$$Z = Z_{vib} Z_{int} Z_p$$

is equivalent to Z given by Eq. (13.97). However, for simplicity we are ignoring E_{vib} and E_{int} in

$$N_2 = \frac{N}{Z}\exp\left(-\frac{\varepsilon_2}{k_B T}\right)$$

(= $N_1 + N_2$) denotes the total number of atoms.

Thus, the excess of atoms aligned along **B** over those aligned anti-parallel to **B** is

$$N_1 - N_2 = \frac{N}{Z}\left[\exp\left(-\frac{\varepsilon_1}{k_B T}\right) - \exp\left(-\frac{\varepsilon_2}{k_B T}\right)\right]$$

$$= \frac{N}{Z}\left[\exp\left(\frac{\mu_B B}{k_B T}\right) - \exp\left(-\frac{\mu_B B}{k_B T}\right)\right]$$

$$= \frac{2N}{Z}\sinh\left(\frac{\mu_B B}{k_B T}\right)$$

On substituting for Z from Eq. (13.97), we get

$$N_1 - N_2 = N\tanh\left(\frac{\mu_B B}{k_B T}\right) \tag{13.98}$$

The net magnetic moment of the system is therefore given by

$$M = \mu_B\,(N_1 - N_2) = N\mu_B\tanh\left(\frac{\mu_B B}{k_B T}\right) \tag{13.99}$$

This expression relates the magnetic moment M to B and T. In fact, M is determined by $/T$, rather than their individual values.

You may now like to answer a practise problem.

Problem 13.7 Using Eq. (13.97), obtain an expression for F and hence derive Eq. (13.99). Show that for strong fields and at low temperatures ($\mu_B B >> k_B T$), $M \rightarrow N\mu_B$ This is referred to as the saturation magnetic moment, M_{sat}.

For weak fields and at high temperatures, where $\mu_B B << k_B T$, $x(= \mu_B B/k_B T)$ will be much less than one. Then, we can write

$$\tanh x = \frac{e^x - e^{-x}}{e^x + e^{-x}} \approx \frac{(1+x)-(1-x)}{(1+x)-(1-x)}$$

where we have retained terms only up to the first power in x. This result shows that

$$M = N\mu_B\frac{\mu_B B}{k_B T} = N\frac{\mu_B^{\,2}\mu_0 H}{k_B T} \tag{13.100}$$

$\mu_0 H$; μ_0 is permeability of free space and H denotes intensity of the magnetic

We know that magnetic susceptibility is given by

$$\chi = \frac{M}{H} = N\frac{\mu_B^2 \mu_0}{k_B T}$$

i.e., the magnetic susceptibility

$$\chi = \frac{C}{T} \tag{13.101}$$

$\dfrac{N\mu_B^2 \mu_0}{k_B}$ is known as the *Curie constant*. Eq. (13.101) is known as *Curie's law*. It states that in weak fields and at high temperatures, magnetic susceptibility is inversely proportional to temperature.

Again using the result, $M_{sat} = N\mu_B$, we can rewrite Eq. (13.99) as

$$\frac{M}{M_{sat}} = \tanh\left(\frac{\mu_B B}{k_B T}\right) \tag{13.102}$$

Figure 13.17 shows the plot of M/M_{sat} as a function of $\mu_B B/k_B T$. This curve is referred *magnetisation curve*.

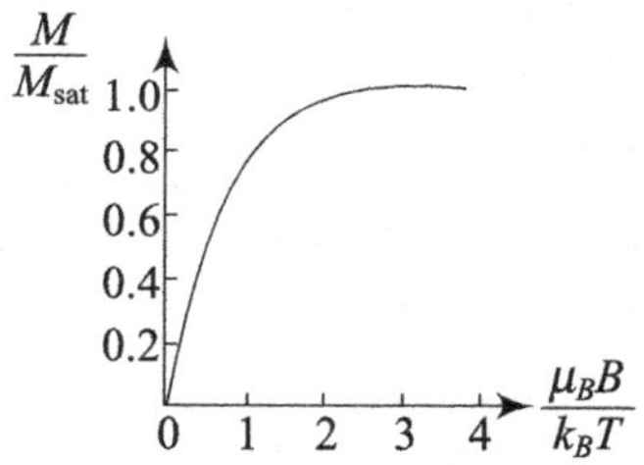

Fig. 13.17 Plot of M/M_{sat} vs. $\mu_B B/k_B T$ for a paramagnetic substance

During magnetisation, the magnetic field tends to align the atomic magnets along its own direction, whereas an increase in temperature causes thermal agitation, which tends to disrupt the alignment. (The former is an ordering effect whereas the latter is a disordering effect.) At any temperature, the magnetisation curve indicates how a balance is struck between these two opposing effects. For low values of H, the two levels are almost evenly populated and the resultant magnetic moment is very small. On the other hand, for high values of H, N_1 tends to dominate N_2 and nearly all the atomic magnets occupy the lower energy level.

Before proceeding further, you may like to answer a practise problem. It deals with other thermodynamic properties of interest for this system.

Problem 13.8 For a two-level system with magnetic potential energy levels, show that the total energy, specific heat capacity and entropy are given by

$$U \quad N\mu_B B \tanh\left(\frac{\mu_B B}{k_B T}\right),$$

$$C_B = \left(\frac{\partial U}{\partial T}\right)_B = Nk_B\left(\frac{\mu_B B}{k_B T}\right)^2 \text{sech}^2\left(\frac{\mu_B B}{k_B T}\right) \text{ and}$$

$$= \quad _B\left[\ln\left\{2\cosh\left(\frac{\mu_B B}{k_B T}\right)\right\} - \left(\frac{\mu_B B}{k_B T}\right)\tanh\left(\frac{\mu_B B}{k_B T}\right)\right]$$

Figures 13.18a, b and c show plots of C_B/Nk_B) and (S/Nk_B) as a function B. Owing to the existence of only two permitted energy levels, the nature of the variation of C_B is very different from that of the specific heat capacity of a collection of harmonic oscillators. In this connection, a comparison between the values of C_B of an assembly of magnetic ions and that of C_V of the entire paramagnetic substance would be quite interesting.

Let us first consider a typical paramagnetic crystal of chromium potassium sulphate, $K_2SO_424H_2O$. Only the chromium atoms in the crystal are responsible for the paramagnetic behaviour of the substance. They exist as ions, Cr^{+++}, with three uncompensated electron spins and a magnetic moment of $3\mu_B$. Note that for each chromium ion, there are 2 sulphur atoms, 1 potassium atom, 20 oxygen atoms and 24 hydrogen atoms. It means that for every chromium ion, there are 47 non-magnetic atoms.

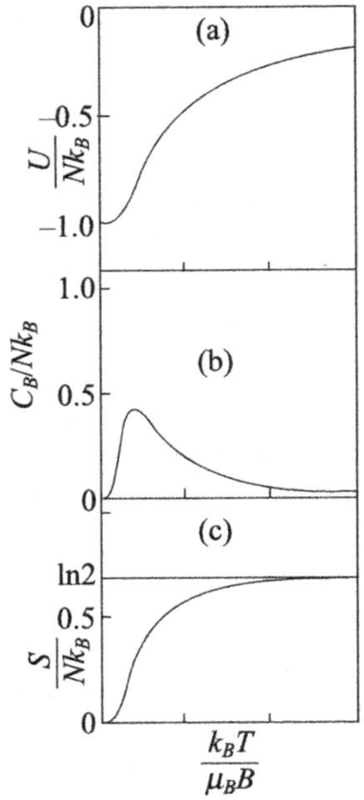

Fig. 13.18 Plot of (a) U/Nk_B vs k_BT/μ_BB, (b) C_B/Nk_B vs k_BT/μ_BB and

= 1 K and B = 1 tesla, $\dfrac{k_BT}{\mu_B B} \approx 1.5$

sech $\left(\dfrac{k_BT}{\mu_B B}\right) \approx 0.8$. Hence, using the result of Problem 13.8b, we get

$$C_B = 0.36\, Nk_B$$

Now, let us assume that there are 50 non-magnetic particles for every magnetic ion. Then considering a Debye temperature of 300 K as a typical value, we have from the -law (see Eq. (13.92):

$$C_V = Nk_B \times 50 \times \frac{12\pi^4}{5}\left(\frac{1}{300}\right)^3$$

$$= 0.5 \times 10^{-5}\ Nk_B$$

That is, at a temperature of 1 K, the heat capacity of magnetic ions is about 10^5 times the vibrational heat capacity of the entire crystal lattice. Hence, the energy required to orient magnetic ions is very high compared to that required for increasing the vibrational energy of the molecules of the lattice. This energy (of orientation) is responsible for cooling of a paramagnetic salt by adiabatic demagnetisation (Sec 10.7).

Figure 13.18c depicts the graph of S/Nk_B versus $\dfrac{k_BT}{\mu_B B}$. At a particular value of B, $S \to 0$

0. At this temperature, all dipoles occupy the lower energy state, i.e., there is only one possible microstate ($\Omega = 1$) and since $S = k_B \ln \Omega$ (where $\Omega = 1$), $S \to 0$.

For high temperatures, i.e., $k_BT >> \mu_BB$, cosh $\left(\dfrac{\mu_B B}{k_BT}\right) \to 1$ and tanh $\left(\dfrac{\mu_B B}{k_BT}\right) \to 1$. So, ln2. It can be seen that entropy is also a function of $\frac{B}{T}$. A reversible adiabatic demagnetisation is an isentropic process, so that $\frac{B}{T}$ remains constant. Thus, T decreases decreases.

Problem 13.9 In the high temperature/low field limit, the entropy can be expressed ln(2^N). Explain the significance of the factor 2^N in this expression.

Negative Temperatures

The concept of negative absolute temperature seems physically unacceptable. But we will now show that it is possible in principle to realise a state where increase in energy is accompanied by decrease in entropy. Such a state is said to correspond to a negative temperature.

To appreciate the point, let us consider a freely moving molecule of a perfect gas or a harmonic oscillator characterised by an infinite number of energy levels. As temperature of such a system is increased, more molecules will gradually go to higher energy levels. So the system becomes more energetic and greater disorder will prevail. We can therefore say that the entropy associated with the ordinary translational and vibrational degrees of freedom of a body increases without limit as energy increases. In such cases, the ratio $\frac{dU}{dS}$ will always be positive definite. We may therefore conclude that a system with infinite energy levels always stays in positive temperature states.

Let us now again refer to the two-level system of magnetic dipoles with parallel and anti-parallel spins. In an external magnetic field **B**, these levels have energies $\varepsilon_1 = -\mu_B B$ and , respectively . Suppose that the number of atoms whose moments are parallel and anti-parallel to **B** are N_1 and N_2, respectively. In the equilibrium state at a temperature , the average occupation numbers of these levels are given by

$$N_1 = \frac{N}{Z}\exp\left(-\frac{\varepsilon_1}{k_B T}\right)$$

$$N_2 = \frac{N}{Z}\exp\left(-\frac{\varepsilon_2}{k_B T}\right)$$

$N_1 + N_2$ denotes the total number of atoms. On combining these expressions, we can write

$$\frac{N_1}{N_2} = \exp\left(\frac{\varepsilon_2 - \varepsilon_1}{k_B T}\right) \tag{13.103}$$

On taking natural logarithm and rearranging terms, we can define temperature in terms of the energies of the levels and their atomic occupancy as

$$T = \frac{1}{k_B}\left(\frac{\varepsilon_2 - \varepsilon_1}{\ln N_1 - \ln N_2}\right) \tag{13.104a}$$

ε_1, we note that T will be positive if $N_1 > N_2$. This situation is depicted schematically in Fig. 13.19a. The lengths of the vertical lines signify level of occupancy

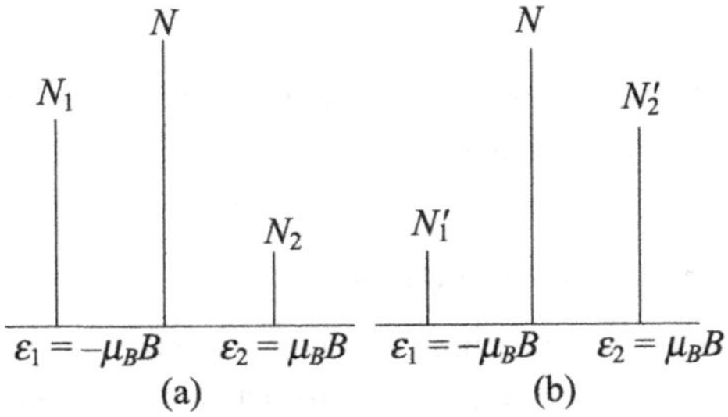

(a) In thermal equilibrium, $N_1 > N_2$ and (b) Reversal of the direction of **B** leads to population inversion, $N_2 > N_1$.

Next we suddenly reverse the direction of magnetic field. The dipoles oriented parallel to the field and having energy ε_1 will be oriented anti-parallel to the field and have a higher energy. But the dipoles originally anti-parallel to the field and having energy ε_2 will now be oriented parallel to the field and have a lower energy. (Eventually, of course, the dipoles in the higher energy state will flip over to the new lower-energy state.) The situation immediately after the reversal of the magnetic field is shown in Fig. 13.19b. Let the average occupation numbers of the new lower and higher energy states be N_1' and N_2', respectively.

$N_1' = N_2$ and $N_2' > N_1'$. This process is referred to as *population inversion*. The temperature corresponding to the situation illustrated in Fig. 13.19b can be mathematically represented by the relation

$$T' = \frac{1}{k_B}\left(\frac{\varepsilon_2 - \varepsilon_1}{\ln N_1' - \ln N_2'}\right) \qquad (13.104b)$$

$N_2' > N_1'$, this state will correspond to negative T.

Now refer to Fig. 13.20a. It shows the variation of entropy with energy.

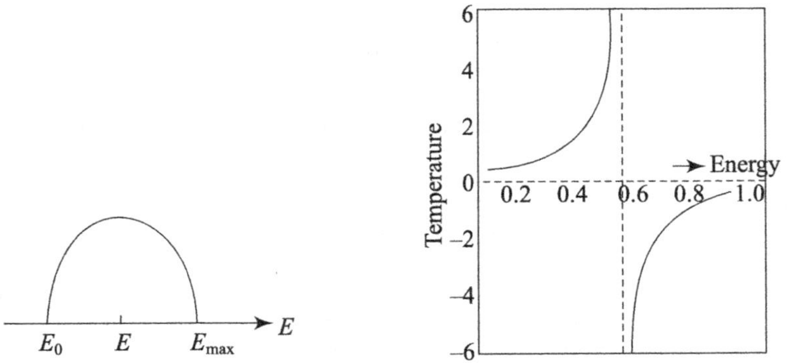

(a) Plot of entropy as a function of internal energy for a two-level system: $\varepsilon_1 = 0$ and $\varepsilon_2 = \varepsilon$ and (b) Plot of temperature as a function of internal energy.

From Eq. (13.93a), we recall that for $T = 0$, the probability of occupancy P_1 will be one; i.e., all the atoms will be in the energy state $\varepsilon_1 = 0$: $N_1 = N$ and $N_2 = 0$. The internal energy of the system will be zero. This is a state of minimum disorder and corresponds to $S = 0$.

increases, the occupancy in the higher level begins to rise. As $T \to \infty$, the probabilities of occupancy P_1 and P_2 will be one-half, i.e., both levels are equally occupied: $N_1 = N_2 = \frac{N}{2}$, and internal energy will be $\frac{N\varepsilon}{2}$. This corresponds to state of maximum disorder and hence maximum entropy. In the region $0 < U \le \frac{N\varepsilon}{2}$, the derivatives of S with respect to U will be positive and we can define a positive temperature.

If more atoms tend to occupy the higher energy state such that $N_2 > N_1$, we say that population inversion has been attained. In the particular case, if $N_2 = N$, the internal energy will be $N\varepsilon$. This corresponds to the state of minimum disorder and zero entropy. Eq. (13.104a) then implies that T will be negative. That is, in the region

$\le U \le N\varepsilon$, the slope of entropy-energy curve is negative. This corresponds to the state of *negative temperature.*

We can understand the concept of negative temperatures in yet another way. This is illustrated in Fig. 13.20b, where we have plotted T versus U. Note that as energy increases T increases from zero and reaches a maximum of ∞ when $U = \frac{N\varepsilon}{2}$. As U is infinitesimally greater than $\frac{N\varepsilon}{2}$, T abruptly changes to $-\infty$ and when $U = N\varepsilon$, we say that it gradually goes to -0 K. We thus arrive at a paradoxical result: *a system at a negative temperature is hotter than infinite absolute temperature!* This means that -0 K is the hottest temperature possible. However, note that the negative temperature is 'not below absolute zero' but above infinite absolute temperature.

Experimental Demonstration of Negative Temperatures In paramagnetic substances, the interactions between atomic magnetic moments (the unpaired electrons) and the crystal lattice are very strong and it is not possible for a substance to exist in the state of population inversion for appreciable time. However, in 1951, Pound, Purcell and Ramsey established the existence of negative absolute temperature by using nuclear magnetic resonance techniques. They observed that the nuclear magnetic moments of the Li atoms in LiF interacted extremely slowly with the lattice and quite a long time, up to several minutes, elapsed before they attained equilibrium with the lattice. As a result, the system existed in the state of population inversion, i.e., the state of negative temperature, for quite some time.

In 1917, Einstein had predicted that population inversion can occur due to transitions between different energy states in an atom with finite number of energy levels when he gave the theory of stimulated emission. It was realised in 1954 by C. H. Townes and his co-workers for *microwaves*. But Maiman operated the first successful laser in 1960 using a ruby crystal.

Transitions between States: Einstein's Formulation of Spontaneous and Stimulated Emission of Radiation

In your school atomic physics classes, you have learnt Bohr's theory of hydrogen spectrum. It stipulates as follows:

Electrons revolve around the nucleus in stationary orbits characterised by discrete energy levels.

The transition of an atom from one energy level to another level occurs in quantum jumps.

When an atom makes a transition from an excited state to a lower energy state, the frequency ν of the emitted radiation is given by

$$h\nu = E_i - E_f, \tag{13.105}$$

is Planck's constant, and E_i and E_f, respectively denote energies of higher (excited) and lower (ground) states.

To understand the physics of atomic processes involved in emission and absorption of light, we consider a two energy level system, as shown in the Fig. 13.21a. Suppose that the energies of the lower and upper energy levels are E_1 and E_2, respectively. An atom in the lower energy level can *absorb* a photon of energy $h\nu(=E_2 - E_1)$ and get excited to the higher energy level E_2. This is shown as *absorption* or *excitation* in Fig. 13.21b.

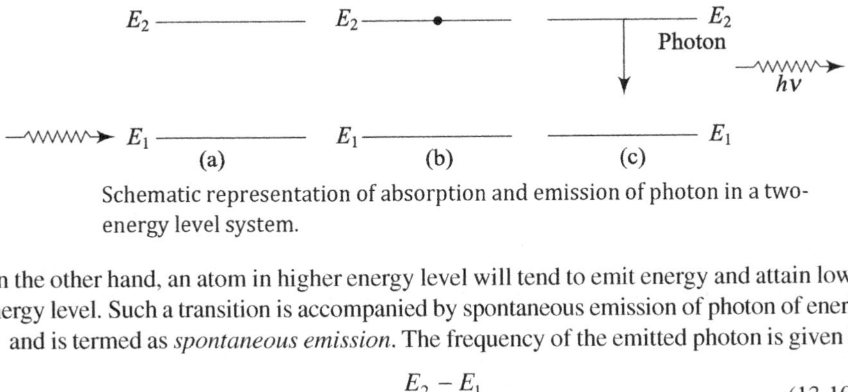

Schematic representation of absorption and emission of photon in a two-energy level system.

On the other hand, an atom in higher energy level will tend to emit energy and attain lower energy level. Such a transition is accompanied by spontaneous emission of photon of energy and is termed as *spontaneous emission*. The frequency of the emitted photon is given by

$$\nu = \frac{E_2 - E_1}{h} \tag{13.106}$$

Note that due to spontaneous transition, photons are emitted randomly, i.e., there is no definite phase relationship amongst them. The emitted light is *incoherent* and has a broad spectrum. This is the basic mechanism of emission of light by ordinary sources.

Next let us suppose that an atom in an excited state is irradiated by a photon of energy . Instead of being absorbed, this photon may trigger the atom to de-excite and emit radiation. As illustrated in Fig. 13.22, two photons, each of energy $h\nu$, will be emitted simultaneously. Such a transition is said to be *stimulated* or *induced.*

Note that the inducing as the well as the induced photon has the same energy and both are in phase. That is, these are *coherent* as well as co-directional and therefore add to amplify the incident beam. These unidirectional photons can de-excite two other atoms in their path producing four other photons and this cascading process of photon multiplication will continue. It means that when a large number of atoms are involved, the stimulated emission generates an intense, highly *coherent* and *directional* beam. In optical range, this

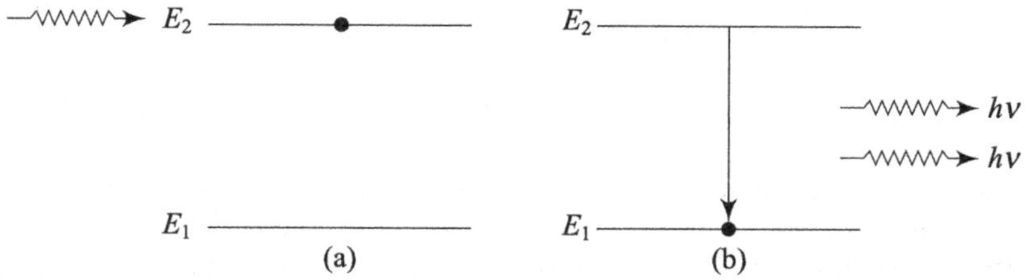

Schematic depiction of stimulated emission of radiation.

process is facilitated by a laser. (The device which facilitates stimulated emission in the microwave range is known as maser.) Note that stimulated emission is the fundamental process responsible for the development of a laser.

Before proceeding further, we give a simple numerical example.

Example 13.4 A sodium lamp emits a spectral line of wavelength 590 nm. Consider it as a two level system so that the spectral line corresponds to a transition from the first excited state (3*p*) to the ground state (3*s*). Calculate the corresponding energy.

From Eq. (13.106), we recall that

$$E_2 - E_1 = h\nu = \frac{hc}{\lambda}$$

On substituting the values of $h(=6.6\times10^{-34}\ \text{Js}), c(=3\times10^{8}\ \text{ms}^{-1})$ and λ = 590 × m, we get

$$E_2 - E_1 = \frac{(6.6\times10^{-34}\ \text{Js})\times(3\times10^{8}\ \text{ms}^{-1})}{590\times10^{-9}\ \text{m}} = 3.4\times10^{-19}\ \text{J} = 2.1\ \text{eV}$$

This corresponds to the energy of the first excited state of sodium.

Suppose that the number of atoms in the upper and lower non-degenerate energy levels N_1, respectively. Absorption of photons by atoms in the lower energy level and emission of photons in the upper energy level is known to be proportional to the intensity of incident light. Moreover, when such an assembly of atoms is in thermal equilibrium at temperature *T*, it can be assumed to obey Maxwell–Boltzmann distribution law and we

$$N_1 \propto \exp(-E_1/k_BT)$$

$$N_2 \propto \exp(-E_2/k_BT), \tag{13.107}$$

is Boltzmann constant.

On combining these expressions, we can write

$$\frac{N_2}{N_1} = \exp\left[-\frac{E_2 - E_1}{k_BT}\right] = \exp\left[-\frac{h\nu}{k_BT}\right] \tag{13.108a}$$

For degenerate energy levels, this expression will modify as

$$\frac{N_2}{N_1} = \frac{g_2}{g_1} \exp\left[-\frac{h\nu}{k_B T}\right] \tag{13.108b}$$

You may now like to know: How will this ratio (of the populations in energy levels) change, if the radiation of energy $h\nu$ is introduced in the system? The answer to this important question was provided by Einstein, who argued that for the given system and the radiation to be in thermal equilibrium, the rate of transitions due to spontaneous and stimulated emission (downward transitions) must be equal to the rate of transitions due to absorption (upward transitions). Before we discuss the implications of this argument further, go through the following example carefully.

Example 13.5 For the sodium lamp considered in Example 13.4, calculate the fraction of sodium atoms in the first excited state, if the lamp is at 327°C. Assume it to be non-degenerate.

In this case, we use Eq. (13.108a):

$$\frac{N_2}{N_1} = \exp\left[-\frac{h\nu}{k_B T}\right] = \exp\left[-\frac{hc}{k_B T\lambda}\right].$$

= 590 nm, and $T = (273 + 327)$ K = 600 K. On substituting the numerical values of various physical constants, we get

$$\frac{N_2}{N_1} = \exp\left[-\frac{(6.6\times10^{-34}\ \text{Js})\times(3\times10^{8}\ \text{ms}^{-1})}{(1.38\times10^{-23}\ \text{JK}^{-1})\times(590\times10^{-9}\ \text{m})\times(600\ \text{K})}\right]$$

$$= \exp(-40.53)$$

$$= 2.5\times10^{-18}$$

This result shows that ordinarily, the number of atoms in a higher energy state is extremely

---

You may now like to answer a practise problem.

Problem 13.10 A two-level ruby laser emits light of wavelength 693 nm. If temperature of the system is 27°C, calculate the ratio of populations in the two levels.

$\times 10^{-31}$

Following Einstein, let us suppose that the energy density of the incident radiation of ν is u_ν. Then the rate of spontaneous emission will be proportional to N_2; the number of atoms in the higher energy state. However, the rate of stimulated emission will be determined by the energy density of the incident radiation as well as the number of atoms in the higher energy state. Hence, we can express the rate of spontaneous and stimulated emissions as

$$P_{21} = A_{21} N_2 + B_{21} N_2 u_\nu . \tag{13.109}$$

and B_{21}, respectively denote constants of proportionality for spontaneous and stimulated emission of radiation from higher state (2) to lower state (1). Note that u_ν does not appear in the first term on the right hand side of Eq. (13.109) because no photon is needed for spontaneous emission.

Similarly, we can write the expression for the rate of absorption as

$$P_{12} = B_{12} N_1 u_\nu, \tag{13.110}$$

is constant of proportionality for absorption of radiation.

The conditions represented by Eqs. (13.109) and (13.110) are depicted schematically in

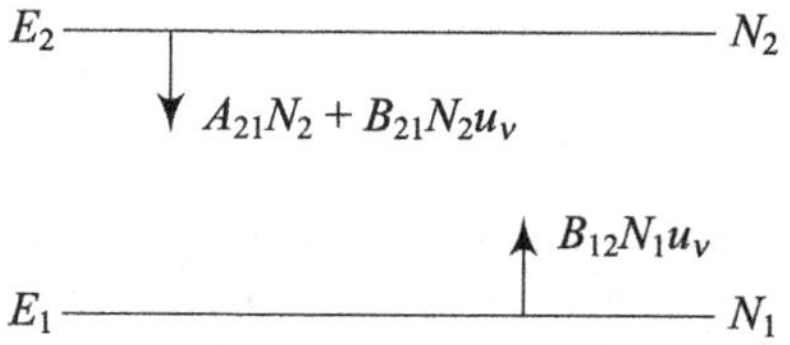

Fig. 13.23 Schematic representation of spontaneous and stimulated emission: Einstein's *A* and *B* coefficients.

If the lower and higher energy levels correspond to respective single non-degenerate states, we take constant $A_{21} = A$, and $B_{12} = B_{21}$. These are known as Einstein's *A* and *B* coefficients. However, for generality, we will continue to work with A_{21}, B_{12} and B_{21}.

For equilibrium, we equate Eqs. (13.109) and (13.110) to obtain

$$N_2 \,[A_{21} + B_{21} u_\nu] = N_1 B_{12} u_\nu$$

$$\frac{N_2}{N_1} = \frac{B_{12} u_\nu}{A_{21} + B_{21} u_\nu} \tag{13.111}$$

On combining this result with Eq. (13.108a), we can write

$$\frac{B_{12} u_\nu}{A_{21} + B_{21} u_\nu} = \exp\left[-\frac{h\nu}{k_B T}\right]$$

On cross-multiplication and rearrangement of terms, we can solve for energy density to

$$u_\nu = \frac{A_{21}}{B_{12}} \frac{1}{\exp(h\nu/k_B T) - B_{21}/B_{12}}. \tag{13.112}$$

From Chapter 11, you may recall Planck's law for blackbody radiation (Eq. (11.44)):

$$u_\nu = \frac{8\pi h\nu^3}{c^3} \frac{1}{\exp(h\nu/k_B T) - 1}. \tag{13.113}$$

As may be noted, Eqs. (13.112) and (13.113) are strikingly similar, though these have been obtained from completely different arguments. (This indicates the soundness of both thought processes.) On comparing these expressions, we obtain

$$B_{21} = B_{12} \tag{13.114a}$$

$$\frac{A_{21}}{B_{12}} = \frac{8\pi h\nu^3}{c^3} \tag{13.114b}$$

This result shows that the proportionality coefficient for stimulated emission, B_{21}, is inversely proportional to the third power of frequency. (This suggests that laser action will be more difficult at higher frequencies.)

For degenerate energy levels, Eqs (13.114a, b) respectively modify as

$$\frac{B_{21}}{B_{12}} = \frac{g_1}{g_2} \tag{13.114c}$$

$$\frac{A_{21}}{B_{12}} = \frac{g_1}{g_2}\frac{8\pi h\nu^3}{c^3}. \tag{13.114d}$$

Note that we would not have obtained the expression for energy density similar to Planck's law without considering stimulated emission. In fact, Einstein predicted stimulated emission in 1917 but it could be realised in 1954 when Townes and his co-workers developed a microwave amplifier using ammonia. In 1960, Maiman succeeded in constructing a Ruby laser. This highlighted the importance of *stimulated emission for the development of a* . This however is very uncommon in nature.

Let us pause for a minute and reflect on what we have achieved so far. From Eq. (13.114a), we note that the probabilities of absorption and stimulated emission are the same. It means that when an atomic system is in thermal equilibrium, absorption and emission occur simultaneously. As shown in Example 13.5, $N_2 < N_1$ in general. It means that from physics point of view, under normal conditions, absorption dominates stimulated emission. However, if we devise a mechanism to ensure that $N_2 > N_1$, stimulated emission may begin to dominate absorption. Such a condition in an atomic system is known as *population inversion.*

Proceeding further, we combine Eqs. (13.112) and (13.114a) and obtain

$$\frac{A_{21}}{B_{21}u_\nu} = \exp\left[\frac{h\nu}{k_B T}\right] - 1 \tag{13.115}$$

Do you recognise this result? Physically, it gives the ratio of the number of spontaneous emissions to stimulated emissions for a system in thermal equilibrium. When $h\nu >> k_B T$, spontaneous emission is so much more probable that stimulated emission may be ignored. This works rather well for electronic transitions in atoms and molecules as also in the case of radiative transitions in nuclei.

To appreciate this point, go through the following solved example.

Example 13.6 In a tungsten filament lamp, thermionic emission takes place at 10^3 K. Calculate the ratio of spontaneous emission to stimulated emission for non-degenerate energy levels. Take $\lambda = 550$ nm, $k_B = 1.38 \times 10^{-23}$ JK^{-1} and $h = 6.67 \times 10^{-34}$ Js.

From Eq. (13.115), we recall that the ratio of spontaneous emissions to stimulated emissions is given by

$$\frac{A_{21}}{B_{21}u_\nu} = \exp\left[\frac{h\nu}{k_B T}\right] - 1 \tag{i}$$

= 550 nm, the corresponding frequency emitted by the tungsten lamp is given by

$$\nu = \frac{c}{\lambda} = \frac{3 \times 10^8 \text{ ms}^{-1}}{550 \times 10^{-9} \text{ m}} = 5.45 \times 10^{14} \text{ Hz.}$$

On inserting this value of frequency along with the required data in Eq. (i), we get

$$\frac{A_{21}}{B_{21} u_\nu} = \exp\left(\frac{(6.62 \times 10^{-34} \text{ Js}) \times (5.45 \times 10^{14} \text{ s}^{-1})}{(1.38 \times 10^{-23} \text{ JK}^{-1}) \times 1200}\right) - 1$$

$$= [\exp(21.79)] - 1$$

$$= 2.91 \times 10^9$$

From this result, we can conclude that at optical frequencies, emission is predominantly due to spontaneous transitions.

$\nu << k_B T$, as in the microwave region of the spectrum, it will suffice to retain only the linear term in the expansion of exponential function in Eq. (13.115). Then it simplifies to

$$\frac{A_{21}}{B_{21} u_\nu} = \frac{h\nu}{k_B T} \tag{13.116}$$

This result shows that for $\frac{h\nu}{k_B T} << 1$, stimulated emission will dominate spontaneous emission. Recall that stimulated emission takes place as a result of the action of the incoming radiation on atoms or molecules of a substance. That is, the resulting emission is due to forced atomic oscillations and bears a constant phase relation to the incoming radiation. As a result, all atoms (or molecules) radiate in phase. That is why *stimulated emission is coherent.* (On the other hand, in spontaneous emission there is no definite phase relation with the result that spontaneous emission is incoherent.) This remarkable characteristic of stimulated emission is utilised in Masers and Lasers.

Laser Action

Refer to Fig. 13.24a, which shows a two-level system, where the lower energy level E_1 is more populated than the upper energy level E_2. It depicts the equilibrium condition, *net* absorption or emission occurs, i.e., when the total number of absorption and emission transitions per unit time is the same. In general, when the system has not attained equilibrium, we have

$$\frac{\text{Rate of emission}}{\text{Rate of absorption}} = \frac{[A_{21} + B_{21} u_\nu] N_2}{B_{12} u_\nu N_1} = \left[1 + \frac{A_{21}}{B_{21} u_\nu}\right] \frac{N_2}{N_1} \tag{13.117}$$

represents the energy of the incoming radiation. For $h\nu << k_B T$, Eq. (13.116) implies that $\frac{A_{21}}{B_{21} u_\nu} \rightarrow 0$. On using this result in Eq. (13.117), we get

$$\frac{\text{Rate of emission}}{\text{Rate of absorption}} \simeq \frac{N_2}{N_1}$$

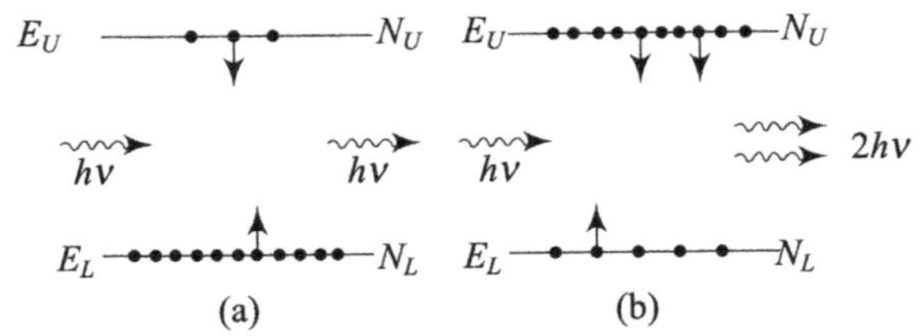

Fig. 13.24 Basic principle of operation of a maser and a laser: (a) Normal state, and (b) State of population inversion

We know that natural tendency of every system is to go to the state of minimum energy, i.e., ground state. As a result of this characteristic, at thermal equilibrium $N_2 < N_1$. That is, the occupancy of higher energy levels is significantly less than that of the ground state. So the rate of emission is smaller than the rate of absorption. But we know that the number of atoms that get de-excited in a certain time is more than those that get excited. Therefore, we expect that the upper level will get depleted with time. Thus, the rate of amplification should decrease gradually. Hence, to obtain a sustained and uniform amplification, it is necessary to continuously replenish the upper level. This is achieved by transferring atoms from the lower to the upper level by some means.

We may now conclude that for the lasing action to be initiated, more atoms/molecules should be in an excited state than in a lower energy state. That is, *the pre-condition for amplification of light intensity by stimulated emission is population inversion*, i.e., $N_2 > N_1$. Then the rate of emission will exceed the rate of absorption. In other words, if radiation of energy density u_ν passes through this system, the outgoing radiation will have more photons of frequency ν than the incoming radiation (Fig. 13.24b). That is, the incident radiation gets amplified.

It may be mentioned here that population inversion is an artificial condition. The mechanism used to attain population inversion is known as *pumping*. (A mechanical analogue of population inversion is pumping of water from the ground floor to the top of a high rise, i.e., from a state of lower potential energy to a state of higher potential energy.) It is important to point out here that population inversion can not be achieved with just two levels. This is because the probabilities of absorption and spontaneous emission are exactly equal.

A number of methods are used to achieve population inversion. These include

Optical pumping or excitation by photons using a light source,
Electrical pumping using gas discharge,
Inelastic atomic collisions (semiconductor laser), and
Chemical pumping using energy from chemical reactions.

For simplicity, we will confine our discussion to optical pumping only.

In optical pumping, a high energy light source such as a Xenon flash lamp or another laser can be used to supply energy to the active medium. This energy comes in the form of short flashes of light and atoms are raised to the appropriate excited state by selective absorption of radiation in an atomic/molecular system. In this process, strong deviations from thermal equilibrium populations of selected states may take place. This method is particularly suited for solid state or liquid lasers whose absorption bands are wide enough.

As such, optical pumping is a resonant phenomenon; the energy of the incident photon must be equal to the difference of energies of excited and normal states. If X respectively denote a normal and excited atom, we can express optical pumping symbolically as

$$X + h\nu \rightarrow X^*$$

If we have a laser which gives out light whose wavelength lies in the absorption bands of the active medium (solid, liquid or a gas), we can use it for pumping. Since the bandwidth of a laser is very narrow, its pumping efficiency will be very high. It may be mentioned here that this method was first used by Maiman for Ruby Laser but even today, it is widely used in solid-state lasers.

As mentioned earlier, population inversion can not be achieved with just two levels. An efficient inversion mechanism uses three level pumping scheme as in ruby laser, or four levels pumping scheme as in Nd: YAG laser. Let us learn about these pumping schemes

Pumping Schemes To discuss pumping schemes, it is convenient to indicate the pumping transition by upward arrow, the lasing transition by downward arrow and non-radiative decay by slanted arrow. (In non-radiative transition, the excess energy does not appear in the form of electromagnetic radiation. It is, in general, utilised in raising the vibrational energy of the host medium, resulting in its heating.)

1. Three-Level Pumping Scheme Refer to Fig. 13.25 (a). It shows a three-level pumping scheme. Note that the ground state is represented as 1, the lasing state as 2 and the pumping state as 3. Consider an assembly of N atoms, each of which can exist in any of the three energy states having energies E_0, E_1 and E_2, and populations N_0, N_1 and N_2, respectively. We assume that to begin with, the system is in thermal equilibrium so that most of the atoms will be in the ground state, i.e., $N_0 \approx N$ and $N_2 \approx N_1 \approx 0$. If a radiation of frequency E_1 is now made to fall of the system, the atoms will begin to get excited from the ground state to the pumping state due to optical absorption. In due course of time, an appreciable number of atoms may be excited to the pumping level.

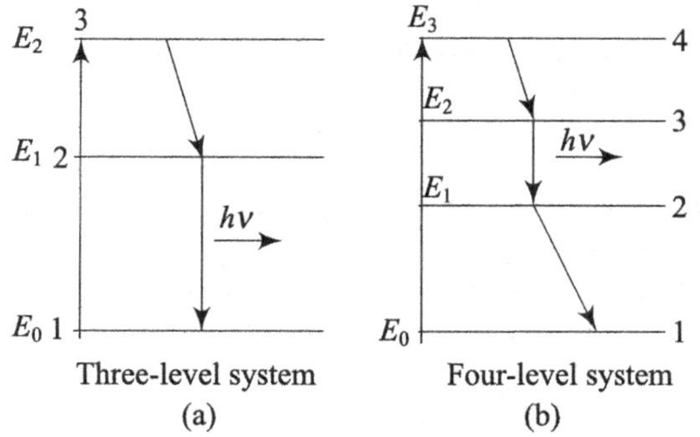

(a) Three-level and (b) Four-level pumping schemes.

In a medium suited for laser operation, the pumped atoms will be expected to decay quickly to the lasing level 2 through a non-radiative transition. (This decay is very fast; the lifetime is of the order of 10^{-8} s and is shown by slanted arrow in Fig. 13.25a. You may recall that lifetime signifies the average time that an atom spends in an excited state.) This state is generally meta-stable. (A meta-stable state is an excited state of an atom/nucleus/molecule that has a lifetime longer than the ordinary excited state, which, in turn, has lifetime shorter than the lowest and the most stable ground energy state. From quantum mechanical point of view, a meta-state is not a truly stationary state but we can say that is almost stationary.

For example, the lifetime of a lasing state $\approx 10^{-3}$ s or even more, and is much greater than the lifetime of an excited or pumping state 3. An atom in state 2 may decay to the ground state by spontaneously emitting a photon of frequency $\nu_L = E_1 - E_0$. This is shown by downward arrow in Fig. 13.25a. If the lifetime of this transition is much greater than the lifetime of the non-radiative transition, the population of the pumping level 3 will drop very rapidly and become essentially zero. As a result, a population of excited atoms will crowd level 2 in a three level pumping scheme. If more than half the total population accumulates in level 2, it will exceed the population of the ground state. We then say that population inversion has occurred between the meta-stable lasing state 2 and the ground state 1 and this may lead to optical amplification at the frequency ν_L.

Note that in three level pumping scheme, more than half the atoms have to be excited from the ground state to obtain population inversion. This feature proves too demanding because ordinarily occupancy of the ground state is higher and in the instant case, the laser medium has to be pumped very strongly. This makes a three level pumping scheme rather inefficient and unattractive.

2. Four-Level Pumping Scheme Refer to Fig. 13.25b. It shows a four-level pumping scheme. The energy levels correspond to energies E_0, E_1, E_2 and E_3 and populations N_0, N_3, respectively. Note that in this scheme, we have two upper levels, instead of one as in three-level pumping scheme. As before, atoms are excited from the ground state to the pumping state (level 4). From level 4, atoms undergo fast non-radiative transition to the upper lasing (meta-state or level 3). Since lifetime of meta-state is longer compared to the pumping state, atoms tend to accumulate in level 3, which may relax by spontaneous or stimulated emission into level 2 (lower lasing level). Atoms in this level can also decay by non-radiative transitions and reach the ground state.

Due to non-radiative transitions, population in the pumping state (level 4) depletes fast, as in the case of three-level pumping scheme. Similarly, atoms in the lower lasing level (level 2) also de-excite rapidly and the number of atoms drops to a negligible number 0). As a result, any appreciable accumulation of atoms in the upper lasing level leads to population inversion between levels 2 and 3. That is, if $N_2 > 0$, N_2 will be greater . Thus, optical amplification and laser operation will take place at a frequency $E_1)/h$.

You may now like to know: Which pumping scheme is more efficient? Since excitation of only a few atoms into the upper level helps achieve population inversion, a four-level scheme is more efficient than a three-level scheme. Primarily for this reason, four-level pumping scheme is used in a wide variety of practical lasers. But a serious drawback in four level scheme arises due to the loss of a lot of energy in non-radiative transitions between the pumping state 4 and the upper lasing state 3, as also the lower lasing state 2 and ground state 1. We may therefore conclude that each pumping scheme has its own limitations but the ultimate choice is dictated by the active medium, type of usage, and such other factors.

So far we have considered systems at room temperature. But at sufficiently high temperatures such as occurring in the interior of stars, most substances exist in an ionised state. This is known as *thermal ionisation* (Ionisation in stars can also occur due to high pressure prevailing there. The theory of pressure ionisation was worked out by Indian physicist D.S. Kothari.) We now illustrate evaluation of the partition function for such systems. For the first time M. N. Saha, an eminent Indian physicist, gave the theory of thermal ionisation to explain ionisation in the solar chromosphere. We now discuss it in some detail.

AHA'S IONISATION FORMULA

We now know that in the interior of stars, temperatures are extremely high and the elements present there are mostly in the atomic state. Saha argued that under such conditions, atoms move very rapidly and undergo frequent collisions. In the process, they are stripped of their valence electrons. This is referred to as *thermal ionisation* and is accompanied by electron recapture to form neutral atoms. Saha's ionisation formula relates the temperature, pressure and ionisation potential of atoms to their degree of ionisation.

To arrive at Saha's formula, we assume that

the contribution of nuclear spin is negligible,
only single ionisation occurs and
dynamical equilibrium between ionisation and recombination (due to electron capture) is reached at a given temperature and pressure so that

$$A \rightleftharpoons A^+ + e \tag{13.118}$$

is a singly ionised positive ion.

If we treat this as a reaction between different components of an ideal gas in equilibrium, we can write

$$\sum_k \zeta_k \mu_k = 0 \tag{13.119}$$

is the component index for A, A^+ or e^-, ζ_k is the stoichiometric coefficient† of the th component, and μ_k is chemical potential of the kth atom.

On combining Eqs. (13.7) and (13.21) for this case, we can write the chemical potential

$$\mu = -k_B T \ln\left[\left(\frac{2\pi m}{h^2}\right)^{3/2} \tilde{Z}(T)\frac{k_B T}{p}\right] \tag{13.120}$$

) refers to the contributions of the internal modes (rotational, vibrational and electronic) to the partition function. Note that it is a function of T but not of pressure.

This configuration is equivalent to the presence of several gases contained in a vessel. Let us assume that p_a, p_i and p_e are the partial pressures exerted by the atom, the ion and the electron, respectively. Let their masses be m_a, m_i and m_e, respectively. Hence, the chemical potential of the kth component in the system is given by

$$\mu_k = -k_B T \ln\left[\left(\frac{2\pi M_k}{N_A h^2}\right)^{3/2} \tilde{Z}_k(T)(k_B T)^{5/2}\right] + k_B T \ln p_k \tag{13.121}$$

is Avogadro's number.

It is important to point out here that for most reactions of the type (13.118), we measure energies from the ground state. The partition functions in this case are given by

$$\tilde{Z}_a = Z_a, \tilde{Z}_e = Z_e, \text{ and } \tilde{Z}_i = Z_i \exp(-\varepsilon^* / k_B T)$$

is the ionisation energy. At equilibrium Eq. (13.119) implies that

$$\mu_a \nu_a - \mu_i \nu_i - \mu_e \nu_e = 0$$

If the number of A, A^+ and e^- components taking part in the reaction be n, n^+ and n^-, respectively, then

Assumption (b) implies that $\nu_a = \nu_i = \nu_e = 1$. Therefore, on substituting for the chemical potentials in this result with appropriate partition functions, we get

$$-\ln\left[\left(\frac{2\pi M_a}{N_A h^2}\right)^{3/2} Z_a(T)(k_B T)^{5/2}\right] + \ln p_a + \ln\left[\left(\frac{2\pi M_i}{N_A h^2}\right)^{3/2} Z_i(T)(k_B T)^{5/2}\right] - \ln p_i$$

$$+\ln\left[\left(\frac{2\pi M_e}{N_A h^2}\right)^{3/2} Z_e(T)\exp(-\varepsilon^*/k_B T)(k_B T)^{5/2}\right] - \ln p_e = 0$$

On rearranging terms and taking antilog of both sides, we can write

$$\frac{p_i p_e}{p_a} = \left(\frac{2\pi}{N_A h^2}\right)^{3/2} (k_B T)^{5/2} \left(\frac{M_i M_e}{M_a}\right)^{3/2} \frac{Z_i Z_e}{Z_a} \exp(-\varepsilon^*/k_B T) \qquad (13.122)$$

$M_i/M_a \approx 1$ as the mass of a hydrogen atom is 10^4 times the mass of an electron. Further, for free atoms and ions, the possible internal states are electronic states. For all practical purposes, these may be taken to be the ground state with degeneracies g_a, g_i and . For an electron, $g_e = 2$ and $m_e = \dfrac{M_e}{N_A}$. Then, we can rewrite Eq. (13.122) as

$$\frac{p_i p_e}{p_a} = \left(\frac{2\pi m_e}{h^2}\right)^{3/2} (k_B T)^{5/2} \frac{2 g_i}{g_a} \exp(-\varepsilon^*/k_B T) \qquad (13.123)$$

A system of ions and electrons will be electrically neutral when the number of ions is equal to the number of electrons. So we can assume $p_i = p_e$ so that $p_i p_e = p_e^2$. Also, you may recall from Dalton's law of partial pressures that the total pressure is sum of the partial pressure and we can write $p = p_a + p_i + p_e$ and we can write

$$p_a = p - 2p_e$$

On combining these results, we can write

$$\frac{p_i p_e}{p_a} = \frac{p_e^2}{p - 2p_e} = \frac{p(p_e/p)^2}{1 - 2(p_e/p)} = \frac{px^2}{1 - 2x} \qquad (13.124)$$

$= (p_e/p)$ denotes the mole fraction of electrons, i.e., the fraction of electrons in the system.

On combining Eqs. (13.123) and (13.124), we get

$$\frac{x^2}{1 - 2x} = \left(\frac{2\pi m_e}{h^2}\right)^{3/2} (k_B T)^{5/2} \frac{1}{p} \frac{2 g_i}{g_a} \exp(-\varepsilon^*/k_B T) \qquad (13.125)$$

This equation expresses the degree of ionisation as a function of temperature, pressure and the ionisation potential of the atoms present in the interior of a star. This is commonly referred to as *Saha's ionisation formula*. It implies that the degree of ionisation will be more if temperature is high, or pressure and ionisation potential are low. Eq. (13.125) also shows that in the limit $T \to 0$, $x \to 0$. However, this result should be taken with caution since gases do not exhibit ideal gas behaviour at such low temperatures.

Saha's ionisation formula finds several applications in diverse areas of physics and astrophysics as well as chemistry. It has been successfully used to explain the Fraunhofer spectrum of the sun, and the spectrum of stars, formation of ionosphere, electrical conductivity of flames and to estimate the temperature of the atmosphere. It also finds applications in plasma studies as well as ionisation processes in semi-conductors. The most spectacular success of ionisation formula is in categorisation of the vast number of stars into a few groups: *B*, *A*, *G*, *F*, *K*, *M*, which form a continuous and linear series.

To illustrate the use of Eq. (13.125), we consider the ionisation of a hydrogen atom: $^{-}$. In this reaction, the process proceeds from the state of predominantly neutral hydrogen to the state dominated by protons. Hence, we can assume that

$$p_a = p_e = p_i$$

= 1/3. We know that the ionisation energy of hydrogen is 13.6 eV, the proton has two spin states, i.e., $g_i = 2$ and for the hydrogen atoms $g_a = 4$, because of 2 electron and 2 proton spin states. Let the number density of all the particles be n. Then, from the equation of state we can write $p = nk_BT$. Hence, from Eq. (13.125), we get

$$\frac{(1/3)^2}{1-2(1/3)} = \left(\frac{2\pi m_e}{h^2}\right)^{3/2} (k_BT)^{5/2} \frac{1}{nk_BT} \frac{2\times 2}{4} \exp(-\varepsilon^*/k_BT)$$

On rearranging terms, we obtain expression for the number density for electrons:

$$n = 3\left(\frac{2\pi m_e k_B}{h^2}\right)^{3/2} T^{3/2} \exp(-\varepsilon^*/k_BT) \tag{13.126}$$

Problem 13.11 Substitute the values of m_e, k_B, h and ε^* in Eq. (13.126) and show

$$\log_{10} n = 15.86 + 1.5 \log_{10} T - \frac{3.63\times 10^5}{T}$$

is in cm^{-3} and T is absolute temperature.

The relation between n and T based on Eq. (13.126) is shown in Fig. 13.26. Neutral hydrogen dominates the radiation at high atomic densities and low temperatures while the protons and electrons dominate at high temperatures and low densities. For n = 1 $n = 10^{25}$ cm^{-3}, we observe from Fig. 13.26 that the corresponding temperatures are given by $\log_{10}$ T = 3.8 and $\log_{10}$ = 6. This means that the transition occurs at $10^{3.8}$K 7500 K) and 10^6 K.

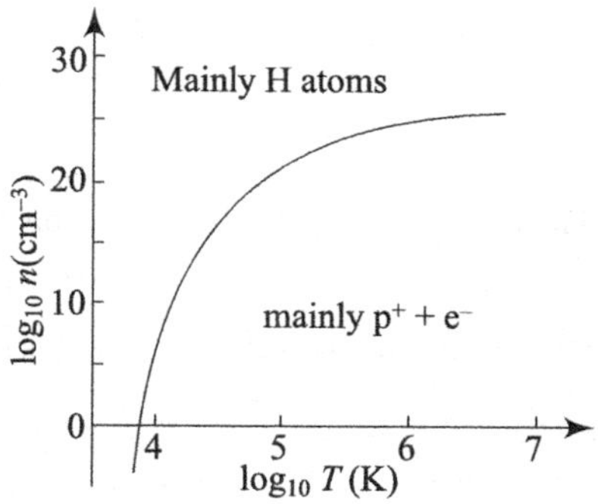

Fig. 13.26 Temperature variation of number density for ionisation equilibrium of hydrogen in a star.

DDITIONAL EXAMPLES

Example 13.7 The molecules of a gas have two states of internal energy with statistical weights g_1, g_2 and energies 0 and ε, respectively. Calculate the contribution of these states to the specific heat capacity of the gas.

The molecular partition function in this case can be expressed as

$$Z = g_1 + g_2 \exp(-\varepsilon/k_B T)$$

Suppose that n_1 molecules are in state 1 and n_2 molecules are in state 2. Then, we can write

$$\frac{n_2}{n_1} = \frac{g_2}{g_1} \exp(-\varepsilon/k_B T) \qquad \text{(i)}$$

If the total number of particles $N = n_1 + n_2$, we can rewrite Eq. (i) as

$$\frac{N - n_1}{n_1} = \frac{g_2}{g_1} \exp(-\varepsilon/k_B T)$$

$$\frac{N}{n_1} - 1 = \frac{g_2}{g_1} \exp(-\varepsilon/k_B T)$$

$$n_1 = \frac{N}{1 + \frac{g_2}{g_1} \exp(-\varepsilon/k_B T)} \qquad \text{(ii)}$$

$$n_2 = N - n_1 = N - \frac{N}{1 + \frac{g_2}{g_1} \exp(-\varepsilon/k_B T)}$$

$$= \frac{N \frac{g_2}{g_1} \exp(-\varepsilon/k_B T)}{1 + \frac{g_2}{g_1} \exp(-\varepsilon/k_B T)} \qquad \text{(iii)}$$

The internal energy of the gas can be expressed as

$$U = n_1 \varepsilon_1 + n_2 \varepsilon_2 = n_2 \varepsilon$$

On substituting the value of n_2 in this expression from Eq. (iii), we get

$$U = \frac{N\varepsilon \frac{g_2}{g_1} \exp(-\varepsilon/k_B T)}{1 + \frac{g_2}{g_1} \exp(-\varepsilon/k_B T)}$$

Hence, specific heat capacity of the gas is given by

$$C = \left(\frac{\partial U}{\partial T}\right)_V = N\varepsilon \frac{g_2}{g_1}\frac{\partial}{\partial T}\left(\frac{\exp(-\varepsilon/k_B T)}{1+\frac{g_2}{g_1}\exp(-\varepsilon/k_B T)}\right)$$

$$N\varepsilon\frac{g_2}{g_1}\frac{\left(1+\frac{g_2}{g_1}\exp(-\varepsilon/k_BT)\right)\frac{\mathrm{d}}{\mathrm{d}T}\left(\exp(-\varepsilon/k_BT)\right) - \exp(-\varepsilon/k_BT)\frac{\mathrm{d}}{\mathrm{d}T}\left(1+\frac{g_2}{g_1}\exp(-\varepsilon/k_BT)\right)}{\left(1+\frac{g_2}{g_1}\exp(-\varepsilon/k_BT)\right)^2}$$

$$= N\varepsilon\frac{g_2}{g_1}\cdot\frac{\left(1+\frac{g_2}{g_1}\exp(-\varepsilon/k_BT)\right)(\exp(-\varepsilon/k_BT)\left(\frac{\varepsilon}{k_BT^2}\right) - \exp(-\varepsilon/k_BT)\frac{g_2}{g_1}\exp(-\varepsilon/k_BT)\left(\frac{\varepsilon}{k_BT^2}\right)}{\left(1+\frac{g_2}{g_1}\exp(-\varepsilon/k_BT)\right)^2}$$

$$= N\varepsilon\frac{g_2}{g_1}\cdot\frac{\exp(-\varepsilon/k_BT)\left(\frac{\varepsilon}{k_BT^2}\right)}{\left(1+\frac{g_2}{g_1}\exp(-\varepsilon/k_BT)\right)^2}$$

$$= \frac{\frac{N\varepsilon^2}{k_BT^2}g_1 g_2\exp(-\varepsilon/k_BT)}{[g_1+g_2\exp(-k_BT)]^2}$$

$$= \frac{N\varepsilon^2}{k_BT^2}\frac{g_1 g_2\exp(-\varepsilon/k_BT)}{[g_1\exp(-\varepsilon/k_BT)+g_2]^2}$$

Example 13.8 The atoms of thallium in the vapour state have a doublet ground state energy separation $2_{S_{1/2}} - 2_{S_{3/2}}$ equal to 1.55×10^{-19} J. Calculate the entropy at 604°C and normal atmospheric pressure. Compare it with the measured value of 203 J mol^{-1} K^{-1}.

The doublet separation in terms of temperature can be expressed as

$$\frac{1.55\times10^{-19}\ \text{J}}{k_B} = \frac{1.55\times10^{-19}\ \text{J}}{1.38\times10^{-23}\ \text{JK}^{-1}} = 11{,}200\ \text{K}$$

Using the result $S = Nk_B \ln[(2\pi mk_BT/h^2)^{3/2}\,(k_BT/P)e^{5/2}]$, we get

$$S = 197.5 \text{ J mol}^{-1}\text{ K}^{-1}.$$

The electronic excitation is negligible at $T = 1000$ K, but it gives degeneracy $g = 2$. So we may add $Nk_B \ln 2$ to the above value of S:

$$Nk_B \ \ln 2 = 6.023 \times 10^{23} \text{ mol}^{-1} \times (1.38 \times 10^{23} \text{ JK}^{-1}) \times 0.693$$

$$= 5.76 \text{ J mol}^{-1}\text{K}^{-1}$$

So, the modified values of S is given by

$$(197.5 + 5.76)\text{ J mol}^{-1}\text{K}^{-1} = 203.26 \text{ J mol}^{-1}\text{K}^{-1}$$

which is in excellent agreement with the observed result.

Example 13.9 1g diamond is heated at constant volume from 30 K to 50 K. Calculate its specific heat capacity, change in entropy and the highest lattice frequency. Take Debye's temperature for diamond as 2230 K and assume that the asymptotic expression of Debye's law is valid at low temperatures.

From Eq. (13.92), we recall that in the asymptotic range, specific heat capacity of a solid is given by

$$C_V = \frac{12\pi^4}{5} R\left(\frac{T}{\theta_D}\right)^3 \qquad \text{for } T << \theta_D.$$

$$= \frac{12}{5} \times (3.1417)^4 \times (8.31 \text{ JK}^{-1}\text{mol}^{-1}) \times \left(\frac{30 \text{ K}}{2230 \text{ K}}\right)^3$$

$$= 4.73 \times 10^{-3} \text{ JK}^{-1}\text{mol}^{-1}$$

The change in entropy is given by

$$\Delta S = \int_{T_1}^{T_2} \frac{C_V\,dT}{T} = \frac{12\pi^4}{5}\frac{R}{(\theta_D)^3}\int_{T_1}^{T_2} T^2\,dT$$

$$= \frac{12\pi^4}{3\times 5}\frac{R}{(\theta_D)^3}(T_2^3 - T_1^3)$$

On substituting the given values, we get

$$\Delta S = \frac{4\times(3.1417)^4}{5} \times (8.31 \text{ JK}^{-1}\text{ mol}^{-1})\frac{(50^3 - 30^3)\text{K}^3}{(2230 \text{ K})^3}$$

$$= 647.66 \times (88.37 \times 10^{-7} \text{ JK}^{-1}\text{mol}^{-1})$$

$$= 5.72 \times 10^{-3} \text{ JK}^{-1}\text{mol}^{-1}$$

The maximum lattice frequency

$$\nu_m = \frac{k_B\theta_D}{h} = \frac{(1.38\times10^{-23} \text{ JK}^{-1})(2230 \text{ K})}{6.62 \times 10^{34} \text{ Js}}$$

$$= 4.65 \times 10^{13} \text{ Hz}.$$

Example 13.10 For HCl molecule, successive rotational energy levels have a spacing of 20.8 cm^{-1}. Calculate the bond length, using the following data: mass of proton is $^{-27}$ kg, $h = 6.62 \times 10^{-34}$ Js and atomic weight of chlorine = 35.5.

The energy spacing between successive levels:

$$\Delta E = E_1 - E_2$$
$$= Bch\, j(j+1) - Bch(j-1)j$$
$$= 2\,Bch\, j$$

$\dfrac{h}{8\pi^2 IC}$ is rotational constant.

The wave number corresponding to this difference in energy is given by

$$k = \frac{\Delta E}{hC} = 2\,Bj$$

This result shows that in terms of the wave number, spacing between successive levels is $2B$.

$$2\,B = 20.8\ \text{cm}^{-1}$$

$$B = 10.4\ \text{cm}^{-1}$$

Also, since

$$B = \frac{h}{8\pi^2 IC} = \frac{h}{8\pi^2 \mu r^2 C}$$

where reduced mass of HCl molecule.

$$\mu = \frac{35.5 \times 1}{36.5} \times (1.67 \times 10^{-24}\ \text{g}) = 1.62 \times 10^{-24}\ \text{g}$$

Hence, bond length

$$= \left[\frac{h}{8\pi^2 \mu BC}\right]^{1/2} = \left[\frac{6.62\times10^{-27}\ \text{erg s}}{8\times(3.1417)^2\times(1.62\times10^{-24}\ \text{g})\times(10.4\,\text{cm}^{-1})\times(3\times10^{10}\ \text{ms}^{-1})}\right]^{1/2}$$

$$= \left[\frac{6.62 \times 10^{-27}}{3.99 \times 10^{-11}}\right]^{1/2} \text{cm} = 1.29 \times 10^{-8}\ \text{cm} = 1.29\ \text{Å}$$

Example 13.11 **(a)** State which statistics would be appropriate in the following cases

He gas at room temperature and atmospheric pressure.

(ii) Electrons in silver at room temperature.

(iii) Electrons and holes in semi-conducting Ge, which has band gap of 1eV at room temperature.

State the basic difference in the fundamental assumption in the following cases:

(i) M–B statistics and F–D statistics

(ii) M–B statistics and B–E statistics

(iii) F–D statistics and B–E statistics

(a) (i) For ^{4}He at room temperature

$$\frac{p}{k_B T}\left(\frac{h^2}{2\pi m k_B T}\right)^{3/2}$$

$$\frac{1.01\times 10^5 \text{ Nm}^{-2}}{(1.38\times 10^{-23} \text{ JK}^{-1})\times(300 \text{ K})}\left(\frac{(6.62\times 10^{-34} \text{ Js})^2}{2\times 3.1417\times(6.68\times 10^{-27} \text{ kg})\times (300 \text{ K})\times(1.38\times 10^{-23} \text{ JK}^{-1})}\right)^{3/2}$$

$$=(0.244\times 10^{+26} \text{ m}^{-3})(0.1266\times 10^{-30} \text{ m}^3)$$

$$=0.0309\times 10^{-4}$$

$$=3.09\times 10^{-6}$$

Obviously, this is much less than 1. Therefore, M–B statistics will apply.
(ii) For electrons in silver

$$n\lambda_{\text{dB}}^3 \approx 0.6146$$

This is comparable to unity implying that electron gas is highly degenerate even at room temperature. Moreover, electrons are fermions. Therefore in this case, F–D will apply.
(iii) In this case, M–B statistics will apply because at room temperature, the electrons and holes do not have sufficient energy to jump over the 1eV band gap in sufficient

(i) The F–D statistics has two additional assumptions:

indistinguishability principle
Pauli Exclusion principle; not more than one particle can occupy the same quantum state.
(ii) In B–E statistics, the principle of indistinguishability applies and energy is quantised.
(iii) In F–D statistics, Pauli Exclusion principle implies that a given quantum state cannot be occupied by any two particles with all the four quantum numbers same, whereas no such restriction applies in B-E statistics.

Example 13.12 The entropy of an ideal paramagnet in a magnetic field is given by

$$S = S_0 - CU^2$$

is energy of the spin system and C is a constant. **(a)** Calculate the energy as a function of T, **(b)** Plot U versus T for $-\infty < T < \infty$ and **(c)** discuss the physical significance of the curves obtained in **(b)**.

(a) By definition, temperature is given by

$$T = \left(\frac{\partial U}{\partial S}\right)_V = -\frac{1}{2CU}$$

so that for an ideal paramagnet, internal energy is inversely proportional to temperature:

$$U = -\frac{1}{2CT} \qquad \text{(i)}$$

> 0, plot of U versus T is shown in Fig. 13.27.

Under normal conditions, the number of particles decreases as we go to higher energy states. However, the significance of negative temperature is that under such conditions, the number of particles in higher energy states is greater than that in the ground state. It means that there are more particles with magnetic moments anti-parallel to the magnetic field than those with magnetic moments parallel to the magnetic field.

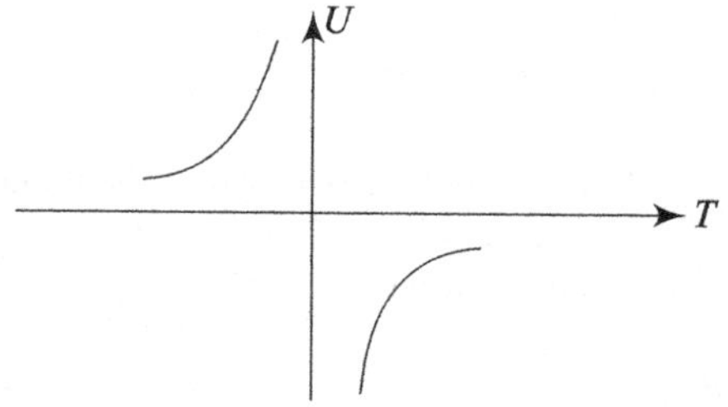

Fig. 13.27 Plot of energy versus T for an

Example 13.13 3,000 particles are distributed among three states having energies $_2 = 2k_BT$ and $\varepsilon_3 = 3k_BT$, respectively. Calculate the total equilibrium energy of the system in terms of k_BT, if the particles obey Maxwell–Boltzmann distribution law.

According to Maxwell–Boltzmann distribution law, the number of particles $_i(\varepsilon_i)$ is proportional to $\exp(-\varepsilon_i / k_BT)$, where n_i is the number of particles in the $= 1, 2, 3$). Therefore, we can write

$$n_1 : n_2 : n_3 = \exp(-k_BT/k_BT) : \exp(-2k_BT/k_BT) : \exp(-3k_BT/k_BT)$$

$$n_1 : n_2 : n_3 = e^{-1} : e^{-2} : e^{-3}$$

$$\frac{n_1}{\frac{1}{e}} : \frac{n_2}{\frac{1}{e^2}} : \frac{n_3}{\frac{1}{e^3}} = \frac{n_1 + n_2 + n_3}{\frac{1}{e} + \frac{1}{e^2} + \frac{1}{e^3}} = \frac{N}{\frac{1}{e} + \frac{1}{e^2} + \frac{1}{e^3}} = \frac{N}{0.5530}$$

$$n_1 = \frac{N}{0.5530} \times \frac{1}{e} = \frac{N}{1.503} = 0.665N = 1995$$

$$n_2 = \frac{N}{0.5530} \times \frac{1}{e^2} = \frac{N}{4.085} = 0.245N = 735$$

$$n_3 = \frac{N}{0.5530} \times \frac{1}{e^3} = \frac{N}{11.107} = 0.090N = 270$$

The value of equilibrium energy

$$U = n_1\varepsilon_1 + n_2\varepsilon_2 + n_3\varepsilon_3$$

$$= 1995k_BT + 735 \times 2k_BT + 270 \times 3k_BT$$

$$= 4275\, k_BT$$

We now summarise what you have learnt in this chapter.

MMARY

For a system of N distinguishable non-interacting particles obeying Maxwell–Boltzmann distribution, the thermodynamic functions can be expressed in terms of the partition function as

$$U = Nk_B T^2 \frac{\partial(\ln Z)}{\partial T}$$

$$S = Nk_B \ln Z + \frac{U}{T}$$

$$F = -Nk_B T \ln Z$$

$$p = \frac{Nk_B T}{Z}\left(\frac{\partial Z}{\partial V}\right)_T$$

The partition function of an ideal monatomic gas consisting of N non-interacting particles is given by

$$Z_N = \frac{V^N}{h^{3N}}(2\pi m k_B T)^{3N/2}$$

The thermodynamic functions of ideal monatomic gas are given by

$$U = \frac{3}{2}RT$$

$$p = \frac{nRT}{V}$$

$$C_V = \frac{3}{2}R$$

$$S = Nk_B \ln\left[\frac{V(2\pi m k_B Te)^{3/2}}{h^3}\right]$$

The expression for entropy obtained by treating gas particles as distinguishable and energy as continuous variable leads to Gibbs Paradox, which implies that entropy can not be function of the thermodynamic state only.

The Sackür–Tetrode equation accounts for indistinguishability of gas molecules. It is given by

$$S_N^c = Nk_B \ln\left[\left(\frac{V}{N\lambda_{dB}^{\;3}}\right)\pi^{3/2}\, e^{5/2}\right]$$

To explain temperature variation of molar specific heat capacity of (hetro-nuclear as well as homo-nuclear) diatomic gases, we have to consider quantisation of rotational as well as vibrational motions of molecules.

The rotational partition function for a hetro-nuclear molecule is given by

$$Z_{\text{rot}} = \frac{T}{\theta_{\text{rot}}}\left[1 + \frac{1}{3}\frac{\theta_{\text{rot}}}{T} + \frac{1}{15}\left(\frac{\theta_{\text{rot}}}{T}\right)^2 + \ldots\right] \qquad T >> \theta_{\text{rot}}$$

$$= 1 + 3e^{-2\theta_{\text{rot}}/T} + 5e^{-6\theta_{\text{rot}}/T} + \ldots \qquad T << \theta_{\text{rot}}$$

The vibrational partition function is given by

$$Z_{\text{vib}} = \frac{1}{2\sinh\left(\frac{\theta_{\text{vib}}}{2T}\right)}$$

$$\theta_{\text{vib}} = \frac{h\nu}{k_B}.$$

The rotational and vibrational molar specific heat capacities are given by

$$(C_V)_{\text{rot}} = R\left[1 + \frac{1}{45}\left(\frac{\theta_{\text{rot}}}{T}\right)^2 + \ldots\right] \qquad T >> \theta_{\text{rot}}$$

$$= 12R\left(\frac{\theta_{\text{rot}}}{T}\right)^2 \exp\left(-\frac{2\theta_{\text{rot}}}{T}\right) \qquad T << \theta_{\text{rot}}$$

$$(C_V)_{\text{vib}} = N_A k_B f\left(\frac{\theta_{\text{vib}}}{T}\right)$$

$$f\left(\frac{\theta_{\text{vib}}}{T}\right) = \left(\frac{\theta_{\text{vib}}}{T}\right)^2 \frac{\exp(\theta_{\text{vib}}/T)}{[\exp(\theta_{\text{vib}}/T) - 1]^2}$$

$>> \theta_{\text{vib}}$, $f\left(\frac{\theta_{\text{vib}}}{T}\right) \to 1$ and we recover the classical result: $(C_V)_{\text{vib}} \to N_A k_B$.

On the other hand, for $T << \theta_{\text{vib}}$, $(C_V)_{\text{vib}} \to 0$.

For hydrogen, the rotational specific heat capacity is given by

$$(C_V)_{\text{rot}} = \frac{1}{4}(C_V)_{\text{para}} + \frac{3}{4}(C_V)_{\text{ortho}}$$

$$(C^{\text{rot}})_{\text{para}} = \frac{\partial}{\partial T}\left\{N_A k_B T^2 \frac{\partial(\ln Z_{\text{rot}})_{\text{para}}}{\partial T}\right\} = 180\,R\left(\frac{\theta_{\text{rot}}}{T}\right)^2 \exp(-6\theta_{\text{rot}}/T) + \ldots$$

$$(C^{\text{rot}})_{\text{ortho}} = \frac{\partial}{\partial T}\left\{N_A k_B T^2 \frac{\partial(\ln Z_{\text{rot}})_{\text{ortho}}}{\partial T}\right\} = \frac{700}{3}R\left(\frac{\theta_{\text{rot}}}{T}\right)^2 \exp(-10\theta_{\text{rot}}/T) + \ldots$$

Dulong–Petit law states that for crystalline solids at room temperature, $C_V \approx 3R$. However, it fails to explain the observed temperature variation of specific molar heat capacity.

Einstein assumed that each atom in a solid vibrates with the same angular frequency and may be treated as an independent oscillator. He obtained the following expression for molar heat capacity of solids:

$$C_V = 3N_A k_B \left(\frac{\theta_E}{T}\right)^2 \frac{\exp(\theta_E/T)}{\left[\exp\left(\frac{\theta_E}{T}\right) - 1\right]^2}$$

$\theta_E = \dfrac{h\nu_E}{k_B}$ is Einstein temperature of the solid.

Einstein's theory fails to explain observed temperature variation of specific heat capacity at low temperatures. Debye replaced the single frequency model of Einstein by a model where a spectrum of vibrational frequencies is considered. He obtained the following expression for molar heat capacity of solids:

$$C_V = 3R\left[12\left(\frac{T}{\theta_D}\right)^3 \int_0^{\theta_D/T} \frac{x^3}{e^x - 1}dx - \frac{3(\theta_D/T)}{e^{\theta_D/T} - 1}\right]$$

$\theta_D = h\nu/k_B$ is Debye temperature.

At high temperatures ($T >> \theta_D$), the molar specific heat capacity of a solid equals . On the other hand, at low temperatures ($T << \theta_D$)

$$C_V = 12\frac{\pi^4}{5}R\left(\frac{T}{\theta_D}\right)^3$$

When a paramagnetic substance having N magnetic atoms per unit volume is placed in an external magnetic field **B**, only two energy states defined by $\varepsilon_1 = -\mu_B B$ and $\mu_B B$ will be available. If we assume that these states are non-degenerate and the system is in thermal equilibrium at temperature T, the net magnetic moment of the system is given by

$$M = \mu_B(N_1 - N_2) = N\mu_B \tanh\left(\frac{\mu_B B}{k_B T}\right)$$

Other thermodynamic functions of the paramagnetic substance are given by

$$U = -N\mu_B B \tanh\left(\frac{\mu_B B}{k_B T}\right),$$

$$C_B = Nk_B\left(\frac{\mu_B B}{k_B T}\right)^2 \text{sech}^2\left(\frac{\mu_B B}{k_B T}\right)$$

$$S = Nk_B\left[\ln\left\{2\cosh\left(\frac{\mu_B B}{k_B T}\right)\right\} - \left(\frac{\mu_B B}{k_B T}\right)\tanh\left(\frac{\mu_B B}{k_B T}\right)\right]$$

When a two-level system of magnetic dipoles with parallel and anti-parallel spins is place in a magnetic field, it is possible to achieve population inversion by reversing the magnetic field. We then obtain to a state of lower entropy. This corresponds to a negative temperature state. The negative temperatures are higher than positive absolute temperatures and -0 K is the highest attainable temperature.

In 1951, Pound, Purcell and Ramsey established the existence of negative absolute temperatures by using nuclear magnetic resonance techniques.

Based on the theory of stimulated emission, Einstein predicted that population inversion can occur due to transitions between different energy states in an atom with finite number of energy levels. It was realised in 1960 by Maiman using a ruby crystal. Three and four level pumping schemes are more efficient.

In the core of a star, temperatures are extremely high and the elements present there are mostly in the atomic state. Under such conditions, atoms move very rapidly and undergo frequent collisions. In the process, they are stripped of valence electrons. This is referred to as *thermal ionisation*. The expression for the number density for electrons in a star is given by Saha's ionisation formula:

$$n = 3\left(\frac{2\pi m_e k_B}{h^2}\right)^{3/2} T^{3/2} \exp(-\varepsilon^* / k_B T)$$

ERCISES

Monatomic molecules absorbed on a surface are free to move on this surface and can be treated as a classical ideal two-dimensional gas. At absolute temperature T, what is the heat capacity per mole (C_A) of molecules thus absorbed on a surface of fixed area? (*Ans:* $C_A = N_A k_B$)

A system consisting of two particles, each of which can be in any one of three quantum states of respective energies 0, ε and 3ε is in thermal equilibrium at temperature T. Write expressions for the partition function of the system if the particles obey **(a)** B–E statistics and **(b)** F-D statistics.

(*Ans:* **(a)** $Z = 1 + e^{-\beta\varepsilon} + e^{-2\beta\varepsilon} + e^{-3\beta\varepsilon} + e^{-4\beta\varepsilon} + e^{-6\beta\varepsilon}$

(b) $Z = e^{-\beta\varepsilon} + e^{-3\beta\varepsilon} + e^{-4\beta\varepsilon}$)

Consider a mixture of ideal gases consisting of N_1 diatomic molecules of type 1 N_2 diatomic molecule of type 2. Obtain the expression for the total partition function, where Z_1 and Z_2 are the single particle partition functions of molecules of types 1 and 2 respectively. Also obtain the equation of state for this system.

(*Ans:* $Z = (Z_1^{N_1} / N_1 !)(Z_2^{N_2} / N_2 !)$; Equation of state is $pV = Nk_BT$ where $p = p_1 + p_2$ and $N = N_1 + N_2$)

The partition function of a system is given by

$$\ln Z = \alpha T^4 V$$

where α is a constant, T is the absolute temperature and V is the volume. Calculate the internal energy, the pressure and the entropy.

(*Ans:* $E = 4k_B\alpha T^5 V$; $p = k_B\alpha T^5$; $S = 5k_B T^4 V$)

Prove that

$$C_p - C_V = \frac{k_B\,[(\partial/\partial T)\{T(\partial \ln Z/\partial V)_T\}]^2}{(\partial^2 \ln Z/\partial V^2)_T} > 0$$

Verify that the value of $(C_p - C_V)$ for an ideal gas is Nk_B

Show that the partition function of a relativistic gas consisting of N monatomic molecules and having the energy-momentum relation : $\varepsilon = pc$, is given by

$$Z = \frac{1}{N!}\left\{8\pi^2 V\left(\frac{k_B T}{hc}\right)^3\right\}^N$$

Show that the entropy of an assembly of N Einstein oscillators is given by

$$S = 3Nk_B\left\{\frac{\theta_E/T}{e^{\theta_E/T} - 1} - \ln[1 - e^{-\theta_E/T}]\right\}$$

Show that $S \to 0$ as $T \to 0$

Show that for large T, S approaches $3Nk_B\,[1 + \ln(T/\theta_E)]$.

The magnetic moment of nuclei is of the order of 10^{-26} Am^2. Estimate the magnetic field required at 0.01 K to produce appreciable alignment of such nuclei.

(*Ans:* B = 14 tesla)

Consider a system of N distinguishable particles distributed in two non-degenerate levels separated by an energy ε and in equilibrium with a reservoir at a temperature . Calculate **(a)** the internal energy, **(b)** the entropy and **(c)** C_V of the system.

$$\left(\textit{Ans:}\ \text{(a)}\ E = N\varepsilon\,[1 + e^{\varepsilon/k_B T}]^{-1}\right.$$

$$\text{(b)}\ S = Nk_B \ln[1 + e^{\varepsilon/k_B T}] + N\varepsilon\{T[1 + e^{\varepsilon/k_B T}]\}^{-1}$$

$$\left.\text{(c)}\ C_V = Nk_B\left(\frac{\varepsilon}{k_B T}\right)^2 e^{\varepsilon/k_B T}\,(1 + e^{\varepsilon/k_B T})^{-2}\right)$$

A paramagnetic salt contains 10^{25} magnetic ions per cubic metre. Each has a magnetic moment of 1 Bohr magneton. Calculate the difference between the number of ions aligned parallel and anti-parallel to a field of strength 1 tesla at **(a)** 300 K and **(b)** 4 K, if the volume of the sample is 100 cm^3. Calculate the magnetic moment of the sample at these two temperatures.

(*Ans:* (a) 2.24×10^{18}, 2.08×10^{-6} A m^2

(b) 1.66×10^{20}, 1.54×10^{-4} A m^2)

4

FERMI–DIRAC STATISTICS

ing Objectives

hapter, you will learn how to:

- ain the concept of zero point energy;
- n expressions for Fermi energy, Fermi pressure and other modynamic functions of a completely degenerate and a strongly nerate F–D system;
- ain temperature dependence of electronic specific heat capacity;
- y F–D statistics to electrons in metals and derive Richardson–mann equation for thermionic emission;
- e an expression for photoelectric current density; and
- uss the constitution of a white dwarf star and discuss Chandrasekhar s limit formula.

NTRODUCTION

In Chapter 13, we considered Maxwell–Boltzmann statistics and obtained the thermodynamic functions of a gas made up of non-interacting distinguishable particles. In so doing, we came across a paradox in that self-diffusion of a gas causes increase in its entropy. The genesis of the paradox was traced to the treatment of identical molecules of a gas as distinguishable and thereby over-estimation of the number of accessible states. Sackür–Tetrode equation provided a way out and successfully resolved the paradox. The study of thermodynamic behaviour of a two-level paramagnetic substance in a magnetic field led to the realisation of the condition of population inversion and negative temperatures. Population inversion is also the basic requirement for operation of a laser—a highly directional, intense source of monochromatic light.

From Chapter 13, you may recall that classical and quantum results tend to be identical $<< 1$, where n is number density and λ_{dB} is de Broglie wavelength. This happens at high temperatures and low densities. However, when $n\lambda_{\text{dB}} \approx 1$, quantum effects begin to become important. This happens at low temperatures and high densities and the classical results derived in the preceding chapter are not expected to hold. A physical system is then *degenerate*. The familiar situations where quantum effects become particularly important include the atoms of helium at low temperatures, an assembly of photons, conduction electrons in metals and white dwarf stars.

In this chapter, we will begin our discussion by obtaining expressions for a weakly degenerate quantum system. We have worked out expressions for total number of particles and total internal energy in Sec. 14.2. You will discover that the total internal energy of a fermion system is greater than that predicted classically. This is a consequence of the exclusion principle which forbids more than one fermion from occupying the already filled lower energy levels. On the other hand, a boson system tends to have less energy than that predicted by classical statistics because bosons prefer low energy states. It may be mentioned here that the thermodynamic properties of weakly degenerate Bose–Einstein and Fermi–Dirac systems can be discussed simultaneously. To save teaching time as well as space in the book, we have illustrated this in Sec. 14.3. However, when deviation from classical behaviour is very large, i.e., a system is strongly degenerate, we have to treat them separately. Whereas a fermion system is quite alive even at absolute zero, bosons exhibit a spectacular effect – *Bose-Einstein condensation* – at low temperatures. These effects are purely quantum mechanical, completely unknown in tenets of classical physics. A detailed discussion of strongly degenerate Fermi–Dirac systems is presented in Sec. 14.4. The applications of F–D statistics to thermionic emission, photoelectric emission and white dwarf stars are given in Sec. 14.5. Fermi–Dirac statistics can also be used to explain Pauli paramagnetism and calculate the electron distribution in a heavy atom. However, we will not go into the details of these systems here.

DEAL QUANTUM GASES: PARTICLE DISTRIBUTION ND INTERNAL ENERGY

From Eq. (12.65), we recall that the occupation number for a state of energy ε_s is given by

$$n_s = \frac{1}{\exp[\beta(\varepsilon_s - \mu)] + \kappa} \tag{14.1a}$$

$= -1$ for bosons and $+1$ for fermions. For a macroscopic sample we replace n_s by a continuous distribution $n(\varepsilon)$:

$$n(\varepsilon) = \frac{1}{\exp[\beta(\varepsilon - \mu)] + \kappa} \tag{14.1b}$$

The number of particles within the element of phase space $d^3\mathbf{r}\, d^3\mathbf{p}$ is the product of the number of quantum states and the occupation number. The number of quantum states of within this element is given by $(d^3\mathbf{r}\, d^3\mathbf{p}_\ell/h^3)$ so that*

$$dN = \left(\frac{d^3\mathbf{r} d^3\mathbf{p}_\ell}{h^3}\right) n(\varepsilon) \tag{14.2}$$

From Chapter 12, we recall that a system of one type of particles can often be separated into several subsystems distinguishable by their internal properties, such as intrinsic angular momentum. If we denote this number by $\boldsymbol{G}$, then Eq. (14.2) can be rewritten as

$$dN = \boldsymbol{G}\frac{d^3\mathbf{r}\, d^3\mathbf{p}_\ell}{h^3} n(\varepsilon) \tag{14.3}$$

---

*Although the standard notation for linear momentum is p, we have put the subscript ℓ to avoid confusion with pressure.

When the energy of interaction between particles making up the system is negligibly small, the energy of the state, ε, is independent of position. Then integration over $d^3\mathbf{r}$ gives the volume V in the Cartesian coordinate system. Similarly, if the energy of the state does not depend on the direction in which the particle is moving, we can replace d^3p_ℓ by . Then Eq. (14.3) takes the form

$$dN = G\left(\frac{4\pi V p_\ell^2\, dp_\ell}{h^3}\right) n(\varepsilon) \tag{14.4}$$

The momentum and energy of a non-relativistic particle are connected by the relation

so that $p_\ell^2\, dp_\ell = \sqrt{2} m^{3/2} \varepsilon^{1/2} d\varepsilon$ and the number of particles in phase space

can be written as

$$dN = \left(\frac{2\pi V (2m)^{3/2}}{h^3} G\right) \frac{\varepsilon^{1/2}\, d\varepsilon}{\exp[\beta(\varepsilon - \mu)] + \kappa}$$

$$= CV \frac{\varepsilon^{1/2}\, d\varepsilon}{\exp[\beta(\varepsilon - \mu)] + \kappa} \tag{14.5}$$

$\left(\frac{2\pi (2m)^{3/2}}{h^3} G\right)$ is constant for a given system.

Equation (14.5) expresses the number of particles as a function of energy for a non-relativistic quantum gas. In the low-energy limit, the exponential factor in the denominator will be of little consequence, so that

$$\frac{dN}{d\varepsilon} \propto \varepsilon^{1/2} \tag{14.6}$$

On the other hand, in the high-energy limit, the exponential term will dominate and we have

$$\frac{dN}{d\varepsilon} \propto \exp(-\beta\varepsilon) \tag{14.7}$$

Before proceeding further, you may like to answer a practise problem.

Problem 14.1 For a relativistic quantum gas, show that Eq. (14.5) can be expressed $\left(\frac{4\pi V}{h^3 c^3} G\right) \frac{\varepsilon^2\, d\varepsilon}{\exp[\beta(\varepsilon - \mu)] + \kappa}$. Discuss the limiting behaviour of the modified

$\propto \varepsilon^2$ in the low energy limit and $\frac{dN}{d\varepsilon} \propto \exp(-\beta\varepsilon)$ in the high energy limit

In Fig. 14.1 and 14.2, we have plotted $D(\varepsilon)$, $n(\varepsilon)$ and $dN/d\varepsilon$ as a function of energy for non-relativistic fermions and bosons at low and high temperatures.

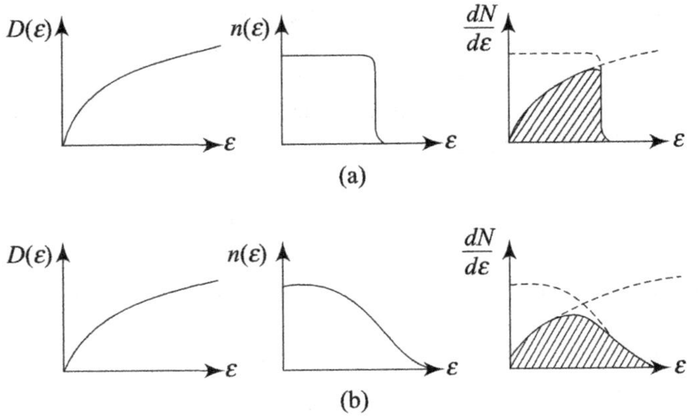

Plot of density of states, occupation number and distribution function of particles in a non-relativistic Fermi gas at (a) low temperatures and (b) high temperatures.

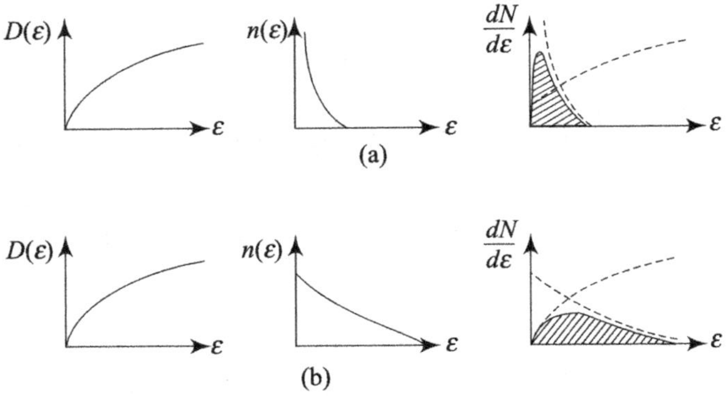

Plot of density of states, occupation number and distribution function of particles in a non-relativistic Bose gas at (a) low temperatures and (b) high temperatures.

The total number of particles making up the system is obtained by integrating Eq. (14.5) over energy:

$$N = CV \int_0^\infty \frac{\varepsilon^{1/2}\, \mathrm{d}\varepsilon}{\exp[\beta(\varepsilon - \mu)] + \kappa} \tag{14.8}$$

The energy of a system whose particles have energies in the range ε and $\varepsilon + \mathrm{d}\varepsilon$ is equal to the product of the number of such particles and the energy of each particle:

$$\mathrm{d}E = \varepsilon\, \mathrm{d}N$$

Using the result given in Eq. (14.5), we can write

$$\mathrm{d}E = CV \frac{\varepsilon^{3/2}\, \mathrm{d}\varepsilon}{\exp[\beta(\varepsilon - \mu)] + \kappa} \tag{14.9}$$

The total internal energy of the system is obtained by integrating this expression over ε in the range 0 to ∞. Hence,

$$E = CV \int_0^\infty \frac{\varepsilon^{3/2}\, d\varepsilon}{\exp[\beta(\varepsilon-\mu)]+\kappa} \tag{14.10}$$

You will note that Eqs. (14.8) and (14.10) for the total number of particles and total internal energy are very similar. So it should be possible to relate them. To this end, we integrate Eq. (14.8) by parts. This gives

$$N = CV\left[\frac{2}{3}\frac{\varepsilon^{3/2}}{\exp[\beta(\varepsilon-\mu)]+\kappa}\right]_0^\infty + \frac{2CV}{3}\int_0^\infty \frac{\varepsilon^{3/2}\, e^{\beta(\varepsilon-\mu)}\, \beta d\varepsilon}{\{\exp[\beta(\varepsilon-\mu)]+\kappa\}^2}$$

The first term vanishes at both limits and the expression for N reduces to

$$N = \frac{2}{3}CV\beta \int_0^\infty \frac{\varepsilon^{3/2}\, e^{\beta(\varepsilon-\mu)}\, d\varepsilon}{[e^{\beta(\varepsilon-\mu)}+\kappa]^2} \tag{14.11}$$

To evaluate this integral, we write

$$\frac{e^{\beta(\varepsilon-\mu)}}{[e^{\beta(\varepsilon-\mu)}+\kappa]^2} = \frac{1}{(e^{\beta(\varepsilon-\mu)}+\kappa)}\left[1 - \frac{\kappa}{e^{\beta(\varepsilon-\mu)}+\kappa}\right]$$

$$N = \frac{2}{3}CV\beta\left[\int_0^\infty \frac{\varepsilon^{3/2}\, d\varepsilon}{[e^{\beta(\varepsilon-\mu)}+\kappa]} - \kappa\int_0^\infty \frac{\varepsilon^{3/2}\, d\varepsilon}{[e^{\beta(\varepsilon-\mu)}+\kappa]^2}\right]$$

By rearranging terms, we get

$$\frac{3}{2}Nk_BT = CV\int_0^\infty \frac{\varepsilon^{3/2}\, d\varepsilon}{e^{\beta(\varepsilon-\mu)}+\kappa} - CV\kappa\int_0^\infty \frac{\varepsilon^{3/2}\, d\varepsilon}{[e^{\beta(\varepsilon-\mu)}+\kappa]^2} \tag{14.12}$$

$k_BT)^{-1}$.

From Eq. (14.12), we note that the first term on the right-hand side is just the total internal energy (Eq. (14.10)) of the system. So we can express the total internal energy of a non-relativistic quantum gas as

$$E = \frac{3}{2}Nk_BT \pm \xi(\mu, T) \tag{14.13}$$

Note that the plus sign holds for fermions and minus sign holds for bosons. The function) is positive definite and is given by

$$\xi(\mu, T) = CV\int_0^\infty \frac{\varepsilon^{3/2}\, d\varepsilon}{[e^{\beta(\varepsilon-\mu)}+\kappa]^2} = CV\int_0^\infty [n(\varepsilon)]^2\, \varepsilon^{3/2}\, d\varepsilon \tag{14.14}$$

From this expression, we note that the correction term will be small when the occupation ε) is small. However, it becomes increasingly important when particle densities become higher, i.e., in the range of energies where occupation numbers are large.

Problem 14.2 Show that the total internal energy of a relativistic gas is given by

$$E = 3Nk_BT \pm \eta(\mu, T)$$

$$\eta(\mu, T) = \left(\frac{4\pi VG}{h^3 c^3}\right) \int_0^\infty \frac{\varepsilon^3 \, d\varepsilon}{[e^{\beta(\varepsilon-\mu)} + \kappa]^2}$$

Equation (14.13) shows that the total internal energy of a non-relativistic fermion system is greater than that predicted classically. This is a consequence of the exclusion principle, which forbids more than one fermion from occupying the already filled lower energy levels. On the other hand, a boson system tends to have less energy than that predicted by classical statistics because more bosons prefer low energy states, as there is no constraint like the Pauli principle on them.

We know that pressure and energy are directly related. So similar considerations will apply to the pressure exerted by a gas and we expect that $p_{F-D} > p_{M-B} > p_{B-E}$. Since

we can write

$$pV = Nk_B T \pm \frac{2}{3} \xi(\mu, T) \tag{14.15}$$

You will note that Eqs. (14.13) and (14.15) respectively give us the relative magnitudes of energy and pressure exerted by fermions and bosons. However, these equations do not enable us to determine the exact values of these parameters. This suggests that we have to characterise quantum effects, which impact the behaviour of the system. We now discuss how we can do so.

For mathematical convenience in obtaining expressions for thermodynamic properties of a system, we define the *degeneracy* or *fugacity* parameter as

$$A = \exp(\mu\beta) \tag{14.16}$$

A system is said to be *weakly degenerate* if $A << 1$. You will soon discover that in the high temperature limit, a quantum system resembles a classical system. But as temperature decreases, the behaviour of a quantum system differs from a classical system significantly. In fact, even Fermi and Bose gases behave dramatically differently as we approach quantum region. We will now evaluate thermodynamic properties of a weakly degenerate quantum system, i. e., in the classical neighbourhood.

Weakly Degenerate Quantum Systems

In terms of fugacity, we can rewrite Eq. (14.8) as

$$N = CV \int_0^\infty \frac{\varepsilon^{1/2} \, d\varepsilon}{(1/A)\, e^{\beta\varepsilon} + \kappa} \tag{14.17}$$

To evaluate this integral, we introduce a change of variable by putting

$$\beta\varepsilon = x \tag{14.18}$$

$\varepsilon = dx$ and

$$\varepsilon^{1/2} \, d\varepsilon = \frac{x^{1/2} \, dx}{\beta^{3/2}} = (k_B T)^{3/2} \, x^{1/2} \, dx$$

On making these substitutions in Eq. (14.17), we get

$$N = CV(k_B T)^{3/2} \int_0^\infty \frac{x^{1/2}\,dx}{(1/A)e^x + \kappa}$$

$$= CV(k_B T)^{3/2} \int_0^\infty A\,e^{-x} x^{1/2} [1 + \kappa A\,e^{-x}]^{-1}\,dx$$

Using binomial expansion, we can write

$$N = CV(k_B T)^{3/2} \int_0^\infty A e^{-x} x^{1/2} [1 - \kappa A\,e^{-x} + \kappa^2 A^2\,e^{-2x} - \ldots]\,dx$$

$$= CV(k_B T)^{3/2} \int_0^\infty x^{1/2}\,dx \sum_{s=1}^\infty (-\kappa)^{s-1} A^s\,e^{-sx}$$

On changing the order of summation and integration, we obtain

$$N = CV(k_B T)^{3/2} \sum_{s=1}^\infty (-\kappa)^{s-1} A^s \int_0^\infty e^{-sx} x^{1/2}\,dx$$

The integral on the right-hand side is a standard gamma function

$$\int_0^\infty e^{-sx} x^{1/2}\,dx = \frac{\Gamma(3/2)}{s^{3/2}}$$

so that the total number of particles is given by

$$N = CV(k_B T)^{3/2}\,\Gamma\left(\frac{3}{2}\right) \sum_{s=1}^\infty (-\kappa)^{s-1} \frac{A^s}{s^{3/2}}$$

$$= CV(k_B T)^{3/2}\,\Gamma\left(\frac{3}{2}\right)\left[A - \frac{\kappa A^2}{2^{3/2}} + \frac{A^3}{3^{3/2}} - \ldots\right] \qquad (14.19)$$

Using the same procedure for internal energy, we can readily show that

$$E = CV(k_B T)^{5/2}\,\Gamma\left(\frac{5}{2}\right)\left[A - \frac{\kappa A^2}{2^{5/2}} + \frac{A^3}{3^{5/2}} - \ldots\right] \qquad (14.20)$$

Equations (14.19) and (14.20) indicate the dependence of the total number of particles and internal energy of a weakly degenerate quantum system on fugacity A (or chemical) and absolute temperature T.

Proceeding further, we put

$$A_0 = \frac{N}{CV(k_B T)^{3/2}\,\Gamma(3/2)} \qquad (14.21)$$

This shows that A_0 is much less than one when the number density (N/V) is small and/or temperature is high. That is a rarefied system at high temperatures will be weakly degenerate. We can correlate A_0 to the ratio of the de Broglie wavelength and average intermolecular distance. To this end, we substitute for C to obtain

$$A_0 = \frac{Nh^3}{V(2\pi m k_B T)^{3/2}\,G}$$

$$\Gamma(3/2) = \frac{\sqrt{\pi}}{2}.$$

To interpret this result physically, we rewrite it as

$$A_0 = \frac{1}{\mathbf{G}}\left[\frac{h}{(V/N)^{1/3}\,(2\pi m k_B T)^{1/2}}\right]^3$$

By definition, $\dfrac{h}{(2\pi m k_B T)^{1/2}} = \lambda_{dB}$ is the de Broglie wavelength associated with a molecule/atom. Similarly, $(V/N)^{1/3}(= l)$ is a measure of the mean distance between molecules. So, the expression for A_0 can be rewritten as

$$A_0 = \frac{1}{\mathbf{G}}\left(\frac{\lambda_{dB}}{l}\right)^3 \qquad (14.22)$$

and the condition for a system to be weakly degenerate takes the form

$$A_0 = \frac{1}{\mathbf{G}}\left(\frac{\lambda_{dB}}{l}\right)^3 << 1 \qquad (14.23)$$

That is, a system will be weakly degenerate when the de Broglie wavelength associated with a molecule is much less than the mean intermolecular separation. For most gases, A_0 is much smaller than one at STP ($A_0(\text{A}) = 2 \times 10^{-6}$, $A_0(\text{Ne}) = 10^{-4}$, $A_0(H_2) = 10^{-12}$). However, for helium gas at 273 K, $\mathbf{G} = 1$ and if the number density is 10^{20} molecules cm^{-3}, we get cm and $\lambda_{dB} = 0.8 \times 10^{-8}$ cm, i.e., $A_0 = 0.6 \times 10^{-4}$. This means that under normal conditions, quantum effects are not important for most gases. Physically, this corresponds to the spacing between molecules being too large for wave-mechanical interference effects between de Broglie waves of different molecules to be significant. That is, Eq. (14.23) holds when the wave nature of molecules is not important.

$n\lambda_{dB}$ is comparable to 1, interference effects tend to become important. For example, for liquid helium $\dfrac{V}{N} = 5 \times 10^{-23}\ \text{cm}^3$ per atom, so that $l = 4 \times 10^{-8}$ cm. The de Broglie wavelength of 4×10^{-8} cm corresponds to a temperature $T = 10$ K. So, we expect quantum effects to become important for liquid helium below about $T = 10$ K.

Let us now consider conduction electrons in a metal, which constitute the most important physical system obeying F–D statistics. In metals, the electrons in the outermost orbit of the atoms may be detached from the atom and roam around freely. Drude suggested that the *electron gas* behaved like a free gas moving inside the metal. He used classical statistics to explain electrical and thermal conductivities, thermionic emission and photoelectric effect, among others. However, he failed to explain the observed results. To understand this, let us assume that every atom has one conduction electron. Then number density

$= \quad = 5 \times 10^{-23}\ \text{cm}^3$ per atom and $l = 2 \times 10^{-8}$ cm. With $m_e = 9.1 \times 10^{-28}$ g, the de Broglie wavelength $\lambda_{dB} = 2 \times 10^{-8}$ cm corresponds to a temperature $T = 3 \times 10^5$ K.

means that we should use quantum statistics to describe the conduction electrons in a metal even at room temperature. In fact, this suggestion was first made by Sommerfeld who argued that electrons should be regarded as an ideal *fermion gas* of free electrons.

From this discussion we can conclude that the value of degeneracy parameter varies over a wide range for different physical systems.

Proceeding further, we combine Eqs. (14.19), and (14.21) to obtain

$$A_0 = A - \frac{\kappa A^2}{2^{3/2}} + \frac{A^3}{3^{3/2}} - \ldots + \ldots \tag{14.24}$$

A in terms of A_0, let us write

$$A = A_0 + (\kappa) a A_0^2 + \ldots \tag{14.25}$$

is an unknown constant.

To determine a, we substitute for A from Eq. (14.25) in Eq. (14.24). This gives

$$A_0 = (A_0 + \kappa a A_0^2 + \ldots) - \frac{\kappa}{2^{3/2}} (A_0 + \kappa a A_0^2 + \ldots)^2 + \ldots$$

On retaining terms only up to A_0^2, a little rearrangement gives

$$a = \frac{1}{2^{3/2}} \tag{14.26}$$

On putting this value of a in Eq. (14.25), we get the desired relation for the degeneracy

$$A = A_0 + \frac{\kappa}{2^{3/2}} A_0^2 + \ldots \tag{14.27}$$

Using this result in Eq. (14.20), we find that the total internal energy of a weakly degenerate quantum gas is given by

$$E = CV (k_B T)^{5/2}\, \Gamma\left(\frac{5}{2}\right)\left[\left(A_0 + \frac{\kappa A_0^2}{2^{3/2}} + \ldots\right) - \frac{\kappa}{2^{5/2}} A_0^2 + \ldots\right]$$

$$= CV\, (k_B T)^{5/2}\, \Gamma\left(\frac{5}{2}\right)\left[A_0 + \frac{\kappa A_0^2}{2^{5/2}} + \ldots\right]$$

$$= CV (k_B T)^{5/2}\, \Gamma\left(\frac{5}{2}\right) A_0 \left(1 + \frac{\kappa a A_0}{2} + \ldots\right)$$

$\frac{1}{2^{3/2}}$. On substituting for A_0 from Eq. (14.21) and using the relation $\Gamma\left(\frac{5}{2}\right) = \frac{3}{2}\Gamma\left(\frac{3}{2}\right)$, a little simplification leads to the result

$$E = \frac{3}{2} N k_B T \left(1 + \frac{\kappa a A_0}{2}\right) \tag{14.28}$$

Note that this expression has the same form as Eq. (14.13). Moreover for $\kappa = 0$, the result of classical theory is reproduced.

The heat capacity at constant volume is defined as

$$C_V = \left(\frac{\partial E}{\partial T}\right)_V$$

A_0, which occurs in Eq. (14.28), is proportional to $T^{-3/2}$. Therefore, to obtain an expression for heat capacity at constant volume we first rewrite Eq. (14.28) as

$$E = \frac{3}{2} N k_B T + \frac{3}{4} \kappa a N k_B \alpha T^{-1/2}$$

$$C_V = \left(\frac{\partial E}{\partial T}\right)_V = \frac{3}{2} N k_B - \frac{3}{8} \kappa a N k_B \alpha T^{-3/2}$$

$$= \frac{3}{2} N k_B - \frac{3}{8} \kappa a N k_B A_0$$

$$= \frac{3}{2} N k_{\text{B}} \left(1 - \frac{\kappa a}{4} A_0 + \ldots\right) \tag{14.29}$$

This result shows that for weak degeneracy, $(C_V)_{\text{BE}} > (C_V)_{\text{Classical}} > (C_V)_{\text{FD}}$. In words, the specific heat capacity of a weakly degenerate Bose–Einstein system is more than that for a classical system, which, in turn, is more than that for a weakly degenerate Fermi–Dirac system. However, the differences are, in general, not very large.

So far we have confined ourselves to systems of weakly interacting particles for which the separation between neighbouring states is rather thin. For such systems, we could treat the energy as continuous and replaced the summation over states by integration. But when the spacings between neighbouring states are *not* small compared to $k_B T$, we cannot do so. At sufficiently low temperatures, the system is confined to the configuration of the lowest possible energy and deviations from classically predicted behaviour are large. The system is then said to be *strongly degenerate* and it exhibits phenomena unknown in tenets of classical physics. A strongly degenerate system of bosons would have all N particles in the single lowest state, while a strongly degenerate fermion system would have the lowest states filled and higher states vacant. This is because there can be only one fermion per state. Even at $T = 0$ K, a F–D system is alive!

For some systems, the spacings between neighbouring states compared to $k_B T$ are small at STP but large at low temperatures. Then, such systems exhibit strong degeneracy because there is a negligible probability of excitation to even the first excited state. We will now study the behaviour of strongly degenerate Fermi systems.

TRONGLY DEGENERATE QUANTUM SYSTEMS

The total number of particles in a quantum system is given by the sum over all states of the occupation number of each state.

$$N = \sum_s n_s = \sum_s \left[\frac{1}{e^{\beta(\varepsilon_s - \mu)} + \kappa}\right] = \sum_s \frac{1}{A^{-1} \exp(\beta \varepsilon_s) \pm 1}$$

signifies the fugacity or degeneracy, the minus sign holds for B–E systems and the plus sign holds for F - D systems.

For a B–E system, when all particles are confined to the lowest level $\varepsilon_1 = 0$, the occupation number will be positive definite only for $A \leq 1$. If A were greater than one, the factor -1 in the denominator of the above equation would imply that N would become negative, which makes no sense. On the other hand, because of the factor $+1$ for F–D system we do not have to restrict A. In fact $n \to 1$ for $A \to \infty$. This difference in the values of the degeneracy parameter suggests that the behaviour of strongly degenerate B–E and F–D systems will be vastly different and we have to analyse them separately. We therefore now consider a completely degenerate F–D system.

Completely Degenerate (T = 0K) F–D Systems

For a Fermi–Dirac system, we can write the occupation number for the sth energy level as

$$n_{FD} = \frac{1}{e^{(\varepsilon_s - \mu)/k_B T} + 1} \tag{14.30}$$

$k_B T)^{-1}$.

While dealing with a macroscopic system, we replace n_{FD} by a continuous function) because energy levels may be spaced so closely that their discrete nature is not observed. For a continuous distribution, we can rewrite Eq. (14.30) as

$$n_{FD}(\varepsilon) = \frac{1}{e^{(\varepsilon - \mu)/k_B T} + 1} \tag{14.31}$$

We can evaluate μ, called the *chemical potential*, for a given system at any temperature by applying the condition

$$N = \int_0^\infty n_{FD}(\varepsilon) \frac{d^3\mathbf{r}\, d^3\mathbf{p}_l}{h^3}$$

$d^3\mathbf{p}_l / h^3$ gives the density of states with energy in the interval ε to $\varepsilon + d\varepsilon$ and is $CV\varepsilon^{1/2}\, d\varepsilon$. Then, we can write

$$N = CV \int_0^\infty n_{FD}(\varepsilon) \varepsilon^{1/2} d\varepsilon \tag{14.32}$$

At absolute zero, we expect all particles to be in their lowest possible energy states since only one particle can occupy one state in view of the exclusion principle. So the occupation number should be 1 for the N lowest energy states and 0 for all higher states.* Equations (14.30) and (14.31) bear this expectation out. At T = 0 K, the exponent in Eq. (14.31) becomes $+\infty$ for $\varepsilon > \mu$ and $-\infty$ for $\varepsilon < \mu$. Since $\lim_{x\to\infty} e^{-x} \to 0$ and $\lim_{x\to\infty} e^{x} \to \infty$, we find that

$$n_{FD}(\varepsilon) = \begin{cases} 1 & \text{for } \varepsilon < \mu \\ 0 & \quad\;\; \varepsilon > \mu \end{cases} \tag{14.33}$$

This is illustrated in Figs. 14.3a and b. The value of chemical potential at T = 0 K corresponding to the energy of the highest occupied energy level is called the *Fermi*

$$\varepsilon_F = \mu(T = 0) \tag{14.34}$$

This suggests that we can replace the upper limit of integration in Eq. (14.32) by ε_F. Eq. (14.32) takes a very compact form

$$N = CV \int_0^{\varepsilon_F} \varepsilon^{1/2} d\varepsilon$$

You can imagine a highly degenerate fermion system to be like a partially filled glass of water; full up to

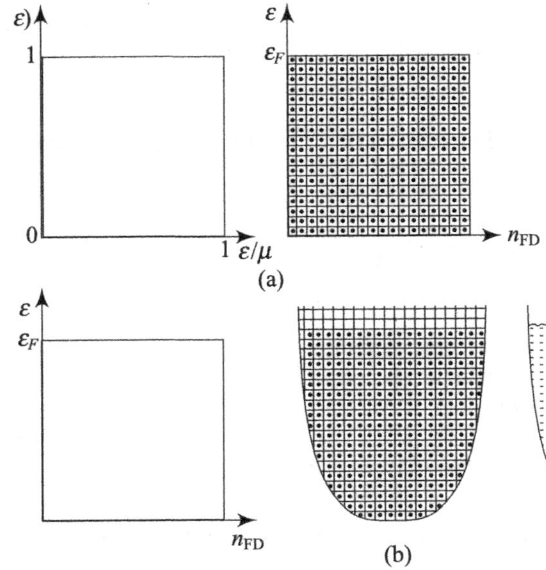

Plot of mean occupation number as a function of energy for (a) a completely degenerate Fermi gas. All states upto the Fermi level are occupied and (b) The

This integral can be readily evaluated to give

$$N = \frac{2}{3} CV \varepsilon_F^{3/2} = \left(\frac{8\pi V\boldsymbol{G}}{3} \right) \left(\frac{2m}{h^2} \right)^{3/2} \varepsilon_F^{3/2} \tag{14.35}$$

We can invert this expression to solve for the Fermi energy:

$$\varepsilon_F = \left(\frac{3}{2} \frac{N}{CV} \right)^{2/3} = \frac{h^2}{2m} \left(\frac{3N}{4\pi \boldsymbol{G} V} \right)^{2/3} \tag{14.36}$$

To get an estimate of the highest occupied state for a metal at absolute zero, we assume that the valence electrons form a Fermi gas. Since an electron has two independent spin states, the multiplicity factor $\boldsymbol{G}(= 2s + 1)$ will be equal to 2, since $s = 1/2$. Hence, the expression for Fermi energy simplifies to

$$\varepsilon_F = \frac{h^2}{8m} \left(\frac{3N}{\pi V} \right)^{2/3} \tag{14.37}$$

This result shows that the Fermi energy depends on the mass of the fermions (here electrons) as well as the number density (N/V) of valence electrons. It is independent of temperature. For monovalent metals (alkali and noble), V/N denotes the atomic volume.

The Fermi energies for a few typical metals are given in Table 14.1. The third column of this table gives the corresponding Fermi temperatures, defined as

$$T_F = \frac{\varepsilon_F}{k_B} = \frac{h^2}{8mk_B} \left(\frac{3N}{\pi V} \right)^{2/3} \tag{14.38}$$

Table 14.1 *The Fermi energies and Fermi temperatures for some metals.*

Metals	ε_F (eV)	T_F(K) $\times 10^{-4}$
Lithium	4.72	5.47
Beryllium	14.14	16.39
Sodium	3.10	3.59
Aluminium	11.66	13.52
Potassium	2.12	2.46
Cesium	1.58	1.83
Copper	7.00	8.12
Silver	5.49	6.37
Gold	5.51	6.39

Note that the Fermi energy

varies in a wide range; from 1.58 eV to 14.14 eV;
is lower for monovalent metals than for trivalent metals; and
corresponds to very high Fermi temperatures in all cases.

This is consistent with our earlier statement that electrons must be treated a quatum system at all temperatures.
You should now go through the following example:

Example 14.1 In silver, we take one conduction electron per atom at room temperature and $N/V = 5.86 \times 10^{28}$ m^{-3}. Show that the electron gas is strongly degenerate.
$= 6.62 \times 10^{-34}$ Js, $k_B = 1.38 \times 10^{-23}$ JK^{-1} and $m_e = 9.11 \times 10^{-31}$ kg.

To show that the electron gas is strongly degenerate, we have to calculate Fermi temperature, which, in turn, requies knowledge of Fermi energy. From Eq. (14.37), we recall that

$$\varepsilon_F = \frac{h^2}{8m}\left(\frac{3N}{\pi V}\right)^{2/3}$$

On substituting the given values, we get

$$\varepsilon_F = \frac{(6.62 \times 10^{-34}\ \text{Js})^2}{8 \times (9.11 \times 10^{-31}\ \text{kg})}\left(\frac{3 \times 5.86 \times 10^{28}\ \text{m}^{-3}}{\pi}\right)^{2/3}$$
$$= 0.60 \times 10^{-37}\ (\text{Js})^2\ \text{kg}^{-1} \times 14.63 \times 10^{18}\ \text{m}^{-2}$$
$$= 8.78 \times 10^{-19}\ \text{J} = 5.49\ \text{eV}$$

Let us now calculate Fermi temperature for silver. From Eq. (14.38), we can write

$$T_F = \frac{\varepsilon_F}{k_B}$$
$$= \frac{8.78 \times 10^{-19}\ \text{J}}{1.38 \times 10^{-23}\ \text{JK}^{-1}}$$
$$= 6.36 \times 10^4\ \text{K}$$

>> T, we can conclude that electrons in silver behave as a strongly degenerate

Proceeding further we calculate the total internal energy of a completely degenerate F–D system. It is given by

$$E_0 = CV \int_0^{\varepsilon_F} \varepsilon^{3/2}\, d\varepsilon$$

$$= \frac{2}{5} CV \varepsilon_F^{5/2} \tag{14.39}$$

On combining Eqs. (14.35) and (14.39), we can express zero-point energy of a completely degenerate F–D system in terms of the number of fermions as

$$E_0 = \frac{3}{5} N \varepsilon_F \tag{14.40a}$$

The mean energy per fermion for a completely degenerate electron gas is given by

$$\bar{\varepsilon} = \frac{E_0}{N} = \frac{3}{5} \varepsilon_F \tag{14.40b}$$

This result shows that unlike a classical particle, a fermion has appreciable energy even at absolute zero. This is a completely unexpected result as far as the domain of classical physics is concerned. For conduction electrons in silver

$$\bar{\varepsilon} = \frac{3}{5} \times 5.49 \text{ eV} = 3.29 \text{ eV}$$

This shows that closely packed electrons in silver would have energy that a Maxwellian gas would possess at a temperature T of about 4×10^4 K. This large difference clearly shows that a fermion system is very active even at absolute zero! Obviously, this is a quantum effect and has arisen out of the Pauli principle. This clearly brings out the inadequacy of classical statistics in describing the behaviour of a completely degenerate F–D system.

$C_V = (\partial E/\partial T)_V$, Eq. (14.39) implies that the heat capacity of a fermion system drops to zero at absolute zero. Similarly, we can show that the entropy of a F–D system also vanishes at absolute zero. This is consistent with the third law of thermodynamics. Now you may ask: Does pressure of a fermion system also drop to zero at absolute zero? To discover answer to this question, we recall that $p = (2E / 3V)$. Using Eq. (14.39) in this expression, we can show that the pressure exerted by a fermion system at absolute zero

$$p_F = \frac{4C}{15} \varepsilon_F^{5/2}$$

On combining this result with Eq. (14.35), we can express pressure in terms of the number of fermions as

$$p_F = \frac{2}{5}\left(\frac{N}{V}\right) \varepsilon_F \tag{14.41}$$

This result shows that if electrons in a metal were neutral, they would exert a pressure of atmospheres! Using the data given in Example 14.1, you can readily convince yourself that the pressure of electrons in silver is 20.82×10^9 N m^{-2} = 20.82×10^4 atm. Do we experience such an enormous pressure when we hold a piece of silver? If not, why?

The answer to this question is that the pressure exerted by electrons is counterbalanced by the Coulomb attraction of electrons by ions.

It may be mentioned here that if we draw a sphere of radius p_F, all particles will be found inside the sphere. This is called the *Fermi surface*. In case of the alkali and noble atoms, the surface is a sphere. In other cases, the Fermi surface can be quite complicated.

You should now go through the following examples carefully.

Example 14.2 Calculate the Fermi energy and the pressure in aluminium at absolute zero. The density of Al is 2.7×10^3 kg m^{-3}, Atomic weight of Al is 26.98 kg (kmol)$^{-1}$, $\times 10^{26}$ atoms (kmol)$^{-1}$, $h = 6.62 \times 10^{-34}$ Js, and $m_e = 9.11 \times 10^{-31}$ kg.

From Eq. (14.37), we recall that Fermi energy is given by

$$\varepsilon_F = \frac{h^2}{8m}\left(\frac{3N}{\pi V}\right)^{2/3}$$

To calculate the Fermi energy, we first determine the number density of electrons:

$$= \frac{N_A d}{\text{At.Wt}} = \frac{(3 \text{ electron atom}^{-1}) \times (6.02 \times 10^{26} \text{ atom kmol}^{-1}) \times (2.7 \times 10^3 \text{ kg m}^{-3})}{26.98 \text{ kg kmol}^{-1}}$$

$$= 18.07 \times 10^{28} \text{ electron m}^{-3}$$

$$\varepsilon_F = \frac{(6.62 \times 10^{-34} \text{ Js})^2}{8 \times 9.11 \times 10^{-31} \text{ kg}}\left(\frac{3 \times 18.07 \times 10^{28} \text{ m}^{-3}}{\pi}\right)^{2/3}$$

$$= 18.66 \times 10^{-19} \text{ J}$$

$$= 11.66 \text{ eV}$$

The pressure exerted by the electrons

$$p_F = \frac{2}{5}\left(\frac{N}{V}\right)\varepsilon_F$$

$$= \frac{2}{5} \times (18.07 \times 10^{28} \text{ m}^{-3}) \times (18.66 \times 10^{-19} \text{ J})$$

$$= 134.87 \times 10^9 \text{ N m}^{-2}$$

$$= 1.35 \times 10^6 \text{ atm.}$$

Before proceeding further, you may like to answer a practise problem.

Problem 14.3 Show that the velocity distribution of a completely degenerate fermion system is given by

$$\frac{dN_v}{dv} = \frac{8\pi m^3 V}{h^3} v^2 \qquad \text{for } v < v_F$$

$$= 0 \qquad \text{for } v > v_F$$

is the velocity of particles corresponding to the Fermi energy.

Work Function We now know that the Fermi energy is the kinetic energy of electrons in the highest occupied state. We can relate this to the work function of a metal. Refer to Fig. 14.4, which shows a potential well in which electrons reside and all states up to ε_F are filled up. If the depth of the well is W, we can write

$$\varepsilon_F = W - \phi$$

is work function. It is defined as the minimum energy required for removing an electron form the metal. Any energy in excess of ϕ is converted into kinetic energy of the ejected electron.

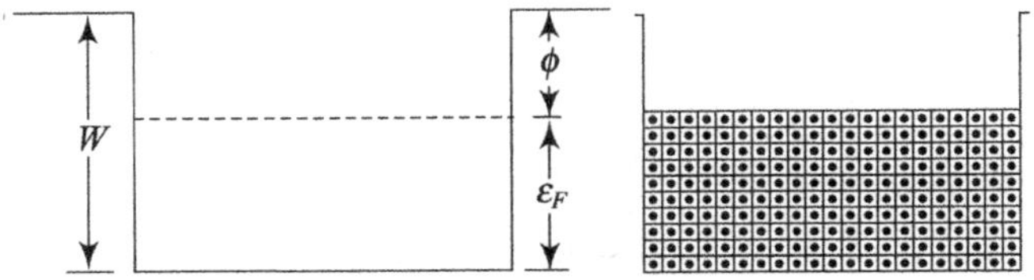

Relation between the depth of potential well, Fermi energy and work function for electrons in a metal. Horizontal lines depict filled energy levels.

So far we have confined our discussion to a completely degenerate Fermi gas (at absolute zero). To know how heat capacity and entropy of a F–D system vary with temperature, we must extend this study to temperatures above absolute zero. In this analysis, the Fermi temperature plays a very significant role because for $T << T_F$, the mean occupation number) and density of states do not differ much from their respective values at absolute zero, as shown in Fig. 14.3a. Such a fermion system is said to be *strongly degenerate.*

Strongly Degenerate F–D System ($T << T_F$)

From Eq. (15.31), we note that for finite values of T, the exponential term will be unity . This means that the value of the mean occupation number $n(\varepsilon)$ drops to (1/2) at . As the temperature is raised above absolute zero, electrons are excited from single particle states with $\varepsilon < \mu$ to states with $\varepsilon > \mu$. This appears in the form of a little tail in) versus ε plot. However, most of this transfer is confined to electrons with energy near the Fermi level. This is illustrated in Fig. 14.5, which shows the mean occupation number as a function of ε/μ for a low as well as a high temperature. The dashed lines are the corresponding curves for $T = 0$ K. You will note that the finite temperature curve is symmetrical about $\varepsilon = \varepsilon_F$. Moreover, the electrons in the tail have some freedom to move into unoccupied states. It is observed that μ depends on temperature. However, for $T << T_F$, the dependence is very weak. Moreover, $k_BT << \varepsilon_F$ for every metal. So we usually set in our discussion.

We now proceed to derive the expression for the total number of particles at a finite temperature. To this end, we express Eq. (14.8) in terms of a dimensionless variable

$$N = CV(k_B T)^{3/2} \int_0^\infty \frac{x^{1/2}\, dx}{e^{x - x_0} + 1} \tag{14.42}$$

$\mu/k_BT.$

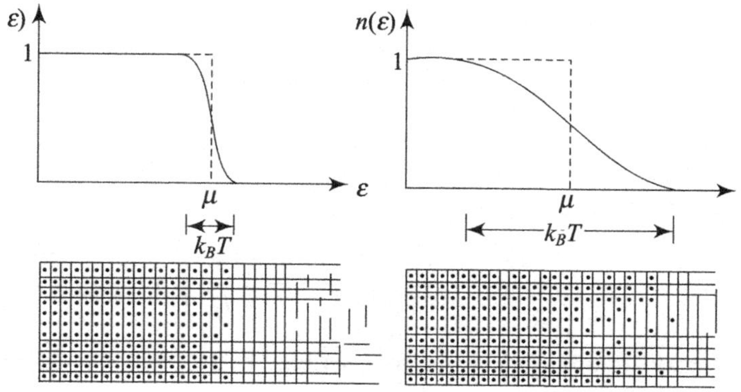

Mean occupation number for a strongly degenerate Fermi gas at two different temperatures. The width of the tail of the Fermi distribution is k_BT.

To evaluate this integral, we use Sommerfeld's lemma:

$$\int_0^\infty f(x-x_0)x^s \mathrm{d}x = \frac{x_0^{s+1}}{s+1}\left[1+\frac{\pi^2}{6}\frac{s(s+1)}{x_0^2}+\ldots\right] \tag{14.43}$$

$$f(x-x_0)=\frac{1}{e^{x-x_0}+1} \text{ and } s=\frac{1}{2}, \text{ we get}$$

$$\int_0^\infty \frac{x^{1/2}\,\mathrm{d}x}{e^{x-x_0}+1}=\frac{2}{3}x_0^{3/2}\left[1+\frac{\pi^2}{8x_0^2}+\ldots\right]$$

$$=\frac{2}{3}\left(\frac{\mu}{k_BT}\right)^{3/2}\left[1+\frac{\pi^2}{8}\left(\frac{k_BT}{\mu}\right)^2+\ldots\right]$$

Using this result in Eq. (14.42), the expression for the total number of particles takes the

$$N=\frac{2}{3}CV\mu^{3/2}\left[1+\frac{\pi^2}{8}\left(\frac{k_BT}{\mu}\right)^2+\ldots\right]$$

On combining this result with Eq. (14.35), we can relate μ and ε_F:

$$\varepsilon_F^{3/2}=\mu^{3/2}\left[1+\frac{\pi^2}{8}\left(\frac{k_BT}{\mu}\right)^2+\ldots\right] \tag{14.44}$$

This shows that for small T, we can write in the first order approximation

$$\varepsilon_F=\mu(T) \tag{14.45}$$

On substituting this result in Eq. (14.44) and inverting the resulting expression, we get

$$\mu(T)=\varepsilon_F\left[1+\frac{\pi^2}{8}\left(\frac{k_BT}{\varepsilon_F}\right)^2+\ldots\right]^{-2/3}$$

$$= \varepsilon_F \left[1 + \frac{\pi^2}{8} \left(\frac{T}{T_F} \right)^2 + \ldots \right]^{-2/3}$$

is the Fermi temperature.

Using binomial expansion and retaining terms only up to $(T/T_F)^4$, we get temperature dependence of the chemical potential:

$$\mu(T) = \varepsilon_F \left[1 - \frac{\pi^2}{12} \left(\frac{T}{T_F} \right)^2 + \frac{\pi^4}{80} \left(\frac{T}{T_F} \right)^4 + \ldots \right] \tag{14.46}$$

Note that for $T = 0$, this equation reproduces Eq. (14.34). However, as temperature increases, chemical potential lowers somewhat. For $\frac{T}{T_F} = 0.2$, the chemical potential changes by about 4%. Temperature variation of chemical potential for fixed density is shown in Fig. 14.6. For comparion, the classical case is also depicted.

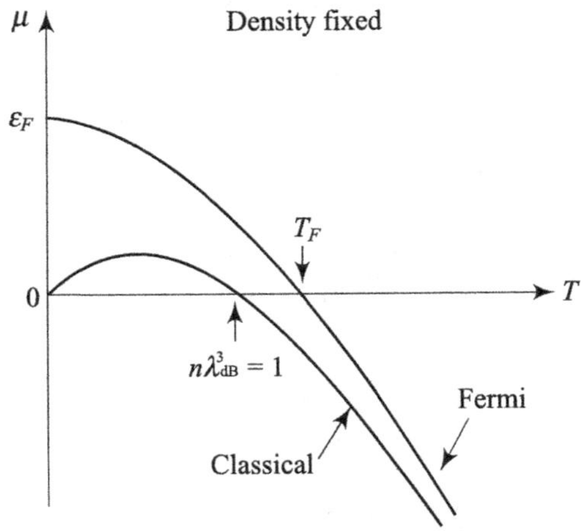

Temperature variation of chemical potential of a Fermi gas. For comparison, the classical case is also shown.

To obtain an expression for constant volume electronic heat capacity and show that it varies linearly with temperature, we have to use F–D statistics. To derive an expression for it, we must first know the total internal energy. To this end, we rewrite Eq. (14.10) in terms of dimensionless variable $x = \varepsilon/k_B T$ as

$$E = CV\,(k_B T)^{5/2} \int_0^\infty \frac{x^{3/2}\, dx}{e^{x - x_0} + 1} \tag{14.47}$$

$\mu/k_B T$.

Using Sommerfeld's lemma [Eq. (14.43)], we can readily evaluate this integral to obtain

$$E = \frac{2}{5} CV \mu^{5/2} \left[1 + \frac{5\pi^2}{8} \left(\frac{T}{T_F} \right)^2 + \ldots \right] \tag{14.48a}$$

where we have put $\mu = \varepsilon_F$ in the second term on the right-hand side of this expression.

On combining this result with Eq. (14.46), we obtain

$$E = \frac{2}{5} C V \varepsilon_F^{5/2} \left[1 - \frac{\pi^2}{12}\left(\frac{T}{T_F}\right)^2 + \ldots\right]^{5/2} \left[1 + \frac{5\pi^2}{8}\left(\frac{T}{T_F}\right)^2 + \ldots\right]$$

Retaining terms only up to the order of $\left(\frac{T}{T_F}\right)^2$ in binomial expansion, we get

$$E = \frac{2}{5} C V \varepsilon_F^{5/2} \left[1 - \frac{5\pi^2}{24}\left(\frac{T}{T_F}\right)^2 + \ldots\right] \left[1 + \frac{5\pi^2}{8}\left(\frac{T}{T_F}\right)^2 + \ldots\right]$$

On simplification, we get

$$E = \frac{2}{5} C V \varepsilon_F^{5/2} \left[1 + \frac{5\pi^2}{12}\left(\frac{T}{T_F}\right)^2 - \frac{\pi^4}{16}\left(\frac{T}{T_F}\right)^4 + \ldots\right]$$

On combining this result with Eq. (14.39), we can express the total internal energy of a strongly degenerate fermion system ($T << T_F$) in terms of the total zero point energy (of a completely degenerate system):

$$E = E_0 \left[1 + \frac{5\pi^2}{12}\left(\frac{T}{T_F}\right)^2 - \frac{\pi^4}{16}\left(\frac{T}{T_F}\right)^4 + \ldots\right] \tag{14.48b}$$

In terms of total number of particles and Fermi energy, we can rewrite it as

$$E = \frac{3}{5} N \varepsilon_F \left[1 + \frac{5\pi^2}{12}\left(\frac{T}{T_F}\right)^2 - \frac{\pi^4}{16}\left(\frac{T}{T_F}\right)^4 + \ldots\right] \tag{14.48c}$$

As expected, this equation predicts that the total internal energy of a fermion system increases with temperature. This is illustrated in Fig. 14.7. Physically this corresponds to the net transfer of particles to higher energy states. The internal energy of an ideal monatomic classical gas $\left(= \frac{3}{2} N k_B T\right)$ is also shown (dashed curve). As may be noted, for given values of *V*, T , the energy of an ideal fermion gas is greater than that of a classical ideal gas for $T < T_F$. However, for , the energy of a fermion gas approaches the classical value. The heat capacity of a fermion system

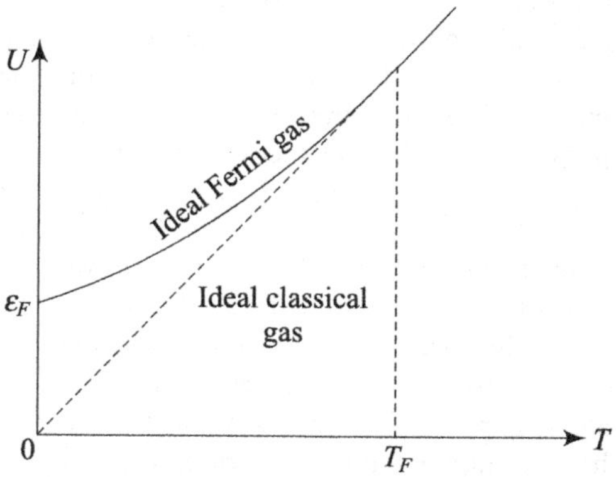

Fig. 14.7 Plot of the internal energy of a strongly degenerate Fermi gas as a

$$C_V = \left(\frac{\partial E}{\partial T}\right)_V = \frac{\partial}{\partial T}\left[\frac{3}{5} N \varepsilon_F \left\{1 + \frac{5\pi^2}{12}\left(\frac{T}{T_F}\right)^2 + \ldots\right\}\right]$$

$$= \frac{3}{5} N\varepsilon_F \left(\frac{5\pi^2}{12} \right) 2 \frac{T}{T_F^{\,2}}$$

$$= \frac{\pi^2}{2} N\varepsilon_F \frac{T}{T_F^2} = \frac{\pi^2}{2} \left(\frac{Nk_B}{T_F} \right) T \tag{14.49a}$$

In writing this expression, we have used Eqs. (14.38). In terms of the gas constant, we can rewrite Eq. (14.49a) for one mole of the gas as

$$C_V = \frac{\pi^2}{2} \left(\frac{k_B T}{\varepsilon_F} \right) R \tag{14.49b}$$

Equation (14.49a) shows that for $T << T_F$, the constant volume heat capacity of fermions varies linearly with temperature:

$$C_V = \alpha T \tag{14.49c}$$

$\dfrac{\pi^2}{2} \dfrac{Nk_B}{T_F}$ is constant for a given system.

This result shows that $C_V \to 0$ as $T \to 0$. For silver, let us calculate C_V at room temperature (300 K). Using the values obtained in Example 14.1, we can write

$$\frac{k_B T}{\varepsilon_F} = \frac{(1.38 \times 10^{-23}\ \text{JK}^{-1}) \times (300\ \text{K})}{8.78 \times 10^{-19}\ \text{J}} \approx 4.72 \times 10^{-3}$$

On using this result in Eq. (14.49b), we get

$$C_V = 2.33 \times 10^{-2}\ R \tag{14.49d}$$

We know that the constant volume heat capacity of a monatomic pefect gas is 1.5R. So we can say that the coefficient of R in Eq. (14.49d) is much less than 1.5. It means that electrons contribute very little to constant volume heat capacity at ordinary temperatures. However, electronic contribution dominates at low temperatures. This is in complete agreement with experimental findings for a number of metals and alloys. We can therefore conclude that at low temperatures, the constant volume heat capacity of a metal is made up of two parts—electronic and lattice—and write:

$$C_V = \alpha T + \beta T^3 \tag{14.49e}$$

Now refer to Fig. 14.8, which shows the constant volume heat capacities of a Maxwellian gas and a F–D gas as a function . Note that heat capacity of a F–D gas approaches the classical value (dashed line) asymptotically. The variation at low temperatures is however linear (dotted line). $T < 0.1\ T_F$, the heat capacity of a F–D gas is much less than the corresponding value for a Maxwellian gas.

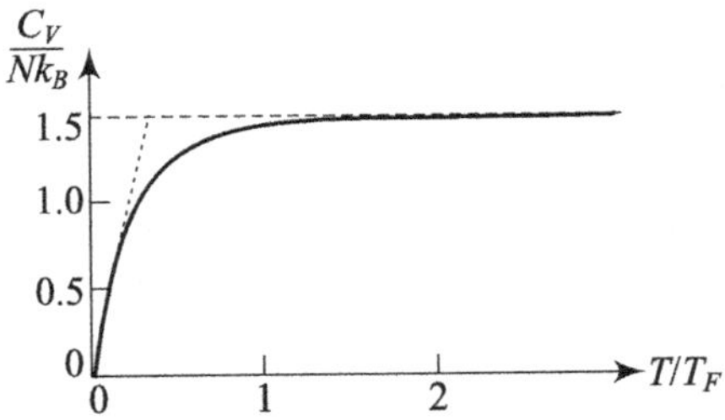

Fig. 14.8 Variation of constant volume heat capacity of an ideal Fermi

If we use Eq. (14.48c) where we have retained terms up to $(T/T_F)^4$ in the expression for energy, Eq. (14.49a) modifies to

$$C_V = \frac{\pi^2}{2} \left(\frac{Nk_B}{T} \right) T \left[1 - \frac{3\pi^2}{10} \left(\frac{T}{T} \right)^2 + \dots \right] \tag{14.50}$$

Before proceeding further, go through the following example carefully.

Example 14.3 Calculate the Fermi energy and the electronic contribution to constant volume heat capacity of copper at 300 K. Take $N/V = 8.5 \times 10^{28}$ m^{-3}, $h = 6.62 \times 10^{-34}$ Js, $\times 10^{-23}$ JK^{-1}, $R = 8.314$ J K^{-1} mol^{-1} and $m_e = 9.11 \times 10^{-31}$ kg.

We know that

$$\varepsilon_F = \frac{h^2}{8m_e}\left(\frac{3N}{\pi V}\right)^{2/3}$$

On substituting the given values, we get

$$\varepsilon_F = \frac{(6.62\times 10^{-34}\text{ Js})^2}{8\times(9.11\times 10^{-31}\text{ kg})}\left(\frac{3\times 8.5\times 10^{28}\text{ m}^{-3}}{\pi}\right)^{2/3}$$

$$= 11.26\times 10^{-19}\text{ J}$$

$$= 7.04\text{ eV}$$

Let us now calculate electronic contribution to constant volume heat capacity for copper. From Eq. (14.49b), we can write

$$C_V = \frac{\pi^2}{2}\left(\frac{k_B T}{\varepsilon_F}\right)R$$

$$= \frac{\pi^2}{2}\left(\frac{(1.38\times 10^{-23}\text{ JK}^{-1})\times 300\text{ K}}{11.26\times 10^{-19}\text{ J}}\right)\times(8.314\text{ JK}^{-1}\text{mol}^{-1})$$

$$= 0.15\text{ JK}^{-1}\text{mol}^{-1}$$

Let us now calculate the entropy of a strongly degenerate F–D gas at a finite temperature. We use the relation

$$S = \int_0^T \frac{C_V}{T}\,dT \tag{14.51}$$

On substituting the value of C_V from Eq. (14.49a) and carrying out the integration, we get

$$S = \left(\frac{\pi^2}{2}\frac{Nk_B}{T_F}\right)T \tag{14.52a}$$

This result shows that as $T \to 0$, $S \to 0$. That is, the entropy of strongly degenerate F–D gas drops to zero at the absolute zero of temperature. This is consistent with the third law of thermodynamics.

If we use Eq. (14.48c) for energy, Eq. (14.52a) modifies to

$$S = \frac{\pi^2}{2}\left(\frac{Nk_B}{T_F}\right)T\left[1 - \frac{\pi^2}{10}\left(\frac{T}{T_F}\right)^2 + \ldots\right] \tag{14.52b}$$

The Helmholtz function F is defined as

$$F = E - TS$$

On substituting for E and S from Eqs. (15.48c) and (15.52b), we get

$$F = \frac{3}{5} N\varepsilon_F \left[1 - \frac{5\pi^2}{12}\left(\frac{T}{T_F}\right)^2 + \dots \right] \tag{14.53}$$

The pressure exerted by a fermion system is given by

$$p = -\left(\frac{\partial F}{\partial V}\right)_T$$

$$= \frac{h^2}{8\,m_e}\left(\frac{3N}{\pi V}\right)^{2/3}$$, we note that

$$\left(\frac{\partial \varepsilon_F}{\partial V}\right)_N = K\frac{\partial}{\partial V}\left(\frac{1}{V}\right)^{2/3} = \frac{2}{3}\frac{K}{V^{5/3}} = -\frac{2}{3}\frac{\varepsilon_F}{V}$$

Hence we can write

$$p = \frac{2}{5}\frac{N\varepsilon_F}{V}\left[1 + \frac{5\pi^2}{12}\left(\frac{T}{T_F}\right)^2 + \dots \right] \tag{14.54a}$$

This defines the equation of state of an electron gas. On comparing it with Eq. (14.48c), we note that pressure of an electron gas is two-thirds of the energy density:

$$p = \frac{2E}{3V} \tag{14.54b}$$

You should now go through the following example:

Example 14.4 Using the data given in Example 14.2, calculate the Fermi momentum, Fermi velocity, Fermi temperature, average energy per atom at absolute zero and electronic contribution to heat capacity of aluminium at 100 K.

The Fermi momentum of the electrons is given by

$$M_F = (2m_e\varepsilon_F)^{1/2}$$
$$= [2 \times (9.11 \times 10^{-31}\ \text{kg}) \times (18.66 \times 10^{-19}\ \text{J})]^{1/2}$$
$$= 18.44 \times 10^{-25}\ \text{kg m s}^{-1}$$

Fermi velocity

$$v_F = M_F / m_e$$
$$= (18.44 \times 10^{-25}\ \text{kg m s}^{-1})/(9.11 \times 10^{-31}\ \text{kg})$$
$$= 2.02 \times 10^6\ \text{ms}^{-1}$$

Fermi temperature

$$T_F = \frac{\varepsilon_F}{k_B}$$
$$= \frac{18.66 \times 10^{-19}\ \text{J}}{1.38 \times 10^{-23}\ \text{JK}^{-1}}$$
$$= 1.35 \times 10^5\ \text{K}$$

Average energy per electron

$$\bar{\varepsilon} = \frac{3}{5}\varepsilon_F = \frac{3}{5 \times (1.6 \times 10^{-19}\ \mathrm{J\,eV^{-1}})} \times (18.66 \times 10^{-19}\ \mathrm{J})$$

$$= 7\ \mathrm{eV}$$

Electronic heat capacity

$$= \frac{\pi^2}{2}\left(\frac{T}{T_F}\right) N_A k_B$$

$$= \frac{(3.14)^2}{2}\left(\frac{100\ \mathrm{K}}{1.35 \times 10^5\ \mathrm{K}}\right) \times (6.02 \times 10^{26}\ \mathrm{kmol^{-1}}) \times (1.38 \times 10^{-23}\ \mathrm{JK^{-1}})$$

$$= 30.34\ \mathrm{J\,K^{1}\,kmol^{-1}}.$$

Applications of F–D Statistics

Fermi–Dirac statistics finds applications in several physical problems. One such physical system is that of conduction electrons in metals. Historically, the electron theory of metals was developed by Drude and Lorentz, who assumed that the electron gas inside a metal obeys classical statistics. This model provided a reasonable theoretical basis for qualitative understanding of electrical and thermal properties of metals. However, it failed to provide a quantitative measure of heat capacity of metals. This unsatisfactory explanation was set right by Sommerfeld when he proposed that free electrons in metals be described by Fermi–Dirac statistics. This led to linear variation of heat capacity of a metal at low temperatures.

Another fascinating category of phenomena exhibited by free (conduction) electrons in *thermionic* and *photoelectric emissions*. Since there is practically no spontaneous emission of electrons at ordinary temperatures, we can say that electrons inside a metal find themselves trapped in a potential well created by coulomb attraction by metallic ions. It means that electrons move freely inside the metal. But when any electron tries to escape, it encounters a constant potential barrier of height W, say. And only those electrons whose energy is greater than W can escape this barrier. The minimum energy required to be supplied to an electron to escape is equal to the work function ($\phi = W - \varepsilon_F$). The energy in excess of ϕ will appear as kinetic energy of the electron.

Strictly speaking, the emission of electrons (thermionic as well as photoelectric) is a non-equilibrium phenomenon. However, if the number of electrons escaping in a given time is small compared to their total population in a metal, we can calculate the current density by assuming that the electron gas is in *quasistatic equilibrium*.

Thermionic Emission

When we give thermal energy to a metal, conduction electrons escape from it. This phenomenon is called *thermionic emission*. We can easily calculate the current density using F–D statistics. Recall that at all temperatures above absolute zero, some electrons in the tail of the distribution occupy energy levels above the Fermi level. This means that the minimum energy required to remove an electron from a metal at temperature T is given by

$$\phi = E_c - \mu(T) \tag{14.55}$$

is the energy required to remove a zero energy electron from a metal.

Proceeding further, we note that to calculate the current density of escaping electrons, we should know their number. To do so, let us take the surface of the metal to lie in the plane so that x-axis is normal to the metal surface. We now assume that

the emitting surface is plane, and

only those electrons leave the metal whose momentum along the x-direction satisfies the condition

$$\frac{p_x^2}{2m} > E_c \tag{14.56}$$

Let us represent the thermionic current per unit area long the x-direction by j_{th}. Then, by definition

$$j_{\text{th}} = \frac{2e}{h^3} \int_{-\infty}^{\infty} \mathrm{d}p_{lz} \int_{-\infty}^{\infty} \mathrm{d}p_{ly} \int_{\sqrt{2mE_c}}^{\infty} n_{\text{FD}}(\varepsilon) v_x \, \mathrm{d}p_{lx}$$

The numerical factor 2 is included to account for spin degeneracy of an electron. On substituting for $n(\varepsilon)$ from Eq. (14.30) and changing from momentum to velocity variable, we can write the expression for thermionic current as

$$j_{\text{th}} = \frac{2em^3}{h^3} \int_{-\infty}^{\infty} \mathrm{d}v_z \int_{-\infty}^{\infty} \mathrm{d}v_y \int_{v_{x_c}}^{\infty} \frac{v_x \, \mathrm{d}v_x}{e^{(\varepsilon-\mu)/k_BT} + 1} \tag{14.57}$$

defines the threshold speed corresponding to energy E_c.

To simplify this expression, we assume that energy distribution of escaping electrons is Maxwellian. Let us pause for a while and ask: If electrons are known to obey F–D statistics, what makes them to switch over to classical statistics? This seems intriguing. The confusion however gets cleared when we note that at all temperatures of interest, the tail of the distribution is almost exponential. This follows from the fact that $\exp\left[\frac{\varepsilon - \mu}{k_B T}\right] >> 1$ and the mean occupation number

$$n_{\text{FD}}(\varepsilon) = \frac{1}{e^{(\varepsilon-\mu)/k_BT} + 1} \rightarrow e^{-(\varepsilon-\mu)/k_BT} = \text{constant} \times e^{-\varepsilon/k_BT}$$

which corresponds to a Maxwellian distribution.

Under this approximation, Eq. (14.57) takes the form

$$j_{\text{th}} = \frac{2em^3}{h^3} e^{\mu(T)/k_BT} \int_{-\infty}^{\infty} e^{-mv_z^2/2k_BT} \, \mathrm{d}v_z$$

$$\times \int_{-\infty}^{\infty} e^{-mv_y^2/2k_BT} \, \mathrm{d}v_y \int_{v_{x_c}}^{\infty} e^{-mv_x^2/2k_BT} \, v_x \, \mathrm{d}v_x \tag{14.58}$$

where we have used the relation $\varepsilon = \frac{m}{2}(v_x^2 + v_y^2 + v_z^2)$.

To simplify this expression we note that the integrals over dv_y and dv_z are identical. This means that it is sufficient to evaluate only one of them. Moreover, we note that these are symmetric so that we can write

$$I_y = \int_{-\infty}^{\infty} \exp\left[-\frac{mv_y^2}{2k_B T}\right] dv_y = 2\int_0^{\infty} \exp\left[-\frac{mv_y^2}{2k_B T}\right] dv_y$$

We evaluated this integral in Chapter 1 in terms of the gamma function. You may recall that

$$\int_0^{\infty} e^{-Bv^2} v^n \, dv = \frac{1}{2B^{\frac{n+1}{2}}} \Gamma\left(\frac{n+1}{2}\right)$$

$= \dfrac{m}{2k_B T}$ and $n = 0$, we get

$$I_y = \sqrt{\frac{2k_B T}{m}}\, \Gamma\left(\frac{1}{2}\right)$$

$$= \sqrt{\frac{2\pi k_B T}{m}} \qquad \left(\because \Gamma\left(\frac{1}{2}\right) = \sqrt{\pi}\right)$$

Using this result in Eq. (14.58), we obtain

$$j_{\text{th}} = \left(\frac{4\pi e m^2 k_B T}{h^3}\right) e^{\mu(T)/k_B T} \int_{v_{x_c}}^{\infty} e^{-mv_x^2/2k_B T} v_x \, dv_x \tag{14.59}$$

In terms of energy, this expression can be rewritten as

$$j_{\text{th}} = \left(\frac{4\pi e m k_B T}{h^3}\right) e^{\mu(T)/k_B T} \int_{E_c}^{\infty} e^{-\varepsilon_x/k_B T} \, d\varepsilon_x \tag{14.60}$$

To evaluate this integral, we introduce a change of variable by writing

$$\frac{\varepsilon_x}{k_B T} = u + \frac{E_c}{k_B T} \tag{14.61}$$

$$d\varepsilon_x = k_B T \, du$$

and the limits of integration run from 0 to ∞. Then Eq. (14.60) takes the form

$$j_{\text{th}} = \left(\frac{4\pi m e k_B^2}{h^3}\right) T^2 \exp\left[\frac{\mu(T) - E_c}{k_B T}\right] \int_0^{\infty} e^{-u} \, du$$

$e^{\ u}\, du = \Gamma(1) = 1$, the expression for the density of escaping electrons simplifies

$$j_{\text{th}} = \left(\frac{4\pi m e k_B^2}{h^3}\right) T^2 \exp\left[-\left\{\frac{E_c - \mu(T)}{k_B T}\right\}\right]$$

$$= AT^2 e^{-\phi/k_B T} \tag{14.62}$$

$= \dfrac{4\pi m e k_B^2}{h^3} = 120\text{ A cm}^{-2}\text{ K}^{-2}$, and $\phi = E_c - \mu(T)$ is the work function. (This *minimum* energy required by an electron to get emitted from the metal surface.) It is of the order of a few eV. Equation (14.62) is known as the *Richardson–Dushmann equation.* It shows that the emission current density depends on the work function of the metal and square of the absolute temperature. Since the work function is a characteristic property of a metal, the emission current will also be a characteristic of the specimen.

It may be noted that in this analysis we have not accounted for the possibility of reflection of electrons from the metallic surface. Equation (14.62) provides us a simple method to determine the work function and hence zero point energy. To show this, we rewrite Eq.

$$\frac{j_{\text{th}}}{T^2} = A\exp\left(-\frac{\phi}{k_B T}\right)$$

On taking natural logarithm of both sides, we get

$$\ln\left(\frac{j_{\text{th}}}{T^2}\right) = \ln A - \frac{\phi}{k_B T}$$

So if we plot $\ln(j_{\text{th}}/T^2)$ versus (T^{-1}), we will obtain a straight line whose slope $(= \phi/k_B)$ gives the work function. And once work function is known, using Eq. (14.55), we can estimate the Fermi energy for a metal.

The experimental results on thermionic emission from tungsten are shown in Fig. 14.9. The value of the work function resulting from the slope of the experimental line is about 4.5 eV.

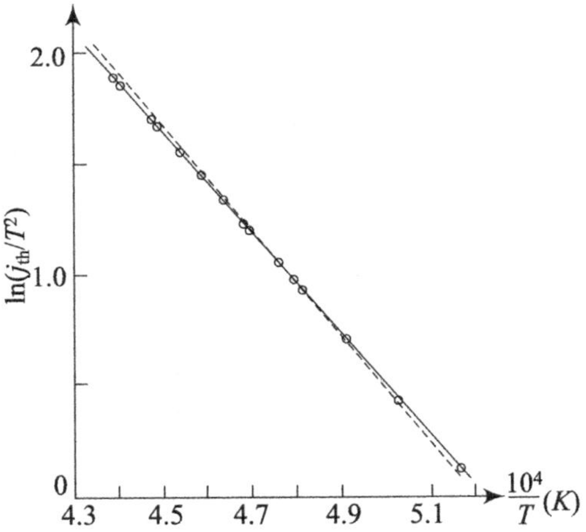

Plot of $\ln(j_{\text{th}}/T^2)$ versus T^{-1} for tungsten.

You may now like to answer a practise problem.

Problem 14.4 If the work function $\phi = E_c - \varepsilon_F$ is independent of temperature, show that the density of escaping electrons is given by

$$j = BT^{1/2}\, e^{-E_C/k_B T}$$

is constant for a given metal. This result is known as the *Richardson equation.*

Photoelectric Emission

When a metal is irradiated with a monoenergetic beam of ultraviolet light, electrons are found to escape from it. This phenomenon is known as *photoelectric emission.* For an electron to escape from the metal now, the condition to be satisfied by the momentum component p_{lx} now modifies to

$$\frac{p_{lx}^2}{2m} + h\nu > E_c \tag{14.63}$$

is the frequency of the incident monochromatic radiation. Proceeding as in the case of thermionic emission, we can write

$$j_{\text{ph}} = \frac{2e}{h^3} \int_{-\infty}^{\infty} dp_{lz} \int_{-\infty}^{\infty} dp_{ly} \int_{\sqrt{\frac{2}{m}(E_c - h\nu)}}^{\infty} \frac{v_x \, dp_{lx}}{e^{(\varepsilon-\mu)/k_B T} + 1}$$

We rewrite it in terms of velocity as

$$j_{\text{ph}} = \frac{2m^3 e}{h^3} \int_{-\infty}^{\infty} dv_z \int_{-\infty}^{\infty} dv_y \int_{v_{x0}}^{\infty} \frac{v_x \, dv_x}{A^{-1} \exp\left[\frac{m}{2k_B T}(v_x^2 + v_y^2 + v_z^2)\right] + 1} \tag{14.64}$$

$= \exp(-\mu/k_B T)$ and v_{x_0} is the threshold velocity below which electrons fail

$$\frac{m v_{x0}^2}{2} + h\nu = E_c \tag{14.65}$$

Next we write v_x, v_y, and v_z in terms of cylindrical coordinates with x–axis as the polar $v_x = v_x$, $v_y = \rho \sin \varphi$ and $v_z = \rho \cos \varphi$ so that $\rho^2 = v_y^2 + v_z^2$ and $dv_y dv_z = \rho d\rho d\varphi$. Then we can rewrite Eq. (14.64) as

$$j_{\text{ph}} = \left(\frac{2m^3 e}{h^3}\right) \int_0^{2\pi} d\varphi \int_0^{\infty} \rho d\rho \int_{v_{x0}}^{\infty} \frac{v_x \, dv_x}{A^{-1} \exp\left[\frac{m}{2k_B T}(v_x^2 + \rho^2)\right] + 1} \tag{14.66}$$

Integration over φ gives 2π so that we can rewrite this expression as

$$j_{\text{ph}} = \frac{4\pi m^3 e}{h^3} \int_0^{\infty} \rho d\rho \int_{v_{x0}}^{\infty} \frac{v_x \, dv_x}{\exp\left[\frac{m}{2k_B T}\left(v_x^2 + \rho^2 - 2\frac{\mu}{m}\right)\right] + 1} \tag{14.67}$$

We now introduce a change of variable by defining

$$t = \frac{m}{2k_B T}\left(v_x^2 + \rho^2 - \frac{2\mu}{m}\right) \tag{14.68}$$

To rewrite Eq. (14.67) in terms of the new variable, we note that

$$t dt = \frac{m}{k_B T} \rho d\rho \Rightarrow \rho d\rho = \frac{k_B T}{m} t dt$$

The limits of integration over *t* will now run from $\frac{m}{2k_BT}\left(v_x^2 - \frac{2\mu}{m}\right)$ to ∞. For mathematical convenience, we write the lower limit as V. On making these substitutions in the expression for photoelectric current density, Eq. (14.67) takes the form

$$j_{\text{ph}} = \frac{4\pi m^3 e}{h^3}\left(\frac{k_B T}{m}\right)\int_{\varsigma}^{\infty} dt \int_{v_{x0}}^{\infty} \frac{v_x\, dv_x}{1+\exp(t)} \tag{14.69}$$

Next we change the order of integration and write

$$j_{\text{ph}} = \frac{4\pi m^3 e}{h^3}\left(\frac{k_B T}{m}\right)\int_{v_{x0}}^{\infty} v_x\, dv_x \int_{\varsigma}^{\infty} dt \frac{1}{1+\exp(t)}$$

By integrating over *t*, we get

$$j_{\text{ph}} = \frac{4\pi m^3 e}{h^3}\left(\frac{k_B T}{m}\right)\int_{v_{x0}}^{\infty} v_x\, dv_x \ln[1+\exp(-\varsigma)]$$

$$= \left(\frac{4\pi m^2 e k_B T}{h^3}\right)\int_{v_{x0}}^{\infty} v_x\, dv_x \ln\left[1+\exp\left(-\frac{mv_x^2}{2k_BT}+\frac{\mu}{k_BT}\right)\right] \tag{14.70}$$

In terms of energy, we can write

$$j_{\text{ph}} = \left(\frac{4\pi m e k_B T}{h^3}\right)\int_{\varepsilon_x = E_c - h\nu}^{\infty} d\varepsilon_x \ln\left[1+\exp\left(-\frac{\varepsilon_x}{k_BT}+\frac{\mu}{k_BT}\right)\right] \tag{14.71}$$

We now introduce a new variable by defining

$$x = \frac{\varepsilon_x - E_c + h\nu}{k_B T} \tag{14.72}$$

= $k_B T\, dx$ and the limits of integration run from 0 to ∞. On making these substitutions, Eq. (14.71) becomes

$$j_{\text{ph}} = \left(\frac{4\pi m e k_B^2}{h^3}\right)T^2 \int_0^{\infty} dx \ln\left[1+\exp\left\{\frac{h(\nu-\nu_0)}{k_BT} - x\right\}\right]$$

$$= \left(\frac{4\pi m e k_B^2}{h^3}\right)T^2 \int_0^{\infty} dx \ln[(1+e^{\xi - x})] \tag{14.73}$$

= $E_c - \mu(T) = E_c - \varepsilon_F = \phi$ and $\xi = h(\nu - \nu_0)/k_B T$.

The characteristic frequency $\nu_0 (= \phi/h)$ is referred to as the *threshold frequency* for photoelectric emission from a metal. On integrating by parts, we find that

$$\int_0^{\infty} dx \ln(1+e^{\xi-x}) = \int_0^{\infty} \frac{x\, dx}{e^{\xi - x}+1}$$

$$j_{\text{ph}} = \left(\frac{4\pi m e k_B^2}{h^3}\right)T^2 f(\xi) \tag{14.74}$$

ν_0 and $h|\nu - \nu_0| >> k_B T$, we can ignore one in comparison to the exponential term. Then the function $f(\xi)$ takes the form

$$f(\xi) = \int_0^\infty \frac{x\,dx}{e^{x-\xi}+1} = e^{\xi} \int_0^\infty e^{-x}\, x\,dx = e^{\xi}\,\Gamma(2) = e^{\xi}$$

Hence, Eq. (14.74) becomes

$$j_{ph} = \left(\frac{4\pi m e k_B^2}{h^3}\right) T^2 \exp\left(\frac{h\nu - \phi}{k_B T}\right) \tag{14.75}$$

On the other hand, for $h(\nu - \nu_0) >> k_B T$, $e^{\xi} >> 1$. Then using Sommerfeld's lemma, we

$$\int_0^\infty \frac{x\,dx}{e^{x-\xi}+1} = \frac{\xi^2}{2}\left[1 + \frac{\pi^2}{3\xi^2} + \dots\right] \tag{14.76}$$

so that the current density of the photoelectric emission is given by

$$j_{ph} = \frac{4\pi m e k_B^2}{h^3} T^2 \xi^2 \left[1 + \frac{\pi^2}{3\xi^2} + \cdots\right]$$

$$\approx \frac{2\pi m e}{h} (\nu - \nu_0)^2 \tag{14.77}$$

This result shows that in the limit $\xi >> 1$, the photoelectric current density is completely independent of temperature. Physically, when energy of the light quantum is much greater than the work function of the metal, the current density is independent of temperature.

At the threshold frequency ($\nu = \nu_0$), $\xi = 0$. Then, integral in Eq. (14.74) gives $\Gamma(2)\ \ (2) = \frac{\pi^2}{6}$ so that

$$j_0 = \frac{2\pi^3 m e k_B^2}{3h^3} T^2 \tag{14.78}$$

Figure (14.10) shows a plot of the experimental results for photoelectric emission from palladium. Note that agreement between theoretical and experimental results is almost

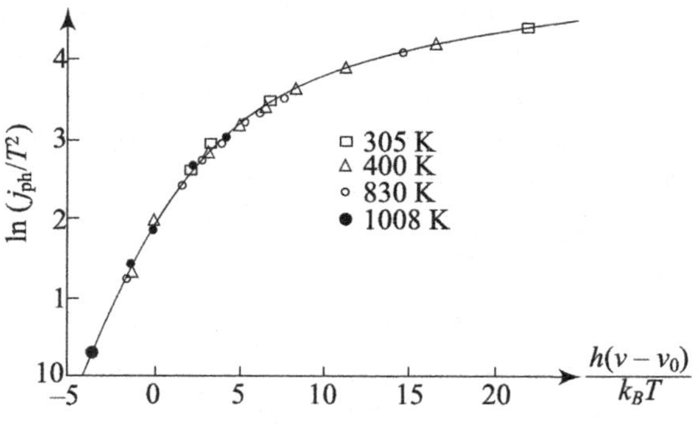

Photoelectric current for palladium as a function of $h(\nu - \nu_0)/k_B T$. Here ν_0 is the threshold frequency.

White Dwarf Stars

Fermi–Dirac statistics finds very important applications in astrophysics. We will apply it to the case of white dwarf stars. White dwarf stars are relatively old stars, almost in the end-phase of their lives. They have a small diameter and are very faint but extremely dense. Typical white dwarf stars are Sirius B, 40 Eridani *B* and Van Maalu and their positions on Hertz–Russel diagram, which is a plot of lumonisity of a star versus temperature, is shown in Fig. 14.11.

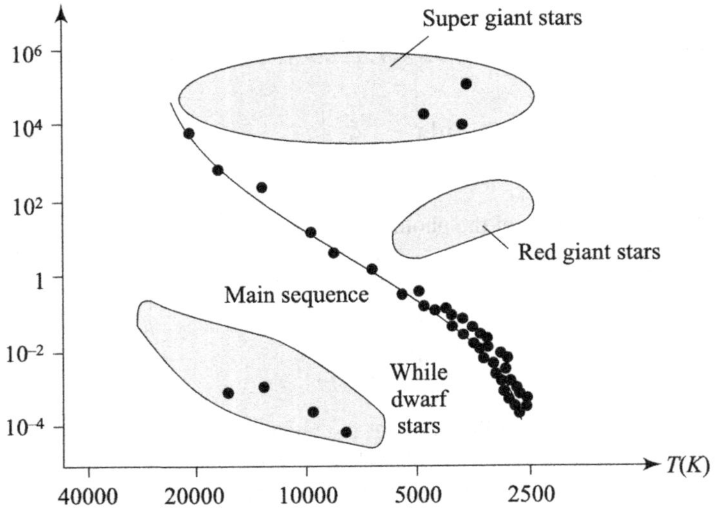

Location of known white dwarf stars on H–R diagram.

As may be noted, white dwarf stars are located below the main sequence. The estimated mass of Sirius *B* is about 2×10^{30} kg, which is roughly the same as the solar mass. However, has a radius of only 5.57×10^3 km, which is less than the radius of our earth. The loss of brightness is due to the fact that their hydrogen content has almost completely burnt out leading to formation of helium. The thermonuclear reactions in white dwarf stars occur at a rather slow pace.

An idealised model of a white dwarf may be constructed along the following lines: A white dwarf star is a mass of helium ($M \approx 10^{30}$ kg) under extreme pressure (density = ^{3}) at a very high temperature (10^7 K). The temperature of 10^7 K corresponds to a mean thermal energy (per particle) of about 10^3 eV, which is much greater than the energy required for ionising a helium atom (~55 eV). This means that the whole of helium in a white dwarf star is completely ionised. The constituents of the star may, therefore, be taken as *N* electrons (each of mass m_e) and *N*/2 helium nuclei (each of mass = $4m_p$). Therefore, the mass of the star is given by

$$M = N m_e + 4m_p \times \frac{N}{2}$$

$$= N(m_e + 2m_p) \approx 2Nm_p \tag{14.79}$$

<< m_p.

So, the number density of electrons in a typical white dwarf star is

$$n = \frac{N}{V} = \frac{M/2m_p}{M/\rho} = \frac{\rho}{2m_p} \tag{14.80}$$

The electron density in Sirius B is about 1.7×10^{36} electrons per m^3. This corresponds to Fermi energy of

$$\varepsilon_F = \frac{h^2}{2m_e}\left(\frac{3N}{8\pi V}\right)^{2/3} \approx 0.33 \text{ MeV}$$

and Fermi temperature

$$T_F = \frac{\varepsilon_F}{k_B} \approx 3.8 \times 10^9 \text{ K}$$

We know that the rest mass energy of an electron is 0.5 MeV. This means that

the dynamics of electrons in a typical white dwarf star is *relativistic*; and statistically speaking, the electron gas is in a state of complete degeneracy. This is because the ratio $(T/T_F) \approx 5 \times 10^{-3}$ (The core of Sirius B is at about 2×10^7 K.)

This highly degenerate electron gas behaves essentially as an electron gas at absolute zero. So the entire analysis of Sec. 14.4.1 applies here.

The pressure exerted by an ideal fermion gas at absolute zero for Sirius B is about Nm^{-2}, which is approximately 1.8×10^{17} atm. Such an enormous pressure is counterbalanced by the binding due to gravitational attraction, which is almost entirely due to the helium nuclei in the star.

For simplicity, we assume that (i) the dynamics of the system is governed only by the electrons and (ii) the electron gas is *uniformly* distributed in the star.

We now have to discuss the ground state properties of an ideal Fermi gas of N relativistic electrons. By inverting Eq. (14.37) we can write

$$N = \frac{8\pi V}{3h^3}(2m_e\varepsilon_F)^{3/2}$$

$p_F = \sqrt{2m_e\varepsilon_F}$, it readily follows that

$$N = \frac{8\pi V}{3h^3}\, p_F^3$$

Hence, Fermi momentum is given by

$$p_F = \left(\frac{3Nh^3}{8\pi V}\right)^{1/3} \tag{14.81}$$

For a relativistic particle, the energy-momentum relation is given by

$$\varepsilon = \sqrt{p_\ell^2 c^2 + m_e^2 c^4}$$

Therefore, the ground state energy of a F–D gas is given by

$$E_0 = \frac{2V}{h^3}\int_0^{p_F} (p_\ell^2 c^2 + m_e^2 c^4)^{1/2}\, 4\pi p_\ell^2\, dp_\ell$$

$$= \frac{4\pi V}{h^3} \int_0^{p_F} m_e c^2 \left[1 + \left(\frac{p_\ell}{m_e c} \right)^2 \right]^{1/2} p_\ell^2 \, dp_\ell \tag{14.82}$$

To evaluate the integral,we introduce a dimensionless variable x by defining

$$\frac{p_\ell}{m_e c} = x$$

$p \ \ dp_\ell = (m_e c)^3 \, x^2 \, dx$. The limits of integration run from 0 to $x_F = p_F/m_e c$. Then, the expression for ground state energy takes the form

$$E_0 = \frac{8\pi V m_e^4 c^5}{h^3} \int_0^{x_F} (1 + x^2)^{1/2} \, x^2 \, dx$$

$$= \frac{8\pi V m_e^4 c^5}{h^3} f(x_F) \tag{14.83}$$

The function $f(x_F)$ can be computed for any value of x. However, there are two limiting cases which can be treated analytically and are sometimes useful. These are given by

$$f(x_F) = \int_0^{x_F} (1 + x^2)^{1/2} \, x^2 \, dx \left[\begin{array}{ll} = \frac{1}{3} x_F^3 \left(1 + \frac{3}{10} x_F^2 + \ldots \right) & \text{for } x_F << 1 \\ = \frac{1}{3} x_F^4 \left(1 + \frac{1}{x_F^2} + \ldots \right) & \text{for } x_F >> 1 \end{array} \right. \tag{14.84}$$

The case $x_F << 1$ corresponds to the non-relativistic case, whereas $x_F >> 1$ defines the relativistic case. We can rewrite x_F in terms of some familiar properties of a star. Note that

$$x_F = \frac{p_F}{m_e c} = \frac{h}{m_e c} \left(\frac{3N}{8\pi V} \right)^{1/3} \tag{14.85}$$

On substituting $V = \frac{4\pi}{3} R^3$ and $M = 2m_p N$, we find that $\frac{N}{V} = \frac{3M}{8\pi m_p R^3}$ so that

$$x_F = \frac{h}{2R\pi m_e c} \left(\frac{9\pi N}{8 m_p} \right)^{1/3} = \frac{\bar{M}^{1/3}}{\bar{R}} \tag{14.86}$$

$$\bar{M} = \frac{9\pi M}{8 m_p} \tag{14.87}$$

is mass of a white dwarf star in terms of the proton mass and

$$\bar{R} = R \left(\frac{2\pi m_e c}{h} \right) \tag{14.88}$$

is radius of a white dwarf star in terms of the Compton wavelength of the electron.

The pressure exerted by the F–D gas is, therefore,

$$p_0 = -\frac{\partial E_0}{\partial V} = -\frac{8\pi m_e^4 c^5}{h^3}\left[f(x_F) + V\frac{\partial f(x_F)}{\partial x_F}\frac{\partial x_F}{\partial V}\right] \tag{14.89}$$

Using the relation connecting x_F and p_F (Eq. (14.85)), we can write

$$\frac{\partial x_F}{\partial V} = \frac{h}{m_e c}\frac{1}{3}\left(\frac{3N}{8\pi V}\right)^{-2/3}\left(-\frac{3N}{8\pi V^2}\right)$$

$$= -\frac{1}{3V}\frac{h}{m_e c}\left(\frac{3N}{8\pi V}\right)^{1/3}$$

$$V\frac{\partial x_F}{\partial V} = -\frac{1}{3}x_F \tag{14.90}$$

Similarly, it follows from Eq. (14.84) that for $x_F << 1$

$$\frac{\partial f(x_F)}{\partial x_F} = x_F^2\left(1 + \frac{3}{10}x_F^2 + \ldots\right) + \frac{1}{3}x_F{}^3\left(\frac{3}{5}\right)x_F + \ldots$$

$$= x_F^2\left(1 + \frac{1}{2}x_F^2 + \ldots\right) \tag{14.91a}$$

>> 1, we can write

$$\frac{\partial f(x_F)}{\partial x_F} = x_F^3\ (1 + x_f^{-2} + \ldots) - \frac{1}{4}x_F^4\left(\frac{2}{x_F{}^3}\right)$$

$$= x_F^3\left(1 + \frac{1}{x_F^2} + \ldots\right) \tag{14.91b}$$

Hence, the expression for pressure exerted by the F–D gas is given by

$$p_0 = \frac{8\pi m_e^4 c^5}{h^3}\begin{cases}\left[\frac{1}{3}x_F^3\left(1 + \frac{1}{2}x_F^2 + \ldots\right) - f(x_F)\right] & \text{for } x_F << 1 \\ \left[\frac{1}{3}x_F^4\left(1 + \frac{1}{2}x_F^{-2} + \ldots\right) - f(x_F)\right] & \text{for } x_F >> 1\end{cases} \tag{14.92}$$

On substituting for $f(x_F)$ from Eq. (14.84) in this expression and simplifying the resultant expression, we get

$$p_0 = \begin{cases}\frac{8\pi m_e^4 c^5}{15h^3}x_F^5 & \text{for } x_F << 1 \\ \frac{2\pi m_e^4 c^5}{3h^3}(x_F^4 - x_F^2) & \text{for } x_F >> 1\end{cases} \tag{14.93}$$

We will now use these results to discuss the mass-radius relation of a white dwarf star within the framework of the model considered here.

Mass-radius relation The pressure exerted by the electron gas in a white dwarf star is counterbalanced by gravitational attraction between helium ions. The gravitational potential energy is given by

$$U_{grav} = -\alpha \frac{GM^2}{R} \tag{14.94}$$

is total mass of the star of radius R, G is the gravitational constant and α is of the order of unity.

The change in the energy of the star due to the work done against pressure p_0 is given by

$$dE_0 = p_0(R)\, 4\pi R^2 dR \tag{14.95}$$

where we have expressed p_0 as $p_0(R)$.

For the configuration to be in equilibrium, the net energy change of the system for an infinitesimal change in its size should be zero. Thus, we must have

$$dU_{grav} = dE_0$$

$$\alpha \frac{GM^2}{R^2} dR = p_0(R)\, 4\pi R^2 dR$$

This can be rewritten as

$$p_0(R) = \frac{\alpha}{4\pi} \frac{GM^2}{R^4} \tag{14.96}$$

On combining this result with Eqs. (14.87) and (14.88), we find that the mass and radius of a white dwarf star are connected through the relation

$$p_0\,(R) = \frac{\alpha G}{8\pi} \left(\frac{8 m_p}{9\pi}\right)^2 \left(\frac{2\pi m_e c}{h}\right)^4 \frac{(\bar{M})^2}{(\bar{R})^4} \tag{14.97}$$

Note that this expression exhibits a remarkable combination of the three tools of theoretical physics developed in the 20th century—quantum mechanics, relativity and gravitation—the holy Triveni in the context of Indian mythology.

R), we now substitute from Eq. (14.93). For the non-relativistic case ($x_F << 1$), using Eq. (14.86) we find that

$$\frac{4k}{5} \frac{(\bar{M})^{5/3}}{(\bar{R})^5} = \frac{k'(\bar{M})^2}{(\bar{R})^4}$$

This can be rewritten as

$$(\bar{M})^{1/3}\ \bar{R} = \frac{4}{5}\frac{k}{k'} = \frac{4}{5} M_0^{2/3}$$

$$\bar{R} = \frac{4}{5}\left(\frac{M_0^2}{\bar{M}}\right)^{1/3} \tag{14.98}$$

$$k = \frac{m_e c^2}{12\pi^2}\left(\frac{2\pi m_e c}{h}\right)^3$$

$$k' = \frac{\alpha G}{4\pi}\left(\frac{8m_p}{9\pi}\right)^2 \left(\frac{2\pi m_e c}{h}\right)^4$$

$$M_0 = \left(\frac{k}{k'}\right)^{3/2} = \left(\frac{27\pi}{64\alpha}\right)^{3/2} \left(\frac{ch}{2\pi G m_p^2}\right)^{3/2} \tag{14.99}$$

For the relativistic case ($x_F >> 1$), Eqs. (14.86), (14.93) and (14.97) give

$$k\left(\frac{(\overline{M})^{4/3}}{(\overline{R})^4} - \frac{(\overline{M})^{2/3}}{(\overline{R})^2}\right) = k'\frac{(\overline{M})^2}{(\overline{R})^4}$$

$$\overline{R} = (\overline{M})^{1/3}\left[1 - \left(\frac{\overline{M}}{M_0}\right)^{2/3}\right]^{1/2} \tag{14.100}$$

Equations (14.98) and (14.100) establish a one-to-one correspondence between the mass and the radius of a white dwarf star. Equation (14.98) tells us that the radius of a white dwarf star and its mass are inversely related: $R \propto M^{-1/3}$. So, smaller the radius (size) of a white dwarf star, the greater will be its mass! This means that the star becomes relatively denser with the passage of time. On the other hand, Eq. (14.100) shows that there exists a limiting mass M_0, given by Eq. (14.99), which corresponds to a very small size of a white dwarf star. We can, therefore, conclude that no white dwarf star can have a mass greater . Physically, it means that for a mass more than this limit, the pressure p_0 of the system is insufficient and hence unable to support the gas against its tendency towards gravitational collapse. If we take $\alpha \approx 1$, the numerical value of this limiting mass comes $\approx 10^{30}$ kg. Detailed investigations by Chandrasekhar on the limiting mass of a white dwarf star led him to the relation

$$M_0 \approx 1.44\Theta \tag{14.101}$$

denotes the mass of the sun ($\approx 2 \times 10^{30}$ kg). These features are depicted in

Equation (14.101) is known as the *Chandrasekhar mass limit* for a white dwarf star.

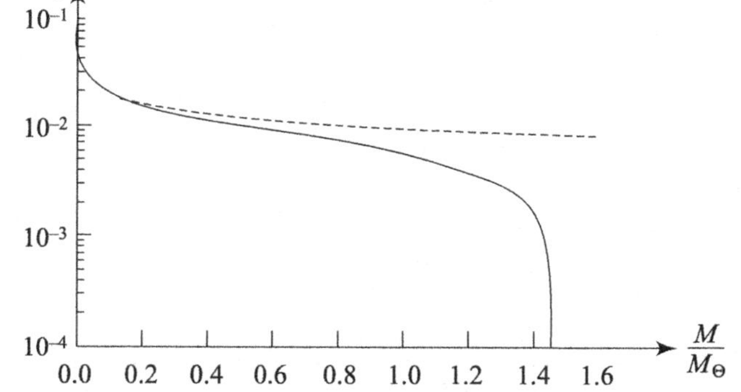

Variation of radius of a white dwarf star with its mass. Both parameters have been expressed in terms of solar radius and solar mass.

Fermi–Dirac statistics can also be used to calculate the electron distribution in a heavy atom or to study other important phenomena. A detailed discussion of these aspects is beyond the scope of this book.

DDITIONAL EXAMPLES

Example 14.5 Prove that for a system at a finite temperature and obeying Fermi–Dirac statistics, the probability that a level lying ΔE below the Fermi level is unoccupied is the same as the probability of occupation of a level lying ΔE above the Fermi level.

We know that according to F–D statistics,

$$n_{\text{FD}} = \frac{1}{\exp[(\varepsilon - \varepsilon_F)/k_B T)] + 1}$$

$+ \Delta E$, we can write $\varepsilon - \varepsilon_F = \Delta E$ so that

$$(n_{\text{FD}})_1 = \frac{1}{\exp(\Delta E / k_B T) + 1}$$

$$\overline{(n_{\text{FD}})}_1 = 1 - (n_{\text{FD}})_1 = 1 - \frac{1}{\exp(\Delta E / k_B T) + 1} = \frac{\exp(\Delta E / k_B T)}{\exp(\Delta E / k_B T) + 1}$$

Similarly, for $\varepsilon = \varepsilon_F - \Delta E$, we can write $\varepsilon - \varepsilon_F = -\Delta E$ so that

$$(n_{\text{FD}})_2 = \frac{1}{\exp(-\Delta E / k_B T) + 1} = \frac{\exp(\Delta E / k_B T)}{\exp(\Delta E / k_B T) + 1}$$

$$\overline{(n_{\text{FD}})}_1 = (n_{\text{FD}})_2$$

Example 14.6 The Fermi energy ε_F for lithium is 4.72 eV at $T = 0$. Calculate the number of conduction electrons per unit volume in lithium. Take $m_e = 9.31 \times 10^{-31}$ kg and 10^{-34} Js.

From Eq. (14.37), we know that Fermi energy is given by

$$\varepsilon_F = \frac{h^2}{8 m_e} \left(\frac{3n}{\pi} \right)^{2/3} \quad \text{where} \quad n = \frac{N}{V}$$

By rearranging terms, we can write

$$\left(\frac{8 m_e \varepsilon_F}{h^2} \right)^{3/2} = \frac{3n}{\pi}$$

$$n = \frac{\pi}{3} \left(\frac{8 m_e \varepsilon_F}{h^2} \right)^{3/2}$$

On substituting the given values, we get

$$n = \frac{\pi}{3} \left(\frac{8 \times (9.1 \times 10^{-31}\ \text{kg}) \times (4.72 \times 1.6 \times 10^{-19}\ \text{J})}{(6.62)^2 \times 10^{-68}\ \text{J}^2\,\text{s}^2} \right)^{3/2}$$

$$= 4.65 \times 10^{28}\ \text{m}^{-3}.$$

Example 14.7 Estimate the Fermi-level for copper and calculate the average electron energy at $T = 0$ K. The molecular weight of copper = 63.57 and density of copper is 8.94 g cm^{-3}. Take $h = 6.62 \times 10^{-34}$ Js, $N_A = 6.023 \times 10^{23}$ mol^{-1} and $m_e = 9.11 \times 10^{-31}$ kg.

From Eq. (14.37), we know that Fermi energy is given by

$$\varepsilon_F = \frac{h^2}{8m}\left(\frac{3N}{\pi V}\right)^{2/3}$$

$$\frac{N}{V} = \frac{2 \times N_A d}{\text{Mol Weight } (M)}.$$ So we can write

$$\varepsilon_F = \frac{h^2}{8m_e}\left(\frac{3}{\pi} \times \frac{2N_A d}{M}\right)^{2/3}$$

On substituting the given values, we get

$$\varepsilon_F = \frac{(6.62 \times 10^{-34}\ \text{Js})^2}{8 \times (9.11 \times 10^{-31}\ \text{kg})} \times$$

$$\left(\frac{3}{\pi} \times \frac{2 \times (6.023 \times 10^{23}\ \text{mol}^{-1}) \times (8.94 \times 10^3\ \text{kg m}^{-3})}{63.57\ \text{kg mol}^{-1}}\right)^{2/3}$$

$$= \frac{(6.62)^2 \times 10^{-68}}{8 \times 9.11 \times 10^{-31}}\left(\frac{3}{\pi} \times \frac{0.2 \times 6.023 \times 8.94}{63.57}\right)^{2/3} \times 10^{18}\ \frac{\text{J}^2\,\text{s}^2}{\text{kg m}^2}$$

$$= 1.79 \times 10^{-20}\ \text{J}$$

$$= 0.112\ \text{eV}.$$

Example 14.8 A flux of 10^{12} neutrons m^{-2} emerges each second from a port in a nuclear reactor. If these neutrons can be characterised by Maxwell–Boltzmann energy distribution corresponding to T = 300 K, calculate the density of neutrons in the beam. = 1.38×10^{-23} JK^{-1} and mass of neutron = 1.67×10^{-27} kg.

Here, density refers to the number density = n

Then, according to M–B distribution law we can write $n = n_0 \exp(-k_B T)$. Also, we recall that

$$\text{Flux} = n_0 \times (\text{velocity})$$

We estimate velocity with $v_{\text{rms}} = \sqrt{\dfrac{3k_B T}{m}}$

$$n_0 = \frac{\text{Flux}}{\sqrt{\dfrac{3k_B T}{m}}} = \sqrt{\frac{m}{3k_B T}}\ (\text{Flux})$$

The required value of n is given by

$$n = n_0 \exp(-k_B T) = \sqrt{\frac{m}{3k_B T}}\ (\text{Flux}) \exp(-k_B T)$$

$$\sqrt{\frac{1.67\times10^{27}\ \text{kg}}{3\times(1.38\times10^{-23}\ \text{JK}^{-1})\times(300\ \text{K})}}(10^{12}\ \text{m}^2\text{s}^{-1})\times\exp(-1.38\times10^{-23}\ \text{JK}^{-1}\times300\ \text{K})$$

$$=0.367\times10^{36}\ \text{m}^{-3}=3.67\times10^{36}\ \text{m}^{-3}.$$

Example 14.9 Calculate the Fermi wavelength and Fermi energy for 4.2 × 10^{21} electrons in a box of volume 1 cm^3. If the electrons are replaced by neutrons, how do Fermi wavelength and Fermi energy change.

From Eq. (14.37), we recall that Fermi energy

$$\varepsilon_F=\frac{p_F^2}{2m}=\frac{h^2}{8m_e}\left(\frac{3N}{\pi V}\right)^{2/3}$$

$$p_F^2=\frac{h^2}{4}\left(\frac{3N}{\pi V}\right)^{2/3}$$

On raising the power of both sides by 3/2, we can write

$$p_F^3=\frac{h^3}{8}\left(\frac{3N}{\pi V}\right)$$

Hence, Fermi wavelength

$$\lambda_F=\frac{h}{p_F}=\left(\frac{8\pi V}{3N}\right)^{1/3}$$

On substituting the given values, we get

$$\lambda_F=\left(\frac{8\times3.14\times1\ \text{cm}^3}{3\times4.2\times10^{21}}\right)^{1/3}$$

$$=(1.995\times10^{-21})^{1/3}\ \text{cm}$$

$$=1.26\times10^{-7}\ \text{cm}$$

$$=126\ \text{nm}$$

Fermi energy

$$\varepsilon_F=\frac{p_F^2}{2m_e}=\frac{1}{2m_e}\left(\frac{h}{\lambda_F}\right)^2=\frac{1}{(2\times9.1\times10^{-31}\ \text{kg})}\left(\frac{6.62\times10^{-34}\ \text{Js}}{1.26\times10^{-9}\ \text{m}}\right)^2$$

$$=1.52\times10^{-19}\ \text{J}$$

Since Fermi wavelength is independent of mass of the particles making up the system, it is not affected when electrons are replaced by neutrons. However, Fermi energy will be

$$\left(\frac{m_e}{m_n}\right)\varepsilon_F:$$

$$(\varepsilon_F)_{\text{neutron}} = \left(\frac{9.1\times10^{-31}\text{ kg}}{1.67\times10^{-27}\text{ kg}}\right)\times1.54\times10^{-19}\text{ J}$$

$$= 8.39\times10^{-23}\text{ J}$$

$$= 5.24\times10^{-4}\text{ eV.}$$

Example 14.10 The work function of a metal is 2.5 eV. Calculate the density of electrons ejected in thermionic emission at 1227°C. Given A = 120A cm^{-2} K^{-2} and × 10^{-23} JK^{-1}.

From Eq. (14.62), we recall that

$$j_{\text{th}} = AT^2 e^{-\phi/k_B T}$$

= 120 A cm^{-2} K^{-2}, T = 1500 K, k_B = 1.38 × 10^{-23} JK^{-1} and ϕ = 2.5 eV = 4 × 10^{-19} J.

$$= (120\text{ Acm}^{-2}\text{K}^{-2})(1500\text{ K})^2\exp\left[-\left(\frac{4\times10^{-19}\text{ J}}{(1.38\times10^{-23}\text{ JK}^{-1})\times1500\text{ K}}\right)\right]$$

$$= (120\,A\text{ cm}^{-2})\times(225\times10^4)\times(4.07\times10^{-9})$$

$$= 1.10\text{ Am}^{-2}$$

Example 14.11 Calculate the current density of photoelectric emission at 27°C at the threshold frequency for silicon. Take m_e = 9.1 × 10^{-31} kg, e = 1.6 × 10^{-19} C, × 10^{-23} JK^{-1} and h = 6.62 × 10^{-34} Js.

From Eq. (14.78), we recall that current density of photoelectric emission at the threshold frequency is given by

$$j_0 = \frac{2\pi^3 me k_B^2}{3h^3}T^2.$$

On substituting the given values, we get

$$j_0 = \frac{[2\times(3.1417)^3\times(9.1\times10^{-31}\text{ kg})]\times(1.6\times10^{-19}\text{ C})\times(1.38\times10^{-23}\text{ JK}^{-1})^2\times(300\text{ K})^2}{3\times(6.62\times10^{-34}\text{ Js})^3}.$$

$$= 17.78\times10^{10}\text{ A m}^{-2}$$

Example 14.12 Show that the single particle occupation number for a fermion at T K with an energy within $\pm k_B T$ of the Fermi energy has an approximate range of 0.46.

For $\varepsilon_1 = \varepsilon_F + k_B T$, we can write

$$n_1 = \frac{1}{\exp[(\varepsilon_F + k_B T - \varepsilon_F)] + 1} = \frac{1}{e+1} = 0.269$$

Similarly, for $\varepsilon_2 = \varepsilon_F - k_B T$, we get

$$n_2 = \frac{1}{\exp[(\varepsilon_F - k_B T - \varepsilon_F)] + 1} = \frac{1}{e^{-1}+1} = \frac{e}{e+1} = 0.731$$

$- n_1 = 0.46.$

We now summarise what you have learnt in this chapter.

MMARY

For a weakly degenerate quantum system, the internal energy and specific heat capacity are respectively given by

$$E = \frac{3}{2} N k_B T \left(1 + \frac{\kappa a A_0}{2} \right)$$

$$C_V = \frac{3}{2} N k_{\mathrm{B}} \left(1 - \frac{\kappa a}{4} A_0 + \ldots \right)$$

For a completely degenerate F–D system, the Fermi distribution function is a step function:

$$n_{\mathrm{FD}}(\varepsilon) = \begin{cases} 1 & \text{for} \quad \varepsilon < \mu \\ 0 & \quad \varepsilon > \mu \end{cases}$$

For a completely degenerate F–D system, the Fermi energy, Fermi temperature, zero-point energy and Fermi pressure are respectively given by

$$\varepsilon_F = \frac{h^2}{8m} \left(\frac{3N}{\pi V} \right)^{2/3}$$

$$T_F = \frac{\varepsilon_F}{k_B} = \frac{h^2}{8 m k_B} \left(\frac{3N}{\pi V} \right)^{2/3}$$

$$E_0 = \frac{3}{5} N \varepsilon_F$$

$$p_F = \frac{2}{5} \left(\frac{N}{V} \right) \varepsilon_F$$

For a strongly degenerate F–D system, the distribution function exhibits a tail. The thermodynamic properties of a strongly degenerate system are given by

$$E = \frac{3}{5} N \, \varepsilon_F \left[1 + \frac{5\pi^2}{12} \left(\frac{T}{T_F} \right)^2 - \frac{\pi^4}{16} \left(\frac{T}{T_F} \right)^4 + \ldots \right]$$

$$F = \frac{3}{5} N \varepsilon_F \left[1 - \frac{5\pi^2}{12}\left(\frac{T}{T_F}\right)^2 + \ldots\right]$$

$$p = \frac{2}{5}\frac{N\varepsilon_F}{V}\left[1 + \frac{5\pi^2}{12}\left(\frac{T}{T_F}\right)^2 + \ldots\right]$$

$$C_V = \frac{\pi^2}{2}\left(\frac{Nk_B}{T_F}\right)T = \frac{\pi^2}{2}\left(\frac{k_B T}{\varepsilon_F}\right)R$$

$$S = \left(\frac{\pi^2}{2}\frac{Nk_B}{T_F}\right)T$$

The expression for density of electrons escaping from a metal surface when it is given thermal energy is

$$j_{\text{th}} = \left(\frac{4\pi m e k_B^2}{h^3}\right)T^2 \exp\left[-\left\{\frac{E_c - \mu(T)}{k_B T}\right\}\right]$$

$$= AT^2\, e^{-\phi/k_B T}$$

where $A = \frac{4\pi m e k_B^2}{h^3} = 120 \text{ A cm}^{-2} \text{ K}^{-2}$ and $\phi = E_c - \mu(T)$ is work function. Knowledge of the work function facilitates determination of zero point energy. The current density of the photoelectric emission is given by

$$j_{\text{ph}} = \frac{4\pi m e k_B^2}{h^3} T^2 \xi^2$$

A white dwarf star is a mass of helium ($M \approx 10^{30}$ kg) under extreme pressure (density kg m^{-3}) at a very high temperature (10^7 K). This means that the dynamics of electrons in a typical white dwarf star is *relativistic*; and statistically speaking, the electron gas is in a state of complete degeneracy. The mass and radius of a white dwarf star are connected through the relation

$$\bar{R} = \frac{4}{5}\left(\frac{M_0^2}{\bar{M}}\right)^{1/3}$$

Chandrasekhar mass limit for a white dwarf star is given by

$$M_0 \approx 1.44\Theta$$

Θ denotes the mass of the sun ($\approx 2 \times 10^{30}$ kg).

ERCISES

A system consisting of two particles, each of which can be in any one of three quantum states of respective energies 0, ε and 3ε is in thermal equilibrium at temperature T. Write expressions for the partition function of the system if the particles obey F–D statistics. (*Ans*: $Z = e^{-\beta\varepsilon} + e^{-3\beta\varepsilon} + e^{-4\beta\varepsilon}$)

Sodium is monovalent. Compute its Fermi energy (in eV) using the following data: $= 9.70$ kg m^{-3}, $A = 23$, $N_A = 6.02 \times 10^{26}$ per kilomole, $m_e = 9.11 \times 10^{-31}$ kg and $= 6.62 \times 10^{-34}$ Js. (*Ans:* 3.14 eV)

$T = 300$ K, the energy of a metal falls to 1.2×10^{-4} eV below the Fermi energy (at 0 K). Calculate ε_F. (*Ans:* 4.58 eV)

Each potassium atom contributes one electron to the electron gas. Calculate **(a)** the Fermi energy, **(b)** the Fermi momentum, **(c)** the Fermi temperature, and **(d)** the contribution of free electrons to the specific heat capacity at 100 K. Given $(N/V) =$ $\times 10^{28}$ m^{-3}. (*Ans:* (a) 2.05 eV, (b) 7.73×10^{-25} kg ms^{-1}, (c) 2.38×10^4 K, (d) 1.72×10^6 J kmol^{-1} K^{-1})

The sun is composed of atomic hydrogen and its mass is 2×10^{30} kg. Calculate the number of electrons in the Sun.
In a white dwarf star of one solar mass, the atoms are completely ionised and contained in a sphere of radius 2×10^7 m. Calculate the Fermi energy of electrons in eV.
If the temperature of the white dwarf is 10^7 K, will the electrons be weakly or strongly degenerate? (*Ans:* 1.2×10^{57}; 4×10^4 eV; $T_F = 4 \times 10^{10}$ K Strongly degenerate)

4.2×10^{21} electrons are confined in a box of volume 10^{-3} m^3. Calculate Fermi wavelength and Fermi energy.
If the electrons are substituted by neutrons, how will these quantities change? (*Ans:* (a) 1.25 nm, 1.54×10^{-19} J; (b) 1.25 nm, 8.34×10^{-23} J)

The Fermi energy in silver is 5.5 eV. Calculate the temperature at which the probability will be 10% that electrons have energy 1% above the Fermi energy. (*Ans:* 290 K)

Calculate the temperature at which the probability that an electron occupies a state with energy 0.5 eV above the Fermi energy will be 1%. (*Ans:* 1264 K)

15

BOSE–EINSTEIN STATISTICS

arning Objectives

In this chapter, you will learn how to

obtain expressions for thermodynamic functions of a strongly degenerate B–E system;

explain Bose–Einstein condensation;

apply B–E statistics to an assembly of photons and derive Planck's law for blackbody radiation; and

explain the behaviour of liquid helium at low temperatures.

NTRODUCTION

In the previous chapter, we considered weakly degenerate quantum systems and strongly degenerate F–D systems. We discovered that F–D systems are quite alive even at absolute zero; a purely quantum mechanical effect completely unknown in the realm of classical physics. On the other hand, a weakly degenerate boson system has less energy and exerts lower pressure than a classical assembly kept under similar conditions. This is essentially because bosons prefer to stick together in lower energy states and there is no restriction on their number in an energy state. However, the specific heat capacity of a weakly degenerate B–E system is greater than that of a classical system.

In this chapter, we begin by discussing the behaviour of an assembly of ideal (non-interacting) bosons at low temperatures. Such a system of bosons has to be treated as strongly degenerate. At extremely low temperatures, bosons exhibits one of the most remarkable quantum effects: a phase change somewhat resembling condensation of a vapour into a liquid. This phenomenon is known as *Bose–Einstein condensation.* It was first predicted by Einstein in 1925. F. London suggested that B-E condensation can be the basis for microscopic understanding of very curious properties of ^{4}He. The two fluid theory of Tisza and Landau's quantum excitations theories of liquid He-II are also discussed in brief in Sec. 15.2, followed by derivation of expressions for thermodynamic properties of a strongly degenerate boson gas.

While obtaining the most probable distribution for an assembly of B–E particles in Chapter 12, it was pointed out that photons are bosons, have zero rest mass and Pauli Exclusion Principle does not apply. Moreover, their number is not conserved so that = 1. Such a system has to be treated as completely degenerate. In Sec. 15.3, we have discussed how Bose used statistical method to derive Planck's law of radiation. We have also evaluated the partition function for a photon gas and used it to obtain expressions for thermodynamic functions.

TRONGLY DEGENERATE BOSON GAS: -E CONDENSATION

The total number of particles in an assembly of bosons is given by Eq. (14.17) with $\kappa = -1$:

$$N = CV \int_0^\infty \frac{\varepsilon^{1/2}\, d\varepsilon}{A^{-1}\, e^{\varepsilon/k_B T} - 1} \tag{15.1}$$

The density of states around a given energy ε is given by

$$\frac{d^3\mathbf{r}\, d^3\mathbf{p}_l}{h^3} = CV\varepsilon^{1/2}\, d\varepsilon$$

You will note that by replacing summation by integration, we are giving zero weightage to the lowest energy state, i.e., $\varepsilon = 0$ energy level. This is not quite justified in quantum mechanical treatment because for any non-degenerate single particle state, a statistical weight other than one makes no sense physically. You may now ask: What are its implications? To know these, we make the substitution $\varepsilon/k_B T = x$. Then Eq. (15.1) takes

$$N = CV(k_B T)^{3/2} \int_0^\infty \frac{x^{1/2}\, dx}{A^{-1}\exp(x) - 1} \tag{15.2}$$

For a completely degenerate boson gas, $A = 1$ ($\mu = 0$); values greater than unity are not admissible because these will correspond to negative number of bosons in the lowest energy state. In this limit, the integral

$$\int_0^\infty \frac{x^{1/2}\, dx}{\exp(x) - 1} = \Gamma\left(\frac{3}{2}\right)\zeta\left(\frac{3}{2}\right)$$

$\Gamma\left(\frac{3}{2}\right) = \frac{\sqrt{\pi}}{2}$ is gamma function of order (3/2) and $\zeta\left(\frac{3}{2}\right) = 2.612$ is the Riemann zeta function of order (3/2).

Using this result in Eq. (15.2), we get

$$N = 1.306\sqrt{\pi}\, CV(k_B T)^{3/2} \tag{15.3}$$

This result shows that the particle density (*N*/*V*) is a function of temperature. Obviously, something is missing in the argument that led us to Eq. (15.3). As expected, the trouble stems from the fact that the energy level corresponding to $\varepsilon = 0$ has been completely neglected in Eq. (15.1). We correct this by treating this state separately in the expression . We therefore write

$$N = N_g + CV \int_0^\infty \frac{\varepsilon^{1/2}\, d\varepsilon}{A^{-1}\, e^{\varepsilon/k_B T} - 1} \tag{15.4}$$

$$N_g = \frac{1}{(1/A) - 1} = \frac{1}{e^{-\mu/k_B T} - 1} \tag{15.5}$$

gives the number of particles in the ground state ($\varepsilon = 0$, $p = 0$) and

$$N_{ex} = CV \int_0^\infty \frac{\varepsilon^{1/2}\, d\varepsilon}{A^{-1} e^{\varepsilon/k_B T} - 1} \tag{15.6}$$

gives the number of particles with non-zero energy and momentum ($\varepsilon > 0$, $p > 0$). This means that for given values of *V* and *T*, the total number of particles in all the excited states > 0) taken together is bounded, since $A \leq 1$:

$$N_{ex} \le 1.306\sqrt{\pi}CV(k_B T)^{3/2}$$

$$\le 2.612\,V\left(\frac{2\pi m k_B T}{h^2}\right)^{3/2} \tag{15.7}$$

Therefore, so long as the actual number of particles is less than this limiting value, everything is well and all the particles in the system are distributed over the excited states. But as soon as the number exceeds this limiting value, the number of particles in the excited states, N_{ex}, is given by

$$N_{ex} = 2.612\,V\left(\frac{2\pi m k_B T}{h^2}\right)^{3/2} \tag{15.8}$$

and the excess particles are pushed *en masse* into the ground state, which, of course, can possess practically an unlimited number of particles. Thus, the number of particles in the ground state, N_g is given by

$$N_g = N - \left[2.612\,V\left(\frac{2\pi m k_B T}{h^2}\right)^{3/2}\right] \tag{15.9}$$

The phenomenon in which bosons gradually accumulate in the lower energy levels and a large number of these occupy the ground state ($\varepsilon = 0$) is referred to as *Bose–Einstein condensation.* It is not condensation in ordinary space like condensation of a vapour into a liquid. Moreover, no spatial separation into two phases with different properties occurs in Bose–Einstein condensation. As we show, the pressure exerted by the condensed Bose gas, like the pressure of a saturated vapour, depends only on its temperature—it is independent of volume. If anything, B–E condensation is a purely quantum mechanical effect, which is completely new and unknown in the realm of classical physics. It was reported by Einstein , a few months after he had forwarded Bose's paper for publication. As he put it:

> *A number of molecules steadily growing with increasing density go over into the first quantum state (which has zero kinetic energy) while the remaining molecules distribute themselves according to the parameter A. A separation is thus effected: one part condenses while the rest remains as a 'saturated ideal gas.'*

The condition for the onset of Bose–Einstein condensation is written as

$$N > N_{ex} = 2.612\,V\left(\frac{2\pi m k_B T_c}{h^2}\right)^{3/2} \tag{15.10}$$

is known as the *Bose–Einstein condensation temperature.*

and V are kept constant but temperature is made to vary, we can express this condition in terms of Bose–Einstein condensation temperature as

$$T < T_c = \frac{h^2}{2\pi m k_B}\left[\frac{N}{2.612\,V}\right]^{2/3} \tag{15.11}$$

Note that Bose–Einstein condensation temperature depends on the particle mass and particle density (N/V) of the system. Above T_c, the number of particles in the ground state

*The first gaseous Bose–Einstein condensate was produced in 1995 by Eric Cornell and Carl Wieman at the University of Colorado at Boulder NIST–JILA laboratory using a gas of rubidium atoms cooled to 170 nanokelvin. For this accomplishment, they received the 2001 Nobel Prize in physics, along with Wolf gang Ketterle. In November 2010, the first photon B-E condensate was observed.

is almost negligible but below T_c, particles begin to accumulate in the quantum state with = 0. Accordingly, for $T < T_c$ a boson gas may be looked upon as a mixture of two phases:

a gaseous phase consisting of N_{ex} particles distributed over the excited states ($\varepsilon > 0$, > 0), and

a condensed phase, consisting of $N_g (= N - N_{ex})$ particles concentrated in the energy state with $\varepsilon = 0 = p_\ell$.

In the framework of this model, we can express N_{ex} and N_g in terms of total number N in the assembly. To this end, we first note from Eqs. (15.10) and (15.11) that

$$\frac{N_{ex}}{N} = \left(\frac{T}{T_c}\right)^{3/2}$$

$$N_{ex} = N\left(\frac{T}{T_c}\right)^{3/2} \tag{15.12}$$

This shows that the number of atoms in excited states increases as $T^{3/2}$. Moreover, at , all the particles are in the excited states.

The number of particles in the ground state N_g, is given by

$$N_g = N - N_{ex} = N\left[1 - \frac{N_{ex}}{N}\right] = N\left[1 - \left(\frac{T}{T_c}\right)^{3/2}\right] \tag{15.13}$$

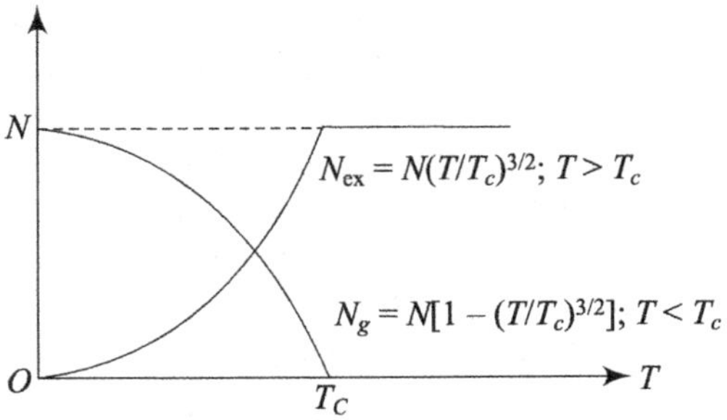

Distribution of bosons as a function of temperature below and above the Bose-Einstein condensation temperature.

Figure 15.1 shows the temperature variation of $\frac{N_{ex}}{N}$ and $\frac{N_g}{N}$ based on Eqs. (15.12) and (15.13). Note that

= 0, all bosons condense into the lowest energy state ($\varepsilon = 0$).

increases, the number of particles in the ground state decreases rapidly and at $_c$, their number is almost zero.

> T_c, all particles are in the excited states.

Figure 15.2 shows the qualitative behaviour of chemical potential, μ as a function of temperature for fixed density. (For comparison, chemical potentials of an ideal M-B and an ideal Fermi gas are also shown.) As may be noted, μ is almost zero for $T \leq T_c$ but drops rapidly for $T > T_c$.

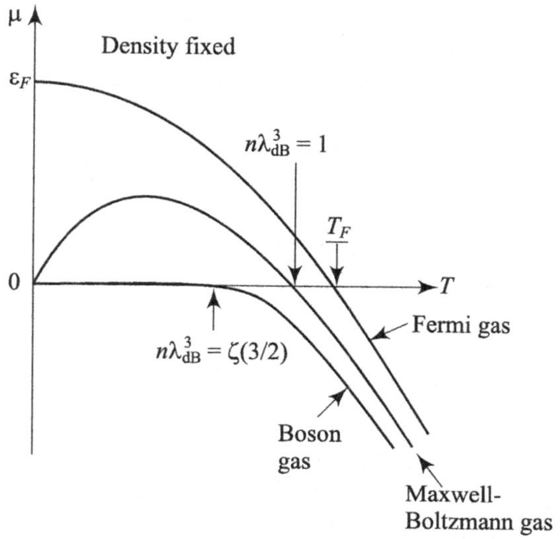

Temperature variation of chemical potential of bosons. For comparison, chemical potentials of Fermi and M-B gases are also shown.

You may now ask: Why does Bose–Einstein condensation occur in a (non-interacting) ideal boson gas? The question is quite logical because condensation of a gas into a liquid occurs because of the existence of forces of attraction (intermolecular potential) between the molecules/atoms. B–E condensation occurs because bosons prefer to stick together in the same state. (Alternatively, we can say that an effective interaction comes into action through the symmetry requirement of the N-body wave function.) In general, it is impossible to observe B–E condensation because every gas, with the exception of helium, crystallises at temperatures much above T_c.

It may be mentioned here that Maxwell–Boltzmann statistics also predicts total occupation of the lowest energy state at absolute zero, no increase in the occupancy of the lowest state at a non-zero temperature occurs within its framework.

We now derive expressions for thermodynamic functions of an ideal boson gas.

Thermodynamic Functions of an Ideal B–E Gas

We now know that the properties of an ideal boson gas above and below T_c are vastly different. Therefore, we have to discuss the thermodynamic functions in these temperature ranges separately. We begin with the calculation of internal energy.

The internal energy of B–E gas is given by Eq. (14.10) with $\kappa = -1$.

$$E = CV \int_0^\infty \frac{\varepsilon^{3/2}\, d\varepsilon}{(1/A)\exp(\varepsilon/k_B T) - 1}$$

= $\exp(\mu/k_BT)$ is degeneracy parameter.

As before, to evaluate this integral we introduce a dimensionless variable by defining . Then $d\varepsilon = k_BT dx$ and $\varepsilon^{3/2} d\varepsilon = (k_BT)^{5/2} x^{3/2} dx$. Note that the limits of integration do not change. Hence, we can write the expression for internal energy as

$$E = (k_B T)^{5/2}\, CV \int_0^\infty \frac{x^{3/2}}{(1/A)\, \exp(x) - 1} dx \tag{15.14}$$

T_c, $A = 1$ so that Eq. (15.14) takes the form

$$E_g = CV(k_B T)^{5/2} \int_0^\infty \frac{x^{3/2}\,dx}{e^x - 1}$$

$$= CV(k_B T)^{5/2}\, \Gamma\left(\frac{5}{2}\right)\zeta\left(\frac{5}{2}\right) \tag{15.15}$$

$\left(\frac{5}{2}\right)$ and $\zeta\left(\frac{5}{2}\right)$ are respectively gamma function and Riemann zeta function $\left(\frac{5}{2}\right)$.

On substituting $C = \dfrac{2\pi(2m)^{3/2}}{h^3}$ and $\Gamma\left(\frac{5}{2}\right) = \frac{3}{2} \times \frac{\sqrt{\pi}}{2}$ in Eq. (15.15), we get

$$E_g = \frac{3}{2}\left(\frac{2\pi m k_B}{h^2}\right)^{3/2} V k_B T^{5/2}\, \zeta\left(\frac{5}{2}\right), T < T_c \tag{15.16}$$

We now intend to express internal energy of particles in the ground state in terms of the Bose–Einstein condensation temperature. To do so, we note from Eq. (15.11) that

$$V\left(\frac{2\pi m k_B}{h^2}\right)^{3/2} = \frac{N}{\zeta(3/2)} \times \frac{1}{T_c^{3/2}}$$

On combining this result with Eq. (15.16), we get

$$E_g = \frac{3}{2} N k_B \left(\frac{T^{5/2}}{T_c^{3/2}}\right) \frac{\zeta(5/2)}{\zeta(3/2)}$$

$$= 0.5134 \times \frac{3}{2} N k_B \left(\frac{T^{5/2}}{T_c^{3/2}}\right)$$

$$= 0.7701\, N k_B \left(\frac{T^{5/2}}{T_c^{3/2}}\right) \quad \text{for } T < T_c \tag{15.17}$$

This result shows that the internal energy of a boson gas below T_c varies as $T^{5/2}$ and that it becomes zero as $T \to 0$, i.e., the condensed phase settles down into the ground state. T_c ($A < 1$), the boson gas will be weakly degenerate and the internal energy of such a system is obtained by following the procedure used to arrive at Eq. (14.20). Therefore, we combine Eqs. (14.19) and (14.20) to obtain

$$E_{ex} = \frac{3}{2} N k_B T\left[1 + \frac{A}{2^{5/2}} + \ldots\right]\left[1 - \frac{A}{2^{3/2}} + \ldots\right]$$

$$= \frac{3}{2} N k_B T\left[1 - \frac{A}{2^{5/2}} - \frac{A^2}{3^{5/2}} + \ldots\right] \quad T > T_c$$

On substituting for $A = \zeta\left(\frac{3}{2}\right)\left(\frac{T_c}{T}\right)^{3/2}$ in this expression, we can express the total internal energy of a weakly degenerate boson gas in terms of T and T_c as

$$E_{ex} = \frac{3}{2} N k_B T \left[1 - 0.1767 \times 2.612 \left(\frac{T_c}{T} \right)^{3/2} + \ldots \right]$$

$$= \frac{3}{2} N k_B T \left[1 - 0.4618 \left(\frac{T_c}{T} \right)^{3/2} + \ldots \right] \quad T > T_c \qquad (15.18)$$

This result shows that for $T >> T_c$, E_{ex} approaches the classical value.

We know that the internal energy and pressure of a gas are directly related. So, for , we can write

$$p(T) = \frac{2}{3} \frac{E_g}{V} = \left(\frac{2\pi m}{h^2} \right)^{3/2} (k_B T)^{5/2} \zeta\left(\frac{5}{2} \right), \quad T < T_c \qquad (15.19)$$

This result shows that pressure of the condensate is proportional to $T^{5/2}$ but is *completely independent of volume* or *number density*. This is very surprising because from kinetic theory we know that pressure exerted by a gas is directly proportional to number density. However, this situation is analogous to the co-existence of liquid-vapour state wherein addition of any number of particles at constant temperature does not lead to increase in pressure; they go to the liquid and vapour phases. Similarly, in B–E condensation, particles go to the zero momentum state without altering pressure.

At the transition point, the pressure is

$$p(T_c) = \left(\frac{2\pi m}{h^2} \right)^{3/2} (k_B T_c)^{5/2} \zeta\left(\frac{5}{2} \right) \qquad (15.20a)$$

Combining this result with Eq. (13.12), we can write

$$p(T_c) = \left(\frac{N}{V} k_B T_c \right) \frac{\zeta(5/2)}{\zeta(3/2)} = 0.5134 \left(\frac{N}{V} k_B T_c \right) \qquad (15.20b)$$

From this we may conclude that an ideal boson gas at the transition temperature exerts pressure which is nearly one-half of what a classical gas would exert under similar

We can write $p(T)$ in terms of $p(T_c)$ as

$$p(T) = p(T_c) \left(\frac{T}{T_c} \right)^{5/2} = 0.5134 \left(\frac{N_{ex} k_B T}{V} \right), T \leq T_c \qquad (15.21)$$

From this, we can say that particles in the condensed phase do not exert any pressure at all—the particles in the excited states are about half as effective as those of a Maxwellian

T_c, the pressure is given by

$$p(T) = \left(\frac{N k_B T}{V} \right) \left[1 - 0.4618 \left(\frac{T_c}{T} \right)^{3/2} + \ldots \right] \qquad (15.22)$$

$> T_c$, pressure approaches the classical value of Nk_BT/V. These features are illustrated in Fig. 15.3a. For a B-E system, we have plotted p as a function of V at different

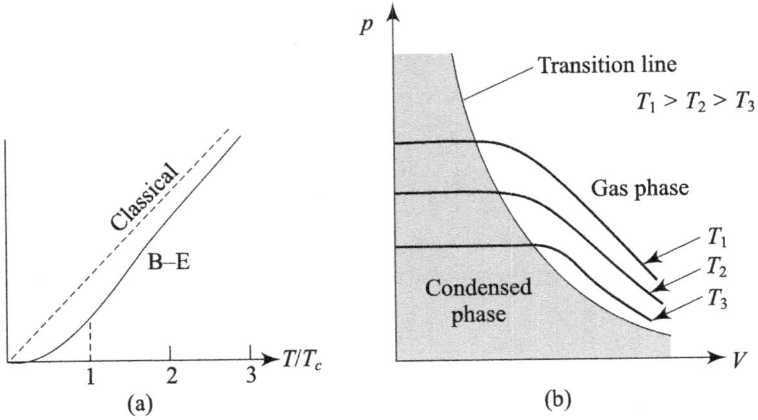

(a) Plot of (a) pV/Nk_BT_c as a function of T/T_c, and (b) p versus V for a strongly degenerate Bose gas.

temperatures in Fig. 15.3b. The horizontal portion in Fig. 15.3b corresponds to the condensate defined by Eq. (15.19). The curved portion follows the equation

$$p = \frac{N k_B T}{V}\left[1 - 0.4618\left(\frac{V_c}{V}\right)\right] \tag{15.23}$$

signifies the volume occupied at condensation temperature. The transition line qualitatively resembles an isotherm for a strongly degenerate boson gas.
Before proceeding further, you may like to answer a practice problem.

Problem 15.1 Consider an ideal classical gas, an ideal Fermi gas or an ideal Bose gas. Which of these gases will exert the maximum and the minimum pressure at a given temperature and why?

While answering this problem, we expect you to argue that an ideal Fermi gas will exert the maximum pressure and an ideal Bose gas will exert minimum pressure. This is because F–D distribution has larger proportion of higher energy particles than the other two gases. On the other hand, the B–E distribution has larger proportion of lower energy particles than the other two gases.

The heat capacity of an ideal Boson gas at constant volume for $T < T_c$ can be calculated using Eq. (15.17):

$$C_V = \left(\frac{\partial E_g}{\partial T}\right)_V = \frac{15}{4}\times 0.5134\, N k_B \left(\frac{T}{T_c}\right)^{3/2} = 1.925\, N k_B \left(\frac{T}{T_c}\right)^{3/2} \quad T < T_c \tag{15.24a}$$

This equation shows that at the condensation temperature, the heat capacity of an ideal Boson gas varies as $T^{3/2}$ and at any temperature just below T_c, it exceeds the classical value . Moreover, $C_V \to 0$ as $T \to 0$.

T_c, the heat capacity of an ideal Boson gas at constant volume can be calculated using Eq. (15.18):

$$C_V = \left(\frac{\partial E_{ex}}{\partial T}\right)_V = \frac{3}{2} N k_B \left[1 + 0.2309\left(\frac{T_c}{T}\right)^{3/2}\right] \tag{15.24b}$$

This implies that for $T > T_c$, $C_V \rightarrow (3/2)\,R$.

T_c, Eqs. (15.24a) and (15.24b) lead to the same value of $1.925Nk_B$. This suggests that the specific heat capacity is continuous at the condensation temperature. These features are illustrated in Fig. 15.4, where we have plotted C_V/Nk_B versus temperature. The plot shows a kink at $T = T_c$ but it is continuous. Also, the curve resembles the symbol *lambda*. To facilitate comparison of heat capacities of bosons and fermions with M–B gas, we have also plotted the values of C_V/Nk_B corresponding to a strongly degenerate F–D gas. As may be noted, the two curves are quite different, particularly in the region $T < T_c$. Above T_c, both curves approach the classical value asymptotically.

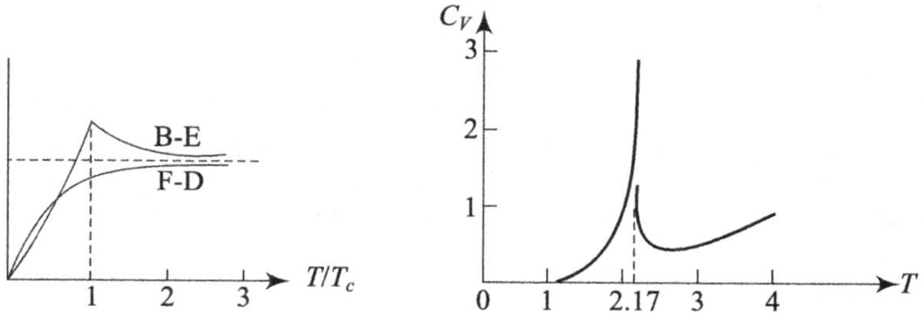

(a) (b)

Plot of (a) C_V/Nk_B as a function of T/T_c for strongly degenerate B–E and F–D gases and (b) C_V as a function of T for a B-E gas around the transition temperature.

In Fig. 15.4b, we have depicted the observed variation of C_V versus T for helium around its transition temperature. As may be noted, (a) the observed curve resembles the symbol lambda -point curve is very sharp compared to the ideal boson gas model curve. Moreover, Eq. (15.24a) suggests that in the region $T < T_c$, the heat capacity of an ideal boson gas varies , whereas the experimentally measured heat capacity of liquid He-II is known to obey -law below 0.6 K. Similarly, the dependence of transition temperature on pressure is not correctly predicted by the ideal boson gas model. This is not surprising as the ideal boson gas model does not account for the weak interatomic forces present in liquid He-II. He also shows superfluid behaviour below 0.003 K due to formation of cooper pairs, which behave like bosons. However, one finds it extremely difficult to observe it.)

Another important thermodynamic parameter of interest, particularly at low temperatures, is the entropy. We can obtain an expression for entropy using the relation

S T below and above T_c with C_V given by Eqs. (15.24a) and (15.24b), respectively. A simple calculation leads to the following result

$$S(T) = \begin{cases} 1.28\,Nk_B \left(\dfrac{T}{T_c}\right)^{3/2} & T < T_c \\[2ex] S(T_c) + \dfrac{3}{2} Nk_B \left[\ln\left(\dfrac{T}{T_c}\right) + 0.1536 \left\{ 1 - \left(\dfrac{T_c}{T}\right)^{3/2} \right\} \right] & T > T_c, \end{cases} \tag{15.25}$$

It may be mentioned here that like heat capacity, entropy also shows a sudden drop for . This amounts to an increase in order in the system. Note that this change in order

That is, the condensed phase has no entropy at absolute zero. This is in conformity with the third law of thermodynamics.

15.2.2 Liquid Helium

At atmospheric pressure, helium condenses into a normal liquid at 4.2 K. As the temperature is lowered further, another phase transition is observed at 2.17 K. One would have expected helium to solidify at this transition. Instead, liquid helium changes into another liquid of vastly different, in fact, amazing properties. (Some of these—superfluidity, fountain effect, second sound and superconductivity—were discussed qualitatively in Chapter 10.) To differentiate these phases, we term the phase above 2.17 K as Helium I and that below 2.17 K as Helium II. It consists of two components, a *normal fluid* and a *superfluid*. The superfluid is characterised by remarkable properties like apparently zero viscosity and infinite thermal conductivity. Moreover, the transition temperature is accompanied by an anomalous behaviour of the heat capacity; the C_V versus T curve exhibits a shape similar to the Greek letter lambda (Fig. 15.4b). The density of helium increases by about 20% as temperature drops from 4.2 K to 2.17 K and attains a maximum value at the transition temperature. However, no abrupt change in vapour pressure, latent heat surface tension and dielectric constant is observed.

If we use $N/V = 2.2 \times 10^{28}$ m^{-3} and $m = 6.65 \times 10^{-27}$ kg for atoms of liquid helium, Eq. (15.11) gives $T_c = 3.15$ K, which is not very far from the observed value of 2.17 K for the onset of condensation in liquid helium. This led London to suggest that λ–transition of He is a form of B–E condensation and has some relevance to its low-temperature behaviour. (However, there were differences in details.)

In Chapter 10, we have highlighted the low temperature behaviour of helium, which has two isotopes: ${}_2^3$He and ${}_2^4$He. (Of these, ${}_2^4$He is more abundant.) These behave quite differently at low temperatures and a logical explanation follows from quantum statistics He is a boson (spin 0) and ${}_2^3$He is a fermion (spin 1/2).

The 2.17 K phase transition from He-I to He-II tells us that condensation into the ground state is a *necessary* condition for the occurrence of superfluid behaviour. But it is *sufficient* condition.

Now you have some idea about Bose–Einstein condensation and its relevance to low temperature properties of ^{4}He. Recall that below the λ-point, ^{4}He is called He-II, which is characterised by very peculiar properties. Some of these properties can be understood in terms of the *two fluid model* given by Tisza. We now consider it for brevity and completeness.

Tisza model In two-fluid model, it is assumed that

Below the λ-point, liquid Helium-II behaves as a mixture of two fluids: a normal fluid and a superfluid, which exhibit no viscous interaction between them. We can express the total number of particles and density as

$$N = N_n + N_s$$

$$\rho = \rho_n + \rho_s \qquad T < T_\lambda$$

The proportion of the superfluid component increases from zero at the λ-point to unity at the absolute zero. In the language of mathematics, we express this condition as

$$\rho_n = \rho \quad \text{for} \quad T \geq T_\lambda$$

$$= 0 \quad \text{for } T = 0.$$

Using this model, Tisza calculated the expression for the speed of second sound. We quote his result without going into details:

$$u_2 = 26\left(\frac{T}{T_\lambda}\right)\left[1-\left(\frac{T}{T_\lambda}\right)^{5.5}\right]^{1/2} \tag{15.26}$$

The calculated as well as observed values of second sound are plotted as a function of temperature in Fig. 15.5. As may be noted, there is significant deviation below about 1 K. It means that two fluid theory does not provide complete understanding of He-II. It was modified by Landau and we will now

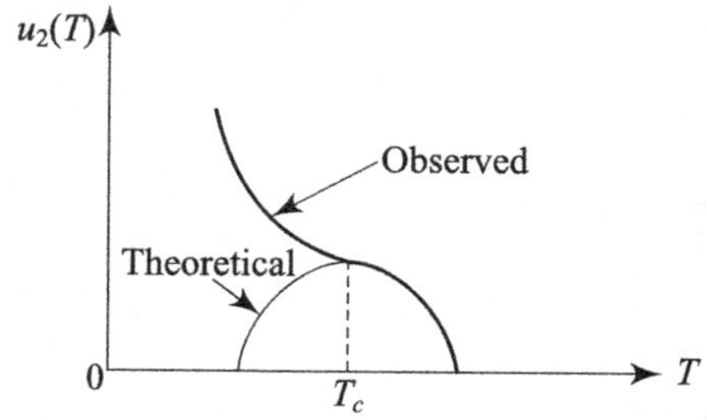

Fig. 15.5 Temperature variation of second sound.

Landau's Theory of Liquid He-II

Landau found the explanation of London and Tisza based on B–E condensation of an ideal Bose gas for the peculiar properties of liquid He-II to be unsatisfactory. In particular, the predictions about second sound and viscosity were not supported by experiments, particularly for $T \to 0$. He was of the view that around absolute zero, every substance should exhibit solid-like behaviour. This was confirmed by the fact that the observed value of specific heat capacity of He-II follows Debye's T^3-law as $T \to 0$. Landau gave a simple theoretical model which fairly well explains the behaviour of liquid He-II in the $T < 2$ K.

Landau argued that at absolute zero, liquid He is devoid of all motion but elementary excitations begin to appear, though discontinuously, as temperature rises. In the range < 0.5 K, these excitations were termed as *phonons* or quanta of sound. In the range < 2 K, Landau assumed the existence of the second group of excitations, called . (The nomenclature roton is due to Landau, who thought that these excitations might in some way represent local disturbances of a rotational character in the liquid.)

Note that both phonons and rotons together determine the behaviour of liquid He-II in the temperature range $0.5 < T < 1$ K, and the contribution of phonons tends to be relatively small and less important above 1 K. That is, above 1 K we have to consider only rotons. It is natural to expect that like phonons, rotons also obey B–E statistics. Moreover, their number in the system is not fixed. Can you say as to why is it so? It is because the energy of quasi-particles such as phonons is a linear function of momentum only for small momenta and elementary excitations with higher momenta are unstable and decouple into several excitations of smaller momenta.

Guided by purely empirical considerations, Landau proposed the following dispersion

$$\varepsilon(p) = \begin{cases} u_1 p & \text{for} \quad p << p_0 \\ \Delta + \dfrac{(p-p_0)^2}{2\sigma} & \text{for} \quad p \approx p_0 \end{cases} \tag{15.27}$$

is velocity of first sound, p_0 is the lowest roton level, Δ denotes the energy gap between the lowest phonon level and the lowest roton level (i.e., minimum energy required to excite a roton at rest) and σ is the effective mass of the roton. These parameters were adjusted to obtain the best fit to specific heat capacity measurements.

If we plot $\varepsilon(p)$ versus p, we note that it increases linearly initially and attains a maximum value. Thereafter, it begins to decrease and goes through a minimum, which defines the lowest roton level. The dispersion curve increases again thereafter. These features are depicted in Fig. 15.6.

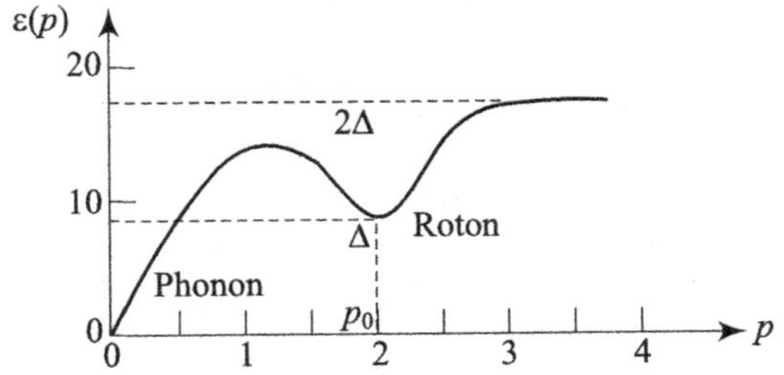

Dispersion curve of quasi-particles in liquid ^{4}He.

The nature of the empirical spectrum proposed by Landau was reproduced by several experimentalists based on neutron scattering/diffraction studies in liquid He-II.

Evaluation of thermodynamic functions of liquid He-II is now a straightforward exercise. The phonon and roton contributions can be readily calculated using Eq. (15.27). We leave it as an exercise for you.

Problem 15.2 Using Eq. (15.27), show that phonon and roton contributions to specific heat capacity and entropy of liquid He-II are respectively given by

$$(C_V)_{\text{phonon}} = \frac{16\pi^5 V k_B{}^4}{15(u_1 h)^3} T^3$$

$$(S)_{\text{phonon}} = \frac{16\pi^5 V k_B{}^4}{45(u_1 h)^3} T^3$$

$$(C_V)_{\text{roton}} = N_r k_B \left[\frac{3}{4} + \frac{\Delta}{k_B T} + \left(\frac{\Delta}{k_B T}\right)^2\right]$$

$$(S)_{\text{roton}} = N_r k_B \left[\frac{3}{2} + \frac{\Delta}{k_B T}\right]$$

denotes the speed of sound and N_r is number of rotons.

From the expression for $(C_V)_{\text{phonon}}$, you will note that specific heat capacity of liquid He-II ^{3} at extremely low temperatures ($0 < T < 0.5$ K). This is in agreement with the observed temperature variation. In the temperature range $0.5 < T < 1$ K, the specific heat capacity will be made up of phonon and roton contributions:

$(C_V)_{\text{He-II}} = (C_V)_{\text{phonon}} + (C_V)_{\text{roton}}$. However, $(S)_{\text{phonon}} \to 0$ as $T \to 0$.

LACKBODY RADIATION

In Chapter 11, we discussed how Rayleigh–Jeans and Planck treated radiation as electromagnetic waves and oscillators, respectively to calculate the number of modes. As we now know, one could instead consider radiation as an assembly of photons in a chamber and use the results of statistical mechanics directly. Such an assembly of photons is called *photon gas*. Since photons have integral spins, B–E statistics finds direct application in the study of radiation emitted and absorbed by objects. That is, it enables us to investigate the equilibrium properties of black body radiation.

Historically, the behaviour of radiation in a cavity was studied along two essentially identical but conceptually different lines:

As an assembly of *harmonic oscillators*, each having energy $\left(n_s + \frac{1}{2}\right)\hbar\omega_s$, where $= 0,1,2,\ldots$ and ω_s is the angular frequency.

As a gas of indistinguishable photons, each with energy $\hbar\omega_s$.

The first of these viewpoints reflects Planck's ideas whereas the second one originated with Bose and was put on a firm theoretical basis by Einstein when he extended Bose's theory to material particles. We now illustrate how B–E statistics can be used to derive Planck's radiation formula.

Bose's derivation of Planck's law Consider an assembly of photons inside a cavity in a heat reservoir at temperature T. Photons are continually emitted and absorbed by the walls of the container so that their number is not conserved. However, their energy remains constant. We know that photons are relativistic particles with zero rest mass and do not interact at all amongst themselves. Moreover they are indistinguishable. Therefore, their chemical potential is zero.* Then the occupation number for a photon of energy ε is given by

$$n_{ph}(\varepsilon) = \frac{1}{e^{\beta\varepsilon} - 1} \tag{15.28}$$

The number of photons in the energy range from ε to $\varepsilon + d\varepsilon$ is given by the product of the density of states in this range and the occupation number of each state (Eq. (14.4)). Hence, we can write

$$dN = \left(\frac{d^3r\, d^3p_\ell}{h^3}\right) n_{ph}(\varepsilon)$$

$$= 2\left(\frac{4\pi V p_\ell^2}{h^3} dp_\ell\right)\frac{1}{e^{\beta\varepsilon} - 1} \tag{15.29}$$

where we have substituted for $n_{ph}(\varepsilon)$ from (15.28). The factor '2' accounts for the two polarization states (clockwise and anti-clockwise) of the photon spin.

To simplify this expression, we recall that energy and momentum of a photon are connected through the relation

$$\varepsilon = p_\ell c$$

is velocity of light. Then, we can write

---

*Alternatively, we can say that the number of photons in an assembly is not conserved, i.e., N is not fixed. This means that the Lagrange multiplier $\alpha = 0$. However, the total energy of the system is constant. This means that the total energy has no direct relationship to the number of particles in this case and therefore the chemical potential of a photon gas is zero.

$$p_\ell^2 \, \mathrm{d}p_\ell = \frac{\varepsilon^2 \, \mathrm{d}\,\varepsilon}{c^3}$$

With these substitutions, the number of photons having energies in the range ε and $\varepsilon + \mathrm{d}\varepsilon$ in the cavity can be expressed as

$$\mathrm{d}N = \left(\frac{8\pi V}{h^3 c^3}\right) \frac{\varepsilon^2 \, \mathrm{d}\varepsilon}{\mathrm{e}^{\beta\varepsilon} - 1} \tag{15.30}$$

Problem 15.3 Integrate Eq. (15.30) and show that

$$N = 8\pi V \left(\frac{k_B T}{hc}\right)^3 \times 2.44 = 2.02 \times 10^7 \, T^3$$

Hence, prove that a cube of size $1 \mathrm{m}^3$ contains nearly 4×10^{14} photons at 273 K.

This result shows that the number of photons per cubic metre is proportional to third power of temperature.

From kinetic theory of gases, we recall that an ideal gas contains 3×10^{25} molecule m^{-3}

The distribution of photon energies in the system is obtained by multiplying the number of photons and the energy ε of a photon. So we can write

$$\mathrm{d}E = \varepsilon \mathrm{d}N = \left(\frac{8\pi V}{h^3 c^3}\right) \frac{\varepsilon^3 \, \mathrm{d}\,\varepsilon}{\mathrm{e}^{\beta\varepsilon} - 1}$$

Hence, the energy density defined as $u = E/V$, is given by

$$u(T, \varepsilon)\, \mathrm{d}\varepsilon = \left(\frac{8\pi}{h^3 c^3}\right) \frac{\varepsilon^3 \, \mathrm{d}\,\varepsilon}{\mathrm{e}^{\beta\varepsilon} - 1} \tag{15.31}$$

This is the famous *Planck's formula* for the distribution of energy over the black body

In terms of photon frequency $\nu(=\varepsilon/h)$, we can express Planck's formula as

$$u(\nu, T)\, \mathrm{d}\nu = \left(\frac{8\pi h}{c^3}\right) \frac{\nu^3 \, \mathrm{d}\nu}{\mathrm{e}^{\beta h\nu} - 1} \tag{15.32}$$

In terms of wavelength $\lambda(= c/\nu)$, the spectral distribution of blackbody radiation in a cavity can be expressed as

$$u(\lambda, T)\, \mathrm{d}\lambda = \left(\frac{8\pi hc}{\lambda^5}\right) \frac{\mathrm{d}\lambda}{\mathrm{e}^{\beta hc/\lambda} - 1} \tag{15.33}$$

You will note that in writing this expression we have assumed that the wavelength region $\lambda + \mathrm{d}\lambda$ corresponds to the frequency region ν to $\nu + \mathrm{d}\nu^*$.

---

is inversely related to ε, we find that

$$|\mathrm{d}\lambda| = \frac{hc}{\varepsilon^2} \mathrm{d}\varepsilon$$

It may be pointed out here that Eqs. (15.31), (15.32) and (15.33) are equivalent representations of Planck's law.

The total energy density at temperature T is obtained by integrating Eq. (15.33) over all wavelengths from 0 to ∞. We evaluated this integral in Chapter 11 and obtained the result

$$u(T) = \frac{8\pi^5 k_B{}^4}{15h^3 c^3} T^4 = aT^4 \tag{15.34}$$

$$= \frac{8\pi^5 k_B{}^4}{15h^3 c^3} = 7.56 \times 10^{-16} \text{ Jm}^3 \text{ K}^{-4}.$$

The mean energy of the photon in the cavity is given by

$$\frac{u(T)}{n} = 2.7 k_B T \tag{15.35}$$

To obtain expressions for thermodynamic functions of a photon gas, we write the expression for partition function.

Partition function of a photon gas We know that photons are bosons and there is no restriction on their number in any state, i.e., their number can vary from 0 to ∞. Let us assume for simplicity that there are $n_1, n_2, n_3 \ldots$ photons in single particle states having $\nu_1, h\nu_2, h\nu_3 \ldots$, respectively. Note that $n_1, n_2, n_3, \ldots$ vary from 0 to ∞. Then we

$$Z_{\text{ph}} = \sum_{n_i=0}^{\infty} \exp\left[-\frac{n_1 h\nu_1}{k_B T} - \frac{n_2 h\nu_2}{k_B T} - \cdots \frac{n_i h\nu_i}{k_B T} - \cdots\right]$$

We can factorise the expression on the RHS of this expression and write

$$Z_{\text{ph}} = \left(\sum_{n_1=0}^{\infty} \exp(-n_1 h\nu_1 / k_B T)\right)\left(\sum_{n_2=0}^{\infty} \exp(-n_2 h\nu_2 / k_B T)\right)\ldots \tag{15.36}$$

You should practise this by giving three values each to $n_1, n_2, n_3, \ldots$ and writing out the sum.

We can evaluate each term in the parentheses in Eq. (15.36) using the binomial theorem:

$x \ = (1-x)^{-1}$. Therefore,

$$\sum_{n_i=0}^{\infty} \exp(-n_i h\nu_i / k_B T) = \left[1 - \exp\left(-\frac{h\nu_i}{k_B T}\right)\right]^{-1}$$

Using this result in Eq. (15.36), we can write

$$_{\text{ph}} = \left[1 - \exp\left(-\frac{h\nu_1}{k_B T}\right)\right]^{-1} \left[1 - \exp\left(-\frac{h\nu_2}{k_B T}\right)\right]^{-1} \cdots \left[1 - \exp\left(-\frac{h\nu_i}{k_B T}\right)\right]^{-1}$$

$$= \prod_{i=1}^{\infty} \frac{1}{1 - \exp(-h\nu_i / k_B T)} \tag{15.37}$$

Taking natural logarithm, we can write

$$\ln Z_{\text{ph}} = -\sum_{i=1}^{\infty} \ln[1 - \exp(-h\nu_i / k_B T)$$

Note that the summation is over all single photon states. Therefore, we have to multiply the expression on the right hand side by the number of states in volume V and in the frequency

$\nu + d\nu$. You may recall that this number is $\dfrac{8\pi V\nu^2\,d\nu}{c^3}$. Hence, on replacing

summation by integration and multiplying by the number of states, we get

$$\ln Z_{\text{ph}} = -\frac{8\pi V}{c^3}\int_0^\infty \nu^2 \ln[1 - \exp(-h\nu/k_B T)\,d\nu$$

To evaluate the integral, we introduce a change of variable by defining $x = h\nu/k_BT$ so that

$h)dx$ and $\nu^2 d\nu = (k_BT/h)^3 x^2 dx$. On making these substitutions, we can write

$$\ln Z_{\text{ph}} = -8\pi V\left(\frac{k_B T}{hc}\right)^3 \int_0^\infty x^2 \ln[1 - \exp(-x)]\,dx$$

On integrating by parts, we get

$$\int_0 x^2 \ln[1 - \exp(-x)]\,dx = -\frac{1}{3}\int_0^\infty \frac{x^3 \exp(-x)}{1 - \exp(-x)}dx = -\frac{1}{3}\int_0^\infty \frac{x^3}{\exp(x) - 1}dx$$

$$= -\frac{1}{3}\Gamma(4)\zeta(4) = -\frac{\pi^4}{45}$$

Hence the expression for partition function simplifies to

$$\ln Z_{\text{ph}} = \frac{8\pi^5}{45}\left(\frac{k_B T}{hc}\right)^3 V \tag{15.38}$$

Thermodynamic functions of a photon gas Using the partition function for a photon gas, we can easily obtain expressions for various thermodynamic quantities. For instance, the chemical potential

$$\mu = -k_B T\left(\frac{\partial \ln Z_{\text{ph}}}{\partial N}\right)_{V,T} = 0 \tag{15.39}$$

Similarly, Helmholtz free energy is given by

$$F = -k_B T \ln Z_{\text{ph}} = -k_B T\,\frac{8\pi^5}{45}\left(\frac{k_B T}{hc}\right)^3 V$$

$$= -\left(\frac{8\pi^5 k_B{}^4 T^4}{45 c^3 h^3}\right)V \tag{15.40}$$

On combining this result with Eq. (15.34), we can express Helmholtz free energy in terms of total energy density u as

$$F = -\frac{uV}{3} \tag{15.41}$$

From this result, it readily follows that radiation pressure

$$p = -\left(\frac{\partial F}{\partial V}\right)_T = \frac{u}{3} = \left(\frac{8\pi^5 k_B{}^4 T^4}{45 c^3 h^3}\right) \tag{15.42}$$

Similarly, we can arrive at the expression for entropy of the photon as

$$S = -\left(\frac{\partial F}{\partial T}\right)_V = \frac{32\pi^5 k_B{}^4 T^3}{45 h^3 c^3} V. \tag{15.43}$$

This result shows that entropy is directly proportional to third power of temperature and tends to zero as $T \rightarrow 0$.

Recall that we derived these results in Chapter 11 also but here we have applied B–E statistics to arrive at the same results. You will agree that this approach is more elegant and powerful.

If radiation is made to undergo an adiabatic change, we know that

$$VT^3 = \text{Constant} \tag{15.44}$$

This result shows that entropy remains constant in a reversible adiabatic expansion. We all know that the universe is pervaded by 3 K blackbody radiation. If the volume of the universe increases by a factor of two in the next 10^{10} years, Eq. (15.44) tells us that the temperature will decrease by a factor of $2^{-1/3}$. So after 10^{10} years, the temperature of the blackbody radiation will be T = (3 K)/$2^{1/3}$ = 2.38 K.

In terms of pressure and volume, the equation for the adiabatics of the system takes

$$pV^{4/3} = \text{Constant} \tag{15.45}$$

From this you may be tempted to conclude that the ratio of heat capacities at constant pressure to that at constant volume is 4/3. But this is not true.

DDITIONAL EXAMPLES

Example 15.1 Calculate the average number of photons in an enclosure of 22.4 litre at 273 K.

Using Eq. (15.30) we can express the average number of photons between frequencies ν and $\nu + d\nu$ as

$$n(\nu)d\nu = \frac{8\pi V}{c^3} \cdot \frac{\nu^2 d\nu}{e^{h\nu/k_B T} - 1} \tag{i}$$

Hence, the average number of photons with frequencies from 0 to ∞ at temperature T is obtained by integrating (i) over all possible values of ν:

$$N(T) = \int_0^\infty n(\nu)d\nu = \frac{8\pi V}{c^3}\int_0^\infty \frac{\nu^2 d\nu}{e^{h\nu/k_B T} - 1} \tag{ii}$$

To evaluate the integral, we introduce a change of variable by defining

$$x = \frac{h\nu}{k_B T}$$

$d = \frac{h}{k_B T} d\nu$ and $\nu^2 d\nu = \left(\frac{k_B T}{h}\right)^3 x^2\, dx$. Using these results in Eq. (ii), we get

$$N(T) = \frac{8\pi V}{c^3}\left(\frac{k_B T}{h}\right)^3 \int_0^\infty \frac{x^2 dx}{e^x - 1} \tag{iii}$$

The integral occurring in (iii) is a standard Riemann zeta function of order 3:

$$\int_0^{\infty} \frac{x^2\,dx}{e^x - 1} = \zeta(3)\Gamma(3)$$

Hence, the average number of photons with all possible frequencies in an enclosure at a temperature T is given by

$$N(T) = 16\pi V\left(\frac{k_B T}{hc}\right)^3 \zeta(3)$$

$$= 16\pi V\left(\frac{k_B T}{hc}\right)^3 \times 1.162$$

= 22.4 litre = 22. $\times$ 10^{-3}m^3 and T = 273 K. Using the standard values of k_B, h and

$$(\quad) = 16 \times 3.1417 \times (22.4 \times 10^{-3}\ \text{m}^3) \times \left(\frac{(1.38 \times 10^{-23}\ \text{JK}^{-1}) \times (273\text{K})}{(6.62 \times 10^{-34}\ \text{Js}) \times (3 \times 10^8\ \text{ms}^{-1})}\right)^3 \times 1.162$$

$$= 1308.39 \times 10^{-3} \times (18.97 \times 10^3)^3$$

$$= 1308.39 \times 6826.33 \times 10^6$$

$$= 8.93 \times 10^{12}.$$

Example 15.2 Blackbody radiation in an enclosure is made to undergo adiabatic change at a pressure of 1.49 $\times$ 10^{-2} Nm^{-2}. Calculate the final pressure, if the volume of the enclosure **(a)** increases by a factor of 2, and **(b)** decreases by a factor of 2.

From Eq. (15.45), we recall that when blackbody radiation is made to undergo adiabatic change, its pressure and volume are related as

$$pV^{4/3} = \text{Constant}$$

When volume increases by a factor of 2, we can write

$$p_f V_f^{4/3} = p_i V_i^{4/3}$$

$$p_f = p_i \left(\frac{V_i}{V_f}\right)^{4/3}$$

$$= (1.49 \times 10^{-2}\ \text{Nm}^{-2}) \times \left(\frac{1}{2}\right)^{4/3}$$

$$= 0.59 \times 10^{-2}\ \text{Nm}^{-2}.$$

When volume decreases by a factor of 2, we get

$$p_f = (1.49 \times 10^{-2}\ \text{Nm}^{-2}) \times 2^{4/3}$$

$$= 3.75 \times 10^{-2}\ \text{Nm}^{-2}.$$

Example 15.3 Consider a non-interacting, non-relativistic, ideal boson gas. Discuss whether the B-E condensation observed with a 3-*D* gas will occur for a 2-*D* gas. Justify your answer.

For a 3-*D* gas, B-E condensation occurs when $A = 1$ or $\mu = 0$. Moreover, the number of particles is given by

$$N(3-D) = \frac{2\pi(2m)^{3/2}}{h^3} V \int_0^{\infty} \frac{\varepsilon^{1/2}\, d\varepsilon}{\exp(\varepsilon/k_B T) - 1}$$

D gas, this expression modifies as

$$N(2-D) = \frac{2\pi mA}{h^2} \int_0^{\infty} \frac{d\varepsilon}{\exp[(\varepsilon - \mu)/k_B T] - 1}$$

We can rewrite it as

$$N(2-D) = \frac{2\pi mA}{h^2} \int_0^{\infty} \left(\sum_{\ell=1}^{\infty} \exp[-\ell(\varepsilon - \mu)/k_B T] \right) d\varepsilon$$

$$= \frac{2\pi mA}{h^2} \sum_{\ell=1}^{\infty} \exp(\mu\ell/k_B T) \int_0^{\infty} \sum_{\ell=1}^{\infty} \exp(-\varepsilon/k_B T)\, d\varepsilon$$

$$= \frac{2\pi mA}{h^2} k_B T \sum_{\ell=1}^{\infty} \exp(\mu\ell/k_B T) \int_0^{\infty} \sum_{\ell=1}^{\infty} \frac{\exp(-\ell x)}{\ell}\, dx$$

$$= \frac{2\pi mA}{h^2} k_B T \sum_{\ell=1}^{\infty} e^{\mu\ell/k_B T} \zeta(2)$$

$= 0$ the RHS diverges. However, since $\mu \neq 0$, B-E condensation does not occur in B-E gas.

Example 15.4 Discuss B-E condensation in the high density limit at a fixed temperature as *V* is reduced.

As *V* is reduced, μ decreases and at an initial value $V = V_c$, $\mu = 0$, say. Then we can write

$$N = \frac{4\pi V_c}{h^3} \left(\int_0^{\infty} \frac{p^2\, dp}{\exp(p^2/2mk_B T) - 1} \right) + n_0$$

The number of particles in the ground state is

$$n_0 = N\left(1 - \frac{V}{V_c}\right).$$

Example 15.5 Show that

$$P - \frac{NkT}{V} = \begin{cases} \leq 0 & \text{for a B.E gas} \\ \geq 0 & \text{for a F.D gas} \end{cases}$$

We know that pressure exerted by a quantum gas is given by

$$p^{\text{FD}}_{\text{BE}} = \pm\frac{k_B T}{V}\sum_i \ln\left\{1 \pm e^{\beta(\mu-\varepsilon_i)}\right\}$$

Similarly, the number of particles is given by

$$N^{\text{FD}}_{\text{BE}} = \sum_i n_i = \sum\left\{e^{\beta(\varepsilon_i-\mu)} \pm 1\right\}^{-1}$$

For a F D gas,

$$p - \frac{Nk_B T}{V} = -\frac{k_B T}{V}\sum\left\{\ln(1-n_i) + n_i\right\}$$

$0 \le \quad \le 1$, the quantity within the bracket is –ve and hence, we note that

$$p - \frac{Nk_B T}{V} \ge 0.$$

Similarly, for a BE gas we can write

$$p - \frac{Nk_B T}{V} = \frac{k_B T}{V}\sum\left\{\ln(1+n_i) - n_i\right\}$$

≥ 0, the quantity within the bracket is –ve

$$p - \frac{Nk_B T}{V} \le 0.$$

MMARY

The phenomenon in which bosons gradually accumulate in the lower energy levels and a large number of them may occupy the ground state ($\varepsilon = 0$) is referred to as *Bose–Einstein condensation*.

The condition for the onset of Bose–Einstein condensation is written as

$$N > N_{ex} = 2.612\,V\left(\frac{2\pi m k_B T_c}{h^2}\right)^{3/2}$$

T_c is known as the *Bose–Einstein condensation temperature.*

The number of particles in the ground state N_g, is given by

$$N_g = N - N_{ex} = N\left[1 - \left(\frac{T}{T_c}\right)^{3/2}\right]$$

The thermodynamic functions of the condensate are given by

$$E_g = 0.7701 N k_B\left(\frac{T^{5/2}}{T_c^{\,3/2}}\right) \text{ for } T < T_c$$

$$p(T) = \frac{2}{3}\frac{E_g}{V} = \left(\frac{2\pi m}{h^2}\right)^{3/2} (k_B T)^{5/2} \zeta\left(\frac{5}{2}\right),$$

$$C_V = 1.925\, N k_B \left(\frac{T}{T_c}\right)^{3/2}$$

$$S = 1.28\, N k_B \left(\frac{T}{T_c}\right)^{3/2}$$

Landau gave a simple theoretical model to explain the behaviour of liquid He-II in the range $0 < T < 2$ K. According to him, liquid He-II is devoid of all motion at absolute zero but elementary excitations begin to appear, though discontinuously, for > 0. In the range $0 < T < 0.5$ K, these excitations are termed as *phonons* whereas in the range $0.5 < T < 2$ K, the excitations are termed as *rotons*.

The contributions of phonons and rotons to specific heat capacity and entropy of liquid He-II are respectively given by

$$(C_V)_{\text{phonon}} = \frac{16\pi^5 V k_B{}^4}{15(u_1 h)^3} T^3$$

$$(S)_{\text{phonon}} = \frac{16\pi^5 V k_B{}^4}{45(u_1 h)^3} T^3$$

$$(C_V)_{\text{roton}} = N_r k_B \left[\frac{3}{4} + \frac{\Delta}{k_B T} + \left(\frac{\Delta}{k_B T}\right)^2\right]$$

$$(S)_{\text{roton}} = N_r k_B \left[\frac{3}{2} + \frac{\Delta}{k_B T}\right]$$

u_1 denotes the speed of sound and N_r is number of rotons.

The partition function of phonons is given by

$$Z_{\text{ph}} = \prod_{i=1}^{\infty} \frac{1}{1 - \exp(-h\nu_i / k_B T)}$$

The Helmholtz free energy of photons is given by

$$F = -\left(\frac{8\pi^5 k_B{}^4 T^4}{45 c^3 h^3}\right) V$$

ERCISES

Imagine the universe to be a spherical cavity with a radius of 10^{28} cm and impenetrable walls. If the temperature of the cavity is 3 K, calculate the total number of photons in the universe and their energy content. (*Ans:* 2.5×10^{87}; 2.6×10^{65} J)

Show that chemical potential of an assembly of bosons is given by $\mu = -k_B T/N$ with ground state at $\varepsilon = 0$. Also calculate μ at 10^{-3} K, if $N = 10^{22}$. (*Ans:* -1.4×10^{-34} J)

Appendix I

ANALYSIS OF EXPERIMENT OF ZARTMANN AND KO

Refer to Fig. I.1, which is a schematic representation of the equipment used by Zartmann and Ko to verify Maxwell's law of distribution of speeds. Suppose that the circular drum D makes μ revolutions in one second. If the diameter of the chamber is d_0, its circumference will be πd_0. So in one second the total distance covered by the drum is $\pi d_0 \mu$. The distance from the centre of the glass plate at which a molecule moving with speed v will strike it is obtained by multiplying the total distance covered in one second by the time taken to go from the entrance to the glass plate. Hence, $s = \pi d_0 \mu (d_0/v)$.

We know that the number of molecules crossing a hole is proportional to their speed as well as to the number per unit volume. Therefore, we can write

$$dN^* = Kv\,dN_v \qquad (I.1)$$

where K is the constant of proportionality, and $dn_v = 4\pi n A^3 \exp(-Bv^2) v^2\, dv$.

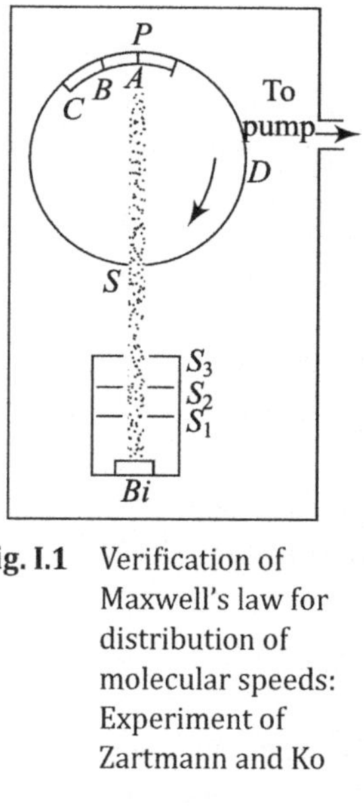

Fig. I.1 Verification of Maxwell's law for distribution of molecular speeds: Experiment of Zartmann and Ko

To calculate the constant of proportionality, we integrate Eq. (I.1) for all values of v and note that $\int dN^* = N^*$. Then we obtain

$$N^* = 4\pi n K A^3 \int_0^\infty \exp(-Bv^2)\, v^3\, dv$$

From Chapter 1, you may recall that the integral

$$\int_0^\infty e^{-Bv^2} v^3\, dv = \frac{1}{2B^2}\Gamma(2) = \frac{1}{2B^2}$$

so that

$$N^* = \frac{4\pi n K A^3}{2B^2}$$

and the constant of proportionality is given by

$$K = \frac{B^2}{2\pi n A^3} N^*$$

On substituting this value in Eq. (I.1), we get

$$dN^* = \frac{4\pi n A^3 B^2}{2\pi n A^3} N^* e^{-Bv^2} v^3\, dv$$

$$= 2B^2 N^* e^{-Bv^2} v^3\, dv \qquad (I.2)$$

We are interested in the number of molecules deposited along the glass plate P. Let us assume that the molecules having speeds in the range v to $v + dv$ will be deposited at distances between s and $s + ds$ from the centre of the plate where s and v are connected through the relation

$$s = \frac{\pi d_0^2 \mu}{v}$$

From this, we can write

$$\left|\frac{ds}{dv}\right| = \frac{\pi d_0^2 \mu}{v^2}$$

Also

$$\frac{dN^*}{ds} = \frac{dN^*}{dv}\frac{dv}{ds} = \frac{v^2}{\pi d_0^2 \mu}\frac{dN^*}{dv}$$

Hence, the number of particles deposited along the plate can be written as a function of v:

$$\frac{dN^*}{ds} = \frac{2B^2 N^* e^{-Bv^2} v^5}{\pi d_0^2 \mu}$$

Since $v_p^2 = \frac{1}{B}$, we can write this expression in terms of the most probable speed v_p as

$$\frac{dN^*}{ds} = \frac{2N^*}{\pi d_0^2 \mu}\frac{v^5}{v_p^4} e^{-(v^2/v_p^2)}$$

$$= \frac{2N^*}{\pi d_0^2 \mu}\left(\frac{v}{v_p}\right)^5 v_p\, e^{-(v^2/v_p^2)} \tag{I.3}$$

If molecules moving with speeds v and v_p are deposited at distances s and s_p respectively from the centre of the glass plate, we can express the number of particles depositing on the plate per unit distance (i.e., the intensity of deposit) as

$$\frac{dN^*}{ds} = \frac{2N^*}{s_p}\left(\frac{s_p}{s}\right)^5 e^{-(s_p/s)^2} \tag{I.4}$$

To evaluate the number of particles corresponding to a given range of speeds, i.e., deposited in the distance range from $s - a$ to $s + a$ on the plate, we integrate Eq. (I.4) with respect to s. This gives

$$N_s^* = \frac{2N^*}{s_p}\int_{s-a}^{s+a}\left(\frac{s_p}{s}\right)^5 \exp[-(s_p/s)^2]\, ds$$

To evaluate this integral, we introduce a change of variable by putting

$$(s_p/s)^2 = t$$

so that

$$ds = -\frac{s^3}{2s_p^2}dt$$

Note that the limits of integration change to $\left(\dfrac{s_p}{s-a}\right)^2$ and $\left(\dfrac{s_p}{s+a}\right)^2$ .Using these in the expression for the number of particles corresponding to a given velocity range, we get

$$N_s^* = N^* s_p \int_{(s_p/s-a)^2}^{(s_p/s+a)^2} t\exp[-t)dt$$

This can be evaluated by method of integration by parts. The result is

$$N_s^* = N^* s_p \left[\left\{ \left(\frac{s_p}{s-a} \right)^2 + 1 \right\} e^{-\left(\frac{s_p}{s-a}\right)^2} - \left\{ \left(\frac{s_p}{s+a} \right)^2 + 1 \right\} e^{-\left(\frac{s_p}{s+a}\right)^2} \right] \tag{I.5}$$

Appendix

II

MEAN FREE PATH: MAXWELLIAN GAS

To calculate $\overline{v}_r$, we first determine the magnitude of relative velocity $(v_r)_{12}$ of one molecule of velocity $\mathbf{v}_1$ with respect to all other molecules of velocity $\mathbf{v}_2$. This expression is averaged for all values of $\mathbf{v}_2$ from 0 to ∞. Next, this mean value is multiplied by the probability that the molecule has velocity $\mathbf{v}_1$ and then averaged for all values of $\mathbf{v}_1$ from 0 to ∞. So, let us consider two molecules of a Maxwellian gas whose velocity vectors ($\mathbf{v}_1$ and $\mathbf{v}_2$) are inclined at an angle θ. Then we can write

$$(v_r)_{12} = \sqrt{v_1^2 + v_2^2 - 2 v_1 v_2 \cos\theta} \tag{II.1}$$

We know that all directions are equally probable for $\mathbf{v}_2$. Therefore, to calculate the average value of $(v_r)_{12}$, we must multiply it by the probability that v_2 lies within a solid angle dΩ. Hence,

$$\overline{(v_r)_{12}} = \int_{\Omega} \frac{d\Omega}{4\pi}(v_r)_{12}$$

We assume azimuthal symmetry and write $d\Omega = 2\pi \sin\theta\, d\theta$. Hence, on substituting the value of $(v_r)_{12}$ from Eq. (II.1), we can write

$$\overline{(v_r)_{12}} = \frac{1}{2}\int_0^{\pi} d\theta \sin\theta (v_1^2 + v_2^2 - 2 v_1 v_2 \cos\theta)^{1/2} \tag{II.2}$$

To evaluate this integral, we introduce change of variable and put $\cos\theta = x$ so that $\sin\theta\, d\theta = -dx$. Hence, Eq. (II.2) takes the form

$$\overline{(v_r)_{12}} = \frac{1}{2}\int_1^{-1} dx (v_1^2 + v_2^2 - 2 v_1 v_2 x)^{1/2}$$

Note that the limits of integration have changed from $(0, \pi)$ to $(1, -1)$. This integral can readily be evaluated giving

$$\overline{(v_r)_{12}} = -\frac{1}{2}\left(\frac{2}{3}\right) \left. \frac{(v_1^2 + v_2^2 - 2 v_1 v_2 x)^{3/2}}{2 v_1 v_2} \right|_{-1}^{+1}$$

or

$$\overline{(v_r)_{12}} = \frac{1}{6 v_1 v_2}(-(v_1^2 + v_2^2 - 2 v_1 v_2)^{3/2} + (v_1^2 + v_2^2 + 2 v_1 v_2)^{3/2}$$

$$= \frac{1}{6 v_1 v_2}[(v_1 + v_2)^3 - |v_1 - v_2|^3] \tag{II.3}$$

You will note that we have written $|v_1 - v_2|^3$ for $\left((v_1 - v_2)^2\right)^{3/2}$ instead of $(v_1 - v_2)^3$. This is because of the fact that $\overline{(v_r)_{12}}$ should always be positive. Therefore, irrespective of the values that v_1 and v_2 take, $\left((v_1 - v_2)^2\right)^{3/2}$ must also be positive. So for $v_1 > v_2$, $|v_1 - v_2|^3 = (v_1 - v_2)^3$ and the angle averaged value of $\overline{(v_r)_{12}}$ is given by

$$\overline{(v_r)_{12}} = \frac{v_2^2 + 3\,v_1^{\,2}}{3\,v_1} \tag{II.4a}$$

Similarly, for $v_2 > v_1$, $|v_1 - v_2|^3 = (v_2 - v_1)^3$ and $\overline{(v_r)_{12}}$ is given by

$$\overline{(v_r)_{12}} = \frac{v_1^2 + 3\,v_2^2}{3\,v_2} \tag{II.4b}$$

The average relative velocity of a molecule having velocity $\mathbf{v}_1$ with respect to all other molecules moving with velocity $\mathbf{v}_2$ lying in the range 0 to ∞ is

$$\overline{(v_r)_1} = \int_0^\infty P(v_2)\,\overline{(v_r)_{12}}\,\mathrm{d}v_2 \tag{II.5}$$

where

$$P(v_2)\,\mathrm{d}v_2 = 4\pi A^3 \exp(-Bv_2^2)\,v_2^2\,\mathrm{d}v_2$$

gives the probability that there exists a molecule with velocity in the range v_2 to $v_2 + \mathrm{d}v_2$.

Note that $\overline{(v_r)_{12}}$ has two distinct values depending upon whether $v_1 > v_2$ or $v_2 > v_1$. Therefore, we break the range of integration from 0 to ∞ in two parts: 0 to v_1 and v_1 to ∞. Then we can rewrite Eq. (II.5) as

$$\overline{(v_r)_1} = 4\pi A^3 \left[\int_0^{v_1} \mathrm{d}v_2 \exp(-Bv_2^2)\left(\frac{3\,v_1^2 + v_2^2}{3\,v_1}\right) v_2^2 + \int_{v_1}^{\infty} \mathrm{d}v_2 \exp(-Bv_2^2)\,v_2^2 \left(\frac{3\,v_2^2 + v_1^2}{3\,v_2}\right)\right] \tag{II.6}$$

To arrive at the average relative velocity of any molecule with respect to any other molecule in the gas, we multiply Eq. (II.6) by the probability that a molecule has velocity components in the range v_1 to $v_1 + \mathrm{d}v_1$. Hence, we can write

$$\overline{v_r} = \int_0^\infty P(v_1)\,\overline{(v_r)_1}\,\mathrm{d}v_1$$

$$= (4\pi A^3)^2 \left[\int_0^\infty \mathrm{d}v_1 \exp(-Bv_1^2)\,v_1^2 \left(\int_0^{v_1} \mathrm{d}v_2 \exp(-Bv_2^2)\,v_2^2 \left(\frac{3\,v_1^2 + v_2^2}{3\,v_1}\right) + \int_{v_1}^{\infty} \mathrm{d}v_2 \exp(-Bv_2^2)\,v_2^2 \left(\frac{3\,v_2^2 + v_1^2}{3\,v_2}\right)\right)\right] \tag{II.7}$$

where we have substituted $P(v_1)\mathrm{d}v_1 = 4\pi A^3\,v_1^2 \exp(-Bv_1^2)\mathrm{d}v_1$.

This expression may be rewritten as

$$\bar{v}_r = (4\pi A^3)^2 \left(\int_0^\infty dv_1 \exp(-Bv_1^2)v_1^2 \int_0^{v_1} e^{-Bv_2^2} v_2^2 \left(\frac{3v_1^2 + v_2^2}{3v_1} \right) dv_2 \right.$$

$$\left. + \int_0^\infty dv_1 \exp(-Bv_1^2)v_1^2 \int_{v_1}^\infty e^{-Bv_2^2} v_2^2 \left(\frac{3v_2^2 + v_1^2}{3v_2} \right) dv_2 \right) \quad \text{(II.8)}$$

$$= (4\pi A^3)^2 (I_1 + I_2)$$

where

$$I_1 = \int_0^\infty dv_1 \exp(-Bv_1^2)v_1^2 \int_0^{v_1} \exp(-Bv_2^2) v_2^2 \left(\frac{3v_1^2 + v_2^2}{3v_1} \right) dv_2 \quad \text{(II.9a)}$$

and

$$I_2 = \int_0^\infty dv_1 \exp(-Bv_1^2) v_1^2 \int_{v_1}^\infty \exp(-Bv_2^2) v_2^2 \left(\frac{3v_2^2 + v_1^2}{3v_2} \right) dv_2 \quad \text{(II.9b)}$$

It may be mentioned here that the integrals I_1 and I_2 are equal. To show this, we note that in Eq. (II.9a), v_1 runs from 0 to ∞ and v_2 runs from 0 to v_1. That is, the upper limit of v_2 is v_1. So, if we interchange the order of integration in Eq. (II.9a) and then correspondingly change the limits of integration over v_1 from v_2 to ∞, the minimum value of v_1 remains v_2 (which varies from 0 to ∞). Moreover, the resultant expression for I_1 will be same as that for I_2. Hence, the expression for relative velocity simplifies to

$$\bar{v}_r = 2(4\pi A^3)^2 I_2 \quad \text{(II.10)}$$

To evaluate I_2, let us first consider the integral

$$I = \int_{v_1}^\infty \exp(-Bv_2^2) v_2^2 \left(\frac{3v_2^2 + v_1^2}{3v_2} \right) dv_2 \quad \text{(II.11)}$$

and break it into two parts:

$$I = \int_{v_1}^\infty \exp(-Bv_2^2) v_2^3 \, dv_2 + \frac{v_1^2}{3} \int_{v_1}^\infty \exp(-Bv_2^2) v_2 \, dv_2$$

Now we introduce a change of variable and put $Bv_2^2 = x$, So that $2Bv_2 dv_2 = dx$ and the lower limit of integration changes from v_1 to Bv_1^2. Then, we can write

$$I = \frac{1}{2B^2} \int_{Bv_1^2}^\infty \exp(-x)\, x\, dx + \frac{v_1^2}{6B} \int_{Bv_1^2}^\infty \exp(-x)\, dx$$

Since $\int_a^b x e^{-x} dx = -(x+1)\, e^{-x} \Big|_a^b$, the expression for integral I simplifies to

$$I = \frac{1}{2B^2}(Bv_1^2 + 1)\exp(-Bv_1^2) + \frac{v_1^2}{6B}\exp(-Bv_1^2)$$

$$= \frac{1}{2B^2}\exp(-Bv_1^2) + \frac{2}{3}\frac{v_1^2 \exp(-Bv_1^2)}{B} \quad \text{(II.12)}$$

On substituting this result in Eq. (II.9b), we get

$$I_2 = \int_0^\infty dv_1 \exp(-2Bv_1^2)\, v_1^2 \left(\frac{1}{2B^2} + \frac{2}{3}\frac{v_1^2}{B}\right)$$

$$= \frac{1}{2B^2}\int_0^\infty dv_1 \exp(-2Bv_1^2)\, v_1^2 + \frac{2}{3B}\int_0^\infty dv_1 \exp(-2Bv_1^2)\, v_1^4 \qquad \text{(II.13)}$$

Now we use the standard integral

$$\int_0^\infty \exp(-Bv^2) v^n\, dv = \frac{1}{2B^{(n+1)/2}}\Gamma\left(\frac{n+1}{2}\right)$$

and simplify the resultant expression to obtain a compact value of I_2:

$$I_2 = \frac{1}{(2B)^{7/2}}\Gamma(3/2) + \frac{1}{(2B)^{7/2}}\Gamma(3/2)$$

$$= \frac{\sqrt{\pi}}{(2B)^{7/2}} \qquad \text{(II.14)}$$

since $\Gamma(3/2) = \sqrt{\pi}/2$.

Hence, the magnitude of average relative speed of any one Maxwellian gas molecule with respect to any other molecule is

$$\overline{v_r} = 2(4\pi A^3)^2 \frac{\sqrt{\pi}}{(2B)^{7/2}}$$

Since $A^6 = (B/\pi)^3$, this expression simplifies to

$$\overline{v_r} = \frac{2\sqrt{2}}{\sqrt{\pi B}}$$

We know that average speed of any molecule of a Maxwellian gas at temperature T is given by $\overline{v} = \sqrt{8k_B T/m\pi}$ and $B = m/(2k_B T)$. Hence, we note that

$$\sqrt{\pi B} = \frac{2}{\overline{v}}$$

so that

$$\frac{\overline{v}}{\overline{v_r}} = \frac{1}{\sqrt{2}}$$

Using this result in Eq. (2.3), we get the expression for mean free path for a Maxwellian gas molecule:

$$\lambda = \frac{1}{\sqrt{2}}\frac{1}{\pi d^2 n} \qquad \text{(II.15)}$$

Appendix III

RIGOROUS PROOF OF KIRCHHOFF'S LAW

Consider a body of emissive power e_λ and absorptive power a_λ placed in an enclosure at temperature T, as shown in Fig. III.1. The enclosure will be filled with radiation emitted by its walls and in due course of time, the body will attain the same temperature as that of the walls. This happens because in the equilibrium state, the radiation inside the enclosure is isotropic.

Let the emissive power of the walls of the enclosure be E_λ. Then the amount of energy emitted by a small element of area ds' of the enclosure in the direction of an elementary area ds of the body per second is given by

$$dQ_\lambda = E_\lambda\, d\lambda ds' \cos\theta' \frac{ds\cos\theta}{r^2}$$

$$= E_\lambda\, d\lambda ds \cos\theta d\Omega'$$

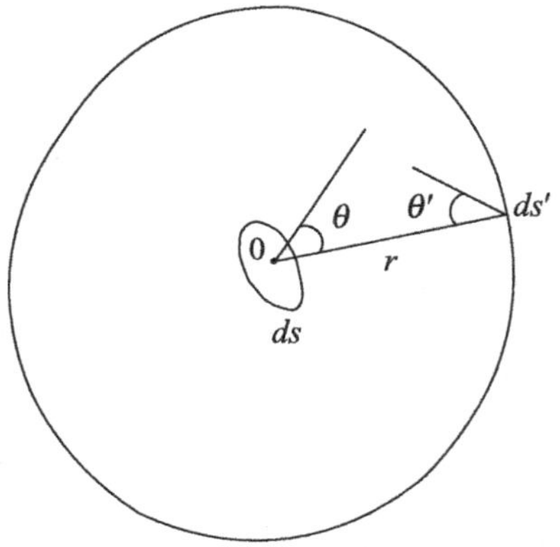

Fig. III.1 A body of emissive power e_λ placed in an enclosure of arbitrary shape.

where $d\Omega'$ is the solid angle subtended by ds' at ds. If we denote the azimuthal angle with respect to origin by ϕ, we can rewrite the expression for the amount of energy emitted by a small element of area ds' as

$$dQ_\lambda = E_\lambda\, d\lambda ds \cos\theta \sin\theta d\theta d\phi \tag{III.1}$$

Of this, $a_\lambda dQ_\lambda$ will be absorbed by ds and we can write

$$a_\lambda\, dQ_\lambda = a_\lambda E_\lambda\, d\lambda ds \cos\theta \sin\theta d\theta d\phi$$

The total energy absorbed by this elementary area of the body is obtained by integrating the expression on the right-hand side over all possible values of θ, ϕ and λ. This gives

$$Q_{ds} = \int_0^{\pi/2} d\theta \int_0^{2\pi} d\phi \int_0 d\lambda\, a_\lambda E_\lambda\, ds\cos\theta\sin\theta$$

$$= ds\int_0^{\infty} E_\lambda a_\lambda d\lambda \int_0^{\pi/2} \cos\theta\sin\theta d\theta \int_0^{2\pi} d\phi$$

$$= ds\left(\int_0^{\infty} E_\lambda a_\lambda d\lambda\right)\times\left(\frac{1}{2}\right)\times(2\pi)$$

$$= \pi ds\int_0^{\infty} E_\lambda a_\lambda d\lambda \tag{III.2}$$

Hence, the expression for radiation absorbed by the whole body is obtained by summing over all elements such as d*s*:

$$Q = \sum_i \pi \mathrm{d}s_i \int_0^\infty E_\lambda a_\lambda \,\mathrm{d}\lambda \qquad \text{(III.3)}$$

Let us now calculate the amount of energy emitted by the elementary area d*s* of the body in the direction θ and $\theta + \mathrm{d}\theta$ and ϕ and $\phi + \mathrm{d}\phi$. It is given by

$$\mathrm{d}I_\lambda = e_\lambda \,\mathrm{d}\lambda \mathrm{d}s \cos\theta \sin\theta \mathrm{d}\theta \mathrm{d}\phi$$

As before, we can obtain an expression for total energy emitted by this elementary area by integrating the expression on the right hand side over all possible values of θ, ϕ and λ. This gives

$$I_{\mathrm{d}s} = \int_0^{\pi/2} \mathrm{d}\theta \int_0^{2\pi} \mathrm{d}\varphi \int_0^\infty \mathrm{d}\lambda\, e_\lambda \,\mathrm{d}s \cos\theta \sin\theta$$

$$= \mathrm{d}s \int_0^\infty e_\lambda \,\mathrm{d}\lambda \int_0^{\pi/2} \cos\theta \sin\theta \mathrm{d}\theta \int_0^{2\pi} \mathrm{d}\phi$$

$$= \mathrm{d}s \left(\int_0^\infty e_\lambda \,\mathrm{d}\lambda \right) \times \left(\frac{1}{2} \right) \times (2\pi)$$

$$= \pi \mathrm{d}s \int_0^\infty e_\lambda \,\mathrm{d}\lambda \qquad \text{(III.4)}$$

Hence, the radiation emitted by the whole body is obtained, as before, by summing over all elements such as d *s*:

$$I = \sum_i \pi \mathrm{d}s_i \int_0^\infty e_\lambda \,\mathrm{d}\lambda \qquad \text{(III.5)}$$

When the system is in thermal equilibrium, the energy gained by the body equals the energy emitted by it. Hence, we can write

$$\sum_i \pi \mathrm{d}s_i \int_0^\infty E_\lambda a_\lambda \,\mathrm{d}\lambda = \sum_i \pi \mathrm{d}s_i \int_0^\infty e_\lambda \,\mathrm{d}\lambda$$

or

$$\int_0^\infty E_\lambda a_\lambda \,\mathrm{d}\lambda = \int_0^\infty e_\lambda \,\mathrm{d}\lambda \qquad \text{(III.6)}$$

We can rewrite it as

$$\int_0^\infty (e_\lambda - E_\lambda a_\lambda)\, d\lambda = 0$$

Since λ is arbitrary, this relation will hold only if the integrand vanishes identically. Hence, we get

$$\frac{e_\lambda}{a_\lambda} = E_\lambda \qquad \text{(III.7)}$$

This is the mathematical statement of Kirchhoff's law.

Proceeding further, we note that the left hand side of Eq. (III.6) is a measure of the total energy absorbed by the body for all wavelengths per unit area per unit time. We denote it by $E_1 a_1$, where E_1 and a_1, respectively denote the emissive power of the enclosure and the absorptive power of the body for all wavelengths. Similarly, we can interpret the right-hand side as the total emission for all wavelengths from unit area in unit time. We denote it by e_1. Hence, we can write

$$\frac{e_1}{a_1} = E_1 \tag{III.8}$$

If you place another body in another enclosure at the same temperature, you will obtain a similar relation:

$$\frac{e_1}{a_1} = E_2 , \tag{III.9}$$

where E_2 is emissive power of the second enclosure.

On comparing the results contained in Eqs. (III.8) and (III.9), we get

$$E_1 = E_2 \tag{III.10}$$

That is, the radiation at the same temperature is independent of the nature or shape of the enclosure.

If we place a blackbody inside the enclosure, $a_1 = 1$ and $e_1 = E_1$. It means that the radiation in a hollow enclosure is identical to the radiation from a blackbody kept at the same temperature. So we can conclude that *radiation inside a hollow enclosure is independent of the nature or shape of its walls and is identical to the blackbody radiation at the same temperature*. Mathematically, we express this result as

$$\frac{e_1}{a_1} = E_1 = e_{BB} \tag{III.11}$$

where e_{BB} is emissive power of the blackbody.

Appendix IV

TRANSMISSION OF HEAT BY CONDUCTION

We know that heat is a form of energy in transit. As mentioned in Chapter 11, transmission of thermal energy can occur through conduction, convection and radiation. Of these modes, conduction is dominant in solids. That is why if you hold one end of a metallic rod in your hand and the other end is heated, you feel warmth. In fact, even when there is difference in temperature between two parts of the same substance, transfer of heat continues till such time that the entire body is at the same temperature. The mechanism of heat transfer by conduction is very simple: The layer of particles (atoms/molecules) in contact with the source of thermal energy gets energised and the particles begin to vibrate about their respective mean equilibrium positions. The vibrating particles part with some of their energy and give it to the particles in the adjacent layer. These particles are, in turn, set into vibrations and the process continues leading to heat transfer without any observable movement of the constituent particles. So we say that *transfer of heat from a higher temperature to a lower temperature in a substance without any observable motion of the constituent particles is known as conduction.*

In your earlier classes, you have learnt about good and bad conductors; while metals are good conductor of heat, wood is a bad conductor of heat. That is, different substances have different abilities to conduct heat. The qualitative measurement of this ability is expressed in terms of *thermal conductivity*, K. Thermal conductivity of a substance is defined as *the rate of flow of heat normally through a slab of unit thickness and unit area of cross-section when temperature difference between the opposite faces is unity.* Mathematically, we can write

$$K = \frac{Q}{At}\frac{dx}{dT} \tag{IV.1}$$

where Q denotes the heat flowing through a slab of thickness dx and area of cross-section A in time t and the two ends of the slab are at a temperature difference dT.

Good conductors are characterised by high K and vice versa. For a perfect conductor, $K = \infty$, whereas for a non-conductor, $K = 0$. The SI unit of thermal conductivity is J m^{-1} K^{-1} s^{-1}. We can also express it as W $m^{-1}K^{-1}$. It may be mentioned here that Eq. (IV.1) holds when flow of heat is in steady state.

By rearranging terms and noting that rate of flow of heat can be expressed as $\frac{dQ}{dt}$, we get

$$\frac{dQ}{dt} = -KA\frac{dT}{dx} \tag{IV.2}$$

From calorimetry, we know that

$$\frac{dQ}{dt} = ms\frac{dT}{dt} \tag{IV.3}$$

By combining Eqs. (IV.2) and (IV.3), we can write

$$ms\frac{dT}{dt} = -KA\frac{dT}{dx} \tag{IV.4}$$

Since density ρ of a substance is mass per unit volume, the LHS may be expressed as volume integral $\iiint_V \rho s \frac{\partial T}{\partial t} dV$, where V is the volume enclosed by a closed surface, S, as shown in Fig. IV.1. Note that time derivative inside the volume integral is partial because the process of differentiation does not intrude with the space coordinates of the volume.

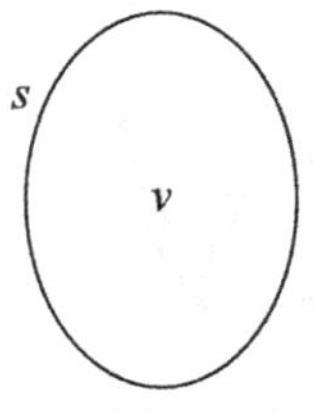

Fig. IV.1 Volume element V surrounded by surface S.

Next, we express the temperature gradient as $\nabla T \left(= \hat{\mathbf{i}}\frac{\partial T}{\partial x} + \hat{\mathbf{j}}\frac{\partial T}{\partial y} + \hat{\mathbf{k}}\frac{\partial T}{\partial z} \right)$ and extend the area of cross-section A to that closed surface. Note that dV is an elementary part of V and is enclosed by **dS**, which points aling the outward drawn normal. In the 1-D case, $\frac{\mathrm{d}T}{\mathrm{d}x}$ is the gradient along x-direction and A is an area normal to the x-direction. For 3-D case, we can write

$$\iiint_V \rho s \frac{\partial T}{\partial t} \mathrm{d}V = \oiint_S (-K\nabla T) \cdot \mathbf{dS}$$

or

$$\iiint_V \frac{\partial T}{\partial t} \mathrm{d}V = -\frac{K}{\rho s} \oiint_S \nabla T \cdot \mathbf{dS} \tag{IV.5}$$

Now we write $\nabla T \cdot \mathbf{dS}$ in component form:

$$\nabla T \cdot \mathbf{dS} = \frac{\partial T}{\partial x}\mathrm{d}S_x + \frac{\partial T}{\partial y}\mathrm{d}S_y + \frac{\partial T}{\partial z}\mathrm{d}S_z \tag{IV.6}$$

Note that $A\frac{\mathrm{d}T}{\mathrm{d}x}$ appearing on the RHS of Eq. (IV.4) is effectively the first term of the RHS of Eq. (IV.6). To simplify Eq. (IV.5), we now use Gauss' Divergence Theorem, which relates the volume and surface integrals as

$$\oiint_S (-\nabla T) \cdot \mathbf{dS} = \iiint_V \nabla \cdot \nabla T \mathrm{d}V = \iiint_V \nabla^2 T \mathrm{d}V$$

Here ∇^2 is the Laplacian: $\nabla^2 = \frac{\partial^2}{\partial x^2} + \frac{\partial^2}{\partial y^2} + \frac{\partial}{\partial z^2}$. Hence, Eq. (IV.5) can be rewritten as

$$\iiint_V \frac{\partial T}{\partial t} \mathrm{d}V + \frac{K}{\rho s} \iiint_V \nabla^2 T \mathrm{d}V = 0$$

or

$$\iiint_V \left(\frac{\partial T}{\partial t} + \frac{K}{\rho s} \nabla^2 T \right) dV = 0$$

Since V is arbitrary, this relation will be satisfied only if the integrand is zero:

$$\frac{\partial T}{\partial t} + \frac{K}{\rho s} \nabla^2 T = 0$$

or

$$\frac{\partial T}{\partial t} = -\frac{K}{\rho s}\nabla^2 T \qquad \text{(IV.7a)}$$

This equation is known as *Fourier heat flow equation.* Note that ρs defines thermal capacity per unit volume and the ratio $\frac{K}{\rho s}$ is called *diffusivity*. It is generally denoted by h. Then we can rewrite Eq.(IV.7a) as

$$\frac{\partial T}{\partial t} = -h\nabla^2 T \qquad \text{(IV.7b)}$$

Heat Conduction through a Bar Refer to Fig. IV.2. It shows a bar AC of thickness x as made up of two bars AB and BC having the same area of cross section but different thicknesses. Let their respective thicknesses be x_1 and x_2 and thermal conductivities K_1 and K_2. Suppose that the temperatures at the end A, junction at B and end C are T_1, T and T_2 respectively. Since the same quantity of heat Q flows end-to-end through the bar in time t, we can write

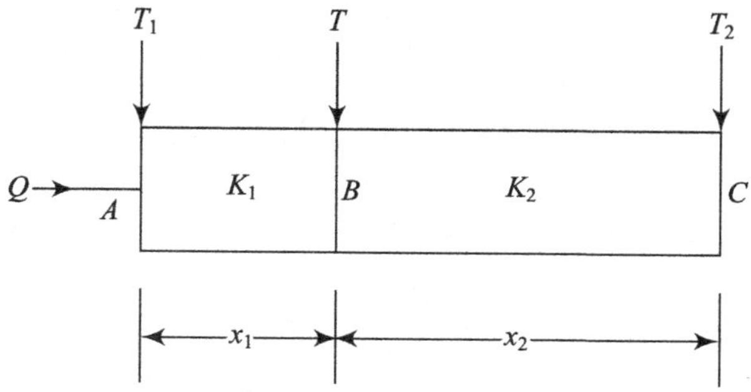

Fig. IV.2 A composite bar.

$$Q = \frac{K(T_1 - T_2)A}{x} \qquad \text{(IV.8a)}$$

$$Q = \frac{K_1(T_1 - T)A}{x_1} \qquad \text{(IV.8b)}$$

and

$$Q = \frac{K_2(T - T_2)A}{x_2} \qquad \text{(IV.8c)}$$

From Eqs. (IV.8b) and (IV.8c), we can write

$$\frac{Q}{A} = \frac{K_1(T_1 - T)}{x_1} = \frac{K_2(T - T_2)}{x_2} \qquad \text{(IV.9)}$$

We can rewrite these as

$$\frac{Q}{A} = \frac{(T_1 - T)}{\frac{x_1}{K_1}} = \frac{(T - T_2)}{\frac{x_2}{K_2}} = \frac{T_1 - T + T - T_2}{\frac{x_1}{K_1} + \frac{x_2}{K_2}}$$

On combining this result with Eq. (IV.8a), we can write

$$\frac{T_1 - T_2}{\dfrac{x_1 + x_2}{K}} = \frac{T_1 - T_2}{\dfrac{x_1}{K_1} + \dfrac{x_2}{K_2}}$$

or

$$\frac{x_1 + x_2}{K} = \frac{x_1}{K_1} + \frac{x_2}{K_2} \tag{IV.10}$$

We can rewrite it as

$$K = \frac{x_1 + x_2}{\dfrac{x_1}{K_1} + \dfrac{x_2}{K_2}} \tag{IV.11}$$

We can express the junction temperature in terms of K_1, K_2, x_1, x_2, T_1 and T_2. To this end, we first write Eq. (IV.9) as

$$K_1 x_2 (T_1 - T) = K_2 x_1 (T - T_2)$$

By multiplying through out and rearranging terms, we get

$$K_1 x_2 T_1 + K_2 x_1 T_2 = (K_1 x_2 + K_2 x_1) T$$

so that

$$T = \frac{K_1 x_2 T_1 + K_2 x_1 T_2}{K_1 x_2 + K_2 x_1} \tag{IV.12}$$

For a composite slab comprising n component slabs, we can generalise Eq. (IV.11) as

$$K = \frac{x_1 + x_2 + x_3 + \ldots + x_n}{\dfrac{x_1}{K_1} + \dfrac{x_2}{K_2} + \ldots + \dfrac{x_n}{K_n}} \tag{IV.13}$$

Let us now consider two bars of length x but cross-sectional areas A_1 and A_2 placed one above the other. If their thermal conductivities are K_1 and K_2 we can show that

$$K = \frac{K_1 A_1 + K_2 A_2}{A_1 + A_2} \tag{IV.14}$$

Similarly, for n bars having conductivities $K_1, K_2, K_3, \ldots K_n$ and cross-sectional areas A_1, $A_2, A_3 \ldots A_n$, we can write

$$K = \frac{K_1 A_1 + K_2 A_2 + K_3 A_3 + \ldots K_n A_n}{A_1 + A_2 + A_3 + \ldots A_n} \tag{IV.15}$$

The Variable and The Steady States of Heat Flow Refer to Fig. IV.3. It shows a rectangular bar whose one end is placed in a constant temperature bath C. Transmission of heat takes place by conduction across the layers normal to the length of the bar. Three such adjacent layers, L, M and N are shown in the diagram. The spaces between the layers have been shown for the sake of convenience of drawing.

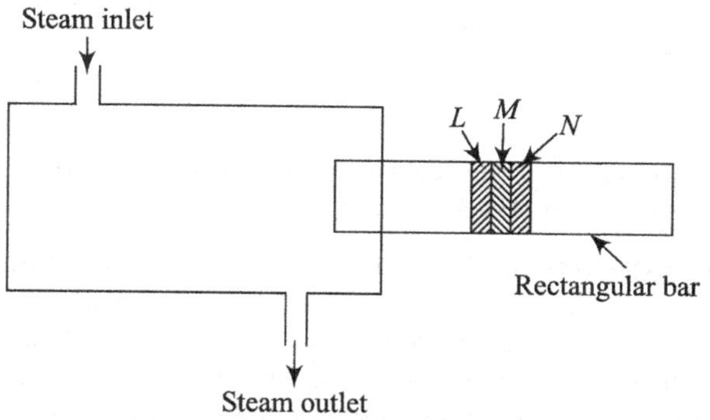

Fig. IV.3 Heat flow through a bar whose on end is placed in a constant temperature bath.

Now, let us concentrate on the middle layer M. You will agree that it receives heat from the immediate preceding layer L and spends it in three ways:

1. A fraction of the heat is absorbed by M. It increases its temperature.
2. A part of the heat is transferred to the subsequent layer N.
3. A fraction of the heat is lost by radiation from its surface. Some heat may also be lost by convection of the surrounding air.

This process is true for every such layer leading to rise in temperature of each layer of the bar gradually. Then the bar is said to be in the *variable state* of heat flow.

With the passage of time, a stage is reached when the rate of absorption of heat by any layer equals the rate of heat loss. At this stage, the temperature of each layer attains a constant value in descending order in respect of its distance from the hot end to the cold end. In other words, if the temperature of the hot end is θ_1 and that of the cold end is θ_2, the temperatures of the intermediate layers lie between these values decreases continuously but do not change with time. Then the bar is said to be in the *steady state* of heat flow. It may be mentioned here that in the steady state, no layer absorbs any heat and the rate of heat flow depends only on the thermal conductivity of the material of the bar and is described by Eq. (IV.2).

Note that in the variable state, the increase in temperature of each layer is essentially determined by two competing phenomena: transmission and absorption of heat. It means that rise in temperature depends on thermal conductivity as well as the heat capacity of the material. For this reason, all measurements of thermal conductivity are made when the steady state has been reached.

From Eq. (IV.7b) we recall that the temporal and spatial rates of change of temperature are connected through diffusivity, which is defined as the ratio of thermal conductivity and thermal capacity per unit volume. To give you an idea about the role of diffusivity in attaining steady state, we compute it for iron and lead. For iron, $\rho = 7.8 \times 10^3$ kg m^{-3}, $s = 0.11$ and $K = 57.2$ J m^{-1}s^{-1}K^{-1} so that $h = 7.83 \times 10^{-2}$ m^{3}s^{-3}K^{-1}. However, for lead $\rho = 11.4 \times 10^3$ kg m^{-3}, $s = 0.03$, $K = 33.6$ J m^{-1}s^{-1}K^{-1} and $h = 9.82 \times 10^{-2}$ m^{3}s^{-3}K^{-1}. It means that

1. Although thermal conductivity of iron is greater than that of lead, its diffusivity is less than that of lead.
2. Under identical conditions of heat flow, the rise of temperature of different sections in a lead bar will be more rapid than in an iron bar. This fact is well illustrated in the Ingen–Hausz's experiment.

The solution of Fourier heat flow equation is given by

$$T = T_0 - \frac{T_0 - T_\ell}{\ell} x \tag{IV.16}$$

where T_0 and T_1 respectively denote temperatures at the ends of a bar of length ℓ.

We now discuss the Ingen–Hausz's Experiment.

Ingen–Hausz's Experiment Refer to Fig. IV.4. It shows an oil bath B which can be put in contact with a source of heat, H. Four identical bars P, Q, R and S of different metals say iron, copper, lead and tin, are inserted in the grooves in the oil bath. These bars are equally coated with wax. It is observed that as metallic bars are heated, the wax first melts along the lead bar. However, when steady state is attained, the length of melted wax will be more for iron than lead; the lengths solely depend on thermal conductivity.

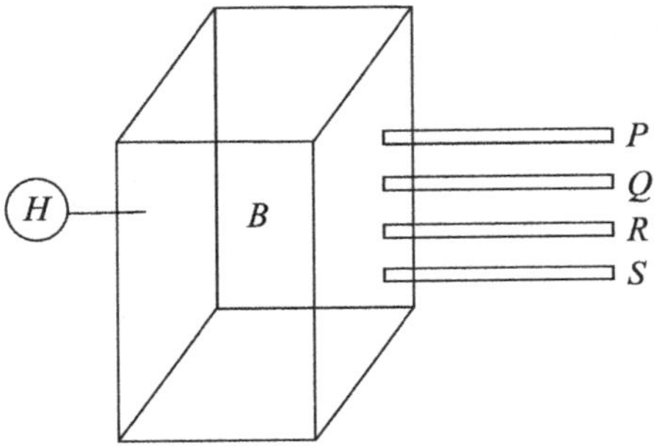

Fig. IV.4 Ingen–Hausz's Experiment.

Freezing of a Lake Thermal conductivity finds a fascinating application in freezing of ponds and lakes during winter in extremely cold conditions. This natural phenomenon is of great significance in that the ice becomes rock hard and helps aquatic life to survive extreme cold. You may have learnt about freezing of Dal Lake in Srinagar, India or Great lakes in U.S.

To understand formation of ice and how its thickness increases with time, we first recall that latent heat of fusion of ice is 336 kJ kg^{-1}. It means that 336 kJ kg^{-1} energy must be given out by one kg of water under STP. And when some layer of ice has been formed, the heat given out in the fusion is conducted through it. To calculate the time required for formation of a particular thickness, let us assume that a layer of thickness x is formed in time t and it takes time dt to increase by dx. Suppose that an amount of heat Q is released when thickness of layer of ice increases from x to x + dx. Hence, we can write

$$Q = A\rho L \mathrm{d}x \tag{IV.17}$$

where A is area of the slab, ρ is density of ice and L is latent heat of fusion.

Suppose that the heat so released flows from one face of the ice slab to another in time dt at temperature $-\theta$°C. So we can write

$$Q = \frac{KA[0-(-\theta)]\mathrm{d}t}{x} = \frac{KA\theta \mathrm{d}t}{x} \tag{IV.18}$$

On comparing Eqs. (IV.17) and (IV.18), we can write

$$A\rho L \mathrm{d}x = \frac{KA\theta}{x}\mathrm{d}t$$

so that

$$x\mathrm{d}x = \frac{K\theta}{\rho L}\mathrm{d}t \tag{IV.19}$$

If the time taken for increase in thickness from x_1 to x_2 is τ, we can integrate this expression to obtain

$$\int_{x_1}^{x_2} x\mathrm{d}x = \frac{K\theta}{\rho L}\int_0^{\tau} \mathrm{d}t$$

Hence,

$$\frac{x_2^2 - x_1^2}{2} = \frac{K\theta}{\rho L}\tau$$

or

$$\tau = \frac{\rho L}{2K\theta}(x_2^2 - x_1^2) \tag{IV.20}$$

If the layer of ice in a lake is 0.2 m and temperature of air is –40°C, using Eq. (IV.20) you can easily convince yourself that it will take more than three days for the ice layer to double. So far we have confined our discussion to heat flow and the role thermal conductivity plays in it. Now we present a few methods for verification of some of these aspects and experimental determination of thermal conductivity of good as well as bad conductors of heat in a physics laboratory.

Searle's Method Searle's method is used for determination of thermal conductivity of a good conductor in the laboratory at fairly high temperatures. The apparatus is shown in Fig. IV.5. The specimen is taken in the form of a thick bar (*XY*) and it is well lagged on the sides by a non-conducting material such as cotton or wool so as to minimise loss of heat due to radiation. End *X* of the bar is kept inside a steam bath, which has inlet and outlet for steam. Mercury is kept in the pockets created in the middle of the apparatus to hold mercury thermometers. This ensures good thermal contact with the specimen. At the

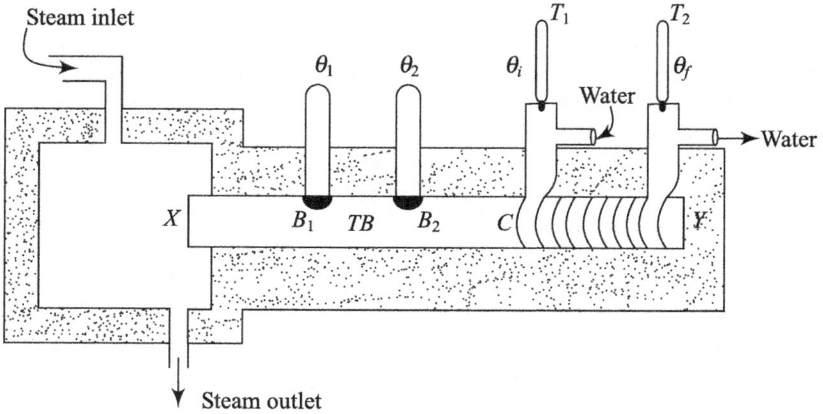

Fig. IV.5 Searle's apparatus for determination of thermal conductivity of a good conductor.

other end (Y) of the specimen, a copper tube C is wound and soldered. A steady flow of water is circulated through tube C, which absorbs heat from the specimen. This water is collected from the outlet. The temperatures of water at the inlet and the outlet are recorded by thermometers T_1, and T_2, respectively.

When steady state is attained, the quantity of heat that passes from section B_1 to B_2 is constant and the same amount of heat flows through the section at Y. Suppose that it causes the temperature of water circulating in tube C to rise from θ_i to θ_f. Therefore, the rate at which heat is conducted from B_1 to B_2 is given by

$$Q = \frac{KA(\theta_1 - \theta_2)}{d}$$

where d is the separation between B_1 and B_2, and A is area of cross section of the specimen.

The heat required to raise the temperature of m gram of water from θ_i to θ_f is $ms(\theta_i - \theta_f)$, where s is specific heat capacity of water. Therefore, by equating these two expressions, we can write

$$\frac{KA(\theta_1 - \theta_2)}{d} = ms(\theta_i - \theta_f)$$

By rearranging terms, we get the expression for thermal conductivity:

$$K = \frac{md(\theta_i - \theta_f)s}{A(\theta_1 - \theta_2)} \quad \text{(IV.21)}$$

From this expression, we note that we can determine thermal conductivity of a metallic specimen by knowing A, s, d, m, $(\theta_i - \theta_f)$ and $(\theta_1 - \theta_2)$.

It may be mentioned here that thermal conductivity of a metal varies with temperature and Searle's method is unsuited at low temperatures. We use Lee's method for this purpose and discuss it now.

Lee's Method for Determination of K of a Metal at Low Temperatures Refer to Fig. IV.6, which shows Lee's apparatus for determination of thermal conductivity of a metal, particularly at low temperatures. It consists of a specimen bar XY fixed to a copper frame CF. A heating coil H_1 is used to heat the end X of the specimen. Another heating coil H_2 is wound over the copper frame so as to attain the desired temperature. P_1 and P_2 are two platinum resistance thermometers and measure steady state temperatures at two points a distance d apart on the specimen. The entire set up is kept in a Dewar flask D so that there are no radiation losses.

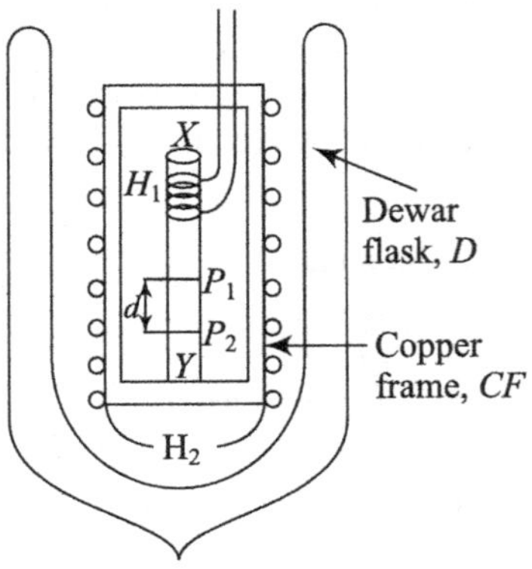

Fig. IV.6 Lee's apparatus for determination of K of a metal.

The flask is filled with liquid air so that the temperature of the apparatus goes down to a value below the temperature at which thermal conductivity is to be measured. Next, the liquid air is poured out of D and current is passed through the heating element H_2 so as to get the desired value of temperature. The end X of the bar is heated by passing current i under a given potential difference V. This process is continued till such time that the temperatures recorded by two platinum resistance thermometers show steady values θ_1 and θ_2. Then, we can write

$$\frac{Eit}{J} = Q = \frac{KA(\theta_1 - \theta_2)}{d} t$$

$$\therefore \qquad K = \frac{Eid}{JA(\theta_1 - \theta_2)} \qquad \text{(IV.22)}$$

where J is Mechanical equivalent of heat and t denotes the time for which current is passed to attain steady state.

From the known values of the quantities on the RHS of Eq. (IV.22), we can easily determine thermal conductivity of the given metallic bar at any temperature. However, it is important to mention here that when current is passed through the coil H_1, some heat is conducted by the copper frame and its temperature rises. This heat backflows and reduces the temperature difference $(\theta_1 - \theta_2)$ somewhat. Therefore, a correction should be applied to obtain the correct value. This can be done as follows before starting the main experiment:

The resistances of coils H_1 and H_2 are made equal and current i is passed through the coil H_1 only. The difference of temperatures recorded by platinum resistance thermometers is recoded once the steady state has been attained. Let us denote it as $d\theta_1$. This step is repeated and same current i is now passed through the coil H_2 only. The difference of temperatures recorded by platinum resistance thermometers is recoded again after the steady state has been attained. Let it be $d\theta_2$. The correct value of temperature difference is taken as $(\theta_1 - \theta_2) + \frac{d\theta_1 + d\theta_2}{2}$.

Forbes' Method Forbe's method is one of the earliest methods used to measure absolute thermal conductivity. You may get an opportunity to perform this experiment in your physics laboratory. It is based on a simple principle but the process is somewhat cumbersome. The experiment is performed in two parts: (i) Static, and (ii) dynamic.
In the static part, the specimen bar is heated at one end till such time that the steady state is reached. In the dynamic part, an identical bar (same material and cross-sectional area) but shorter than the experimental bar, is allowed to cool under similar conditions of temperature and pressure. In order to achieve 'similar conditions' in a laboratory, the experiment should be performed in a part of the laboratory which is not affected by thermal disturbances like wind flow, radiant heating, etc.

Static Experiment: The bar B normally has a curved end E which dips into a convenient molten metal MM in an iron crucible IC on one side of a screen S (Fig. IV.7). A series of pockets H_1, H_2, H_3, H_4 are created on the bar at regular intervals. Mercury is kept in these pockets and thermometers T_1, T_2, T_3, T_4 are dipped in each one.

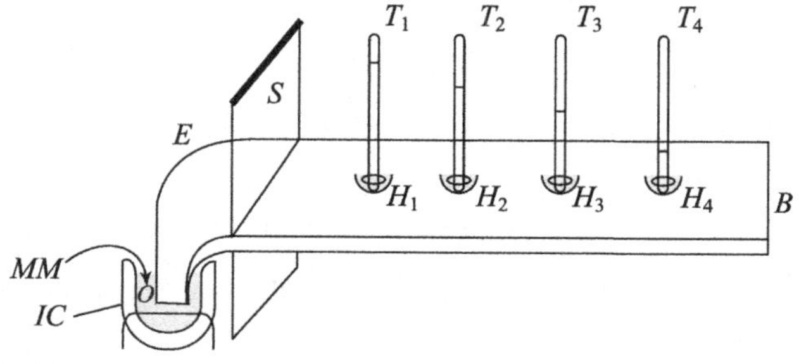

Fig. IV.7 Schematics of the Forbes' apparatus.

Once the steady state has been attained, the temperatures of different thermometers are recorded. If we now plot the temperatures at different points versus distance, we obtain a curve which can be characterised by a decaying exponential (Fig. IV.8). Note that the temperature θ is taken as that in excess over the surroundings and so the curve meets the x–axis at $x = \ell$. This is because the end of the bar is supposed to be at the room temperature and the length of the bar, ℓ is sufficiently large, about 2 m. So the temperature gradient $\left(\frac{d\theta}{dx}\right)_x$ at any distance x can be calculated from this graph.

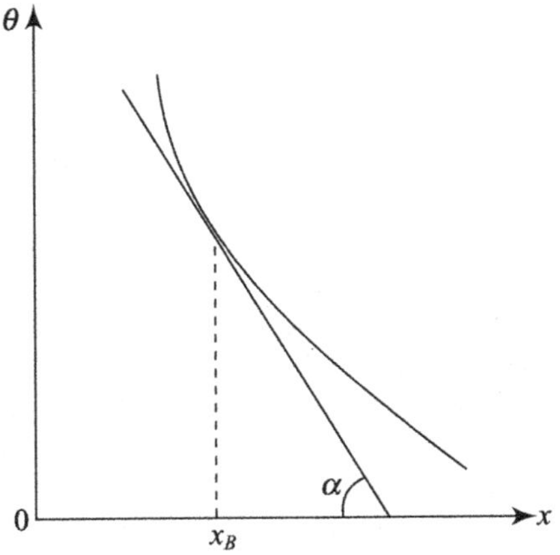

Fig. IV.8 Spatial variation of temperature in the steady state.

In the steady state, all the heat that passes through a section at $x = x_B$ will escape from the region of the surface located between the positions at $x = x_B$ and $x = \ell$. The quantity of heat that passes across the section at x per unit time is given by $-KA\left.\frac{d\theta}{dx}\right|_{x=x_B}$.

The temperature gradient $\left.\frac{d\theta}{dx}\right|_{x=x_B}$ at $x = x_B$ can be obtained from the graph by drawing a tangent at this point and measuring the slope:

$$\left[\frac{d\theta}{dt}\right]_{x_B} = \tan\alpha \qquad \text{(IV.23)}$$

The heat lost over the surface between x and $x + dx$ per second is $A\rho c dx\frac{d\theta}{dt}$, where $\left(-\frac{d\theta}{dt}\right)$ defines the time rate of change of temperature and the negative sign implies cooling. So the total heat lost due to radiation from the surface between $x = x_B$ and $x = \ell$ is given by $-\int_{x_B}^{\ell}\frac{d\theta}{dt}A\rho c dx$. Hence, we can write

$$-KA\left(\frac{d\theta}{dx}\right)_{x_B} = -\int_{x_B}^{\ell}\frac{d\theta}{dt}A\rho c dx$$

Using Eq. (IV.23), we can rewrite it as

$$K = \frac{\rho c\int_{x_B}^{\ell}\left(\frac{d\theta}{dt}\right)dx}{\tan\alpha} \qquad \text{(IV.24)}$$

You must have noted that $\left.\frac{d\theta}{dx}\right|_{x=x_B}$ can be obtained from the static part of the experiment by using the graph shown in Fig. IV.8. Now, we think of the way to determine the time variation of temperature.

Dynamic Part: A second bar identical in all respect with the experimental bar is heated to the temperature of steam. The bar is exposed to the atmosphere and temperatures attained by this bar are recorded at successive time intervals, say 1 minute. Care should however be taken to make sure that the time of cooling should include in the range, the values of temperature that figured in the static part of the experiment. The idea is to determine $\frac{d\theta}{dt}$ by allowing the bar to cool. With the above arrangement, θ is plotted against t to obtain what is called the cooling curve (Fig. IV.9). Now, we have to evaluate the definite integral on the RHS of Eq. (IV.24).

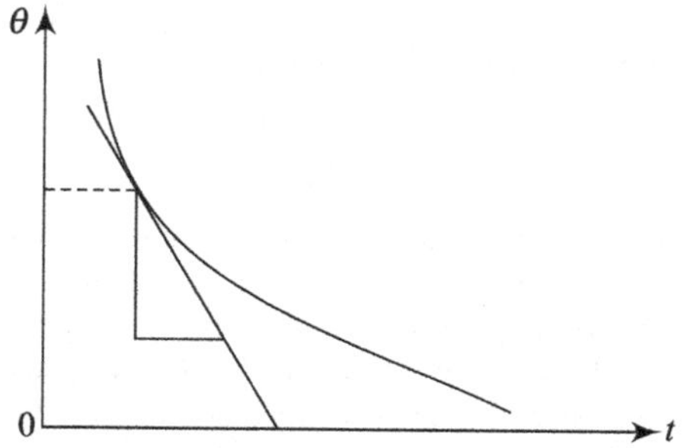

Fig. IV.9 Plot of θ versus t: Dynamical part.

To do so, you may recall that the definite integral $\int_a^b f(x)\,dx$ is the area under the curve $y = f(x)$ between the ordinates $x = a$ and $x = b$. We now apply this principle to evaluate the definite integral on the RHS of Eq. (IV.24). But remember, we need to have a curve of $\frac{d\theta}{dt}$ versus x. For this, the following steps are followed:

- Using the θ versus x plot (Fig IV.8), we obtain a series of values of x corresponding to different values of θ.
- Next we calculate $\frac{d\theta}{dt}$ values corresponding to these values of θ and x.
- Using the one-to-one correspondence between x and θ (as obtained in the first step, we plot $\frac{d\theta}{dt}$ versus x, as shown in Fig. IV.10

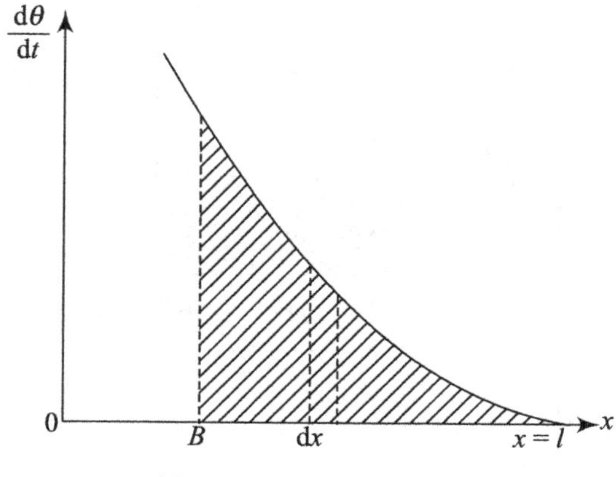

Fig. IV.10 Spatial variation of $\frac{d\theta}{dt}$ in the steady state.

$\frac{d\theta}{dt}$ versus x curve meets the x-axis at the point where $\frac{d\theta}{dt}$ becomes zero. This point on the bar corresponds to the temperature of the surroundings. Can you guess which this point is on the bar? It is the end point of the bar, $x = \ell$.

Note that the shaded portion in Fig. IV.10 represents some area under the curve $\frac{d\theta}{dt}$ versus x. This is equal to the value of the definite integral $\int_x^\ell \left(\frac{d\theta}{dt}\right) dx$. Using this result in Eq. (IV.24), we obtain absolute value of thermal conductivity from the relation

$$K = \frac{\rho c \,(\text{area of the shaded portion})}{\tan \alpha} \tag{IV.25}$$

It may be mentioned here that this method is quite cumbersome and time consuming. Moreover, the specific heat capacity may show some variation with temperature and the distribution of heat may be somewhat different in the two experiments. Therefore, it is quite possible that in a physics laboratory, you may not obtain a very accurate value. Moreover, Forbes' method, as also Searle's method, cannot be used for bad conductors. We now discuss Lees and Charlton method to determine K of a bad conductor in the form of a thin sheet such as a cardboard, asbestos sheet, etc.

Lees and Charlton's Method The experimental set up used in Lee and Charlton's method for determination of K is shown in Fig. IV.11. A brass disc D is supported by strong but light chains tied to a large ring on a retort stand. The experimental specimen M is taken in the form of a circular slab of the same diameter as that of D and is made to rest on it. A hollow brass cylinder C of the same diameter with side tubes for steam inlet and outlet is placed on M. Mercury thermometers T_1 and T_2 are inserted into holes drilled in the base B of C as well as in D.

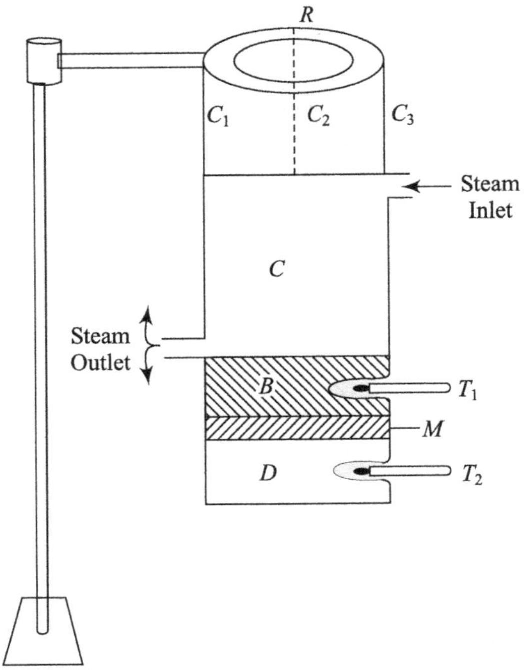

Fig. IV.11 Lee and Charlton's apparatus.

When steam is passed through *C*, heat is conducted through *M* and *D* gets heated up. After some time, steady state is attained and the temperatures recorded by the thermometers T_1 and T_2 attain a constant value. In this condition, the amount of heat conducted through *M* in unit time is totally radiated from the exposed surface of *D*; its temperature does not change despite heat conduction. The amount of heat conducted per unit time is given by

$$\frac{dQ}{dt} = \frac{KA(\theta_1 - \theta_2)}{d}$$

where θ_1 and θ_2 are steady temperatures recorded respectively by T_1 and T_2 and *d* is thickness of the material *M*.

Now we wish to determine the heat radiated by the exposed surface of *D*. To this end, the steam chamber and sheet *M* are removed from the top of *D*. Thereafter *D* is heated with a burner to a temperature a few degrees above θ_1. Thereafter, *M* and the steam chamber are placed again in their previous position and the flow of steam is continued through the chamber. The lower slab, being at a higher temperature, will cool due to radiation. When its temperature falls to θ_1, the steady state temperature, supply of steam is cut off. The lower slab continues to cool and its temperature is recorded at regular intervals of time (say every 30 s) till its temperature is a few degrees below the lower fixed temperature θ_2. As a matter of fact, we need to obtain $\left|\left(\frac{d\theta}{dt}\right)_{\theta_2}\right|$. To obtain this, we plot θ versus *t* and the expected nature of curve is shown in Fig. IV.12.

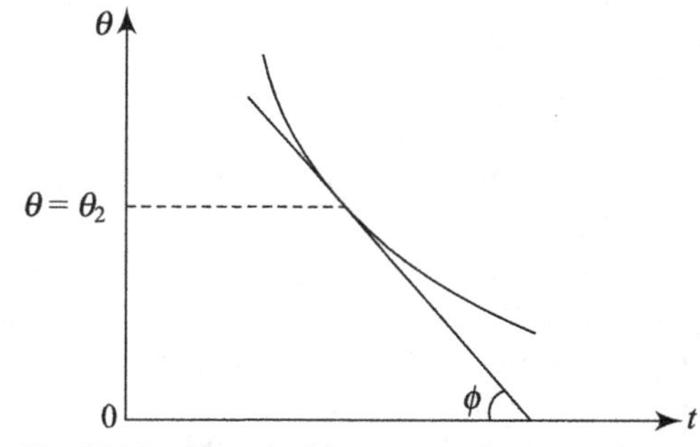

Fig. IV.12 Temporal variation of temperature.

The desired value of $\left|\left(\frac{d\theta}{dt}\right)_{\theta_2}\right|$ is obtained by calculating ϕ, where ϕ is the angle made with the negative direction of *t*–axis by the tangent to the curve at $\theta = \theta_2$. This is because, we need the absolute value of the slope; the angle made with the negative direction of *t*–axis is acute and its tangent is + ve.

Now, if the mass of *D* is *m*, its specific heat capacity *s*, the heat lost by radiation from the exposed surface is given by

$$\frac{dQ}{dt} = ms\left|\left(\frac{d\theta}{dt}\right)_{\theta_2}\right|$$

$$\therefore \quad \frac{KA(\theta_1 - \theta_2)}{d} = ms\left|\left(\frac{d\theta}{dt}\right)_{\theta_2}\right|$$

or

$$K = \frac{msd\left|\left(\frac{d\theta}{dt}\right)_{\theta_2}\right|}{A(\theta_1 - \theta_2)} \tag{IV.26}$$

Since all quantities occurring in the RHS of this expression are known, *K* can be determined easily.

BIBLIOGRAPHY

1. P. B. Pal, *An Introductory Course of Statistical Mechanics*, Narosa, New Delhi, 2009.
2. K. Huang, 2nd Edition, *Statistical Mechanics,* Wiley Eastern, New Delhi, 2008.
3. E. Guha, *Basic Thermodynamics*, Narosa, New Delhi, 2000.
4. Yung-Kuo Lim (Ed), *Problems and Solutions on Thermodynamics and Statistical Mechanics*, World Scientific, 1990.
5. K. Wark, Jr., *Thermodynamics*, McGraw-Hill, New York, 1989.
6. Y.A. Cengel and M.A. Boles, *Thermodynamics*, McGraw-Hill, New York, 1989.
7. F.W. Sears and G.L. Salinger, *Thermodynamics, Kinetic Theory and Statistical Thermodynamics*, Narosa Publishing House, New Delhi, 1988.
8. J.P. Holman, *Thermodynamics*, 3rd edn. McGraw-Hill, 1985.
9. H.B. Callen, *Thermodynamics and an Introduction to Thermostatistics*, 2nd edn. Wiley, New York, 1985.
10. K. Stowe, *Introduction to Statistical Mechanics and Thermodynamics,* Wiley, New York, 1984.
11. C.J. Adkins, *Equilibrium Thermodynamics*, 3rd edn. Cambridge University Press, Cambridge, 1983.
12. W.G.V. Rosser, *An Introduction to Statistical Physics,* Wiley Eastern, New Delhi, 1982.
13. F. Mendle, *Statistical Physics*, Wiley, New York, 1982.
14. M.W. Zemansky and R.H. Dittman, *Heat and Thermodynamics*, 6th edn. McGraw-Hill, New York, 1981.
15. L.D. Landau and E.M. Lifshitz, *Statistical Physics*, 3rd edn. part 1 (revised and enlarged by E.M. Lifshitz and L.D. Pitaevskiii). Pergamon, Oxford, 1980.
16. C. Kittel/H. Kroemer, *Thermal Physics,* 2nd edn. W.H. Freeman, San Francisco, 1980.
17. E.S.R. Gopal, *Statistical Mechanics and Properties of Matter,* Ellis Horwood, Chichester, 1974.
18. M.N. Saha and B.N. Srivastava, *A Treatise on Heat*, 5th edn. The Indian Press (Publications) Pvt. Ltd. Allahabad, 1969.
19. F. Reif, *Fundamentals of Statistical and Thermal Physics*, McGraw-Hill, New York, 1965.

INDEX

Made in the USA
Monee, IL
04 November 2026

67384909R10380